Handbook of Experimental Pharmacology

Continuation of Handbuch der experimentellen Pharmakologie

Vol. 57

Tissue Growth Factors

Contributors

J. Abrahm · R. Baserga · G. Carpenter · D. R. Clemmons
M. J. Cline · V. J. Cristofalo · D. D. Cunningham
P. Datta · Ch. M. Dollbaum · J. H. Fitchen · F. Grinnell
P. M. Gullino · R. G. Ham · L. Harel · O. H. Iversen
M. J. Koroly · D. Metcalf · B. A. Rosner · R. Ross
G. Rovera · H. S. Smith · J. J. Van Wyk · M. Young

Editor

R. Baserga

Springer-Verlag Berlin Heidelberg New York 1981

Professor RENATO BASERGA, M.D.
Senior Investigator, Fels Research Institute and Department of Pathology
Temple University School of Medicine, Health Sciences Center
Philadelphia, PA 19140/USA

With 47 Figures

ISBN 3-540-10623-5 Springer-Verlag Berlin Heidelberg New York
ISBN 0-387-10623-5 Springer-Verlag New York Heidelberg Berlin

Library of Congress Cataloging in Publication Data. Main entry under title: Tissue growth factors. (Handbook of experimental pharmacology; v. 57) Bibliography: p. Includes index. 1. Tissues – Growth. 2. Growth promoting substances. 3. Cell proliferation. I. Abraham, J. (Janet) II. Baserga, Renato. III. Series. [DNLM: 1. Cell division. 2. Growth substances. 3. Peptides. W1 HA51L v. 57/QH 511 T616]   QP905.H3 vol. 57 [QP88]   615'.1s [599.08'2]   80-14574AACR2   ISBN 0-387-10623-5 (U.S.).

Typesetting, printing, and bookbinding: Brühlsche Universitätsdruckerei, Giessen.
2122/3130-543210

List of Contributors

Dr. JANET ABRAHM, Hematology-Oncology Division, Department of Internal Medicine, University of Pennsylvania, Philadelphia, PA 19140/USA

Professor R. BASERGA, M.D., Senior Investigator, Fels Research Institute and Department of Pathology, Temple University School of Medicine, Health Sciences Center, Philadelphia, PA 19140/USA

Professor G. CARPENTER, Departments of Biochemistry and Medicine, Vanderbilt University, School of Medicine, Nashville, TN 37232/USA

Professor D. R. CLEMMONS, M.D., University of North Carolina, Department of Medicine and Pediatrics, Clinical Sciences Building 229H, Chapel Hill, NC 27514/USA

Professor M. J. CLINE, M.D., Department of Medicine, Center for the Health Sciences, University of California, Los Angeles, CA 90024/USA

Professor V. J. CRISTOFALO, Ph.D., The Wistar Institute, 36th Street at Spruce, Philadelphia, PA 19104/USA

Professor D. D. CUNNINGHAM, Department of Microbiology, College of Medicine, University of California, Irvine, CA 92717/USA

Professor P. DATTA, Department of Biological Chemistry, University of Michigan, Medical School, P.O.B. 034, Ann Arbor, MI 48109/USA

Professor CH. M. DOLLBAUM, Peralta Cancer Research Institute, 3023 Summit Street, Oakland, CA 94609/USA

J. H. FITCHEN, M.D., Assistant Professor of Medicine, Department of Medicine, UCLA School of Medicine, Center for the Health Sciences, Los Angeles, CA 90024/USA

Professor F. GRINNELL, Department of Cell Biology, University of Texas, Health Sciences Center at Dallas, Southwestern Medical School, Dallas, TX 75235/USA

Dr. P. M. GULLINO, Department of Health, Education, and Welfare, Public Health Service, National Institutes of Health, Bethesda, MD 20014/USA

Professor R. G. HAM, University of Colorado, Department of Molecular, Cellular, and Developmental Biology, Biosciences Building, Boulder, CO 80309/USA

Dr. LOUISE HAREL, Institut de Recherches, Scientifiques Sur Le Cancer, 7, Rue Guy-Mocquet, F-94800 Villejuif

Professor O. H. IVERSEN, Universitet I Oslo, Institutt for Patologi, Rikshospitalet, N-Oslo 1. McArdle Laboratory for Cancer Research, University of Wisconsin, Madison, WI 53706/USA

Professor MARY J. KOROLY, Department of Biochemistry and Molecular Biology, University of Florida, College of Medicine, Gainesville, FL 32610/USA

D. METCALF, M.D., Head Cancer Research Unit, The Walter and Eliza Hall, Institute of Medical Research, Royal Melbourne Hospital, AUS-Victoria 3050

Dr. B. A. ROSNER, The Wistar Institute of Anatomy and Biology, 36th Street at Spruce, Philadelphia, PA 19104/USA

Professor R. ROSS, University of Washington, School of Medicine, Department of Pathology and Biochemistry, Seattle, WA 98195/USA

Professor G. ROVERA, The Wistar Institute of Anatomy and Biology, 36th Street at Spruce, Philadelphia, PA 19104/USA

HELENE S. SMITH, Ph.D., Assistant Director, Peralta Cancer Research Institute, 3023 Summit Street, Oakland, CA 94609/USA

Professor J. J. VAN WYK, Departments of Medicine and Pediatrics, University of North Carolina, School of Medicine, Clinical Sciences, Building 229H, Chapel Hill, NC 27514/USA

Professor M. YOUNG, Department of Biochemistry and Molecular Biology, University of Florida, College of Medicine, Gainesville, FL 32610/USA

Preface

From a logical point of view, cell division is regulated by the environment and by the ability of the cell to respond to the environmental signals. The terminology of the cell cycle, the elaborate mathematical models, and the kinetic analyses are all convenient notations and descriptions of the behavior of populations of cells. However, they tell us very little about the fundamental molecular mechanisms that control cell proliferation. Stated in other terms, what controls cell reproduction are growth factors in the environment and genes and gene products inside the cell or at its surface.

This book examines the aforementioned growth factors, the study of which has made very rapid progress in the past few years. The selection of topics has been influenced by logistic considerations, but the book, as a whole, gives a broad survey of the state of the art of this exciting field. For this, thanks are due to the contributors, who have given much time to the preparation of the manuscripts and have met the deadline with a punctuality that is uncommon among biomedical scientists.

I would also like to thank Ms. NORA PERRETT and the staff of Springer-Verlag for their help in editing the manuscripts and in preparing the production of the book.

Philadelphia

R. BASERGA

Contents

CHAPTER 3

Epidermal Growth Factor. G. CARPENTER. With 3 Figures

CHAPTER 4

The Platelet-Derived Growth Factor. R. Ross. With 14 Figures

CHAPTER 5

Somatomedin: Physiological Control and Effects on Cell Proliferation
D. R. CLEMMONS and J. J. VAN WYK. With 13 Figures

CHAPTER 6

Glucocorticoid Modulation of Cell Proliferation
V. J. CRISTOFALO and B. A. ROSNER. With 2 Figures

CHAPTER 7

Proteases as Growth Factors. D. D. CUNNINGHAM

CHAPTER 10

Membrane-Derived Inhibitory Factors. P. DATTA. With 4 Figures

CHAPTER 11

Diffusible Factors in Tissue Cultures. LOUISE HAREL

CHAPTER 12

Hemopoietic Colony Stimulating Factors. D. METCALF
With 6 Figures

CHAPTER 13

Inhibition of Hematopoietic Cell Proliferation. J. H. FITCHEN and M. J. CLINE
With 3 Figures

CHAPTER 14

Inducers and Inhibitors of Leukemic Cell Differentiation in Culture
JANET ABRAHM and G. ROVERA

CHAPTER 15

Angiogenesis Factor(s). P. M. GULLINO. With 2 Figures

CHAPTER 16

Growth of Human Tumors in Culture. HELENE S. SMITH and
CH. M. DOLLBAUM. With 1 Figure

CHAPTER 17

The Chalones. O. H. IVERSEN. With 4 Figures

CHAPTER 1

Introduction to Cell Growth: Growth in Size and DNA Replication

R. BASERGA

A. Introduction

This book is devoted to growth factors, which can be defined as those hormones or hormone-like substances that regulate the growth of cells, either in the intact animal or in tissue cultures. Arbitrarily, the book has been restricted to growth factors of *animal* cells. I have no justification to offer for this decision except those logistic considerations that necessarily limit the size of a book, and force some decisions.

There was a time, only a few years ago, when growth factors were looked upon with a mixture of interest and incredulity (BASERGA, 1976). Leaving aside the nutrients and the ionic milieu (amino acids, vitamins, pH, etc.) necessary for the balanced maintenance and growth of cells, tissue growth factors were poorly characterized and had a short half-life in the literature. All this has changed, as a number of growth factors have been purified, their purification allowing a reproducible analysis of their biological action.

Several growth factors are described in this book, some quite different from each other and yet all having the ability of stimulating cell growth. What is the common mechanism, inside the cell, by which many different molecules in the outside environment bring about the same end result? It is this question that I would like to examine briefly in this introduction and, to begin with, I would like to focus the question by making a hypothesis. The hypothesis states that in order to grow a cell must receive *two signals* that are separate and distinct: a signal to double its size and a signal to replicate DNA. This statement requires the demonstration of three separate substatements, namely: 1) that growing cells double in size; 2) that growing cells replicate their DNA (or, better, their genetic material); and 3) that the two processes can be independent of each other. This hypothesis is not new. It has been brilliantly proposed in 1971 by MITCHISON for the yeast cell cycle, but it has been neglected in the field of mammalian cells. Yet, I believe it is the key to our understanding of cell growth. To begin with, let us examine separately the three arguments of the hypothesis.

B. Doubling of Size in Growing Cells

When a cell goes from one mitosis to the next one it doubles its size. This is almost intuitive since if dividing cells were not to double their size from G_1 to M, they would become progressively smaller, and eventually vanish. In fact, size progressively increases from G_1 to M (GAZITT et al., 1978), so much so that it is sometimes

Table 1. Evidence for increased ribosomal RNA synthesis in cells during the transition from a resting to a growing stage

Cells or tissue	Stimulus	Methodology	Reference
Rabbit kidney cells	Explant	Inhibition of Actinomycin D	LIEBERMANN et al. (1963)
Liver	Partial hepatectomy	Incorporation of precursors	TSUKADA and LIEBERMANN (1964)
Salivary glands	Isoproterenol	Incorporation of precursors	NOVI and BASERGA (1972)
WI-38 cells	Serum	Incorporation of precursors	ZARDI and BASERGA (1974)
3T6 cells	Serum	Nuclear monolayers	MAUCK and GREEN (1973)
Lymphocytes	Phytohemag- glutinin	RNA pol I activity	JAEHNING et al. (1975)
CV-1 fibroblasts	Serum	Nuclear monolayers	ROVERA et al. (1975)
Fibroblasts in culture	Serum	Inhibition by Actinomycin D	EPIFANOVA et al. (1975)
Liver	Partial hepatectomy	RNA pol I activity	ORGANTINI et al. (1975)
Uterus	Estrogens	Isolated nucleoli	NICOLETTE and BABLER (1974)
Liver	Partial hepatectomy	Isolated nucleoli	SCHMID and SEKERIS (1975)
Mouse kidney cells	Polyoma virus	Incorporation of precursors	BENJAMIN (1966)
Mouse kidney cells	Polyoma virus	Amount of RNA	WEIL et al. (1975)
Mouse kidney cells	SV 40	Incorporation of precursors	MAY et al. (1976)
tsAF 8	Serum	Isolated nucleoli	ROSSINI and BASERGA (1978)
Liver cells	SV 40 T-antigen	Isolated nucleoli	WHELLY et al. (1978)
Human cells	Growth factors	Ag-staining of NORs	SCHMIADY et al. (1979)
S. cerevisiae	Zinc salts	Incorporation of precursors	JOHNSTON and SINGER (1978)
S. cerevisiae	Methionine	Incorporation of precursors	SINGER et al. (1978)

used to obtain pure populations of synchronized cells (MITCHELL and TUPPER, 1977).

Proteins and nucleic acids are a good indicator of the size of the cell (COHEN and STUDZINSKI, 1967; WHATLEY and HILL, 1979), not only because they constitute more than 50% of the dry weight of cell, but also because from them (especially from proteins and RNA) depend the amounts of other components, such as sugars and lipids. There are several references showing that growing cells, in going from early G_1 to M, double in size (SKOG et al., 1979, see also review by HARTWELL, 1978) and double their protein content (SKOG et al., 1979; BAXTER and STANNERS,

1978; LEE and ENGELHARDT, 1977). There are also several references showing that the amount of RNA/cell also doubles as dividing cells go from one mitosis (or from quiescence) to the next mitosis (ASHIHARA et al., 1978b; DARZYNKIEWICZ et al., 1979). The nucleus also doubles in size (YEN and PARDEE, 1979), and although there are exceptions (see below), one can say that under ordinary circumstances dividing cells will double in size from the immediate post-mitotic stage to the next mitosis.

Once we accept the notion that cells must double in size before division, one is not surprised to find that ribosomal RNA (rRNA) genes are one of the targets of growth stimulating factors. rRNA constitutes about 85% of total cellular RNA and it provides the machinery of protein synthesis. Often, when faced with the evidence that rRNA synthesis and accumulation are increased in cells stimulated to proliferate, some of my colleagues have exclaimed "Why should rRNA synthesis be stimulated?" I would find it more surprising if rRNA synthesis were *not* stimulated. After all, if the cell is to double in size, the rRNA genes are a good place to start from, since it would increase both the RNA content/cell and the framework for the doubling of the protein content. The evidence for increased rRNA synthesis (or accumulation) in cells stimulated to proliferate is overwhelming and is summarized in Table 1.

Please notice: 1) The Table is (with one exception) limited to rRNA synthesis. The other references dealing with an increased amount of RNA/cell were given above: 2) because it is mostly limited to rRNA synthesis, the references deal only with G_0 cells stimulated to proliferate. This is because quiescent cells which, for brevity, we call G_0 cells (see below), have decreased rRNA synthesis. In actively growing cells, RNA synthesis is pretty much constant, except during mitosis when it is virtually nonexistent (BASERGA, 1962), 3) the evidence is sustained by a variety of techniques, and by diverse systems. Since there is a pecking order in cell biology, the evidence from tissue cultures is considered more impressive than that obtained from intact animals, but not as impressive as that obtained from yeasts. For this reason, the group that reported in 1978 the vital role of rRNA synthesis in yeast cell growth could majestically ignore 15 previous years of experience with mammalian cells.

In conclusion, cells double in size before dividing, and the rRNA genes are an ideal target for those growth factors that must signal the cell to double in size.

C. Replication of DNA in Growing Cells

In most cases there is a temporal gap between completion of mitosis and onset of DNA synthesis, a gap which has been called G_1 (HOWARD and PELC, 1951) and which is markedly lengthened in quiescent cells stimulated to proliferate. Periodically we are faced in the literature with theoretical papers questioning the existence of G_1 or, for that matter, of G_0. These are cosmic questions which end up generating a lot of arguments but very few data. That certain cells can be without a G_1 has been known for a long time, in fact, since our own observation in 1963 on Ehrlich ascites tumor cells (BASERGA, 1963). It was confirmed repeatedly in subsequent years (LALA and PATT, 1966; ROBBINS and SCHARFF, 1967) and G_1 cells have been the object of elegant studies by PRESCOTT and his collaborators (LISKAY and PRESCOTT, 1978; LISKAY, 1978). So, everybody knows that rapidly growing cells can do without a G_1, just as everybody knows that embryo cells can condense the whole cell cycle in 15 min (GRAHAM and MORGAN, 1966) with an S phase of less than 10 min and a mitosis lasting 2 min as in stage 7 of *X. laevis* embryo cleavage. But most cells, after mitosis, go through a phase during which no DNA synthesis occurs, which we call G_1. Using Cartesian logic, if this phase does not exist, these cells do not exist either. The fact is that we should be careful not to take for absolute truths our convenient notations. There is, in most cells, an interval between mitosis and S. Something happens in this interval that is a prerequisite for entry into S – obviously some cells can carry out these prerequisite processes in the previous cycle. A similar situation is clearly illustrated in yeast, where *S. cerevisiae* controls its rate of growth in G_1, while *S. pombe* uses G_2 to regulate its growth (HARTWELL, 1978). A simple model to explain the variable length (or even the absence) of G_1, could be based on the scarcity of mRNAs necessary for the entry into S. Unequal partition at mitosis of mRNAs present in 1–2 copies per cell could easily account for the variability in G_1. Indeed, BROOKS (1979) suggested that the variability of the length of G_1 in mammalian cells is of cytoplasmic origin.

We have said "Something happens". What is the evidence that something must happen in G_1 before the cell can enter S? Formal evidence derives from the existence of ts mutants that specifically arrest in G_1 at the nonpermissive temperature (BASILICO, 1977; SIMINOVITCH and THOMPSON, 1978). It should be made clear that these G_1 ts mutants, when shifted to the nonpermissive temperature in late G_1, S, G_2 or M, flow regularly through the cycle until they reach a point in G_1. However, they arrest in mid-G_1 even when shifted up after mitosis (BURSTIN et al., 1974; SMITH and WIGGLESWORTH, 1974; ASHIHARA et al., 1978), or, if made quiescent, after serum stimulation (ASHIHARA et al., 1978). Therefore, it is clear that a number of functions must be completed before cells can enter into S, and that, in most cells, at least some of these functions are carried out in the interval between M and S, conveniently called G_1. Supporting evidence to the existence of these functions is given by experiment with drugs that inhibit RNA or protein synthesis (for reviews see BASERGA, 1976; PARDEE et al., 1978), and by the existence of a number of G_1 markers, i.e., of detectable biochemical changes that make their appearance in G_1. These markers are summarized in Table 2.

It has been said that nothing happens in G_1, and that everything starts in the previous S phase (COOPER, 1979). Possible, but there is no evidence for it, and there will not be unless we can identify and study specific gene products like some of those studied by PARDEE and collaborators (ROSSOW et al., 1979; RIDDLE et al., 1979). But there are already two observations that indicate that some things start in G_1: 1) the doubling in RNA and protein amounts between early G_1 and M, mentioned above; and 2) the fact that cells arrested in mitosis by Colcemid, when released, enter S at the same speed as untreated cells (ASHIHARA et al., 1978a). If things started in S, and G_1 exists only to complete previous processes, some of these processes should be carried out toward completion during the 2–3 h of mitotic arrest.

Some of these G_1 markers appear after quiescent cells are stimulated to proliferate by the addition of growth factors. Which brings up the next point: is G_0 different from G_1? This controversy has also generated many more arguments than data. The fact is that serum-deprived cells *are* biochemically different from G_1 cells

Table 2. Biological markers of the G_1 phase of the cell cycle

Marker	Reference
Synthesis of actin	RIDDLE et al. (1979)
Polypeptide synthesized in mid-G_1	GATES and FRIEDKIN (1978)
Last serum requirement before S	YEN and PARDEE (1978)
Requirements for plasma, serum and isoleucine	YEN and RIDDLE (1979)
Phosphorylation of ribosomal protein S6	HASELBACHER et al. (1979)
Deciliation and duplication of centrioles	TUCKER et al. (1979)
Synthesis of labile protein in early G_1	NOVI and BASERGA (1972)
	SCHNEIDERMAN et al. (1971)
	ROSSOW et al. (1979)
	SHILO et al. (1979)
Synthesis of nonhistone chromosomal protein(s)	TSUBOI and BASERGA (1972)
ts function of ts 13	FLOROS et al. (1978)
Requirement for RNA polymerase II	ROSSINI et al. (1980)
Ribosome content	BECKER et al. (1971)

(see reviews by BASERGA, 1976; EPIFANOVA, 1977; PARDEE et al., 1978; also YEN et al., 1978; and also some data in Table 2). Again, using the term G_0 is a convenient notation to distinguish them from G_1 cells, just as we distinguish epithelial cells of the jejunum from those of the colon, although the two cell types have a lot of characteristics in common. A formal demonstration that serum-deprived cells are different from G_1 cells comes from our own experiments (JONAK and BASERGA, in preparation), fusing cytoplasts to whole cells. Cytoplasts from G_0 cells have a different informational content than cytoplasts from nonquiescent cells (JONAK and BASERGA, 1979; JONAK and BASERGA, 1980). Significantly, one of the growth factors, PDGF, has been shown to inhibit the entry of cells into G_0 (SCHER et al., 1979 b). One cannot help repeating the statement that STANNERS and co-workers made in 1971, that their data supported "the existence of a quiescent state that is distinct from the G_1 phase of the cell cycle" (BECKER et al., 1971).

Whether in G_1 cells in an actively growing population, or in G_0 cells stimulated by growth factors, DNA replication is the *conditio sine qua non* of dividing cells. In fact, DNA synthesis has become synonymous with cell proliferation, and it is considered evidence of cell growth, to the exclusion, some time, of other important parameters, such as growth in size or growth in number.

As far as I know there is only a single exception. In animals treated with azathioprine, partial hepatectomy causes cell division in tetraploid hepatocytes without previous DNA synthesis (MALAMUD et al., 1972). Clearly, this is a special case, essentially of cells blocked in G_2 indefinitely that are now stimulated to return to the diploid state.

How many genes are necessary for the G_0 (or G_1)→S transition? Impossible to say, at the present. But we know that a minimum of 32 gene products are necessary for the yeast cell cycle (EDWARDS et al., 1978). Mammalian cells contain about 100 times the amount of DNA of a diploid Saccharomyces, so that 3,000 genes could be a possible number for the mammalian cell cycle. But a lot of the mammalian DNA is non-coding, and the figure is probably lower by an order of magnitude.

D. Independence of Signal to Grow in Size from Signal to Replicate DNA

It is clear from the studies of Ross and collaborators (Ross et al., 1978; and see Chap. 4 in this book) and of PLEDGER et al. (STILES et al., 1979; PLEDGER et al., 1977), that there are at least two signals for the entry of cells into S. The first signal comes from the PDGF that "primes" the cell (PLEDGER et al., 1977), stimulates RNA synthesis (ABELSON et al., 1979), but does not, by itself, cause cells to enter into S. A second signal comes from a factor in platelet-poor plasma that makes cells, made competent by PDGF, enter into S (for review see SCHER et al., 1979 a). The simplest explanation is that a plasma factor (somatomedin?) interacts with a PDGF-induced cellular product to initiate DNA synthesis. We put forward the hypothesis that PDGF gives the signal for doubling the size of the cell, while the signal for DNA replication comes from the plasma factor plus a PDGF-induced cellular product.

Let us see how available data can fit this hypothesis.

1. It is possible to dissociate an increase in rRNA synthesis (presumably a marker of increase in size) from DNA replication. Thus, nucleolar RNA synthesis increases in ts AF 8 serum-stimulated at the nonpermissive temperature, i.e., blocked in mid-G_1 (ROSSINI and BASERGA, 1978), and PDGF, by itself, can stimulate RNA but not DNA synthesis (ABELSON et al., 1979). It is also possible to dissociate the rate of mRNA synthesis from the timing of DNA replication, at least in yeast (FRASER and NURSE, 1978).

2. 422 E cells, a mutant of BHK, that fail to accumulate ribosomes at the nonpermissive temperature (TONIOLO et al., 1973) will enter DNA synthesis even under nonpermissive conditions (GRUMMT et al., 1979), although they fail to divide (MORA et al., 1980).

3. Adenovirus 2 stimulates DNA synthesis in infected cells (LAUGHLIN and STROHL, 1976; ROSSINI et al., 1979), but does not cause an increase in rRNA synthesis (ELICEIRI, 1973; SOPRANO et al., 1980; POCHRON et al., in preparation). Infection by Adenovirus also fails to increase the accumulation of cellular RNA (POCHRON et al., in preparation). This is at variance with SV 40, which stimulates rRNA synthesis (SOPRANO et al., 1979) and accumulation of cellular RNA (WEIL et al., 1975; PETRALIA et al., in preparation). This difference will become important later on.

4. The SV 40 T-antigen stimulates cellular DNA synthesis (TJIAN et al., 1978) even in ts 13 cells at the nonpermissive temperature (FLOROS et al., in preparation). Ts 13 cells are a ts mutant of BHK cells that, like ts AF 8, arrest in G_1 at the nonpermissive temperature (FLOROS et al., 1978). Adenovirus 2 can also bypass the ts block of these G_1 mutants (ROSSINI et al., 1979), but the gene or genes responsible for the stimulation of cellular DNA synthesis have not yet been identified. Polyoma-transformed ts AF 8 to through one round of DNA synthesis at the nonpermissive temperature, but then they die (BURSTIN and BASILICO, 1975).

As already mentioned serum growth factors cannot overcome the G_1 ts block.

This information is summarized in Table 3. All of this information comes from cells in culture, which may not reflect the reality of the living animal. But then science is concerned with truth, not with reality. Our interpretation of these results is the following:

Two signals exist for cell growth, one to double the cellular size, one to replicate DNA. PDGF gives the first signal only, Adenovirus gives only the second signal. SV 40 T antigen gives both signals, while platelet poor plasma plus a PDGF-induced cellular factor can give the signal for DNA replication. The availability of cloned genes and the technique of microinjection developed by GRAESSMANN should allow us to test this hypothesis. To give an example, if the hypothesis is correct, cells in platelet-poor plasma should go into DNA synthesis if microinjected with mRNA from PDGF-treated quiescent cells.

In this view, certain virally-coded proteins can be seen as replacing different growth factors, a sort, so to speak, of internal growth factors.

Table 3. Effect of serum and DNA viruses on cellular RNA and DNA

	Stimulation of cellular		Accumulation of cellular RNA	In G$_1$ mutants at the nonpermissive temperature	
	rRNA synthesis	DNA synthesis		RNA synthesis	DNA synthesis
Serum growth factors	+	+	+	+	−
PDGF	+	¬	+	ND	ND
SV 40 T-antigen	+	+	+	ND	+
Adenovirus 2	−	+	−	−	+

ND, not done

E. Mechanism of Action of Growth Factors

Probably, most growth factors act at the membrane level, as discussed in the following chapters. What we are concerned with here is how the growth factor-induced cellular proteins or the viral proteins act on the cellular genome. Let us take as an illustration the SV 40 T-antigen (or, I should say, the product of the SV 40 A gene), which has been most extensively studied and characterized.

1. Purified SV 40 T-antigen, micro-injected into quiescent cells, stimulates cellular DNA synthesis (TJIAN et al., 1978). So does the SV 40 A gene (FLOROS et al., in preparation) that codes for large T (WEIL, 1978), and so does SV 40 mRNA (FLOROS et al., in preparation).

2. SV 40 T-antigen reactivates silent rRNA genes (SOPRANO et al., 1979; SO-PRANO et al., 1980) in human > mouse hybrids, where mouse rRNA genes are repressed.

3. At least in our hands, small t is not required for cellular DNA synthesis (FLOROS et al., in preparation), nor for reactivation of silent rRNA genes (SOPRANO et al., 1980).

4. There is essentially a complete overlapping in the DNA sequences of the SV 40 genome that is necessary and sufficient for the reactivation of silent rRNA genes and the stimulation of cellular DNA synthesis. These sequences map between 0.38 and 0.27 of the conventional SV 40 map (SOPRANO et al., in preparation).

This means that SV 40 uses the same 600 base pairs (about 200 aminoacids) to stimulate transcription *and* DNA replication. This is somewhat puzzling, but it is not without a precedent. After all, SV 40 T-antigen is known to be required for both viral DNA replication *and* transcription of late viral genes (for reviews see LEVINE, 1976; and WEIL, 1978). If large T can have this double function on viral genes, why can it not have the same dual effect on the mammalian genome?

Nothing is known about the mechanism of action of SV 40 T-antigen on rRNA genes. The in vitro experiments of WHELLY et al. (1978) suggest a direct effect, either on the genes or on RNA polymerase I. One way to look at T-antigen could be to consider it a protein that increases the affinity of promoters for the RNA polymerases, an equivalent of the CAP protein in bacteria (DE CROMBUGGHE et al., 1971) that increases transcription in a nonspecific way. If this were true, then transcription by RNA polymerases II and III should also be increased, but, for the moment there is no strong evidence that this may be the case, with the exception, perhaps, of *Drosophila*, where the synthesis of different types of RNA is co-ordinated by a gene on the Y chromosome (CLARK and KIEFER, 1977). However, a number of investigators have reported a modest increase in RNA polymerase II activity in cells stimulated to proliferate (ORGANTINI et al., 1975; COX, 1976; HARDIN et al., 1976; BENZ et al., 1977; ROSSINI and BASERGA, 1978).

Since at this point I am committing the sin, reproached to others above, of abandoning myself to cosmic questions, let me continue into a final speculation for which I cannot see yet a clear experimental way.

In bacteria, the enzyme Primase synthesizes an RNA primer that is necessary for the initiation of DNA replication (see review by KORNBERG, 1979). Primase has been shown to have ATPase activity (HILLENBRAND et al., 1979) and ATPase activity has also been found in DNA helicase, a DNA unwinding enzyme (KUHN et al., 1979) and in purified SV 40 T-antigen (TJIAN and ROBBINS, 1970). Could it be that T-antigen simply stimulates transcription? When it makes rRNA, it could give the signal for growth in size, and when it makes primer RNA, the signal for DNA replication. In this respect, the suggestion of BRUN and WEISSBACH (1978) that RNA polymerase I may be involved in the initiation of DNA synthesis in mammalian cells is of considerable importance.

It is to early to say, but in the next few years, the many growth factors described in this book will have to be tested for these two functions: growth in size and DNA replication. Anyway, the time is past when we can be satisfied with general comments on RNA synthesis and protein synthesis. Both the field of growth factors and of genes regulating cell proliferation have come of age. Purified growth factors and specific genes and gene products can now be used to understand the mechanisms that regulate cell growth.

References

Abelson, H.T., Antoniades, H.N., Scher, C.D.: Uncoupling of RNA and DNA synthesis after plasma stimulation of G_0-arrested cells. Biochim. Biophys. Acta *561*, 269–275 (1979)

Ashihara, T., Chang, S.D., Baserga, R.: Constancy of the shift-up point in two temperature-sensitive mammalian cell lines that arrest in G_1. J. Cell. Physiol. *96*, 15–22 (1978a)

Ashihara, T., Traganos, F., Baserga, R., Darzynkiewicz, Z.: A comparison of cell cycle related changes in post mitotic and quiescent AF 8 cells as measured by cytofluorimetry after acridine orange staining. Cancer Res. *38*, 2514–2518 (1978 b)

Baserga, R.: A study of nucleic acid synthesis in ascites tumor cells by two-emulsion autoradiography. J. Cell Biol. *12*, 633–637 (1962)

Baserga, R.: Mitotic cycle of ascites tumor cells. Arch. Path. *75*, 156–161 (1963)

Baserga, R.: Multiplication and division in mammalian cells. New York: Marcel Dekker 1976

Basilico, C.: Temperature-sensitive mutations in animal cells. In: Advances in cancer research. Kline, G., Weinhouse, S. (eds.), vol. 24, pp. 223–266. New York: Academic Press 1977

Baxter, G.C., Stanners, C.P.: The effect of protein degradation on cellular growth characteristics. J. Cell. Physiol. *96*, 139–146 (1978)

Becker, H., Stanners, C.P., Kudlow, J.E.: Control of macromolecular synthesis in proliferating and resting Syrian hamster cells in monolayer culture. J. Cell. Physiol. *77*, 43–50 (1971)

Benjamin, T.L.: Virus specific RNA in cells productively infected or transformed by Polyoma virus. J. Mol. Biol. *16*, 359–373 (1966)

Benz, E.W. jr., Getz, M.J., Wells, D.J., Moses, H.L.: Nuclear RNA polymerase activities and poly (A) containing mRNA accumulation in cultured AKR mouse embryo cells stimulated to proliferate. Exp. Cell Res. *108*, 157–165 (1977)

Brooks, R.F.: The cytoplasmic origin of variability in the timing of S phase in mammalian cells. Cell Biol. Int. Rep. *3*, 707–716 (1979)

Brun, G., Weissbach, A.: Initiation of HeLa cell DNA synthesis in a subnuclear system. Proc. Nat. Acad. Sci. *75*, 5931–5935 (1978)

Burstin, S.J., Basilico, C.: Transformation by Polyoma virus alters expression of a cell mutation affecting cycle traverse. Proc. Natl. Acad. Sci. U.S.A. *72*, 2540–2544 (1975)

Burstin, S.J., Meiss, J.I., Basilico, C.: A temperature-sensitive cell cycle mutant of the BHK cell line. J. Cell. Physiol. *84*, 397–408 (1974)

Clark, S.H., Kiefer, B.I.: Genetic modulation of RNA metabolism in Drosophila. II. Coordinate rate change in 4 S, 5 S, and poly-A-associated RNA synthesis. Genetics *86*, 801–811 (1977)

Cohen, L.S., Studzinski, G.P.: Correlation between cell enlargement and nucleic acid and protein content of HeLa cells in unbalanced growth produced by inhibitors of DNA synthesis. J. Cell. Physiol. *69*, 331–340 (1976)

Cooper, S.: A unifying model for the G_1 period in prokaryotes and eukaryotes. Nature *280*, 17–19 (1979)

Cox, R.F.: Quantitation of elongating form A and B RNA polymerases in chick oviduct nuclei and effects of estradiol. Cell *7*, 455–465 (1976)

Darzynkiewicz, Z., Evanson, D.P., Staiano-Coico, L., Sharpless, T.K., Melamed, M.L.: Correlation between cell cycle duration and RNA content. J. Cell Physiol. *100*, 425–438 (1979)

De Crombrugghe, B., Chen, B., Anderson, W., Nissley, S., Gottesman, M., Pastan, I., Pearlman, R.: Lac DNA, RNA polymerase and cyclic AMP lac repressor and inducer are the essential elements for controlled lac transcription. Nature New Biol. *231*, 139–142 (1971)

Edwards, D.L.W., Taylor, J.B., Wakeling, W.F., Watts, F.C., Johnston, I.R.: Studies on the prereplicative phase of the cell cycle in Saccharomyces Cerivisiae. Cold Spring Harbor Symp. Quant. Biol. *43*, 577–586 (1978)

Eliceiri, G.L.: Ribosomal RNA synthesis after infection with Adenovirus type 2. Virology *56*, 604–607 (1973)

Epifanova, O.I.: Mechanism underlying the differential sensitivity of proliferating and resting cells to external factors. Int. Rev. Cytol. Suppl. *5*, 303–335 (1977)

Epifanova, O.I., Abuladze, M.K., Zosimovksa, A.I.: Effect of low concentrations of Actinomycin D on the initiation of DNA synthesis in rapidly proliferating and stimulated cell cultures. Exp. Cell Res. *92*, 25–30 (1975)

Floros, J., Ashihara, T., Baserga, R.: Characterization of ts 13 cells a temperature-sensitive mutant of the G_1 phase of the cell cycle. Cell Biol. Int. Rep. *2*, 259–269 (1978)

Fraser, R.S.S., Nurse, P.: Novel cell cycle control of RNA synthesis in yeast. Nature *271*, 726–730 (1978)

Gates, B.J., Friedkin, M.: Mid-G_1 marker proteins in 3T3 mouse fibroblast cells. Proc. Natl. Acad. Sci. U.S.A. *75*, 4959–4961 (1978)

Gazitt, Y., Deitch, A.D., Marks, P.A., Rifkind, R.A.: Cell volume changes in relation to the cell cycle of differentiating erythroleukemic cells. Exp. Cell Res. *117*, 413–420 (1978)

Graham, C.F., Morgan, R.W.: Changes in the cell cycle during early amphibian development. Dev. Biol. *14*, 439–460 (1966)

Grummt, F., Grummt, I., Mayer, E.: Ribosome biosynthesis is not necessary for initiation of DNA replication. Eur. J. Biochem. *97*, 37–42 (1979)

Hardin, J.W., Clark, J.H., Glasser, S.R., Peck, E.J. jr.: RNA polymerase activity and uterine growth: differential stimulation by estradiol, estriol, and nafoxidine. Biochemistry *15*, 1370–1374 (1976)

Hartwell, L.: Cell division from a genetic perspective. J. Cell Biol. *77*, 627–637 (1978)

Haselbacher, G.K., Humble, R.E., Thomas, G.: Insulin like growth factor: insulin or serum increase phosphorylation of ribosomal protein S6 during transition of stationary chick embryo fibroblasts into early G_1 phase of the cell cycle. FEBS Letters *100*, 185–190 (1979)

Hillenbrand, G., Morelli, G., Lanka, E., Scherzinger, E.: Cold Spring Harbor Symp. Quant. Biol. *43*, 449–459 (1979)

Howard, A., Pelc, S.R.: A nuclear incorporation of P^{32} as demonstrated by autoradiographs. Exp. Cell Res. *2*, 178–187 (1951)

Jaehning, J.A., Stewart, C.C., Roeder, R.C.: DNA dependent RNA polymerase levels during the response of human peripheral lymphocytes to phytohemagglutinin. Cell *4*, 51–57 (1975)

Johnston, G.C., Singer, R.A.: RNA synthesis in control of cell division in the yeast S. cerevisiae. Cell *14*, 951–958 (1978)

Jonak, G., Baserga, R.: The cytoplasmic appearance of three functions expressed during the $G_1 \rightarrow G_1 \rightarrow S$ transition is nucleus-dependent. J. Cell. Physiol. *105*, 347–354 (1980)

Kornberg, A: Aspects of DNA replication. Cold Spring Harbor Symp. Quant. Biol. *43*, 1–9 (1979)

Kuhn, B., Abdel-Monem, M., Hoffmann-Berling, H.: DNA helicases. Cold Spring Harbor Symp. quant. Biol. *43*, 63–67 (1979)

Lala, P.K., Patt, H.M.: Cytokenetic analysis of tumor growth. Proc. Natl. Acad. Sci. U.S.A. *56*, 173–1742 (1966)

Laughlin, C., Strohl, W.A.: Factors regulating cellular DNA synthesis induced by Adenovirus infection. 2. The effects of Actinomycin D on productive virus cell systems. Virology *75*, 44–56 (1976)

Lee, G.T.Y., Engelhardt, D.L.: Protein metabolism during growth of Vero cells. J. Cell. Physiol. *92*, 293–302 (1977).

Levine, A.J.: SV40 and Adenovirus early functions involved in DNA replication and transformation. Biochim. Biophys. Acta *458*, 213–241 (1976)

Lieberman, I., Abrams, R., Ove, P.: Changes in the metabolism of ribonuclic acid preceding the synthesis of deoxyribonucleic acid in mammalian cells cultured from the animal J. Biol. Chem. *238*, 2141–2149 (1963)

Liskay, R.M.: Genetic analysis of a Chinese hamster cell line lacking a G_1 phase. Exp. Cell Res. *114*, 69–77 (1978)

Liskay, R.M. Prescott, D.M.: Genetic analysis of the G_1 period: isolation of mutants (or variants) with a G_1 period from a Chinese hamster cell line lacking G_1. Proc. Natl. Acad. Sci. U.S.A. *75*, 2873–2877 (1978)

Malamud, D., Gonzalez, E.M., Chiu, H., Malt, R.: Inhibition of cell proliferation by Azathioprine. Cancer Res. *32*, 1226–1229 (1972)

Mauck, J.C., Green, H.: Regulation of RNA synthesis in fibroblasts during transition from resting to growing state. Proc. Natl. Acad. Sci. U.S.A. *70*, 2819–2822 (1973)

May, P., May, E., Bordé, J.: Stimulation of cellular RNA synthesis in mouse kidney cell cultures infected with SV40 virus. Exp. Cell Res. *100*, 433–436 (1976)

Mitchell, B.F., Tupper, J.T.: Synchronization of mouse 3T3 and SV40 3T3 cells by way of centrifugal elutriation. Exp. Cell Res. *106*, 351–355 (1977)

Mitchison, J.M.: The biology of the cell cycle, pp. 313. Cambridge Univ. Press 1971

Mora, M., Darzynkiewicz, Z., Baserga, R.: DNA synthesis and cell division in mammalian cell mutant temperature-sensitive for the processing of ribosomal RNA. Exp. Cell Res. 125, 241–249 (1980)

Nicolette, A. A., Babler, M.: The selective inhibitor effect of NH_4Cl on estrogen stimulated rat uterine in vitro RNA synthesis. Arch. Biochem. Biophys. 163, 656–665 (1974)

Novi, A.M., Baserga, R.: Correlation between synthesis of ribosomal RNA and stimulation of DNA synthesis in mouse salivary glands. Lab. Invest. 26, 540–547 (1972)

Organtini, J.E., Joseph, C.R., Farber, J.L.: Increase in the activity of the solubilized rat liver nuclear polymerases following partial hepatectomy. Arch. Biochem. Biophys. 170, 485–491 (1975)

Pardee, A.B., Dubrow, R., Hamlin, J.L., Kletzien, R.F.: Animal cell cycle. Ann. Rev. Biochem. 47, 715–750 (1978)

Pledger, W.J., Stiles, C.D., Antoniades, H.N., Scher, C.D.: Induction of DNA synthesis in Balb C 3 T 3 cells by serum components: re-evaluation of the commitment process. Proc. Natl. Acad. Sci. U.S.A. 74, 4481–4485 (1977)

Riddle, V.G.H., Dubrow, R., Pardee, A.B.: Changes in the synthesis of Actin and other cell proteins after stimulation of serum arrested cells. Proc. Natl. Acad. Sci. U.S.A. 76, 1298–1302 (1979)

Robbins, E., Scharff, M.D.: The absence of a detectable G_1 phase in a cultured strain of Chinese hamster lung cells. J. Cell Biol. 34, 684–686 (1967)

Ross, R., Nist, C., Kariya, B., Rivest, M.J., Raines, E., Callis, J.: Physiological guiescence in plasma derived serum. Influence of platelet derived growth in culture. J. Cell. Physiol. 97, 497–508 (1978)

Rossini, M., Baserga, R.: RNA synthesis in a cell cycle specific temperature-sensitive mutant from a hamster cell line. Biochemistry 17, 858–863 (1978)

Rossini, M., Weinmann, R., Baserga, R.: DNA synthesis in temperature-sensitive mutants of the cell cycle infected by Polyoma virus and Adenovirus. Proc. Natl. Acad. Sci. U.S.A. 76, 4441–4445 (1979)

Rossini, M., Baserga, S., Huang, C.H., Ingles, C.J., Baserga, R.: Changes in RNA polymerase II in a cell-cycle specific temperature-sensitive mutant of hamster cells. J. Cell. Physiol. 103, 97–103 (1980)

Rossow, P.W., Riddle, V.G.H., Pardee, A.B.: Synthesis of labile serum dependent protein in early G_1 controls animal cell growth. Proc. Natl. Acad. Sci. USA 76, 4446–4450 (1979)

Rovera, G., Mehta, S., Maul, G.: Ghost monolayers in the study of the modulation of transcription in cultures of CV 1 fibroblasts. Exp. Cell Res. 89, 295–305 (1975)

Scher, C.D., Shepherd, R.C., Antoniades, H.N., Stiles, C.D.: Platelet derived growth factor and the regulation of the mammalian fibroblast cell cycle. Biochim. Biophys. Acta 560, 217–241 (1979 a)

Scher, C.D., Stone, M.E., Stiles, C.D.: Platelet derived growth factor prevents G_0 growth arrest. Nature 281, 390–392 (1979 b)

Schmiady, H., Munke, M., Sperling, K.: Ag staining of nucleolus organizer regions on human prematurely condensed chromosomes from cells with different ribosomal RNA genes. Exp. Cell Res. 121, 425–428 (1979)

Schmid, W., Sekeris, C.E.: Nuclear RNA synthesis in the liver of partially hepatectomized and Cortisol treated rats. Biochim. Biophys. Acta 402, 244–252 (1975)

Schneiderman, M.H., Dewey, W.C., Highfield, D.P.: Inhibition of DNA synthesis in synchronized Chinese hamster cells treated in G_1 with cycloheximide. Exp. Cell Res. 67, 147–155 (1971)

Shilo, B., Riddle, V.G.H., Pardee, A.B.: Protein turnover and cell cycle initiation in yeast. Exp. Cell Res. 123, 221–227 (1979)

Siminovitch, L., Thompson, L.H.: The nature of conditionally lethal temperature-sensitive mutations in somatic cells. J. Cell. Physiol. 95, 361–366 (1978)

Singer, R.A., Johnston, G.C., Bedard, D.: Methionine analogs and cell division in the yeast S. cerevisiae. Proc. Natl. Acad. Sci. U.S.A. 75, 6083–6087 (1978)

Skog, S., Eliasson, E., Eliasson, Eva: Correlation between size and position within the division cycle in suspension cultures of Chang liver cells. Cell Tissue Kinet. 12, 501–511 (1979)

Smith, B.J., Wigglesworth, N.M.: Studies on the Chinese hamster line that is temperature-sensitive for the commitment to DNA synthesis. J. Cell. Physiol. *84*, 127–134 (1974)

Soprano, K.J., Dev, V.G., Croce, C.M., Baserga, R.: Reactivation of silent rRNA genes by Simian virus 40 in human-mouse hybrid cells. Proc. Natl. Acad. Sci. U.S.A. *76*, 3885–3889 (1979)

Soprano, K.J., Rossini, M., Croce, C., Baserga, R.: The role of large T antigen in SV 40 induced reactivation of silent rRNA genes in human-mouse hybrid cells. Virology *102*, 317–326 (1980)

Stiles, C.D., Capone, G.T., Scher, C.D., Antoniades, H.N., Van Wyke, J.J., Pledger, W.J.: Dual control of cell growth by somatomedins and platelet derived growth factor. Proc. Natl. Acad. Sci. U.S.A. *76*, 1279–1283 (1979)

Tjian, R., Robbins, A.: Enzymatic activities associated with a purified simian virus 40 T antigen-related protein. Proc. Natl. Acad. Sci. U.S.A. *76*, 610–614 (1979)

Tjian, R., Fey, G., Graessmann, A.: Biological activity of purified simian virus 40 T antigen proteins. Proc. Natl. Sci. U.S.A. *75*, 1279–1283 (1978)

Toniolo, D., Meiss, H.K., Basilico, C.: A temperature-sensitive mutation affecting 28 S ribosomal RNA production in mammalian cells. Proc. Natl. Acad. Sci. U.S.A. *70*, 1273–1277 (1973)

Tsuboi, A., Baserga, R.: Synthesis of nuclear acidic proteins in density inhibited fibroblasts stimulated to proliferate. J. Cell. Physiol. *80*, 107–118 (1972)

Tsukada, K., Lieberman, I.: Synthesis of ribonucleic acid by liver nuclear and nucleolar preparations after partial hepatectomy. J. Biol. Chem. *239*, 2952–2956 (1964)

Tucker, R.W., Pardee, A.B., Fujiwara, K.: Centriole ciliation is related to quiescence and DNA synthesis in 3 T 3 cells. Cell *17*, 527–535 (1979)

Weil, R.: Viral tumor antigens. A novel type of mammalian regulator protein. Biochim. Biophys. Acta *516*, 301–388 (1978)

Weil, R., Salomon, C., May, E., May, P.: A simplifying concept in tumor virology: Virus specific pleiotropic effectors. Cold Spring Harbor Symp. Quant. Biol. *39*, 381–395 (1975)

Whatley, S.A., Hill, B.T.: The relationship between RNA content, cell volume and growth potential in ageing human embryonic mesenchymal cells. Cell Biol. Int. Rep. *3*, 671–683 (1979)

Whelly, S., Ide, T., Baserga, R.: Stimulation of RNA synthesis in isolated nucleoli by preparations of simian virus 40 T antigen. Virology *88*, 82–91 (1978)

Yen, A., Pardee, A.B.: Exponential 3T3 cells escape in mid-G_1 from their high serum requirement. Exp. Cell Res. *116*, 103–113 (1978)

Yen, A., Pardee, A.B.: Role of nuclear size in cell growth initiation. Science *204*, 1315–1317 (1979)

Yen, A., Riddle, V.G.H.: Plasma and platelet associated factors act in G_1 to maintain proliferation and to stabilize arrested cells in a viable, quiescent state. Exp. Cell Res. *120*, 349–357 (1979)

Yen, A., Warrington, R.C., Pardee, A.B.: Serum stimulated 3T3 cells undertake a histidinol-sensitive process which G_1 cells do not. Exp. Cell Res. *114*, 458–462 (1978)

Zardi, L., Baserga, R.: Ribosomal RNA synthesis in WI 38 cells stimulated to proliferate. Exp. Mol. Path. *20*, 69–77 (1974)

CHAPTER 2

Survival and Growth Requirements of Nontransformed Cells

R. G. HAM

A. Definitions

This chapter describes the environmental requirements for survival and growth of nontransformed cells and the development of assay systems for studying those requirements. As a starting point, it is useful to define as precisely as possible a number of terms that will be used repeatedly.

I. Normal, Nontransformed, and Transformed Cells

As used in this chapter, "normal" refers only to cells that do not differ in any significant way from cells found in a healthy intact organism. By definition, a normal cell has an unaltered euploid karyotype (usually diploid). In addition, normal cells typically undergo cellular senescence after a finite number of doublings in culture. They also exhibit all of the properties described below for the broader category of "nontransformed" cells, of which they are a subset.

At the other extreme, "transformed" cells are those that exhibit malignant properties. Precise definition of the boundary between cells that are transformed and those that are not is difficult due to the existence of some types of cells that exhibit intermediate properties. As a first approximation, transformed cells can be regarded as those that form malignant tumors in appropriate test animals, such as nude mice (FREEDMAN and SHIN, 1974). In most cases, the ability to multiply in culture without anchorage dependence and the absence of density-dependent inhibition are also indicative of transformation (KAHN and SHIN, 1979; BOREK, 1979; CIFONE and FIDLER, 1980). However, there are some exceptions, such as normal chondrocytes and normal lymphocytes.

The term "nontransformed" is used in this chapter to describe all cells that are characterized by the absence of any obviously malignant properties. In addition to unaltered normal cells, the term also refers to cell lines that are aneuploid and continue to multiply indefinitely in culture, but that do not exhibit transformed properties under conventional testing conditions. Mouse 3T3 cells are an example. They are aneuploid and immortal, but exhibit anchorage dependence and density dependent inhibition of multiplication. In addition, they acquire "transformed" properties after exposure to oncogenic viruses. When injected into test animals they usually do not form tumors. However, tumor formation can be obtained with 3T3 cells by first attaching the cells to small glass beads and then injecting them into test animals (BOONE, 1975), or by injecting unusually large numbers of cells (KAHN and SHIN, 1979).

II. Survival and Survival Requirements

The term "survival" refers specifically to the maintenance of viability. In most cases, survival implies retention of the ability to respond by multiplication when all growth requirements (defined below) are satisfied. The assay for survival requirements described later in this chapter is based primarily on the ability to resume active multiplication at the end of the assay period. However, there are special cases, such as survival of senescent cells long after cessation of active multiplication (BELL et al., 1978), in which survival must be measured in terms of maintenance of the structural and metabolic integrity of the cells, rather than maintenance of the ability to multiply.

The term "survival requirement" refers to any member of the set of minimal environmental conditions that must be provided in order for the cells in question to remain fully viable. It is generally assumed that the growth requirements of nontransformed cells are more complex than their survival requirements, and that under minimal survival conditions nontransformed cells are fully quiescent and do not attempt to enter the cell cycle. Nontransformed cells, however, do not necessarily cease attempting to initiate cell cycles when their growth requirements are not fully satisfied, and may fail to survive in a deficient medium for that reason (SCHIAFFONATI and BASERGA, 1977).

III. Cell Cycle, Growth, Division, Multiplication, and Proliferation

Actively multiplying cells move repeatedly through a replicative cell cycle consisting of four stages designated M (mitosis), G_1 ("gap" between M and S), S (DNA synthesis), and G_2 ("gap" between S and M). A variety of environmental conditions can cause nontransformed cells to withdraw from the cell cycle into a quiescent G_1-like state that is usually designated G_0 on the basis of differences between normal G_1 events and the events that occur during reactivation of G_0 cells (BASERGA, 1976; PRESCOTT, 1976).

The terminology used to describe progress of cells through the cell cycle and the accumulation of increased numbers of cells in a culture can be quite confusing. When used to describe a process that continues over a long period of time, "growth" generally implies an increase in total mass and volume, accompanied by a proportional increase in number of cells. However, on a short term basis, "growth" can also describe an increase in cell size (mass plus volume) with no change in cell number. This occurs, for example, during the formation of "giant" cells in lethally irradiated feeder layers (PUCK and MARCUS, 1955) and in the "growth" of an oocyte as it matures.

The term "division" is also ambiguous, in that it refers primarily to the mitotic process itself, with no assurance of continued cycling or net increase in mass and volume. "Multiplication" and "proliferation" both imply a net increase in cell number, with a corresponding increase in total mass and volume, such that both daughter cells become essentially identical to their parental cell. These terms are far less ambiguous than "growth" or "division," and are used throughout this chapter despite their slightly cumbersome nature.

IV. Nutrient, Growth Requirement, Growth Factor, Mitogen, and Hormone

The terminology used to describe substances and conditions that promote cellular multiplication is also confusing. "Nutrient" is a highly restrictive term that refers specifically to a chemical substance that is taken into a cell and utilized as a substrate in biosynthesis or energy metabolism, or else as a catalyst in one of those processes. Substances that promote cellular multiplication by other means, such as regulatory interactions with cell surface receptors should not be referred to as "nutrients." However, many nutrients also have regulatory roles that appear to be distinct from their roles as metabolic substrates or catalysts.

"Growth requirement" is an extremely broad term, which can be used to refer to virtually anything that has a positive effect on cellular multiplication. In addition to nutrients and regulatory factors, cellular growth requirements include highly diverse environmental variables such as temperature and other physiological conditions, quantitative balance relationships, and the nature of the surface on which the cells are grown. The total set of growth requirements of nontransformed cells is discussed in greater detail in Sect. C of this chapter.

"Growth factor" is a particularly ambiguous term. Historically it was used by nutritional biochemists to refer to any chemical substance that promoted cellular multiplication, including common nutrients such as the B vitamins. However, current usage in cell culture is much more restrictive. A distinction is generally made between "nutrients" and "growth factors," with growth factors being defined quite specifically as non-nutritive substances that do not participate in biosynthesis, metabolism or catalysis, but instead control proliferation in a regulative manner. All such growth factors that have been studied in detail have been shown to interact with specific cellular receptors, although it has not yet been demonstrated conclusively that the growth factor-receptor interactions are the controlling step in the proliferative response.

In the current literature, the term "growth factor" is widely used, both in the naming of individual multiplication-promoting substances such as "epidermal growth factor" and "fibroblast growth factor" and as a collective generic term for all such substances (GOSPODAROWICZ and MORAN, 1976). This usage has become so widespread that it must be accepted, as it has been in this chapter. However, it is important to avoid confusion with older definitions, particularly when reading the older literature.

"Mitogen" refers rather specifically to substances that stimulate quiescent cells to leave G_0 and enter the cell cycle. Although the term "mitogen" literally means something that causes mitosis, the actual response that is measured is frequently DNA synthesis rather than cell division *per se*. As used by most investigators, "mitogen" refers to agents that evoke short term responses, with no assurance of continued long term proliferation of the affected cells.

"Hormones" are classically defined as chemical substances that are transported through body fluids and affect target cells at locations remote from the cells that produce them. Many of the regulatory growth factors described above are similar in their mechanisms of action to classical peptide hormones, and there is a growing tendency by investigators who replace serum with mixtures of growth factors and classical hormones to refer to the growth factors as "hormones" (BOTTENSTEIN et al., 1979; Sato and Ross, 1979; Rizzino et al., 1979; Scher et al., 1979 a). Although this practice may ultimately prove to be valid, the in vivo functions of many of the growth factors are not yet well understood. Therefore, in this chapter, the term "hormone" will be used only to refer to substances with well-defined endocrinological roles.

Finally, a purist might be tempted to reject the terms "growth factor" and "growth requirement" in favor of terms that incorporate the words "multiplication" or "proliferation." However, with the exception of "multiplication-stimulating activity" or "MSA" (Dulak and Temin, 1973), such terms are not generally used, and will not be introduced here.

B. Assay Systems for the Measurement of Growth Requirements

I. Background

The history of studies of cellular growth requirements can be subdivided into three relatively distinct eras. The first, which extended from the earliest beginnings of cell culture to the early 1950's, was characterized by the use of assays based on growth or survival responses of normal cells in primary culture. Substantial advances were made during that period, including the use of dialyzed supplements to permit precise analysis of small molecular growth requirements (Fischer, 1941), and the development of medium 199 (Morgan et al., 1950), which is still widely used 30 years later.

The second era began when permanent cell lines such as mouse L (Earle, 1943; Sanford et al., 1948) and HeLa (Gey et al., 1952) became generally available. In terms of convenience and reproducibility, cellular multiplication assays based on the established cell lines were far superior, and they almost totally displaced primary cultures in the study of cellular growth requirements. In addition, the aneuploid permanent lines possessed a genetic instability that permitted them to undergo substantial evolutionary adaptation and simplification of their growth requirements in response to selective pressure.

Many important assay techniques that are still widely used were developed during this period. Replicate assay techniques were developed that permitted precise comparison of experimental culture media under strictly controlled conditions (Evans et al., 1951). Cellular dispersal with trypsin (Moscona, 1952) greatly simplified the preparation of suspensions of viable cells for use in such assays. Initially, cellular multiplication assays could only be done with relatively large numbers of cells, and the first media that were developed had to be "conditioned" by the cells that were inoculated into them before they would support satisfactory multiplication. However, within a short time, clonal growth assays were developed, in which the multiplication of isolated individual cells could be studied (Puck and Marcus, 1955; Puck et al., 1956). Such assays permitted a more critical measurement of the adequacy and freedom from toxicity of test media in the absence of conditioning and with minimal carryover in the cellular inoculum of nutrients from the previous culture.

Spectacular progress was made toward an understanding of the growth requirements of established cell lines, and by the early 1960's, protein-free synthetic media had been developed that would support continuous multiplication of many different types of permanent cell lines with transformed properties (reviewed by HIGUCHI, 1973; KATSUTA and TAKAOKA, 1973). However, these media would support satisfactory multiplication of nontransformed cells only when supplemented with large amounts of serum or other undefined additives. Initially, it was generally assumed that the additional requirements of nontransformed cells would be relatively trivial (EAGLE, 1963). However, it gradually became evident that the permanent lines had undergone very extensive adaptation and simplification of their growth requirements, and that major new studies would be needed to evaluate the growth requirements of nontransformed cells (HAM, 1974a).

The third era in the study of cellular growth requirements, which is still gaining momentum, is characterized by the use of nontransformed cells to measure cellular multiplication responses. For a period of nearly 20 years, study of the growth requirements of nontransformed cells could legitimately be classified as a "neglected area of modern biology" (HAM, 1974a). However, since the early 1970's, more and more laboratories have turned their attention to nontransformed cells, and impressive results are beginning to accumulate.

One of the major conclusions that is emerging is that the classical assay systems that were developed to analyze the growth requirements of highly adapted established lines are inadequate for analysis of the more complex growth requirements of nontransformed cells. These inadequacies are discussed in the next section, together with a discussion of means to improve the assay systems.

II. Inadequacy for Nontransformed Cells of Classic Assay Systems

Cellular multiplication requires a complex mixture of nutrients and growth factors, together with an acceptable physiological environment. In order to analyze specific multiplication responses, it is necessary to devise assay systems that satisfy cellular needs for all except one of the essential substances or conditions, and that yield a positive multiplication response when the missing substance or condition is provided.

For many (but not all) types of nontransformed cells, reasonably good proliferation was obtained by adding relatively large amounts of serum to synthetic media that had been developed for highly adapted permanent lines of transformed cells. Therefore it was assumed that definition of the growth requirements of the nontransformed cells would be a relatively simple matter of identifying and supplying in pure form the growth-promoting substances found in serum. Unfortunately, this has not proven to be as simple a task as was once anticipated. When viewed retrospectively, two major reasons can be identified: 1) adaptive simplification of the growth requirements of the permanent cell lines used in the initial studies of growth requirements; and 2) the multiplicity of different functions that serum serves in promoting multiplication of nontransformed cells.

1. Evolutionary Adaptation of Permanent Lines

It is now clear that much of the difficulty that has been encountered in attempts to develop defined media for nontransformed cells has been due to failure to appreciate the extent of the evolutionary adaptation and simplification of growth requirements that occurred during the culture histories of the permanent lines that were used for the development of synthetic media in the late 1950's and early 1960's. Since these media are still widely used as starting points for contemporary studies of the growth requirements of nontransformed cells, it is appropriate to explore this point in some detail.

Major selection and adaptation occurred at least twice during the development of the mouse L cell line, which has had a greater influence on medium development than any other single line (reviewed by HAM, 1974a). The first period of adaptation occurred during initial establishment of the culture. It was only after prolonged maintenance as a slowly growing explant cultured in a totally undefined natural medium (consisting of a mixture of serum, embryo extract, and saline), followed by extended treatment with a carcinogen, that the rapidly proliferating mouse L cell line emerged (EARLE, 1943). Later, mouse L cells were adapted to growth in protein-free synthetic media in a number of different laboratories. This was almost invariably achieved by progressive reduction of the serum concentration in a manner that provided a strong selective advantage to cells that were able to grow with smaller amounts of serum and ultimately with no serum at all.

At a slightly later time, many other cell lines were shown to have very similar growth requirements. However, the cultures whose requirements were analyzed were almost invariably initiated in media that had been previously developed for mouse L cells (or HeLa cells, whose early history was somewhat the same). Such cells also underwent adaptation, and erroneous conclusions concerning the simplicity and uniformity of cellular growth requirements were reached. These conclusions are still plaguing investigators seeking to analyze the growth requirements of nontransformed cells.

2. Role of Serum in the Multiplication of Nontransformed Cells

Serum is an extremely complex mixture that contains substances released into it by virtually every type of cell in the body. In addition, it undergoes further modifications and additions during the clotting process. The components of serum, many of which are present at very low concentrations, appear to interact with many different cellular growth requirements, either directly or indirectly. Experience in this and other laboratories is making it clear that the number of different functions that serum performs in promoting cellular multiplication is far larger than has been realized until quite recently. (For detailed discussions of this point from two very different perspectives, cf. BARNES and SATO, 1980; HAM and McKEE-HAN, 1978b).

In retrospect, it is clear that most of the difficulties encountered in early attempts to analyze the growth requirements of nontransformed cells were due to experimental designs that did not take into account the complexity and diversity of roles that serum plays in promoting their multiplication. The number of different molecular species from serum that contribute to the overall growth response observed in classical assay systems is so great that the individual activities are not clearly separated by conventional fraction procedures. This results in small amounts of biological activity appearing to be "smeared" across many different fractions due to fortuitous combinations of the various essential molecules in the impure fractions and makes it very difficult to develop a highly specific assay system for any one of the activities.

3. Problems Awaiting Solution for Nontransformed Cells

As a minimum, contemporary studies seeking to replace serum and define fully the growth requirements of nontransformed cells must take into account all of the following:

a) The growth requirements of the cell lines that were used to develop most of the media that are currently in widespread use are not representative of cells in general or even of non-adapted transformed cells. Recent viral transformants and freshly established tumor lines that have not been subjected to selective pressure generally will not multiply in conventional media in the total absence of serum, and at least superficially, they appear to have growth requirements that are almost as complex as those of nontransformed cells.

b) The growth requirements of diverse types of nontransformed cells are far less uniform than those of adapted established lines. Recent studies of various types of cells, both transformed and nontransformed, that have not undergone adaptation have revealed extensive individuality of growth requirements. Studies on the replacement of serum with various mixtures of hormones and growth factors have shown that a different mixture is required for serum-free growth of each cell type tested (BOTTENSTEIN et al., 1979; BARNES and SATO, 1980). Studies in the author's laboratory have shown that fibroblasts from four different avian and mammalian species have very different quantitative growth requirements (HAM and MCKEEHAN, 1978a, b) and that even human skin and lung fibroblasts have slightly different requirements (HAM, 1980). Also the growth requirements of human epidermal keratinocytes and human fibroblasts are very different from each other, so much so that a medium optimized for growth of either type of cell is actually selective against growth of the other type at low serum protein concentrations (PEEHL and HAM, 1980b; HAM, 1980). This concept is developed further in Sects. C.V.2 and C.VII.2.

c) The total set of growth requirements of nontransformed cells goes far beyond nutrients and growth factors in the culture medium. Recent studies with nonadapted cells have shown serum-replacing effects of changes in the culture environment and culture technique that have nothing directly to do with the composition of the culture medium. Major benefits have been obtained by modifying the culture surface on which the cells are grown and by use of a gentle low temperature trypsinization technique. Growth requirements related to subculturing technique and cellular attachment to the substrate are discussed in detail in Sect. C.II.

d) Complex quantitative interactions occur among components of the synthetic medium and components of serum (or the hormones and growth factors that are used to replace serum). The amount of serum protein needed for cellular multiplication is clearly related to concentrations of components of the culture medium such as divalent cations (MCKEEHAN and HAM, 1978a) and pyruvate or other 2-oxocarboxylic acids (MCKEEHAN and MCKEEHAN, 1979a). In serum replacement studies, the amounts and kinds of hormones and other growth factors needed to replace serum are clearly affected by the composition of the basal medium that is used (WU and SATO, 1978). This concept is developed further in Sect. B.III and C.V.

e) Many of the biologically active components of serum are present at concentrations that are too low to detect directly by any means other than their biological activity. Although serum contains on the order of 50 mg/ml of macromolecular constituents, many of the multiplication-promoting substances that it contains appear to be present at only nanogram per ml concentrations (less than one part per million of the total protein). As an example, purified platelet-derived growth factor has a specific activity 20 million times that of whole serum (ANTONIADES et al., 1979). Many of the early attempts at purification of serum growth-promoting activity were doomed to failure simply because they assumed that the activity would be associated with a specific major serum protein such as fetuin or serum albumin.

f) Although serum masks a multitude of different deficiencies in assay systems, there are some very serious deficiencies that are not overcome even by very high concentrations of serum. As a historical accident, current cell culture technology and cell growth assays are organized primarily around wound-healing responses rather than around multiplication of the types of cells that cycle most frequently in vivo. The prototype cells that were initially studied were mostly fibroblastic, and the growth supplement that was used with the greatest frequency was serum, which occurs in vivo only in extremely small amounts and in direct response to wounding and blood clotting. Rapidly cycling epithelial cells multiply poorly, if at all, in most currently available culture systems. A detailed study of the multiplication requirements of human epidermal keratinocytes in the author's laboratory has revealed that their quantitative growth requirements differ greatly from those of human fibroblasts. However, in an appropriately optimized medium (MCDB 151), the keratinocytes can be grown quite well with small amounts of dialyzed serum as their only supplement (PEEHL and HAM, 1980 a, b; HAM, 1980).

Preliminary studies with human prostatic epithelial cells appear to show that they require a quantitatively different medium from the optimum medium for either fibroblasts or keratinocytes (LECHNER et al., 1980; HAM, 1980). It is likely that other types of glandular and epithelial cells that are currently very difficult to culture will also require media so different from conventional media that serum alone cannot fully compensate for the inadequacies of the current media. The fact that such cells proliferate so well in vivo clearly indicates that the primary problem lies in current culture media and technology, and not in any inherent inability of the cells to continue cycling.

III. Holistic Approach to Cellular Growth Requirements

The total set of variables that affect multiplication has not yet been defined with certainty for any type of nontransformed cell. However, substantial progress is currently being made toward that goal in several laboratories. As emphasized in the preceding discussion of deficiencies of classical assay systems, the growth requirements of nontransformed cells are proving to be significantly more complex than originally anticipated, and also to include many complex interactions. As a minimum, the significant variables that affect multiplication of nontransformed cells include, but are not necessarily limited to the following:

1) The maintenance of a suitable physiological environment, including the correct temperature, balanced concentrations of essential inorganic ions, suitable pH and osmolarity, adequate buffering, an acceptable redox potential, partial pressures of oxygen and carbon dioxide within acceptable ranges, and freedom from toxicity;

2) The availability of all nutrients needed for biosynthesis, energy metabolism, and biochemical catalysis, including appropriate inorganic trace elements;

3) The availability of a culture surface that satisfies the need for anchorage exhibited by most nontransformed cells, and that has an optimal distribution of surface charges;

4) Utilization of gentle procedures to minimize cellular damage during subculturing, or else provision in the cellular environment of "factors" whose only functions appear to be to neutralize the agents used for cellular dispersal and to help the cells overcome the deleterious effects of the subculturing process;

5) The availability of appropriate carriers for hormones, trace elements and lipids in order to provide adequate supplies of these substances without toxicity;

6) The availability of appropriate attachment factors for cells that do not produce adequate amounts of such factors for themselves;

7) The availability of appropriate hormones and hormone-like macromolecular growth factors, whose roles appear to be primarily regulatory rather than nutritional;

8) Precise quantitative balance of the concentrations of all nutrients, ions, hormones and other growth factors (this has repeatedly proven to be as important as the qualitative presence or absence of specific components of the culture medium);

9) The occurrence of complex regulatory interactions among the various components of culture media and the culture environment, such that it is often possible to satisfy specific growth requirements by more than one means.

The multiplicity of different ways in which cellular multiplication can be affected and the complexity of the apparent interactions that occur among the environmental variables that affect multiplication make it useful to take a holistic approach to cellular growth requirements (HAM and McKEEHAN, 1978a, b, 1979). In this approach, all variables, qualitative or quantitative, that affect cellular multiplication, no matter how diverse they may seem, are viewed as a single interacting set, so constituted that a change in any one member of the entire set can potentially alter cellular growth responses to any other member of the set.

When viewed from this perspective, seemingly conflicting data on improvement of cellular growth by highly diverse means begin to make sense. Thus, for example, a cellular requirement for a particular macromolecular factor from serum might be altered by each of the following: 1) quantitative changes in the amounts of specific nutrients in the culture medium; 2) addition of specific hormones to the culture medium; 3) alteration of the concentration of divalent cations in the culture medium; 4) and treatment of the cells with reagents that alter intracellular levels of cyclic nucleotides or sodium ion.

IV. Development of Specific Assays for Individual Growth Requirements

While viewing the total set of growth requirements holistically, it is also necessary to develop a highly specific assay for every member of that set. The ideal assay system for the measurement of a cellular growth requirement is one that is totally defined and fully satisfies every growth requirement of the cell in question except the one being measured. Unfortunately, such an assay system becomes available only after the total set of growth requirements is already known. Thus, in the design of assay systems that are to be used to direct the isolation and purification of specific growth-promoting substances from natural sources, such as serum or organ extracts, some degree of improvising is always necessary.

Since the chemical identity of the active substance is not known until after it has been purified and analyzed chemically, the assay system must be based on biological activity. The main problem that is usually encountered in assays based on cellular multiplication is dependence of the multiplication on more than one component of the crude starting material. There are several different approaches that can be taken to overcome this problem.

1. Background Media Lacking a Single Growth-Promoting Substance

If only two or three components are needed, it is often possible to separate them completely enough during preliminary fractionation so that adequate specificity can be achieved by adding fractions containing all but one of the activities to the background medium used in the assay for the remaining activity. Alternately, it is sometimes possible to employ a naturally-occurring undefined supplement that contains all except one of the activities found in the more complete supplement. An example of this is the use of plasma-derived serum in the basal medium used to assay for platelet derived growth factor (Ross, this volume).

2. First Limiting Factor

In cases where the complexity of the crude mixture of growth-promoting substances is high and no alternate supplements are available that lack a single activity, it is often useful to use the first limiting factor approach. This approach is based on the assumption that reduction of the overall concentration of a complex undefined mixture results in one particular component of that mixture becoming rate-limiting for cellular multiplication before any of the others. When this happens, the other components are still at concentrations that will support some multiplication (or DNA synthesis) when the rate-limiting component is added from alternate sources, or from fractions of the crude mixture. The growth response that is obtained is often sufficiently specific so that the first limiting factor can be isolated and identified.

Once this has been accomplished, the newly identified growth factor is added to the medium in pure form and the concentration of the crude mixture is reduced further until another component of it becomes rate-limiting for cellular multiplication. The entire process is then repeated, leading to isolation and identification of the second limiting factor. In theory, this process can be continued until every component of the original undefined supplement that is needed for cellular multiplication has been identified and added to the medium in a pure form.

3. Replaceable Requirements

The first limiting factor approach is also very useful in analyzing alternate ways of satisfying the same requirement. For example, with the concentration of serum protein reduced to a level that supports suboptimal growth, it is possible to search for serum-sparing effects of qualitative and quantitative adjustments in the composition of the synthetic portion of the medium and of other modifications in the culture system. Experience in the author's laboratory has shown that very substantial reductions in the total amount of serum protein needed for optimal growth can often be achieved in this manner (McKeehan et al., 1977; Peehl and Ham, 1980 b; Shipley and Ham, 1981; Agy et al., 1981). The functions of serum protein that can be replaced by optimizing the culture medium and the culture system have been designated operationally as "replaceable" functions of serum (cf. Ham and McKeehan, 1978 b, for a detailed discussion of such requirements).

4. Sequential Depletion

If a very sensitive measurement that will detect small amounts of cellular multiplication or DNA synthesis is available, it is possible to employ a variation of the first-limiting factor approach. This variation assumes that rapidly multiplying cells have small intracellular stores of everything that they need for multiplication, and that they continue to multiply for a short time after the undefined supplement has been removed. In addition, the stores are assumed to be depleted unequally so that only one, or a few, of the components supplied by the undefined supplement initially become rate limiting. If this is the case, a limited amount of additional multiplication (or DNA synthesis) can be achieved by supplying the first limiting component, or components, from alternative sources or from fractions of the supplement. Such a multiplication response can then be used as the basis for an assay system leading to isolation and identification of the first-limiting factor(s). In theory, this approach can also be used to identify each additional factor that becomes rate limiting after the previous one is added back in pure form to the test system. An example of this approach is the use of serum-starved chicken embryo fibroblasts as the assay system for the isolation of multiplication stimulating activity (MSA) from rat liver cell conditioned medium (Dulak and Temin, 1973).

5. Systematic Testing of Suspected Growth-Promoting Substances

Another approach that is sometimes used in conjunction with the depletion technique described above is systematic testing of mixtures of defined substances that are known or suspected to have growth-promoting activity for at least some types of cells. This approach has been used particularly successfully in Gordon Sato's laboratory, where serum-free multiplication of many different types of cells has been achieved through the addition of mixtures of hormones, growth factors, carrier proteins, attachment factors, and trace elements to conventional culture media (Hayashi and Sato, 1976; Bottenstein et al., 1979; Rizzino et al., 1979; Barnes and Sato, 1980).

V. Types of Measurement and Analysis of Data

If an adequate degree of specificity is achieved, it is normally quite easy to tell by visual inspection which cultures are multiplying well and which are not. However, visual impressions alone are not adequate for quantification of results and comparison of specific activities of various multiplication-promoting fractions. Many different approaches to quantitative measurement are possible. Short term or long term responses can be measured, sparse or dense cultures can be used, and the actual measurements can be based on DNA synthesis or increase in cell number (or in special cases on percentage of inoculated cells that form colonies). In addition, measurements can either be done only at the end of the assay to yield data on total multiplication per assay period, or repeatedly to yield data on rate of multiplication. The advantages and disadvantages of these various approaches are discussed below.

1. Short-Term Vs Long-Term Multiplication

The only true test of the adequacy of a culture medium is its ability to support sustained cellular multiplication over a long period of time. Short term tests do not rule out the possibility that small amounts of an essential growth requirement could have been present in the cellular inoculum, thereby permitting the cells to complete a limited number of multiplication cycles before the required material is totally depleted. Thus, as a final test of adequacy, it is very important to perform long term multiplication assays.

For routine testing, however, long term assays have some distinct disadvantages, including the length of time that they require for completion and the extra labor that is involved in changing media and/or subculturing during the assay period. Short term testing generates data much more quickly and with a minimum of effort. However, there is a substantial risk that certain growth requirements may not be detected due to failure to exhaust multiplication-promoting activity that was contained within the initial cellular inoculum.

A third option, which is currently used in the author's laboratory, allows the cells a period of 1–3 days to deplete carryover nutrients and growth factors and to finish cycles that were in progress or committed at the time of inoculation, and then measures continuing multiplication in terms of incorporation of ^{3}H-thymidine into DNA during a 24 h labeling period. This delayed thymidine incorporation assay yields results that closely parallel those of longer term clonal growth or serial monolayer passage assays in a much shorter time. Only the most significant results from the delayed thymidine incorporation assays need to be verified with long term cellular multiplication assays.

2. Sparse Vs Dense Cultures

The requirements for multiplication of nontransformed cells in very sparse (clonal) cultures and in very crowded (dense monolayer) cultures both tend to be more complex than these for multiplication of subconfluent monolayer cultures. These differences make it desirable to test the adequacy of culture media at several different cellular densities, including clonal growth assays, in which well-separated single cells (generally about 100 per 5 ml of medium in a 60 mm Petri dish) are tested for their ability to form large colonies, and crowded monolayer assays, in which the parameter that is measured is the highest density (cells/cm^2) that can be achieved.

Under clonal growth conditions, the volume of medium per cell is very large, such that the cells are totally unable to "condition" the medium. Thus, growth is obtained only when the medium as formulated, and without modification by the cells, is totally adequate to support cellular multiplication. In addition to satisfying all of the growth requirements of subconfluent monolayer cultures, media for clonal growth must be free from toxicity and must provide intermediary metabolites such as carbon dioxide, pyruvate (or other 2-oxocarboxylic acids), and non-essential amino acids that tend to be lost into the surrounding medium (cf. Sect. C.VIII.1).

At the other extreme, most nontransformed cells are quite sensitive to density-dependent inhibition of multiplication and require high levels of serum supplementation, or else elevated levels of special growth factors, such as platelet derived growth factor (PDGF), to multiply at high cell densities (Ross, this volume). Depending on the purpose that a particular medium is to be used for, assays with a clonal, subconfluent, or very crowded inoculum may be the most important. However, whenever feasible it is desirable to develop culture media that will support optimal multiplication over the full range of population densities, including both extremes.

3. DNA Synthesis Vs Increase in Cell Number

The final test of whether or not net cellular multiplication is occurring is whether or not the total number of cells is increasing. However, measurements based on numbers of cells tend either to be tedious and time-consuming or else to require very expensive specialized equipment such as Coulter counters or computerized TV scanning. For laboratories that are already equipped to do isotope counting, it is often much more convenient to measure incorporation of ^{3}H-thymidine into cold TCA insoluble material (presumably DNA).

Although convenient and very useful for exploratory studies, measurement of thymidine incorporation as a criterion of cellular multiplication has some severe limitations. First of all, anything that shifts the balance relationship between *de novo* synthesis of thymidine nucleotides and utilization of exogenously supplied thymidine can falsely appear to affect cellular multiplication. Thus, for example, an inhibitor of thymidylate biosynthesis, such as 5-fluorodeoxyuridine or aminopterin could appear to be a potent mitogen in an appropriate background medium. In addition, DNA synthesis is only one part of the cell cycle. Cells stimulated to enter DNA synthesis may or may not complete the cycle that they have started, and those that do so may or may not initiate a second cycle. Also, it is possible to visualize a steady state condition (due either to contact inhibition or a restricted supply of a stable nutrient) in which cellular proliferation exactly matches cellular death and a constant total population is maintained. In such a system, anything that increases the rate of cellular death would also increase the rate of cellular multiplication based on ^{3}H-thymidine incorporation, even though the steady state total population would remain unchanged. Finally, DNA-damaging agents, such as ultraviolet irradiation, that stimulate "unscheduled" repair synthesis of DNA can increase thymidine incorporation into non-cycling cells.

^{3}H-thymidine incorporation is used extensively in many laboratories, including that of the author, for preliminary surveys. However, important results should always be verified with assays based on actual cellular multiplication. The clonal growth assay (PUCK and MARCUS, 1955; PUCK et al., 1956; HAM and PUCK, 1962;

Ham, 1972) is particularly useful for final verification of the adequacy of a culture medium (Ham and McKeehan, 1978 a, b). The cellular inoculum is small, such that there is very little carryover of growth-promoting substances from previous cultures and virtually no conditioning of the test medium. Production of a large colony requires at least 10 population doublings, which is sufficient for thorough depletion of any residual growth-promoting substances that are introduced with the cellular inoculum. In most cases, this number of divisions is also sufficient to demonstrate the ability of the test medium to support "long term" growth. The plating efficiency (percentage of inoculated cells that develop into macroscopically visible colonies) is a sensitive index of the freedom from toxicity of the initial medium. The number of cells per colony gives an accurate measurement of total growth, and also permits approximate calculation of exponential growth rate. If extreme precision is not needed, photometrically measured colony area can be used in place of total number of cells per colony for convenient comparison of growth responses to various media (Ham, 1963; McKeehan et al., 1977).

One limitation of the clonal growth assay, however, is that it does not effectively measure the ability of test media to support multiplication of dense cultures. Media that are found to be optimal in clonal growth assays may fail to support dense cultures either because of depletion of growth-promoting substances by the larger number of cells, or because such media lack specific factors that are needed to overcome density-dependent inhibition of multiplication. Thus, any medium that is being designed for use with dense cultures needs to be tested for its ability to support continuing cellular multiplication through several sequential monolayer passages at high cell densities.

4. Analysis of Data Based on Total Multiplication

Precise analysis of a cellular multiplication response requires testing the variable that is responsible for the response over as wide a range of values as is practical. In order to keep the terminology manageable, the following discussion is limited to specific growth-promoting substances but the same principles also apply to all other variables that affect cellular multiplication. A typical positive multiplication response can be divided into three distinct parts: 1) the positive response itself, in which cellular multiplication is positively correlated with the concentration of the stimulatory substance; 2) a plateau region, in which the multiplication response is saturated and essentially unaffected by changes in the concentration of the substance (zero order response); and 3) an inhibitory range, in which there is a negative correlation between cellular multiplication and the concentration of the test substance.

Depending on circumstances, the concentration selected for use in future media can be at any of several points along the response curve. If a general purpose medium is being designed, it is usually desirable to select a concentration that is as far separated as possible both from deficiency and from inhibitory effects (i.e., the geometric mean of the range of concentrations that support optimal multiplication). This can be achieved conveniently by selecting the midpoint of the plateau of optimal response on a semilog plot with concentration on the logarithmic scale and multiplication response on the linear scale (Ham and McKeehan, 1978 a).

In cases where the medium must support a very large cellular population, it may be desirable to select a value near the upper end of the optimum range, particularly if the optimum range has been determined in a clonal growth assay. If the substance that is being tested is expensive or scarce, or if there is a potential problem of undesirable interactions with other components of the overall test system, it may be preferable to select a value near the lower end of the optimum range. Finally, when studying replacement of the substance or its interaction with other components of the medium, one must select a concentration located in the ascending portion of the response curve where multiplication is suboptimal and responsive to changes in the concentration of the substance in question. This permits additive effects of other variables that may interact with the variable being tested to be detected.

5. Analysis Based on Rate of Multiplication

In cases where cellular multiplication has not stopped or slowed due to causes such as density-dependent inhibition, accumulation of inhibitory substances, or depletion of essential growth-promoting substances, total multiplication during the assay period can be used to determine approximate multiplication rate. However, because of the exponential nature of unrestricted cellular multiplication, it is necessary to convert the data to number of doublings ($\log_2$ multiplication) in order to obtain an accurate picture of multiplication rate.

For precise analysis of multiplication rate, it is desirable to take several measurements during the multiplication period and to eliminate the deviations from exponential that occur early in the assay period due to slow resumption of multiplication after the cultures are inoculated, and late in the assay period, due to general depletion of the culture medium, decomposition of unstable nutrients and growth factors, or the beginnings of density dependent inhibition.

Cellular multiplication responses often appear to be quite different when viewed from the perspective of multiplication rate rather than total amount of multiplication, particularly in long term assays that involve many cell cycles. For example, in a clonal growth assay that involves ten population doublings and yields an average colony size of 1,024 cells, a 10% increase in multiplication rate will permit eleven population doublings in the same time period and result in colonies that are twice as large (2,056 cells/colony). Similarly, reduction of multiplication rate to one half the original value will reduce colony size by a factor of 32 (from large colonies containing an average of 1,024 cells per colony to minute, barely visible colonies containing only 32 cells per colony). Thus, most of the responses that are dealt with in conventional clonal growth assays actually involve quite small changes in overall rate of cellular multiplication.

Thymidine incorporation data can also provide information on rate of multiplication in certain cases. Such data are subject to all of the reservations expressed above about short-term assays and effects of factors other than multiplication rate on thymidine incorporation. In addition, it is necessary to assume that there is no synchrony of DNA synthesis, that length of the S phase is unaltered by different growth rates, and that all cells in the population are responding reasonably uniformly to the variable whose effect on multiplication rate is being analyzed. If all of these conditions are met, and if the total labeling period is short enough so that no cell is labelled during more than one cycle, the total incorporation of tritiated thymidine will be approximately a linear function of the rate of population doubling and therefore usable as input data for the kinetic analysis of multiplication rate effects described below (ELLEM and GIERTHY, 1977).

6. Multiplication Rate Kinetics

Recent studies in several laboratories have shown that it is possible to apply the principles of Michaelis-Menton enzyme rate kinetic analysis to the analysis of cellular multiplication rate responses (ELLEM and GIERTHY, 1977; MCKEEHAN and HAM, 1978a; MCKEEHAN and MCKEEHAN, 1979a, b; LECHNER and KAIGHN, 1979a, b, 1980; LECHNER et al., 1980). Lineweaver-Burke double reciprocal plots of multiplication rate versus concentration of growth-promoting substance yield straight lines that can be interpreted in terms of classical Michaelis-Menton formulations. Extrapolation of the lines yields the concentration of the growth-promoting substance that supports a half maximal rate of multiplication (equivalent to K_m in enzyme rate kinetics) and a theoretical maximum multiplication rate (equivalent to V_{max} in enzyme rate kinetics).

Kinetic analysis of multiplication rates is particularly useful for analyzing complex interactions among growth-promoting substances. For example, if a small stimulation is observed by adding a particular growth factor to a medium containing an amount of serum protein that is rate limiting for cellular multiplication, it is not immediately clear whether the growth factor partially replaces the need for serum protein or enhances multiplication independently of the amount of serum protein that is present. This distinction can be made by determining K_m and V_{max} values for cellular multiplication responses to serum protein in the presence and absence of the growth factor. If part of the need for serum has been replaced, the K_m value for the remaining essential functions of serum protein will be smaller and the V_{max} that is approached with high concentrations of serum protein will be unchanged. On the other hand, if the enhanced multiplication is independent of serum replacement, K_m values for serum will be unchanged, but multiplication rates will be greater at all levels of serum protein and the V_{max} value for serum protein will be increased by the presence of the factor. This type of analysis has been used to study interactions among calcium, magnesium, and serum protein (MCKEEHAN and HAM, 1978a), 2-oxocarboxylic acids and serum protein (MCKEEHAN and MCKEEHAN, 1979a), EGF and serum protein (LECHNER and KAIGHN, 1979a), EGF and calcium (LECHNER and KAIGHN, 1979b; LECHNER et al., 1980; MCKEEHAN and MCKEEHAN, 1979b); and EGF and phorbol esters (LECHNER and KAIGHN, 1980). Kinetic analysis of thymidine uptake rates has been used to analyze growth responses involving different kinds and amounts of serum and different cell densities (ELLEM and GIERTHY, 1977).

VI. Systematic Analysis of Growth Requirements

With currently available assay systems and cell culture technology, it is now feasible to undertake a total analysis of the growth requirements of virtually any type of nontransformed cell that has retained the capacity to multiply. The approaches that have been taken in different laboratories vary significantly in their areas of primary emphasis (e.g., optimization of basal medium versus replacement of serum requirements with hormones and growth factors). However, there is a distinct trend toward a more global approach at present (Wu and Sato, 1978; Rizzino et al., 1979; Shipley and Ham, 1980; Walthall and Ham, 1980; Phillips and Cristofalo, 1980a, b).

The total set of growth requirements is not yet known with certainty for any type of nontransformed cell, although that knowledge appears to be imminent for a few types of cells currently being studied intensively, including human diploid fibroblasts and mouse 3 T 3 cells. The suggested sequence of steps presented below is a composite of approaches that have been used and are projected in the future for a variety of different cell types, both in the author's laboratory and in other laboratories.

1. Obtain growth in vitro of the cell type of interest by any means necessary. A variety of different kinds of media and different kinds of supplements, including sera from diverse species, pituitary extract, embryo extract, etc., should be tested in view of the extensive individuality of cellular growth requirements discussed earlier in this chapter. This is particularly true for epithelial cells that cycle very readily in vivo, but cannot be grown well with existing cell culture technology.

Innovative approaches may be needed for some cell types. For example, it may be necessary to use an irradiated feeder layer (Puck and Marcus, 1955; Rheinwald and Green, 1975). Alternately, it may be necessary to provide an acceptable substitute for the basement membrane that the particular type of cell normally grows on, such as a floating collagen raft (Michalopoulos and Pitot, 1975; Emerman and Pitelka, 1977; Reid and Rojkind, 1979).

The literature contains many references to novel means of stimulating multiplication of cells that are difficult to culture. These include the use of agents such as cholera toxin that enhance intracellular cyclic AMP levels (Green, 1978; Pruss and Herschman, 1979), alterations of intracellular levels of ions (Cone and Cone, 1976; Koch and Leffert, 1979a, Rozengurt, 1979), and enhancement of effects of growth factors with tumor promoters such as phorbal esters (Frantz et al., 1979). While the agents used in such studies are not naturally occurring growth-promoting substances, they may be useful in providing an initial level of multiplication that is sufficient to permit assays for the natural mitogens. Such agents are discussed in greater detail in Sect. C.IX.

2. Modify media, supplements, and culture techniques as needed to obtain clonal growth. This is normally relatively easy to achieve once adequate monolayer growth has been accomplished. Special attention should be given to metabolic intermediates such as carbon dioxide, pyruvate, and other 2-oxocarboxylic acids, and non-essential amino acids that are frequently needed by sparse but not crowded cultures (cf. Sect. C.VIII.1). If nothing else works, it is generally possible to obtain clonal growth through the use of media conditioned by monolayers of the cells in question. This can be done as a temporary expedient until Step 5 below.

3. Survey readily available media and supplements and select the combination that supports the best clonal growth. Since quantitative growth requirements vary widely from one cell type to another, it is highly desirable to test as diverse a selection of culture media as possible during this phase of the investigation. Measurement of continuing thymidine incorporation after an initial period of depletion of the medium (Sect. B.V.3) can be used as a preliminary assay tool in this and subsequent steps, but it is highly desirable to verify each major conclusion with a full clonal growth assay.

4. If a feeder layer has been used in Steps 1–3, find a means to replace it. This can often be achieved by use of a medium better suited to the cell type or else by adding complex extracts such as pituitary extract or embryo extract. If necessary, the first step toward replacement of the feeder layer can be the use of a medium that has been conditioned by a feeder layer.

5. If a conditioned medium is in use, find a means to replace it. Further testing of metabolic intermediates that promote clonal growth (Step 2) should be undertaken. In addition, attempts can be made to replace conditioning with hormones, growth factors, and tissue extracts. Mixtures of conditioned and unconditioned medium can be tested to determine whether the effect of conditioning is detoxification or the release of the positive-acting growth factor into the medium. If necessary, growth-promoting activity can be partially purified from the conditioned medium by techniques such as acetone precipitation. If a small amount of unconditioned medium appears to be inhibitory when added to a large amount of conditioned medium, concentrations of individual components in the unconditioned medium can be reduced, more pure starting materials can be tested, and detoxification by techniques such as exposure to charcoal can be tried.

6. Replace all undefined supplements that are required for multiplication (e.g., serum, embryo extract, pituitary extract, conditioned medium factors) with dialyzed supplements. If necessary, undefined low molecular weight supplements (e.g., ultrafiltrates, hydrolysates, etc.) can also be added to the medium to supply small molecules removed during dialysis. The purposes of this step are a) to separate the sources of undefined small and large molecules that are added to the medium, so that each type of requirement can be analyzed independently, and b) to verify that none of the required growth promoting activities have been inactivated or "lost" during the separation of large and small molecules.

7. Identify all low molecular weight substances (presumably nutrients) needed for clonal growth in the presence of dialyzed supplements. In some cases, this can be accomplished by trial and error substitution of biochemical substances considered likely to be required for cellular multiplication. If this approach fails, it will be necessary to isolate from natural sources and identify chemically the low molecular weight substances that are essential for cellular multiplication in the presence of the dialyzed supplements. As the low molecular weight growth requirements are identified, they are added as pure compounds to the synthetic portion of the medium in place of the ultrafiltrates, hydrolysates, etc., that were previously needed. The objective of this step is to obtain a synthetic medium that will support reasonably good clonal growth with dialyzed supplements as the only undefined additives.

8. Reduce the amount of the dialyzed supplement(s) to a level that supports less than optimal growth. As discussed previously, it is necessary for the supplements to be made rate-limiting for multiplication in order to detect partial replacement of them in the following steps.

9. Adjust all components of the synthetic portion of the medium to experimentally determined optimum concentrations for clonal growth with minimal amounts of the dialyzed supplement(s). Because of the first-limiting factor principle, it is usually necessary to perform these adjustments in a sequential manner, beginning with the nutrient whose current concentration in the medium is found to be the most limiting to cellular multiplication. A preliminary assay in which each component is tested at 0.1, 1.0, and 10 x its normal concentration is very helpful in identifying those components whose concentrations are marginally too high or too low.

Starting with the component that is found to be the farthest from optimum in the preliminary assay, measure the cellular multiplication response over the entire range of positive stimulation, optimal growth, and inhibition. Adjust the concentration of that component in the synthetic medium to the geometric mean (the midpoint on a semilog plot) of its optimum concentration range. Then proceed to the next most limiting nutrient, and continue adjusting until no further benefits can be obtained and all components of the synthetic medium are at optimum values. As growth is improved in this and subsequent steps, it will be necessary to reduce the amount of the dialyzed supplement, such that it is always rate-limiting for multiplication (cf. Step 14). Quantitative optimization of nutrient concentrations has proven to be one of the most important steps in reducing the amount of dialyzed supplement that is needed for several different types of cells (McKEEHAN et al., 1977; HAM and McKEEHAN, 1978 a, b; PEEHL and HAM, 1980 b; SHIPLEY and HAM, 1981; AGY et al., 1981).

10. Systematically test low molecular weight compounds not already in the basal medium that are considered likely to have multiplication-promoting activity (e.g., trace elements and nutrients known to be required by other types of cells or organisms).

11. Test complex low molecular weight mixtures, such as hydrolysates, extracts, and ultrafiltrates, for multiplication-promoting activity. If such activity is found, return to Step 7.

12. Attempt to separate low molecular weight multiplication-promoting activity from the dialyzed supplements through the use of dissociating conditions. Again, if low molecular weight multiplication-promoting activity is found, return to Step 7.

13. Attempt to improve cellular proliferation through refinement of the culture techniques that are employed. Examples include the use of very gentle trypsinization procedures to minimize cellular damage (McKEEHAN, 1977) and coating of culture surfaces with various substances that permit multiplication with less serum protein in the medium (McKEEHAN and HAM, 1976; BARNES and SATO, 1980).

14. Whenever cellular multiplication is improved significantly by any of Steps 9–13, reexamine the multiplication response to the dialyzed supplement and, whenever necessary, reduce the concentration of the dialyzed supplement so that the replacement assays are always performed at a concentration of the supplement that is rate-limiting for multiplication.

15. Keep repeating Steps 9–14 until no further benefit can be obtained. One of the difficulties created by the first-limiting factor principle is that it is often not possible to detect multiplication responses caused by the second and subsequent limiting components of an undefined supplement until after the first-limiting component has been identified and replaced. Frequently, it is necessary to reduce the overall concentration of the undefined supplement in order to reduce the amount of its next most limiting component to a level that becomes rate-limiting for multiplication (cf. HAM and McKEEHAN, 1978b for a detailed discussing of this principle). When no additional improvement in multiplication with rate-limiting amounts of the dialyzed supplement can be achieved by further manipulation of the defined portion of the medium, that medium is assumed to be "optimized" and, it becomes necessary to seek alternate ways of replacing the remaining requirement for the supplement.

16. With the dialyzed supplement either absent or at a rate-limiting level for multiplication, test all hormones or hormone-like growth factors with known or suspected multiplication-promoting activity for their ability to stimulate multiplication of the cell type in question. These substances should be tested both individually and in combinations. If a dialyzed supplement is used initially, its level must be kept low enough so that it is always rate-limiting for multiplication.

For many types of cells, mixtures of hormones and growth factors have sufficient serum-replacing activity so that it is possible to obtain serum-free multiplication without optimizing the synthetic portion of the medium as described in Steps 8–15 (BOTTENSTEIN et al., 1979; BARNES and SATO, 1980). However, optimization clearly reduces the amounts of hormones and growth factors needed for multiplication (WU and SATO, 1978; SHIPLEY, G.D., WALTHALL, B.J. and HAM, R.G., unpublished work). In cases where effective replacement of serum with hormones and growth factors can be achieved without prior optimization of the basal medium, it may be desirable to do the replacement first (after Step 7) and then to optimize the synthetic medium (Steps 8–15) in the presence of rate-limiting amounts of hormones and growth factors, rather than with limiting amounts of the more complex dialyzed serum.

17. If complete replacement of the dialyzed supplements is achieved in Step 16, proceed directly to Step 19. If not, test various organ extracts and also media conditioned by various types of cells for ability to replace partially or fully the multiplication-promoting activities of the dialyzed supplements. Such tests should be done in the presence of a mixture of hormones and growth factors with optimum supplement-replacing activity, as determined in Step 16. If one of the extracts or conditioned media has a higher specific activity than the original supplement, determine whether the activity is macromolecular or of low molecular weight by procedures such as dialysis, ultrafiltration and gel filtration. If low molecular weight activities are present, return to Step 7. If all activity is of high molecular weight, proceed to Step 18.

18. Determine which of the macromolecular supplements (either the original ones from Step 6 or new ones introduced in Step 17) have the highest specific activity. Isolate and identify the specific multiplication-stimulating substances that they contain, using an assay system consisting of the optimized synthetic medium from Steps 8–15 with an optimum mixture of hormones and growth factors from Step 16, so that the requirement for additional

factors will be held to a minimum. Replace the undefined macromolecular supplements with the newly identified multiplication-promoting substances in purified form and determine whether additional improvement of multiplication can be achieved by further addition of undefined supplements. If further stimulation of multiplication is detected, isolate and identify the additional multiplication-stimulating substances that are involved and add them to the medium in purified form.

19. Identify all "contaminants," both in the synthetic medium and in the "purified" macromolecular substances that are required for multiplication of the cell in question. If any such contaminants appear to stimulate multiplication, deliberately add them to the synthetic medium at their experimentally determined optimum concentrations. If this is not done, multiplication will be dependent on the presence of chance contaminants that could be eliminated from future batches of chemicals or growth factors with no warning, thereby causing unexpected and difficult-to-trace failures of multiplication. If inhibitory contaminants are present, they should be eliminated or reduced to harmless levels whenever feasible. If not, they should be monitored carefully and medium components with excessive levels of them should be rejected.

20. Make any changes that may be needed to bring every component of the fully defined medium to its experimentally determined optimum concentration. Because of holistic interactions, any of the changes made in Steps 16–20 can potentially alter responses to components of the optimized medium developed in Steps 8–15. Therefore, it is desirable to verify experimentally that all concentrations in the "final" medium are indeed optimal after all of the adjustments have been made.

21. After all requirements for clonal growth of isolated cells have been identified and optimized, it may be necessary to make additional adjustments for dense monolayer growth and/or for maximum expression of differentiated properties by the cells in question.

The procedures described in this sequence are long and tedious, but they are also highly effective. Their effectiveness has been demonstrated particularly well by the major progress that has recently been made toward defining the growth requirements of human epidermal keratinocytes, which traditionally have been extremely difficult to grow in culture. In abbreviated form, the sequence described above was applied to human keratinocytes as follows:

Steps 1 and 2. A feeder layer of irradiated 3T3 cells yielded reasonably good clonal growth of human keratinocytes in Dulbecco's modified Eagle's medium with 20% whole fetal bovine serum (RHEINWALD and GREEN, 1975).

Steps 3 and 4. Substitution of medium 199 for DME eliminated the need for a feeder layer and made possible clonal growth in medium conditioned by 3T3 cells (PEEHL and HAM, 1980a).

Step 5. Addition of pituitary extract and very high concentrations of hydrocortisone made possible clonal growth in medium 199 without conditioning (PEEHL and HAM, 1980a).

Steps 6 and 7. A further survey of media revealed that limited clonal growth could be obtained without conditioning in medium F12 supplemented with dialyzed fetal bovine serum and high concentrations of hydrocortisone (PEEHL and HAM, 1980b).

Steps 8–15. A process of quantitative optimization, starting from medium F12, revealed that the quantitative growth requirements of human keratinocytes are very different from those of human fibroblasts. A new medium, MCDB 151, was developed, which will support clonal growth of human epidermal keratinocytes with greatly reduced concentrations of dialyzed fetal bovine serum (equivalent in total protein concentration of 2% whole serum). MCDB 151 is also somewhat selective against the growth of human diploid fibroblasts (PEEHL and HAM, 1980b).

Steps 16 and 17. These steps are still in progress at the time that this is being written. However, substantial further reduction in the total amount of undefined supplementation needed for clonal growth has been achieved. Small amounts of pituitary extract replace larger amounts (but not all) of the requirement for serum protein. Supplementation with a mixture of hormones and growth factors (WALTHALL, B.J. and HAM, R.G., unpublished work) and further adjustments in the composition of the synthetic medium have also been beneficial in reducing the requirement for undefined supplements.

Experience in this laboratory with a variety of other cell lines has been similar, although not as dramatic, since the cells that were studied could already be grown quite well in available media with sufficient serum supplementation (MCKEEHAN et al., 1977; HAM and MCKEEHAN, 1978a, b, SHIPLEY and HAM, 1980, 1981; AGY et al., 1981). Innovative procedures for obtaining some degree of growth (Step 1) of many types of cells that have traditionally been difficult to grow in culture are correctly being undertaken in many laboratories (HARRIS et al., 1980). Once that initial barrier has been passed, it currently appears to be a relatively straightforward process to develop assays for and to identify the total set of growth requirements of virtually any type of nontransformed (or transformed) cell. The only major limitation is the time and effort that is involved (Sects. E.III and E.IV).

C. Requirements for Survival and Growth of Nontransformed Cells

The overall set of growth requirements for nontransformed cells was outlined briefly in Sect. B.III. In the sections that follow, these requirements are explored in greater detail and in a slightly different organizational framework. The approach that has been taken is highly inclusive, with essentially everything that stimulates cellular multiplication under any conditions (other than obvious reversal of toxicity) classified as a growth requirement. Because of the large number of different types of requirements that are discussed, it has been necessary to abbreviate the discussion of each type and to rely heavily on references to review papers for the details.

I. Transformed Vs Nontransformed Cells

Although substantial information is available concerning the growth requirements of a few types of nontransformed cells, the total set of growth requirements is not yet understood completely for any one of them. In addition, there are still some types of normal cells that cannot yet be grown well enough in culture to undertake a detailed study of their growth requirements. One of the assumptions that has been made in preparing this summary of growth requirements is that virtually anything that stimulates multiplication of any type of transformed cell is also likely to be required by some type of nontransformed cell. At present, there are a few apparent exceptions, such as the requirement of leukemic cells for asparagine under conditions where it is not needed by comparable normal cells (HALEY et al., 1961; OHNUMA et al., 1971), and the requirement of transformed 3T3 cells for a specific macromolecular factor that is not needed by nontransformed 3T3 cells (LIPTON et al., 1979). However, transformed cells exhibit altered patterns of gene expression and it is not unusual for them to express traits that are normally observed only in fetal stem cells or in cells with different types of differentiation (MARKERT, 1968; URIEL, 1979). Because of this, an unusual growth requirement that is exhibited by a transformed cell could easily reflect a requirement of a normal cell quite different than the immediate precursor of the transformed cell. Thus, in the absence of definitive information to the contrary, everything that has been reported to stimulate multiplication of any type of transformed cell is considered to be a potential growth requirement of some type of nontransformed cell.

The reciprocal question – whether or not nontransformed cells have unique growth requirements that are not shared by transformed cells has proven to be surprisingly difficult to answer. Initially, it seems very clear that there exist many circumstances in which transformed cells multiply and nontransformed cells do not, and that nontransformed cells therefore must have more complex growth requirements. However, on more careful analysis, the answer is not that straightforward. First of all, it is necessary to eliminate highly adapted cell lines that have undergone evolutionary selection in vitro for simplified growth requirements, as discussed earlier in this chapter. In the absence of such selection, transformed cells generally require serum protein, although often in smaller amounts than nontransformed cells. Next, it is necessary to consider the effects of density-dependent inhibition. Much of the requirement of nontransformed cells for high concentrations of serum protein appears to be to overcome the effect of crowding on their multiplication (TODARO et al., 1965; HOLLEY and KIERNAN, 1968). Another important consideration is the composition of the basal defined medium. When nontransformed human diploid fibroblasts are grown at clonal density in optimized media and under optimized culture conditions, their requirement for dialyzed serum is no larger than that of their transformed counterparts (MCKEEHAN et al., 1977; HAM and MCKEEHAN, 1978a; MCKEEHAN and HAM, 1978a). This comparison is not entirely fair, since a comparable optimization of media was not done for the transformed cells. Nevertheless, it is clear that for many different types of nontransformed cells, the amount of serum protein needed for multiplication can be reduced greatly by use of optimized media and culture conditions (HAM and MCKEEHAN, 1978a, b; SHIPLEY and HAM, 1981; PEEHL and HAM, 1980b; HAM, 1980).

In the limited number of cases where replacement of serum protein with hormones and growth factors has been relatively successful, no special growth requirements appear to be emerging for the nontransformed cells (BARNES and SATO, 1980; PHILLIPS and CRISTOFALO, 1980a; WALTHALL and HAM, 1980; SHIPLEY and HAM, 1980). There are reports that transformation of specific cell types reduces or eliminates their requirements for specific growth factors associated with initiation of the cell cycle, such as EGF and PDGF (SCHER et al., 1978; CHERINGTON et al., 1979). However, serum replacement studies with a number of different cell lines have provided evidence for EGF requirements by transformed as well as nontransformed cells (BARNES and SATO, 1980).

It is still too early to rule out the possibility that unique factors required only for the multiplication of nontransformed cells may be discovered.

However, for those types of nontransformed cells whose growth requirements are already relatively well understood, it currently appears likely that any differences that are found will prove to be quantitative rather than qualitative in nature. If this prediction is confirmed by studies now in progress in several laboratories, it will be necessary to examine higher levels of regulation in the search for differences in control of multiplication between nontransformed and transformed cells. Possible mechanisms of control that must be considered include density-dependent inhibition and the factors that overcome it (Ross, this volume), specific inhibitors of multiplication (STECK et al., 1979; DATTA, this volume; IVERSON, this volume), and special factors such as those that are involved in control of liver regeneration (SEKAS et al., 1979). Such higher orders of control are clearly beyond the scope of the current chapter, which is limited to the immediate requirements of nontransformed cells for survival and multiplication.

II. Requirements Related to Subculturing and Cellular Attachment

1. Anchorage Dependence

Although there are some exceptions, most types of nontransformed cells must attach to a solid substrate in order to multiply (MACPHERSON and MONTAGNIER, 1964; STOKER et al., 1968). One of the problems encountered with such cells is the necessity of releasing the parental cells from the substrate when subculturing. It is often possible to shake limited numbers of mitotic cells loose from the substrate. However, the yield is small, and the more commonly used mass subculturing procedures all require release of the cells from the substrate by enzymatic digestion or treatment with chelating agents. A significant subset of the growth requirements of nontransformed cells is directly related to neutralization of the dispersing agent and to cellular attachment and spreading on the new culture surface. Many of the requirements can be regarded as artificial, since it is possible to eliminate them through the use of very gentle and nondamaging techniques of subculture. However, limited recent evidence suggests that for some cell types, specific attachment and spreading factors are needed even when the most gentle subculturing procedures that are currently feasible are used.

2. Neutralization of the Dispersing Agent

In order to achieve cellular reattachment and activation of anchorage-dependent multiplication, it is first necessary to stop any further action of the dispersing agent. As a first step, it is generally desirable to minimize the amount of the agent that is introduced into the subcultures. One means of achieving this is gentle centrifugation of the cells, followed by resuspension in fresh medium that does not contain the agent. In most cases this is distinctly beneficial, but the value must be weighed against the damaging effects of the centrifugation. Some types of cells suffer a significant loss of viability during even very gentle centrifugation, particularly if they have already suffered membrane damage during a dispersal procedure. For cells that are capable of clonal growth, plating efficiencies before and after centrifugation can be used to evaluate the effect of centrifugation. Such tests must, of course, be done in a medium that will adequately neutralize the effects of the dispersing agent in the absence of its removal by centrifugation.

In cases where chelating agents are used as the dispersing agent, they are generally neutralized by the divalent cations of the culture medium. In terms of allowing normal reattachment this works quite well, provided that the amount of chelating agent introduced with the cellular inoculum is not excessive, and that the test medium contains an adequate level of calcium and magnesium. However, there remains a more subtle danger that appears not to have been examined adequately. Many chelating agents have higher affinities for inorganic trace elements that are essential for cellular multiplication than they do for calcium and magnesium. Even when they are fully neutralized with respect to calcium and magnesium such agents may preferentially bind certain of the trace elements and reduce their concentrations as free ions in the culture medium below the minimum levels that are needed for satisfactory cellular multiplication.

When proteolytic enzymes such as trypsin are used to disperse the cells, specific steps must be taken to neutralize the enzymes in the subculture. For trypsin-dispersed cells inoculated into media containing substantial amounts of serum, potent trypsin inhibitors in the serum neutralize the trypsin so effectively that many investigators routinely perform monolayer subcultures without removal of the trypsin from the cellular inoculum. In low serum and serum-replacement experiments, this is not possible and the trypsin is generally removed by centrifugation and resuspension of the cells in fresh medium. However, this may not be adequate, since trypsin tends to remain associated with the cells, and under some conditions is actually taken into the interiors of the cells (HODGES et al., 1973). In certain cases, multiple washing of cells after trypsinization will permit serum-free multiplication under conditions where serum or serum fractions with anti-tryptic activity are otherwise needed (WALLIS et al., 1969). The use of soybean trypsinization inhibitor may also be useful at times (BUSH and SHODELL, 1977; BARNES and SATO, 1980).

3. Low Temperature Trypsinization

The effects of conventional trypsinization on cellular growth requirements appear to go far beyond the need for neutralization of the trypsin. A substantial reduction in the amount of serum protein needed for clonal growth of various types of non-transformed cells was achieved through the use of a special low temperature trypsinization technique (MCKEEHAN, 1977; MCKEEHAN et al., 1977; HAM and MCKEEHAN, 1979). In this technique, the cells are chilled to ice bath temperature before the start of the trypsinization, and are kept at this temperature throughout the entire set of manipulations until the daughter cultures in their final media are returned to the incubator. Below 15 °C, cell membranes undergo a phase transition that essentially freezes them and makes them less susceptible to injury. Cellular metabolism is greatly reduced and internalization of the trypsin is essentially prevented.

When compared to control cultures trypsinized at room or incubator temperatures, cells trypsinized at ice bath temperatures exhibit a higher plating efficiency and form larger colonies with the same amount of serum protein. In addition, cells trypsinized at ice bath temperatures are able to form colonies in media supplemented with smaller amounts of serum protein than those trypsinized at conventional temperatures. For a number of types of cells, low temperature trypsinization, in combination with the use of minimal volumes and concentrations of trypsin and minimal times of exposure to trypsin, yields a cellular inoculum that will attach, spread and survive for prolonged times in the complete absence of serum protein. This is true even for cells that are dependent on serum protein for their multiplication (HAM and McKEEHAN, 1978 a; McKEEHAN et al., 1978; SHIPLEY and HAM, 1981). It is highly desirable to use such an inoculum for experiments seeking to replace serum proteins with mixtures of hormones and growth factors.

4. Attachment and Spreading Factors

Many types of cells exhibit specific macromolecular requirements for attachment and spreading after trypsinization that appear to go beyond simple neutralization of trypsin. The history of such factors is long and confusing (reviewed by TEMIN et al., 1972; GRINNELL, 1978), but it currently appears that many such requirements reflect a specific need for fibronectins (reviewed by YAMADA and OLDEN, 1978; GRINNELL, this volume). Fibronectins occur both as cell surface antigens, where they were often referred to as large external transformation sensitive (LETS) protein, and in plasma and serum, where they have been known by such names as cold-insoluble globulin and spreading factor. In retrospect, it seems likely that other attachment and spreading factors, such as fetuin (FISHER et al., 1958) and protein flattening factor (LIEBERMAN and OVE, 1957) contained fibronectins as contaminants. It should be noted, however, that some spreading factor activity may reside in molecules other than fibronectin (BARNES and SATO, 1980; KNOX and GRIFFITHS, 1979).

Nontransformed cells generally produce more fibronectin than transformed cells do (YAMADA and OLDEN, 1978) and some types of nontransformed cells, such as human diploid fibroblasts appear to synthesize and secrete fibronectins in amounts sufficient to take care of their needs for attachment and stretching in the complete absence of serum or other exogenous sources of fibronectins (GRINNELL and FELD, 1979). Fibronectin has been reported to be required for serum-free multiplication of the F 9 embryonal carcinoma line (RIZZINO and CROWLEY, 1980) and also for a nontransformed rat ovarian follicle cell line (ORLY and SATO, 1979). Mouse 3 T 3 cells grown in medium MCDB 402 with FGF, insulin and dexamethasone continue to multiply for a longer period without detachment in the presence of fibronectin (SHIPLEY and HAM, 1980), although it is not needed for initial attachment and spreading of low temperature trypsinized cells on polylysine-coated surfaces (SHIPLEY and HAM, 1981).

It is not entirely clear at present that the only role of fibronectin is in cellular attachment and spreading. Cytokinesis fails to occur normally in rat follicular cells (RF-1) in the absence of fibronectin, leading to the formation of binucleate cells (ORLY and SATO, 1979). In addition, data have been reported that appear to show that fibronectin is not confined to the cell surface and that it can be found in the nucleus in association with chromatin (ZARDI et al., 1979). Also, it is well known that fibronectin interacts with intracellular microfilament bundles, and that it can alter the morphology of transformed cells to make them more like nontransformed cells (reviewed by YAMADA and OLDEN, 1978). Thus, it is quite possible that the fibronectins may have effects on multiplication of some types of cells that have nothing to do with attachment and spreading.

Finally, it must be noted that fibronectins are associated with only a limited number of cell types in normal development (YAMADA and OLDEN, 1978). Other cell types that do not have cell surface fibronectin are also capable of adhesiveness, both in vivo and in vitro, and it is quite likely that they will prove to have other types of molecules with functions similar to those of the fibronectins on their surfaces. Very recently there has been a report of such a factor for chondrocytes, which has been given the name "chondronectin" (HEWITT et al., 1980). Other molecules that mediate cellular adhesion, variously referred to by terms such as "cell adhesion molecule" (RUTISHAUSER et al., 1978) and "cognins" (HAUSMAN and MOSCONA, 1979) are also likely to prove to be important when the growth requirements of diverse types of cells become more clearly understood.

5. The Culture Surface

Historically, cell culture was done "in vitro" in culture vessels made of glass. More recently the glass has been replaced with specially treated polystyrene plastic with a similar negative surface charge. For satisfactory multiplication of attachment-dependent cells, the culture surface must be wettable and it must adsorb proteins (GRINNELL, 1978). In this respect, treated polystyrene is clearly superior to untreated bacteriological polystyrene. However, it is not clear that the negative charge density of commonly used culture surfaces is optimal (MACIEIRA-COELHO and AVRAMEAS, 1972).

Multiplication of nontransformed fibroblast-like cells with minimal amounts of serum protein can often be improved dramatically by coating the culture surface with a positively-charged polymer such as polylysine (McKEEHAN and HAM, 1976). In extreme cases at very low serum concentration, colony formation can be totally dependent on the use of polylysine coating (McKEEHAN and HAM, 1976; SHIPLEY and HAM, 1981). Polylysine-coated surfaces are also beneficial for serum-free growth of some cell types (BARNES and SATO, 1980; SHIPLEY and HAM, 1980; WALTHALL and HAM, 1980) The surface charge appears to effect both attachment and subsequent multiplication, since a polylysine coat also results in the formation of larger colonies.

The possibility that small amounts of the basic polymers are slowly leached from the culture surface and act as a component of the medium cannot be fully ruled out at the present time. Growth of Yoshida sarcoma cells in suspension culture has been reported to be improved by addition of small amounts of basic polymers directly to the medium (MURAKAMI and YAMANE, 1976). Modification of substrate adhesiveness by applying different concentrations of poly-(2-hydroxyethyl methacrylate) is also reported to affect multiplication of nontransformed cells. The more flattened the cells are on the substrate, the more they multiply (FOLKMAN and MOSCONA, 1978).

Improvement in cellular multiplication can also be achieved for some types of cells by coating the culture surface with various other substances, including fibronectin, collagen, and insulin (BARNES and SATO, 1980; AGY et al., 1981). In some cases, serum can be replaced by coating the culture dishes with serum and then rinsing off all material that is not tightly bound (ORLY and SATO, 1979; BARNES and SATO, 1980).

6. Artificial Substitutes for the Basement Membrane

Many types of epithelial cells multiply in vivo only when they are in the basal layer of the epithelium and directly in contact with a collagen-rich basement membrane. Collagen coated culture vessels have been used for many years as a means of improving cellular multiplication (EHRMANN and GEY, 1956). For example, muscle differentiation in culture is greatly enhanced by a collagen substrate (HAUSCHKA and KONIGSBERG, 1966).

More recently, a floating collagen raft technique has been developed in which the cells are separated from the bulk of the medium by a thin collagen film (MICHALOPOULOS and PITOT, 1975; EMERMAN and PITELKA, 1977). This approach is said to offer at least four distinct advantages over conventional attachment of cells to a collagen-coated dish: 1) access of nutrients to the basolateral cell surfaces; 2) close proximity of the cells to the medium surface and the gas phase; 3) interaction of epithelial cells with stromal elements; and 4) substrate flexibility, permitting cell shape change as the cultures mature (EMERMAN et al., 1979).

Artificial basement membranes have been refined somewhat further by REID and ROJKIND (1979) who add fibronectin (LETS protein) and proteoglycans or glycosaminoglycans to a floating collagen raft to simulate more closely the composition of naturally occurring basement membranes.

III. Inorganic Ions, Physical Chemistry, and Cell Physiology

The culture environment in which nontransformed cells are grown must satisfy a variety of physicochemical and physiological requirements of the cells. Such requirements are often assumed to be standardized and tend to be dismissed rather lightly during the development of culture conditions and media for growth of various types of cells. However, the vertebrate body is far from a strictly uniform environment. Substantial differences occur among its specialized tissues and organs with regard to direct or indirect access to the circulatory system, with resulting differences in oxygenation, removal of waste products, access to nutrients, exposure to plasma proteins, etc. There also exist substantial differences in pH, osmolarity, and patterns of aerobic and anaerobic metabolism, and even significant differences in temperature between surface and internal structures. Thus, it is not reasonable to assume that these parameters will all be the same for optimal growth of each type of cell that is placed in culture.

Many of the physicochemical and physiological requirements of cultured cells are intimately related to the concentrations of six major inorganic ions that are strictly required for cellular multiplication in vitro, sodium, potassium, calcium, magnesium, chloride, and phosphate, plus a seventh, bicarbonate, which is formed from metabolically generated carbon dioxide in crowded cultures, but is an essential growth requirement when the cellular inoculum is small. These seven inorganic ions play a multitude of different roles in cellular multiplication and must be viewed from at least four different perspectives: 1) as substances that have a major effect on the physicochemical environment that cultured cells are maintained and grown in; 2) as substances with major roles in cellular physiology that fall outside the narrow definition of nutrients given at the beginning of this chapter; 3) as true nutrients that participate in cellular metabolism as substrates and/or catalysts; and 4) as potential regulators of cellular multiplication. Since these roles tend to be intermingled in a manner that defies precise classification, all aspects of growth requirements for these seven inorganic ions are considered in this section, together with discussions of the physicochemical and physiological parameters that they affect.

1. Bicarbonate, Carbon Dioxide, pH, and Buffering

Historically, a bicarbonate-CO_2 system has served two totally different functions in cellular multiplication: 1) it has been used to control the pH of the culture medium; and 2) it has served as a substrate for a number of essential biosynthetic reactions, including synthesis of purines, pyrimidines, long chain fatty acids, and four-carbon intermediates of the tricarboxylic acid cycle. In crowded cultures, the amount of metabolically generated CO_2, which at neutral pH is in rapid equilibrium with bicarbonate ion, is more than adequate for biosynthesis and tends to contribute to gradual acidification of the culture medium. However, in sparsely inoculated cultures, CO_2 diffuses out of the individual cells so rapidly and becomes so extremely diluted that it is unavailable for biosynthesis. Because of this, it is essential to supply bicarbonate and/or CO_2 from an exogenous source in order to obtain satisfactory biosynthesis and cellular multiplication (HAM, 1972, 1974b; HAM and McKEEHAN, 1978b; McLIMANS, 1972).

As a buffer, the bicarbonate-CO_2 system has some serious drawbacks including lack of precise control, large changes in pH that occur when the cultures are removed from the CO_2 incubator, and the difficulty of maintaining a precisely controlled gas mixture in the cell culture incubator. These problems have led to the development of alternative buffering systems and media designed for use in equilibration with air or reduced concentrations of carbon dioxide (e.g., 2%). High concentrations of amino acids in combination with phosphate have been used for buffering in a number of media including L 15 (LEIBOVITZ, 1963), and CMRL 1415-ATM (HEALY and PARKER, 1966). Other buffers that have been widely used include β-glycerophosphate (reviewed by WAYMOUTH, 1978, 1981) and various synthetic organic buffers, the most popular of which is HEPES (reviewed by EAGLE, 1974). Such media generally work well without carbon dioxide incubation when the cellular population density is kept reasonably high. For clonal growth, it is necessary to provide an exogenous source of carbon dioxide and/or CO_2. However, the amount that is needed appears to be substantially less than the 5–10% CO_2 traditionally used in buffer systems (reviewed by WAYMOUTH, 1978). In the author's laboratory, clonal growth media for human diploid fibroblasts are buffered with 30 mM HEPES and equilibrated with an atmosphere of 2% CO_2. No bicarbonate is added as such, but a substantial amount is formed as the buffered medium equilibrates with the CO_2 in the incubator (HAM and McKEEHAN, 1978b; McKEEHAN et al., 1977). The combined use of HEPES buffering and a minimal amount of CO_2 greatly reduces the problem of pH change as cultures are moved in or out of the cell culture incubator.

The range of hydrogen ion concentrations that will support optimal cellular multiplication is quite narrow. In addition, the optimum value differs somewhat from one type of cell to another (reviewed by EAGLE, 1974). For clonal growth of WI-38 human diploid fibroblasts in an optimized medium with minimal amounts of serum protein, the optimum pH range is 7.30 ± 0.15 (measured under actual incubation conditions). For chicken embryo fibroblasts, the optimum is approximately 7.12 ± 0.18. Although these optima overlap, the range for chicken cells is appreciably more acidic than that for human cells. Optima for nontransformed lines tend to be relatively more alkaline than those of typical transformed permanent lines (EAGLE, 1974). As more cell types are studied there is a distinct possibility that major differences in optimum pH may be discovered. One recent report claims that growth of human epidermal keratinocytes in Eagle's minimum essential medium is optimal at pH 5.6–5.8 and that no growth occurs above pH 7.0 (EISINGER et al., 1979). This report raises the interesting possibility that the pH optimum may be related to the basal medium used, since other studies have shown that good growth of human epidermal keratinocytes can be obtained at pH 7.4 in a medium that has been specifically optimized for keratinocytes, but not in conventional media (PEEHL and HAM, 1980b).

2. Sodium, Chloride, Osmolarity, and Humidification of Incubators

Sodium and chloride ions make by far the largest contributions to the total osmolarity of the culture medium. For most types of cells, optimal growth occurs only in a relatively narrow range of osmolarities. This topic was reviewed in detail by WAYMOUTH (1970) and only a few additional observations will be included here. For clonal growth with minimal amounts of serum protein, human diploid fibroblasts have an optimal range of osmolarity of 285 ± 40 milliosmoles per kg and chicken embryo fibroblasts a range of 300 ± 20 mOsm/kg (HAM and McKEEHAN, 1978b). These values are fairly typical, but preliminary evidence suggests that some types of cells may have osmotic requirements that are quite different. Several investigators have reported that lymphocytes survive best at very low osmolarities in the range of 228–237 mOsm/kg (reviewed by WAYMOUTH, 1970). At the other extreme, an osmolarity of 375 ± 5 mOsm/kg is reported to be optimal for mouse pancreatic epithelium (WAYMOUTH, 1978).

The narrowness of the optimum range of osmolarity for individual types of cells makes it necessary to consider carefully the osmotic effect of additions to a culture medium. For example, the addition of 50 mM HEPES plus enough NaOH to adjust its pH would shift the osmolarity of a culture medium outside of acceptable limits if compensating adjustments were not made in the sodium chloride concentration. High concentrations of HEPES are often considered toxic, but when adequate adjustments of osmolarity are made, clonal growth of human diploid fibroblasts is unaffected by 50 mM HEPES (McKEEHAN et al., 1977).

Adequate humidification of cell culture incubators is very important, particularly in dry climates where partial evaporation of the culture medium can raise osmolarity to inhibitory levels very quickly. Clonal growth experiments whose medium must be allowed to equilibrate with the carbon dioxide in the atmosphere of the incubator are the most sensitive to this problem. An easy way to test for the adequacy of humidification is to measure the osmolarity of uninoculated control dishes of media at the beginning and again at the end of clonal growth experiments.

Incubator humidification is adequate only when the osmolarity remains within acceptable limits throughout the entire experiment, which may require two weeks or longer. (This problem is discussed in greater detail in HAM and McKEEHAN, 1979.)

Sodium and chloride ions also have major physiological roles related to the maintenance of membrane potential and active transport across cellular membranes. Both ions are largely excluded from the intracellular space, and although both are absolutely essential for cellular survival and multiplication, it is somewhat doubtful that either fully satisfies the definition of a cellular nutrient given at the beginning of this chapter.

Current research suggests that sodium ion may be directly involved in the regulation of cellular multiplication. Sodium influx into cells has recently been implicated as one of the very early events involved in the initiation of DNA synthesis (CONE and CONE, 1976; KOCH and LEFFERT, 1979 a, b; ROZENGURT, 1979). Pharmacological agents that stimulate sodium entry into cells have been reported to cause short term increases in cellular multiplication (ROZENGURT et al., 1979).

3. Potassium and $Na^+:K^+$ Ratios

Potassium serves as the major intracellular cation and also has many important cofactor roles in cellular metabolism. The amount of potassium that is optimal varies substantially from one type of cell to another (reviewed by WAYMOUTH, 1972). Since KCl can contribute significantly to total osmolarity in high potassium media, the relative amounts of potassium in various media are often described in terms of the $Na^+:K^+$ ratio at constant osmolarity. All members of a group of 27 different media whose molar concentrations[1] are summarized in a recent review (HAM and McKEEHAN, 1979) have $Na^+:K^+$ ratios between 24:1 and 50:1, and it is often assumed that this range is optimal for multiplication of all types of cells. However, there are many indications that such an assumption is not uniformly valid. Medium MCDB 151, which was optimized for human epidermal keratinocytes (PEEHL and HAM, 1980b) has an unusually low potassium level (1.5 mM) and a very high $Na^+:K^+$ ratio of 100:1. Ratios on the order of 100:1 are also reported to be optimal for continued beating of cultured heart cells (DeHAAN, 1970). WAYMOUTH (1978) reports that mouse liver and prostate cells grow better in media with relatively high $Na^+:K^+$ ratios from 40:1 to 48:1, whereas mouse pancreatic epithelium grows better with a low $Na^+:K^+$ ratio of 15:1. Potassium-rich media with $Na^+:K^+$ ratios as low as 5:1 have been reported to be optimal for mouse strain L cells (KUCHLER, 1967), and ratios greater than 25:1 are reported to be inhibitory to protein synthesis in mouse 3 T 6 cells (POLLACK and FISHER, 1976). The wide range of ratios that appear to be optimal for different types of cells make the $Na^+:K^+$ ratio an important experimental variable that should not be overlooked in the development of media for types of nontransformed cells whose quantitative growth requirements are not yet fully understood.

---

1 The concentration of K^+ in MCDB 411 (AGY et al., 1981) is 3.0×10^{-3} M, and not 1×10^{-3} M as listed in HAM and McKEEHAN (1978b and 1979)

4. Calcium, Magnesium, and Regulatory Roles of Divalent Cations

Calcium and magnesium both have a myriad of essential roles in cellular multiplication. Like potassium and sodium, they distribute asymmetrically across the cellular membrane with a high intracellular concentration of magnesium and a relatively low intracellular concentration of calcium. Both serve as cofactors in enzymatic reactions and therefore qualify as true nutrients. They are also of major importance in cellular attachment and spreading reactions. Cells can be released from substrate attachment by treatment with chelating agents for divalent cations. However, current evidence suggests that the mechanisms of cellular attachment are far more complex than the simple bridging between negatively charged substrates and negatively charged cell surfaces that was once thought to be involved (GRINNELL, 1978; DAMLUJI and RILEY, 1979).

Calcium and magnesium both appear to have major roles in the control of cellular multiplication, and there is currently a lively debate in the literature concerning which interacts at the most fundamental level with the mechanisms that ultimately control proliferation. Since the outcome of that debate has no direct effect on the fact that both ions are essential for cellular multiplication, only a few of the key points will be summarized here.

Intracellular magnesium ion concentration appears to be closely related to the coordinated control of a number of different biochemical events that are involved in initiation of the cell cycle (RUBIN, 1975a, 1976; RUBIN and KOIDE, 1976; RUBIN and SANUI, 1979). Normal calcium dependence can be bypassed by raising extracellular magnesium levels abnormally high (RUBIN et al., 1979). When intracellular levels of monovalent and divalent cations are artifically manipulated, it is claimed that only the magnesium level accurately parallels the rate of initiation of DNA synthesis (MOSCATELLI et al., 1979).

There is also evidence suggesting that calcium ion has a regulatory role, probably in conjunction with cyclic AMP (WHITFIELD et al., 1976; MACMANUS et al., 1978; WHITFIELD et al., 1979). Calcium ionophores have been reported to be mutagenic to lymphocytes (JENSEN et al., 1977). Nontransformed fibroblasts tend to require very high concentrations of calcium and in the case of human diploid fibroblasts the amount of calcium required for multiplication has been shown to increase sharply as the concentration of serum protein is reduced (MCKEEHAN and HAM, 1978a). Similar interactions have also been observed between calcium and ectodermal growth factor, both for human diploid fibroblasts (MCKEEHAN and MCKEEHAN, 1979b) and for normal human prostatic epithelium (LECHNER and KAIGHN, 1979b).

Many of the regulatory functions of calcium are mediated by an intracellular regulatory protein known as calmodulin or calcium dependent regulatory protein, including cyclic nucleotide metabolism (KRETSINGER, 1976; RASMUSSEN and GOODMAN, 1977; PERRY et al., 1979; CHEUNG, 1980). One of the late events before the start of DNA synthesis is an increase in intracellular cyclic AMP, which appears to be mediated by calcium ion concentration (WHITEFIELD et al., 1976; BOYNTON and WHITEFIELD, 1979). Human diploid fibroblasts maintained with a low concentration of calcium become arrested late in G_1 (BOYNTON et al., 1977), and supplying such cells with increased amounts of calcium has been described as a means of generating synchronous cultures (ASHIHARA and BASERGA, 1979). Recent studies of the time dependence of Ca^{2+} and Mg^{2+} requirements during G_1 in human diploid fibroblasts suggest that there are two critical periods of Ca^{2+} dependence, one at the beginning of G_1 and the other near the end of G_1, whereas the requirement for Mg^{2+} appears to be extended throughout the entire G_1 period (HAZELTON et al., 1979). 3 T 3 cells also have a Ca^{2+} sensitive restriction point in G_1 (PAUL et al., 1978 b). It has also been suggested that a transient increase in intracellular calcium levels may be closely associated with the transient deciliation of the centriole that occurs early in the initiation of a new cell cycle (TUCKER et al., 1979).

The addition of abnormally high concentrations of calcium to the culture medium causes 3 T 3 cells in G_0 to initiate DNA synthesis (DULBECCO and ELKINGTON, 1975; BOYNTON and WHITFIELD, 1976 a). However, it currently appears that the formation of a calcium phosphate precipitate is responsible for the mitogenic effect (BARNES and COLOWICK, 1977). Calcium pyrophosphate crystals have a similar effect (RUBIN and SANUI, 1977; RUBIN and BOWEN-POPE, 1979) as do various other fine particulates, including polystyrene beads, kaolin and barium sulfate (BARNES and COLOWICK, 1977). Recent studies have shown that such particulates act early in the initiation of the cell cycle and have an effect similar to that of platelet-derived growth factor (STILES et al., 1979 a; cf. Sect. C.VII.2).

Although it is clear that both calcium and magnesium are essential requirements for cellular multiplication, it does not appear to be safe to assume that the same concentrations of them should be used for all types of cells. Transformed fibroblasts tend to be able to multiply with less Ca^{2+} and Mg^{2+} than nontransformed (BOYNTON and WHITEFIELD, 1976 b; BOYNTON et al., 1977; MCKEEHAN and HAM, 1978 a; BALK et al., 1979; VAN DER BOSCH et al., 1979). In addition, caution must be exercised in generalizing from the calcium requirements of fibroblast-like cells to those of other cell types. Keratinocytes, both from mice (HENNINGS et al., 1980) and from humans (PEEHL and HAM, 1980 b) multiply best in media with rather low concentrations of calcium and tend to withdraw from the proliferative pool and enter terminal differentiation when the calcium concentration is raised. Adhesion of keratinocytes to their substrate has been reported to be independent of calcium and strongly dependent on magnesium, unlike that of other cell types, which tend to be strongly calcium dependent (FRITSCH et al., 1979). A low level of calcium also appears to be essential for proliferation of rat urinary bladder epithelial cells in organ culture (REESE and FRIEDMAN, 1978). In view of these differences in requirements of various types of cells, careful attention should always be given to quantitative requirements for Ca^{2+} and Mg^{2+} during the development of culture media for cell types that have not previously been well studied.

5. Phosphate

Phosphate has many roles in cellular metabolism, including functioning as a component of catalytic coenzymes, as a component of high energy metabolic intermediates such as ATP, and as a substrate for many types of biosynthesis, such as that of nucleic acids. It also contributes to the buffering capacity of culture media.

Typical culture media generally contain about 1 mM phosphate. However, its concentration in a series of media that have been optimized for clonal growth of various types of cells ranges from 0.1 mM in medium MCDB 501, which was developed for duck embryo fibroblasts, to 3.0 mM in medium MCDB 105, which was developed for human diploid fibroblasts (HAM and McKEEHAN, 1978 b).

6. Other Physicochemical Parameters

In addition to the effects on cellular multiplication related to the seven major inorganic ions, discussed above, other aspects of the physicochemical environment can also have important effects. Included among these are temperature, illumination, oxygen tension, and atmospheric pressure. These variables have been discussed in great detail elsewhere (HAM and McKEEHAN, 1978 b, 1979) and will only be summarized briefly here.

Temperatures approximating mammalian deep body temperature are usually utilized for cell culture, generally quite successfully. However, it is important to remember that different species have different body temperatures and that different parts of the body of any given species are normally at slightly different temperatures. It is well known, for example, that mammalian testes do not function well at normal body temperature. Similarly, skin normally has a lower temperature. One recent report suggests that best results can be obtained with human skin cells by keeping the temperature in the range of 35–37 °C (EISINGER et al., 1979).

Exposure of cell culture media to short wavelength visible light has been shown to generate toxic photoproducts (WANG, 1976), including peroxides (WANG and NIXON, 1978). Because of this it is generally considered best to incubate cell culture experiments and store culture media in total darkness. The effects of light on cultured cells, including chromosomal damage have recently been reviewed by SANFORD et al. (1978, 1979). Paradoxically, brief exposure to light sometimes appears to be beneficial to cellular multiplication in culture.

Some oxygen is essential for multiplication of most types of cells and total oxygen depletion, which is most likely to be encountered when working with crowded cultures, must be avoided (McLIMANS et al., 1968; RADLETT et al., 1972). However, oxygen tensions greater than those in the ambient atmosphere are generally considered toxic to cultured cells (BALIN et al., 1976), and it is frequently reported that reduced partial pressures of oxygen are beneficial (RADLETT et al., 1972; SANFORD et al., 1978, 1979). Experience in the author's laboratory confirms the deleterious effects of high oxygen tension, but also suggests that media can be designed around ambient oxygen tension so that it is not necessary to use reduced oxygen tension in the cell culture incubator. In particular, an adequate amount of the trace element selenium in the culture medium is important to optimize intracellular glutathione peroxidase levels (McKEEHAN et al., 1976, 1977; HAMILTON and HAM, 1977). The fact that the author's laboratory is at an altitude of 1650 m above sea level and has an ambient oxygen tension that is only about 82% of that at sea level may have influenced the results, however.

Addition of specific reducing agents to culture media has been practiced by some investigators for many years, while others suggest that it is not necessary.

Many factors, including type of cell, population density, oxygen tension and overall composition of the culture medium, are probably involved. Addition of a reducing agent remains an important option to try when cultures are not growing well. The requirement of a mouse teratocarcinoma line for a feeder layer was eliminated through the addition of β-mercaptoethanol to the culture medium (OSHIMA, 1978). Growth of a similar embryonal carcinoma line in serum-free medium was also improved by the addition of β-mercaptoethanol (RIZZINO and SATO, 1978). The addition of dithiothreitol or high concentrations of cysteine to the culture medium is reported to be beneficial to cultures of human diffuse histiocytic lymphoma cells (EPSTEIN and KAPLAN, 1979). There are also reports of benefits from the addition of vitamin E, which may be due to its function as a reducing agent (Sect. C.IV.3).

IV. Qualitative Nutrient Requirements

The qualitative nutrient requirements of cultured vertebrate cells have been reviewed extensively (MORGAN, 1958; LEVINTOW and EAGLE, 1961; WAYMOUTH, 1965, 1972; SWIM, 1967; HIGUCHI, 1973; HAM and McKEEHAN, 1979; RIZZINO et al., 1979) and will therefore be summarized only briefly here. Nutrient aspects of the major inorganic ions have already been discussed in Sect. C.I.3 and lipid requirements will be discussed separately in Sect. C.VI. All other medium components that fit the definition given at the beginning of this chapter for "nutrients" are discussed here.

1. Components of Eagle's Minimum Essential Medium

After an extensive study of the growth requirements of Mouse L, HeLa, and various other established cell lines in the presence of small amounts of dialyzed serum, Eagle developed a "minimum essential medium" (MEM) that contained only those components that could be shown to be necessary for cellular multiplication with moderate amounts of dialyzed serum (EAGLE, 1959). In addition to the bulk inorganic ions discussed above, MEM contains 13 amino acids (arginine, cystine, glutamine, isoleucine, leucine, lysine, methionine, phenylalanine, threonine, tryptophan, tyrosine, and valine), six vitamins (folic acid, niacinamide, pantothenic acid, pyridoxine, riboflavin, and thiamine), one sugar (glucose), and two other organic nutrients (choline and inositol). Although some types of cells can be grown under specialized conditions without some of the components of MEM (EAGLE and LEVINTOW, 1965; NAYLOR et al., 1976; YASUMURA et al., 1978), requirements for all of the components of MEM can be viewed as sufficiently generalized so that they should be included in all culture media except those designed for very special purposes.

Despite the name "minimum essential medium," MEM was never intended for use as a synthetic medium without dialyzed serum. Because of this, it lacks several essential nutrients that are difficult to remove from dialyzed serum, including biotin, vitamin B_{12}, iron and all other inorganic trace elements. Cellular requirements for nutrients in addition to the components of MEM are discussed in the following sections.

2. Other Amino Acids

Seven of the twenty amino acids that participate in protein synthesis are not included in MEM (alanine, asparagine, aspartic acid, glutamic acid, glycine, proline, serine). When cells are grown in media containing only the thirteen "essential" amino acids in MEM, the remaining seven must be synthesized *de novo*. Since cultured cells do not utilize inorganic nitrogen sources, the nitrogen for the biosynthesis of these seven "nonessential" amino acids must be obtained by catabolism of the essential amino acids. In view of the energy required for this process and the increased amounts of essential amino acids that must be added to supply nitrogen for the nonessential amino acids, many investigators prefer to supply all 20 of the amino acids needed for protein synthesis directly to the cells.

In cases where the nonessential amino acids are not supplied as nutrients, they must be viewed as metabolic intermediates in the process of protein biosynthesis from the 13 "essential" amino acids that are supplied. Like other biosynthetic intermediates, the nonessential amino acids tend to diffuse out of the cells. When the cellular inoculum is low, certain of the nonessential amino acids (particularly serine, asparagine, and glycine) tend to diffuse out of the cells and become so diluted in the extracellular volume that their intracellular levels fall below the minimum needed for protein biosynthesis and cellular multiplication. It is therefore necessary to supply some or all of the "nonessential" amino acids from exogenous sources in order to keep intracellular levels high enough to support cellular multiplication in clonal cultures (LOCKART and EAGLE, 1959; EAGLE, 1959; EAGLE and PIEZ, 1962; HAM, 1972, 1974b). Also, if media are deficient in vitamins that have cofactor roles in biosynthesis of the "nonessential" amino acids, such as pyridoxine (SWIM and PARKER, 1958) or folic acid (EAGLE, 1959; McKEEHAN et al., 1977), cellular multiplication will not occur in the absence of exogenous supplies of the appropriate amino acids. Thus, although the seven "nonessential" amino acids can be left out of culture media under some conditions, all 20 amino acids that are involved in protein biosynthesis are being included in medium formulations with increasing frequency.

3. Other Vitamins

Two water soluble vitamins, biotin and vitamin B_{12}, that are not contained in MEM are now rather generally accepted as growth requirements for cultured cells (HIGUCHI, 1973). Both interact strongly with serum proteins (VALLOTTON et al., 1965; ALLEN, 1975) and requirements for them tend to be difficult to demonstrate in media containing dialyzed serum. Growth responses to biotin have been demonstrated through use of serum free media (HAM, 1967; HAGGERTY and SATO, 1969; MOSKOWITZ and CHENG, 1979); and also when dialyzed serum levels are kept low (MESSMER and YOUNG, 1977). Biotin has also been reported to be needed for transferrin mediated stimulation of SV3T3 cells (YOUNG et al., 1979) and for adipose conversion of 3T3 cells (KURI-HARCUCH et al., 1978). Stimulation of multiplication by vitamin B_{12} has also been reported by a number of investigators (PRICE et al., 1966, 1967; ROTHERHAM et al., 1971; HIGUCHI, 1973; MIERZEJEWSKI and ROZENGURT, 1976, 1977a; RUDLAND et al., 1977; O'FARRELL et al., 1979). Vitamin B_{12} is normally taken into cells in vivo as a complex with a binding protein known as transcobalamin II (FRIEDMAN et al., 1977; YOUNGDAHL-TURNER et al., 1978). The possibility remains that some types of cultured cells may in the future be found to be dependent on this protein. Since both of these vitamins are precursors of cofactors that function in a very limited number of biosynthetic reactions, there is at least a theoretical possibility of complete replacement of the requirements by providing appropriate end products from the biosynthetic reactions.

The literature contains many conflicting reports concerning possible multiplication-promoting effects of ascorbic acid (vitamin C). Ascorbic acid is extremely unstable in culture media (FENG et al., 1977), which makes negative results difficult to interpret. In addition, under oxidizing conditions ascorbic acid may have toxic effects, which appear to be due to peroxide formation in the culture medium (PETERKOFSKY and PRATHER, 1977; KOCH and BIAGLOW, 1978). The best known biochemical role of ascorbic acid is in the hydroxylation of proline, and there are a number of reports of enhanced hydroxypyroline synthesis in ascorbic acid enriched cultures (e.g., LEVENE and BATES, 1975). Multiplication of a mouse plasmacytoma cell line has been reported to require ascorbic acid (PARK et al., 1971) and there are scattered reports of improved growth of human diploid fibroblasts in the presence of ascorbic acid (LEMBACH, 1976; ROWE et al., 1977). Further study of the effects of ascorbic acid on cultured cells is clearly needed. There is also a need for the development of means of stabilizing ascorbic acid in culture media.

The fat soluble vitamins (A, D, E, K) are somewhat of an enigma in that they are clearly established as growth requirements for intact animals, but generally appear not to be required for cultured cells with the possible exception of vitamin E, whose status remains uncertain. Vitamin E has been reported to be beneficial to growth of guinea pig aortic smooth muscle cells, but several non-related antioxidants have similar activity such that a specific vitamin role is questionable (CORNWELL et al., 1979). Vitamin E also appears to be slightly stimulatory to 3 T 3 cells grown with delipidated serum (GIASUDDIN and DIPLOCK, 1979). The widely cited extension of lifespan of cultured human diploid fibroblasts by vitamin E (PACKER and SMITH, 1974) has proven not to be reproducible, even in the hands of the investigators who originally reported it (PACKER and SMITH, 1977; BALIN et al., 1977; SAKAGAMI and YAMADA, 1977). One possible explanation for the lack of apparent response in cultured cells to the fat soluble vitamins is that they may be involved only in highly specialized functions in the intact animal, rather than general cellular metabolism. It is also not clear whether the serum protein that is still required for multiplication of most types of nontransformed cells may be supplying small amounts of bound fat soluble vitamins. An answer to that question should become available in the near future for a few types of nontransformed cells that are currently close to being grown in the total absence of serum proteins.

4. Carbohydrates and Intermediates of Energy Metabolism

All cell culture media must contain a sugar or closely related substance as the major energy source. Glucose is by far the most widely used sugar. Mannose can generally be substituted for glucose, and fructose or galactose usually work reasonably well if pyruvate is also added (EAGLE et al., 1958; BURNS et al., 1976). In culture media designed for use without a CO_2-bicarbonate buffer system, part or all of the glucose is often replaced with a mixture of galactose and pyruvate to reduce the rate of anerobic metabolism and acid production (LEIBOVITZ, 1963; LING et al., 1968; WAYMOUTH, 1978).

Pyruvate or other 2-oxocarboxylic acids are required for growth of small inocula of many different types of cells (NEUMAN and McCOY, 1958; EAGLE, 1959; HERZENBERG and ROOSA, 1960; HAM, 1962). Recent data suggest that any of several 2-oxocarboxylic acids are interchangeable in promoting the multiplication of human diploid fibroblasts (McKEEHAN and McKEEHAN, 1979a). The amount of 2-oxocarboxylic acid that is needed to promote cellular multiplication rises sharply as the amount of serum protein is reduced, leading to the speculation that the 2-oxocarboxylic acids may be involved in a fundamental multiplication-regulating mechanism, possibly mediated by the intracellular $NAD^+/NADH$ ratio (McKEEHAN and McKEEHAN, 1979a). A possible regulatory role of 2-oxocarboxylic acids has also been postulated by GROELKE et al. (1979).

Acetate is a component of several culture media. Although its position in metabolism of mammalian cells remains obscure, acetate was found to be somewhat beneficial for clonal growth of human epidermal keratinocytes with minimal amounts of serum protein and was included in the optimized medium for those cells (PEEHL and HAM, 1980b).

5. Nucleic Acid Components

With an adequate supply of essential vitamins, most cultured cells are capable of synthesizing all needed nucleic acid components. However, many types of nontransformed cells are relatively inefficient in uptake and utilization of folic acid and thus often exhibit deficiencies in synthesis of purines and thymidine (HUENNEKENS et al., 1978; BALK et al., 1978; NEUGUT and WEINSTEIN, 1979). About half of the widely used media whose compositions are summarized by HAM and McKEEHAN (1979) contain one or more purine or pyrimidine components. Even with the availability of folinic acid, which is utilized more efficiently than folic acid, many modern media continue to be supplemented with purines and thymidine. For example, during the course of quantitative optimization for clonal growth of human diploid fibroblasts, a medium supplemented with adenine and thymidine and containing only a very low level of folinic acid was found to be slightly more effective than media containing larger amounts of folic or folinic acid but no purines or thymidine (McKEEHAN et al., 1977). Clonal growth of human epidermal keratinocytes with minimal amounts of serum protein is improved by adding 1.8×10^{-4} M adenine to the medium. The very high level of adenine in the keratinocyte medium (MCDB 151) is somewhat inhibitory to fibroblasts and contributes to the selective growth of keratinocytes in preference to fibroblasts in that medium (PEEHL and HAM, 1980b).

A novel purine derivative, 6,8-dihydroxypurine, which was isolated from peptone, has been reported to stimulate multiplication of some types of cells (YAMANE and MURAKAMI, 1973). However, it is not included in most culture media and many types of cells appear to be able to multiply without it.

6. Other Organic Nutrients

Putrescine or other polyamines are required for growth of Chinese hamster ovary cells in protein free media (HAM, 1964; HAMILTON and HAM, 1977) and are also reported to be beneficial for growth of various other types of cells (POHJANPELTO and RAINA, 1972; CLO et al., 1976, 1979; JÄNNE et al., 1978; BOTTENSTEIN and SATO, 1979; BARNES and SATO, 1980). Even in cells that appear not to require an exogenous source of putrescine for multiplication, one of the early events in the activation of multiplication is an increase in the level of ornithine deoxycarboxylase, the enzyme responsible for putrescine synthesis (reviewed by RUSSELL, 1973; TABOR and TABOR, 1976).

Many other organic substances have been observed to stimulate cellular multiplication under certain circumstances. However, in many cases the responses are marginal or potentially attributable to contaminants in the supposedly active growth-promoting substances. Since the growth requirements of most types of nontransformed cells are not fully understood, it is extremely important to keep an open mind about the possibility of growth stimulation by such substances. However, a detailed review of isolated reports of multiplication responses that have not been verified by other laboratories will not be undertaken here.

7. Inorganic Trace Elements

One of the most serious deficiencies of MEM is its total lack of inorganic trace elements. The presence of bound trace elements makes it almost impossible to demonstrate trace element requirements in the presence of large amounts of dialyzed serum. Even under serum-free conditions, contaminants in the chemicals and water used for medium preparation make trace element studies difficult. It has not yet been possible to demonstrate cellular growth responses for many of the trace elements that are currently known to be needed for whole animal nutrition (UNDERWOOD, 1977). However, several trace elements are clearly needed by cultured cells and others can be shown to be at least somewhat beneficial under certain conditions. In addition, it is likely that still other elements will prove to be essential when more highly purified chemicals and growth factors are available for medium preparation. Some investigators feel that it is desirable to add as many trace elements as possible to current culture media in order to avoid possible future growth failure if improvements in the purity of chemicals are made quietly by manufacturers without calling attention to the levels of impurity that were previously present.

Iron is the most clearly established and generally accepted of the trace element requirements of cultured cells (reviewed by HIGUCHI, 1973). In view of the major roles played by iron in energy metabolism, it is unlikely that any type of cultured cell could multiply in the total absence of iron. The ferric form of iron is quite insoluble at physiological pH values and is likely to be lost from culture media during sterilization by filtration (BIRCH and PIRT, 1970). Ferrous iron is far more soluble, and is a more effective way to supply iron to culture media, although ferric iron can also be used if it is complexed (MESSMER, 1973; YOUNG et al., 1979). However, ferrous iron becomes oxidized to the ferric form rather quickly and thus can also be lost during filtration (BIRCH and PIRT, 1970). Hence, it is desirable not to add ferrous iron to the culture medium until just before the medium is to be sterilized.

For many types of cells grown in the complete absence of serum protein, multiplication is better when the iron transport protein transferrin is included in the culture medium (reviewed by BARNES and SATO, 1980). In at least some cases, there appears to be a specific interaction between transferrin and the B vitamin biotin (YOUNG et al., 1979). A siderophore-like iron-binding factor that solubilizes iron is found in medium conditioned by cells that have been adapted to grow in the presence of picolinic acid. Similar factors are postulated to be involved in normal cell multiplication (FERNANDEZ-POL, 1978 a, b).

Zinc and selenium are two other trace elements whose cell culture requirement can be considered to be relatively well established. Zinc tends to be an ubiquitous contaminant, such that maximum responses are often not seen unless special steps are taken to purify the basal medium (THOMAS and JOHNSON, 1967; cf. also review by HIGUCHI, 1973). Selenium is required by a number of different types of cells (MCKEEHAN et al., 1976; HAMILTON and HAM, 1977; GUILBERT and ISCOVE, 1976; GIASUDDIN and DIPLOCK, 1979; BOTTENSTEIN and SATO, 1979; BOTTENSTEIN et al., 1979; BARNES and SATO, 1980; AGY et al., 1981). Selenium is a component of the enzyme glutathione peroxidase which helps to remove metabolically generated peroxides from cells. Selenium deficient cells are more sensitive to oxygen toxicity than cells grown with sufficient amounts of selenium (MCKEEHAN et al., 1977). Selenium is an essential nutrient for at least 40 different animal species (FROST and LISH, 1975; GRIFFIN, 1979), and it is likely that it will ultimately prove to be required by all types of cultured cells.

Several other trace elements are also strongly implicated as possible cell growth requirements, although in most cases background media have not been prepared that are clean enough to demonstrate absolute requirements. At least marginal stimulation of growth can be obtained by the addition of copper, manganese, molybdenum and vanadium to the media for human diploid fibroblasts (MCKEE-HAN et al., 1977). A tripeptide (glycylhistidylserine) that stimulates multiplication of certain types of cells is thought to function as a chelating agent that acts synergistically with iron and copper (PICKART and THALER, 1979).

Excess manganese is quite toxic to cultured cells (THOMAS and JOHNSON, 1967; BARNES and SATO, 1980; AGY et al., 1980), although at slightly lower concentrations it appears to be at least somewhat beneficial for cellular multiplication (MCKEEHAN et al., 1977; SHIPLEY and HAM, 1980a; AGY et al., 1980). It has been reported that the toxicity of manganese can be reversed by increasing the amount of iron in the medium (THOMAS and JOHNSON, 1967). There are a few reports that cadmium may stimulate cellular multiplication (RUBIN and KOIDE, 1973; RUBIN, 1975b; BARNES and SATO, 1980). Cadmium is generally viewed as a highly toxic substance with no known beneficial effects, and the possibility remains that the benefit that is observed may be only the paradoxical stimulation that is often observed with toxic substances at levels just below their toxic thresholds (cf. Sect. C.IX.1). However, not too many years ago other elements now viewed as essential trace elements, such as chromium and selenium, were viewed only as toxic substances.

There remain a number of confirmed or suspected essential trace elements in animal nutrition that have not yet been shown to be beneficial to cultured cells. The number of elements to be included in such a list varies from one trace element "expert" to another. However, in addition to the elements discussed above, a reasonably complete list of trace elements should probably include arsenic, chromium, fluorine, iodine, nickel, silicon, and tin (UNDERWOOD, 1977; NIELSEN, 1981). Cobalt could also be added, but it is probably supplied in sufficient amounts in most media as a structural component of vitamin B_{12}.

In addition to these elements with known or suspected essential roles, there are a number of other elements that accumulate in biological systems for which no essential roles are known at the present time (UNDERWOOD, 1977). These elements should also be kept in mind as media of higher and higher purity are prepared. It will not be known with certainty whether or not they have essential roles until contamination of culture media with these elements can be reduced to a level too low to reasonably expect biological activity.

V. Quantitative Optimization of Synthetic Media

1. Species and Cell Type Individuality of Quantitative Requirements

There have been hints in the literature for a long time that quantitative nutrient balance is very important in the design of culture media and that quantitative requirements differ from one cell type to another. For example, the optimum amino acid concentrations for HeLa and Mouse L cells were found to be different and Eagle actually published two different versions of his "basal medium," one for HeLa cells and one for Mouse L cells (EAGLE, 1955). SWIM (1967) lists differing optimum concentrations of amino acids for a number of different types of established cell lines. The "optimum" concentrations of at least seven relatively hydrophobic amino acids for clonal growth of Chinese hamster ovary cells in protein free media are strongly interdependent, suggesting that balance is more important than absolute concentration (HAM, 1974a).

During studies seeking to minimize the amount of serum protein required for clonal growth of various types of cells, members of the author's research group have found repeatedly that quantitative adjustment of nutrient concentrations is of major importance (HAM and MCKEEHAN, 1978a, b, 1979; HAM, 1980). Optimized media have been developed for human fibroblasts (MCKEEHAN et al., 1977), human keratinocytes (PEEHL and HAM, 1980b), Chinese hamster ovary lines (HAM, 1965; HAMILTON and HAM, 1977), mouse 3T3 cells (SHIPLEY and HAM, 1981), mouse neuroblastoma cells (AGY et al., 1981), and chicken and duck fibroblasts (HAM and MCKEEHAN, 1978b). Qualitatively, these media are all quite similar. However, quantitatively each is unique.

The compositions of all except the keratinocyte medium are compared in the summary table of medium compositions in HAM and MCKEEHAN (1979) and optimized media for human fibroblasts and keratinocytes are compared in PEEHL and HAM (1980b). Table 1 summarizes the differences in the molar ratio of histidine to tryptophan in the various optimized media. Although the 200-fold difference in ratios seen among the media in Table 1 is more dramatic than is the case for most other nutrients, very distinctive quantitative patterns have emerged for every cell type studied thus far. Human epidermal keratinocytes require an unusually high concentration of adenine and grow well with an unusually low concentration of calcium. Concentrations of the branched chain of amino acids are very high in the

Table 1. Histidine: tryptophan molar ratios in optimized media

Medium	Cell type optimized for	Histidine moles/liter[a]	Tryptophan moles/liter[a]	Ratio his:tryp
MCDB 501	Duck embryo fibroblast	1.0E-5	1.0E-5	1.0
MCDB 411	Mouse neuroblastoma	3.0E-5	1.0E-5	3.0
MCDB 202	Chicken embryo fibroblast	1.0E-4	3.0E-5	3.3
MCDB 151	Human epidermal keratinocyte	8.0E-5	1.5E-5	5.3
MCDB 301	Chinese hamster ovary	1.0E-4	1.0E-5	10.0
MCDB 105	Human diploid fibroblast	1.0E-4	1.0E-5	10.0
MCDB 402	Mouse 3T3 (Swiss)	2.0E-3	1.0E-5	200.0

[a] Computer style exponential notation is used. 1.0E-5 means $1.0 \times 10^{-5} M$

medium for mouse 3 T 3 cells and significantly lower in the medium for mouse neuroblastoma cells. Niacinamide concentrations are very high in the optimized media for several of the nontransformed cell types. Quantitative differences in $Na^+ : K^+$ ratios among various optimized media have already been discussed in Sect. C.III. Quantitative balance relationships among the inorganic trace elements are also of major importance. For example, zinc "deficiency" can be induced in L-cells by increasing the concentration of iron, manganese, or cobalt in the medium (ZOMBOLA et al., 1979) and manganese "toxicity" can be reversed with iron (THOMAS and JOHNSON, 1967).

2. Reduction of Requirements for Serum and Growth Factors

The amount of reduction of serum protein requirement that can be achieved by quantitative optimization of the culture medium is dramatic (McKEEHAN et al., 1977; HAM and McKEEHAN, 1978 a; PEEHL and HAM, 1980 b; SHIPLEY and HAM, 1981; AGY et al., 1981). Although the total number of different cell types from different species for which optimized media have been developed is still rather small, currently available information suggests that virtually every cell type from every species will prove to have individualized quantitative growth requirements. Serum proteins have the ability to compensate for media that have not been optimized, apparently in many different ways. (This topic is explored in detail in HAM and McKEEHAN, 1978 b). The number of different growth-promoting activities from dialyzed serum that are needed for cellular multiplication appears to be reduced substantially through the use of an optimized culture medium. Preliminary studies on the replacement of serum with hormones and growth factors in optimized and non-optimized media also appear to indicate that a less complex mixture is needed to replace serum for HeLa cells when an optimized medium is used (WU and SATO, 1978). There is also an older report that HeLa cells can be grown in yet another basal medium with insulin as the only macromolecular supplement (BLAKER et al., 1971), which has not been reconciled with the more recent studies. In extreme cases, such as Chinese hamster ovary cells (HAM, 1965; HAMILTON and HAM, 1977) and mouse neuroblastoma cells (AGY et al., 1981), the need for serum protein has been totally eliminated by optimizing the medium and adding a few nutrients that were not previously present.

Any time that growth factor assays are performed in a medium that has not been optimized for the cell type in question, it is likely that additional "growth factors" will be found to be needed above and beyond those that are needed in an optimized medium. It is not quite fair to call such requirements "spurious," since they represent alternative ways of satisfying the same basic cellular growth requirements. As emphasized in Sect. B.III, many complex interactions are involved in satisfying the growth requirements of cultured cells, such that a holistic approach is necessary. Thus, for example, cellular multiplication can be improved by increasing the extracellular concentration of a rate-limiting nutrient, by increasing the efficiency of transport of that nutrient into the cell, or by increasing efficiency of utilization of that nutrient within the cell. Therefore, a growth factor that stimulates pinocytosis, as has been reported for epidermal growth factor (HAIGLER et al., 1979) and platelet-derived growth factor (DAVIES and ROSS, 1978), or that stimulates active transport (HOCHSTADT et al., 1979), could promote cellular multiplication at least partially by indirect mechanisms involving changes in intracellular concentrations of nutrients. Similarly, a hormone that increases the efficiency of utilization of the nutrient after it is in the cell could also increase the multiplication response to that nutrient even though its extracellular and intracellular concentrations remain unchanged.

Since defined nutrients are generally easier to work with than specialized macromolecular growth factors or totally undefined serum factors, it is usually desirable to optimize the basal medium and minimize the amount of supplement needed irrespective of the nature of the supplement that is being used. Unfortunately, the process of quantitative optimization is long, tedious and labor intensive. However, in every case where it has been carried to completion in the author's laboratoy, it has proven to be well worthwhile. In the most extreme case, it made possible growth with 2% dialyzed serum of human epidermal keratinocytes, which previously could be grown only through the use of a feeder layer in a medium supplemented with 20% whole serum (PEEHL and HAM, 1980b).

VI. Lipids and Related Substances

A substantial fraction of the total mass of typical cultured cells is composed of lipids. When grown in the presence of serum, which even after dialysis is a rich source of lipids, cultured cells utilize lipids derived from the serum in preference to *de novo* synthesis. Free fatty acids in serum are bound to serum albumin, while all other lipids, including cholesterol, cholesteryl esters, triglycerides, and phospholipids, are carried by the serum lipoproteins (reviewed by SPECTOR, 1972; KING and SPECTOR, 1981).

1. Fatty Acids

When deprived of an exogenous source of lipids, cultured cells are able to satisfy most of their lipid requirements by *de novo* synthesis. However, mammalian cells are unable to synthesize linoleic acid and related polyunsaturated fatty acids, which are often referred to as "essential fatty acids." The essential fatty acids are required nutrients for intact animals. However, cultures of highly adapted established lines can be grown indefinitely in serum-free and lipid-free media with no apparent source of linoleic acid or other essential fatty acids (reviewed by KATSUTA and TAKAOKA, 1973, 1978; KING and SPECTOR, 1981). In such cases, the essential fatty acid of the cellular lipids appears to drop to zero (KAGAWA et al., 1970; LENGLE and GEYER, 1972; BAILEY and DUNBAR, 1973).

Not all types of cultured cells will multiply in the total absence of lipids, however. There are many reports that linoleic acid, oleic acid, or mixtures of both stimulate multiplication of cultured cells (reviewed by YAMANE, 1978; KING and SPECTOR, 1981; BARNES and SATO, 1980). Because of the severe toxicity of free fatty acids, delipidated serum albumin is often added as a carrier (JENKIN and ANDERSON, 1970; YAMANE et al., 1976; KING and SPECTOR, 1981; BARNES and SATO, 1980). β-lactoglobulin can also be used as a carrier for fatty acids (NILAUSEN, 1978).

Biotin and pantothenic acid are both intimately involved in the biosynthesis of fatty acids. Since biotin appears to have relatively few other roles, biotin deficiency is sometimes used to inhibit synthesis of fatty acids in cultured cells (CORNELL et al., 1977; MESSMER and YOUNG, 1977). When cells are grown in synthetic media that contain only minimal amounts of linoleic acid, optimal levels of both biotin and pantothenic acid must be provided to facilitate *de novo* synthesis of other lipids, including saturated and monounsaturated fatty acids and steroids.

The fatty acid composition of the membrane phospholipids of cultured cells can be manipulated extensively by varying the types of fatty acids supplied in the culture medium. The fatty acids can be supplied in any of several different ways, including complexes bound to delipidated serum albumin, Tween esters of fatty acids, lipids added back to delipidated lipoproteins, or phospholipid vesicles (reviewed by KING and SPECTOR, 1981). Growth rate of Yoshida sarcoma cells is reported to be proportional to membrane fluidity produced by various mixtures of fatty acids (YAMANE and TOMIOKA, 1979).

2. Prostaglandins

In addition to serving as a source of fatty acids for membrane phospholipids, linoleic acid also serves as a precursor for the biosynthesis of prostaglandins, a family of hormone-like substances with multiple biological roles. There are a number of reports in the literature suggesting that prostaglandins are involved in cellular proliferation, probably in regulatory roles. Quiescent mouse fibroblasts can be stimulated to undergo DNA synthesis by prostaglandin $F_{2\alpha}$ (JIMENEZ DE ASUA et al., 1975; O'FARRELL et al., 1979). Detailed analysis of this stimulation suggests that it is similar in many respects to the effects of fibroblast growth factor (RUDLAND and JIMENEZ DE ASUA, 1979). Thymic lymphocytes are stimulated by prostaglandin E_1 and also to a lesser extent by prostaglandin A_1 (MACMANUS and WHITFIELD, 1974). Canine kidney line MCDK requires prostaglandin E_1 or dibutyryl cyclic AMP for growth in a synthetic medium supplemented only with purified hormones and growth factors (TAUB et al., 1979). Guinea pig aortic smooth muscle cells are stimulated by prostaglandins $F_{1\alpha}$ and $F_{2\alpha}$ (CORNWELL et al., 1979). Mammary carcinoma line MCF requires prostaglandin $F_{2\alpha}$ for multiplication in a hormone and growth factor supplemented medium (BARNES and SATO, 1979). Thus, although the prostaglandins have not yet been fully integrated into most cell culture systems, there are many indications in the literature that they may be of major importance, along with the essential fatty acids that they are derived from.

3. Phospholipids

Choline and myo-inositol, which are two of the basic building blocks for synthesis of phospholipids, are rather widespread requirements for multiplication of cultured cells, and both are included in Eagle's MEM. Very recently, phosphoethanolamine, a basic building block for a third class of phospholipids, has been shown to be a specific requirement for multiplication of a rat mammary carcinoma line (KANO-SUEOKA et al., 1979 b).

Sonication of phospholipid suspensions results in the formation of membranous phospholipid vesicles often referred to as liposomes (POSTE et al., 1976). Liposomes appear to be a very promising method for supplying lipids without toxicity to cultured cells in protein-free media (MCKEEHAN and HAM, 1978 b; ISCOVE and MELCHERS, 1978). In addition to supplying lipids directly, liposomes surround an internal aqueous phase which can be used to "inject" almost anything into cells (PAGANO and WEINSTEIN, 1978). Both cholesterol (ROTHBLAT et al., 1978) and free fatty acids (HOSICK, 1979) have been supplied to cultured cells contained within liposomes.

4. Cholesterol

Cholesterol and related steroids can be synthesized *de novo* in adequate amounts by most types of cultured cells. However, as was the case with the fatty acids, *de novo* synthesis is suppressed when an exogenous supply is available. There are scattered reports in the literature that cholesterol is required by or improves growth of certain types of cultured cells (reviewed by HIGUCHI, 1973; CHEN and KANDUTSCH, 1981). Oxidized derivatives of cholesterol such as 25-hydroxycholesterol inhibit the synthesis of cholesterol and make cells dependent on cholesterol supplied from exogenous sources (SINENSKY, 1979; CHEN and KANDUTSCH, 1981).

5. Synthetic Media

Both for the essential fatty acids and for cholesterol, there is a critical need for more precise studies of the requirements of nontransformed cells grown under lipid-free conditions. Delipidated serum is not fully satisfactory as a substrate for such studies, since the apo-lipoproteins may attract lipids so avidly that they disrupt normal lipid-containing structures in the cells grown in such media. In the near future, as nontransformed cells are grown in media supplemented only with hormones and purified growth factors, it will become possible to evaluate their lipid requirements more precisely. At present, all classes of lipids should be viewed as potential growth requirements for nontransformed cells.

VII. Hormones, Hormone-Like Growth Factors, and Carrier Proteins

1. Hormones

As defined at the beginning of this chapter, the term "hormone" is being used here only to refer to substances with well-defined endocrinological roles. Although it is just now coming to full fruition, the use of hormones in culture media has a long history. Insulin was first reported to be beneficial for cultured cells even before it was completely purified (GEY and THALHIMER, 1924) and available hormones have been used quite extensively in cell culture experiments ever since (reviewed by WAYMOUTH, 1954; LASNITZKI, 1965; HIGUCHI, 1973; BARNES and SATO, 1980; CRISTOFALO, this volume).

The boundary line between classically defined hormones and the cellular growth factors described in the following section of this chapter (and elsewhere in this volume) is virtually imperceptible, and will probably disappear entirely in the near future. Perhaps the key test for a true hormone should be clear demonstration of an in vivo endocrine function. In the case of somatomedin (once known as sulfation factor) such a relationship had already been established. Many authors assume that all cellular growth factors that do not function as nutrients will ultimately prove to be true hormones and use the term "hormone" freely when refering to them (e.g., RIZZINO et al., 1979; SCHER et al., 1979a). However, others have remained more cautious and continue to refer to "hormone-like growth factors" (e.g., BARNES and SATO, 1980).

2. Growth Factors

Growth factors were defined at the beginning of this chapter as substances that promote cellular multiplication without participating in cellular metabolism as nutrients. Although detailed mechanisms of action remain poorly understood, it is generally assumed that the effects of growth factors on cellular multiplication are strictly regulatory. Since growth factors are the major topic of this volume, they will be discussed here only briefly and in general terms.

Tissue specific growth factors have been known for a long time, but until relatively recently they have been studied primarily in terms of cellular differentiation, rather than multiplication (RUTTER et al., 1973). Epidermal growth factor (EGF) (CARPENTER and COHEN, 1979; CARPENTER, this volume) is an example of a factor originally isolated on the basis of its developmental effects that has proven to be of major importance as a cellular growth factor. Similarly, the somatomedins (LUFT and HALL, 1975; VAN WYK, this volume), which are now of major importance as cellular growth factors, can be traced back to "sulfation factor" activity for chondrocyte growth and differentiation (DAUGHADAY et al., 1972). Nerve growth factor (YOUNG, this volume) is an example of a well known differentiation-promoting factor that has thus far achieved only limited importance as a cellular growth factor (BARNES and SATO, 1980).

There are also a number of cellular growth factors that have been isolated through use of assay systems based on cellular multiplication. By the early 1970's, it had become clear that serum was a difficult starting material for the isolation of multiplication-promoting substances for cultured cells, both because of the extreme complexity of serum, and because of the low concentrations of the individual factors in it. The search for alternate sources of multiplication-promoting activities led ultimately to the isolation and characterization of two growth factors that are now of major importance. Crude luteinizing hormone preparations from the pituitary gland were found to contain a potent growth-promoting activity for 3 T 3 cells (ARMELIN, 1973), which was later identified by GOSPODAROWICZ (1974) as fibroblast growth factor (FGF). In other studies, plasma was found not to contain growth-promoting activity comparable to that of serum (BALK, 1971). This led to the isolation from blood platelets of platelet-derived growth factor (PDGF) (ROSS, this volume), which is released from platelets into serum during clotting.

Many other growth factors have been reported in the literature (reviewed by GOSPODAROWICZ and MORAN, 1976; RUDLAND and JIMENEZ DE ASUA, 1979), and new ones are constantly appearing. A recent tabulation by RUDLAND and JIMENEZ DE ASUA (1979) lists 19 different peptide factors with multiplication-promoting activity for vertebrate cells, and it would not be difficult to construct a substantially longer list from the current literature. For example, a partially characterized growth factor has been prepared from extracts of bovine hypothalamus that enhances multiplication of human endothelial cells and that appears to be distinct from FGF (MACIAG et al., 1979). However, since many of the reports of multiplication-promoting activity are very recent, and since many of the factors have not yet been fully purified or adequately compared to known factors from other sources, such a listing would be premature. One recent example of a "growth factor" that later proved to be a nutrient is "mammary growth factor" (KANO-SUEOKA et al., 1979a), which was later identified as phosphoethanolamine, a structural component of phospholipids (KANO-SUEOKA et al., 1979b).

Recent reports of growth factor production by tumor cell lines suggest that one of the major differences between normal and transformed cells could be autonomous production of essential regulatory factors by the latter (DE LARCO and TODARO, 1978; TODARO et al., 1979; HELDIN et al., 1979).

The biological roles of most of the growth factors are still incompletely understood. Analysis of temperature sensitive cell cycle mutants of BHK cells suggests at least two separate genes whose mutants can complement each other are involved in initiation of the cell cycle (JONAK and BASERGA, 1979). Recent studies of the initiation of proliferation in quiescent 3 T 3 cells reveal the presence of at least two, and probably three distinct control points in the sequence that leads from G_0 to the initiation of DNA synthesis (PLEDGER et al., 1977, 1978; VOGEL et al., 1978; RUDLAND and JIMENEZ DE ASUA, 1979; ROSSOW et al., 1979; BROOKS et al., 1980). Analysis of the growth factor requirements for these transitions (STILES et al., 1979a; SCHER et al., 1979a) indicates that the earliest event, which is referred to as acquisition of "competence" is promoted by platelet growth factor, fibroblast growth factor, or the presence of a calcium phosphate precipitate in the medium; and that the subsequent steps, which are referred to a "progression," are mediated by somatomedin-like factors that are present in plasma as well as serum. During active proliferation, competence for the next cycle is acquired during G_1 (YEN and PARDEE, 1978; YEN and RIDDLE, 1979) or M (SCHER et al., 1979b) of the previous cycle. Once the mitogen-induced competence has been achieved, the cells will proceed autonomously through the cell cycle without further need for the mitogen. Similarly, temperature-sensitive mutants that arrest in G_1 at nonpermissive temperatures proceed autonomously through the rest of the cycle and retain the ability to reactivate nuclei in heterokaryons when they are transferred to a nonpermissive temperature after initiation at a permissive temperature (FLOROS and BASERGA, 1980).

In addition to specific growth factors, many other substances, including sodium, calcium and magnesium ions, polyamines, prostaglandins, cyclic nucleotides, 2-oxocarboxylic acids and nutrients have been implicated in the overall process of controlling whether or not nontransformed cells proliferate. Many complex theories have been proposed for control of the cell cycle, but at present there is no generally agreed upon regulatory mechanism and a detailed discussion of the competing theories is clearly beyond the scope of this chapter.

3. Replacement of Serum with Hormones and Growth Factors

One of the most rapidly moving areas of cell culture at the present time is the replacement of serum requirements with mixtures of hormones, growth factors, and other highly purified proteins (PAPACONSTANTINOU and RUTTER, 1978; SANFORD, 1978; SATO and ROSS, 1979; BOTTENSTEIN et al., 1979; WAYMONTH et al., 1981; BARNES and SATO, 1980). A definitive hypothesis that the primary role of serum in promoting cellular multiplication is to provide hormones and hormone-like factors was stated in a theoretical paper published in 1975 (SATO, 1975), and was followed the next year by a practical demonstration of the complete replacement of serum with mixtures of hormones and growth factors (HAYASHI and SATO, 1976). Since then there has been a veritable flood of papers extending these results to many other types of cells (reviewed by BOTTENSTEIN et al., 1979; BARNES and SATO, 1980).

The review by BARNES and SATO (1980) list twenty different hormones and hormone-like growth factors that have been found to be beneficial for multiplication of one or more types of cells. These include: 1) steroid hormones (testosterone, estrogen, progesterone, hydrocortisone); 2) thyroid hormone (triiodothyronine); 3) prostaglandins (PGE_1, $PGF_{2\alpha}$); 4) classical peptide hormones (insulin, glucagon, follicle stimulating hormone, luteinizing hormone, parathyroid hormone, growth hormone, somatomedin C, thyrotropin releasing hormone, and luteinizing hormone releasing hormone); and 5) hormone-like growth factors (epidermal growth factor, fibroblast growth factor, gimmel factor, and nerve growth factor). Another class of proteins found stimulatory to multiplication are those that function primarily as carriers of essential substances (transferrin, which carries iron, and bovine serum albumin, which carries fatty acids). A third class of substances reported to be beneficial are the attachment and spreading factors which were discussed earlier in this chapter. Those listed by BARNES and SATO (1980) include cold insoluble globulin, serum spreading factor, fetuin, collagen gel, and polylysine coating of the culture dishes.

4. Individuality of Cellular Requirements

Although the total list is weighty, each individual cell type that has been tested has been found to require only a small subset of hormone and growth factor supplements for serum-free growth in a defined nutrient medium. One of the striking aspects of these studies has been the extent of differences in hormone and growth factor requirements that have been observed from one type of cell to another (BARNES and SATO, 1980; RIZZINO et al., 1979). For growth in the background media that have been employed, insulin and transferrin are rather uniformly required. However, the requirements for the various other hormones and growth factors vary widely from cell to cell, such that each general type of cell appears to have an unique set of requirements. These studies strongly reinforce the concept of individuality of cellular growth requirements that has also been observed during medium optimization studies in the author's laboratory (Sect. C.V).

The combined impact of these two diverse studies of cellular growth requirements is that very few predictions can be made about specific growth requirements for cell types that have not yet been studied. In essence, it is necessary to start over each time that the growth requirements of a new kind of cell are studied. Failure to recognize the extent of individuality of cellular growth requirements is probably one of the major reasons why investigators who are attempting to culture cells that proliferate continually in vivo have had so much difficulty getting them to do so in vitro. Many of our current prejudices about what cells *should* require for multiplication, both qualitatively and quantitatively, may have to be discarded in order to develop media and culture conditions that will support optimal proliferation in culture of rapidly cycling stem cells and other curently difficult-to-grow cells. It will be necessary to try things that do not "make sense" in terms of conventional growth requirements in order to discover the unconventional requirements of the cells that do not grow well under conventional conditions.

5. Nontransformed Cells

At present, the prospects for extending serum replacement by hormones and growth factors to nontransformed cells seem to be quite good, particularly if the more stringent requirements of nontransformed cells for precisely optimized basal media are also taken into account. In a number of cases, serum-free media that were developed for permanent lines of differentiated cells have also proven to be relatively effective for normal cells of the same type in primary culture, often with the added advantage that they do not support fibroblastic overgrowth (MATHER and SATO, 1979; BARNES and SATO, 1980). Two laboratories have recently reported growth of human diploid fibroblasts in optimized media supplemented with hormones and growth factors (PHILLIPS and CRISTOFALO, 1980 a, b; WALTHALL and HAM, 1980). Balb/c and Swiss 3 T 3 cells have also been grown in media supplemented with hormones and growth factors in at least two different laboratories (BOTTENSTEIN et al., 1979; SERRERO et al., 1979; SHIPLEY and HAM, 1980). A novel growth factor prepared from female rat submaxillary glands, designated gimmel factor, has been found by Sato's group to be beneficial for growth of both types of 3 T 3 cells (SERRERO et al., 1979; BOTTENSTEIN et al., 1979; BARNES and SATO, 1980). However, gimmel factor does not appear to be uniquely a factor for nontransformed cells, since it is also beneficial to the C 6 rat glioma cell lines (BOTTENSTEIN et al., 1979; BARNES and SATO, 1980). In addition, when optimized medium MCDB 402 is used, there appears not to be a specific requirement for gimmel factor for multiplication of 3 T 3 cells with hormones and growth factors (SHIPLEY and HAM, 1980; G. D. SHIPLEY, personal communication).

6. Relationship to In Vivo Growth Requirements

One of the major criticisms of hormonal substitution studies is the fact that in many cases there is no obvious rationale for the multiplication-stimulating activity of particular hormones or growth factors. The whole area of study is still very new, and most of the serum-replacing activities have been discovered as the result of trial-and-error screening programs in which essentially all hormones and hormone-like growth factors that are available for testing have been tested. There is a significant risk that the multiplication-promoting activities that are identified in this manner may not be the same as the components of serum that support multiplication. However, there is also a risk that the ways in which serum supports cellular multiplication may differ significantly from normal in vivo conditions that promote and control cellular proliferation.

For nontransformed cells in general, it is highly probable that future investigations will reveal that there are multiple ways of satisfying many of their individual growth requirements. It is already clear, for example, that the set of hormones and growth factors that are required for cellular multiplication can be altered by other aspects of the culture system. When corneal epithelial cells are grown directly on plastic surfaces, they have a flattened morphology and respond to fibroblast growth factor. However, when grown on collagen, they assume a tall columnar morphology and respond primarily to epithelial growth factor (GOSPODAROWICZ et al., 1978, 1979). As discussed in Sect. C.V.2, the requirement of HeLa cells for hormones and growth factors is also altered by the culture environment (WU and SATO, 1978).

For each of the major growth factor activities that have been discovered there appear to exist many different molecular species with at least some degree of biological activity. Many apparently different molecules cross react in varying degrees with insulin, including the somatomedins, multiplication-stimulating activity, and insulin-like growth factors IGF-I and IGf-II (FROESCH et al., 1979; NISSLEY et al., 1979; VAN WYK, this volume). Epidermal growth factor was originally isolated from male mouse salivary glands, but is also found in a slightly different molecular form in human urine (CARPENTER and COHEN, 1969; CARPENTER, this volume). The inhibitor of gastric secretion known as urogastrone appears to be identical with human urinary EGF (GREGORY, 1975). Cells transformed by murine sarcoma viruses also produce EGF-like peptides (TODARO et al., 1979). Platelet derived growth factor or closely related molecules have been isolated from several sources besides platelets, including serum and pituitary (ANTONIADES and SCHER, 1978). Fibroblast growth factor also shares many of the activities of platelet-derived growth factor, although in some systems responses to the two are not the same (SCHER et al., 1979 a).

In many cases, little is known concerning the precise molecular form of specific growth-promoting activities either in serum or in the interstitial fluids surrounding undisturbed cells in vivo. For each hormone or growth factor that is found to be stimulatory to multiplication of any kind of cell in culture, studies should be undertaken to identify the molecular forms that are active in serum and in interstitial fluids. In many cases, it is likely that the naturally active form in vivo will prove to be somewhat different from the storage form that has been isolated from tissue extracts. In some cases, complex interactions or substitutions may be involved, such that the natural in vivo form is quite different from the easier-to-isolate forms that are now being studied.

There still remain many cell types that can be grown only through the use of grossly undefined supplements. Because of the complexities that are encountered when an attempt is made to isolate multiple-growth promoting activities directly from serum or other unidentified supplements, it is highly desirable to replace as much of the requirement for undefined supplements as possible by optimizing the synthetic portion of the medium and by using hormonal and growth factor supplementation whenever possible. At the present time, direct isolation of growth-promoting activities from natural sources should be viewed as a last resort. However, as discussed in the previous paragraph, it is desirable after a full set of hormonal growth factor substitutions has been achieved to isolate and characterize any alternative forms of the individual growth-promoting activities that may be found in serum or interstitial fluids. The advantage of waiting until a defined hormone and growth factor supplemented medium is available is that a highly specific assay system for the alternative forms can then be generated by deleting one component at a time from the defined medium.

The use of hormones and growth factors is a major step forward in the development of defined media for cultured cells. The fact that many of the observed hormonal or growth factor replacements of serum activities do not "make sense" from our limited current perspective does not invalidate the importance of cellular multiplication under highly defined conditions. Also, as discussed above, innovations that do not "make sense" from a conventional viewpoint are almost certainly going to be required to obtain proliferation in culture of cell types that do not grow well in conventional culture systems.

Many years of detailed studies of intracellular regulatory mechanisms will be required to gain a full understanding of the reasons for growth stimulation by various hormones and growth factors and the relationship of such growth stimulation to events that occur naturally in vivo. The availability of serum-free media will make such studies much easier. Furthermore, when the true mechanisms responsible for growth simulation by the hormones and growth factors are fully understood, we will almost certainly discover that the observed responses failed to "make sense" only because our knowledge of the biological roles of the hormones and growth factors that were involved was far too limited.

VIII. Special Requirements Related to Cellular Density

As discussed briefly in Sect. B.V.2, most nontransformed cells exhibit special growth requirements when grown either at extremely low or extremely high cell densities. Since each condition results in a different set of special growth requirements, the two will be discussed separately.

1. Requirements for Clonal Growth

In moderately crowded cultures where the volume of medium per cell is not too large, complex interactions occur between the cells and the medium, which collectively are known as "conditioning" of the medium, a process that often is essential for cellular multiplication. The two main features of conditioning are detoxification and release into the medium of substances that are synthesized by the cells.

When the cellular inoculum is very small, the volume of medium per cell is so large that essentially no conditioning can be accomplished. Cellular multiplication can occur under these conditions only if the medium as formulated is free from toxicity and satisfies all of the growth requirements of the cells. Included in the "nutrients" that must be provided in the medium for clonal growth is a set of metabolic intermediates that are synthesized by the cells in sufficient amounts, but that are not retained in the intracellular space against a concentration gradient. In more crowded cultures, such substances accumulate in the conditioned medium at concentrations high enough to maintain adequate intracellular levels by equilibration. However, under clonal conditions, they fail to accumulate due to near infinite dilution and it becomes necessary to provide them from exogenous sources to support multiplication.

Most such requirements for clonal growth are relatively straightforward. The metabolic intermediates that most frequently must be provided in the culture medium for clonal growth have already been discussed in previous sections of this chapter. They are the nonessential amino acids (Sect. C.IV.2), pyruvate or other 2-oxocarboxylic acids (Sect. C.IV.4), and carbon dioxide (Sect. C.II.1). Clonal cultures also tend to be very sensitive to inadequacies in procedures used to prepare the cellular inoculum (Sect. C.II.3). The special requirements for clonal growth have been reviewed extensively in other publications (EAGLE and PIEZ, 1962; HAM, 1972, 1974b; HAM and McKEEHAN, 1978b).

The possibility also exists that special macromolecular factors may be needed for clonal growth in some cases. For example, mouse neuroblastoma Cl300 cells can be grown at monolayer densities in a totally protein-free synthetic medium, designated MCDB 411. However, at clonal density, no growth occurs in unsupplemented MCDB 411. Clonal growth can be restored through the use of medium previously conditioned by a monolayer of mouse neuroblastoma cells. A conditioned medium factor other than the common intermediary metabolites mentioned above appears to be involved. Addition of small amounts of insulin to the culture medium or simply coating the culture dishes with insulin and then washing off all that is not tightly bound before the cells or synthetic medium are placed in the dishes eliminates the need for conditioning and permits clonal growth of the neuroblastoma cells in the total absence of serum supplementation (AGY et al., 1981).

2. Requirements for Multiplication of Dense Cultures

At the other extreme, cellular crowding also generates special growth requirements. Nontransformed cells are highly sensitive to density-dependent inhibition of multiplication and it has been known for many years that the cellular density at which multiplication stops is determined by the amount of serum protein that is present (TODARO et al., 1965; HOLLEY and KIERNAN, 1968, 1971). The factor needed for dense growth is missing from plasma (BALK, 1971), and several research groups have demonstrated that it is released from blood platelets during clotting (Ross et al., 1974; KOHLER and LIPTON, 1974; WESTERMARK and WASTESON, 1976). The factor is currently known as platelet-derived growth factor (Ross and VOGEL, 1978; Ross et al., 1979; SCHER et al., 1979a; Ross, this volume) and appears to be required by sparse as well as dense cultures, as will be discussed below. It has recently been purified to homogeneity, with a specific activity of 2×10^7 times that of whole unfractionated serum (ANTONIADES et al., 1979).

Fibroblast growth factor, which is isolated from pituitary glands, has been reported to be capable of supporting multiplication of at least some types of dense cultures (GOSPODAROWICZ et al., 1975; STILES et al., 1979a). There are also reports that epidermal growth factor will overcome density-dependent inhibition of multiplication (CARPENTER and COHEN, 1976; WESTERMARK, 1976; MIERZEJEWSKI and ROZENGURT, 1977b). However, STILES et al. (1979a) report that EGF has no competence activity and only weak progression activity for 3T3 cells, and a recent abstract claims that PDGF is needed in addition to EGF for dense growth of human diploid fibroblasts in a synthetic medium supplemented only with hormones and growth factors (PHILLIPS and CRISTOFALO, 1980a).

Although definitive data are rather limited, it currently appears that the special requirements for growth of dense cultures of 3T3 cells are quantitative, rather than qualitative. Rapid depletion of the serum factors that are needed for multiplication of dense cultures has been described (HOLLEY and KIERNAN, 1971; cf. also comments by HOLLEY and BALDWIN, 1979). In early studies involving partial replacement of serum with hormones and growth factors (HOLLEY and KIERNAN, 1974a), FGF was found to be needed by sparse as well as dense cultures in the presence of 0.2% serum. Most current assays for PDGF activity are done with density-inhibited cultures. However, cultures of 3T3 cells have been held at a population density of about 4×10^3 cells/cm^2 in medium containing 5% "plasma-derived serum" (which lacks PDGF activity) for over a month with no loss of viability, but with less than one doubling during the entire period (Ross et al., 1978). Thus, it currently appears that PDGF is required by 3T3 cells at all cell densities for the initiation of the replicative cycle, and that the requirement for PDGF to overcome density-dependent inhibition of multiplication is quantitative in nature. The concentrations of nutrients in the culture medium also have an effect on the cell density at which multiplication of nontransformed cells stops (HOLLEY et al., 1978).

The continuing controversy over the exact mechanisms involved in density dependent inhibition of multiplication (WESTERMARK, 1977; WHITTENBERGER and GLASER, 1978; HOLLEY and BALDWIN, 1979; LIEBERMAN et al., 1979) make it difficult to evaluate the exact significance of the reversal of that inhibition by PDGF and related factors. However, investigators studying the growth requirements of nontransformed cells must remain aware that they are likely to encounter special requirements (which may be purely quantitative) when their cultures become crowded. Also, since PDGF is specifically associated with platelet damage, clotting, and wound healing, it is not safe to assume that PDGF has anything to do with density-related requirements of nonfibroblastic cells without performing specific confirmatory experiments.

IX. Regulatory Interactions and Artificial Stimulation of Multiplication

Although there still exist major gaps in our total understanding of the intracellular mechanisms that are responsible for the control of cellular multiplication, sufficient knowledge has been gained so that it is frequently possible to stimulate the cell cycle artificially on a short term basis. Thus, there are many substances that are capable of initiating a cellular multiplication response or at least DNA synthesis that cannot reasonably be classified as growth requirements for the cells in question. No attempt is made here to establish rigid definitions that will completely separate natural from artificial stimulation of multiplication. However, the basic principle that is involved is that true "growth factors" should be at least closely related to the circumstances that lead to multiplication of the type of cell under consideration within the intact organism. Even this is not entirely straightforward, however, since many of the types of cells that are studied in vitro in a state of rapid multiplication multiply only rarely in vivo. Nevertheless, there exist numerous techniques for grossly artificial stimulation of multiplication in culture. Selected examples are given in this section, with no attempt to cover the topic in a comprehensive manner or to pass judgement on borderline cases between normal and artificial stimulation of multiplication.

1. Borderline Toxicity

Paradoxical stimulation of multiplication by marginal levels of toxic substances has been known to microbial physiologists for a long time (LAMANNA and MALLETTE, 1965). There is evidence that comparable phenomena may also occur in cultured cells. For example, certain metal ions such as zinc, cadmium, and mercury and a variety of other toxic substances stimulate DNA synthesis in serum-deprived chicken embryo fibroblasts. In each case, the stimulation occurs at levels only slightly below those that are clearly toxic (RUBIN and KOIDE, 1973, 1975; RUBIN, 1975 b). Subtoxic concentrations of 9,10-dimethyl-1,2-benzanthracene have a similar effect. Control experiments appear to indicate that the stimulation is not due to release of stimulatory material by cells damaged by the toxic agents. The effects of zinc can also be separated from stimulatory effects on zinc as a trace element, which occur at lower concentrations.

2. Proteolytic Enzymes

Treatment of nontransformed cells with various types of proteases often leads to a temporary stimulation of DNA synthesis and multiplication. Since this phenomenon is the subject of another chapter (CUNNINGHAM, this volume), it will not be described in detail here.

3. Tumor Promoters

Tumor promoters are defined as agents which, although not themselves carcinogenic, induce tumors in animals previously exposed to a subthreshold dose of a carcinogen (DRIEDGER and BLUMBERG, 1980). Although the mechanisms are not yet fully understood, there are a number of reports in the recent literature that tumor promoters, and particularly the phorbol esters, stimulate DNA synthesis in serum deprived quiescent cultures of various types of nontransformed cells (DIAMOND et al., 1974; BOYNTON et al., 1976; SIVAK, 1977; FRANTZ et al., 1979; LECHNER et al.,

1980). There are also reports that the proliferative responses of Balb/c 3T3 cells to hormone-like growth factors are significantly enhanced by 12-0-tetradecanoyl-phorbol-13-acetate (DICKER and ROZENGURT, 1978; FRANTZ et al., 1979). Growth factors whose effects are enhanced included insulin, FGF, EGF, and a crude platelet extract with PDGF activity.

The exact role of phorbol esters in promoting cellular multiplication is not fully understood, but they appear to interact with highly specific receptor sites that are present both in chicken embryo fibroblasts and in intact mouse skin (DRIEDGER and BLUMBERG, 1980). There is also evidence for interaction between phorbol ester and EGF receptors and a modulation of EGF binding by active phorbol ester (SHOYAB et al., 1979). Multiplication rate kinetic analysis with mouse prostate epithelial cells has shown that phorbol esters both inhibit the multiplication-stimulating effects of EGF and have a lesser multiplication-promoting effect of their own (LECHNER and KAIGHN, 1980). It has also been reported that phorbol esters, like EGF, stimulate pinocytosis (HAIGLER et al., 1979).

4. Inorganic Ions

In recent years, inorganic ions, which were once viewed primarily as uniform constituents of the cellular environment, have been recognized as important control agents. Dramatic examples of control of biological processes by calcium, sodium and hydrogen ions can be seen in the acrosome reaction and egg activation during fertilization in sea urchins and many other species (EPEL, 1978). Evidence is accumulating that sodium, calcium, and magnesium ions all play important roles in control of cellular multiplication. Apart from any attempt to resolve the current controversies over which of these three ions are closest to the ultimate mechanisms that control cellular proliferation (discussed in Sect. C.III), it has become quite clear that artificial manipulation of the intracellular concentrations of any one of them can induce at least a temporary increase in DNA synthesis and cellular multiplication. Examples of such experiments include the following: 1) Mitosis has been induced in neurons that normally do not divide by sustained depolarization and increased intracellular concentrations of sodium (CONE and CONE, 1976); 2) Vasopressin, which promotes sodium influx, has a potent mitogenic effect on 3T3 cells (ROZENGURT, 1979; ROZENGURT et al., 1979); 3) Lithium stimulates DNA synthesis and cell multiplication in mouse mammary gland explants (HORI and OKA, 1979); 4) A calcium ionophore is mitogenic for human peripheral lymphocytes (JENSEN et al., 1977); 5) Ultrahigh concentrations of magnesium added in the absence of calcium stimulate DNA synthesis in chicken embryo fibroblasts (RUBIN et al., 1979; RUBIN and SANUI, 1979).

5. Cyclic Nucleotides

The literature on the role of cyclic nucleotides in cellular proliferation is extensive and sometimes contradictory. Although cyclic AMP was at one time viewed primarily as inhibitory to cellular proliferation, there is beginning to accumulate substantial evidence that a surge of cyclic AMP production may be an important event in the multiplication of a variety of types of cells (WHITFIELD et al., 1976; BOYNTON et al., 1978; GREEN, 1978; WHITFIELD et al., 1979). This viewpoint has been sub-stantiated by artificial stimulation of DNA synthesis and cellular multiplication by direct addition of dibutyryl cyclic AMP to the medium and by addition of agents that artificially increase intracellular cyclic AMP levels (PAWELEK et al., 1975; PA-WELEK, 1979; GREEN, 1978; PRUSS and HERSCHMAN, 1979; BOYNTON and WHIT-FIELD, 1979; TAUB et al., 1979). Among the agents used to stimulate such growth, cholera toxin has been found to be particularly effective (GREEN, 1978; PRUSS and HERSCHMAN, 1979) and there is a possibility that its effects may go beyond en-hancement of intracellular cyclic AMP levels. The prostaglandins, which were dis-cussed under the lipid category (Sect. C.VI.2), have hormone-like regulatory roles that appear to be related to the cyclic nucleotides. Not enough is yet known of their roles to be certain whether they should be treated as natural growth-promoting substances, as they have been in this chapter, or as artificial growth stimulants.

6. Defining a Genuine Growth Requirement

These selected examples make it adequately clear that there are many ways to stim-ulate short term DNA synthesis and cellular multiplication that have little to do with natural growth requirements. In view of such responses, it is necessary to ex-amine carefully each report of the discovery of a new growth factor for cultured cells to determine its relationship to natural multiplication-promoting substances. In most cases, sustained multiplication over a long period of time is probably a good test. However, in cases where regulatory, rather than nutritional mechanisms are involved, it might well be possible to maintain an artificial growth stimulus over a long period of time. In addition, there is the problem of adaptation of growth requirements over a period of time. As an extreme example, cells that have been adapted gradually to the replacement of thymidine with 5-bromodeoxyuridine can become totally dependent on 5-bromodeoxyuridine as an essential nutrient (DAVIDSON and BICK, 1973).

It is probably not possible to formulate a single definition for a "genuine" growth requirement that will adequately cover all circumstances. However, in gen-eral, the aspect that should be emphasized the most strongly is the relationship be-tween the apparent "requirement" and the conditions that are essential for multi-plication of the cell in question in its natural habitat in vivo. Beyond that, anything that is needed for multiplication in vitro can be classified as a "growth require-ment" in some sense of the word, although many such "requirements" should probably be viewed as artificial.

D. Requirements for Survival of Nontransformed Cells Without Proliferation

The requirements for prolonged survival of nontransformed cells in a quiescent state are still being worked out. It has been known for a long time that the require-ments for survival are less complex than those for proliferation, and current re-search is making it appear that survival requirements may be even simpler than generally appreciated.

I. Early Studies

Information concerning survival requirements has been accrued in several steps. Length of cellular survival was routinely used as the assay system during the development of medium 199 by MORGAN et al. (1950). Average survival in medium 199 with no serum or protein supplementation was 33 days and some cultures survived beyond 70 days. Thus, it was evident very early that, at least for primary cultures, prolonged survival without multiplication in the total absence of macromolecular supplementation was possible.

II. Survival Factor

Somewhat later, when established cell lines became widely available for nutritional studies, a variety of completely synthetic media were developed to support their multiplication (reviewed by HIGUCHI, 1973; KATSUTA and TAKAOKA, 1973). In these media not only survival, but also multiplication continued indefinitely in the total absence of undefined supplementation. However, when these media were utilized for growth of nontransformed cells, large amounts of serum supplementation were needed for multiplication. When attempts were made to replace whole serum with serum fractions or mixtures of hormones and growth factors, it was observed that 3 T 3 cells required a "survival" factor from serum in order to remain viable under conventional culture conditions, even when they were in a completely quiescent state (PAUL et al., 1971; LIPTON et al., 1972; GOSPODAROWICZ and MORAN, 1975; PAUL et al., 1978 a). Even in experiments seeking to replace the serum requirements of 3 T 3 cells with hormones and mixtures of incompletely defined growth factors, it was found necessary to use a serum-containing medium for initial establishment of the test cultures before washing them and changing to serum-free medium for the actual multiplication experiments (SERRERO et al., 1979).

III. Elimination of the Need for Survival Factor

The exact role of the serum "survival factor" has always been poorly defined, and recent experiments suggest that it is at least partially an artifact of the use of inadequate media and subculturing conditions. As described earlier in this chapter, the total requirement for serum protein for multiplication of several types of nontransformed cells has been reduced greatly through use of a combination of optimized media, polylysine-coated culture vessels, and gentle low temperature trypsinization techniques. This combination of improvements in the culture medium and techniques has also completely eliminated the "survival factor" requirement of human diploid fibroblasts and has at least greatly reduced the requirement of Swiss/c 3 T 3 cells. Human diploid fibroblasts that have been trypsinized at low temperature can be inoculated at clonal density on polylysine coated dishes in medium MCDB 105 and kept in the cell culture incubator for at least a week in the total absence of serum protein. Addition of serum protein at the end of the survival period results in prompt resumption of multiplication with virtually no loss in plating efficiency (McKEEHAN et al., 1978; HAM and McKEEHAN, 1978 a). In the case of Swiss/c 3 T 3 cells, survival of low temperature trypsinized cells on polylysine-coated dishes in medium MCDB 402 for a period of 6 days was approximately 50%, compared to virtually no survival in Dulbecco's modified Eagle's medium on dishes that had not been coated with polylysine (SHIPLEY and HAM, 1981).

IV. Nutrients and Survival

Complete data are not yet available on survival in the absence of nutrients that are essential for cellular multiplication. However, in many cases G_1/G_0 arrest of the cell cycle can be achieved by withholding specific nutrients including various amino acids, glucose, or phosphate ion (HOLLEY and KIERNAN, 1974b; HOLLEY et al., 1978; STILES et al., 1979b). In at least some cases, prolonged survival without such nutrients is possible, and multiplication can be quickly reactivated by restoring the nutrients. Calcium ion, which is needed in substantial amounts for multiplication of human diploid fibroblasts with minimal amounts of serum protein, appears not to be required for survival without multiplication (MCKEEHAN and HAM, 1978a). Cells maintained with low concentrations of calcium accumulate near the G_1/S boundary line and enter S rapidly when the calcium concentration is raised to a higher level (BOYNTON et al., 1977; ASHIHARA and BASERGA, 1979). Amino acid deprivation arrests multiplication of 3T3 cells at a restriction point about 6 h before the initiation of DNA synthesis (STILES et al., 1979b).

V. Differences in Survival Between Nontransformed and Transformed Cells

Nontransformed human diploid fibroblasts, when they reach a saturation density determined by the amount of serum protein in the culture medium, cease multiplying and remain viable for long periods of time. However, transformed human fibroblasts do not go into a quiescent state when nutritional conditions are inadequate and as a result they die quite rapidly (KRUSE et al., 1969; SCHIAFFONATI and BASERGA, 1977). Thus, nontransformed cells often exceed their transformed counterparts in their ability to survive under non-optimal conditions. It should be noted, however, that this applies primarily to growth arrests due to depletion of essential serum factors. Growth arrest due to amino acid deprivation tends to occur equivalently in transformed and nontransformed cells (LEY and TOBEY, 1970; HOLLEY, 1975).

In a fully rigorous analysis of the differences of survival and growth requirements of nontransformed and transformed cells, it would be desirable first to define the minimal set of requirements for survival without multiplication for each type of cell and then to define additional requirements, both qualitative and quantitative, that must be satisfied for multiplication, but not for survival. Such an analysis has not yet been carried to completion for any type of nontransformed cell or for transformed cells that require serum protein for survival. However, as soon as the growth requirements of specific types of nontransformed and serum-dependent transformed cells are completely understood, such an analysis will become relatively easy to do.

E. Future Studies of the Growth and Survival Requirements of Nontransformed Cells

I. Integration of New Findings

At a workshop held in November, 1973, little could be said about the specific growth requirements of normal cells in primary culture other than the fact that they represented a neglected area of modern biology (HAM, 1974a). Today that is not true. The lag phase has ended and we are now well into a period of exponential growth of knowledge about the growth requirements of normal and other nontransformed cells. A vast wealth of information has been published in the last few years and new reports are now accumulating so fast that it is becoming difficult to integrate all of the new data into a coherent picture. In addition, new approaches and concepts are currently evolving so rapidly that in most cases each of the pioneering laboratories has not yet incorporated into its assay systems all of the advances made in the other laboratories. Thus, nearly all of the assay systems still incorporate some of the defects of conventional assays that could now be eliminated. Clearly, one of the urgent priorities of the immediate future is to correct this problem.

II. Questions Remaining to be Solved After Development of Synthetic Media

Despite the current rapid progress, however, there is no indication that we are yet anywhere near to the end of the exponential phase. Quite soon (probably by the time this is published) we can expect to have available serum-free synthetic media for a few cell types that are currently being studied intensively – particularly mouse 3 T 3 cells, human diploid fibroblasts, and perhaps also human epidermal keratinocytes and chicken embryo fibroblasts. Although such media will represent major accomplishments, three major classes of questions will remain to be answered for each cell type.

1. What role are contaminants playing in such growth? A truly defined medium is an abstract ideal that can never be fully realized, since all real chemicals and growth factor preparations contain at least some degree of impurity (HAM and MCKEEHAN, 1979). Thus, even after a synthetic medium is available, extensive studies of the impurities that are present will be required before we can say that we understand all of the growth requirements that are being satisfied by the medium. This is particularly true for inorganic trace elements and for potential contaminants in any macromolecular growth factors that are needed by the cell type in question.

2. How do the growth requirements observed in the synthetic medium relate to natural in vivo growth requirements? On the basis of current trends, it is likely that long-term multiplication of nontransformed cells in a synthetic medium will first be achieved through the use of mixtures of hormones and hormone-like growth factors and that these will be identified through large scale screening programs rather than by systematic isolation and identification of the "factors" from serum that promote multiplication. Because of the complex holistic interactions that can occur among growth-promoting substances, there is no assurance that the growth-promoting components of the synthetic medium will be identical to the

components of serum that promote multiplication in less defined media, or that either set of "growth requirements" will match all of the conditions that promote multiplication in vivo. Thus, ultimately there must be a detailed analysis of the microenvironment that supports multiplication in vivo for each type of cell whose growth requirements have been "defined" in vitro.

3. What regulates multiplication in vivo? For many types of research, and particularly for those related to cancer, this is the most important question of all. As was discussed in Sect. C.I, current data suggest that we will probably not find anything to be qualitatively unique about the requirements for multiplication of nontransformed cells in synthetic media. If this prediction is confirmed, it will be necessary to seek mechanisms that are more complex than the simple availability of specific growth-promoting substances for nontransformed cells to explain the fact that many types of nontransformed cells are normally quiescent in vivo. It is possible that for some cell types the explanation could be as simple as density-dependent inhibition of multiplication and the absence of platelet-derived growth factors in all circumstances except wound healing. However, a detailed analysis of the normal microenvironment of each cell type that can be grown in a synthetic medium will be necessary to determine the nature of the regulatory interactions that keep that particular type of cell from proliferating inappropriately. Related studies will also be needed for stem cells that normally multiply rapidly in vivo in order to determine the nature of control mechanisms that differentiate between their normal proliferation and invasive malignant growth that ultimately destroys the host organism.

III. Magnitude of Remaining Work

In addition to these types of continuing study with each cell type for which a synthetic medium has been developed, there also remains a vast amount of work to be done with other cell types, many of which cannot currently be grown well in any type of medium, no matter how heavily it is supplemented. The number of experiments that ultimately must be performed is staggering. For each species, there are probably at least a hundred different normal cell types whose growth in culture is of interest to investigators in some field of research. Currently available data suggest that growth requirements are likely to be significantly different, at least quantitatively, for each new type of cell that is studied, and also for each species that the cell is taken from. If we consider 100 different cell types from 10 different species and assume that a total of at least 100 different nutrients, hormones, growth factors, etc. must be tested for effects on the cell in question and adjusted to an optimal concentration if they are found to be required, the total is on the order of 100,000 different combinations of cell types, species and medium components that must be worked with to carry such studies to completion, just for nontransformed cells.

In addition, if cancer research is also considered, new dimensions are added by the sequential changes that can occur in growth requirements during tumor cell evolution through various pre-malignant and malignant stages of development. The study of differentiation also adds complexities, both because of potential changes in growth requirements associated with the expression of specific differentiated properties and because of the potential need for specific environmental factors (in some cases already clearly defined) that are essential for full phenotypic expression of differentiated properties (ROVERA, this volume).

IV. Selective Allocation of Resources

In view of the magnitude of such an undertaking, it is unlikely that the growth requirements of all cell types and species that are of experimental interest will be worked out very rapidly. For selected cell types, the prognosis is excellent. All of the methodology that is needed for complete definition of growth requirements is already available. With an adequately equipped laboratory, an appropriate mixture of skills in cell culture and biochemistry, and an investment of two or three person-years of labor, the qualitative and quantitative growth requirements of virtually any cell type can be worked out in detail. Even for cell types that are currently very difficult to culture, the prognosis is potentially good, thanks to techniques such as the use of feeder layers and conditioned medium. Current progress with human epidermal keratinocytes illustrates well the potentialities that exist. However, the sheer bulk of the work that must be done with each cell type makes it necessary to select cell types for study very carefully and insures that it will be a long time before complete information on growth requirements is available for all of the cell types that are of potential interest in a wide variety of biological and biomedical studies.

F. Summary and Conclusions

Conventional assay systems, based on the methods used to identify the growth requirements of permanent lines of transformed cells, have generally proven to be inadequate for comparable studies with nontransformed cells. Serum serves so many different functions in promoting the multiplication of nontransformed cells in conventional media and the individual growth-promoting substances are present in such small amounts that it is extremely difficult to isolate and characterize the individual active components. In addition, complex interactions among growth-promoting substances require a holistic view of the entire set of growth requirements for each type of cell.

For a few selected types of nontransformed cells, optimized media and improved assay systems for growth factors have been developed in the last few years, and essentially all of their growth requirements have been identified. At present, it appears likely that there will be nothing qualitatively unique about the growth requirements of these cells. Thus, it will apparently be necessary to seek an understanding of the differences between growth control mechanisms in transformed and nontransformed cells at higher levels of regulatory interaction than the minimal set of requirements for multiplication of isolated individual cells.

An overview has been presented of the total set of growth requirements for various types of nontransformed cells as they are currently understood. One of the major patterns that is emerging is the high degree of individuality of growth requirements for each type of cell and each species of origin, both in terms of quantitative optimization of nutrient concentrations in the synthetic medium and in terms of the combinations of hormones and growth factors that replace the requirement for serum.

The overall theme of this volume is growth factors. This chapter has emphasized three aspects of the relationship between growth factors and the growth requirements of nontransformed cells: 1) growth factors are only one small subset of the total set of environmental variables that affect cellular multiplication, 2) cellular responses to specific growth factors cannot be adequately measured and evaluated unless careful attention is given to all of the other variables that affect cellular multiplication, and 3) holistic interactions with other aspects of the total cellular environment can alter cellular responses to specific growth factors. Thus, it is essential that the material that follows on cellular growth factors be considered in the context of the overall cellular environment and the total set of growth requirements for each type of cell that is studied.

G. Note Added in Proof

Since this chapter was written, multiplication of several types of normal cells in media that contain no deliberately added undefined supplements has been achieved. For lack of a more appropriate adjective, the term "defined" is used to describe these media, although the possibility that trace contaminants in the chemicals, water, or sterilizing filters used in their preparation could be contributing to the multiplication that support cannot yet be ruled out completely. Rapid clonal growth of human diploid fibroblasts can be obtained in an optimized nutrient medium (MCDB 110) supplemented with insulin, EGF, dexamethasone, mixed phospholipids, cholesterol, sphingomyelin, vitamin E, prostaglandins, reducing agents, and phosphoenolpyruvate (WALTHALL and HAM, 1981; BETTGER et al., 1981). Clonal growth of human epidermal keratinocytes occurs in a very different optimized basal medium (MCDB 152) supplemented with insulin, EGF, hydrocortisone, ethanolamine, phosphoethanolamine, transferrin and progesterone (TSAO et al., 1981). Clonal growth of rabbit ear chondrocytes occurs in MCDB 110 supplemented with insulin, FGF, and the lipid mixture that is used with human fibroblasts (JENNINGS and HAM, 1981). Functional kidney epithelial cells from baby mice and various other mamalian species can be grown in medium K-1, which consists of a 50:50 mixture of DME and F 12, supplemented with insulin, hydrocortisone, transferrin, triiodothyronine, and prostaglandin E_1 (TAUB and SATO, 1980). Adult human prostatic epithelial cells have been reported to grow in RPMI 1640 supplemented with polyvinylpyrollidone, vitamin A, EGF, dexamethasone, transferrin and putrescine or spermine (CHAPRONIERE-RICKENBERG and WEBBER, 1980). Although a total analysis of growth requirements has not yet been reported for any of these systems, the ability to grow a number of different cells in media that are quite highly defined represents significant progress toward that goal, which should be achieved for several types of normal cells in the very near future.

References

Agy, P.C., Shipley, G.D., Ham, R.G.: Growth of mouse neuroblastoma cells in a protein-free medium. In Vitro, in press. (1981)

Allen, R.H.: Human vitamin B_{12} transport proteins. Prog. Hematol. 9, 57–84 (1975)

Antoniades, H.N., Scher, C.D.: Growth factors derived from human serum, platelets, and pituitary: Properties and immunologic cross-reactivity. Natl. Cancer Inst. Monogr. No. 48, 137–140 (1978)

Antoniades, H.N., Scher, C.D., Stiles, C.D.: Purification of human platelet-derived growth factor. Proc. Natl. Acad. Sci. U.S.A. 76, 1809–1813 (1979)

Armelin, H.A.: Pituitary extracts and steroid hormones in the control of 3 T 3 cell growth. Proc. Natl. Acad. Sci. U.S.A. 70, 2702–2706 (1973)

Ashihara, T., Baserga, R.: Cell synchronization. Methods Enzymol. 58, 248–262 (1979)

Bailey, J.M., Dunbar, L.M.: Essential fatty acid requirements of cells in tissue culture. Exp. Mol. Pathol. 18, 142–161 (1973)

Balin, A.K., Goodman, D.B.P., Rasmussen, H., Cristofalo, V.J.: The effect of oxygen tension on the growth and metabolism of WI-38 cells. J. Cell. Physiol. 89, 235–249 (1976)

Balin, A.K., Goodman, D.B.P., Rasmussen, H., Cristofalo, V.J.: The effect of oxygen and vitamin E on the lifespan of human diploid cells in vitro. J. Cell Biol. 74, 58–67 (1977)

Balk, S.D.: Calcium as a regulator of the proliferation of normal, but not of transformed, chicken fibroblasts in a plasma-containing medium. Proc. Natl. Acad. Sci. U.S.A. 68, 271–275 (1971)

Balk, S.D., LeStourgeon, D., Mitchell, R.S.: 5-methyltetrahydrofolic acid, 5-formyltetrahydrofolic acid (folinic acid), and folic acid requirements of normal and Rous sarcoma virus-infected chicken fibroblasts. Cancer Res. 38, 3966–3968 (1978)

Balk, S.D., Polimeni, P.I., Hoon, B.S., LeStourgeon, D.N., Mitchell, R.S.: Proliferation of Rous sarcoma virus-infected, but not of normal, chicken fibroblasts in a medium of reduced calcium and magnesium concentration. Proc. Natl. Acad. Sci. U.S.A. 76, 3913–3916 (1979)

Barnes, D.W., Colowick, S.P.: Stimulation of sugar uptake and thymidine incorporation in mouse 3 T 3 cells by calcium phosphate and other extracellular particles. Proc. Natl. Acad. Sci. U.S.A. 74, 5593–5597 (1977)

Barnes, D., Sato, G.: Growth of a human mammary tumor cell line in a serum-free medium. Nature 281, 388–389 (1979)

Barnes, D., Sato, G.: Methods for growth of cultured cells in serum-free medium: Review. Anal. Biochem. 102, 255–270 (1980)

Baserga, R.: Multiplication and division in mammalian cells. New York: Marcel Dekker Inc. 1976

Bell, E., Marek, L.F., Levinstone, D.S., Merrill, C., Sher, S., Young, I.T., Eden, M.: Loss of division potential in vitro: Aging or differentiation? Science 202, 1158–1163 (1978)

Bettger, W.J., Boyce, S., Walthall, B.J., Ham, R.G.: Rapid clonal growth and serial passage of human diploid fibroblasts in a lipid-enriched, synthetic medium supplemented with EGF, insulin and dexamethasone. Proc. Natl. Acad. Sci. U.S.A., in press

Birch, J.R., Pirt, S.J.: Improvements in a chemically defined medium for the growth of mouse cells (strain LS) in suspension. J. Cell Sci. 7, 661–670 (1970)

Blaker, G.J., Birch, J.R., Pirt, S.J.: The glucose, insulin and glutamine requirements of suspension cultures of HeLa cells in a defined culture medium. J. Cell Sci. 9, 529–537 (1971)

Boone, C.W.: Malignant hemangioendotheliomas produced by subcutaneous inoculation of Balb/3 T 3 cells attached to glass beads. Science 188, 68–70 (1975)

Borek, C.: Malignant transformation in vitro: Criteria, biological markers, and application in environmental screening of carcinogens. Radiation Res. 79, 209–232 (1979)

Bottenstein, J.E., Sato, G.H.: Growth of a rat neuroblastoma cell line in serum-free supplemented medium. Proc. Natl. Acad. Sci. U.S.A. 76, 514–517 (1979)

Bottenstein, J., Hayashi, I., Hutchings, S., Masui, H., Mather, J., McClure, D.B., Ohasa, S., Rizzino, A., Sato, G., Serrero, G., Wolfe, R., Wu, R.: The growth of cells in serum-free hormone-supplemented media. Methods Enzymol. 58, 94–109 (1979)

Boynton, A.L., Whitfield, J.F.: The different actions of normal and supranormal calcium concentrations on the proliferation of BALB/c 3 T 3 mouse cells. In Vitro 12, 479–484 (1976 a)

Boynton, A.L., Whitfield, J.F.: Different calcium requirements for proliferation of conditionally and unconditionally tumorigenic mouse cells. Proc. Natl. Acad. Sci. U.S.A. 73, 1651–1654 (1976 b)

Boynton, A.L., Whitfield, J.F.: The cyclic AMP-dependent initiation of DNA synthesis by T 51 B rat liver epithelial cells. J. Cell. Physiol. 101, 139–148 (1979)

Boynton, A.L., Whitfield, J.F., Isaacs, R.J.: Calcium-dependent stimulation of BALB/c 3 T 3 mouse cell DNA synthesis by a tumor-promoting phorbol ester (PMA). J. Cell. Physiol. 87, 25–32 (1976)

Boynton, A.L., Whitfield, J.F., Isaacs, R.J., Tremblay, R.: The control of human WI-38 cell proliferation by extracellular calcium and its elimination by SV-40 virus-induced proliferative transformation. J. Cell. Physiol. 92, 241–247 (1977)

Boynton, A.L., Whitfield, J.F., Isaacs, R.J., Tremblay, R.G.: An examination of the roles of cyclic nucleotides in the initiation of cell proliferation. Life Sci. 22, 703–710 (1978)

Brooks, R.F., Bennett, D.C., Smith, J.A.: Mammalian cell cycles need two random transitions. Cell 19, 493–504 (1980)

Burns, R.L., Rosenberger, P.G., Klebe, R.J.: Carbohydrate preferences of mammalian cells. J. Cell. Physiol. 88, 307–316 (1976)

Bush, H., Shodell, M.: Cell cycle changes in transformed cells growing under serum-free conditions. J. Cell. Physiol. 90, 573–583 (1977)

Carpenter, G., Cohen, S.: Human epidermal growth factor and the proliferation of human fibroblasts. J. Cell. Physiol. 88, 227–237 (1976)

Carpenter, G., Cohen, S.: Epidermal growth factor. Annu. Rev. Biochem. 48, 193–216 (1979)

Chaproniere-Rickenberg, D., Webber, M.M.: A synthetic medium for the growth of primary cultures of adult human prostatic epithelium. In Vitro 16, 214 (Abstract No. 39) (1980)

Chen, H.W., Kandutsch, A.A.: Cholesterol requirement for cell growth: Endogenous synthesis vs. exogenous sources. In: The growth requirements of vertebrate cells in vitro. Waymouth, C., Ham, R.G., Chapple, P.J. (eds.). New York: Cambridge University Press, in press 1981

Cherington, P.V., Smith, B.L., Pardee, A.B.: Loss of epidermal growth factor requirement and malignant transformation. Proc. Natl. Acad. Sci. U.S.A. 76, 3937–3941 (1979)

Cheung, W.Y.: Calmodulin plays a pivotal role in cellular regulation. Science 207, 19–27 (1980)

Cifone, M.A., Fidler, I.J.: Correlation of patterns of anchorage-independent growth with in vivo behavior of cells from a murine fibrosarcoma. Proc. Natl. Acad. Sci. U.S.A. 77, 1039–1043 (1980)

Clo, C., Orlandini, G.C., Casti, A., Guarniere, C.: Polyamines as growth stimulating factors in eukariotic cells. Ital. J. Biochem. 25, 94–114 (1976)

Clo, C., Caldarera, C.M., Tantini, B., Benalal, D., Bachrach, U.: Polyamines and cellular adenosine 3′:5′-cyclic monophosphate. Biochem. J. 182, 641–649 (1979)

Cone, C.D. jr., Cone, C.M.: Induction of mitosis in mature neurons in central nervous system by sustained depolarization. Science 192, 155–158 (1976)

Cornell, R., Grove, G.L., Rothblat, G.H., Horwitz, A.F.: Lipid requirement for cell cycling. The effect of selective inhibition of lipid synthesis. Exp. Cell Res. 109, 299–307 (1977)

Cornwell, D.G., Huttner, J.J., Milo, G.E., Panganamala, R.V. Sharma, H.M., Geer, J.C.: Polyunsaturated fatty acids, vitamin E, and the proliferation of aortic smooth muscle cells. Lipids 14, 194–207 (1979)

Damluji, R., Riley, P.A.: On the role of calcium in adhesion of cells to solid substrates. Exp. Cell Biol. 47, 226–237 (1979)

Daughaday, W.H., Hall, K., Raben, M.S., Salmon, W.D. jr., van den Brande, J.S., van Wyk, J.J.: Somatomedin: Proposed designation for sulphation factor. Nature 235, 107 (1972)

Davidson, R.L., Bick, M.D.: Bromodeoxyuridine dependence – A new mutation in mammalian cells. Proc. Natl. Acad. Sci. U.S.A. 70, 138–142 (1973)

Davies, P.F., Ross, R.: Mediation of pinocytosis in cultured arterial smooth muscle and endothelial cells by platelet-derived growth factor. J. Cell Biol. 79, 663–671 (1978)

DeHaan, R.L.: The potassium-sensitivity of isolated embryonic heart cells increases with development. Dev. Biol. 23, 226–240 (1970)

De Larco, J.E., Todaro, G.J.: Growth factors from murine sarcoma virus-transformed cells. Proc. Natl. Acad. Sci. USA 75, 4001–4005 (1978)

Diamond, L., O'Brien, S., Donaldson, C., Shimizu, Y.: Growth stimulation of human diploid fibroblasts by the tumor promoter, 12-0-tetradecanoylphorbol-13-acetate. Int. J. Cancer 13, 721–730 (1974)

Dicker, P., Rozengurt, E.: Stimulation of DNA synthesis by tumor promoter and pure mitogenic factors. Nature 276, 723–726 (1978)

Driedger, P.E., Blumberg, P.M.: Specific binding of phorbol ester tumor promoters. Proc. Natl. Acad. Sci. U.S.A. 77, 567–571 (1980)

Dulak, N.C., Temin, H.M.: Multiplication-stimulating activity for chicken embryo fibroblasts from rat liver cell conditioned medium: A family of small polypeptides. J. Cell. Physiol. 81, 161–170 (1973)

Dulbecco, R., Elkington, J.: Induction of growth in resting fibroblastic cell cultures by Ca^{++}. Proc. Natl. Acad. Sci. U.S.A. 72, 1584–1588 (1975)

Eagle, H.: Nutrition needs of mammalian cells in tissue culture. Science 122, 501–504 (1955)

Eagle, H.: Amino acid metabolism in mammalian cell cultures. Science 130, 432–437 (1959)

Eagle, H.: The nutritional requirements and metabolic activities of human cells in vivo and in vitro. Proc. of Symposium on the characterization and uses of human diploid cell strains. Opatija, Yugoslavia, pp. 143–159 (1963)

Eagle, H.: Some effects of environmental pH on cellular metabolism and function. In: Control of proliferation in animal cells. Clarkson, B., Baserga, R. (eds.), pp. 1–11. Cold Spring Harbor, New York: Cold Spring Harbor Laboratory 1974

Eagle, H., Levintow, L.: Amino acid and protein metabolism. I. The metabolic characteristics of serially propagated cells. In: Cells and tissues in culture. Willmer, E.N. (ed.), pp. 277–296. New York: Academic Press 1965

Eagle, H., Piez, K.: The population-dependent requirement by cultured mammalian cells for metabolites which they can synthesize. J. Exp. Med. 116, 29–43 (1962)

Eagle, H., Barban, S., Levy, M., Schulze, H.O.: The utilization of carbohydrates by human cell cultures. J. Biol. Chem. 233, 551–558 (1958)

Earle, W.R.: Production of malignancy in vitro. IV. The mouse fibroblast cultures and changes seen in the living cells. J. Natl. Cancer Inst. 4, 165–212 (1943)

Ehrmann, R.L., Gey, G.O.: The growth of cells on a transparent gel of reconstituted rat-tail collagen. J. Natl. Cancer Inst. 16, 1375–1403 (1956)

Eisinger, M., Lee, J.S., Hefton, J.M., Darzynkiewicz, Z., Chiao, J.W., De Harven, E.: Human epidermal cell cultures: Growth and differentiation in the absence of dermal components or medium supplements. Proc. Natl. Acad. Sci. U.S.A. 76, 5340–5344 (1979)

Ellem, K.A.O., Gierthy, J.F.: Mechanism of regulation of fibroblastic cell replication. IV. An analysis of the serum dependence of cell replication based on Michaelis-Menten kinetics. J. Cell. Physiol. 92, 381–400 (1977)

Emerman, J.T., Pitelka, D.R.: Maintenance and induction of morphological differentiation in dissociated mammary epithelium on floating collagen membranes. In Vitro 13, 316–328 (1977)

Emerman, J.T., Burwen, S.J., Pitelka, D.R.: Substrate properties influencing ultrastructural differentiation of mammary epithelial cells in culture. Tissue Cell 11, 109–119 (1979)

Epel, D.: Mechanisms of activation of sperm and egg during fertilization of sea urchin gametes. Curr. Top. Dev. Biol. 12, 185–246 (1978)

Epstein, A.L., Kaplan, H.S.: Feeder layer and nutritional requirements for the establishment and cloning of human malignant lymphoma cell lines. Cancer Res. 39, 1748–1759 (1979)

Evans, V.J., Earle, W.R., Sanford, K.K., Shannon, J.E., Waltz, H.K.: The preparation and handling of replicate tissue cultures for quantitative studies. J. Natl. Cancer Inst. 11, 907–927 (1951)

Feng, J., Melcher, A.H., Brunette, D.M., Moe, H.K.: Determination of L-ascorbic acid levels in culture medium: Concentrations in commercial media and maintenance of levels under conditions of organ culture. In Vitro 13, 91–99 (1977)

Fernandez-Pol, J.A.: Siderophore-like growth factor synthesized by SV 40-transformed cells adapted to picolinic acid stimulates DNA synthesis in cultured cells. FEBS Lett. 88, 345–348 (1978 a)

Fernandez-Pol, J.A.: Isolation and characterization of a siderophore-like growth factor from mutants of SV 40-transformed cells adapted to picolinic acid. Cell *14*, 489–499 (1978 b)

Fischer, A.: Die Bedeutung der Aminosäuren für die Gewebezellen in vitro. Acta Physiol. Scand. *2*, 143–188 (1941)

Fisher, H.W., Puck, T.T., Sato, G.: Molecular growth requirements of single mammalian cells: The action of fetuin in promoting cell attachment to glass. Proc. Natl. Acad. Sci. U.S.A. *44*, 4–10 (1958)

Floros, J., Baserga, R.: Reactivation of G_0 nuclei by S-phase cells. Cell Biol. Int. Rep. *4*, 75–82 (1980)

Folkman, J., Moscona, A.: Role of cell shape in growth control. Nature *273*, 345–349 (1978)

Frantz, C.N., Stiles, C.D., Scher, C.D.: The tumor promoter 12-0-tetradecanoyl-phorbol-13-acetate enhances the proliferative response of Balb/c-3 T 3 cells to hormonal growth factors. J. Cell. Physiol. *100*, 413–424 (1979)

Freedman, V.H., Shin, S.-I.: Cellular tumorigenicity in nude mice: Correlation with cell growth in semi-solid medium. Cell *3*, 355–359 (1974)

Friedman, P.A., Shia, M.A., Wallace, J.K.: A saturable high affinity binding site for transcobalamin II-vitamin B_{12} complexes in human placental membrane preparations. J. Clin. Invest. *59*, 51–58 (1977)

Fritsch, P., Tappeiner, G., Huspek, G.: Keratinocyte substrate adhesion is magnesium-dependent and calcium-independent. Cell Biol. Int. Rep. *3*, 593–598 (1979)

Froesch, E.R., Zapf, J., Rinderknecht, E., Morell, B., Schoenle, E., Humbel, R.E.: Insulin-like growth factor (IGF-NSILA): Structure, function, and physiology. In: Hormones and cell culture Book A. Sato, G.H., Ross, R. (eds.), pp. 61–77. Cold Spring Harbor, New York: Cold Spring Harbor Laboratory 1979

Frost, D.V., Lish, P.M.: Selenium in biology. Annu. Rev. Pharmacol. *15*, 259–284 (1975)

Gey, G.O., Thalhimer, W.: Observations on insulin introduced into the medium of tissue culture. J. Am. Med. Assoc. *82*, 1609 (1924)

Gey, G.O., Coffman, W.D., Kubicek, M.T.: Tissue culture studies of the proliferative capacity of cervical carcinoma and normal epithelium. Cancer Res. *12*, 264–265 (1952)

Giasuddin, A.S.M., Diplock, A.T.: The influence of vitamin E and selenium on the growth and plasma membrane permeability of mouse fibroblasts in culture. Arch. Biochem. Biophys. *196*, 270–280 (1979)

Gospodarowicz, D.: Localisation of a fibroblast growth factor and its effect alone and with hydrocortisone on 3 T 3 cell growth. Nature *249*, 123–127 (1974)

Gospodarowicz, D., Moran, J.: Optimal conditions for the study of growth control in BALB/c 3 T 3 fibroblasts. Exp. Cell Res. *90*, 279–284 (1975)

Gospodarowicz, D., Moran, J.S.: Growth factors in mammalian cell culture. Annu. Rev. Biochem. *45*, 531–558 (1976)

Gospodarowicz, D., Greene, G., Moran, J.: Fibroblast growth factor can substitute for platelet factor to sustain the growth of Balb/3 T 3 cells in the presence of plasma. Biochem. Biophys. Res. Commun. *65*, 779–787 (1975)

Gospodarowicz, D., Greenburg, G., Birdwell, C.R.: Determination of cellular shape by the extracellular matrix and its correlation with the control of cellular growth. Cancer Res. *38*, 4155–4171 (1978)

Gospodarowicz, D., Vlodavsky, I., Greenburg, G., Johnson, L.K.: Cellular shape is determined by the extracellular matrix and is responsible for the control of cellular growth and function. In: Hormones and cell culture Book B. Sato, G.H., Ross, R. (eds.), pp. 561–592. Cold Spring Harbor, New York: Cold Spring Harbor Laboratory 1979

Green, H.: Cyclic AMP in relation to proliferation of the epidermal cell: A new view. Cell *15*, 801–811 (1978)

Gregory, H.: Isolation and structure of urogastrone and its relationship to epidermal growth factor. Nature *257*, 325–327 (1975)

Griffin, A.C.: Role of selenium in the chemoprevention of cancer. Adv. Cancer Res. *29*, 419–449 (1979)

Grinnell, F.: Cellular adhesiveness and extracellular substrata. Int. Rev. Cytol. *53*, 65–144 (1978)

Grinnell, F., Feld, M.K.: Initial adhesion of human fibroblasts in serum-free medium: Possible role of secreted fibronectin. Cell *17*, 117–129 (1979)

Groelke, J.W., Baseman, J.B., Amos, H.: Regulation of the $G_1 \to S$ phase transition in chick embryo fibroblasts with α-keto acids and L-alanine. J. Cell. Physiol. *101*, 391–398 (1979)

Guilbert, L.J., Iscove, N.N.: Partial replacement of serum by selenite, transferrin, albumin, and lecithin in haemopoietic cell cultures. Nature *263*, 594–595 (1976)

Haggerty, D.F., Sato, G.H.: The requirement for biotin in mouse fibroblast L-cells cultured on serumless medium. Biochem. Biophys. Res. Commun. *34*, 812–815 (1969)

Haigler, H.T., McKanna, J.A., Cohen, S.: Rapid stimulation of pinocytosis in human carcinoma cells A-431 by epidermal growth factor. J. Cell Biol. *83*, 82–90 (1979)

Haley, E.E., Fischer, G.A., Welch, A.D.: The requirement for L-asparagine of mouse leukemia cells L 5178 Y in culture. Cancer Res. *21*, 532–536 (1961)

Ham, R.G.: Clonal growth of diploid Chinese hamster cells in a synthetic medium supplemented with purified protein fractions. Exp. Cell Res. *28*, 489–500 (1962)

Ham, R.G.: An improved nutrient solution for diploid Chinese hamster and human cell lines. Exp. Cell Res. *29*, 515–526 (1963)

Ham, R.G.: Putrescine and related amines as growth factors for a mammalian cell line. Biochem. Biophys. Res. Commun. *14*, 34–38 (1964)

Ham, R.G.: Clonal growth of mammalian cells in a chemically defined, synthetic medium. Proc. Natl. Acad. Sci. U.S.A. *53*, 288–293 (1965)

Ham, R.G.: Discussion after Somatic cells in vitro: Their relationship to progenitive cells and to artificial millieux. In: Second decennial review conference on cell tissue and organ culture. Westfall, B.B. (ed.), pp. 19–21. Natl. Cancer Inst. Monogr. 26 (1967)

Ham, R.G.: Cloning of mammalian cells. Methods Cell Physiol. *5*, 37–74 (1972)

Ham, R.G.: Nutritional requirements of primary cultures. A neglected problem of modern biology. In Vitro *10*, 119–129 (1974a)

Ham, R.G.: Unique requirements for clonal growth. J. Natl. Cancer Inst. *53*, 1459–1463 (1974b)

Ham, R.G.: Dermal fibroblasts. Methods Cell Biol. *21 A*, 255–276 (1980)

Ham, R.G., McKeehan, W.L.: Development of improved media and culture conditions for clonal growth of normal diploid cells. In Vitro *14*, 11–22 (1978a)

Ham, R.G., McKeehan, W.L.: Nutritional requirements for clonal growth of nontransformed cells. In: Nutritional requirements of cultured cells. H. Katsuta (ed.), pp. 63–115. Tokyo, Japan: Japan Scientific Societies Press (1978b)

Ham, R.G., McKeehan, W.L.: Media and growth requirements. Methods Enzymol. *58*, 44–93 (1979)

Ham, R.G., Puck, T.T.: Quantitative colonial growth of isolated mammalian cells. Methods Enzymol. *5*, 90–119 (1962)

Hamilton, W.G., Ham, R.G.: Clonal growth of Chinese hamster cell lines in protein-free media. In Vitro *13*, 537–547 (1977)

Harris, C.C., Trump, B.F., Stoner, G.D. (eds.): Normal human tissue and cell culture. Methods Cell Biol. *21 A and 21 B* (1980)

Hauschka, S.D., Konigsberg, I.R.: The influence of collagen on the development of muscle clones. Proc. Natl. Acad. Sci. U.S.A. *55*, 119–126 (1966)

Hausman, R.E., Moscona, A.A.: Immunologic detection of retina cognin on the surface of embryonic cells. Exp. Cell Res. *119*, 191–204 (1979)

Hayashi, I., Sato, G.H.: Replacement of serum by hormones permits growth of cells in a defined medium. Nature *259*, 132–134 (1976)

Hazelton, B., Mitchell, B., Tupper, J.: Calcium, magnesium, and growth control in the WI-38 human fibroblast cell. J. Cell Biol. *83*, 487–498 (1979)

Healy, G.M., Parker, R.C.: An improved chemically defined basal medium (CMRL-1415) for newly explanted mouse embryo cells. J. Cell Biol. *30*, 531–538 (1966)

Heldin, C.-H., Westermark, B., Wasteson, A.: Purification and characterization of human growth factors. In: Hormones and cell culture Book A. Sato, G.H., Ross, R. (eds.), pp. 17–31. Cold Spring Harbor, New York: Cold Spring Harbor Laboratory 1979

Hennings, H., Michael, D., Cheng, C., Steinert, P., Holbrook, K., Yuspa, S.H.: Calcium regulation of growth and differentiation of mouse epidermal cells in culture. Cell *19*, 245–254 (1980)

Herzenberg, L.A., Roosa, R.A.: Nutritional requirements for growth of a mouse lymphoma in cell culture. Exp. Cell Res. *21*, 430–438 (1960)

Hewitt, A.T., Kleinman, H.K., Pennypacker, J.P., Martin, G.R.: Identification of an adhesion factor for chondrocytes. Proc. Natl. Acad. Sci. U.S.A. *77*, 385–388 (1980)

Higuchi, K.: Cultivation of animal cells in chemically defined media, a review. Adv. Appl. Microbiol. *16*, 111–136 (1973)

Hochstadt, J., Quinlan, D.C., Owen, A.J., Cooper, K.O.: Regulation of transport upon interaction of fibroblast growth factor and epidermal growth factor with quiescent (G_0) 3T3 cells and plasma-membrane vesicles isolated from them. In: Hormones and cell culture Book B. Sato, G.H., Ross, R. (eds.), pp. 751–771. Cold Spring Harbor, New York: Cold Spring Harbor Laboratory 1979

Hodges, G.M., Livingston, D.C., Franks, L.M.: The localization of trypsin in cultured mammalian cells. J. Cell Sci. *12*, 887–902 (1973)

Holley, R.W.: Control of growth of mammalian cells in cell culture. Nature *258*, 487–490 (1975)

Holley, R.W., Baldwin, J.H.: Cell density is determined by a diffusion-limited process. Nature *278*, 283–284 (1979)

Holley, R.W., Kiernan, J.A.: "Contact inhibition" of cell division in 3T3 cells. Proc. Natl. Acad. Sci. U.S.A. *60*, 300–304 (1968)

Holley, R.W., Kiernan, J.A.: Studies of serum factors required by 3T3 and SV3T3 cells. In: Symposium on growth control in cell cultures, London 1970. Growth control in cell cultures, Wolstenholme, G.E.W., Knight, J. (eds.), pp. 3–15. London: Churchill, Livingstone 1971

Holley, R.W., Kiernan, J.A.: Control of the initiation of DNA synthesis in 3T3 cells: Serum factors. Proc. Natl. Acad. Sci. U.S.A. *71*, 2908–2911 (1974a)

Holley, R.W., Kiernan, J.A.: Control of the initiation of DNA synthesis in 3T3 cells: Low-molecular weight nutrients. Proc. Natl. Acad. Sci. U.S.A. *71*, 2942–2945 (1974b)

Holley, R.W., Armour, R., Baldwin, J.H.: Density-dependent regulation of growth of BSC-1 cells in cell culture: Control of growth by low molecular weight nutrients. Proc. Natl. Acad. Sci. U.S.A. *75*, 339–341 (1978)

Hori, C., Oka, T.: Induction by lithium ion of multiplication of mouse mammary epithelium in culture. Proc. Natl. Acad. Sci. U.S.A. *76*, 2823–2827 (1979)

Hosick, H.L.: Uptake and utilization of free fatty acids supplied by liposomes to mammary tumor cells in culture. Exp. Cell Res. *122*, 127–136 (1979)

Huennekens, F.M., Vitols, K.S., Henderson, G.B.: Transport of folate compounds in bacterial and mammalian cells. Adv. Enzymol. *47*, 313–346 (1978)

Iscove, N.N., Melchers, F.: Complete replacement of serum by albumin, transferrin, and soybean lipid in cultures of lipopolysaccharide-reactive B lymphocytes. J. Exp. Med. *147*, 923–931 (1978)

Jänne, J., Poso, H., Raina, A.: Polyamines in rapid growth and cancer. Biochim. Biophys. Acta *473*, 241–293 (1978)

Jenkin, H.M., Anderson, L.E.: The effect of oleic acid on the growth of monkey kidney cells (LLC-MK$_2$). Exp. Cell Res. *59*, 6–10 (1970)

Jennings, S.D., Ham, R.G.: Clonal growth of primary cultures of rabbit chondrocytes in a defined medium. In Vitro, abstract in press

Jensen, P., Winger, L., Rasmussen, H., Nowell, P.: The mitogenic effect of A 23187 in human peripheral lymphocytes. Biochim. Biophys. Acta *496*, 374–383 (1977)

Jimenez de Asua, L., Clingan, D., Rudland, P.S.: Initiation of cell proliferation in cultured mouse fibroblasts by prostaglandin $F_{2\alpha}$. Proc. Natl. Acad. Sci. U.S.A. *72*, 2724–2728 (1975)

Jonak, G.J., Baserga, R.: Cytoplasmic regulation of two G_1-specific temperature-sensitive functions. Cell *18*, 117–123 (1979)

Kagawa, Y., Takaoka, T., Katsuta, H.: Absence of essential fatty acids in mammalian cell strains cultured in lipid- and protein-free chemically defined synthetic media. J. Biochem. *68*, 133–136 (1970)

Kahn, P., Shin, S.-I.: Cellular tumorigenicity in nude mice. Test of associations among loss of cell-surface fibronectin, anchorage independence, and tumor-forming ability. J. Cell Biol. *82*, 1–16 (1979)

Kano-Sueoka, T., Errick, J.E., Cohen, D.M.: Effects of hormones and a novel mammary growth factor on a rat mammary carcinoma in culture. In: Hormones and cell culture Book A. Sato, G.H., Ross, R. (eds.), pp. 499–512. Cold Spring Harbor, New York: Cold Spring Harbor Laboratory 1979 a

Kano-Sueoka, T., Cohen, D.M., Yamaizumi, Z., Nishimura, S., Mori, M., Fujiki, H.: Phosphoethanolamine as a growth factor of a mammary carcinoma cell line of rat. Proc. Natl. Acad. Sci. U.S.A. 76, 5741–5744 (1979 b)

Katsuta, H., Takaoka, T.: Cultivation of cells in protein- and lipid-free synthetic media. Methods Cell Biol. 6, 1–42 (1973)

Katsuta, H., Takaoka, T.: Protein- and lipid-free synthetic media for the cultivation of mammalian cells in tissue culture. In: Nutritional requirements of cultured cells. Katsuta, H. (ed.), pp. 257–275. Tokyo, Japan: Japan Scientific Societies Press 1978

King, M.E., Spector, A.A.: Lipid metabolism in cultured cells. In: The growth requirements of vertebrate cells in vitro. Waymouth, C., Ham, R.G., Chapple, P.J. (eds.). New York: Cambridge University Press in press 1981

Knox, P., Griffiths, S.: A cell spreading factor in human serum that is not cold-insoluble globulin. Exp. Cell Res. 123, 421–424 (1979)

Koch, C.J., Biaglow, J.E.: Toxicity, radiation sensitivity modification, and metabolic effects of dehydroascorbate and ascorbate in mammalian cells. J. Cell. Physiol. 94, 299–306 (1978)

Koch, K.S., Leffert, H.L.: Ionic landmarks along the mitogenic route. Nature 279, 104–105 (1979 a)

Koch, K.S., Leffert, H.L.: Increased sodium ion influx is necessary to initiate rat hepatocyte proliferation. Cell 18, 153–163 (1979 b)

Kohler, N., Lipton, A.: Platelets as a source of fibroblast growth-promoting activity. Exp. Cell Res. 87, 297–301 (1974)

Kretsinger, R.H.: Evolution and function of calcium-binding proteins. Int. Rev. Cytol. 46, 323–393 (1976)

Kruse, P.F. jr., Whittle, W., Miedema, E.: Mitotic and nonmitotic multiple-layered perfusion cultures. J. Cell Biol. 42, 113–121 (1969)

Kuchler, R.J.: The role of sodium and potassium in regulating amino acid accumulation and protein synthesis in LM-strain mouse fibroblasts. Biochim. Biophys. Acta 136, 473–483 (1967)

Kuri-Harcuch, W., Wise, L.S., Green, H.: Interruption of the adipose conversion of 3 T 3 cells by biotin deficiency: Differentiation without triglyceride accumulation. Cell 14, 53–59 (1978)

Lamanna, C., Mallette, M.F.: Basic bacteriology: Its biological and chemical background. 3 rd ed., pp. 897–901. Baltimore, Maryland: Williams and Wilkens Company 1965

Lasnitzki, I.: The action of hormones on cell and organ cultures. In: Cell and tissues in culture. Willmer, E.N. (ed.), Vol. 1, pp. 591–658. New York: Academic Press 1965

Lechner, J.F., Kaighn, M.E.: Application of the principles of enzyme kinetics to clonal growth rate assays: An approach to delineating interactions among growth promoting agents. J. Cell. Physiol. 100, 519–530 (1979 a)

Lechner, J.F., Kaighn, M.E.: Reduction of the calcium requirement of normal human epithelial cells by EGF. Exp. Cell Res. 121, 432–435 (1979 b)

Lechner, J.F., Kaighn, M.E.: EGF growth promoting activity is neutralized by phorbol esters. Cell Biol. Int. Rep. 4, 23–28 (1980)

Lechner, J.F., Babcock, M.S., Marnell, M., Narayan, K.S., Kaighn, M.E.: Normal human prostate epithelial cell cultures. Methods Cell Biol. 21 B, 195–225 (1980)

Leibovitz, A.: The growth and maintenance of tissue-cell cultures in free gas exchange with the atmosphere. Am. J. Hyg. 78, 173–180 (1963)

Lembach, K.J.: Induction of human fibroblast proliferation by epidermal growth factor (EGF): Enhancement by an EGF-binding arginine esterase and by ascorbate. Proc. Natl. Acad. Sci. U.S.A. 73, 183–187 (1976)

Lengle, E., Geyer, R.P.: Comparison of cellular lipids of serum-free strain L mouse fibroblasts. Biochim. Biophys. Acta 260, 608–616 (1972)

Levene, C.I., Bates, C.J.: Ascorbic acid and collagen synthesis in cultured fibroblasts. Ann. NY Acad. Sci. 258, 287–306 (1975)

Levintow, L., Eagle, H.: Biochemistry of cultured mammalian cells. Annu. Rev. Biochem. *30*, 605–640 (1961)

Ley, K.D., Tobey, R.A.: Regulation of initiation of DNA synthesis in Chinese hamster cells. II. Induction of DNA synthesis and cell division by isoleucine and glutamine in G_1-arrested cells in suspension culture. J. Cell Biol. *47*, 453–459 (1970)

Lieberman, I., Ove, P.: Purification of a serum protein required by a mammalian cell in tissue culture. Biochim. Biophys. Acta *25*, 449–450 (1957)

Lieberman, M.A., Whittenberger, B., Raben, D., Glaser, L.: Whittenberger and Glaser reply. Nature *278*, 284 (1979)

Ling, C.T., Gey, G.O., Richters, V.: Chemically characterized concentrated corodies for continuous cell culture (The 7 C's culture media). Exp. Cell Res. *52*, 469–489 (1968)

Lipton, A., Paul, D., Henahan, M., Klinger, I., Holley, R.W.: Serum requirements for survival of 3 T 3 and SV 3 T 3 fibroblasts. Exp. Cell Res. *74*, 466–470 (1972)

Lipton, A., Schuler, M.F., Musselman, E., Kepner, N., Feldhoff, R.C., Jefferson, L.S.: Liver as a source of transformed-cell growth factor. In: Hormones and cell culture Book A. Sato, G.H., Ross, R. (eds.), pp. 461–475. Cold Spring Harbor, New York: Cold Spring Harbor Laboratory 1979

Lockart, R.Z. jr., Eagle, H.: Requirements for growth of single human cells. "Nonessential" amino acids, notably serine, are necessary and sufficient nutritional supplements. Science *129*, 252–254 (1959)

Luft, R., Hall, K. (eds.): Advances in metabolic disorders, Vol. 8, Somatomedins and some other growth factors. New York: Academic Press 1975

Maciag, T., Cerundolo, J., Ilsley, S., Kelley, P.R., Forand, R.: An endothelial cell growth factor from bovine hypothalamus: Identification and partial characterization. Proc. Natl. Acad. Sci. U.S.A. *76*, 5674–5678 (1979)

Macieira-Coelho, A., Avrameas, S.: Modulation of cell behavior in vitro by the substratum in fibroblastic and leukemic mouse cell lines. Proc. Natl. Acad. Sci. U.S.A. *69*, 2469–2473 (1972)

MacManus, J.P., Whitfield, J.F.: Cyclic AMP, prostaglandins, and the control of cell proliferation. Prostaglandins *6*, 475–487 (1974)

MacManus, J.P., Boynton, A.L., Whitfield, J.F.: Cyclic AMP and calcium as intracycle regulators in the control of cell proliferation. Adv. Cyclic Nucleotide Res. *9*, 485–491 (1978)

Macpherson, I., Montagnier, L.: Agar suspension culture for the selective assay of cells transformed by polyoma virus. Virology *23*, 291–294 (1964)

Markert, C.L.: Neoplasia: A disease of cell differentiation. Cancer Res. *28*, 1908–1914 (1968)

Mather, J.P., Sato, G.H.: The use of hormone-supplemented serum-free media in primary cultures. Exp. Cell Res. *124*, 215–221 (1979)

McKeehan, W.L.: The effect of temperature during trypsin treatment on viability and multiplication potential of single normal human and chicken fibroblasts. Cell Biol. Int. Rep. *1*, 335–343 (1977)

McKeehan, W.L., Ham, R.G.: Stimulation of clonal growth of normal fibroblasts with substrata coated with basic polymers. J. Cell Biol. *71*, 727–734 (1976)

McKeehan, W.L., Ham, R.G.: Calcium and magnesium ions and the regulation of multiplication in normal and transformed cells. Nature *275*, 756–758 (1978a)

McKeehan, W.L., Ham, R.G.: Phospholipid vesicles as synthetic, defined carriers for lipids in aqueous culture media. In Vitro *14*, 353 (Abstract No. 77) (1978b)

McKeehan, W.L., McKeehan, K.A.: Oxocarboxylic acids, pyridine nucleotide-linked oxidoreductases and serum factors in regulation of cell proliferation. J. Cell. Physiol. *101*, 9–16 (1979a)

McKeehan, W.L., McKeehan, K.A.: Epidermal growth factor modulates extracellular Ca^{2+} requirement for multiplication of normal human skin fibroblasts. Exp. Cell Res. *123*, 397–400 (1979b)

McKeehan, W.L., Hamilton, W.G., Ham, R.G.: Selenium is an essential trace nutrient for growth of WI-38 diploid human fibroblasts. Proc. Natl. Acad. Sci. U.S.A. *73*, 2023–2027 (1976)

McKeehan, W.L., McKeehan, K.A., Hammond, S.L., Ham, R.G.: Improved medium for clonal growth of human diploid fibroblasts at low concentrations of serum protein. In Vitro *13*, 399–416 (1977)

McKeehan, W.L., Genereux, D.P., Ham, R.G.: Assay and partial purification of factors from serum that control multiplication of human diploid fibroblasts. Biochem. Biophys. Res. Commun. *80*, 1013–1021 (1978)

McLimans, W.F.: The gaseous environment of the mammalian cell in culture. In: Growth, nutrition and metabolism of cells in culture. Rothblat, G.H., Cristofalo, V.J. (eds.), Vol. 1, pp. 137–170. New York: Academic Press 1972

McLimans, W.F., Crouse, E.J., Tunnah, K.V., Moore, G.E.: Kinetics of gas diffusion in mammalian cell culture systems. I. Experimental. Biotechnol. Bioeng. *10*, 725–740 (1968)

Messmer, T.O.: Nature of the iron requirement for Chinese hamster V 79 cells in tissue culture medium. Exp. Cell Res. *77*, 404–408 (1973)

Messmer, T.O., Young, D.V.: The effects of biotin and fatty acids on SV 3 T 3 cell growth in the presence of normal calf serum. J. Cell. Physiol. *90*, 265–270 (1977)

Michalopoulos, G., Pitot, H.C.: Primary culture of parenchymal liver cells on collagen membranes. Exp. Cell Res. *94*, 70–78 (1975)

Mierzejewski, K., Rozengurt, E.: Stimulation of DNA synthesis and cell division in a chemically defined medium: Effect of epidermal growth factor, insulin, and vitamin B_{12} on resting cultures of 3 T 6 cells. Biochem. Biophys. Res. Commun. *73*, 271–278 (1976)

Mierzejewski, K., Rozengurt, E.: Vitamin B_{12} enhances the stimulation of DNA synthesis by serum in resting cultures of 3 T 6 cells. Exp. Cell Res. *106*, 394–397 (1977 a)

Mierzejewski, K., Rozengurt, E.: Density-dependent inhibition of fibroblast growth is overcome by pure mitogenic factors. Nature *269*, 155–156 (1977 b)

Morgan, J.F.: Tissue culture nutrition. Bact. Rev. *22*, 20–45 (1958)

Morgan, J.F., Morton, H.J., Parker, R.C.: Nutrition of animal cells in tissue culture. I. Initial studies on a synthetic medium (17557). Proc. Soc. Exp. Biol. Med. *73*, 1–8 (1950)

Moscatelli, D., Sanui, H., Rubin, A.H.: Effects of depletion of K^+, Na^+, or Ca^{2+} on DNA synthesis and cell cation content in chick embryo fibroblasts. J. Cell. Physiol. *101*, 117–128 (1979)

Moscona, A.: Cell suspensions from organ rudiments of chick embryos. Exp. Cell Res. *3*, 535–539 (1952)

Moskowitz, M., Cheng, D.K.S.: Growth and G 1 arrest of Sarcoma virus transformed cells in serum free media. J. Cell. Physiol. *100*, 589–602 (1979)

Murakami, O., Yamane, I.: Effects of basic polymers on growth of tumor cells in suspension cultures with "serum-free" medium. Cell Structure Function *1*, 285–290 (1976)

Naylor, S.L., Busby, L.L., Klebe, R.J.: Biochemical selection systems for mammalian cells: The essential amino acids. Somatic Cell Genet. *2*, 93–111 (1976)

Neugut, A.I., Weinstein, I.B.: Growth limitation of BHK-21 cells and its relation to folate metabolism. In Vitro *15*, 363–367 (1979)

Neuman, R.E., McCoy, T.A.: Growth-promoting properties of pyruvate, oxalacetate and α-ketoglutarate for isolated Walker carcinosarcoma 256 cells. Proc. Soc. Exp. Biol. Med. *98*, 303–306 (1958)

Nielsen, F.H.: Consideration of trace element requirements for preparation of chemically defined media. In: The growth requirements of vertebrate cells in vitro. Waymouth, C., Ham, R.G., Chapple, P.J. (eds.). New York: Cambridge University Press, in press 1981

Nilausen, K.: Role of fatty acids in growth-promoting effect of serum albumin on hamster cells in vitro. J. Cell. Physiol. *96*, 1–13 (1978)

Nissley, S.P., Rechler, M.M., Moses, A.C., Eisen, H.J., Higa, O.Z., Short, P.A., Fennoy, I., Bruni, C.B., White, R.M.: Evidence that multiplication-stimulating activity, purified from the BRL-3 A rat liver cell line, is found in rat serum and fetal liver organ culture. In: Hormones and cell culture Book A. Sato, G.H., Ross, R. (eds.), pp. 79–94. Cold Spring Harbor, New York: Cold Spring Harbor Laboratory 1979

O'Farrell, M.K., Clingan, D., Rudland, P.S., Jimenez de Asua, L.: Stimulation of the initiation of DNA synthesis and cell division in several cultured mouse cell types. Exp. Cell Res. *118*, 311–321 (1979)

Ohnuma, T., Waligunda, J., Holland, J.F.: Amino acid requirements *in vitro* of human leukemic cells. Cancer Res. *31*, 1640–1644 (1971)

Orly, J., Sato, G.: Fibronectin mediates cytokinesis and growth of rat follicular cells in serum-free medium. Cell *17*, 295–305 (1979)

Oshima, R.: Stimulation of the clonal growth and differentiation of feeder layer dependent mouse embryonal carcinoma cells by β-mercaptoethanol. Differentiation *11*, 149–155 (1978)

Packer, L., Smith, J.R.: Extension of the lifespan of cultured normal human diploid cells by vitamin E. Proc. Natl. Acad. Sci. U.S.A. *71*, 4763–4767 (1974)

Packer, L., Smith, J.R.: Extension of the lifespan of cultured normal human diploid cells by vitamin E: A reevaluation. Proc. Natl. Acad. Sci. U.S.A. *74*, 1640–1641 (1977)

Pagano, R.E., Weinstein, J.N.: Interactions of liposomes with mammalian cells. Annu. Rev. Biophys. Bioeng. *7*, 435–468 (1978)

Papaconstantinou, J., Rutter, W.J.: Molecular control of proliferation and differentiation. New York: Academic Press Inc. 1978

Park, C.H., Bergsagel, D.E., McCulloch, E.A.: Ascorbic acid: A culture requirement for colony formation by mouse plasmacytoma cells. Science *174*, 720–722 (1971)

Paul, D., Lipton, A., Klinger, I.: Serum factor requirements of normal and Simian virus 40-transformed 3T3 mouse fibroblasts. Proc. Natl. Acad. Sci. U.S.A. *68*, 645–648 (1971)

Paul, D., Ristow, H.J., Rupniak, H.T., Messmer, T.O.: Growth control by serum factors and hormones in mammalian cells in culture. In: Molecular control of proliferation and differentiation. Papaconstantinou, J., Rutter, W.J. (eds.), pp. 65–79. New York: Academic Press 1978a

Paul, D., Brown, K.D., Rupniak, H.T., Ristow, H.J.: Cell cycle regulation by growth factors and nutrients in normal and transformed cells. In Vitro *14*, 76–84 (1978b)

Pawelek, J.M.: Evidence suggesting that a cyclic AMP-dependent protein kinase is a positive regulator of proliferation in Cloudman S91 melanoma cells. J. Cell. Physiol. *98*, 619–626 (1979)

Pawelek, J., Halaban, R., Christie, G.: Melanoma cells which require cyclic AMP for growth. Nature *258*, 539–540 (1975)

Peehl, D.M., Ham, R.G.: Growth and differentiation of human keratinocytes without a feeder layer or conditioned medium. In Vitro *16*, 516–525 (1980a)

Peehl, D.M., Ham, R.G.: Clonal growth of human keratinocytes with small amounts of dialyzed serum. In Vitro *16*, 526–540 (1980b)

Perry, S. V., Grand, R.J.A., Nairn, A.C., Vanaman, T.C., Wall, C.M.: Calcium-binding proteins and the regulation of contractile activity. Biochem. Soc. Trans. *7*, 619–622 (1979)

Peterkofsky, B., Prather, W.: Cytotoxicity of ascorbate and other reducing agents towards cultured fibroblasts as a result of hydrogen peroxide formation. J. Cell. Physiol. *90*, 61–70 (1977)

Phillips, P.D., Cristofalo, V.J.: Growth regulation of WI-38 cells in a serum-free hormone-supplemented medium. In Vitro *16*, 250 (abstract No. 193) (1980a)

Phillips, P.D., Christofalo, V.J.: The growth of young and old WI-38 cells in chemically defined medium. In: Neural regulatory mechanisms during aging. Adelman, R.C., Roberts, J., Cristofalo, V.J., (eds.). pp. 199–202 New York: Alan R. Liss, Inc. (1980b)

Pickart, L., Thaler, M.M.: Growth-modulating human plasma tripeptide: Relationship between molecular structure and DNA synthesis in hepatoma cells. FEBS Lett. *104*, 119–122 (1979)

Pledger, W.J., Stiles, C.D., Antoniades, H.N., Scher, C.D.: Induction of DNA synthesis in BALB/c3T3 cells by serum components: Reevaluation of the commitment process. Proc. Natl. Acad. Sci. USA *74*, 4481–4485 (1977)

Pledger, W.J., Stiles, C.D., Antoniades, H.N., Scher, C.D.: An ordered sequence of events is required before BALB/c-3T3 cells become committed to DNA synthesis. Proc. Natl. Acad. Sci. USA *75*, 2839–2843 (1978)

Pohjanpelto, P., Raina, A.: Identification of a growth factor produced by human fibroblasts in vitro as putrescine. Nature New Biology *235*, 247–249 (1972)

Pollack, M., Fisher, H.W.: Dissociation of ribonucleic acid and protein synthesis in mammalian cells deprived of potassium. Arch. Biochem. Biophys. *172*, 188–190 (1976)

Poste, G., Papahadjopoulos, D., Vail, W.J.: Lipid vesicles as carriers for introducing bio-logically active materials into cells. Methods Cell Biol. *14*, 33–71 (1976)

Prescott, D.M.: Reproduction of eukaryotic cells. New York: Academic Press 1976

Price, F.M., Kerr, H.A., Andresen, W.F., Bryant, J.C., Evans, V.J.: Some in vitro require-ments of cells of C 3 H mouse, Chinese hamster, green monkey, and human origin. J. Natl. Cancer Inst. *37*, 601–617 (1966)

Price, F.M., Rotherham, J., Evans, V.J.: Pyrimidine nucleoside requirements of a neoplastic C 3 H mouse cell strain. J. Natl. Cancer Inst. *39*, 529–538 (1967)

Pruss, R.M., Herschman, H.R.: Cholera toxin stimulates division of 3 T 3 cells. J. Cell. Physiol. *98*, 469–473 (1979)

Puck, T.T., Marcus, P.I.: A rapid method for viable cell titration and clone production with HeLa cells in tissue culture: The use of X-irradiated cells to supply conditioning factors. Proc. Natl. Acad. Sci. USA *41*, 432–437 (1955)

Puck, T.T., Marcus, P.I., Cieciura, S.J.: Clonal growth of mammalian cells in vitro. Growth characteristics of colonies from single HeLa cells with and without a "feeder" layer. J. Exp. Med. *103*, 273–284 (1956)

Radlett, P.J., Telling, R.C., Whitside, J.P., Maskell, M.A.: The supply of oxygen to sub-merged cultures of BHK 21 cells. Biotechnol. Bioeng. *14*, 437–445 (1972)

Rasmussen, H., Goodman, D.B.P.: Relationships between calcium and cyclic nucleotides in cell activation. Physiol. Rev. *57*, 421–509 (1977)

Reese, D.H., Friedman, R.D.: Suppression of dysplasia and hyperplasia by calcium in or-gan-cultured urinary bladder epithelium. Cancer Res. *38*, 586–592 (1978)

Reid, L.M., Rojkind, M.: New techniques for culturing differentiated cells: Reconstituted basement membrane rafts. Methods Enzymol. *58*, 263–278 (1979)

Rheinwald, J.G., Green, H.: Serial cultivation of strains of human epidermal keratinocytes: The formation of keratinizing colonies from single cells. Cell *6*, 331–344 (1975)

Rizzino, A., Crowley, C.: Growth and differentiation of embryonal carcinoma cell line F_9 in defined media. Proc. Natl. Acad. Sci. USA *77*, 457–461 (1980)

Rizzino, A., Sato, G.: Growth of embryonal carcinoma cells in serum-free medium. Proc. Natl. Acad. Sci. USA *75*, 1844–1848 (1978)

Rizzino, A., Rizzino, H., Sato, G.: Defined media and the determination of nutritional and hormonal requirements of mammalian cells in culture. Nutr. Rev. *37*, 369–378 (1979)

Ross, R., Vogel, A.: The platelet-derived growth factor. Cell *14*, 203–210 (1978)

Ross, R., Glomset, J., Kariya, B., Harker, L.: A platelet-dependent serum factor that stimu-lates the proliferation of arterial smooth muscle cells in vitro. Proc. Natl. Acad. Sci. USA *71*, 1207–1210 (1974)

Ross, R., Nist, C., Kariya, B., Rivest, M.J., Raines, E., Callis, J.: Physiological quiescence in plasma-derived serum: Influence of platelet-derived growth factor on cell growth in culture. J. Cell. Physiol. *97*, 497–508 (1978)

Ross, R., Vogel, A., Davies, P., Raines, E., Kariya, B., Rivest, M.J., Gustafson, C., Glom-set, J.: The platelet-derived growth factor and plasma control cell proliferation. In: Hor-mones and cell culture Book A. Sato, G.H., Ross, R. (eds.), pp. 3–16. Cold Spring Har-bor, New York: Cold Spring Harbor Laboratory 1979

Rossow, P.W., Riddle, V.G.H., Pardee, A.B.: Synthesis of labile, serum dependent protein in early G_1 controls animal cell growth. Proc. Natl. Acad. Sci. USA *76*, 4446–4450 (1979)

Rothblat, G.H., Arbogast, L.Y., Ray, E.K.: Stimulation of esterified cholesterol accumula-tion in tissue culture cells exposed to high density lipoproteins enriched in free choles-terol. J. Lipid Res. *19*, 350–358 (1978)

Rotherham, J., Price, F.M., Otani, T.T., Evans, V.J.: Some aspects of the role of vitamin B_{12} and folic acid in DNA-thymidine synthesis in a neoplastic C 3 H mouse cell strain. J. Natl. Cancer Inst. *47*, 277–287 (1971)

Rowe, D.W., Starman, B.J., Fujimoto, W.Y., Williams, R.H.: Differences in growth re-sponse to hydrocortisone and ascorbic acid by human diploid fibroblasts. In Vitro *13*, 824–830 (1977)

Rozengurt, E.: Biochemical basis of the early events stimulated by serum and mitogenic factors in cultures of quiescent cells. In: Hormones and cell culture Book B. Sato, G.H., Ross, R. (eds.), pp. 773–788. Cold Spring Harbor, New York: Cold Spring Harbor Lab-oratory 1979

Rozengurt, E., Legg, A., Pettican, P.: Vasopressin stimulation of mouse 3T3 cell growth. Proc. Natl. Acad. Sci. USA *76*, 1284–1287 (1979)

Rubin, H.: Central role for magnesium in coordinate control of metabolism and growth in animal cells. Proc. Natl. Acad. Sci. USA *72*, 3551–3555 (1975a)

Rubin, H.: Nonspecific nature of the stimulus to DNA synthesis in cultures of chick embryo cells. Proc. Natl. Acad. Sci. USA *72*, 1676–1680 (1975b)

Rubin, H.: Magnesium deprivation reproduces the coordinate effects of serum removal or cortisol addition on transport and metabolism in chick embryo fibroblasts. J. Cell. Physiol. *89*, 613–626 (1976)

Rubin, A.H., Bowen-Pope, D.F.: Coordinate control of Balb/c 3T3 cell survival and multiplication by serum or calcium pyrophosphate complexes. J. Cell. Physiol. *98*, 81–94 (1979)

Rubin, H., Koide, T.: Stimulation of DNA synthesis and 2-deoxy-D-glucose transport in chick embryo cultures by excessive metal concentrations and by a carcinogenic hydrocarbon. J. Cell. Physiol. *81*, 387–396 (1973)

Rubin, H., Koide, T.: Early cellular responses to diverse growth stimuli independent of protein and RNA synthesis. J. Cell. Physiol. *86*, 47–58 (1975)

Rubin, H., Koide, T.: Mutual potentiation by magnesium and calcium of growth in animal cells. Proc. Natl. Acad. Sci. USA *73*, 168–172 (1976)

Rubin, H., Sanui, H.: Complexes of inorganic pyrophosphate, orthophosphate, and calcium as stimulants of 3T3 cell multiplication. Proc. Natl. Acad. Sci. USA *74*, 5026–5030 (1977)

Rubin, A.H., Sanui, H.: The coordinate response of cells to hormones and its mediation by the intracellular availability of magnesium. In: Hormones and cell culture Book B. Sato, G.H., Ross, R. (eds.), pp. 741–750. Cold Spring Harbor, New York: Cold Spring Harbor Laboratory 1979

Rubin, A.H., Terasaki, M., Sanui, H.: Major intracellular cations and growth control: Correspondence among magnesium content, protein synthesis, and the onset of DNA synthesis in BALB/c 3T3 cells. Proc. Natl. Acad. Sci. USA *76*, 3917–3921 (1979)

Rudland, P.S., Jimenez de Asua, L.: Action of growth factors in the cell cycle. Biochim. Biophys. Acta *560*, 91–133 (1979)

Rudland, P.S., Durbin, H., Clingan, D., Jimenez de Asua, L.: Iron salts and transferrin are specifically required for cell division of cultured 3T6 cells. Biochem. Biophys. Res. Commun. *75*, 556–562 (1977)

Russell, D.H.: The roles of the polyamines, putrescine, spermidine, and spermine in normal and malignant tissues. Life Sci. *13*, 1635–1647 (1973)

Rutishauser, U., Gall, W.E., Edelman, G.M.: Adhesion among neural cells of the chick embryo. IV. Role of the cell surface molecule CAM in the formation of neurite bundles in cultures of spinal ganglia. J. Cell Biol. *79*, 382–393 (1978)

Rutter, W.J., Pictet, R.L., Morris, P.W.: Toward molecular mechanisms of developmental processes. Annu. Rev. Biochem. *42*, 601–646 (1973)

Sakagami, H., Yamada, M.-A.: Failure of vitamin E to extend the life span of a human diploid cell line in culture. Cell Structure Function *2*, 219–227 (1977)

Sanford, K.K. (ed.): Third Decennial review conference: Cell, tissue, and organ culture. Gene expression and regulation in cultured cells. Natl. Cancer Inst. Monograph 48 (1978)

Sanford, K.K., Earle, W.R., Likely, G.D.: The growth in vitro of single isolated tissue cells. J. Natl. Cancer Inst. *9*, 229–246 (1948)

Sanford, K.K., Parshad, R., Gantt, R.: Light and oxygen effects on chromosomes, DNA and neoplastic transformation of cells in culture. In: Nutritional requirements of cultured cells. Katsuta, H. (ed.), pp. 117–148. Tokyo, Japan: Japan Scientific Societies Press 1978

Sanford, K.K., Parshad, R., Jones, G., Handleman, S., Garrison, C., Price, F.: Role of photosensitization and oxygen in chromosome stability and "spontaneous" malignant transformation in culture. J. Natl. Cancer Inst. *63*, 1245–1255 (1979)

Sato, G.H.: The role of serum in cell culture. In: Biochemical actions of hormones. Litwack, G. (ed.), Vol. III, pp. 391–396. New York: Academic Press 1975

Sato, G.H., Ross, R. (eds.): Hormones and cell culture Books A and B. Cold Spring Harbor, New York: Cold Spring Harbor Laboratory 1979

Scher, C.D., Pledger, W.J., Martin, P., Antoniades, H., Stiles, C.D.: Transforming viruses directly reduce the cellular growth requirement for a platelet derived growth factor. J. Cell. Physiol. *97*, 371–380 (1978)

Scher, C.D., Shepard, R.C., Antoniades, H.N., Stiles, C.D.: Platelet-derived growth factor and the regulation of the mammalian fibroblast cell cycle. Biochim. Biophys. Acta *560*, 217–241 (1979 a)

Scher, C.D., Stone, M.E., Stiles, C.D.: Platelet-derived growth factor prevents G_0 growth arrest. Nature *281*, 390–392 (1979 b)

Schiaffonati, L., Baserga, R.: Different survival of normal and transformed cells exposed to nutritional conditions nonpermissive for growth. Cancer Res. *37*, 541–545 (1977)

Sekas, G., Owen, W.G., Cook, R.T.: Fractionation and preliminary characterization of a low molecular weight bovine hepatic inhibitor of DNA synthesis in regenerating rat liver. Exp. Cell Res. *122*, 47–54 (1979)

Serrero, G.R., McClure, D.B., Sato, G.H.: Growth of mouse 3T3 fibroblasts in serum-free, hormone-supplemented media. In: Hormones and cell culture Book B. Sato, G.H., Ross, R. (eds.), pp. 523–530. Cold Spring Harbor, New York: Cold Spring Harbor Laboratory 1979

Shipley, G.D., Ham, R.G.: Attachment and growth of Swiss and Balb/c 3T3 cells in a completely serum-free medium. In Vitro *16*, 218 (abstract No. 57) (1980)

Shipley, G.D., Ham, R.G.: Improved medium and culture conditions for clonal growth with minimal serum protein and for enhanced serum-free survival of Swiss 3T3 cells. In Vitro, in press (1981)

Shoyab, M., De Larco, J.E., Todaro, G.J.: Biologically active phorbol esters specifically alter affinity of epidermal growth factor membrane receptors. Nature *279*, 387–391 (1979)

Sinensky, M.: A nutritional test for the low density lipoprotein receptor in Chinese hamster fibroblasts. FEBS Lett. *106*, 129–131 (1979)

Sivak, A.: Induction of cell division in BALB/c-3T3 cells by phorbol myristate acetate or bovine serum: Effects of inhibitors of cyclic AMP phosphodiesterase and Na^+-K^+-ATPase. In Vitro *13*, 337–343 (1977)

Spector, A.A.: Fatty acid, glyceride, and phospholipid metabolism. In: Growth, nutrition, and metabolism of cells in culture. Rothblat, G.H., Cristofalo, V.J. (eds.), Vol. 1, pp. 257–296. New York: Academic Press 1972

Steck, P.A., Voss, P.G., Wang, J.L.: Growth control in cultured 3T3 fibroblasts. Assays of cell proliferation and demonstration of a growth inhibitory activity. J. Cell Biol. *83*, 562–575 (1979)

Stiles, C.D., Capone, G.T., Scher, C.D., Antoniades, H.N., Van Wyk, J.J., Pledger, W.J.: Dual control of cell growth by somatomedins and platelet-derived growth factor. Proc. Natl. Acad. Sci. USA *76*, 1279–1283 (1979 a)

Stiles, C.D., Isberg, R.R., Pledger, W.J., Antoniades, H.N., Scher, C.D.: Control of the BALB/c-3T3 cell cycle by nutrients and serum factors: analysis using platelet-derived growth factor and platelet-poor plasma. J. Cell. Physiol. *99*, 395–406 (1979 b)

Stoker, M., Piggott, D., O'Neill, M.C., Berryman, S., Waxman, V.: Anchorage and growth regulation in normal and virus-transformed cells. Int. J. Cancer *3*, 683–693 (1968)

Swim, H.E.: Nutrition of cells in culture – a review. In: Lipid metabolism in tissue culture cells. Rothblat, G.H., Kritchevsky, D. (eds.), pp. 1–16. Philadelphia, Pennsylvania: The Wistar Institute Press 1967

Swim, H.E., Parker, R.F.: Vitamin requirements of uterine fibroblasts, strain U 12–79; their replacement by related compounds. Arch. Biochem. Biophys. *78*, 46–53 (1958)

Tabor, C.W., Tabor, H.: 1,4-diaminobutane (putrescine), spermidine, and spermine. Annu. Rev. Biochem. *45*, 285–306 (1976)

Taub, M., Chuman, L., Saier, M.H., Jr., Sato, G.: Growth of Madin-Darby canine kidney epithelial cell (MDCK) line in hormone-supplemented, serum-free medium. Proc. Natl. Acad. Sci. USA *76*, 3338–3342 (1979)

Taub, M., Sato, G.: Growth of functional primary cultures of kidney epithelial cells in defined medium. J. Cell. Physiol. *105*, 369–378 (1980)

Temin, H.M., Pierson, R.W., Jr., Dulak, N.C.: The role of serum in the control of multiplication of avian and mammalian cells in culture. In: Growth, nutrition, and metabolism of cells in culture. Rothblat, G.H., Cristofalo, V.J. (eds.), Vol. 1, pp. 49–81. New York: Academic Press 1972

Thomas, J.A., Johnson, M.J.: Trace-metal requirements of NCTC clone 929 strain L cells. J. Natl. Cancer Inst. *39*, 337–345 (1967)

Todaro, G.J., Lazar, G.K., Green, H.: The initiation of cell division in a contact-inhibited mammalian cell line. J. Cell. Physiol. *66*, 325–333 (1965)

Todaro, G.J., De Larco, J.E., Marquardt, H., Bryant, M.L., Sherwin, S.A., Sliski, A.H.: Polypeptide growth factors produced by tumor cells and virus-transformed cells: A possible growth advantage for the producer cells. In: Hormones and cell culture Book A. Sato, G.H., Ross, R. (eds.), pp. 113–127. Cold Spring Harbor, New York: Cold Spring Harbor Laboratory 1979

Tsao, M., Walthall, B.J., Ham, R.G.: Clonal growth of normal human epidermal keratinocytes in a defined medium. J. Cell. Physiol. in press

Tucker, R.W., Scher, C.D., Stiles, C.D.: Centriole deciliation associated with the early response of 3T3 cells to growth factors but not to SV40. Cell *18*, 1065–1072 (1979)

Underwood, E.J.: Trace elements in human and animal nutrition, 4th ed. New York: Academic Press 1977

Uriel, J.: Retrodifferentiation and the fetal patterns of gene expression in cancer. Adv. Cancer Res. *29*, 127–174 (1979)

Vallotton, M., Hess-Sander, U., Leuthardt, F.: Fixation spontanée de la biotine à une protéine dans le sérum humain. Helv. Chim. Acta *48*, 126–133 (1965)

Van der Bosch, J., Sommer, I., Maier, H., Rahmig, W.: Density-dependent growth adaptation kinetics in 3T3 cell populations following sudden (Ca^{2+}) and temperature changes. A comparison with SV40-3T3 cells. Z. Naturforsch. *34*, 279–283 (1979)

Vogel, A., Raines, E., Kariya, B., Rivest, M.-J., Ross, R.: Coordinate control of 3T3 cell proliferation by platelet-derived growth factor and plasma components. Proc. Natl. Acad. Sci. USA *75*, 2810–2814 (1978)

Wallis, C., Ver, B., Melnick, J.L.: The role of serum and fetuin in the growth of monkey kidney cells in culture. Exp. Cell Res. *58*, 271–282 (1969)

Walthall, B.J., Ham, R.G.: Growth of human diploid fibroblasts in serum free medium. In Vitro *16*, 250–251 (Abstract No. 194) (1980)

Walthall, B.J., Ham, R.G.: Multiplication of human diploid fibroblasts in a synthetic medium supplemented with EGF, insulin, and dexamethasone. Exp. Cell. Res. **134**, 301–309 (1981)

Wang, R.J.: Effect of room fluorescent light on the deterioration of tissue culture medium. In Vitro *12*, 19–22 (1976)

Wang, R.J., Nixon, B.T.: Identification of hydrogen peroxide as a photoproduct toxic to human cells in tissue-culture medium irradiated with "daylight" fluorescent light. In Vitro *14*, 715–722 (1978)

Waymouth, C.: The nutrition of animal cells. Int. Rev. Cytol. *3*, 1–68 (1954)

Waymouth, C.: Construction and use of synthetic media. In: Cells and tissues in culture. Willmer, E.N. (ed.), pp. 99–142. New York: Academic Press 1965

Waymouth, C.: Osmolality of mammalian blood and of media for culture of mammalian cells. In Vitro *6*, 109–127 (1970)

Waymouth, C.: Construction of tissue culture media. In: Growth, nutrition and metabolism of cells in culture. Rothblat, G.H., Cristofalo, V.J. (eds.), Vol. I, pp. 11–47. New York: Academic Press 1972

Waymouth, C.: Studies on chemically defined media and the nutritional requirements of cultures of epithelial cells. In: Nutritional requirements of cultured cells. Katsuta, H. (ed.), pp. 39–61. Tokyo, Japan: Japan Scientific Societies Press 1978

Waymouth, C.: Major ions, buffer systems, pH, osmolality, and water quality. In: The growth requirements of vertebrate cells in vitro. Waymouth, C., Ham, R.G., Chapple, P.J. (eds.). New York: Cambridge University Press, in press 1981

Waymouth, C., Ham, R.G., Chapple, P.J. (eds.): The growth requirements of vertebrate cells in vitro. New York: Cambridge University Press, in press 1981

Westermark, B.: Density dependent proliferation of human glia cells stimulated by epidermal growth factor. Biochem. Biophys. Res. Commun. *69*, 304–310 (1976)

Westermark, B.: Local starvation for epidermal growth factor cannot explain density-dependent inhibition of normal human glial cells. Proc. Natl. Acad. Sci. USA *74*, 1619–1621 (1977)

Westermark, B., Wasteson, A.: A platelet factor stimulating human normal glial cells. Exp. Cell Res. *98*, 170–174 (1976)

Whitfield, J.F., MacManus, J.P., Rixon, R.H., Boynton, A.L., Youdale, T., Swierenga, S.: The positive control of cell proliferation by the interplay of calcium ions and cyclic nucleotides. A review. In Vitro *12*, 1–18 (1976)

Whitfield, J.F., Boynton, A.L., MacManus, J.P., Sikorska, M., Tsang, B.K.: The regulation of cell proliferation by calcium and cyclic AMP. Mol. Cell. Biochem. *27*, 155–179 (1979)

Whittenberger, B., Glaser, L.: Cell saturation density is not determined by a diffusion-limited process. Nature *272*, 821–823 (1978)

Wu, R., Sato, G.H.: Replacement of serum in cell culture by hormones: A study of hormonal regulation of cell growth and specific gene expression. J. Toxicol. Environ. Health *4*, 427–448 (1978)

Yamada, K.M., Olden, K.: Fibronectins-adhesive glycoproteins of cell surface and blood. Nature *275*, 179–184 (1978)

Yamane, I.: Development and application of a serum-free culture medium for primary culture. In: Nutritional requirements of cultured cells. Katsuta, H. (ed.), pp. 1–21. Tokyo, Japan: Japan Scientific Societies Press 1978

Yamane, I., Murakami, O.: 6,8-dihydroxypurine. A novel growth factor for mammalian cells in vitro, isolated from a commercial peptone. J. Cell. Physiol. *81*, 281–284 (1973)

Yamane, I., Tomioka, F.: The concomitant effect of unsaturated fatty acid supplemented to medium on cellular growth and membrane fluidity of cultured cells. Cell Biol. Int. Rep. *3*, 515–523 (1979)

Yamane, I., Murakami, O., Kato, M.: "Serum-free" culture of various mammalian cells and the role of bovine albumin. Cell Structure Function *1*, 279–284 (1976)

Yasumura, Y., Niwa, A., Yamamoto, K.: Phenotypic requirement for glutamine of kidney cells and for glutamine and arginine of liver cells in culture. In: Nutritional requirements of cultured cells. Katsuta, H. (ed.), pp. 223–255. Tokyo, Japan: Japan Scientific Societies Press 1978

Yen, A., Pardee, A.B.: Exponential 3T3 cells escape in mid-G1 from their high serum requirement. Exp. Cell Res. *116*, 103–113 (1978)

Yen, A., Riddle, V.G.H.: Plasma and platelet associated factors act in G1 to maintain proliferation and to stabilize arrested cells in a viable quiescent state. A temporal map of control points in the G1 phase. Exp. Cell Res. *120*, 349–357 (1979)

Young, D.V., Cox, F.W. III, Chipman, S., Hartman, S.C.: The growth stimulation of SV3T3 cells by transferrin and its dependence on biotin. Exp. Cell Res. *118*, 410–414 (1979)

Youngdahl-Turner, P., Rosenberg, L.E., Allen, R.H.: Binding and uptake of transcobalamin II by human fibroblasts. J. Clin. Invest. *61*, 133–141 (1978)

Zardi, L., Siri, A., Carnemolla, B., Santi, L., Gardner, W.D., Hoch, S.O.: Fibronectin: A chromatin-associated protein? Cell *18*, 649–657 (1979)

Zombola, R.R., Bearse, R.C., Kitos, P.A.: Trace element uptake by L-cells as a function of trace elements in a synthetic growth medium. J. Cell. Physiol. *101*, 57–66 (1979)

CHAPTER 3

Epidermal Growth Factor

G. Carpenter

A. Introduction

During the course of purifying nerve growth factor from the submaxillary gland of the mouse, Cohen (1960) and Levi-Montalcini and Cohen (1960) noticed that daily injections of certain gland extract fractions into newborn mice produced developmental changes that could not be ascribed to nerve growth factor. These changes included precocious opening of the eyelids (7 days compared to the usual 14 days) and a similar early eruption of the incisors. Using these gross anatomical changes as an assay, Cohen (1962) proceeded to isolate the active factor – a polypeptide which he termed epidermal growth factor (EGF).

In the ensuing two decades, and particularly in the last seven years, a considerable number of studies concerning EGF have been published. The most recent and exciting progress has been in understanding the mechanism of action of this growth factor at the cellular level. These advances which have paralleled studies of the polypeptide hormones, for example insulin and the gonadotrophins, may provide a general framework for understanding how extracellular signals regulate cell development and physiology.

Although this article is intended to be comprehensive, the focus will be on the more recent advances at the cell and subcellular levels. Additional information, particularly on earlier studies, can be obtained elsewhere (Cohen, 1965a, 1971, 1972; Cohen and Taylor, 1974; Cohen and Savage, 1974; Cohen et al., 1975; Carpenter and Cohen, 1978a, 1979; Carpenter, 1978; Haigler and Cohen, 1979).

B. Chemical and Physical Properties of EGF

I. Mouse EGF

1. Isolation

Purified EGF can be prepared in essentially a two-step procedure (Savage and Cohen, 1972). This is a rapid purification based on the observation that, at low pH, columns of polyacrylamide (Bio-Gel) are capable of selectively and reversibly adsorbing EGF from crude homogenates of the mouse submaxillary gland. The reason for the anomalous behavior of EGF on Bio-Gel columns at low pH is not known. It is suspected, however, that the two adjacent Trp-Trp residues in the COOH-terminal region of the molecule (Fig. 1) are responsible for the adsorption of EGF. The yield can be further increased by treating the mice with a single injection of testosterone propionate one week before sacrificing the animals. One can

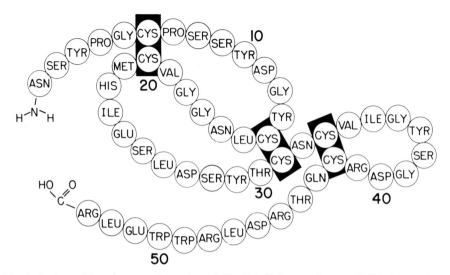

Fig. 1. Amino acid sequence and location of disulfide linkages of mouse EGF. (From Savage et al., 1973)

obtain up to 1.0 mg of EGF per gram wet weight of salivary glands. Rapid isolation procedures yielding milligram quantities of purified growth factor are an important technical aspect in the study of mitogens or hormones. In addition to the time and cost saved, the availability of substantial quantities of the growth factor has allowed detailed physical and chemical studies, the use of ferritin:EGF conjugates for electron microscopy studies, and the use of immobilized EGF for affinity chromatography purification of the receptor.

2. Chemical and Physical Properties

Studies by Taylor et al. (1972) reported many of the basic chemical and physical properties of mouse EGF. The growth factor is a heat-stable, non-dialyzable, single polypeptide chain of 53 amino acid residues and is devoid of alanine, phenylalanine, and lysine residues. No free sulfhydryl groups, hexosamine, or neutral sugar groups were detected. The molecular weight, as determined by amino acid composition, was reported as 6,045, in agreement with molecular weight estimates of 6,400 and 7,000, determined by sedimentation equilibrium and gel filtration, respectively. The extinction coefficient ($E_{1\,cm}^{1\%}$ at 280 nm) was determined to be 30.9, and a sedimentation constant of 1.25 S was reported (Cohen, 1962).

Studies by Savage et al. (1972, 1973) have reported the primary amino acid sequence of mouse EGF and the location of the three intramolecular disulfide bonds (Fig. 1). The disulfide bonds in mouse EGF are required for biological activity (Taylor et al., 1972). Reduction of the disulfides in the presence of mercaptoethanol and urea yielded an inactive polypeptide. Biological activity could be almost completely restored by removal of the mercaptoethanol and urea by dialysis and reoxidation in air.

Circular dichroic examination of the far-ultraviolet spectrum of mouse EGF indicated the absence of significant α-helical structure, the presence of 25% β-helix, and a random coil content of 75% (TAYLOR et al., 1972). These results were confirmed by HOLLADAY et al. (1976) who extended circular dichroic studies to include the nearultraviolet spectrum. Their results indicated that while intact mouse EGF has 22% β-structure, derivates affecting residues 49–53 at the carboxy-terminus and residues 21 and 22 had significantly less β-structure. These residues, therefore, would appear to contribute to the ordered structure of mouse EGF.

The sedimentation constant of 1.25 S (COHEN, 1962) and frictional ratio (f/f_0) of 1.12 (L. HOLLADAY, personal communication) suggest that mouse EGF is a relatively compact and globular polypeptide.

Examination of the circular dichroic spectra of mouse EGF under different conditions indicates that the molecule has a very stable tertiary structure (HOLLA-DAY et al., 1976). Equilibrium studies of the reversible unfolding of mouse EGF in the presence of a denaturant showed a high transition midpoint (6.89 M guanidinium hydrochloride at 25 °C) compared to other proteins. By extrapolations, an estimation of 16 kcal/mole was made for the free energy of unfolding (ΔG °C) in the absence of denaturant. The thermal stability of mouse EGF was also examined and various thermodynamic properties estimated for this polypeptide at 40 °C in the absence of denaturant. These data indicated that the apparent free energy of unfolding (ΔG^0_{app}) is 18 kcal/mole; the apparent enthalpy of unfolding (ΔH_{app}) is 24.4 kcal/mole; and the apparent entropy of unfolding (ΔS_{app}) is 20.4 cal/(mole deg). The absolute magnitude of the apparent heat capacity (ΔC_p) was estimated at less than 0.5 kcal/mole deg^{-1}), substantially less than that reported for other proteins. It was suggested that the small ΔC_p value for mouse EGF is due in part to the relatively high degree of exposure of hydrophobic sites to the solvent. In the absence of completely buried aromatic residues, it is not clear what forces, other than the three disulfide bonds, produce such a highly stable native conformation for mouse EGF. Perhaps the β-structure plays a large role. In summary, the data indicate that mouse EGF is one of the most energetically stable proteins that has been described.

3. Derivatives

Several derivates of mouse EGF have been characterized and are described in this section. In later sections the relationship of these structural derivates to biological activities of EGF will be discussed.

a) EGF-2

The isolation of intact mouse EGF from crude extracts of submaxillary glands was dependent on the pH. If the pH of the extract was lowered from 4.5 to 3.5, a derivative of mouse EGF was produced (termed EGF-2 or EGF$_{1-51}$), presumably by an acidic protease, that lacked the carboxy-terminal Leu-Arg residues (SAVAGE and COHEN, 1972).

b) EGF-5

SAVAGE et al. (1972) showed that exposure of intact mouse EGF to mild tryptic digestion resulted in cleavage of the peptide bond between residues 48 (Arg) and 49

(Trp). Following this procedure a pentapeptide comprised of residues 49–53 and a polypeptide composed of residues 1–48 (referred to as EGF-5 or EGF_{1-48}) were isolated.

c) Cyanogen Bromide EGF (CNBr-EGF)

SAVAGE et al. (1972) treated intact mouse EGF with cyanogen bromide and showed that the peptide bond between residues 21 (Met) and 22 (His) was cleaved.

4. High Molecular Weight Mouse EGF

In crude homogenates of the mouse submaxillary gland prepared at neutral pH, EGF was found as a component of a high molecular weight complex (TAYLOR et al., 1970, 1974). The complex had a molecular weight of approximately 74,000, as judged by sedimentation equilibrium, and was composed of two molecules of mouse EGF (6,045 molecular weight) and 2 molecules of binding protein (29,300 molecular weight). The high molecular weight complex was stable over the pH range 5–8 and had an isoelectric point of 5.4. Other properties of the complex include a sedimentation value of 4.81 S, an extinction coefficient ($E_{1\,cm}^{1\%}$ at 280 nm) of 19.1, and a diffusion constant ($D_{20,\,w}^{0}$) of 6.7×10^{-7} cm^2/s.

The high molecular weight complex could be dissociated into subunits by ion-exchange chromatography at pH 7.5 or gel filtration at pH values below 5 or above 8. Low molecular weight mouse EGF released from the complex by dissociation was identical to the mouse EGF described previously and shown in Fig. 1. The dissociated mouse EGF and binding protein reassociated at a neutral pH to form a high molecular weight complex of approximately the same molecular weight as the native complex. The capacity of mouse EGF to reassociate with its binding protein is dependent on the presence of the carboxy-terminal arginine residue (SERVER et al., 1976). Mouse EGF-5 ($mEGF_{1-51}$), lacking the carboxy-terminal Arg-Leu residues, did not recombine with binding protein. The importance of the carboxy-terminal arginine residue was demonstrated by treating native mouse EGF with carboxypeptidase B to produce a derivative lacking the carboxyterminal arginine residue. This derivative showed no capacity to recombine with the binding protein.

The biological significance of the carboxy-terminal arginine residue of mouse EGF has been indicated by the identification of the binding protein's enzymatic activity toward arginine esters (TAYLOR et al., 1970, 1974). The male mouse submaxillary gland is reported to contain a number of enzymes capable of hydrolyzing arginine esters (ANGELLETTI et al., 1967; CALISSANO and ANGELLETTI, 1968). The γ subunit proteins associated with nerve growth factor (NGF) isolated from the mouse submaxillary gland also possess arginine esterase activity (GREENE et al., 1968). Although the arginine esteropeptidases associated with NGF and mouse EGF have similar molecular weights, amino acid compositions, and substrate specificities, they differ in their electrophoretic properties. Also, while they cross-react immunologically, they are not identical antigenically, and the mouse EGF-binding protein does not substitute for the γ subunits in the formation of 7 S NGF (SERVER and SHOOTER, 1976). The results indicate specific associations between each of these growth factors and particular arginine esteropeptidases. Since both mouse EGF and NGF have carboxy-terminal arginine residues and are associated with arginine esteropeptidases, it is possible that these low molecular weight polypeptides are produced from precursor proteins by the action of the peptidases (TAYLOR et al., 1970; ANGELLETTI and BRADSHAW, 1971; SERVER and SHOOTER, 1976). FREY et al. (1979) have isolated from submaxillary gland extracts a polypeptide precursor molecule for mouse EGF. This preEGF has a molecular weight of approximately 9,000 and is converted to EGF (molecular weight 6,000) by the arginine esteropeptidase found in the high molecular weight EGF complex. Conversion of the preEGF to EGF takes place by the release of material from the carboxy terminus. Interestingly, the arginine esterase protein associated with mouse EGF does exhibit biological activity, enhancing the mitogenic properties of mouse EGF in cell culture systems (BARNES and COLOWICK, 1976; LEMBACH, 1976 a, b).

II. Human EGF

1. Identification and Isolation

STARKEY et al. (1975) identified the presence of an activity in human urine that possessed biological activities identical to those exhibited by mouse EGF. Based on the ability of urine concentrates to compete with ^{125}I-labeled mouse EGF in binding to human fibroblasts, COHEN and CARPENTER (1975) isolated a polypeptide from human urine that demonstrated biological activities previously ascribed to mouse EGF.

2. Chemical and Physical Properties

Although human EGF has not been characterized as well as the mouse-derived polypeptide, the available data indicate that the two polypeptides are very similar, but not identical, with respect to their chemical and physical properties (COHEN and CARPENTER, 1975).

Polyacrylamide disc gel electrophoresis at pH 9.5 shows that human EGF migrates slightly faster than mouse EGF, indicating a greater net negative charge for the human polypeptide at this pH. At a low pH (pH 2.3), both growth factors migrate at approximately the same rate. The different electrophoretic properties of human and mouse EGF are reflected in their similar, but slightly different, amino acid compositions. The two polypeptides also crossreact antigentically. Studies of human EGF by gel filtration chromatography and sedimentation equilibrium indicate molecular weight values of approximately 5,700 and 5,290, respectively. Because mouse EGF and human EGF exhibit identical biological activities, are structurally similar, and have some common antigenic sites, the results support the idea that the human growth factor is an evolved form of the mouse polypeptide.

HIRATA and ORTH (1979 a) examined the molecular size of human EGF in various body fluids by gel exclusion chromatography and radioimmuno- and radioreceptor-assays. While small molecular weight human EGF (molecular weight $\approx 6,000$) was the predominant size found in all fluids, a large form of human EGF (molecular weight $\approx 30,000$) was detected in urine. The size of the high molecular weight form of human EGF was not affected by chromatography in acetic acid or urea. The high molecular weight human EGF was equivalent to the low molecular weight form of the polypeptide in radioimmunoassays, but was less potent in radioreceptor assays.

HIRATA and ORTH (1979 b) demonstrated that the high molecular weight form of human EGF could be converted to the low molecular weight form by incubation with the mouse arginine esterase which is associated with the high molecular weight form of mouse EGF. The low molecular weight EGF generated by esterase treatment cochromatographed with standard human EGF (molecular weight ≈ 6,000), was fully active in both radioreceptor and radioimmunoassays, and was able to stimulate DNA synthesis in cultured fibroblasts as effectively as standard human EGF. These authors conclude the high molecular weight form of human EGF found in urine may represent a precursor molecule which can be cleaved by specific endoproteolytic activity to the fully active small molecular weight human EGF. GREGORY et al. (1977) have noted the presence of a high molecular form of human EGF in serum samples.

3. Relationship of Human EGF and Urogastrone

A new aspect of the chemistry and biology of EGF emerged with the publication by GREGORY (1975) of the amino acid sequence and disulfide linkages of β-urogastrone, a gastric antisecretory hormone isolated from human urine. A comparison of the primary structures of human urogastrone and mouse EGF is shown in Fig. 2. Of the 53 amino acids residues comprising each of the two polypeptides, 37 are common to both molecules, and the three disulfide bonds are formed in the same

									10		
Asn	Ser	Tyr	Pro	Gly	Cys	Pro	Ser	Ser	Tyr	Asp	Gly
Asn	Ser	Asp	Ser	Glu	Cys	Pro	Leu	Ser	His	Asp	Gly
							20				
Tyr	Cys	Leu	Asn	Gly	Gly	Val	Cys	Met	His	Ile	Glu
Tyr	Cys	Leu	His	Asp	Gly	Val	Cys	Met	Tyr	Ile	Glu
				30							
Ser	Leu	Asp	Ser	Tyr	Thr	Cys	Asn	Cys	Val	Ile	Gly
Ala	Leu	Asp	Lys	Tyr	Ala	Cys	Asn	Cys	Val	Val	Gly
		40									
Tyr	Ser	Gly	Asp	Arg	Cys	Gln	Thr	Arg	Asp	Leu	Arg
Tyr	Ile	Gly	Glu	Arg	Cys	Gln	Tyr	Arg	Asp	Leu	Lys
50											
Trp	Trp	Glu	Leu	Arg	———	EGF	(mouse)				
Trp	Trp	Glu	Leu	Arg	———	UROGASTRONE					

Fig. 2. Amino acid sequences of mouse EGF and human urogastrone. (From GREGORY, 1975)

relative positions. Sixteen variable residues occur at intervals along the polypeptide chain; of these, 14 could result from single base changes in the triplet code and most are conservative replacements. Furthermore β-urogastrone and mouse EGF elicit nearly identical biological responses in intact animals, organ cultures and cell cultures. The structural similarities of these polypeptides is also substantiated by the ability of urogastrone to compete with ^{125}I-labeled mouse EGF in highly specific radioreceptor assays (HOLLENBERG and GREGORY, 1976). The available data, therefore, strongly suggest that human EGF and β-urogastrone are identical and the amino acid sequence for β-urogastrone (Fig. 2) is undoubtedly very close to that for human EGF.

III. Rat EGF

MOORE (1978) has reported the isolation and characterization of EGF from rat submaxillary glands. The rat EGF is a single chain polypeptide with a molecular weight of 6,000 and an isoelectric pH of 4.3–4.6. The amino acid composition of rat EGF resembles the compositions of both mouse and human EGF, but is not identical to either.

C. Physiological Aspects of EGF

I. Concentrations in Body Fluids

1. Mouse EGF

In the adult mouse EGF concentrations of approximately 1–2 ng/ml in plasma, 300 ng/ml in milk, and about 1,000 ng/ml in saliva and urine have been reported (BYYNY et al., 1974; HIRATA and ORTH, 1979c). Factors which alter the levels of EGF in these fluids in the mouse are discussed below in Sect. C.IV.

2. Human EGF

Human EGF has been detected in various body fluids as follows: urine (approximately 100 ng/ml), milk (about 80 ng/ml), plasma (about 2 ng/ml), saliva (approximately 12 ng/ml) and amniotic fluid (about 1 ng/ml) (STARKEY and ORTH, 1977; GREGORY et al., 1977; BARKA et al., 1978). Factors which affect the level of EGF in human body fluids are discussed below in Sect. C.V.

II. Localization

1. Mouse EGF

Numerous studies have demonstrated that in the adult male mouse, EGF levels are much higher in the submaxillary gland compared to other sites. In other mouse organs studied, less than 0.5 ng EGF was detected per mg wet weight tissue. In the adult male mouse submaxillary gland the concentration of EGF is approximately 1–3 µg per mg wet weight tissue, while levels of 30–70 ng per mg wet weight tissue are present in the adult female (BYYNY et al., 1972; HIRATA and ORTH, 1979 c). Immunocytochemical studies (TURKINGTON et al., 1971; COHEN and SAVAGE, 1974; VAN NOORDEN et al., 1977; GRESIK and BARKA, 1977) have demonstrated that EGF is synthesized in the submaxillary gland and stored in secretion granules of the granular convoluted tubule cells. There are secondary sites of growth factor synthesis, however, as surgical ablation of the submaxillary gland does not reduce plasma levels of EGF (BYYNY et al., 1974). Recently, the guinea pig prostate has been reported to contain large quantities of EGF (HARPER et al., 1979).

2. Human EGF

The site(s) of EGF synthesis and/or storage in the human is not known. Using an indirect immunofluorescent technique, ELDER et al. (1978) reported the presence of EGF in two human tissues – the submandibular (submaxillary) glands and the glands of Brunner, located near the duodenum. HIRATA and ORTH (1979 d) used immunoaffinity chromatography and radioimmunoassays to extract and quantitate EGF in various human tissues. Their data reported concentrations of 1.3 to 5.5 ng of EGF per g wet weight tissue as follows: submandibular gland (1.3), duodenum (2.3), pancreas (2.8), jejunum (4.8), thyroid (5.3) and kidney (5.5). The quantities of EGF detected in these human tissues are exceedingly low compared to the daily urinary excretion of the growth factor (50–60 µg/24 h), and their physiological significance therefore, is not clear.

III. Control of Submaxillary Gland Content of Mouse EGF

The EGF containing tubular cells of the mouse submaxillary gland exhibit sexual diphorism. The morphology and granule content of these cells are dependent on the hormonal status of the animal. The cells are developed fully in the male only after puberty; castration results in atrophy of the tubular portion of the gland. The injection of testosterone into female mice results in a hypertrophy and hyperplasia of these cells (LACASSAGNE, 1940). The quantity of EGF present in the submaxillary gland closely parallels the development of the tubular system (COHEN, 1965 a; GRESIK and BARKA, 1977). In the immature 15-day old male mouse, the amount of submaxillary gland EGF is very low, about 0.02 ng per mg wet tissue. The concentration increases rapidly with age until maximal levels of about 1 µg per mg wet tissue are reached in adult male mice aged 50 days or more (BYYNY et al., 1972). In contrast to the adult male mouse, the concentration of EGF in the submaxillary gland of the adult female is only about 70 ng per mg wet tissue. VAN NOORDEN et al. (1977) reported that in pregnant females EGF levels increase to a concentration close to that present in males. Subsequently, the growth factor levels decline during lactation.

The androgen dependence of EGF concentrations in the mouse submaxillary gland was shown by the reduction of EGF levels in the castrated male (113 ng/mg wet tissue) and the increased levels of EGF in the testosterone-treated female (2,900 ng/mg wet tissue) (BYYNY et al., 1972). The *tfm/y* mouse has a genetic defect which results in target organ insensitivity to androgens. The concentration of EGF in the submaxillary glands of *tfm/y* mice is low, approximately 2 ng/mg wet tissue, and is not increased by the administration of androgens (BARTHE et al., 1974). The effect of progestins, alone and in combination with testosterone, on the concentrations of EGF in the submaxillary glands of normal female mice have been reported (BULLOCK et al., 1975). Medroxyprogesterone acetate produced a 40-fold increase in submaxillary EGF levels. Cyproterone acetate had no androgenic activity when administered alone and acted as a potent anti-androgen when used in combination with testosterone. The only synandrogenic response was elicited with doses of progesterone caproate.

A circadian periodicity in the mouse submaxillary gland content of EGF, which was abolished by superior cervical ganglionectomy, has been noted (KRIEGER et al., 1976).

IV. Secretion of Mouse EGF from the Submaxillary Gland

Despite the androgen controlled levels of EGF in the mouse submaxillary gland, there are no significant differences in the plasma concentrations of EGF between male and female mice or between immature and adult mice (BYYNY et al., 1974; HIRATA and ORTH, 1979c). The release of submaxillary gland EGF into the circulatory system is controlled, either directly or indirectly, by α-adrenergic agents (BYYNY et al., 1974). α-Adrenergic agents are known to bring about the degranulation of the peritubular cells in the submaxillary gland (JUNQUIERA et al., 1964). The intravenous administration of an α-adrenergic agent, such as phenylephrine, into adult male mice, results in a 65% decrease in submaxillary gland EGF within 60 min and a concomitant increase in plasma EGF levels from 1.2 to 152 ng/ml. MURPHY et al. (1979) have suggested that administration of α-adrenergic agents may increase secretion directly into saliva and not plasma. The α-adrenergic-stimulated saliva, rich in EGF, might then pass into the digestive tract and be absorbed into the circulation. These authors suggest that the mouse submaxillary gland is an exocrine rather than endocrine gland (MURPHY et al., 1977, 1979).

The concentration of EGF in mouse saliva reflects the differences between males and females in submaxillary gland content of EGF. HIRATA and ORTH (1979c) report a difference of approximately 40-fold in the level of EGF in the saliva of adult male (about 2,400 ng/ml) and female (approximately 50 ng/ml) mice; hypersalivation was induced with carbachol prior to collection of the saliva. MURPHY et al. (1979) also demonstrated a similar difference in the EGF content of saliva between male and female mice.

The ability of adrenergic agents to control the release of EGF from the mouse submaxillary gland has been demonstrated in the intact animal (BYYNY et al., 1974; MURPHY et al., 1979) and in vitro with minced submaxillary glands (ROBERTS, 1977, 1978).

V. Factors Affecting Levels of Human EGF

STARKEY and ORTH (1977) and DAILEY et al. (1978) have used heterologous and homologous radioimmunoassays, respectively, to measure levels of EGF in human fluids, primarily urine, under different physiological influences. Twenty four-hour urinary excretion of human EGF from normal adult males and females was about

63 and 52 µg per total volume. No diurnal or postprandial variation was detected; however, an excellent linear correlation was observed between urinary human EGF and urinary creatinine concentrations in each sample. The range of EGF concentrations in human urine was 29–272 ng/ml. Unusually low lovels of urinary human EGF, 5–15 ng/ml, were reported for newborns and high levels were detected in the 24 h urinary excretions from females taking oral contraceptives.

D. Biological Activities of EGF In Vivo

I. Skin

The initial observation which led to the recognition of epidermal growth factor and to its biological assay was that daily subcutaneous injections of extracts of the mouse submaxillary gland into newborn mice resulted in the precocious opening of the eyelids (6–7 days instead of the normal 12–14 days) and precocious eruption of the incisors (COHEN, 1962). Similar effects were demonstrable in newborn rats and dogs. These gross biological changes were ascribed mainly to an enhancement of epidermal growth and keratinization, as ascertained from histological studies (COHEN and ELLIOTT, 1963). A thickening of the epithelial layer also was noted in the oral cavity and esophagus. It is not known whether EGF directly stimulated keratinization or whether accelerated keratin formation was a consequence of the stimulation of basal cell proliferation. Increased epidermal mitotic activity has been found in the skin of newborn mice treated with mouse EGF (BIRNBAUM et al., 1976). Curiously, in somewhat older mice (12–20 days), the daily injection of mouse EGF had little visible effect on dorsal skin but produced a marked thickening of tail and foot-pad epidermis (COHEN, 1971).

The eyelid-opening effect of mouse EGF in newborn rats was demonstrable at a dosage level of 0.1 µg per gram per day. The complete sequence of 53 amino acid residues of native mouse EGF was not essential to elicit this in vivo response (SAVAGE et al., 1972). A derivative of mouse EGF lacking five amino acids at the carboxy terminus, EGF-5 (EGF_{1-48}), was as active as the intact molecule in the eyelid opening assay. In view of these results and the fact that this derivative is produced by mild trypsinization, perhaps the intact form of EGF (EGF_{1-53}) is cleaved by proteases to the EGF_{1-48} derivative when injected into the intact animal. It is not known whether the endogenous EGF found in plasma in the intact molecule or perhaps the slightly smaller derivative.

The histological evidence of epidermal hypertrophy and hyperplasia following the administration of EGF to newborn rats has been augmented by cytological quantitation of increases in the number of mitotic epidermal cells (BIRNBAUM et al., 1976), chemical measurements of increases in dry weight, DNA and RNA content of the epidermis (ANGELETTI et al., 1964), and biochemical assays of enhanced enzyme activities (STASTNY and COHEN, 1970; BLOSSE et al., 1974). The hyperkeratosis caused by EGF was accompanied by an increase in the disulfide group content of the epidermal cells and a decrease in free sulfhydryl groups (C. FRATI et al., 1972).

An effect of EGF on skin that is probably related to its mitogenicity, is the ability of EGF to enhance the carcinogenic potential of methylcholanthrene (REYNOLDS et al., 1965; ROSE et al., 1976). In these studies the growth factor by itself did not produce tumors, but decreased the latency period and increased the numbers of papillomas and carcinomas per animal when methylcholanthrene was topically applied to the skin. An early, persistent alopecia and a distinct pachyderma developed in the area of methylcholanthrene application in only those animals that also received EGF.

II. Corneal Epithelium

L. Frati et al. (1972) showed that when labeled EGF was administered by intraperitoneal injection to rabbits the labeled EGF become highly concentrated in two tissues, the skin and cornea, relative to blood. Studies by two groups (Savage and Cohen, 1973; L. Frati et al., 1972) demonstrated that EGF is a potent mitogen for corneal epithelial tissue in vivo. The corneal epithelium of rabbits was experimentally wounded and EGF was topically applied to one-half the wounded animals. Application of the growth factor to these wounds resulted in a marked hyperplasia of the corneal epithelium and a slight decrease in the time required for wound closure. During regeneration of the epithelia in animals treated with EGF, the epithelium increased from its normal thickness of 4–6 layers to 10–15 cell layers after six days and then returned to the normal thickness by 14 days. Topical application of EGF to control corneas (non-wounded) had no discernible effect. Gospodarowicz et al. (1979) have reported that the slow release of EGF from a polyacrylamide pellet implanted in the rabbit cornea induces neovascularization of the cornea.

A possible therapeutic use for EGF in humans has been advanced by Daniele et al. (1979) who reported that the effect of application of EGF to wounds of the corneal epithelium in humans. An acceleration of the healing process was affected by the growth factor provided the integrity of the corneal stroma was maintained.

III. Respiratory Epithelium

Infusion of fetal lambs in utero with EGF for 3–5 days markedly stimulated the proliferation and development of epithelial tissue in lungs, trachae, and esophagus (Sundell et al., 1980). Also, comparison of lung deflation pressure-volume curves from fetal rabbits showed that fetuses injected with EGF had increased total lung capacity consistent with an increase in alveolar surface active material (Catterton et al., 1979). These workers suggest that EGF may protect the premature fetus from the development of respiratory distress syndromes such as hyaline membrane disease.

IV. Gastrointestinal Tract

The completely unforeseen finding that the amino acid sequence of mouse EGF shows considerable homology with that of human urogastrone led to the observation that mouse EGF has a direct and rapid effect upon gastric acid secretion in both rats and dogs (Bower et al., 1975). Either histamine or pentagastrin were administered to an experimental animal to increase gastric acid secretion. Then an intravenous injection of mouse EGF was given (10 µg/kg in rats and 0.5 µg/kg in dogs) and within 45 minutes gastric acid secretion was reduced. Since mouse EGF had a short half-life in blood (about 4 min), the reduction in acid secretion was transient. GREGORY et al. (1977) have reported that EGF is effective in increasing the healing of experimentally-induced ulcers. Although a role for EGF in controlling the growth of mucosal tissues has not been well documented, FELDMAN et al. (1978) have demonstrated that subcutaneous injection of EGF into young mice produces a significant increase in ornithine decarboxylase activity in the stomach and duodenum. SCHEVING et al. (1979) have shown that within 4–8 h after the administration of EGF to adult mice there was increased (approximately 100%) incorporation of radioactive thymidine in both glandular and nonglandular stomach tissue. These workers also recorded larger increases in DNA synthesis in the tongue (290%) and esophagus (580%) following administration of EGF.

V. Liver

BUCHER et al. (1978) reported that intraperitoneal infusion of EGF increased DNA synthesis in hepatocytes in the intact liver of normal adult rats. This action of EGF was markedly enhanced by the simultaneous administration of glucagon or insulin. Neither insulin nor glucagon alone or in combination increased DNA synthesis. Treatment with EGF and insulin also produced hepatic enlargement, with considerable hypertrophy and hyperplasia.

HEIMBERG et al. (1965) noted that at a high dosage of EGF (2 µg per gram per day), a distinct growth inhibition of the neonatal rat occurred. This effect was accompanied by the accumulation of lipids, mainly triglycerides, in the liver.

E. Organ Culture Studies of EGF

I. Skin

Different lines of evidence demonstrate that EGF has a direct effect on the epidermis. Within 2 h after the intraperitoneal injection of ^{125}I-labeled EGF into rats, the radioactivity was concentrated 3-fold in the epidermis relative to the blood (COVELLI et al., 1972b). Specific binding sites for ^{125}I-EGF in epidermal tissue in vitro have been reported (O-KEEFE et al., 1974). COHEN and co-workers have shown that EGF stimulates the proliferation of epidermal cells in organ cultures of chick embryo skin (COHEN, 1965b) and human fetal head skin (COHEN and SAVAGE, 1974). A significant aspect of this work was the demonstration that EGF stimulated ornithine decarboxylase activity, RNA, protein, and DNA synthesis, cell division and eventually keratinization either in the presence of "killed" dermis or in the absence of any dermis (COHEN, 1965b; HOOBER and COHEN, 1967b). It appears, therefore, that EGF interacts directly with the epidermis in the skin and its action on this target tissue is not mediated by a second hormone.

Biochemical studies of the effects of EGF on chick embryo epidermal tissue in organ culture have been observed. EGF rapidly stimulated the transport of certain metabolites. Within 15 min following the addition of EGF, there were stimulations of approximately 2-fold of the uptakes of radioactive α-aminoisobutyric acid and uridine into the trichloroacetic acid soluble fraction of the cells. These increases in transport were not prevented by inhibitors of protein synthesis, such as cyclo-heximide, indicating that the synthesis of new proteins was not required for these permeability changes (COHEN, 1971).

Within 90 min following the addition of EGF to epidermal cultures, there was a conversion of preexisting ribosomal monomers into functional polysomes. Accompanying this alteration was an increase in protein synthesis and an increase in the synthesis of all classes of cytoplasmic RNA detectable on sucrose gradients. The ribosomal monomer-to-polysome conversion did not appear to require the synthesis of new protein (COHEN and STASTNY, 1968). The total ribosomal population, isolated from cells treated with EGF for 4 h, was more active in a cell-free protein synthesizing system than ribosomes isolated from control cells (HOOBER and COHEN, 1967a). It was suggested that ribosomes prepared from EGF-treated cells had a greater ability to bind messenger RNA in a functional manner.

The similarities between the response of epidermal cells to EGF and the response of many other cell types to a growth-stimulating condition or mitogen prompted an investigation of the effects of EGF on the induction of ornithine decarboxylase. The activity of this enzyme is enhanced under a variety of conditions, the central element of which appears to be growth stimulation. The addition of mEGF to chick epidermal cells induced within 4 h a marked (40-fold) but transient increase of ornithine decarboxylase activity, and led to the intracellular accumulation of putrescine (STASTNY and COHEN, 1970).

In the presence of mEGF a 2-fold increase in total protein and RNA content was detected during the first 48 h of incubation (HOOBER and COHEN, 1967b), as was stimulation of thymidine incorporation into DNA (BERTSCH and MARKS, 1974).

Thus, a series of metabolic alterations which accompany the growth-stimulatory effects of mEGF on epidermal cells have been described. Because many of these changes take place in a variety of cells when a growth stimulus is applied, these biochemical events seem to reflect the inherent program and capabilities of the cells rather than the specific nature of the growth stimulus.

II. Other Tissues

1. Cornea

The hyperplastic effect of EGF on the corneal epithelium was demonstrated in organ cultures of chick and human corneas (SAVAGE and COHEN, 1973). A similar stimulation of the proliferation of bovine corneal epithelium occurred in organ culture (GOSPODAROWICZ et al., 1977c). Since, multiplication of isolated bovine corneal epithelial in cell culture was not affected by EGF, despite the presence of specific binding sites for ^{125}I-EGF, these authors propose that epithelial-mesenchyme are necessary for EGF to have a mitogenic effect on the corneal epithelium. GOSPODAROWICZ et al. (1978b) have suggested that the failure of corneal epithelial cells to respond to EGF in culture is due to their inability to produce their own basement membrane which results in an altered, i.e. flattened, cell morphology. In contrast, when the corneal epithelium is maintained in organ culture its topological relationship with its basement membrane is maintained, the typical tall and columnar shape is kept, and the cells are able to proliferate in response to added EGF. Gospodarowicz and Greenburg (1979) have shown that EGF significantly accelerated the repair process of wounded bovine corneal endothelium maintained in organ culture. Similar results were reported for the corneal endothelium from other species – cats and humans.

2. Palate

Studies of cleft palate fusion in rodents indicates that the epithelium of the palatal shelves is sensitive to EGF. Normal fusion of the secondary palate in rodents requires movement of the palatal shelves into aposition, adhesion, and fusion of the mesenchymal tissue of the joining shelves. During the fusion process, epithelial layers of the adhering shelves degenerate allowing fusion to occur between layers of mesenchymal cells. Hassell (1975) demonstrated that when EGF was added to organ cultures of apposed palatal shelves from rat embryos, the epithelia thickened and keratinized and the adhesion and fusion processes did not occur. Bedrick and Ladda (1978) have reported a similar inhibitory effect of EGF on palatal fusion in vivo.

3. Bone

Canalis and Raisz (1979) have examined the effect of EGF on cultured half-calvaria from fetal rats. In the presence of exogenous EGF, DNA synthesis was increased as judged by the incorporation of labeled thymidine, chemical measurement of total DNA, and measurement of the number mitoses in histological sections. In this system, EGF also decreased the incorporation of labeled proline into collagen. Additional effects of EGF on bone metabolism have been described by Tashjian and Levine (1978). Using cultured neonatal mouse calvaria, these authors report that EGF stimulates bone resorption and increases the production of prostaglandin E_2. These various effects of EGF on bone were all observed at concentrations of the growth factor similar to the levels of EGF present in mouse plasma.

F. Cell Culture Studies of EGF

I. Cell Nutrition

Several studies have demonstrated that in the presence of EGF cellular nutrition is altered in several respects. Perhaps the most profound effect that EGF has on the requirement of cells for extracellular nutrients is to decrease the amount of serum needed to maintain cell growth. Using diploid human fibroblasts, Carpenter and Cohen (1978 b) showed that these cultured cells were able to multiply in media containing a low concentration of serum (1%) and supplemented with EGF. Under these conditions cell proliferation was equal to that obtained in media containing an optimal serum level (10%). Also, the growth of human fibroblasts was severely restricted in the presence of deficient-sera, such as γ-globulin free serum and plasma. However, the addition of EGF to media containing these growth-poor sera resulted in significant cell multiplication. LECHNER and KAIGHN (1979a) also have demonstrated a similar reduction in the serum requirement of normal cells cultured in the presence of EGF. These results suggested that either EGF reduced or eliminated cell requirements for certain serum components or that the growth factor was able to directly substitute for those serum components.

This work has been extended by the experiments of SATO and his co-workers who have demonstrated that the serum requirement of cultured cells can be completely replaced by a carefully designed mixture of hormones tailored to the requirements of individual cell lines. HUTCHINGS and SATO (1978) and BARNES and SATO (1979) showed, respectively, that HeLa cells and mammary tumor cells can be cultured in the absence of serum if EGF and several other hormones are provided.

The presence of EGF has a marked effect on the extracellular level of Ca^{+2} required by normal cells for cell multiplication. McKEEHAN and McKEEHAN (1979) and LECHNER and KAIGHN (1979b) demonstrated that the extracellular Ca^{+2} requirement of normal human fibroblasts or normal human epithelial cells, respectively, was reduced 50- to 80-fold by the presence of EGF.

Certain similarities appear between the nutritional requirements of normal cells grown in the presence of EGF and tumor cells. Both cell types are able to proliferate in the presence of low or deficient sera and there is a similar reduction in the level of extracellular calcium required for cell growth.

II. Types of Cells Affected by EGF

Although studies of EGF action in intact animals and in organ culture systems have identified, nearly exclusively, epidermal and epithelial cells as target cells, investigations of other cell types in cell culture have demonstrated that many cells are responsive to EGF. Cultured cells which respond to the mitogenic activity of EGF include the following; murine and human fibroblasts (ARMELIN, 1973; HOLLENBERG and CUATRECASAS, 1973); human glia (WESTERMARK, 1976); human keratinocytes (RHEINWALD and GREEN, 1977); rabbit chondrocytes (GOSPODAROWICZ and MESCHER, 1977); human and bovine amniotic fluid-derived cells (GOSPODAROWICZ et al., 1977e); bovine and human smooth muscle cells (BHARGAVA et al., 1979; GOSPODAROWICZ et al., 1977d); bovine, porcine and human granulosa cells (GOSPODAROWICZ et al., 1977a; GOSPODAROWICZ and BIALECKI, 1979; OSTERMAN and HAMMOND, 1979); rat hepatocytes (RICHMAN et al., 1976); rabbit lens epithelial cells (HOLLENBERG, 1975); bovine corneal endothelial cells (GOSPODAROWICZ et al., 1977b); human vascular endothelial cells (GOSPODAROWICZ et al., 1978a); rabbit endometrial cells (GERSCHENSON et al., 1979); monkey kidney epithelial cells (HOLLEY et al., 1977); human cervical epithelial cells (STANLEY and PARKINSON, 1979); rodent and human mammary epithelial cells (TURKINGTON, 1969a, b; STOKER et al., 1976; KIRKLAND et al., 1979).

III. EGF and the Growth of Cell Populations

CARPENTER and COHEN (1976a) cultured normal human fibroblasts in media containing an optimal serum concentration (10%) and supplemented some of the cultures with EGF. Addition of the growth factor under these conditions produced a small decrease in the population doubling time, but had a substantial effect on the maximum saturation density achieved by the cultures. In several experiments, cells grown in the presence of EGF reached cell densities 3- to 6-fold higher than that obtained by fibroblasts grown in the absence of exogenous EGF. Similar increases in saturation density due to EGF have been reported for mammary epithelial cells (KIRKLAND et al., 1979), glial cells (WESTERMARK, 1977), and 3T6 mouse fibroblasts (MIERZEJEWSKI and ROZENGURT, 1977). Milk is mitogenic for cultured cells and CARPENTER (1980) has shown that 90% of the growth stimulatory effect of human milk is neutralized when anti-human EGF antibodies are added to the milk.

The EGF-promoted increase in saturation density is accompanied by a marked change in the topology of cell populations. The growth of human fibroblasts is restricted by density-dependent-inhibition of growth and usually a maximal saturation density is reached when the cells have formed a confluent monolayer covering the growth surface. The increased saturation density achieved in the presence of EGF, in contrast, is accompanied by growth of the cells into multiple layers (CARPENTER and COHEN, 1976a). These multi-layered cell populations are not randomly oriented like transformed cells; rather, each layer is arranged in a perpendicular orientation to the layers of cells above and below – resembling the orientation of cells in a tissue. The multi-layered populations of human fibroblasts cannot be dispersed into single cells with trypsin, but require additional treatment with collagenase and either chondroitinase or hyaluronidase (G. CARPENTER, unpublished observation). Human fibroblasts cultured in the presence of EGF would appear to secrete extracellular material which probably allows the cells to proliferate into organized, multilayered structures.

RHEINWALD and GREEN (1977) and GOSPODAROWICZ and BIALECKI (1978) have reported that EGF has a very significant effect on the replicative lifespan of cultured keratinocytes and granulosa cells, respectively. The capacity of EGF to increase the growth potential of these cell populations apparently was due to a delay of terminal differentiation by the growth factor.

IV. Components of the Mitogenic Response

1. Rapid Biological Responses at the Membrane

Following the addition of EGF to cultured cells the most rapid biological response observed are those associated with the plasma membrane.

a) Uridine Uptake

ROZENGURT et al. (1978) have examined the uptake of ^{3}H-uridine in cultured fibroblasts exposed to EGF and other mitogens. After a short lag of approximately 5–10 min, uptake of the exogenous labeled uridine is increased. The data show, however, that this rapid effect is due to the metabolic trapping of intracellular uridine by increased phosphorylation rather than a stimulation transport. The enhanced phosphorylation of intracellular uridine occurs by a mechanism which is not sensitive to cycloheximide, Actinomycin D or ammonium chloride. These authors report that mitogens do produce an increase in the transport of uridine in quiescent cells, but only after a period of approximately 10 h when the cells are beginning to enter the S phase of the cell cycle.

b) Sugar Transport

BARNES and COLOWICK (1976) described an increase in the rate of sugar uptake after the addition of EGF to quiescent fibroblasts. Increased transport, measured by the uptake of radioactive 2-deoxyglucose or 3-0-methylglucose, was observed within 15 min after the addition of EGF and was maximal after approximately 2 h of exposure to the mitogen. The EGF-stimulated transport of glucose analogues could be further increased by addition of EGF-binding arginine esterase (see Sect. B.I.4) which had no stimulatory effect by itself. The increase in sugar transport elicited by EGF was not sensitive to inhibitors of protein or RNA synthesis and was not blocked by ammonium chloride.

c) Cation Fluxes

ROZENGURT and his colleagues have studied the effect of EGF addition on cation fluxes in quiescent fibroblasts. Using $^{86}Rb^+$ as a tracer for K^+, ROZENGURT and HEPPEL (1975) showed that $^{86}Rb^+$ influx was rapidly increased following the addition of EGF. Increases were measurable within 2 min following EGF addition and maximal stimulation was reached after 5 min of exposure to EGF. The EGF-stimulated influx of $^{86}Rb^+$ was not affected by cycloheximide and was accompanied by an increase in the maximum velocity (V_{max}) of entry. This putative effect on K^+ entry was reversible, dependent on K^+ in the media, and could be blocked by ouabain suggesting involvement of the membrane Na^+K^+ pump. SMITH and ROZENGURT (1978) examined the influence of EGF and other mitogens on Na^+ flux in quiescent fibroblasts by using Li^+ as a sodium marker. Mitogens promoted increased rates of Li^+ influx with characteristics similar to those described for $^{86}Rb^+$ entry. With Li^+, however, the mitogens also enhanced the rate of cation efflux. Since the putative Na^+ transport system was sensitive to ouabain and amiloride, apparently both the Na^+K^+ pump and membrane Na^+-specific "porter" are affected. The authors suggest that mitogens stimulate the activity of the Na^+K^+ pump by increasing the intracellular concentration of Na^+. Higher intracellular Na^+ levels presumably lead to increased Na^+ efflux with a concomitant influx of K^+. Increased intracellular levels of K^+ are considered to be important for promoting higher rates of macromolecular synthesis.

d) Putrescine Transport

DIPASQUALE et al. (1978) have reported that the transport of the polyamine putrescine is increased upon the addition of EGF to quiescent fibroblasts. This stimulated transport increased linearly for several hours after EGF addition, was accompanied by a 2-fold increase in the maximum velocity (V_{max}), and resulted in an intracellular putrescine concentration 50- to 150-fold higher than that in the media. Curiously, both the basal and EGF-stimulated rates of putrescine transport were inversely related to the cell density. Although the intracellular effects of polyamines are multifarious, these compounds may play an important role in stabilizing newly synthesized RNA.

e) Alanine Transport

HOLLENBERG and CUATRECASAS (1975) have reported increased alanine transport, measured by the uptake of labeled α-aminoisobutyrate, after the addition of EGF to cultured cells. In these studies the increased transport of the radioactive tracer occurred rapidly (15–30 min) after the addition of EGF, but the level of increase was modest – approximately 1.2- to 2-fold.

f) Membrane Ruffling and Macropinocytosis

BRUNK et al. (1976) showed that membrane ruffling and macropinocytosis, assessed by morphological criteria are enhanced within 4–6 h after the addition of EGF to quiescent glia cells. CHINKERS et al. (1979) and HAIGLER et al. (1979 b) have examined a unique tumor cell line, A-431, to investigate EGF-promoted membrane ruffling and macropinocytosis, respectively. This tumor line is unique as it possesses a very high concentration of EGF receptor, approximately 2×10^6 receptors per cell. This is 20-fold greater than the concentration of receptors on other cell lines and it is thought that A-431 cells may exaggerate responses to EGF which would be more subtle and difficult to detect in normal cells.

CHINKERS et al. (1979) used scanning electron microscopy to demonstrate that very soon after the addition of EGF the A-431 cells showed extensive increases in membrane ruffling and extension of filopodia. These morphological changes were apparent within 5 min after the addition of EGF and were transient in nature subsiding after about 15 min.

HAIGLER et al. (1979 b) examined the effect of EGF on macropinocytosis, assessed by measuring the uptake of horseradish peroxidase, in A-431 cells. This group demonstrated a large (10-fold) increase in the rate of macro- or fluid phase pinocytosis which occurred in temporal coordination with the enhanced membrane ruffling described by CHINKERS et al. (1979).

g) Other Membrane Responses

AHARONOV et al. (1978 b) showed that pretreatment of fibroblasts with EGF increases their binding to Concanavalin A-coated nylon fibers and reduces Concanavalin A-mediated hemadsorption. These authors suggest that EGF may increase the lateral mobility of lectin receptors in plasma membrane.

ANDERSON et al. (1979) have reported that EGF blocks the prostaglandin E_1-stimulation of cyclic AMP synthesis in fibroblastic cells. LEVINE and HASSID (1977) have demonstrated that EGF stimulates the formation of prostaglandins $F_{2\alpha}$ and E_2 by 40% and 25%, respectively, in cultured canine kidney cells. Although cyclic AMP and prostaglandins may be involved in the mitogenesis, there is no clear evidence that the EGF activation of quiescent cells directly involves these molecules.

CHEN et al. (1977) have reported that when 3T3 cells are maintained in low serum, the fibrous network of LETS cell surface protein is lost. Subsequent addition of EGF results in the reappearance of the LETS protein network.

2. Responses to EGF Occurring in the Cytoplasm

Following the activation of membrane processes and preceding increased synthesis of DNA in the nucleus, a series of EGF responses occur in the cytoplasm – generally in a time frame of 2–12 h.

a) Activation of Glycolysis

Activation of quiescent 3 T 3 cells by EGF produces a rapid increase in glycolytic activity as shown by enhanced production of lactic acid (DIAMOND et al., 1978). The level of lactate in the media increased very rapidly (4-fold within 2 h) after addition of EGF, insulin, or serum to quiescent cells. The increases in lactic acid production during the first 3 h of exposure to EGF were not affected by cycloheximide which indicates that the initial activation of glycolysis did not require new protein synthesis. The addition of dibutyryl cyclic AMP, dibutyryl cyclic GMP, or theophylline alone or in combination with EGF did not alter the production of lactic acid. The authors also report that ouabain did not block the activation of glycolysis by EGF which suggests that the Na^+ pump is not crucial. The ability of EGF to stimulate lactate production was markedly decreased by omission of Ca^{2+} from the media. This result may be of interest in view of the proposed role of Ca^{2+} as a hormone second messenger.

The stimulatory effect of EGF on glycolysis could be demonstrated in cell-free homogenates prepared from cells activated with EGF (SCHNEIDER et al., 1978). Addition of EGF to homogenates prepared from quiescent cells did not affect glycolysis. These studies have further shown that the specific activity of phosphofructokinase is significantly increased in cell-free homogenates prepared from quiescent cells exposed to EGF for 3 h. The ability of EGF to stimulate phosphofructokinase activity was dependent on the presence of Ca^{2+} in the medium and was not affected by the presence of cycloheximide. The results of mixing experiments, and the persistence of EGF-activated phosphofructokinase after gel chromatography and at alkaline pH, where the influence of many allosteric effectors of the enzyme is reduced, indicate that diffusible activators or inhibitors are probably not involved in this response to EGF.

b) Synthesis of Extracellular Macromolecules

LEMBACH (1976a) has demonstrated that the addition of EGF to cultured human fibroblasts increases the incorporation of ^3H-glucosamine into both cellular and extracellular glycosaminoglycans. Incorporation of $^{35}SO_4^{2-}$ into these cells was not altered by the presence of EGF. EGF stimulated the incorporation of ^3H-glucosamine into cellular glycosaminoglycans within 4 h and into extracellular material within 8–12 h. The labeled extracellular material was predominately hyaluronic acid. Also, EGF stimulated incorporation of ^3H-glucosamine into both cellular and extracellular glycoproteins within 4 h (LEMBACH, 1976a).

c) Activation of RNA and Protein Synthesis

Relatively few studies have reported alterations in protein and RNA synthesis following the activation of cells by EGF. HOLLENBERG and CUATRECASAS (1973) have presented data that show increased incorporation of [^3H]-uridine into RNA in human fibroblasts approximately 15–20 h after the addition of EGF. This experiment was performed in serum-deficient media which may account for the rather slow response – serum stimulation by itself produced an increased ^3H-uridine incorporation in 5 h. Experiments with HeLa cells (COVELLI et al., 1972a) and chick embryo epidermis (COHEN and STASTNY, 1968) have shown increased (2- to 5-fold) ^3H-uridine incorporation within 5 h following addition of EGF.

Experiments with chick embryo epidermis in organ culture demonstrated a rapid increase in the rate of protein synthesis following the addition of EGF (HOOBER and COHEN, 1967a). Increased polysome formation (COHEN and STASTNY, 1968) and increased activity of ribosomes in vitro (HOOBER and COHEN, 1967b) were evident within 5 h after activation by EGF. Increased incorporation of labeled amino acids at approximately 10 h after addition of EGF to HeLa cells has been noted (COVELLI et al., 1972a). There are no reports to date indicating what effects EGF may have on the process of protein degradation which plays an important role in determining the rate of net protein synthesis.

d) Activation of Ornithine Decarboxylase

Increased activity of the enzyme ornithine decarboxylase has often been associated with the stimulation of cell proliferation. However, detailed studies of the influence of EGF on the activity of this enzyme have not appeared. Preliminary studies have demonstrated that approximately 6 h after the addition of EGF ornithine decarboxylase activity was increased about 6-fold in certain cell lines: skin fibroblasts (DIPASQUALE et al., 1978), pheochromocytoma (HUFF and GUROFF, 1979), and granulosa cells (OSTERMAN and HAMMOND, 1979).

e) Protein Phosphorylation

SMITH et al. (1979) have shown that 5 min after the addition of EGF (or insulin or serum) to differentiated 3T3-L1 preadipocytes a protein of molecular weight 31,000 became heavily phosphorylated. This protein was tentatively identified as the ribosomal protein S6. HUFF and GUROFF (1978) demonstrated that after a 30 min exposure of chick embryo epidermis to EGF, a 30,000 molecular weight protein became heavily phosphorylated. Based on electrophoretic migration in SDS gels, these authors have tentatively identified the phosphorylated protein as f_1 histone. HUFF and GUROFF were also able to demonstrate increased phosphorylation of this protein in vivo by injecting EGF through the shell and into the amniotic sac of chick embryos. As discussed in detail below (see Sect. I), CARPENTER et al. (1978, 1979) have described a subcellular system in which protein phosphorylation is increased by EGF.

3. Stimulation of DNA Synthesis

The addition of EGF to quiescent cells, presumably arrested in G_1, leads to activation of the cells and transition into the S or replicative phase of the cell cycle. This transition is marked by an increased rate of DNA synthesis which in most cell culture systems is detectable no sooner than 12 h after growth factor addition and generally is maximal after approximately 20 h (CARPENTER and COHEN, 1976a). In most cell culture systems the EGF-enhancement of DNA synthesis, measured by the incorporation of radioactive thymidine, is maximal at a growth factor concentration of 5–10 ng/ml (about 1.4×10^{-9} M). In homologous systems, for example, the addition of human EGF to human fibroblasts, slightly lower concentrations (2 ng/ml) may be required and with most all cultured cells concentrations greater than 20 ng/ml may produce less stimulation of DNA synthesis (COHEN et al., 1975;

CARPENTER and COHEN, 1976a). GOSPODAROWICZ et al. (1977a) have reported that bovine granulosa cells can be stimulated to increased rates of DNA synthesis by very low concentrations of EGF. Maximal enhancement of ^{3}H-thymidine incorporation in granulosa cells occurred at an EGF concentration of 3×10^{-11} M, approximately 100-fold less than that required by other cells.

Stimulation of DNA synthesis in cultured cells by EGF usually requires the presence of a low level of serum, 0.5–2.0% (COHEN et al., 1975; LEMBACH, 1976b; CARPENTER and COHEN, 1976a; MIERZEJEWSKI and ROZENGURT, 1978). Although the components in serum which act permissively to promote EGF activity have not been isolated, many workers have reported that certain defined reagents enhance synergistically the effect of EGF on DNA synthesis. These compounds include the following: ascorbic acid (LEMBACH, 1976b; CARPENTER and COHEN, 1976a), the EGF-binding arginine esterase (LEMBACH, 1976b), insulin (RICHMAN et al., 1976; O'FARRELL et al., 1979), thrombin (ZETTER et al., 1977; GOSPODAROWICZ et al., 1978a); vitamin B_{12} (MIERZEJEWSKI and ROZENGURT, 1976), tumor promoter TPA (DICKER and ROZENGURT, 1978, 1979; FRANTZ et al., 1979), vasopressin (DICKER and ROZENGURT, 1979), and antitubulin agents such as colchicine (FRIEDKIN et al., 1979). The ability of each of these reagents to increase the stimulation of DNA synthesis by EGF is dependent on the cell type employed and the experimental conditions.

V. Responses Not Related to Mitogenesis

Several reports have described effects of EGF on cell culture systems which do not appear to be directly related to or a part of the mitogenic response. BENVENISTE et al. (1978) have reported that the addition of EGF to cultured human choriocarcinoma cells increases by 2-fold the amount of human chorionic gonadotropin secreted by these cells. The growth factor did not affect the growth rate of the choriocarcinoma cells nor did it affect the overall rate of protein synthesis. The authors suggest that, in addition to mitogenic action, EGF may act as a trophic modulator of hormone secretion.

KNOX et al. (1978) demonstrated that pretreatment of human fibroblasts with EGF suppressed the infectivity of cytomegalovirus and enhanced the infectivity of herpes simplex type 1 virus. GREEN (1977) has shown that EGF increased the production of squames 2- to 5-fold in cultured human keratinocytes.

G. Growth Factor: Receptor Interactions

I. Receptors for EGF

Specific, saturable receptors for EGF have been demonstrated using ^{125}I-labeled mEGF or hEGF (urogastrone) and a wide variety of cultured cells including corneal cells, human fibroblasts, lens cells, human glial cells, human epidermoid carcinoma cells, 3T3 cells, granulosa cells, human vascular endothelial cells, human choriocarcinoma cells, and a number of other cell types.

For different strains of human fibroblasts it has been estimated that each cell contains 40,000–100,000 binding sites for EGF, and apparent dissociation constants of $2-4 \times 10^{-10}$ M have been calculated (CARPENTER et al., 1975; HOLLEN-BERG and CUATRECASAS, 1975). The human epidermoid carcinoma cell line A-431 has an extraordinary number of binding sites for EGF, $2-3 \times 10^6$ receptors per cell (FABRICANT et al., 1977; HAIGLER et al., 1978).

Receptors for EGF were detected by O'KEEFE et al. (1974) with crude membrane fractions prepared from a variety of mammalian tissues. These authors reported that placental and liver membranes have a high capacity to bind EGF. Specific binding of EGF also has been detected in liver membrane fractions of evolutionarily distant organisms, such as certain teleosts and the dogfish shark (NAFTEL and COHEN, 1978). The specificity and high affinity of cells and membrane preparations for ^{125}I-EGF may be employed as the basis for competitive radioreceptor assays for EGF.

Indirect evidence suggests that the receptor for EGF is a glycoprotein. CAR-PENTER and COHEN (1977) demonstrated reversible inhibition of ^{125}I-EGF binding by a variety of lectins. HOCK et al. (1979) reported that crosslinked EGF:receptor complexes are absorbed by immobilized lectins. PRATT and PASTAN (1978) studied a mutant of 3T3 cells that has a lowered ^{125}I-EGF binding capacity and a decreased content of cell surface carbohydrate due to a block in the acetylation of glucosamine-6-phosphate. The binding was partially restored upon the addition of N-acetyl glucosamine to the media.

II. Internalization and Degradation of EGF

1. Biochemical Evidence

The first experimental data which indicated that EGF and its receptor were internalized resulted from the determination of the time-course of binding of ^{125}I-hEGF to human fibroblasts at 37 °C and 0 °C, shown in Fig. 3 (CARPENTER and COHEN, 1976b). Maximal binding was reached in 45–60 min. On continued incubation of the labeled hormone with fibroblasts at 37 °C, the amount of cell-bound radioactivity decreased until a constant level of 20–25% of the initial maximal amount of cell-bound radioactivity remained associated with the cells. When the cells were incubated with the labeled hormone at 0 °C, there was no net loss of cell-bound radioactivity. CARPENTER and COHEN (1976b) postulated that subsequent to the initial binding of ^{125}I-EGF to specific plasma membrane receptors, the EGF:receptor complex is internalized and the hormone is ultimately degraded in lysosomes. These conclusions were drawn from the following series of observations: 1. Cell-bound ^{125}I-EGF was rapidly degraded to mono [^{125}I]iodotyrosine at 37 °C. 2. At 0 °C cell-bound ^{125}I-EGF was not degraded but slowly dissociated from the cell. 3. When the binding of ^{125}I-EGF was carried out at 37 °C and the cells then incubated at 0 °C, almost no release of cell-bound radioactivity was detected. 4. The degradation, but not the binding, required metabolic energy. 5. The degradation was inhibited by drugs that inhibit lysosomal function, such as chloroquine and ammonium chloride. 6. When ^{125}I-EGF was bound to cells at 0 °C, the hormone was much more accessible to surface reactive agents, such as trypsin and antibodies to EGF, than when the hormone was bound at 37 °C. 7. Ex-

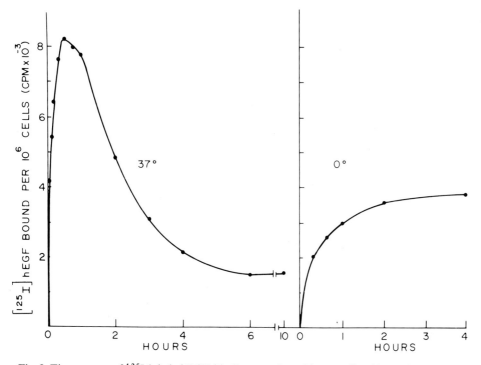

Fig. 3. Time course of ^{125}I-labeled EGF binding to cultured human fibroblasts at 37 °C and 0 °C. ^{125}I-EGF (final concentration 4 ng/ml, 24,400 cpm/ng) was added to each culture dish containing the standard binding medium. At the indicated time intervals, duplicate dishes were selected and the cell-bound radioactivity was determined. (From CARPENTER and CO-HEN, 1976 b)

posure of fibroblasts to EGF resulted in an apparent loss of plasma membrane receptors for EGF, which suggested that the receptor also is internalized.

Internalization and degradation of ^{125}I-EGF have been confirmed in a number of laboratories with a variety of cells.

2. Morphological Evidence

The problem of direct visualization of the internalization of EGF has been approached in a number of laboratories using three procedures: the preparation and tracing of fluorescent derivatives of EGF, the tracing of ^{125}I-EGF by electron microscopy autoradiography, and the preparation and tracing of EGF-ferritin conjugates by electron microscopy. With the exception of ^{125}I-EGF all of the other derivatives were covalent modifications of the single N terminal amino group; all the derivatives retained substantial binding and biological activity.

SCHLESSINGER et al. (1978b) and SHECHTER et al. (1978b) prepared a highly fluorescent rhodamine derivative of EGF and examined the binding and internalization of this derivative to 3 T 3 cells using image-intensified video fluorescent microscopy. HAIGLER et al. (1978) prepared fluorescein-conjugated EGF and examined the binding and internalization of this derivative by direct fluorescence microscopy using human epidermoid carcinoma cells (A-431) which are capable of binding much larger quantities of EGF than fibroblasts, thus rendering visualization possible. Both laboratories found that the initial binding of the derivatives to the cell surface was diffuse (except for a concentration of staining at the cell borders with the A-431 cells). Within 10–30 min at 37 °C the labeled hormone was found within the cells in endocytotic vesicles. An intermediary patching stage was seen in the 3 T 3 cells but was not detected in the A-431 cells; a "microclustering" or receptors was suggested for the latter cells.

GORDEN et al. (1978) used quantitative electron microscopic autoradiography to localize ^{125}I-EGF in human fibroblasts. The initial binding of the labeled hormone was localized to the plasma membrane with some preference to coated pit regions. The membrane-bound ^{125}I-EGF was internalized by the cell in a time- and temperature-dependent fashion. The internalized grains were almost exclusively related to lysosomal structures.

HAIGLER et al. (1979 a) have prepared a conjugate of epidermal growth factor and ferritin that retains substantial binding affinity for cell receptors and is biologically active. Monolayers of human epithelioid carcinoma cells (A-431) were incubated with ferritin-EGF (F-EGF) at 4 °C and processed for transmission electron microscopy. Under these conditions, 6×10^5 molecules of F-EGF bound to the plasma membrane of each cell. In the presence of excess native EGF, the number of bound ferritin particles was reduced by 99%, indicating that EGF:ferritin binds specifically to cellular EGF receptors. At 37 °C, cell-bound F-EGF rapidly redistributed in the plane of the plasma membrane to form small groups that were subsequently internalized into pinocytic vesicles. By 2.5 min at 37 °C, 32% of the cell-bound F-EGF was localized in vesicles. After 2.5 min, there was a decrease in the proportion of conjugate in vesicles with a concomitant accumulation of F-EGF in multivesicular bodies. The authors suggest that the F-EGF-containing pinocytic vesicles evert upon fusion with multivesicular bodies. By 30 min, 84% of the conjugate was located in structures morphologically identified as multivesicular bodies or lysosomes. These results are consistent with other morphological and biochemical studies utilizing ^{125}I-EGF and fluorescein-conjugated EGF.

III. Internalization of the Receptor

1. Indirect Evidence

Since exposure of fibroblasts to EGF resulted in an apparent loss of plasma membrane receptors, "down regulation," for EGF, the possibility was suggested (CARPENTER and COHEN, 1976 b) that receptors were internalized together with the hormone. It was suggested by AHARONOV et al. (1978 a) that low concentrations of EGF result in the down regulation of unoccupied receptors. However, since the initial incubation with low concentrations of EGF were for an extended period of time (h) and internalization occurs in minutes, it seems likely that new hormone-receptor complexes were formed and that more of the receptor was bound than expected. The question of EGF-induced internalization of unoccupied receptors is still open.

2. Chemical Evidence

DAS, FOX, and co-workers have provided information concerning the internalization and degradation of the receptor for EGF, utilizing a different approach. DAS et al. (1977) and DAS and FOX (1978) prepared a photoreactive derivative of ^{125}I-EGF and, following incubation of the derivative with 3T3 cells and photolysis, were able to detect in the cells a radioactive band on SDS polyacrylamide gels with a molecular weight of 190,000, albeit in low yield (1.5–2%). Incubation of affinity-labeled cells at 37 °C resulted in the loss of the radioactive 190,000 molecular weight band and the accumulation of three lower molecular weight bands (MW 62,000, 47,000, and 37,000). These presumed proteolytic products of the receptor were localized to the lysosomal fraction upon subcellular fractionation. Whether the covalent crosslinking of the receptor to EGF influenced its rate of degradation is not clear.

A novel interaction of ^{125}I-EGF (and possibly other ligands) with cell membranes has been reported by BAKER et al. (1979) and LINSLEY et al. (1979). These workers have reported that a small fraction of cell-bound ^{125}I-EGF becomes covalently linked, by an unknown mechanism, to the presumed receptor, identified as a 170,000–190,000 molecular weight protein. Upon prolonged incubation smaller fragments are noted similar to those observed in the studies with photoactivatable ^{125}I-EGF. The biological significance of this interesting direct-linkage phenomenon is not known.

A number of other attempts have been made to label the receptor for EGF. WRANN et al. (1979) and WRANN and FOX (1979) have reported the detection of the EGF receptors as 170,000 molecular weight protein on 3T3 cells and a 175,000 molecular weight protein on A-431 cells by direct surface iodination. SAHYOUN et al. (1978) covalently linked ^{125}I-labeled human EGF to rat liver membranes by a glutaraldehyde procedure and suggested a "subunit" molecular weight of about 100,000 for the receptor. HOCK et al. (1979) reported molecular weights of 160,000 and 180,000 for ^{125}I-labeled human EGF covalently crosslinked to human placental membranes.

3. Morphological Evidence

McKANNA et al. (1979) have developed ultrastructural criteria for identification of the F-EGF:receptor complex, thereby enabling utilization of the F-EGF as an indirect marker to localize the receptor for this peptide hormone. The ferritin cores of bound F-EGF are situated 4–6 nm from the extracellular surface of the membrane. When cells were incubated for up to 30 min at 37 °C, this characteristic spatial relationship was observed in all uptake stages (surface clustering, endocytosis, and incorporation into multivesicular bodies), indicating that the hormone:receptor complex remains intact through these steps. However, when incubation was continued for times sufficient to allow hormone degradation (30–60 min), pools of free ferritin were observed in lysosomes. These authors also have developed a method for viewing the surface of intact cells *en face*, allowing closer scrutiny of the clustering of F-EGF:receptor complexes in the plane of the membrane prior to internalization. The particles in the F-EGF clusters observed by this method are spaced at 12 nm center-to-center, serving to set upper limits on the packing dimensions of the EGF:receptor complex.

Employing fluorescent derivatives of EGF and insulin, Schlessinger et al. (1978 a) were able to determine the lateral diffusion coefficients of the EGF:receptor and insulin:receptor complexes in 3 T 3 cells. Both hormone receptor complexes were mobile on the cell surface, with similar diffusion coefficients, $D = 3-5 \times 10^{-10}$ cm^2/s at 23 °C, a diffusion coefficient much lower than that of a lipid probe. Increasing the temperature to 37 °C resulted in rapid receptor immobilization. The immobilization was attributed to aggregation of hormone-receptor complexes or internalization.

In a very interesting experiment Maxfield et al. (1978), using fluorescent derivatives of EGF, insulin, and α_2-macroglobulin added to 3 T 3-4 cells, concluded that all three polypeptides were internalized within the same vesicles by a common pathway. It would appear that all of these ligand-receptor complexes migrate to specialized regions of the cell surface where pinocytosis occurs. The molecular mechanisms involved in the formation of these clusters and their subsequent pinocytosis are not understood. It has been suggested that these specialized regions are coated and form "coated pits," although it is not at all certain that pinocytosis only occurs in coated regions. In fact the data of Haigler et al. (1979 b) and Chinkers et al. (1979) show clearly that, in A-431 cells, EGF induces ruffling and that although most of the EGF enters the cell in small vesicles, at least some of the membrane bound EGF enters the cell in large vesicles concerned with bulk pinocytosis.

To delineate the biological consequences of receptor clustering and internalization, it would be useful to have an agent which inhibits these processes. It has been reported (Maxfield et al., 1979 a, b) that ammonia and amines block the clustering of EGF receptors on 3 T 3 cells when examined by their fluorescent enhancement methodology. However, Gorden et al. (1978) have reported that ^{125}I-EGF is found in lysosomes even in the presence of ammonia, and McKanna et al. (1979) have reported, using F:EGF and various amines, that internalization of the EGF:receptor complex into multivesicular bodies proceeded, but that hormone-receptor degradation was blocked. The differing reports on amine inhibition may be reconciled if one assumes that these agents slow the rate of internalization but do not prevent internalization.

IV. Recovery of Receptor Activity

In the previous sections we have documented evidence that EGF:receptor complexes are rapidly and extensively removed from the cell surface by an active metabolic process. The results shown in Figure 3 indicate that after incubation of human fibroblasts with ^{125}I-EGF for 6 h or longer a constant level of binding is maintained (Carpenter and Cohen, 1976 b). This binding, which is approximately 25% of the initial binding capacity, does not represent stable receptors that are not being internalized. Rather this reduced level of binding reflects a steady-state in which the loss of hormone:receptor complexes from the cell surface is equalled by the insertion of unoccupied receptors into the plasma membrane (Carpenter and Cohen, 1976 b; Haigler and Carpenter (to be published).

The apparent extent of down regulation, when examined in the continuous presence of labeled ligand as in Fig. 3, can be influenced by factors which increase or decrease either the rate of receptor loss or the rate of receptor replacement. That some cell lines only show a small decrease in binding with increasing periods of time may reflect a higher rate of receptor replacement.

CARPENTER and COHEN (1976b) showed that following down regulation of 75% of the ^{125}I-EGF binding capacity of human fibroblasts, all of the original binding capacity could be recovered within 10 h by the addition of serum to ligand free media. The serum-stimulated recovery process was blocked by cycloheximide or Actinomycin D. This suggests that either the synthesis of receptor molecules occurs during recovery or that the "used" receptors are somehow recycled in a manner requiring macromolecular synthesis. AHARONOV et al. (1978a) also have observed the recovery process and reported that sparse, growing 3T3 cells recovered from down regulation much more quickly than did resting, confluent cells.

H. Relationship of EGF Binding and Metabolism to Biological Activity

In the previous section we have reviewed what is known about the biochemical mechanics by which EGF interacts with cell surface receptors and is subsequently internalized and degraded. An important question is how events involved in hormone recognition and metabolism may be related to biological responses. Various approaches to this problem have been employed, but to date convincing answers have not been forthcoming. Nevertheless, some points are reasonably clear and can form a conceptual framework to discuss this important topic.

I. Rapid Changes in Cell Physiology

Clearly the biological responses produced by EGF are numerous and several separate, but interrelated, mechanisms may be involved. The most rapid effects observed upon the addition of EGF to cultured cells are morphological changes (CHINKERS et al., 1979) and increased rates of transport (see Sect. F.IV.1).

Using scanning electron microscopy, CHINKERS et al. (1979) have demonstrated that within five minutes of exposure of A-431 tumor cells to EGF a large increase in membrane ruffling and extension of filopodia is apparent. These changes can be observed at very short intervals (60 s) after addition of EGF and are transient in nature – subsiding within 15 min. HAIGLER et al. (1979b) measured the effect of EGF on the uptake of horseradish peroxidase (HRP) in A-431 cells and demonstrated a large (10-fold) increase in the rate of fluid-phase pinocytosis in temporal coordination with enhanced ruffling. The ability of EGF to increase the rate of HRP uptake was detectable 30 s after addition of the growth factor and was transient in nature, returning to control levels within 15 min. Similar events probably occur but to a less marked extent in cells having lower concentrations of receptors. Since these uptake and morphological alterations occur very rapidly, degradation of cell-bound EGF is not likely to be necessary. Since increased HRP uptake occurs

very rapidly and requires the occupancy of only a small fraction of available receptors (4% occupancy for half-maximal uptake), the signal for this cellular response
to EGF may occur during clustering or initial internalization of hormone:receptor
complexes. A candidate for such a signal is the EGF-sensitive protein kinase activity present in cell membranes (CARPENTER et al., 1978, 1979). Similar analyses
probably apply to the increases in active transport of ions and low-molecular
weight nutrients.

II. Stimulation of DNA Synthesis

The relationship of responses such as increased DNA synthesis to binding, internalization and degradation is more difficult to assess. The primary complication
is that enhanced DNA synthesis is observed no sooner than 12 h after the addition
of EGF.

Two points relating EGF binding to maximal stimulation of DNA synthesis are
reasonably clear: 1) persistent interactions between EGF and cell surface receptors
must occur for many hours and 2) occupancy of approximately 25% of the available binding sites is necessary. The occupancy ratio has been reported by several
different investigators (HOLLENBERG and CUATRECASAS, 1975; CARPENTER and CO-
HEN, 1976 b; AHARONOV et al., 1978 a). However, the interpretation of this fraction-
al occupancy ratio is difficult as binding analyses are complicated and the possible
functional heterogeneity of binding sites has not been addressed.

To investigate the length of time cells needed to be exposed to EGF before they
became committed to increased DNA synthesis, CARPENTER and COHEN (1976 a)
added antibody to EGF to cultures of human fibroblasts at various times after the
addition of EGF. They then related the length of EGF exposure to the capacity
of the cells to increase DNA synthesis. Removal of EGF from the media by antibody addition at 0–6 h prevented any stimulation of DNA synthesis and not until
12 h was the level of stimulation of DNA synthesis refractory to the antibody.
Similar results have been published by LINDGREN and WESTERMARK (1976, 1977);
AHARONOV et al. (1978 a), and SHECHTER et al. (1978 a). HAIGLER and CARPENTER
(1980) also reported similar results using either antibody to EGF or antibody to
the EGF receptor to interrupt hormone:receptor interactions.

SHECHTER et al. (1978 a) have reported that although the addition of anti-EGF
antibody to cultured fibroblasts at times up to 8 h after the addition of EGF blocks
enhanced DNA synthesis, removal of the growth factor by washing 30 min after
its addition does not completely prevent stimulation of DNA synthesis. They propose that a small fraction of the cell-bound EGF is not removed by washing and
does not dissociate into EGF-free media for at least 8 h. The authors' data further
indicate that this very tightly bound EGF remains on the extracellular surface of
the cells, since it can be inactivated by antibodies to EGF. SHECHTER and his colleagues suggest that this very high affinity binding of EGF is necessary for the induction of DNA synthesis, but represents a negligible fraction of the total binding.
The covalent linkage of EGF to its receptor (BAKER et al., 1979; LINSLEY et al.,
1979) might explain the necessary high affinity binding to support these observations. However, chemical modification of the amino terminus of EGF prevents covalent crosslinking to the receptor, but does not reduce the binding activ-

ity. The conclusion of SHECHTER et al. (1978 a) would also require that the high affinity EGF:receptor complexes are not subject to internalization and degradation. Such a subclass of receptors has not been detected. It should be noted that in experiments reported by other groups (LINDGREN and WESTERMARK, 1976, 1977; AHARONOV et al., 1978 a) removal of EGF from cultured cells by washing did prevent stimulation of DNA synthesis.

In a subsequent publication SHECHTER et al. (1979) have addressed the question of whether the clustering or aggregation of EGF:receptor complexes is required for the stimulation of DNA synthesis. Cultures of fibroblasts were incubated with either biologically inactive cyanogen bromide-treated EGF (which has a 10-fold lower binding affinity compared to native EGF) or with a concentration of native EGF too low to be biologically effective (0.1 pg/ml). Neither of these additions stimulated DNA synthesis; however, addition of small amounts of anti-EGF antibodies shortly (20 min) after either the cyanogen bromide treated EGF or the low amounts of native EGF produced a stimulation of DNA synthesis that was 50%–100% of the maximal stimulation obtained by native EGF at optimal concentrations. Monovalent Fab' antibody was not effective and no enhancement of DNA synthesis was produced by the divalent antibody alone. The authors suggested that the divalent antibody evoked EGF activity by crosslinking and aggregating otherwise inactive hormone:receptor complexes. Although only a very restricted range of antibody dilutions were effective, the authors indicate that the local aggregation of EGF:receptor complexes is a necessary step for the induction of DNA synthesis.

MAXFIELD et al. (1979 a, b) have reported that bacitracin and certain amines (ammonium salts, ethylamine, propylamine, n-butylamine) block the clustering of EGF:receptor complexes on the cell surface and thereby prevent internalization. This group (MAXFIELD et al., 1979 a) reports that these putative inhibitors of clustering do not block stimulation of DNA synthesis by EGF, but actually potentiate the effect of the growth factor. It should be noted that this effect of amines on the EGF stimulation of DNA synthesis was only observed in very restricted conditions – a limited (2½ h) exposure of quiescent cells to EGF in the presence or absence of the amines. The evidence to support the capacity of these inhibitors to block clustering and internalization, however, is derived solely by morphological criteria – by using an image-intensifier television camera to view the interaction of fluorescein or rhodamine conjugated ligands with intact cells (WILLINGHAM and PASTAN, 1978). Although this technology certainly has advantages, resolution in terms of dimensions of the aggregates and their specific location would appear to be difficult.

Using ferritin conjugated EGF and electron microscopy, McKANNA et al. (1979) demonstrated that EGF:receptor complexes are internalized in the presence of amines. Also, GORDEN et al. (1978) performed quantitative electron microscopic autoradiography after incubating cultured fibroblasts with ^{125}I-labeled EGF in the presence and absence of ammonium chloride. Their data also indicate that hormone:receptor complexes are internalized in the presence of this inhibitor. Thus, conclusions regarding the relationship of the internalization of EGF:receptor complexes to enhanced DNA synthesis which are based solely on amine inhibition are not yet warranted.

In summary the available data show that binding of EGF to specific cell surface receptors is obligatory for stimulation of DNA synthesis and that clustering of hormone:receptor complexes is involved. However the possible role of internalization and degradation in the production of biological responses by EGF is not resolved.

Fox and DAS (1979) and HELDIN et al. (1979) have presented indirect evidence to support the idea that internalization is an important step in the activation of DNA synthesis by EGF. Other groups (AHARONOV et al., 1978a; SHECHTER et al., 1979; MAXFIELD et al., 1979a) conclude that internalization is not necessary for stimulation of DNA synthesis and actually may be a mechanism for stopping the generation of intracellular mitogenic signals at the cell surface.

I. Other Controls of Receptor Activity

I. Transforming Agents

Several studies have shown interesting relationships between effects of EGF and the transformed state or between EGF and transforming agents. CARPENTER and COHEN (1976a, 1978b) demonstrated that human fibroblasts grown in the continuous presence of EGF did not exhibit two growth controlling mechanisms otherwise associated with the regulated proliferation of these cells. When grown in the continuous presence of EGF, the cells were not restricted to a tightly packed confluent monolayer but rather formed populations several cell layers thick. Also, the growth of cells in media containing a low concentration of serum or deficient sera was not impeded when EGF was present. It was noted that these behaviors of cells in vitro were more similar to those of transformed cells than normal cells. A notable exception was the inability of EGF to promote anchorage-independent growth, which is the characteristic of cultured cells most closely associated with malignant potential.

Studies of ^{125}I-EGF binding to cultured cells by CARPENTER et al. (1975) showed that normal rat kidney cells bound the growth factor, but the same cells transformed by the Kirsten sarcoma virus did not. They noted that transformation by DNA viruses did not reduce binding capacity and that several other cell lines which were reportedly infected with RNA-transforming viruses exhibited low ^{125}I-EGF binding capacity.

Studies of ^{125}I-EGF binding to pairs of normal and transformed cell lines were reported by TODARO et al. (1976, 1977). Their results showed that transformation by murine or feline sarcoma viruses consistently decreased ^{125}I-EGF binding capacity in different lines of normal cells. In these studies the activity of receptors for other ligands, such as multiplication stimulating activity, was not affected by transformation. Also, cells transformed by DNA viruses, chemicals (with a few exceptions), and spontaneous transformants did not exhibit altered ^{125}I-EGF binding capacity. These authors postulated several possible explanations for the selective reduction of ^{125}I-EGF binding capacity by sarcoma viruses. One explanation was that sarcoma transformation resulted in the production of an EGF-like molecule which would lower the apparent binding capacity for exogenous ^{125}I-EGF by direct competition. Investigation of this possibility has revealed that sarcoma transformed cell lines excrete a polypeptide termed sarcoma growth factor (SGF) which is capable of stimulating cell growth and which apparently interacts with the EGF receptor (DeLarco and Todaro, 1978a; Todaro and DeLarco, 1978; Todaro et al., 1979). In vitro SGF not only stimulates cellular proliferation but is able to promote the anchorage-independent growth of non-transformed cells – an effect not produced by EGF. SGF competes with ^{125}I-EGF in specific radioreceptor assays, but is immunologically distinct from EGF. Guinivan and Ladda (1979) report the presence of a factor in the culture media of Kirsten sarcoma transformed cells which stimulates DNA synthesis in normal cells, but enhances rather than inhibits ^{125}I-EGF binding. Pruss et al. (1978) have demonstrated that 3T3 variants lacking the EGF receptor are readily transformed by the Kirsten sarcoma virus. This result suggests that either interaction of SGF with the EGF receptor is not necessary for transformation or that the sarcoma factor interacts with sites other than the EGF receptor. It will be important to determine the chemical nature of SGF and if this growth factor binds to cellular sites other than the EGF receptor. Experiments concerning the possible activities of SGF in vivo have not been reported. It would be helpful to know whether SGF mimics EGF in the newborn mouse eyelid opening assay which is the most specific index of EGF biological activity.

Todaro and his colleagues (1979) have suggested that the ectopic production of growth factors by cells that also respond to the same factor(s) may offer a selective advantage to such cells and provide means by which some transformation properties are acquired. A reduction in EGF receptor activity has been noted in some instances of chemical transformation – particularly by benzopyrene (Todaro et al., 1976; Hollenberg et al., 1979; Brown et al., 1979b). However, no evidence has been reported for the ectopic production by these cells of factors able to interact with the EGF receptor.

II. Tumor Promoters

EGF has been shown to act as tumor promoter in vivo in studies involving the application of methylcholanthrene to mouse skin (Reynolds et al., 1965; Rose et al., 1976). Another potent tumor promoter is 12-O-tetradecanoyl phorbol-13-acetate (TPA) which is derived from croton oil. In cell culture systems it has been demonstrated that TPA at concentrations of 10^{-8} to 10^{-10} M blocks ^{125}I-EGF binding to cell surface receptors (Lee and Weinstein, 1978, 1979; Shoyab et al., 1979; Brown et al., 1979a; Murray and Fusenig, 1979). Derivatives of TPA which are inactive promoters in vivo do not block ^{125}I-EGF binding. The mechanism by which TPA effects ^{125}I-EGF binding is not clear, but appears to be indirect, i.e., TPA does not compete for ^{125}I-EGF binding in the same manner as unlabeled EGF. At present the indirect effect of TPA is thought to result from perturbations of the phospholipid microenvironment near the EGF receptor or by TPA binding to the receptor at a site other than the EGF binding site. However, even these mechanisms are likely to be simplifications as no effect of TPA on ^{125}I-EGF binding in membrane preparations has been demonstrated. Studies with cultured cells indicate that the TPA effect on EGF binding: 1) is specific (binding of insulin, multiplication stimulating activity, concanavalin A, nerve growth factor, low density lipoprotein, or murine type C ectopic viral glycoprotein [gp 60] is not altered), 2) is strongly temperature dependent, 3) is energy independent, and 4) results in a lowered receptor affinity for EGF without a decrease in receptor number.

EGF and TPA produce a number of similar responses in cultured cells (LEE and WEINSTEIN, 1978, 1979) and TPA acts synergistically with EGF (and other mitogens) to enhance the stimulation of DNA synthesis (DICKER and ROZENGURT, 1979; FRANTZ et al., 1979). In regard to the potentiation of biologic responses to mitogenic agents the action of TPA is not specific for EGF, but seems to produce an enhancement of mitogenic activities in general. TPA also produces mitogenic activities in cells that do not respond to EGF.

III. Differentiation

In what may be an important area of receptor regulation several reports have examined the influence of cell differentiation on ^{125}I-EGF binding. VLODAVSKY et al. (1978) report that cultured bovine granulosa cells respond to EGF by increasing cell numbers and have approximately 23,000 EGF receptors per cell. When these cells spontaneously differentiate in culture to luteal cells, they no longer respond to EGF. Surprisingly, the loss of sensitivity to EGF is accompanied by a 5-fold increase in the number of receptors per cell (105,000 receptors per cell). Both granulosa and luteal cells internalize and degrade ^{125}I-EGF and the receptor K_D was not significantly altered.

HUFF and GUROFF (1979) have studied the rat pheochromocytoma clone PC 12 as a model of neuronal development. This cell line has specific high affinity EGF receptors and ornithine decarboxylase activity is increased by the addition of EGF. When the PC 12 cell line is grown in the presence of nerve growth factor for several days many morphologic, biochemical, and electrical aspects of neuronal differentiation appear. At this time the binding capacity for ^{125}I-EGF is decreased by approximately 80% and the ability of EGF to increase ornithine decarboxylase activity is similarly reduced.

REES et al. (1979) have examined the differentiation of cultured mouse teratocarcinoma stem cells. These undifferentiated embryonal carcinoma cells are pleuripotent and highly tumorgenic. When grown in vitro for several days, the embryonal carcinoma cells form morphologically distinct endoderm-like cells that have a reduced tumorgenic potential. ^{125}I-EGF binding is barely detectable in the embryonal carcinoma cells, but increases approximately 12-fold as the cells develop into endoderm-like cells. At this time the apparently differentiated cells are sensitive to the mitogenic activity of EGF.

IV. Lectins and Glycoprotein Metabolism

The activity of the EGF receptor can be inhibited by interfering with glycoprotein synthesis or by agents that react with glycoprotein. CARPENTER and COHEN (1977) demonstrated that lectins block ^{125}I-EGF binding at 37 °C or at 0 °C and in a reversible manner. PRATT and PASTAN (1978) studied ^{125}I-EGF binding in a 3T3 mutant cell line defective in glycoprotein synthesis due to a partial block in the acetylation of glucosamine-6-phosphate. This block in carbohydrate processing could be overcome by exogenous N-acetyl-glucosamine in the media. In the absence of added N-acetylglucosamine, ^{125}I-EGF binding was reduced approximately 80% in the mutant. Feeding N-acetylglucosamine to the mutant for several days partially restored the ^{125}I-EGF binding capacity.

V. Glucocorticoids

EGF binding capacity can be affected by incubation of cells for several days in media supplemented with glucocorticoids (BAKER et al., 1978; BAKER and CUNNINGHAM, 1978). Cultured human fibroblasts grown for several days in media containing 100 ng/ml dexamethasone exhibited a 50–100% increase in ^{125}I-EGF binding capacity and an increased responsiveness to the mitogenic action of EGF.

VI. Modulation of Protein Synthesis

Modulation of protein synthesis also affords an experimental technique to influence EGF receptor activity. CARPENTER (1979a) reported a half life of approximately 15 h for EGF receptor activity when protein synthesis was inhibited in human fibroblasts. AHARONOV et al. (1978a), using cycloheximide to inhibit protein synthesis, calculated a similar value ($t_{1/2}$) for the EGF receptor in 3T3 cells. In the study carried out by CARPENTER (1979a) protein synthesis was stopped by the removal of histidine from the media and the addition of L-histidinol to create a stringent amino acid starvation. In those experiments protein synthesis could be rapidly reinitiated by the addition of L-histidine. When protein synthesis was reinitiated by this procedure, EGF receptor activity was completely recovered within 10 h and the recovery process was stimulated by the presence of fresh serum. Also, complete recovery took place in the presence of Actinomycin D.

It should be noted that culture conditions can influence the capacity of cells to bind ^{125}I-EGF. Such variables as serum concentration (CARPENTER, unpublished results), cell density (PRATT and PASTAN, 1978; BROWN et al., 1979b; BHARGAVA et al., 1979), and the age of the culture (BHARGAVA et al., 1979) have been noted to influence growth factor binding.

K. A Biochemical Response to EGF in Subcellular Systems

The evidence that EGF forms a complex with plasma membrane receptors, initiating an intricate series of biochemical and morphological events within the cell that includes internalization and degradation of the hormone:receptor complex has now been reviewed. These events ultimately result in cell growth. It is reasonable to assume that the observed biochemical and morphological alterations induced by EGF result from the generation, amplification, and propagation of a series of "signals."

One approach to the problem of the nature of these signals is to develop a cell free system which responds in a measurable way to the presence of EGF. CARPENTER et al. (1978, 1979) have reported that membranes may be prepared from A-431 cells which retain the ability to bind ^{125}I-EGF in vitro and that following the formation of EGF:receptor complexes the capacity of these membranes to phosphorylate endogenous proteins in the presence of γ-labeled [^{32}P]ATP is enhanced. The net incorporation of ^{32}P into endogenous proteins was increased 2- to 3-fold by EGF. The A-431 membrane preparation appeared to have EGF-stimulated protein kinase activity toward exogenous substrates (histone) as well as toward membrane associated proteins. These phosphorylation reactions did not depend on the presence of cyclic AMP or cyclic GMP. Both the endogenous phosphorylation and EGF-stimulated phosphorylation were dependent on the presence of Mg^{2+} or Mn^{2+}; Ca^{2+} was ineffective.

Partial acid hydrolysis and electrophoresis of the phosphorylated membranes showed that the major phosphorylated product, in both the presence and absence of EGF, was phosphotyrosine (USHIRO and COHEN, 1980). SDS polyacrylamide gel electrophoresis and autoradiography indicated that although EGF increased the phosphorylation of a number of membrane proteins, two components which appear to be glycoproteins with molecular weights of 170,000 and 150,000, were primarily affected (KING et al., 1980). A similar but quantitatively much lower, phosphorylation effect of EGF was seen with membranes prepared from human fibroblasts or placenta (CARPENTER et al., 1980).

The activation of the kinase by EGF appears to be a reversible phenomenon since removal of EGF by anti-EGF IgG results in a "deactivation" of the kinase to the original basal level of activity (COHEN et al., 1980).

The biological role of the EGF-enhanced phosphorylation reaction is not known and one can only speculate concerning its possible relationship to nutrient transport, hormone-receptor internalization, modification of cytoskeletal elements, mitogenic signalling or some other aspect of the "pleiotropic" effect of EGF.

L. Prospectus

The most obvious prospect for EGF in the coming years is that there will be an increasing number of publications to be read and, for better or worse, reviewed. Between 1975 and 1979 the number of scientific papers concerning EGF increased six-fold from 22 to 75, respectively. The number of papers published in the last two years of the 1970's was more than the number of all papers published before that time. It is likely that future advances in the biology and chemistry will continue at both the basic and applied research levels.

The EGF receptor has been obtained in an active solubilized form (CARPENTER, 1979 b) and purified to near homogeneity by affinity chromatography (COHEN et al., manuscript submitted). The details of ligand-receptor interactions and the subsequent activation of intracellular biochemical processes, such as protein kinase activity, may soon become known. Important experiments to resolve the question of whether the internalization and/or degradation of EGF and/or its receptor are necessary for biological responses may be realized. Further elucidation of events taking place within the activated cells and leading to increased DNA synthesis and cell division is likely to take place. These studies should give a better understanding of the mechanism of action of EGF.

An equally important question to be answered is: what is the biological function of endogenous EGF in the intact animal? This may prove to be an especially difficult problem to approach experimentally.

It is possible that EGF may become a therapeutic agent. Areas to be considered might include wound healing and the control of gastric acid secretion and ulcers. GREEN et al. (1979) have suggested that human epidermal cells might be cultured in vitro, in the presence of EGF, into multiple epithelia suitable for grafting. Perhaps, EGF or its receptors may be used as markers for certain types of neoplasia. Certainly, EGF would seem to be an ideal compound for experimental studies of growth control and developmental processes in mammalian biology.

Acknowledgements. The author wishes to acknowledge funding from the National Cancer Institute (CA 24071) and American Cancer Society (BC 294) during the preparation of this review.

References

Aharonov, A., Pruss, R.M., Herschman, H.R.: Epidermal growth factor: relationship between receptor regulation and mitogenesis in 3 T 3 cells. J. Biol. Chem. *253*, 3970–3977 (1978 a)

Aharonov, A., Vlodavsky, I., Pruss, R.M., Fox, C.F., Herschman, H.R.: Epidermal growth factor induced membrane changes in 3 T 3 cells. J. Cell. Physiol. *95*, 195–202 (1978 b)

Anderson, W.B., Gallo, M., Wilson, J., Lovelace, E., Pastan, J.: Effect of epidermal growth factor on prostaglandin E_1-stimulated accumulation of cyclic AMP in fibroblastic cells. FEBS Letters *102*, 329–332 (1979)

Angelletti, R.H., Bradshaw, R.A.: Nerve growth factor from mouse submaxillary gland: amino acid sequence. PNAS *68*, 2417–2420 (1971)

Angeletti, P.U., Salvi, M.L., Chesanow, R.L., Cohen, S.: Azione dell' „Epidermal Growth Factor" sulla sintesi di acidi nucleici e proteine dell'epitelio cutaneo. Experientia *20*, 1–6 (1964)

Angeletti, R.A., Angeletti, P.U., Calissano, P.: Testosterone induction of estero-proteolytic activity in the mouse submaxillary gland. Biochim. Biophys. Acta *139*, 372–381 (1967)

Baker, J.B., Cunningham, D.D.: Glucorticoid-mediated alteration in growth factor binding and action: analysis of the binding change. J. Supramol. Struct. *9*, 69–77 (1978)

Baker, J.B., Barsh, G.S., Carney, D.H., Cunningham, D.D.: Dexamethasone modulates binding and action of epidermal growth factor in serum-free cell culture. Proc. Natl. Acad. Sci. USA *75*, 1882–1886 (1978)

Baker, J.B., Simmer, R.L., Glenn, K.C., Cunningham, D.D.: Thrombin and epidermal growth factor become linked to cell surface receptors during mitogenic stimulation. Nature *278*, 743–745 (1979)

Barka, T., Noen, H. van der, Gresik, E.W., Kerenyi, T.: Immunoreactive epidermal growth factor in human amniotic fluid. Mt. Sinai J. Med. *45*, 679–684 (1978)

Barnes, D., Colowick, S.P.: Stimulation of sugar uptake in cultured fibroblasts by epidermal growth factor (EGF) and EGF-binding arginine esterase. J. Cell. Physiol. *89*, 633–640 (1976)

Barnes, D., Sato, G.: Growth of a human mammary tumour cell line in a serum-free medium. Nature *281*, 388–389 (1979)

Barthe, P.L., Bullock, L.P., Mowszowicz, I., Bardin, C.W., Orth, D.N.: Submaxillary gland epidermal growth factor: a sensitive index of biologic androgen activity. Endocrinology *95*, 1019–1025 (1974)

Bedrick, A.D., Ladda, R.L.: Epidermal growth factor potentiates cortisone-induced cleft palate in the mouse. Teratology *17*, 13–18 (1978)

Benveniste, R., Speeg, K.V., Carpenter, G., Cohen, S., Lindner, J., Rabinowitz, D.: Epidermal growth factor stimulates secretion of human chorionic gonadotropin by cultured human choriocarcinoma cells. J. Clin. Endocrinol. Metab. *46*, 169–172 (1978)

Bertsch, S., Marks, F.: Effect of foetal calf serum and epidermal growth factor on DNA synthesis in explants of chick embryo epidermis. Nature *251*, 517–519 (1974)

Bhargava, G., Rifas, L., Markman, M.H.: Presence of epidermal growth factor receptors and influence of epidermal growth factor on proliferation and aging in cultured smooth muscle cells. J. Cell. Physiol. *100*, 365–374 (1979)

Birnbaum, J.E., Sapp, T.M., Moore, J.B. jr.: Effects of reserpine epidermal growth factor, and cyclic nucleotide modulators on epidermal mitosis. J. Invest. Dermatol. *66*, 313–318 (1976)

Blosse, P.T., Fenton, E.L., Henningsson, S., Kahlson, G., Rosengren, E.: Activities of decarboxylases of histidine and ornithine in young mice after injection of epidermal growth factor. Experientia *30*, 22–23 (1974)

Bower, J.M., Camble, R., Gregory, H., Gerring, E.L., Willshire, I.R.: The inhibition of gastric acid secretion by epidermal growth factor. Experientia *31*, 825–826 (1975)

Brown, K.D., Dicker, P., Rozengurt, E.: Inhibition of epidermal growth factor binding to surface receptors by tumor promoters. Biochem. Biophys. Res. Commun. *86*, 1037–1043 (1979a)

Brown, K.D., Yeh, Y.C., Holley, R.W.: Binding, internalization, and degradation of epidermal growth factor by Balb 3T3 and BP3T3 cells: relationship to cell density and the stimulation of cell proliferation. J. Cell. Physiol. *100*, 227–238 (1979b)

Brunk, U., Schellens, J., Westermark, B.: Influence of epidermal growth factor (EGF) on ruffling activity, pinocytosis, and proliferation of cultivated human glia cells. Exp. Cell Res. *103*, 295–302 (1976)

Bucher, N.L.R., Patel, U., Cohen, S.: Hormonal factors and liver growth. Adv. Enzyme Regul. *16*, 205–213 (1978)

Bullock, L.P., Barthe, P.L., Mowszowicz, I., Orth, D.N., Bardin, C.W.: The effect of progestins on submaxillary gland epidermal growth factor: demonstration of androgenic, synandrogenic, and antiandrogenic actions. Endocrinology *97*, 189–195 (1975)

Byyny, R.L., Orth, D.N., Cohen, S.: Radioimmunoassay of epidermal growth factor. Endocrinology *90*, 1261–1266 (1972)

Byyny, R.L., Orth, D.N., Cohen, S., Doyne, E.S.: Epidermal growth factor: effects of androgens and adrenergic agents. Endocrinology *95*, 776–782 (1974)

Calissano, P., Angeletti, P.U.: Testosterone effect on the synthetic rate of two esteropeptidases in the mouse submaxillary gland. Biochim. Biophys. Acta *156*, 51–58 (1968)

Canalis, E., Raisz, L.G.: Effect of epidermal growth factor on bone formation in vitro. Endocrinology *104*, 862–869 (1979)

Carpenter, G.: The regulation of cell proliferation: advances in the biology and mechanism of action of epidermal growth factor. J. Invest. Dermatol. *71*, 283–287 (1978)

Carpenter, G.: Regulation of EGF receptor activity during the modulation of protein synthesis. J. Cell. Physiol. *99*, 101–106 (1979a)

Carpenter, G.: Solubilization of the membrane receptor for epidermal growth factor. Life Sci. *24*, 1691–1698 (1979b)

Carpenter, G., Cohen, S.: Human epidermal growth factor and the proliferation of human fibroblasts. J. Cell. Physiol. *88*, 227–237 (1976a)

Carpenter, G., Cohen, S.: ^{125}I-labeled human epidermal growth factor (hEGF): binding, internalization, and degradation in human fibroblasts. J. Cell Biol. *71*, 159–171 (1976b)

Carpenter, G., Cohen, S.: Influence of lectins on the binding of ^{125}I-EGF to human fibroblasts. Biochem. Biophys. Res. Commun. *79*, 545–552 (1977)

Carpenter, G., Cohen, S.: Epidermal growth factors. In: Biochemical actions of hormones. Litwack, G. (ed.), vol. V, pp. 203–247. New York: Academic Press 1978a

Carpenter, G., Cohen, S.: Biological and molecular studies of the mitogenic effects of human epidermal growth factor. In: Molecular control of proliferation and differentiation. Papaconstantinou, J. (ed.), pp. 13–31. New York: Academic Press 1978b

Carpenter, G., Cohen, S.: Epidermal growth factor. Ann. Rev. Biochem. *48*, 193–216 (1979)

Carpenter, G.C., Lembach, K.J., Morrison, M.M., Cohen, S.: Characterization of the binding of ^{125}I-labeled epidermal growth factor to human fibroblasts. J. Biol. Chem. *250*, 4297–4304 (1975)

Carpenter, G., King, L. jr., Cohen, S.: Epidermal growth factor stimulates phosphorylation in membrane preparations in vitro. Nature *276*, 409–410 (1978)

Carpenter, G., King, L. jr., Cohen, S.: Rapid enhancement of protein phosphorylation in A-431 cell membrane preparations by epidermal growth factor. J. Biol. Chem. *254*, 4884–4891 (1979)

Carpenter, G., Poliner, L., King, L., Jr.: Protein phosphorylation in human placenta: Stimulation by epidermal growth factor. Mol. Cell. Endocrinol. *18*, 189–199 (1980)

Catterton, W.Z., Escobedo, M.B., Sexson, W.R., Gray, M.E., Sundell, H.W., Stahlman, M.T.: Effect of epidermal growth factor on lung maturation in fetal rabbits. Pediat. Res. *13*, 104–108 (1979)

Chen, L.B., Gudor, R.C., Sun, T.T., Chen, A.B., Mosesson, M.W.: Control of a cell surface major glycoprotein by epidermal growth factor. Science *197*, 776–778 (1977)

Chinkers, M., McKanna, J.A., Cohen, S.: Rapid induction of morphological changes in human carcinoma cells A-431 by epidermal growth factor. J. Cell Biol. *83*, 260–265 (1979)

Cohen, S.: Purification of a nerve-growth promoting protein from the mouse salivary gland and its neuro-cytotoxic antiserum. Proc. Natl. Acad. Sci. U.S.A. *46*, 302–311 (1960)

Cohen, S.: Isolation of a mouse submaxillary gland protein accelerating incisor eruption and eyelid opening in the new-born animal. J. Biol. Chem. *237*, 1555–1562 (1962)

Cohen, S.: Growth factors and morphogenic induction. In: Developmental and metabolic control mechanisms and neoplasia, pp. 251–272. Baltimore: Williams and Wilkins Company 1965a

Cohen, S.: The stimulation of epidermal proliferation by a specific protein (EGF). Dev. Biol. *12*, 394–407 (1965b)

Cohen, S.: Studies on the mechanism of action of epidermal growth factor (EGF). In: Hormones in development. Hamburgh, M., Barrington, E.J.W. (eds.), pp. 753–766. New York: Appleton-Century-Crofts 1971

Cohen, S.: Epidermal growth factor. J. Invest. Derm. *59*, 13–16 (1972)

Cohen, S., Carpenter, G.: Human epidermal growth factor: isolation and chemical and biological properties. Proc. Natl. Acad. Sci. USA. *72*, 1317–1321 (1975)

Cohen, S., Elliott, G.A.: The stimulation of epidermal keratinization by a protein isolated from the submaxillary gland of the mouse. J. Invest. Dermatol. *40*, 1–5 (1963)

Cohen, S., Savage, C.R. jr.: Part II – Recent studies on the chemistry and biology of epidermal growth factor. Recent Prog. Horm. Res. *30*, 551–574 (1974)

Cohen, S., Stastny, M.: Epidermal growth factor. III. The stimulation of polysome formation in chick embryo epidermis. Biochim. Biophys. Acta *166*, 427–437 (1968)

Cohen, S., Taylor, J.M.: Epidermal growth factor: chemical and biological characterization. Recent Prog. Horm. Res. *30*, 533–550 (1974)

Cohen, S., Carpenter, G., Lembach, K.J.: Interaction of epidermal growth factor (EGF) with cultured fibroblasts. Adv. Metab. Disord. *8*, 265–284 (1975)

Cohen, S., Carpenter, G., King, L. jr.: Epidermal growth factor (EGF)-receptor-protein kinase interactions: co-purification of receptor and EGF-enhanced phosphorylation activity. J. Biol. Chem. *255*, 4834–4842 (1980)

Covelli, I., Mozzi, R., Rossi, R., Frati, L.: The mechanism of action of the epidermal growth factor. III. Stimulation of the uptake of labeled precursors into RNA, DNA, and proteins induced by EGF in isolated tumor cells. Hormones *3*, 183–191 (1972a)

Covelli, I., Rossi, R., Mozzi, R., Frati, L.: Synthesis of bioactive ^{131}I-labeled epidermal growth factor and its distribution in rat tissues. Eur. J. Biochem. *27*, 225–230 (1972b)

Dailey, G.E., Kraus, J.W., Orth, D.N.: Homologous radioimmunoassay for human epidermal growth factor (urogastrone). J. Clin. Endocrinol. Metab. *46*, 929–935 (1978)

Daniele, S., Frati, L., Fiore, C., Santoni, G.: The effect of the epidermal growth factor (EGF) on the corneal epithelium in humans. Albrecht von Graefes Arch. Klin. Ophthalmol. *210*, 159–165 (1979)

Das, M., Fox, C.F.: Molecular mechanism of mitogen action: processing of receptor induced by epidermal growth factor. Proc. Natl. Acad. Sci. U.S.A. *75*, 2644–2648 (1978)

Das, M., Miyakawa, T., Fox, C.F., Pruss, R.M., Aharonov, A., Herschman, H.R.: Specific radiolabeling of a cell surface receptor for epidermal growth factor. Proc. Natl. Acad. Sci. U.S.A. *74*, 2790–2794 (1977)

DeLarco, J.E., Todaro, G.J.: Growth factors from murine sarcoma virus-transformed cells. Proc. Natl. Acad. Sci. U.S.A. *75*, 4001–4005 (1978a)

Diamond, I., Legg, A., Schneider, J.A., Rozengurt, E.: Glycolysis in quiescent cultures of 3T3 cells. J. Biol. Chem. *253*, 866–871 (1978)

Dicker, P., Rozengurt, E.: Stimulation of DNA synthesis by tumour promoter and pure mitogenic factors. Nature *276*, 723–726 (1978)

Dicker, P., Rozengurt, E.: Synergistic stimulation of early events and DNA synthesis by phorbol esters, polypeptide growth factors, and retinoids in cultured fibroblasts. J. Supramol. Struct. *11*, 79–93 (1979)

DiPasquale, A., White, D., McGuire, J.: Epidermal growth factor stimulates putrescine transport and ornithine decarboxylase activity in cultivated human fibroblasts. Exp. Cell Res. *16*, 317–323 (1978)

Elder, J.B., Williams, G., Lacey, E., Gregory, H.: Cellular localisation of human urogastrone/epidermal growth factor. Nature *271*, 466–467 (1978)

Fabricant, R.N., DeLarco, J.E., Todaro, G.J.: Nerve growth factor receptors on human melanoma cells in culture. Proc. Natl. Acad. Sci. U.S.A. *74*, 565–569 (1977)

Feldman, E.J., Aures, D., Grossman, M.I.: Epidermal growth factor stimulates ornithine decarboxylase activity in the digestive tract of mouse (40357). Proc. Soc. Exp. Med. Biol. *159*, 400–402 (1978)

Fox, C.F., Das, M.: Internalization and processing of the EGF receptor in the induction of DNA synthesis in cultured fibroblasts: the endocytic activation hypothesis. J. Supramol. Struct. *10*, 199–214 (1979)

Frantz, C.N., Stiles, C.D., Scher, C.D.: The tumor promoter 12-O-tetradecanoyl-phorbol-13-acetate enhances the proliferative response of Balb/3T3 cells to hormonal growth factors. J. Cell. Physiol. *100*, 413–424 (1979)

Frati, C., Covelli, I., Mozzi, R., Frati, L.: Mechanism of action of epidermal growth factor: effect on the sulfhydryl and disulfide groups content of the mouse epidermis during keratinization. Cell Differ. *1*, 239–244 (1972)

Frati, L., Daniele, S., Delogu, A., Covelli, I.: Selective binding of the epidermal growth factor and its specific effects on the epithelial cells of the cornea. Exp. Eye Res. *14*, 135–141 (1972)

Frey, P., Forand, R., Maciag, T., Shooter, E.M.: The biosynthetic precursor of epidermal growth factor and the mechanism of its processing. Proc. Natl. Acad. Sci. U.S.A. *76*, 6294–6298 (1979)

Friedkin, M., Legg, A., Rozengurt, E.: Antitubulin agents enhance the stimulation of DNA synthesis by polypeptide growth factors in 3T3 mouse fibroblasts. Proc. Natl. Acad. Sci. U.S.A. *76*, 3909–3912 (1979)

Gerschenson, L.E., Conner, E.A., Yang, J., Andersson, M.: Hormonal regulation of proliferation in two populations of rabbit endometrial cells. Life Sci. *24*, 1337–1344 (1979)

Gorden, P., Carpentier, J.L., Cohen, S., Orci, L.: Epidermal growth factor: morphological demonstration of binding, internalization, and lysosomal association in human fibroblasts. Proc. Natl. Acad. Sci. U.S.A. *75*, 5025–5029 (1978)

Gospodarowicz, D., Bialecki, H.: The effects of the epidermal and fibroblast growth factors on the replicative lifespan of cultured bovine granulosa cells. Endocrinology *103*, 854–865 (1978)

Gospodarowicz, D., Bialecki, H.: Fibroblast and epidermal growth factors are mitogenic agents for cultured granulosa cells of rodent, porcine, and human origin. Endocrinology *104*, 757–764 (1979)

Gospodarowicz, D., Greenburg, G.: The effects of epidermal and fibroblast growth factors on the repair of corneal endothelial wounds in bovine corneas maintained in organ culture. Exp. Eye Res. *28*, 147–157 (1979)

Gospodarowicz, D., Mescher, A.L.: A comparison of the responses of cultured myoblasts and chondrocytes to fibroblasts and epidermal growth factors. J. Cell. Physiol. *93*, 117–128 (1977)

Gospodarowicz, D., Ill, C.R., Birdwell, C.R.: Effects of fibroblast and epidermal growth factors on ovarian cell proliferation in vitro. I. Characterization of the response of granulosa cells to FGF and EGF. Endocrinology *100*, 1108–1120 (1977a)

Gospodarowicz, D., Mescher, A.L., Birdwell, C.R.: Stimulation of corneal endothelial cell proliferation in vitro by fibroblast and epidermal growth factors. Exp. Eye Res. *25*, 75–89 (1977 b)

Gospodarowicz, D., Mescher, A.L., Brown, K.D., Birdwell, C.R.: The role of fibroblast growth factor and epidermal growth factor in the proliferative response of the corneal and lens epithelium. Exp. Eye Res. *25*, 631–649 (1977 c)

Gospodarowicz, D., Moran, J.S., Braun, D.L.: Control of proliferation of bovine vascular endothelial cells. J. Cell. Physiol. *91*, 377–386 (1977 d)

Gospodarowicz, D., Moran, J.S., Owashi, N.D.: Effects of fibroblast growth factor and epidermal growth factor on the rate of growth of amniotic fluid-derived cells. J. Clin. Endocrinol. Metab. *44*, 651–649 (1977 e)

Gospodarowicz, D., Brown, K.D., Birdwell, C.R., Zetter, B.R.: Control of proliferation of human vascular endothelial cells: characterization of the response of human umbilical vein endothelial cells to fibroblast growth factor, epidermal growth factor, and thrombin. J. Cell Biol. *77*, 774–788 (1978 a)

Gospodarowicz, D., Greenburg, G., Birdwell, C.R.: Determination of cellular shape by the extracellular matrix and its correlation with the control of cellular growth. Cancer Res. *38*, 4155–4171 (1978 b)

Gospodarowicz, D., Bialecki, H., Thakral, T.K.: The angiogenic activity of the fibroblast and epidermal growth factor. Exp. Eye Res. *28*, 501–514 (1979)

Green, H.: Terminal differentiation of cultured human epidermal cells. Cell *11*, 405–416 (1977)

Green, H., Kehinde, O., Thomas, J.: Growth of cultured human epidermal cells into multiple epithelia suitable for grafting. Proc. Natl. Acad. Sci. U.S.A. *76*, 5665–5668 (1979)

Greene, L.A., Shooter, E.M., Varon, S.: Enzymatic activities of mouse nerve growth factor and its subunits. PNAS *60*, 1383–1388 (1968)

Gregory, H.: Isolation and structure of urogastrone and its relationship to epidermal growth factor. Nature *257*, 325–327 (1975)

Gregory, H., Bower, J.M., Willshire, I.R.: Urogastrone and epidermal growth factor. In: Growth factors. Kastrup, K.W., Nielsen, J.H. (eds.), pp. 75–84. Elmsford, New York: Pergamon Press 1977

Gresik, E., Barka, T.: Immunocytochemical localization of epidermal growth factor in mouse submandibular gland. J. Histochem. Cytochem. *25*, 1027–1035 (1977)

Guinivan, P., Ladda, R.L.: Decrease in epidermal growth factor receptor levels and production of material enhancing epidermal growth factor binding accompany the temperature-dependent changes from normal to transformed phenotype. Proc. Natl. Acad. Sci. U.S.A. *76*, 3377–3381 (1979)

Haigler, H.T., Carpenter, G.: Production and characterization of antibody blocking epidermal growth factor: receptor interactions. Biochim. Biophys. Acta *598*, 314–325 (1980)

Haigler, H.T., Cohen, S.: Epidermal growth factor-interaction with cellular receptors. TIBS. *June*, 132–134 (1979)

Haigler, H., Ash, J.F., Singer, S.J., Cohen, S.: Visualization by fluorescence of the binding and internalization of epidermal growth factor in human carcinoma cells A-431. Proc. Natl. Acad. Sci. U.S.A. *75*, 3317–3321 (1978)

Haigler, H.T., McKanna, J.A., Cohen, S.: Direct visualization of the binding and internalization of a ferritin conjugate of epidermal growth factor in human carcinoma cells A-431. J. Cell Biol. *81*, 382–395 (1979 a)

Haigler, H.T., McKanna, J.A., Cohen, S.: Rapid stimulation of pinocytosis in human carcinoma cells A-431 by epidermal growth factor. J. Cell Biol. *83*, 82–90 (1979 b)

Harper, G.P., Barde, Y.A., Burnstock, G., Carstairs, J.R., Dennison, M.E., Suda, K., Vernon, C.A.: Guinea pig prostate is a rich source of nerve growth factor. Nature *279*, 160–162 (1979)

Hassell, J.R.: The development of rat palatal shelves in vitro: an ultrastructural analysis of the inhibition of epithelial cell death and palate fusion by the epidermal growth factor. Develop. Biol. *45*, 90–102 (1975)

Heimberg, M., Weinstein, I., Lequire, V.S., Cohen, S.: The induction of fatty liver in neonatal animals by a purified protein (EGF) from mouse submaxillary gland. Life Sci. 4, 1625–1633 (1965)

Heldin, C.H., Westermark, B., Wasteson, Å.: Desensitisation of cultured glial cells to epidermal growth factor by receptor down-regulation. Nature 282, 419–420 (1979)

Hirata, Y., Orth, D.N.: Epidermal growth factor (urogastrone) in human fluids: size heterogeneity. J. Clin. Endocrinol. Metab. 48, 673–679 (1979a)

Hirata, Y., Orth, D.N.: Conversion of high molecular weight human epidermal growth factor (hEGF)/urogastrone (UG) to small molecular weight hEGF/UG by mouse EGF-associated arginine esterase. J. Clin. Endocrinol. Metab. 49, 481–483 (1979b)

Hirata, Y., Orth, D.N.: Concentrations of epidermal growth factor, nerve growth factor, and submandibular gland renin in male and female mouse tissue and fluids. Endocrinology 105, 1382–1387 (1979c)

Hirata, Y., Orth, D.N.: Epidermal growth factor (urogastrone) in human tissues. J. Clin. Endocrinol. Metab. 48, 667–672 (1979d)

Hock, R.A., Nexø, E., Holenberg, M.D.: Isolation of the human placenta receptor for epidermal growth factor-urogastrone. Nature 277, 403–405 (1979)

Holladay, L.A., Savage, C.R. jr., Cohen, S., Puett, D.: Conformation and unfolding thermodynamics of epidermal growth factor and derivatives. Biochemistry 15, 2624–2633 (1976)

Hollenberg, M.D.: Receptors for insulin and epidermal growth factor: relation to synthesis of DNA in cultured rabbit lens epithelium. Arch. Biochem. Biophys. 171, 371–377 (1975)

Hollenberg, M.D., Cuatrecasas, P.: Epidermal growth factor: receptors in human fibroblasts and modulation of action by cholera toxin. Proc. Natl. Acad. Sci. U.S.A. 70, 2964–2968 (1973)

Hollenberg, M.D., Cuatrecasas, P.: Insulin and epidermal growth factor: human fibroblast receptors related to deoxyribonucleic acid synthesis and amino acid uptake. J. Biol. Chem. 250, 3845–3853 (1975)

Hollenberg, M.D., Gregory, H.: Human urogastrone and mouse epidermal growth factor share a common receptor site in cultured human fibroblasts. Life Sci. 20, 267–274 (1976)

Hollenberg, M.D., Barrett, J.C., Ts'o, P.O.P., Berhanu, P.: Selective reduction in receptors for epidermal growth factor-urogastrone in chemically transformed tumorigenic syrian hamster embryo fibroblasts. Cancer Res. 39, 4166–4169 (1979)

Holley, R.W., Armour, R., Baldwin, J.H., Brown, K.D., Yeh, Y.C.: Density-dependent regulation of growth of BSC-1 cells in cell culture: control of growth by serum factors. Proc. Natl. Acad. Sci. U.S.A. 74, 5046–5050 (1977)

Hoober, J.K., Cohen, S.: Epidermal growth factor. II. Increased activity of ribosomes from chick embryo epidermis for cell-free protein synthesis. Biochim. Biophys. Acta 138, 357–368 (1967a)

Hoober, J.K., Cohen, S.: Epidermal growth factor. I. The stimulation of protein and ribonucleic acid synthesis in chick embryo epidermis. Biochim. Biophys. Acta 138, 347–356 (1967b)

Huff, K.R., Guroff, G.: Epidermal growth factor stimulates the phosphorylation of a specific nuclear protein in chick embryo epidermis. Biochem. Biophys. Res. Commun. 85, 464–472 (1978)

Huff, K.R., Guroff, G.: Nerve growth factor-induced reduction in epidermal growth factor responsiveness and epidermal growth factor receptors in PC 12 cells: an aspect of cell differentiation. Biochem. Biophys. Res. Commun. 89, 175–180 (1979)

Hutchings, S.E., Sato, G.H.: Growth and maintenance of HeLa cells in serum-free medium supplemented with hormones. Proc. Natl. Acad. Sci. U.S.A. 75, 901–904 (1978)

Junqueira, L.C.V., Toledo, A.M.S., Saad, A.: Amylase and protease activities in serum, submaxillary gland and submaxillary saliva of rat and mouse: In: Salivary glands and their secretions. Sreebny, L.M., Meyer, J. (eds.), pp. 105–118. New York: the Macmillan Co. 1964

King, L.E. jr., Carpenter, G., Cohen, S.: Characterization by electrophoresis of epidermal growth factor-stimulated phosphorylation using A-431 membranes. Biochemistry 19, 1524–1528 (1980)

Kirkland, W.L., Yang, N.S., Jorgensen, T., Langley, C., Furmanski, P.: Growth of normal and malignant human mammary epithelial cells in culture. J. Natl. Cancer Inst. *63*, 29–39 (1979)

Knox, G.E., Reynolds, D.W., Cohen, S., Alford, C.A.: Alteration of the growth of cytomegalovirus and herpes simplex virus type 1 by epidermal growth factor, a contaminant of crude human chorionic gonadotropin preparations. J. Clin. Invest. *61*, 1635–1644 (1978)

Krieger, D.T., Hauser, H., Liotta, A., Zelenetz, A.: Circadian periodicity of epidermal growth factor and its abolition by superior cervical ganglionectomy. Endocrinology *99*, 1589–1596 (1976)

Lacassange, A.: Dimorphism sexual de la glande sousmaxillaire chez la Sourix. C.R. Soc. Biol. (Paris) *133*, 180–181 (1940)

Lechner, J.F., Kaighn, E.M.: Application of the principles of enzyme kinetics to clonal growth rate assays: an approach for delineating interactions among growth promoting agents. J. Cell Physiol. *100*, 519–530 (1979a)

Lechner, J.F., Kaighn, E.M.: Reduction of the calcium requirement of normal human epithelial cells by EGF. Exp. Cell Res. *121*, 432–435 (1979b)

Lee, L.S., Weinstein, I.B.: Tumor-promoting phorbol esters inhibit binding of epidermal growth factor to cellular receptors. Science *202*, 313–315 (1978)

Lee, L.S., Weinstein, I.B.: Mechanism of tumor promoter inhibition of cellular binding of epidermal growth factor. Proc. Natl. Acad. Sci. U.S.A. *76*, 5168–5172 (1979)

Lembach, K.J.: Enhanced synthesis and extracellular accumulation of hyaluronic acid during stimulation of quiescent human fibroblasts by mouse epidermal growth factor. J. Cell Physiol. *89*, 277–288 (1976a)

Lembach, K.J.: Induction of human fibroblast proliferation by epidermal growth factor (EGF): Enhancement by an EGF-binding arginine esterase and by ascorbate. Proc. Natl. Acad. Sci. U.S.A. *73*, 183–187 (1976b)

Levi-Montalcini, R., Cohen, S.: Effects of extracts of the mouse submaxillary glands on the sympathetic system of mammals. Ann. N.Y. Acad. Sci. *85*, 324–341 (1960)

Levine, L., Hassid, A.: Epidermal growth factor stimulates prostaglandin biosynthesis by canine kidney (MDCK) cells. Biochem. Biophys. Res. Commun. *76*, 1181–1187 (1977)

Lindgren, A., Westermark, B.: Subdivision of the Gl phase of human glia cells in culture. Exp. Cell Res. *99*, 357–362 (1976)

Lindgren, A., Westermark, B.: Reset of the pre-replicative phase of human glia cells in culture. Exp. Cell Res. *106*, 89–93 (1977)

Linsley, P.S., Blifeld, C., Wrann, M., Fox, C.F.: Direct linkage of epidermal growth factor to its receptor. Nature *278*, 745–748 (1979)

Maxfield, F.R., Schlessinger, J., Shechter, V., Pastan, I., Willingham, M.C.: Collection of insulin, EGF and α_2-macroglobulin in the same patches on the surface of cultured fibroblasts and common internalization. Cell *14*, 805–810 (1978)

Maxfield, F.R., Davies, P.J.A., Klempner, L., Willingham, M.C., Pastan, I.: Epidermal growth factor stimulation of DNA synthesis is potentiated by compounds that inhibit its clustering in coated pits. Proc. Natl. Acad. Sci. U.S.A. *76*, 5731–5735 (1979a)

Maxfield, F.R., Willingham, M.C., Davies, P.J.A., Pastan, I.: Amines inhibit the clustering of α_2-macroglobulin and EGF on the fibroblast cell surface. Nature *277*, 661–663 (1979b)

McKanna, J.A., Haigler, H.T., Cohen, S.: Hormone receptor topology and dynamics: morphological analysis using ferritin-labeled epidermal growth factor. Proc. Natl. Acad. Sci. U.S.A. *76*, 5689–5693 (1979)

McKeehan, W.L., McKeehan, K.A.: Epidermal growth factor modulates extracellular Ca^{+2} requirement for multiplication of normal human skin fibroblasts. Exp. Cell Res. *123*, 397–400 (1979)

Mierzejewski, K., Rozengurt, E.: Stimulation of DNA synthesis and cell division in a chemically defined medium: effect of epidermal growth factor, insulin, and vitamin B_{12} on resting cultures of 3T3 cells. Biochem. Biophys. Res. Commun. *73*, 271–278 (1976)

Mierzejewski, K., Rozengurt, E.: Density-dependent inhibition of fibroblast growth is overcome by pure mitogenic factors. Nature *269*, 155–156 (1977)

Mierzejewski, K., Rozengurt, E.: A partially purified serum fraction synergistically enhances the mitogenic activity of epidermal growth factor and insulin in quiescent cultures of 3 T 3 cells. Biochem. Biophys. Res. Commun. *83*, 874–880 (1978)

Moore, J.B. jr.: Purification and partial characterization of epidermal growth factor isolated from the male rat submaxillary gland. Archiv. Biochem. Biophys. *189*, 1–7 (1978)

Murphy, R.A., Saide, J.D., Blanchard, M., Young, M.: Nerve growth factor in mouse serum and saliva: role of the submandibular gland. Proc. Natl. Acad. Sci. U.S.A. *74*, 2330–2333 (1977)

Murphy, R.A., Pantazis, N.J., Papastavros, M.: Epidermal growth factor and nerve growth factor in mouse saliva: a comparative study. Develop. Biol. *71*, 356–370 (1979)

Murray, A.W., Fusenig, N.E.: Binding of epidermal growth factor to primary and permanent cultures of mouse epidermis:inhibition by tumor-promoting phorbol esters. Cancer Letters *7*, 71–77 (1979)

Naftel, J., Cohen, S.: Phylogeny of receptors for epidermal growth factor. J. SC Med. Assoc. *74*, 53 (1978)

O'Farrell, M.K., Clingan, D., Rudland, P.S., DeAsua, L.J.: Stimulation of the initiation of DNA synthesis and cell division in several cultured mouse cell types: effect of growth-promoting hormones and nutrients. Exp. Cell Res. Vol. 118, 311–321 (1979)

O'Keefe, E., Hollenberg, M.D., Cuatrecasas, P.: Epidermal growth factor: characteristics of specific binding in membranes from liver, placenta, and other target tissues. Archiv. Biochem. Biophys. *164*, 518–526 (1974)

Osterman, J., Hammond, J.M.: Effects of epidermal growth factor, fibroblast growth factor and bovine serum albumin on ornithine decarboxylase activity of porcine granulosa cells. Horm. Metab. Res. *11*, 485–488 (1979)

Pratt, R.M., Pastan, I.: Decreased binding of epidermal growth factor to Balb/c 3 T 3 mutant cells defective in glycoprotein synthesis. Nature *272*, 68–70 (1978)

Pruss, R.M., Herschman, H.R., Klement, V.: 3 T 3 variants lacking receptors for epidermal growth factor are susceptible to transformation by Kirsten sarcoma virus. Nature *274*, 272–274 (1978)

Rees, A.R., Adamson, E.D., Graham, C.F.: Epidermal growth factor receptors increase during the differentiation of embryonal carcinoma cells. Nature *281*, 309–311 (1979)

Reynolds, V.H., Boehm, F.H., Cohen, S.: Enhancement of chemical carcinogenesis by an epidermal growth factor. Surg. Forum. *16*, 108–109 (1965)

Rheinwald, J.G., Green, H.: Epidermal growth factor and the multiplication of cultured human epidermal keratinocytes. Nature *265*, 421–424 (1977)

Richman, R.A., Claus, T.H., Pilkis, S.J., Friedman, D.L.: Hormonal stimulation of DNA synthesis in primary cultures of adult rat hepatocytes. Proc. Natl. Acad. Sci. U.S.A. *73*, 3589–3593 (1976)

Roberts, M.L.: The in vitro secretion of epidermal growth factor by mouse submandibular salivary gland. Archiv Pharmacol. *296*, 301–305 (1977)

Roberts, M.L.: Secretion of epidermal growth factor: the role of calcium in stimulus-secretion coupling and structural modification of the growth factor molecule during secretion. Biochim. Biophys. Acta *540*, 246–252 (1978)

Rose, S.P., Stahn, R., Passovoy, D.S., Herschman, H.: Epidermal growth factor enhancement of skin tumor induction in mice. Experientia *32*, 913–915 (1976)

Rozengurt, E., Heppel, L.A.: Serum rapidly stimulates quabain-sensitive ^{86}Rb$^+$ influx in quiescent 3 T 3 cells. Proc. Natl. Acad. Sci. U.S.A. *72*, 4492–4495 (1975)

Rozengurt, E., Mierzejewski, K., Wigglesworth, N.: Uridine transport and phosphorylation in mouse cells in culture: effect of growth-promoting factors, cell cycle transit and oncogenic transformation. J. Cell Physiol. *97*, 241–252 (1978)

Sahyoun, N., Hock, R.A., Hollenberg, M.D.: Insulin and epidermal growth factor-urogastrone: affinity crosslinking to specific binding sites in rat liver membranes. Proc. Natl. Acad. Sci. U.S.A. *75*, 1675–1679 (1978)

Savage, C.R. jr., Cohen, S.: Epidermal growth factor and a new derivative: rapid isolation procedures and biological and chemical characterization. J. Biol. Chem. *247*, 7609–7611 (1972)

Savage, C.R. jr., Cohen, S.: Proliferation of corneal epithelium induced by epidermal growth factor. Exp. Eye Res. *15*, 361–366 (1973)

Savage, C.R. jr., Inagami, T., Cohen, S.: The primary structure of epidermal growth factor. J. Biol. Chem. *247*, 7612–7621 (1972)

Savage, C.R. jr., Hash, J.H., Cohen, S.: Epidermal growth factor: location of disulfide bonds. J. Biol. Chem. *248*, 7669–7672 (1973)

Schechter, Y., Hernaez, L., Schlessinger, J., Cuatrecasas, P.: Local aggregation of hormone-receptor complexes is required for activation by epidermal growth factor. Nature *278*, 835–838 (1979)

Scheving, L.A., Yeh, Y.C., Tsai, T.H., Scheving, L.E.: Circadian phase-dependent stimulatory effects of epidermal growth factor on deoxyribonucleic acid synthesis in the tongue, esophagus, and stomach of the adult male mouse. Endocrinology *105*, 1475–1480 (1979)

Schlessinger, J., Shechter, Y., Cuatrecasas, P., Willingham, M.C., Pastan, I.: Quantitative determination of the lateral diffusion coefficients of the hormone-receptor complexes of insulin and epidermal growth factor on the plasma membrane of cultured fibroblasts. Proc. Natl. Acad. Sci. U.S.A. *75*, 5353–5357 (1978a)

Schlessinger, J., Shechter, Y., Willingham, M.C., Pastan, I.: Direct visualization of binding, aggregation, and internalization of insulin and epidermal growth factor on living fibroblastic cells. Proc. Natl. Acad. Sci. U.S.A. *75*, 2659–2663 (1978b)

Schneider, J.A., Diamond, I., Rozengurt, E.: Glycolysis in quiescent cultures of 3T3 cells: addition of serum, epidermal growth factor, and insulin increases the activity of phosphofructokinase in a protein synthesis-independent manner. J. Biol. Chem. *253*, 872–877 (1978)

Server, A.C., Shooter, E.M.: Comparison of the arginine esteropeptidases associated with the nerve and epidermal growth factors. J. Biol. Chem. *251*, 165–173 (1976)

Server, A.C., Sutter, A., Shooter, E.M.: Modification of the epidermal growth factor affecting the stability of its high molecular weight complex. J. Biol. Chem. *251*, 1188–1196 (1976)

Shechter, Y., Hernaez, L., Cuatrecasas, P.: Epidermal growth factor: biological activity requires persistent occupation of high affinity cell surface receptors. Proc. Natl. Acad. Sci. U.S.A. *75*, 5788–5791 (1978a)

Shechter, Y., Schlessinger, J., Jacobs, S., Chang, K.J., Cuatrecasas, P.: Fluorescent labeling of hormone receptors in viable cells: preparation and properties of highly fluorescent derivatives of epidermal growth factor and insulin. Proc. Natl. Acad. Sci. U.S.A. *75*, 2135–2139 (1978b)

Shoyab, M., DeLarco, J.E., Todaro, G.J.: Biologically active phorbol esters specifically alter affinity of epidermal growth factor membrane receptors. Nature *279*, 387–391 (1979)

Smith, J.B., Rozengurt, E.: Lithium transport by fibroblastic mouse cells: characterization and stimulation by serum and growth factors in quiescent cultures. J. Cell Physiol. *97*, 441–450 (1978)

Smith, C.J., Wejksnora, P.J., Warner, J.R., Rubin, C.S., Rosen, O.M.: Insulin-stimulated protein phosphorylation in 3T3-L1 preadipocytes. Proc. Natl. Acad. Sci. U.S.A. *76*, 2725–2729 (1979)

Stanley, M.A., Parkinson, E.K.: Growth requirements of human cervical epithelial cells in culture. Int. J. Cancer *24*, 407–414 (1979)

Starkey, R.H., Orth, D.N.: Radioimmunoassay of human epidermal growth factor (urogastrone). J. Clin. Endocrinol. Metab. *45*, 1144–1153 (1977)

Starkey, R.H., Cohen, S., Orth, D.N.: Epidermal growth factor: identification of a new hormone in human urine. Science *189*, 800–802 (1975)

Stastny, M., Cohen, S.: Epidermal growth factor. IV. The induction of ornithine decarboxylase. Biochim. Biophys. Acta *204*, 578–589 (1970)

Stoker, M.G.P., Pigott, D., Taylor-Papadimitriou, J.: Response to epidermal growth factors of cultured human mammary epithelial cells from benign tumours. Nature *264*, 764–767 (1976)

Sundell, H., Gray, M.E., Serenius, F.S., Escobedo, M.B., Stahlman, M.T.: Effects of epidermal growth factor on lung maturation in fetal lambs. Am. J. Pathol. *100*, 707–726 (1980)

Tashjian, A.H. jr., Levine, L.: Epidermal growth factor stimulates prostaglandin production and bone resorption in cultured mouse calvaria. Biochem. Biophys. Res. Commun. *85*, 966–975 (1978)

Taylor, J.M., Cohen, S., Mitchell, W.M.: Epidermal growth factor: high and low molecular weight forms. Proc. Natl. Acad. Sci. U.S.A. *67*, 164–171 (1970)

Taylor, J.M., Mitchell, W.M., Cohen, S.: Epidermal growth factor: physical and chemical properties. J. Biol. Chem. *247*, 5928–5934 (1972)

Taylor, J.M., Mitchell, W.M., Cohen, S.: Characterization of the high molecular weight form of epidermal growth factor. J. Biol. Chem. *249*, 3198–3203 (1974)

Todaro, G.J., DeLarco, J.E.: Growth factors produced by sarcoma virus-transformed cells. Cancer Res. *38*, 4147–4154 (1978)

Todaro, G.J., DeLarco, J.E., Cohen, S.: Transformation by murine and feline sarcoma viruses specifically blocks binding of epidermal growth factor to cells. Nature *264*, 26–31 (1976)

Todaro, G.J., DeLarco, J.E., Nissley, S.P., Rechler, M.M.: MSA and EGF receptors on sarcoma virus transformed cells and human fibrosarcoma cells in culture. Nature *267*, 526–528 (1977)

Todaro, G.J., DeLarco, J.E., Marquardt, H., Bryant, M.L., Sherwin, S.A., Sliski, A.H.: Polypeptide growth factors produced by tumor cells and virus-transformed cells: a possible growth advantage for the producer cells. In: Hormones and cell culture. Sato, G., Ross, R. (eds.), pp. 113–127. Cold Spring Harbor, New York: Cold Spring Harbor Laboratory 1979

Turkington, R.W.: Stimulation of mammary carcinoma cell proliferation by epithelial growth factor in vitro. Cancer Res. *29*, 1457–1458 (1969a)

Turkington, R.W.: The role of epithelial growth factor in mammary gland development in vitro. Exp. Cell Res. *57*, 79–85 (1969b)

Turkington, R.W., Males, J.L., Cohen, S.: Synthesis and storage of epithelial-epidermal growth factor in submaxillary gland. Cancer Res. *31*, 252–256 (1971)

Ushiro, H., Cohen, S.: Identification of phosphotyrosine as a product of epidermal growth factor-activated protein kinase in A-431 cell membranes. J. Biol. Chem. *255*, 8363–8363 (1980)

Noorden, S. van, Heitz, P., Kasper, M., Pearse, A.G.E.: Mouse epidermal growth factor: light and electron microscopical localisation by immunocytochemical staining. Histochemistry *52*, 329–340 (1977)

Vlodavsky, I., Brown, K.D., Gospodarowicz, D.: A comparison of the binding of epidermal growth factor to cultured granulosa and luteal cells. J. Biol. Chem. *253*, 3744–3750 (1978)

Westermark, B.: Density dependent proliferation of human glia cells stimulated by epidermal growth factor. Biochem. Biophys. Res. Commun. *69*, 304–309 (1976)

Westermark, B.: Local starvation for epidermal growth factor cannot explain density-dependent inhibition of normal human glial cells. Proc. Natl. Acad. Sci. U.S.A. *74*, 1619–1621 (1977)

Willingham, M.C., Pastan, I.: The visualization of fluorescent proteins in living cells by video intensification microscopy (VIM). Cell *13*, 501–507 (1978)

Wrann, M.M., Fox, C.F.: Identification of epidermal growth factor receptors in a hyperproducing human epidermoid carcinoma cell line. J. Biol. Chem. *254*, 8083–8086 (1979)

Wrann, M., Linsley, P.S., Fox, C.F.: Identification of the EGF receptor on 3T3 cells by surface-specific iodination and gel electrophoresis. FEBS Letters *104*, 415–419 (1979)

Zetter, B.R., Sun, T.T., Chen, L.B., Buchanan, J.M.: Thrombin potentiates the mitogenic response of cultured fibroblasts to serum and other growth promoting agents. J. Cell Physiol. *92*, 233–240 (1977)

CHAPTER 4

The Platelet-Derived Growth Factor

R. Ross

A. Serum, the Platelet-Derived Growth Factor, and Cell Culture

Most diploid cell strains are unable to multiply in cell culture unless a minimal amount of whole blood serum is added to the culture medium. This requirement for serum or its derivatives for growth in culture was recognized quite early. ALEXIS CARREL (1912) noted that a serum-saturated plasma clot could serve as a source of nutrients and also stimulated the growth of embryonic cells in culture. Since those early observations a number of laboratories have examined the role of serum in the culture medium in relation to cell attachment, proliferation and viability.

Numerous attempts have been made to fractionate serum and to determine which of the components present in serum are critical for the successful establishment of different strains or lines of cells in culture (HOLLEY and KIERNAN, 1971; LIPTON et al., 1972; HOLLEY, 1975a). This has been a most difficult problem because serum is an extraordinarily complex mixture of components that have multiple and different effects upon different types of cells in culture. Their interactions and interdependencies remain an important unresolved series of questions. HAM and his collaborators (1965) have made important inroads into the development of defined culture media, and SATO and his colleagues (HAYASHI and SATO, 1976) have developed several completely defined media that will permit the growth of specific non dipoid cell lines in culture. Limited success has thus far been achieved for the growth of diploid cells in completely defined culture media. In general, if diploid cells are to survive and proliferate, relatively small amounts of serum must be added to the culture medium.

By definition, whole blood serum consists of a complex mixture of substances derived from platelets, together with a large variety of low and high molecular weight substances present in the fluid phase of the blood, the plasma. Thrombin is generated during the preparation of whole blood serum as a result of the activation of a cascade of enzymes and cofactors present in plasma. Exposure of the platelets to thrombin results in platelet aggregation and release of material stored in the different cytoplasmic compartments of the platelet. As a result, a large number of substances are released from the different platelet granules. These include low and high molecular weight substances such as serotonin, ADP, ATP, calcium, Platelet Factor 4, beta Thromboglobulin, Fibrinogen, the platelet derived growth factor and several others. These substances can then combine with a number of hormones and nutrients present in the plasma, such as insulin, the somatomedins, transferrin, and other substances, to act upon susceptible cells at sites where platelet release has been stimulated.

Until recently, if one wished to study cell metabolism in a population of quiescent cells in culture, it was necessary to arrest cell growth by markedly reducing the serum concentration (1% or less), and thus expose the cells to a process of relative serum starvation. Under these circumstances, the cells become arrested in a G_0/G_1 state. A number of important observations concerning the cell cycle have resulted from studies of cells inhibited by such serum arrested growth. Since this form of growth arrest resulted from relative serum starvation, BALK (1971) tried a different approach. He noted that chicken fibroblasts were quiescent in culture when the culture medium contained 5% heat inactivated chicken plasma, instead of whole blood serum, in a low calcium containing medium. He later went on to suggest (BALK et al., 1973) that serum might contain a mitogenic factor not present in plasma that might possibly be a "wound" hormone for fibroblasts, that might be derived from the platelets. He was quite correct.

The type of growth regulatory factor present in serum and absent in plasma, represents a newly recognized class of hormones, or growth regulatory factors, that have been shown to be present in serum, but which can also be derived from several other sources such as some lines of sarcoma cells in culture (TODARO et al., 1979), and which have also been found in a number of tissues such as brain and pituitary (ARMELIN, 1973; GOSPODAROWICZ and MORAN, 1976).

The observations by BALK and his colleagues (1971, 1973) were extended and clarified when ROSS et al. (1974) and KOHLER and LIPTON (1974) discovered that the platelet or thrombocyte was the source of the mitogenic activity present in whole blood serum and missing from cell-free, plasma derived serum. This discovery was made in a series of studies in which the mitogenic capacity of whole blood serum was compared with that of cell-free, plasma derived serum, both obtained from the same donor pool of blood. The effects of serum versus plasma were first examined in primate arterial smooth muscle cells (ROSS et al., 1974) and dermal fibroblasts (RUTHERFORD and ROSS, 1976) and in 3T3 Swiss mouse embryo cells (KOHLER and LIPTON, 1974) in culture. Five percent whole blood serum (as it has generally been used in most laboratories) will support typical logarithmic proliferation of these cells in culture, whereas 5% cell free, plasma derived serum from the same donor pool is able to maintain the cells in culture but will not induce their proliferation. The mitogenic capacity of whole blood serum can be restored to plasma derived serum by adding a fraction of material released from the platelets. This fraction is obtained from a purified preparation of platelets, either by exposing the platelets to purified thrombin, or by freeze-thawing. In either case, the supernatent obtained after centrifuging the platelet debris contains all of the activity released from the cells (Fig. 1). These studies were extended by WESTERMARK and his colleagues (WESTERMARK and WASTESON, 1976; BUSCH et al., 1976) who observed that a similar fraction of material derived from platelets also stimulated glial cells in culture.

During this time, ANTONIADES et al. (1975) had partially isolated a growth factor from whole blood serum that was a relatively low molecular weight, cationic protein. Later, in their series of studies, ANTONIADES and SCHER (1977) demonstrated that a partially purified preparation of the mitogenic material derived from platelets, antigenically cross-reacted with the serum growth factor that ANTONIADES et al. (1975) had previously isolated. A similar factor has been isolated

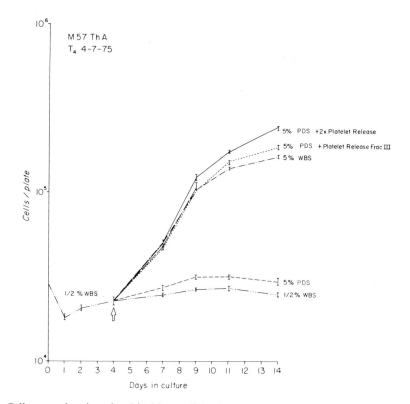

Fig. 1. Cells were plated on day 0 in 35 mm dishes in 0.5% monkey whole blood serum. On day 4, the medium was changed and the various test media were added. Two plates per group were trypsinized and counted each day. Medium was changed every 2 days. The 2 X platelet release is material obtained from thrombin-treated human platelets. Platelet release fraction III is platelet releasate that has been partially purified by carboxymethyl Sephadex chromatography. *WBS*, whole blood serum; *PDS*, plasma-derived serum. (Reproduced with permission of *Cell*)

by CASTOR et al. (1977) that they have termed CTAP (connective tissue activating peptide). These studies provided additional support for the suggestion that the platelet derived growth factor represented the principal source of mitogenic activity in whole blood serum. As will be discussed later in this chapter, this mitogen acts in an interesting and coordinate fashion with other substances and hormones in the plasma to stimulate cells to synthesize DNA and to undergo multiple rounds of cell division.

B. The Platelet

The platelet, or thrombocyte, is an anucleate, disc shaped cell that has for many years been known to play a dominant role in the process of blood coagulation, hemostasis, and thrombosis. Platelets are produced in the bone marrow from a precursor cell, the megakaryocyte, and are the result of a predetermined process of

fragmentation of the cytoplasm of this cell. The megakaryocyte is a multinu-
cleated, large cell that arises from an undifferentiated stem cell, capable of differentiating in a number of pathways. Stem cell differentiation can lead to the formation of erythrocytes, and of the granulocyte cell series as well as the platelet. During
maturation, the megakaryocyte fragments by forming new membranes to provide
approximately 10^3 platelets per megakaryocyte. After fragmentation, the platelets
are released into the circulation. In man they have a lifetime of approximately 8–10
days. The megakaryocyte contains all of the appropriate cytoplasmic machinery
for the synthesis of secretory proteins and the formation of storage granules, including a rough endoplasmic reticulum and Golgi complex. After the mature
thrombocytes are released into the circulation, they contain no rough endoplasmic
reticulum or Golgi complex and few, if any, free ribosomes. The mature platelet
contains at least two populations of granules, the so-called "dense" granules and
the alpha granules, as well as a dense tubular system and aggregates of glycogen.
The cell maintains a discoid shape in the circulation due to the presence of a highly
developed peripheral ring of cytoplasmic microtubules (Fig. 2).

Platelets can be activated to release their granules by exposing them to a number of different substrates. Most commonly, exposure to thrombin, as occurs during serum or clot formation, or exposure to ADP or collagen lead to platelet degranulation or the "platelet release reaction." During this process of activation, the
peripheral microtubular rings disaggregate, the platelet undergoes a marked shape
change and, depending upon the stimulus, release all or part of the material stored
in the two populations of granules. Recently, LINDER et al. (1979) showed that the
platelet derived growth factor and other platelet constituents can be released from
platelets by arachidonic acid via an indomethacin sensitive pathway suggesting
possible approaches to control the release of this growth factor.

The normal level of platelets in the circulation in man is approximately
200,000–300,000 platelets per ml of blood. When platelets are activated in vivo,
they play a complex and critical role in the process of hemostasis. After activation,
they form hemostatic plugs at sites of vessel injury by changing their discoid shape
into a complex convoluted form. At the same time the surface of the platelet becomes "sticky," and they adhere to one another to form large aggregates. Platelets
will also adhere to sites of injury upon exposure to substances such as collagen or
to the surfaces of damaged cells. During platelet activation, they release a membrane phospholipid that accelerates blood coagulation leading to the conversion
of fibrinogen to fibrin. Platelets also appear to play some sort of role in the protection of the endothelium.

C. Platelet-Structure and Function

A number of substances have been localized in the granules of the platelet. The
dense granule contains calcium, serotonin, ADP, and ATP, whereas the alpha
granule contains as least four well characterized proteins. These include platelet
factor 4, beta thromboglobulin, fibrinogen, and the platelet derived growth factor.

The localization of these four constituents in the alpha granule was first suggested in experiments by WITTE et al. (1978) who examined the kinetics of release

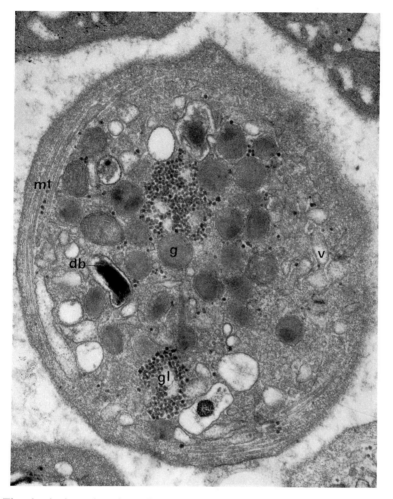

Fig. 2. The platelet is sectioned parallel to the long axis of the disc-shaped cell. The peripheral ring of microtubules (*mt*) and numerous alpha granules (*g*) are visible. Human platelets have fewer dense bodies than other species. One is visible (*db*) in this section. Several elements of the vesicles (*v*) in the platelet are also present in this micrograph, as well as aggregates of glycogen particles (*gl*). Micrograph courtesy of Dr. James G. WHITE. × 33,000

of substances from the two different populations of granules after exposure of the platelets to thrombin. KAPLAN et al. (1979a) and KAPLAN et al. (1979b) induced platelet lysis and studied the contents of each granule fraction by gradiant centrifugation and observed the same distribution of substances in the two classes of granules as had been observed by WITTE et al. (1979).

Observations have recently been obtained from patients with hereditary platelet granule abnormalities that further support the localization of the platelet derived growth factor in the alpha granule of the platelets. WEISS et al. (1977) examined platelets from a patient with a special form of platelet storage pool disease.

This patient had a marked deficiency of alpha granules. When material released from these platelets was studied, it was relatively poor in platelet factor 4, beta thromboglobulin and the platelet derived growth factor. At approximately the same time WHITE (1978) began to study two patients with the Gray Platelet Syndrome.

D. The Gray Platelet Syndrome

The Gray Platelet Syndrome represents a relatively rare inherited disorder in which the platelets are rather large and of nonuniform size. This bestows the name upon the syndrome due to the "Gray" appearance of platelet smears from these patients on microscope slides. Upon examination by electronmicroscopy the platelets appear to be vacuolated and relatively devoid of alpha granules. In a morphometric study, WHITE (1978) examined the fine structure and cytochemistry of platelets from the two patients with this disorder. He observed that their platelets contained normal numbers of mitochondria, dense granules, peroxisomes and lysosomes, but that they specifically lacked alpha granules (Fig. 3).

Examination of these platelets by GERRARD et al. (1980) showed that they contained fewer than 15% of the normal population of alpha granules, whereas the number of dense granules were within normal limits. Furthermore, the Gray platelets had less than 15% of the normal level of platelet factor 4, less than 25% of the normal level of beta thromboglobulin and showed a marked deficiency in thrombin sensitive protein, another protein that is a normal constituent of the alpha granule.

The capacity of the platelets from these patients to provide mitogenic stimulation for cells in culture was essentially absent. A releaseate obtained from platelets of patients with the Gray Platelet Syndrome provided no support for the growth of cells in culture in the presence of cell free plasma derived serum (Fig. 4). In contrast, the levels of lysosomal enzymes, adenine nucleotides, serotonin and catalase, and their capacity to convert arachidonic acid to prostaglandin derivatives were all within normal limits (GERRARD et al., 1980). In addition, functional studies of platelets from patients with the Gray Platelet Syndrome showed marked deficiencies in ADP, thrombin, and collagen stimulated aggregation, suggesting that the alpha granules not only serve as a source of the platelet derived growth factor, but that they also play an important role in normal platelet aggregation.

E. The Megakaryocyte as the Source of Platelet-Derived Growth Factor

Platelets are well known to be capable of taking up low molecular weight substances from the plasma, such as serotonin, and can concentrate these substances in the dense granules. It is important, therefore, to determine whether the platelet derived growth factor is similarly concentrated by the platelet by obtaining it from the plasma, or whether it is formed in the megakaryocyte during its development. It has been suggested that the platelet derived growth factor is similar, if not iden-

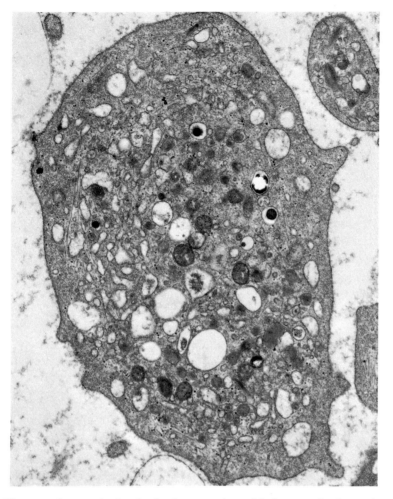

Fig. 3. Electron micrograph of a platelet from a patient with the gray platelet syndrome. α-Granules are missing. A number of vacuoles of varying sizes are apparent together with normal appearing dense granules and mitochondria. Micrograph courtesy of Dr. James G. WHITE. × 30,500

tical to, Fibroblast Growth Factor, and that this pituitary derived hormone is subsequently concentrated in the platelet (GOSPODAROWICZ et al., 1975). This suggestion, however, is incorrect, since the two factors are different chemically.

CHERNOFF et al. (1980) examined lysates of guinea pig platelets and bone marrow megakaryocytes. In a series of quantitative dilution studies, they observed that of all of the cells present in the bone marrow, only the megakaryocyte possessed a quantitatively significant amount of growth factor activity. The amount present in each megakaryocyte was equivalent to that present in 10^3 to 5×10^3 platelets, the approximate number of platelets obtained from a single megakaryocyte. In appropriate control experiments they were unable to obtain any growth factor activ-

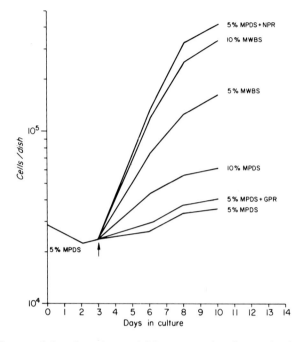

Fig. 4. The influence of the released material from normal and gray platelets on the growth of 3 T 3 cells in culture. At 3 days in culture (*arrow*), monkey whole blood serum (*MWBS*) or monkey plasma-derived serum (*MPDS*), alone or together with the normal platelet release (*NPR*) material or gray platelet release (*GPR*) material, was added

ity from the other populations of cells derived from the marrow. Thus, in their studies, essentially all of the activity was derived from the purified preparations of megakaryocytes. They provided no data for the synthesis of growth factor by megakaryocytes, a necessary observation to prove their postulate; nevertheless, their observations strongly suggest that the platelet derived growth factor has its origin in the megakaryocyte.

F. Purification and Characterization of the Platelet-Derived Growth Factor

Early studies of the platelet derived growth factor (Ross et al., 1974; Antoniades et al., 1975; Westermark and Wasteson, 1976) demonstrated that this material was heat stable (56 °C for 30 min, 100 °C for 10 min) and was a highly cationic (pI, 9.8), trypsin labile, mercaptoethynol sensitive protein.

Most studies have used outdated human platelet rich plasma as a source of the material. Unfortunately, this source contains a great deal of contaminating plasma protein that must be removed during the purification procedure. Nevertheless, progress has been significant. A homogeneous pure protein has been isolated and purification appears to be essentially complete, although there are some differences in the characterization of this protein between the different laboratories that remain to be clarified.

ANTONIADES et al. (1979), and HELDIN et al. (1979) have used different approaches to the purification of the platelet derived growth factor. ANTONIADES et al. (1979), used material derived from lysed outdated human platelets. As a first step in purification, they heated the material to 100 °C for 10 min, centrifuged this material and utilized the supernatent fluid which contained approximately 40% of the starting activity. They subjected this material to a series of chromatographic steps including ion exchange chromatography on carboxymethyl sephadex C-50, followed by gel filtration in Biogel P-150, followed by isoelectric focusing and electrophoresis in 15% polyacrylamide gels containing 2% SDS. In the SDS gels they observed a number of bands in the active fraction. They eluted each band separately and found all of the activity in a band of 35,000–38,000 Dalton molecular weight under non reducing conditions. After disulfide bond reduction, all of the activity was destroyed; the 35,000–38,000 molecular weight band was missing and was replaced by a new band of molecular weight 13,000–16,000 Daltons that contained no activity. The 35,000–38,000 molecular weight material was active at concentrations of approximately 1 ng per ml of culture media.

HELDIN et al. (1977, 1979) have utilized a slightly different approach to the purification of this material. After lysing fresh human platelets, the lysate was taken through a series of chromatographic steps including ion exchange chromatography, followed by hydrophobic chromatography with Cibachrome Blue Sepharose, gel filtration and isoelectric focusing. They radioiodinated the active fraction and found that the electrophoretic behavior of the biologically active fraction occurred in a molecular weight range of approximately 28,000 Daltons. After reduction, the 28,000 Dalton material disappeared and was replaced by two different polypeptides. One of these was 13,000–14,000 Daltons and the other 16,000–17,000 Daltons. The 28,000 Dalton material was active at a level at which approximately 4 ng per ml of culture medium was equivalent to 1% human serum in terms of its mitogenic capacity.

A somewhat different approach has been taken by Ross et al. (1979). They also start with a lysate from outdated human platelet-rich plasma that is subjected to a series of 7 chromatographic steps. These include carboxymethyl sephadex, gel filtration in Biogel P-150, followed by chromatography with DEAE-sephadex. This latter step removes many contaminating acidic proteins. They further purified the active fraction with hydrophobic chromatography using phenyl sepharose and eluted the activity from the hydrophobic gel with guanidine hydrochloride. The active fraction is then filtered on Biogel P-60 to remove the guanidine and is further purified in an activated I-125 silica gel, high pressure liquid chromatography system to obtain a homogenous protein. This material is seen to consist of a species of proteins of molecular weight 28,000–31,000 Daltons when placed on an analytical SDS-PAGE system. Upon reduction of disulfide bonds, this material loses all of its activity, and the 28,000–31,000 bands disappear and are replaced by a band of approximately 13,000 Daltons (Fig. 5; Table 1). This procedure is presently being simplified to reduce the number of steps and to improve the yields.

All of these studies demonstrate that the platelet derived growth factor is a highly cationic, heat stable, disulfide bonded protein of relatively low molecular

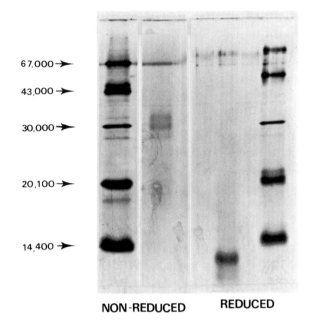

$67,000 \rightarrow$

$43,000 \rightarrow$

$30,000 \rightarrow$

$20,100 \rightarrow$

$14,400 \rightarrow$

NON-REDUCED   REDUCED

Fig. 5. SDS-PAGE of platelet-derived growth factor obtained from the active fraction eluted after high pressure liquid chromatography on a Waters ^{125}I-silica gel system. Before reduction the active material clusters in three bands of molecular weight 28,000–31,000. After reduction the material is inactive and of molecular weight approximately 13,000

weight that is present in relatively small quantities in each platelet, but which is sufficiently potent so that with its hormone-like characteristics only small quantities need to be released to stimulate a large number of cells to synthesize DNA.

G. The Spectrum of Cell Response

A large number of diploid cells are able to respond to the platelet derived growth factor. Included in the list of responsive cells are dermal fibroblasts, arterial adventitial fibroblasts, glial cells, smooth muscle cells, and mouse embryo 3 T 3 cells. At least two classes of diploid cells do not respond to this mitogen. These include diploid arterial endothelial cells (DAVIES and ROSS, 1978; WALL et al., 1978; and THORGEIRSSON and ROBERTSON, 1978), and a relatively large number of clonal lines of virally transformed cells, transformed by SV 40 or retroviruses. These showed similar growth in the presence or absence of the platelet factor (SCHER et al., 1979). In contrast, their nontransformed parent cell lines grow to higher densities in the presence of the platelet derived growth factor, suggesting that the factor is either not necessary or of relatively little importance in the proliferation of the virally transformed cell lines. A loss of the requirement for the platelet derived growth factor may be a characteristic of most transformed cell lines, as shown by VOGEL and POLLACK (1974).

Table 1. Growth promoting activity assayed on Balb/C3T3 cells grown to confluence in either 35 mm plates or 16 mm wells (4–50^4 cells/cm^2)[a]

	Fraction	Specific activity µg/ml medium	Growth promoting		
			Units of activity per mg	% Recovery of starting activity	
				Total	Recovery at individual steps
Standard conditi	5% calf serum	3,250	1.0		
	Human platelet rich plasma frozen/thawed, defibrinogenated	1,075	3.0		
	CM – Sephadex fraction III 0.1 M (NH$_4$)$_2$ CO$_3$, pH 8.9	3.0	1,083	42	42
	Biogel P-150 in 1 N ACOH fraction II MW 10,000–40,000	0.79	4,167	30	71
	DEAE-Sephadex fraction I flow through in 0.05 M Tris, pH 8.9	0.52	6,250	19	64
	Phenyl-Sepharose fraction III 0.05 M Tris, 3 M GuHCl, pH 8.9	0.15	21,667	15	79
	Biogel P-60 in 1 N ACOH 2 fractions	0.05	65,000	12	77
	HPLC Waters I-125 in 1.0 N ACOH – 2 passes	0.008	406,250	2.4	20

[a] All values normalized to growth response to additional 5% calf serum. Protein concentrations were estimated from A$_{280}$ or folins or in the last step. Values represent an average of 2–5 experiments Ag-stained gels

SCHER et al. (1979) examined a number of human and mouse cell lines for their tumorigenic potential by injecting them into athymic nude mice. They found that each of the cell lines that formed tumors in nude mice was capable of growth in culture medium containing plasma and lacking the platelet derived growth factor. In contrast, when a cell line grew poorly in plasma supplemented medium, it also failed to grow tumors after they were innoculated into nude mice. Therefore they concluded that a reduction in the requirement for the platelet derived growth factor may be a necessary although not a sufficient step for tumorigenecity in vivo.

A number of diploid cell types have not yet been examined in terms of their response to the platelet derived growth factor in culture. Cells like arterial endothelium that grow in a uniform monolayer and have already been shown to be independent of this factor may be truly contact inhibited in culture. However, it is also possible that they may be unresponsive due to their incapacity to interact with the mitogen. This is an important question that remains to be resolved.

H. Control of Cell Proliferation by Platelet-Derived Growth Factor and Plasma

Fibroblasts and/or smooth muscle cells will remain quiescent for relatively long periods of time in the presence of 5% cell free plasma derived serum in the culture medium (RUTHERFORD and ROSS, 1976; ROSS et al., 1978). Exposure to the platelet derived growth factor for only one hour results in the commitment of that portion of the cells that are susceptible during the period of exposure, to DNA synthesis and subsequent cell doubling. Thus, only brief exposure to the platelet derived growth factor appears to be required for cell doubling to occur, as long as plasma constituents are present in the culture medium.

VOGEL et al. (1978) demonstrated that if 3T3 cells had been previously maintained in medium containing plasma and were then placed in serum free medium, exposure to semipurified platelet derived growth factor committed approximately 40% of the cells to synthesize DNA and go through one cell doubling (Fig. 6). Multiple cell doublings required the presence of plasma in the culture medium (Fig. 7).

PLEDGER et al. (1977) and STILES et al. (1979) suggest that exposure to the platelet derived growth factor induces cells such as 3T3 cells to become "competent" to synthesize DNA. In their studies the cells remain fixed in the G_0/G_1 phase of the cell cycle in the absence of plasma. Addition of plasma to cells previously exposed to the factor permitted the "competent" cells to traverse from the G_0/G_1 phase of the cell cycle to S. They suggest that this stimulus to traverse the G_1 portion of the cell cycle and to enter into DNA synthesis requires nutrients and hormones in the plasma such as amino acids, and that the nutrient independent, plate-

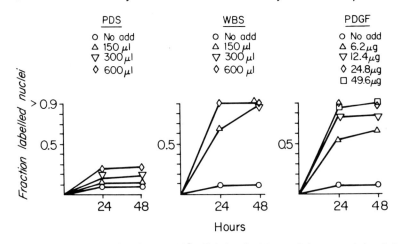

Fig. 6. 3T3 cells were plated at 2×10^5 cells/plate in 35-mm dishes containing 1.5 ml 5% calf serum supplemented medium. Three days later, when the cells had become quiescent, platelet derived growth factor, human PDS or whole blood serum was added. ^{3}H Tdr (2.5 μCi/ml, 6.17 Ci/mm) was added at the time of addition of growth factors. Plates were processed for autoradiography at the indicated times. Platelet factor was partially purified by carboxymethyl Sephadex chromatography

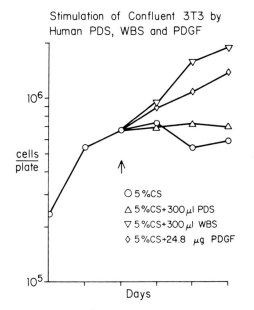

Stimulation of Confluent 3T3 by
Human PDS, WBS and PDGF

cells / plate

10^6

10^5

Days

○ 5%CS
△ 5%CS+300 µl PDS
▽ 5%CS+300 µl WBS
◇ 5%CS+24.8 µg PDGF

Fig. 7. 3 T 3 cells were plated at 2×10^5 cells/plate in 35-mm dishes containing 1.5 ml 5% calf serum supplemented medium. Three days later, when the cells had become quiescent, platelet derived growth factor, human PDS or whole blood serum was added. Plates were processed for determination of cell number at the indicated times. Platelet factor was partially purified by carboxy-methyl Sephadex chromatography

let derived growth factor modulated, growth regulatory event which they have termed "competence," is located 12 h prior to the G_1/S boundary in quiescent, density arrested 3 T 3 cells. On the other hand, the plasma modulated event that appears to be nutrient dependent is located 6 h prior to entry into DNA synthesis and corresponds in their studies, to a plasma dependent arrest point of growth. Clearly, components of plasma are required for the optimal response of susceptible cells to the platelet derived growth factor although the nature of all of these events remains to be clarified (PLEDGER et al., 1977; VOGEL et al., 1978).

Each of the two components of whole blood serum appear to regulate different events in the cell cycle. VOGEL et al. (1980) have further examined the role of the platelet derived growth factor in density dependent inhibition of growth of cells in culture such as 3 T 3 cells. They observed, similar to the studies of PLEDGER et al. (1977), that both plasma and the platelet derived growth factor are not only necessary for an optimal response of cells, such as 3 T 3, for growth stimulation in culture, but in addition, VOGEL et al. (1980) demonstrated that the limiting entity in whole blood serum that determines the density at which cells become density inhibited in culture is the amount of platelet derived growth factor per cell. They arrived at this conclusion as a result of several approaches. In one series of studies, they observed that when a population of 3 T 3 cells becomes density arrested and thus quiescent in 5% whole blood serum, that additional amounts of plasma do not permit the cells to grow to a higher density whereas addition of semipurified platelet derived growth factor stimulates the cells to grow to a newer higher density

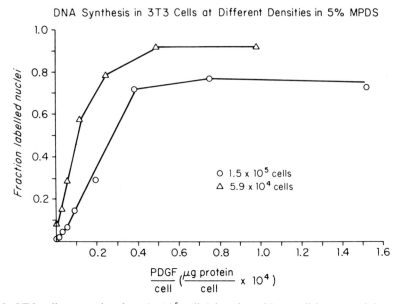

Fig. 8. 3T3 cells were plated at 4×10^5 cells/plate into 35-mm dishes containing 1.5 ml growth medium supplemented with 5% monkey plasma derived serum. Two days later, partially purified platelet factor was added along with 2.5 µCi/ml ^{3}H-Tdr. Cells were processed for autoradiography 28 h later. The platelet factor was partially purified by CM Sephadex chromatography followed by chromatography on Biogel P-100

(Fig. 7). If the cells were plated at different densities, the greater the density, the greater the amount of platelet derived growth factor required per cell, for the same number of cells to synthesize DNA (Fig. 8). The interactions are complex; however, since, as suggested by HOLLEY (1975b) earlier, the density of the cells determines the response of the same numbers of cells to a given amount of platelet derived growth factor per cell (Fig. 9).

One of the ways in which cell density may be important may reflect the amount of exposed cell surface, and thus of putative receptors for growth factors at the surface, that can bind to the hormone and stimulate DNA synthesis. It has been hypothesized that the platelet derived growth factor binds to a high affinity receptor, similar to the one demonstrated for another well characterized growth factor, Epidermal Growth Factor. Cell density, together with the amount of growth factor in the environment, control the growth response, probably by permitting access of the cells to the factor, as well as determining the responsivity of each individual cell. These responses appear to be based upon the localization of growth factor receptors, their ability to bind to the factor and to initiate the sequence of intracellular events that ultimately lead to DNA synthesis and cell division.

I. The Role of Plasma

Plasma is a physiological fluid to which most quiescent cells in the body are normally exposed. Thus, teleologically, it is not surprising that most diploid cells are

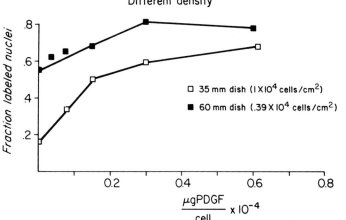

Fig. 9. 3 T 3 cells were plated at 6×10^4 cells in 35-mm or 60-mm dishes containing 1.5 ml growth medium supplemented with 5% monkey plasma derived serum. Two days later, partially purified platelet factor (BG-2) and ^{3}H-Tdr were added and the plates processes for autoradiography 24 h later

quiescent in the presence of 5% or 10% cell free, plasma derived serum in the culture medium. In fact, Ross et al. (1978) have shown that it is possible to maintain cells such as 3 T 3 cells or arterial smooth muscle cells in a quiescent state for up to 8 weeks in culture in medium containing 5% plasma derived serum. Exposure to whole blood serum or to the platelet derived growth factor after 5 or 6 weeks in culture stimulates such quiescent cells into logarithmic growth (Fig. 10). As a result of their studies, they suggested that diploid cells maintained in 5% plasma derived serum might provide a more physiological base for the study of cell quiescence and for the responsivity of cells to various growth factors. In contrast, quiescent cells maintained in low concentrations of serum (1% or less) probably undergo protein starvation. However, in either case, flow microfluorometric studies show that in 1% serum or 5% plasma, the cells appear to be arrested in the G_0/G_1 phase of the cell cycle (Ross et al., 1978; Yen and Riddle, 1979).

The role of plasma in providing support for the platelet derived growth factor in cell culture has also been examined by Stiles et al. (1979). They were particularly interested in the role of the somatomedins and insulin in regulating the growth of fibroblasts in culture. They observed that plasma from hypophysectomized rats was 20-fold less efficient in permitting cells exposed to the platelet derived growth factor to synthesize DNA than was plasma from normal rats. They were able to restore the activity of the hypophysectomized plasma by adding physiological concentrations of pure Somatomedin C. Somatomedin C alone was unable to promote DNA synthesis, suggesting that other materials in the hypophysectomized plasma were also essential for cell growth. However, they interpreted their experiments to suggest that Somatomedin C was a critical component of plasma that was required for platelet factor treated cells to progress into S. In contrast, insulin only in phar-

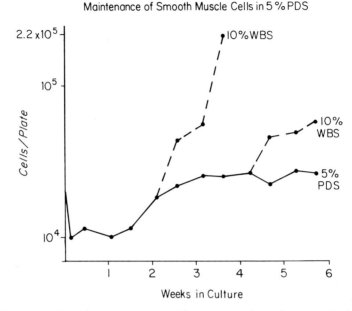

MI26THA

T_2, 4/4/78

Maintenance of Smooth Muscle Cells in 5 % PDS

Fig. 10. Demonstration of quiescence in cell-free, plasma-derived serum. Nonhuman primate arterial smooth muscle cells were plated in DME containing 5% homologous plasma derived serum (PDS). 10% homologous whole blood serum was added to sets of these cultures after 2 and 4 weeks in culture. The cells undergo less than one doubling in PDS over a 6-week interval and respond to WBS at each time of addition

macologic doses provided a stimulus for cells to progress into S, but was more potent than many other hormones they tested such as hydrocortisone or growth hormone.

YEN and RIDDLE (1979) examined the quiescent state of cells in plasma derived serum, lacking the platelet derived growth factor, using flow microfluorimetry. They found that cells were only sensitive to deprivation of the platelet derived growth factor at the beginning of G_1. By depriving the cells of plasma for a prolonged period of time, they became progressively unstable and underwent necrosis. In addition to being required by cycling cells, factors present in plasma are also required to maintain G_1 cells that had been previously arrested in a stable quiescent state by deprivation of the platelet derived growth factor. The observations of YEN and RIDDLE (1979) suggested that two distinct processes were involved in the arrest of cells in culture. The first of these was arrest of proliferation in G_1 by deprivation of the platelet derived growth factor, and the second was a stabilization of the arrested cells in a viable quiescent state by plasma components. Further examination of this process (SCHER et al., 1979) suggested not only that the platelet derived growth factor was important for initiation of DNA synthesis, but that exposure to the factor also prevents replicating cells from entering the G_0/G_1 phase of the cell cycle. Their data suggests that the platelet derived growth factor acts during the late stages of the cell cycle, shortly after mitosis, to prevent cells from returning to G_0 and thus permits them to go into another cycle of division that is stimulated by continuing exposure to the platelet derived growth factor.

K. Endocytosis and the Platelet-Derived Growth Factor

DAVIES and ROSS (1978) examined the effects of the platelet derived growth factor on pinocytosis in culture by monkey arterial smooth muscle cells, bovine aortic endothelial cells, and Swiss 3T3 cells. Markers of endocytosis that they studied included ^{14}C-sucrose and horseradish peroxidase. Arterial smooth muscle cells and 3T3 cells maintained in a quiescent state in 5% cell free, plasma derived serum containing medium demonstrated relatively low levels of pinocytosis. These cells were stimulated to markedly increased rates of endocytosis by exposure to the platelet derived growth factor. This enhancement of the rate of pinocytosis occurred within 4 h after exposure to the growth factor and preceeded the onset of DNA synthesis by 8–10 h.

In contrast, endothelial cells proliferate equally well in the presence or absence of the platelet derived growth factor and do not become quiescent at subconfluent densities in medium containing 5%–20% plasma derived serum. As a consequence, endothelial cells maintain their rate of proliferation and of pinocytosis regardless of whether they are in plasma derived or whole blood serum containing medium. Hence, although endothelial cells do not require the platelet derived growth factor for stimulation of DNA synthesis in culture, if they are actively proliferating they will have an increased rate (2- to 3-fold) of pinocytosis compared to quiescent cells. Enhanced pinocytosis appears to be an early response of cells that are stimulated to the synthesis of DNA that precedes and accompanies such stimulation.

L. Modulation of Receptors for Epidermal Growth Factor by Platelet-Derived Growth Factor

The most extensively studied growth factor that has been studied in many laboratories is Epidermal Growth Factor (EGF) (CARPENTER and COHEN, 1979). This hormone initiates its action by binding to specific receptors on the surface of susceptible cells (CARPENTER and COHEN, 1976; DAS et al., 1977). After EGF is bound to its cell surface receptor, it is internalized by the process of endocytosis and is subsequently degraded in lysosomes. This process of internalization and degradation of epidermal growth factor is associated with a process that has been termed down regulation, meaning a decrease in the binding capacity of the cell surface for EGF. Such internalization of hormone-receptor complexes appears to be a general response to many peptide hormones. It is assumed, although there is little evidence to date, that platelet derived growth factor acts in a similar fashion.

WRANN et al. (1980) have examined the effects of platelet derived growth factor and epidermal growth factor on the receptors for EGF. When they added semipurified platelet factor together with EGF, the platelet factor had no effect on the capacity of [125]-I-EGF to bind to the cell surface. However, if the platelet factor was added two hours before exposing the cells to EGF, there was a transient loss in EGF binding activity which was restored to normal levels within 6 h. Such transient down regulation is different from the down regulation induced in cells by prior exposure to EGF, which remains decreased as long as the cells are exposed to EGF.

In addition to inducing a transient down regulation of EGF binding and increased DNA synthesis 24 h later, the platelet derived growth factor can also act to inhibit down regulation of EGF receptors by EGF itself. This can be demonstrated by preincubating 3 T 3 cells with the platelet derived growth factor and after one additional hour to unlabeled EGF. After an additional 6 h, a sufficient period of time for the cells to recover from platelet factor induced down regulation of EGF receptors, they washed the cells to remove any unassociated platelet factor. They then exposed the cells to ^{125}I-EGF and found decreased ability of EGF to bind to the cells. In earlier experiments they demonstrated that there was no competition by the platelet factor for binding to EGF receptors at 0 °C; hence they interpret these observations to suggest that receptors for the platelet factor and EGF are clustered in similar loci on the cell surface. If this proves to be the case, then exposure to one growth factor could lead to internalization of receptors for both growth factors and make the cell refractory to binding to either factor. There are other possible explanations for their data; however, it is clear that responsiveness to one of these hormones may be modulated by prior exposure to the other, leading to increasing complexity and diversity in growth control by these various hormones.

M. Lipid Metabolism and the Platelet-Derived Growth Factor

The stimulation of pinocytosis by the platelet derived growth factor has also been examined by HABENICHT et al. (1980), and by CHAIT et al. (1980). HABENICHT et al. (1980) confirmed the stimulation of pinocytosis by the platelet derived factor observed by DAVIES and ROSS (1978). They went on to show that when both monkey arterial smooth muscle cells and Swiss 3 T 3 cells are exposed in culture to the platelet derived growth factor in the absence of a source of lipoproteins in the culture medium, they increase their rate of pinocytosis and are stimulated to synthesize cholesterol de novo. HABENICHT et al. (1980) maintained smooth muscle cells in a quiescent state at low density in culture medium lacking both the platelet factor and lipoproteins. Addition of platelet derived growth factor to such quiescent cultures was followed by a sharp increase in the biosynthesis of cholesterol within 3–6 h (Fig. 11). These cells will go on to multiple doublings under these conditions (Fig. 12). They studied inhibition of cholesterol synthesis in the platelet factor stimulated cultures by simultaneous exposure of the cells to a specific inhibitor of HMG-CoA reductase, 25-hydroxycholesterol. Exposure to 25-hydroxycholesterol resulted in a decrease in cholesterol synthesis to 15% of that in quiescent control cultures, but had no effect on DNA synthesis or cell division. If however, they added a different inhibitor of cholesterol synthesis, compactin (ML-236 B), to the culture medium, then high doses (15 μM) decreased cholesterol synthesis to less than

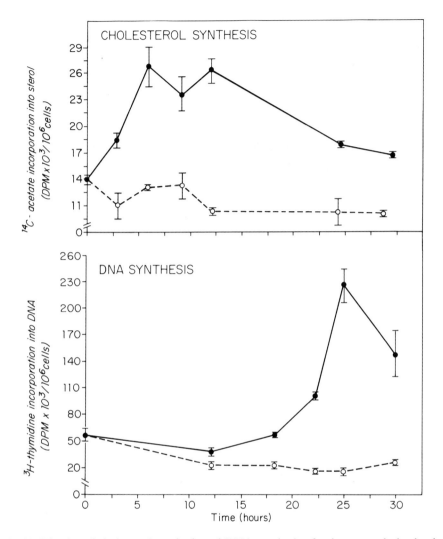

Fig. 11. Kinetics of cholesterol synthesis and DNA synthesis of quiescent and platelet derived growth factor (PDGF)-stimulated smooth muscle cells. Cells were plated at a density of 1×10^5/35-mm plate, maintained and fed every other day with medium containing 5% platelet deficient, lipoprotein deficient serum (PDLDS). After day 5 the dishes were separated into two groups and 8.4 µg PDGF/ml dissolved in 100 µl phosphate buffered saline (PBS) (●) or 100 µl PBS (○) was added to the cultures. Before each data point indicated in the upper part of the figure, cells were incubated for 3 h with 25 µCi/ml (2-^{14}C)-acetate dissolved in 50 µl PBS and thereafter analyzed for content of (^{14}C)-cholesterol. Before each data point indicated in the lower part of the figure, additional cells were incubated for 3 h with 2 µCi(methyl-^{3}H)thymidine/ml dissolved in 100 µl PBS and thereafter analyzed for radioactivity in DNA. Each point represents the mean of three determinations ± 1 S.D. (Reproduced with permission of the Journal of Biological Chemistry)

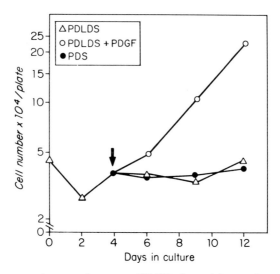

Fig. 12. Maintenance and responsiveness to PDGF of arterial smooth muscle cells in medium containing 5% PDLDS. The cells were seeded in 35 mm Falcon plastic dishes containing Dulbecco's modification of Eagle's minimal medium supplemented with 5% PDLDS (protein concentration 2.5 mg/ml, cholesterol concentration <0.2 µg/ml) and fed every 2 days. After day 4 (*arrow*) the dishes were separated into three groups and refed with 5% PDLDS (△), 5% PDS (●), or 5% PDLDS + 8.4 µg PDGF/ml (○). Each number represents the mean of triplicate counts. (Reproduced with permission of the Journal of Biological Chemistry)

5% of quiescent control cultures and inhibited DNA synthesis as well. Both 25-hydroxycholesterol and compactin inhibit cholesterol synthesis by inhibiting the rate limiting enzyme of this process, HMGCoA Reductase.

If low density lipoproteins, cholesterol or mevalonic acid are added to the culture at the same time as the inhibitor, then DNA synthesis is not inhibited in the presence of the growth factor by lower amounts of compactin (0.25–2.5 µM). However, the inhibition of DNA synthesis observed after high doses (15 µM) of compactin could be prevented, but only by the addition of mevalonic acid (Fig. 13). Mevalonic acid is one of the first intermediate products in the biosynthesis of cholesterol by HMGCoA Reductase. In addition to cholesterol formation, mevalonic acid can lead to the production of substances such as dolichols, ubiquinones, isopentanyl RNA, or other RNA or DNA precursors. Consequently, the addition of mevalonic acid to the cells acts to bypass the inhibitory effects of compactin (Brown et al., 1978). Mevalonic acid is also required for survival of the smooth muscle cells in culture at high concentrations of compactin, since they will undergo necrosis in its absence, even in medium containing 5% monkey serum supplemented with low density lipoproteins. Thus even though cholesterol accumulates in the early phase of the cell cycle, it does not appear to be required for traverse of the cells through the cycle and for DNA synthesis. Rather a different product of mevalonic acid metabolism appears to play an indispensible role in cell division, demonstrating for the first time that DNA synthesis and cholesterol metabolism are separable events in this process, both of which are stimulated by exposure to growth factors such as the platelet derived growth factor.

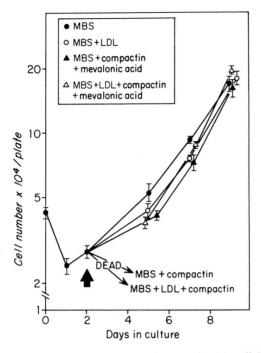

Fig. 13. Growth dependency of smooth muscle cells grown in 5% undialyzed monkey blood serum upon mevalonic acid in the presence of 15 μ*M* compactin. Cells were seeded at a density of 22×10^3/35-mm plate in medium containing 0.5% monkey blood serum. After day 2 (*arrow*) the dishes were separated into six groups and refed every other day with 5% monkey blood serum (●), 5% monkey blood serum plus monkey LDL ($d=1.019–1.063$ g/ml, 15 μg LDL-cholesterol/ml) (○), 5% monkey blood serum plus 15 μ*M* compactin (dead 72 h after addition), 5% monkey blood serum plus monkey LDL plus 15 μ*M* compactin (dead 72 h after addition), 5% monkey blood serum plus 15 μ*M* compactin plus 20 mm mevalonic acid lactone (▲), and 5% monkey blood serum plus monkey LDL plus 15 μ*M* compactin plus 20 mm mevalonic acid lactone (△). Each number represents the mean of two counts ±1 S.D. (Reproduced with permission of the Journal of Biological Chemistry)

In a somewhat related series of studies, CHAIT et al. (1980) observed that the platelet derived growth factor stimulated low density lipoprotein binding and degradation in cultured monkey arterial smooth muscle cells (Fig. 14). Their studies demonstrated that the platelet derived growth factor appears to act by increasing the number of available low density lipoprotein receptors when the factor is added to cells quiescent in serum free medium. Addition of the factor stimulated increased cholesterol synthesis within 2–4 h and prior to lipoprotein binding which occurred within 4–8 h, suggesting that the platelet derived growth factor can serve to provide an increase in the exogenous supply of cholesterol to cells for use during cell proliferation, in addition to de novo stimulation of cholesterol synthesis by these cells in culture.

Another component of lipid metabolism has also been shown to be stimulated by exposure of 3T3 cells to the platelet derived growth factor. SHIER (1979), observed that the platelet derived growth factor stimulates 3T3 cells to synthesize

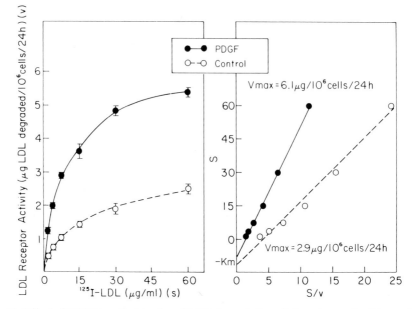

Fig. 14. Effect of PDGF on the kinetics of LDL degradation. Cells quiescent in 5% monkey PDS were changed to serum-free medium. Twenty-four hours later PDGF (●—●) or platelet factor IV (control; o--o) was added. After a further 24 h, (^{125}I)-LDL (specific activity = 25 cpm/ng) was added at the concentrations indicated for determination of LDL degradation during the ensuing 24 h. **A** (^{125}I)-LDL saturation curves, **B** linearization plot (24) of the data in **A**. The slope of the line = apparent "V_{max}", while the point of intersection with the y axis–"-K_m". Mean ± SD of quadruplicates are shown. (Reproduced with permission of the Proceedings of the National Academy of Science)

prostaglandins and to markedly increase the level of phospholipase A-2 activity. Such stimulated cells release up to $^1/_3$ more biosynthetically incorporated arachidonic acid as a hydrolysis product of increased phospholipase activity. The role of increased phospholipase and prostaglandin synthesis stimulated by the platelet derived growth factor in relation to stimulation of cell proliferation in culture remains to be clarified.

N. Platelets and Cell Proliferation In Vivo

A great deal of evidence has accumulated to support the hypothesis that intact platelet function is required for the in vivo development of experimentally induced smooth muscle proliferative lesions of atherosclerosis (ROSS and GLOMSET, 1976). It has also been suggested that platelets play an important role in initiating the early fibroproliferative response that occurs in healing wounds. The data presented in this chapter demonstrate that a hormone-like molecule, the platelet derived growth factor, plays a critical role as one of the principal mitogens in whole blood serum responsible for initiation of DNA synthesis and cell proliferation in culture. These observations, coupled with the in vivo demonstration that platelet function is necessary for the cell proliferation associated with tissue injury suggests that the platelet derived growth factor may be responsible for this phenomenon. However, there is no data available to determine whether the platelet derived growth factor is capable of reaching sites of cell proliferation in vivo and whether this factor is in fact responsible for the fibroproliferative response associated with healing wounds and with atherogenesis. One study has suggested that this may be the case. GOLDBERG et al. (1980) have shown that after removing the endothelium of the aorta of rats with an intraarterial balloon catheter, Platelet Factor 4 is found in the intima and media of the injured artery within 10 min. They demonstrated this by immunofluorescence with an antibody against Platelet Factor 4. Interestingly, they were unable to demonstrate this factor in the tissue beyond a period of 240 min following injury. Since this factor is present in the same granule as the platelet derived growth factor, this provides indirect evidence that the growth factor may also have access to the artery wall after endothelial injury.

Serum, by definition, contains factors derived from platelets and from the coagulation cascade of enzymes. Cells in vivo would only be exposed to the equivalent of serum or to serum constitutents under circumstances that involve injury and blood coagulation. Most cell populations in adults are quiescent in vivo with the exception of those few populations of cells that undergo constant turnover such as bone marrow, skin, epithelia lining various surfaces and a few others. In most instances, cell turnover in the adult is at a relatively low level. Quiescent cells in the adult are generally exposed to the equivalent of plasma or to a filtrate of plasma and would only be exposed to the equivalent of serum during injury. Consequently, to expose such cells to whole blood serum in culture is analogous to exposing them to a fluid milieu that would be present in the response to injury in vivo. This response usually results in connective tissue cell proliferation and scar formation. The quiescence normally observed in mature adult cells in vivo is probably more analogous to the quiescence that can be obtained with such cells in culture when they are maintained in the presence of plasma rather than low concentrations of whole blood serum.

The data suggesting that platelets play a critical role in the initiation of lesions of experimentally induced atherosclerosis is derived from studies in three different animal species. HARKER et al. (1976) induced atherosclerosis in the baboon by intravenously infusing homocysteine so that the level of homocystinemia achieved in these animals was equivalent to that found in individuals with the genetic disease homocystinuria. Such individuals have long been known to have advanced arteriosclerosis at an early age and often have associated infarcts. Observation of the homocystinemic baboons demonstrated a marked decrease in platelet survival (or increased platelet utilization) associated with injury to approximately 10% of the endothelium lining their thoracic and abdominal aortae. When platelet function was interfered with using pharmacologic agents such as dipyridamole (a pyridine nucleotide analogue that inhibits platelet adherence and aggregation and, thus, release) or sulfinpyrazone (a uricosuric agent that partially inhibits platelet adherence and prevents platelet vessel wall interactions by somehow protecting the endothelium), platelet survival was restored to normal levels. After use of either agent, no proliferative lesions of atherosclerosis were found in the homocystinemic baboons. This study was one of the first to suggest that inhibition of platelet function interfered with the intimal smooth muscle proliferative response of atherosclerosis.

MOORE et al. (1976), and FRIEDMAN et al. (1977) induced intimal smooth muscle proliferative lesions, similar to those of atherosclerosis, in rabbits with an indwelling arterial catheter or with a balloon catheter by injuring or removing the endothelium from a given segment of the aorta (BAUMGARTNER and STUDER, 1966; STEMERMAN and ROSS, 1972). When the rabbits were given sufficient antiplatelet serum to induce a thrombocytopenia, in the absence of circulating platelets, they did not respond to these forms of injury with a smooth muscle proliferative response.

FUSTER and his colleagues (1978) examined the role of the platelets in atherogenesis in swine with von Willebrand's disease fed a high fat, high cholesterol diet. In von Willebrand's disease, factor VIII antigen, a factor normally required for platelet adherence and release, is missing from the plasma. These genetically altered swine provide an opportunity to study the role of platelet function in a situation in which atherosclerosis can be induced in matched normal versus genetically altered swine whose platelets will not adhere and undergo the release reaction. Very few proliferative lesions of atherosclerosis were found in the hyperlipemic swine with von Willebrand's disease whereas the normal swine that were made hyperlipemic had extensive proliferative lesions of atherosclerosis.

All of these approaches suggest that functional platelet interactions at sites of injury are critical to the initiation of the proliferative lesions of atherosclerosis. The role that platelets may play in lesion progression or in lesion regression is not yet clear. As noted above, however, data is not yet available to determine whether the role the platelets play in vivo is to release the platelet derived growth factor so that it may act at sites of endothelial injury. This important question must be answered before definitive approaches toward antiplatelet therapy or therapy to inhibit the action of the growth factor itself can be contemplated.

In addition to the data reviewed in this chapter, two reviews have appeared concerning the platelet derived growth factor (ROSS and VOGEL, 1978; SCHER et al., 1979).

O. Summary

Platelets contain a very potent mitogen, the platelet derived growth factor, that appears to be formed in the megakaryocyte and is stored in the alpha granules of the platelet. This growth factor is of relatively low molecular weight (approximately 30,000 Daltons), and is a highly stable, cationic (pI, 9.8) disulfide bonded protein. The factor stimulates cell cycle traverse of sensitive cells into S and appears to act in combination with components in the plasma to stimulate multiple cell doublings. Cells such as fibroblasts, smooth muscle, glial and 3T3 cells are highly sensitive to the factor, whereas arterial endothelial cells and many virally transformed cells do not require the factor to grow in culture. Exposure of susceptible cells to the platelet derived growth factor leads to DNA synthesis within 12–16 h. Three to six hours after exposure to this hormone, cells are markedly stimulated to synthesize cholesterol de novo, to bind increased amounts of low density lipoprotein by forming increased numbers of high affinity LDL receptors; to 3- to 4-fold increased rates of endocytosis; and to marked increases in the synthesis of proteins and glycosaminoglycans. Consequently, exposure to nanogram quantities of this hormone has early multiple complex effects upon cells and their metabolism.

Studies in whole animals clearly demonstrate that platelets play a role in experimentally induced proliferative lesions of atherosclerosis. A specific role for the platelet derived growth factor in this process remains to be established. Clearly much fruitful research on this growth factor as it relates to growth of cells in culture and response of cells and tissue to injury in vivo remains to be done.

This research is supported by grants HL-18645, AM-13970, and RR-00166 from the U.S. Public Health Service.

References

Antoniades, H.N., Scher, C.D.: Radioimmunoassay of a human serum growth factor for Balb/c-3 T 3 cells: derivation from platelets. Proc. Natl. Acad. Sci. U.S.A. *74*, 1973–1977 (1977)

Antoniades, H.N., Stathakos, D., Scher, C.D.: Isolation of a cationic polypeptide from human serum that stimulates proliferation of 3 T 3 cells. Proc. Natl. Acad. Sci. U.S.A. *72*, 2635–2639 (1975)

Antoniades, H.N., Scher, C.D., Stiles, C.D.: Purification of the human platelet-derived growth factor. Proc. Natl. Acad. Sci. U.S.A. *76*, 1809–1813 (1979)

Armelin, H.A.: Pituitary extract and steroid hormones in control of 3 T 3 cell growth. Proc. Natl. Acad. Sci. U.S.A. *70*, 2702–2706 (1973)

Balk, S.D.: Calcium as a regulator of the proliferation of normal but not of transformed, chicken fibroblasts in plasma-containing medium. Proc. Natl. Acad. Sci. U.S.A. *68*, 271–275 (1971)

Balk, S.D., Whittfield, J.F., Youdale, T., Braun, A.C.: Roles of calcium, serum, plasma, and folic acid in the control of proliferation of normal and Rous sarcoma virus-infected chicken fibroblasts. Proc. Natl. Acad. Sci. U.S.A. *70*, 675–679 (1973)

Baumgartner, H.R., Studer, A.: Folgen des GafaSkatheterismus am normo- und hypercholesterinaemischen Kaninchen. Path. Microbiol. *29*, 393–405 (1966)

Brown, M.S., Faust, J.R., Goldstein, J.L., Kaneko, I., Endo, A.: Induction of 3-Hydrox-3-methyl-glutaryl coenzyme A reductase activity in human fibroblasts incubated with compactin (ML-236 B), a competitive inhibitor of the reductase. J. Biol. Chem. *253*, 1121–1128 (1978)

Busch, E., Wasteson, A., Westermark, B.: Release of a cell growth promoting factor from human platelets. Thromb. Res. *8*, 493–500 (1976)

Carpenter, G., Cohen, S.: Epidermal growth factor. Ann. Rev. Biochem *48*, (1979)

Carpenter, G., Cohen, S.: ^{125}I-labelled human epidermal growth factor. Binding, Internalization and Degradation in human fibroblasts. J. Cell Biol. *71*, 159–170 (1976)

Carrel, A.: On the permanent life of tissues outside the organism. J. Exp. Med. *15*, 516–528 (1912)

Castor, C.W., Ritchie, J.C., Scott, M.E., Whitney, S.L.: Connective tissue activation. XI. Stimulation of glycosaminoglycan and DNA formation by a platelet factor. Arthritis Rheumat. *20*, 859–868 (1977)

Chait, A., Ross, R., Albers, J., Bierman, E.: Platelet derived growth factor stimulates low density lipoproteins receptor activity. To be published. Proc. Natl. Acad. Sci. U.S.A. (1980)

Chernoff, A., Goodman, D.S., Levine, R.F.: Origin of platelet derived growth factor in megakaryocytes. J. Clin. Invest. To be published (1980)

Das, M., Miyakawa, T., Fox, F.C., Pruss, R.M., Aharonov, A., Herschman, H.: Specific radiolabeling of a cell surface receptor for epidermal growth factor. Proc. Natl. Acad. Ser. U.S.A. *74*, 2790–2793 (1977)

Davies, P.F., Ross, R.: Mediation of pinocytosis in cultured arterial smooth muscle and endothelial cells by platelet derived growth factor. J. Cell Biol. *79*, 663–671 (1978)

Friedman, R.J., Stemerman, M.B., Wenz, B., Moore, S., Gauldie, J., Gent, M., Tiell, M.I., Spaet, T.H.: The effect of thrombocytopenia on experimental arteriosclerotic lesion formation in rabbits. Smooth muscle cell proliferation and re-endothelialization. J. Clin. Invest. *60*, 1191–1201 (1977)

Fuster, V., Bowie, E.J.W., Lewis, J.C., Fass, D.N., Owen, C.A. jr., Brown, A.L.: Resistance to arteriosclerosis in pigs with von Willebrand's disease. J. Clin. Invest. *61*, 722–730 (1978)

Gerrard, J.M., Phillips, D.R., Rao, G.H.R., Plow, E.F., Walz, D.A., Ross, R., White, J.G.: Biochemical studies of two patients with the gray platelet syndrome – selective deficiency of platelet alpha granules. J. Clin. Invest. (to be published) (1980)

Goldberg, I.D., Stemerman, M.B., Handin, R.I.: Vascular permiation of platelet factor IV following endothelial injury. Science (to be published) (1980)

Gospodarowicz, D., Moran, J.S.: Growth factors in mammalian cell culture. Ann. Rev. Biochem. *45*, 531–558 (1976)

Gospodarowicz, D., Greene, G., Moran, J.: Fibroblast growth factor can substitute for platelet factor to sustain the growth of BALB/3T3 cells in the presence of plasma. Biochem. Biophys. Res. Commun. *65*, 779–787 (1975)

Habenicht, A., Glomset, J., Ross, R.: Relation of cholesterol and mevalonic acid to the cell cycle in smooth muscle and Swiss 3T3 cells stimulated to divide by platelet derived growth factor (to be published). J. Biol. Chem. (1980)

Ham, R.G.: Clonal growth of mammalian cells in a chemically defined medium. Proc. Natl. Acad. Sci. U.S.A. *53*, 288–293 (1965)

Harker, L.A., Ross, R., Slichter, S.J., Scott, C.R.: Homocystine-induced arteriosclerosis: the role of endothelial cell injury and platelet response in its genesis. J. Clin. Invest. *58*, 731–741 (1976)

Hayashi, I., Sato, G.H.: Replacement of serum by hormones permits growth of cells in defined medium. Nature *259*, 132–133 (1976)

Heldin, C.-H., Wasteson, A., WEstermark, B.: Partial purification and characterization of platelet factors stimulating the multiplication of normal human glial cells. Exp. Cell Res. *109*, 429–437 (1977)

Heldin, C.-H., Westermark, B., Wasteson, A.: Purification platelet derived growth factor and partial characterization. Proc. Natl. Acad. Sci. U.S.A. *76*, 3722–3726 (1979)

Holley, R.W.: Factors that control the growth of 3T3 cells and transformed 3T3 cells. In: Proteases and biological control, 2, pp. 455–460. Cold Spring Harbor, New York: Cold Spring Harbor Laboratory 1975 a

Holley, R.W.: Control of growth of mammalian cells in culture. Nature *258*, 487–490 (1975b)

Holley, R.W., Kiernan, J.A.: Studies of serum factors required by 3T3 and SV3T3 cells. In: Ciba Foundation Symposium: Growth Control in Cultures, pp. 5–10. London: Churchill Livingstone 1971

Kaplan, D.R., Chaso, F.C., Stiles, C.D., Antoniades, H.N., Scher, C.D.: Platelet a granules contain a growth factor for fibroblasts. Blood *53*, 1043–1052 (1979 a)

Kaplan, K.L., Brockman, M.J., Chernoff, A., Lesznik, G.R., Drillings, M.: Platelet a granule proteins: studies on release and subcellular localization. Blood *53*, 604–618 (1979 b)

Kohler, N., Lipton, A.: Platelets as a source of fibroblast growth-promoting activity. Exp. Cell Res. *87*, 297–301 (1974)

Linder, B.L., Chernoff, A., Kaplan, K.L., Goodman, D.S.: Release of platelet derived growth factor from human platelets by arachidonic acid. Proc. Natl. Acad. Sci. U.S.A. *76*, 4107–4111 (1979)

Moore, S., Friedman, R.J., Singal, D.P., Gauldie, J., Blajchman, M.S., Roberts, R.S.: Thrombosis and haemostasis. J. Int. Soc. Thromb. Hemost. *35*, 70–81 (1976)

Pledger, W.J., Stiles, C.D., Antoniades, H.N., Scher, C.D.: Induction of DNA synthesis in BALB/c3T3 cells by serum components: Reevaluation of the commitment process. Proc. Natl. Acad. Sci. U.S.A. *74*, 4481–4485 (1977)

Ross, R., Glomset, J.: The pathogenesis of atherosclerosis. New Engl. J. Med. *295*, 369–377, 420–425 (1976)

Ross, R., Vogel, A.: The platelet-derived growth factor. Cell *14*, 203–210 (1978)

Ross, R., Glomset, J., Kariya, B., Harker, L.: A platelet-dependent serum factor that stimulates the proliferation of arterial smooth muscle cells in vitro. Proc. Natl. Acad. Sci. U.S.A. *71*, 1207–1210 (1974)

Ross, R., Nist, C., Kariya, B., Rivest, M., Raines, E., Callis, J.: Physiological quiescence in plasma-derived serum: Influence of platelet-derived growth factor on cell growth in culture. J. Cell Physiol. *97*, 497–508 (1978)

Ross, R., Vogel, A., Davies, P., Raines, E., Kariya, B., Rivest, M., Gustafson, C., Glomset, J.: The platelet-derived growth factor and plasma control cell proliferation. In: Hormones and cell culture, Sato, G.H., Ross, R. (eds.), pp. 5–16. Cold Spring Harbor, NY: Sixth Cold Spring Harbor Conference 1979

Rutherford, R.B., Ross, R.: Platelet factors stimulate fibroblasts and smooth muscle cells quiescent in plasma serum to proliferate. J. Cell Biol. *69*, 196–203 (1976)

Scher, C.D., Shepard, R.C., Antoniades, H.N., Stiles, C.D.: Platelet-derived growth factor and the regulation of the mammalian fibroblast cell cycle. Biochim. Biophys. Acta *560*, 217–241 (1979)

Shier, W.T.: Serum stimulation of phospholipase A_2 and prostaglandin release in 3 T 3 cells is associated with platelet-derived growth promoting activity. Proc. Natl. Acad. Sci. U.S.A. *77*, 137–141 (1981)

Stemerman, M.B., Ross, R.: Experimental arteriosclerosis. I. Fibrous plaque formation in primates, an electron microscope study. J. Exp. Med. *136*, 769–789 (1972)

Stiles, C.D., Isberg, R.R., Pledger, W.J., Antoniades, H.N., Scher, C.D.: Control of the Balb/c-3 T 3 cell cycle by nutrients and serum factors. Analysis using platelet-derived growth factor and platelet-poor plasma. J. Cell Physiol. *99*, 395–405 (1979)

Todaro, G.J., DeLarco, J.E., Marquardt, H., Bryant, M.L., Sherwin, S.A., Sliski, A.H.: Polypeptide growth factors produced by tumor cells and virus transformed cells: a possible growth advantage for the procedure cells. In: Hormones and cell culture. Sato, G.H., Ross, R. (eds.), pp. 113–127. Cold Spring Harbor, N.Y.: Sixth Cold Spring Harbor Conference (1979)

Thorgeirsson, G., Robertson, A.L. jr.: Platelet factor and the human vascular wall. Atherosclerosis *31*, 231–238 (1978)

Vogel, A., Pollack, R.: Methods for obtaining revertants of transformed cells. In: Methods in cell biology, vol. 8. New York: Academic Press 1974

Vogel, A., Raines, E., Kariya, B., Rivest, M., Ross, R.: Coordinate control of 3 T 3 cell proliferation by platelet-derived growth factor and plasma components. Proc. Natl. Acad. Sci. U.S.A. *75*, 2810–2814 (1978)

Vogel, A., Ross, R., Raines, E.: Role of serum components in density-dependent inhibition of growth in 3 T 3 cells: platelet-derived growth factor is the major determinant of saturation density. J. Cell Biol. (1980)

Wall, R.T., Harker, L.A., Quadracci, L.J., Striker, G.E.: Factors influencing endothelial cell proliferation in vitro. J. Cell. Physiol. *96*, 203–214 (1978)

Weiss, H.J., Lages, G.A., Witte, L.D., Kaplan, K.L., Goodman, D.S., Nossel, H.L., Baumgartner, H.R.: Storage pool disease: evidence for clinical and biochemical heterogeneity. Proceedings of the VI Congress, Int. Soc. Thromb. Haemostat., Philadelphia. *38*, 377 (1977)

Westermark, B., Wasteson, A.: A platelet factor stimulating human normal glical cells. Exp. Cell Res. *98*, 170–174 (1976)

White, J.G.: Ultrastructural studies of the Gray Platelet Syndrome. Am. J. Path. *95*, 445–462 (1978)

Witte, L.D., Kaplan, K.L., Nossel, H.L., Lages, G.A., Weiss, H.J., Goodman, D.S.: Studies of the release from human platelets of the growth factor for cultured human arterial smooth muscle cells. Circulation Res. *42*, 402–409 (1978)

Wrann, M., Fox, C.F., Ross, R.: Modulation of epidermal growth factor receptors on 3 T 3 cells by platelet derived growth factor. Science (in Press) (1980)

Yen, A., Riddle, V.G.H.: Plasma and platelet associated factors act in G 1 to maintain proliferation and to stabilize arrested cells in a viable quiescent state. Exp. Cell Res. *120*, 349–357 (1979)

CHAPTER 5

Somatomedin: Physiological Control and Effects on Cell Proliferation*

D. R. CLEMMONS and J. J. VAN WYK

A. Introduction

The pituitary somatotropic hormone has long been believed to be the principal regulator of balanced growth in vivo since growth hormone deficient patients exhibit proportional dwarfism and patients with acromegaly undergo diffuse enlargement of their internal organs and soft tissues. Administration of growth hormone to pituitary dwarfs is followed by dramatic increases in both cell number and cell size (CHEEK and HILL, 1974). Attempts to develop in vitro bioassays based on the supposed effects of growth hormone, however, made it apparent that the magnitude of the in vitro effects produced by growth hormone correlated poorly with its in vivo actions. While searching for an in vitro bioassay system, SALMON and DAUGHADAY (1957) observed that growth hormone itself had no direct effect on cartilage metabolism, but that plasma from normal or growth hormone-treated hypophysectomized rats contained a growth hormone-inducible factor which directly stimulated sulfate uptake by this tissue. This serum factor, initially designated "sulfation factor," subsequently was shown to consist of a family of closely related peptide growth factors which were given the generic designation "somatomedin" (DAUGHADAY et al., 1972; HALL and VAN WYK, 1974). Although the somatomedins originally were postulated to mediate the anabolic effects of growth hormone only on skeletal tissue, they subsequently have been shown to promote cellular proliferation in a wide variety of cell types (VAN WYK and UNDERWOOD, 1978; ZAPF et al., 1978 b). These substances now are believed to form a vital link between growth hormone and the stimulation of the metabolic processes leading to cellular proliferation.

Three major properties characterize the somatomedins: 1) their serum concentrations are growth hormone dependent, 2) they produce insulin-like pleiotypic actions, and 3) they are a mitogenic stimulus for multiple tissues and cell types. The nomenclature of the somatomedins remains confused since several of the peptides were initially isolated and named on the basis of the latter two properties before their growth hormone dependence and metabolic effects in cartilage were recognized. Thus, "non-suppressible insulin-like activity" (NSILA), which was originally isolated on the basis of its insulin-like activity in adipose tissues, was subsequently renamed "insulin-like growth factor" when it was recognized that its predominant effect is to stimulate cell and tissue growth (RINDERKNECHT and HUMBEL, 1976 b).

---

* Supported by NIH Research Grant AM 01022. JUDSON J. VAN WYK is a recipient of Research Career Award No. 4 K 06 AM 14115 from the NIH. DAVID R. CLEMMONS is supported in part by General Clinical Research Center Grant and USPHS grant No. RR 0046

Multiplication stimulating activity (MSA) is the term which TEMIN originally proposed for the mitogenic activity of bovine serum and this term continues to be used for similar mitogenic peptides produced in vitro by specific rat hepatic cell lines (DULAK and TEMIN, 1973 b).

The somatomedins in human plasma which have been best characterized are somatomedin-A (SM-A), somatomedin-C (SM-C) and insulin-like growth factors I and II (IGF-I and IGF-II)[1]. Of these, SM-C and IGF-I are identical in function and clearly different from IGF-II and SM-A in both structure and functional properties. SM-C and IGF-I are basic peptides with more stringent growth hormone dependence and lesser insulin-like activity than IGF-II (VAN WYK et al., 1980). SM-A has been less well characterized, but like IGF-II, it is more nearly neutral than SM-C/IGF-I (FRYKLUND et al., 1974). POSNER et al. (1978) have reported the isolation from human plasma of an insulin-like growth factor (ILAs) which is slightly acidic. The MSA peptides produced by rat hepatocytes, like IGF-II, are neutral and more insulin-like than SM-C and IGF-I (NISSLEY and RECHLER, 1977). The somatomedins, therefore, can be grouped conveniently into two major classes according to their net charge properties. Although such grouping correlates fairly well with certain differences in their growth hormone dependencies and spectrum of biological activities, a more satisfactory classification must await complete chemical characterization of certain of these peptides.

B. Assay Systems Used to Measure Somatomedins

Present concepts concerning the biological role of the somatomedins have been derived from studies utilizing methods of detection which have varied widely in their specificities. Since certain biological assays may be unable to discriminate clearly between the somatomedins and several of the other growth promoting substances in plasma, the development of radioreceptor assays and radioimmunoassays has represented an important advance. There are, however, new difficulties inherent in the latter techniques and there is no single method which faithfully measures all of the somatomedin activity.

I. Biological Assays

The bioassay systems employed to measure the somatomedins have been based on three physiologic actions. These include cartilage assays based on sulfation or thymidine uptake (DAUGHADAY et al., 1975 b; ALMQVIST et al., 1961 b), assays for insulin-like activity in adipose tissue (SCHLUMPF et al., 1976), and fibroblast assays based on the stimulation of DNA synthesis and cell replication (RECHLER et al., 1976 a). While these biological assays are very sensitive, they lack specificity and undoubtedly measure the net effects of a variety of stimulatory and inhibitory substances. In addition to being slow, cumbersome and time-consuming, they are inherently of relatively low precision. In spite of these drawbacks, isolation of the

---

1 The substance described as somatomedin-B has subsequently been identified as an antiprotease lacking several important characteristics common to the other somatomedin peptides. The term somatomedin-B is now considered a misnomer

pure somatomedins would not have been possible without these assays and they continue to provide model systems for studying the mechanisms by which the individual somatomedins act at the cellular level.

II. Radioreceptor Assays

The demonstration by HINTZ et al. (1972a) that somatomedin competed with ^{125}I-insulin for binding to specific cellular receptors made possible the development of a radioreceptor assay using ^{125}-I-insulin as the ligand. Although this receptor assay was not specific for individual somatomedins, it had far greater precision than the bioassay and was less cumbersome and time consuming to perform. This assay was particularly valuable in monitoring purification before specific radioligands became available (VAN WYK et al., 1974).

The discovery that many tissues possess somatomedin receptors distinct from insulin receptors (see Sect. G.II) made possible the development of radioreceptor assays of greater sensitivity and somewhat greater specificity than the receptor assay for insulin-like activity (MARSHALL et al., 1974). Although receptor assays for the various somatomedins have proven to be far more useful than the insulin receptor assay in monitoring purification, these assays possess insufficient specificity to permit distinction between the individual somatomedins (VAN WYK et al., 1980).

III. Protein Binding Assays

The observation by ZAPF et al. (1975c) that ^{125}I-NSILA will bind to carrier proteins in serum has led to the development of competitive protein binding assays which are based on the ability of unlabeled somatomedin to compete with radioiodinated peptides in binding to these carrier proteins (HINTZ and LIU, 1977). As a source of binding protein, these assays use either whole serum itself or relatively crude preparations of binding protein from which somatomedin has been removed (SCHALCH et al., 1978; ZAPF et al., 1977). A major disadvantage of the protein binding assay method is that it depends on removing all traces of binding protein from the sample to be assayed (MOSES et al., 1976; SCHALCH et al., 1978). This is usually accomplished by a chromatographic procedure in acid which not only may result in some loss of somatomedin activity, but more importantly, will fail to measure any acid stable high molecular weight somatomedin species in the preparation (ZAPF et al., 1978a). Although the protein binding technique provides relatively good discrimination between somatomedin values in normal and hypopituitary subjects, the discrimination between normal and patients with acromegaly has been less consistent (ZAPF et al., 1977; HEINRICH et al., 1978).

IV. Radioimmunoassays

During the past several years, highly specific radioimmunoassays have been developed for each of the somatomedins. The assays for SM-C and IGF-I and II are rapid and can be carried out using whole serum, thus eliminating a possible source of loss (FURLANETTO et al., 1977; ZAPF et al., 1978b; BALA and BHAUMICK, 1979). In addition to increasing the sensitivity 10–20 fold over the somatomedin receptor

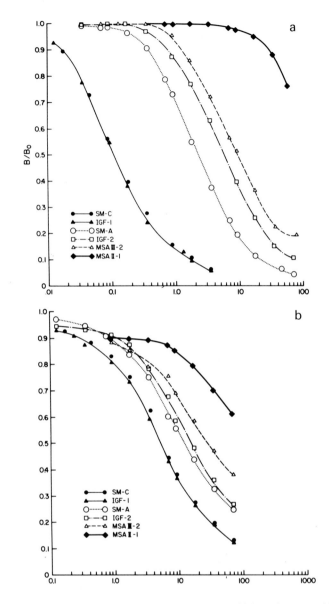

Fig. 1

Table 1. Concentrations of insulin and somatomedin-C which reduce specific binding by 50%

Assay	Insulin (M)	Somatomedin-C (M)	Ratio $\frac{(\text{Insulin})}{(\text{Sm-C})}$
[125]I insulin: RIA[a]	3.3×10^{-11}	$\gg 1 \times 10^{-6\,c}$	$\ll$ 0.00003
[125]I insulin: MBA	1×10^{-9}	5×10^{-8}	0.02
[125]I Sm-C: MBA	1.7×10^{-6}	1×10^{-9}	1,700
[125]I Sm: RIA[b]	5×10^{-5}	1×10^{-10}	500,000

[a] Guinea pig anti-porcine insulin 1:250,000
[b] Rabbit anti-human somatomedin 1:10,000
[c] Highest dose tested

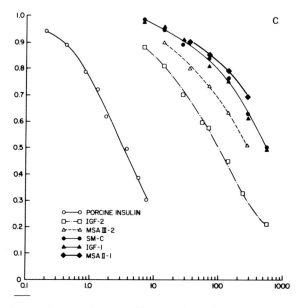

C   NANOGRAMS/0.5ml

o———o PORCINE INSULIN
□--·-□ IGF-2
△---△ MSA Ⅲ-2
●——● SM-C
▲——▲ IGF-1
◆——◆ MSA Ⅱ-1

Fig. 1 a–c. Curves of competition produced by various somatomedin peptides in **a** the radioimmunoassay for somatomedin-C, **b** radioreceptor assay for somatomedin-C, and **c** radioreceptor assay for insulin. (VAN WYK et al., 1980)

assays, the radioimmunoassays have proven to have a high degree of specificity for the individual somatomedins (VAN WYK et al., 1980; MOSES et al., 1979b). This technique should make it possible to determine the physiological significance of each peptide independently. Although, when carried out on unprocessed serum samples, the RIA for somatomedin-C measures only a portion of the total somatomedin-C content (see Sect. E), the fraction measured correlates well with the total somatomedin-C content and accurately reflects pathophysiologic alterations in growth hormone secretion.

Table 1 compares the relative activities of insulin and somatomedin-C in their respective radioimmunoassays and placental membrane receptor assays. Although somatomedin-C is recognized minimally by the insulin antiserum, it is 2% as active as insulin itself in competing for insulin receptors. Similarly, insulin is recognized by the somatomedin receptor better than by the somatomedin-C antibody. By comparing the relative activities of each peptide in the two receptor assays, it can be seen that somatomedin-C is more effective in competing for insulin receptors than is insulin in competing for somatomedin-C receptors.

Figure 1 shows the displacement curves produced by the different somatomedins in the radioimmunoassay for somatomedin-C, radioreceptor assay for somatomedin-C, and placental insulin receptor assay. In the radioimmunoassay for somatomedin-C the curves of competition produced by somatomedin-C and IGF-I are superimposable with 50% inhibition of binding observed at 0.2 ng/ml (Fig. 1 A). Somatomedin-A, IGF-II, and MSA III-2 were respectively 5%, 2.4%, and 1.2% as potent as somatomedin-C/IGF-I. MSA II-1 caused only 24% inhibition of binding at a concentration of 140 ng/ml, the highest concentration tested.

In the placental membrane binding assay for somatomedin-C (Fig. 1 B), somatomedin-C and IGF-I again exhibited essentially identical activity with 50% inhibition of binding of the tracer observed at 10 ng/ml. Somatomedin-A, IGF-II, MSA III-2, and MSA II-1 were respectively 45%, 36%, 21%, and 4.7% as potent as somatomedin-C/IGF-I.

In the placental membrane binding assay for insulin (Fig. 1 C), somatomedin-C and IGF-I again produced identical displacement curves, although 50% inhibition of ^{125}I-insulin binding required 1,400 ng/ml of each peptide. In this assay, the order of potencies was reversed: IGF-II, the most potent, was followed by MSA III-2 and then by somatomedin-C and IGF-I. Somatomedin-C and IGF-I were only 14% as active as IGF-II in this assay and only 43% as potent as MSA III-2. MSA II-1 was now 68% as potent as somatomedin-C and IGF-I. Insufficient somatomedin-A was available for this comparison.

V. Standards Used for the Quantitation of Somatomedin Activity

As yet, there is no universally accepted standard which can be used to express the quantity of somatomedin detected in bioassays and radioligand assays. Porcine insulin has generally been used as a standard for assays based on insulin-like activity and for the measurement of multiplication stimulation activity in chick embryo fibroblast assays. In such assays, somatomedin activity is expressed in insulin equivalents.

Insulin itself is too inactive for use as a standard in cartilage assays based on stimulation of $^{35}SO_4$ or ^{3}H-thymidine uptake and therefore pools of serum or plasma from healthy adults have been used as the standard (PHILLIPS and DAUGHADAY, 1977). In such assays, 1 somatomedin unit is defined as the somatomedin activity *as measured in that assay* contained in 1 ml of the reference serum. Among the many limitations of such standards are the following: a) serum contains multiple species of somatomedin and the ratios between them is probably not constant, b) no two serum pools are identical and c) no two assays have identical specificities, and d) results are influenced by the method in which serum is collected and processed. Although highly purified somatomedin preparations are likely to remain in too short supply to permit their routine use as assay standards, their increasing availability should at least permit calibration of reference sera to ensure greater uniformity between laboratories and to aid in the interpretation of results.

C. Isolation and Properties of the Individual Somatomedins

I. Basic Somatomedins

1. Somatomedin-C (SM-C)

One of the most dramatic consequences of growth hormone administration to hypophysectomized animals is the increased uptake of radiolabeled sulfate by cartilage (ELLIS et al., 1953). While attempting to utilize these in vivo observations to develop an in vitro bioassay, SALMON and DAUGHADAY (1957) observed that sulfate incorporation into cartilage explants was not stimulated directly by growth hormone, although sulfate uptake was markedly stimulated by normal rat serum. Serum from hypophysectomized rats was deficient in this property unless these animals previously had been treated with growth hormone. These observations led to their hypothesis that growth hormone stimulates sulfate uptake by cartilage by inducing a "sulfation factor" which can be detected in serum. Subsequently, these investigators demonstrated that a growth hormone dependent serum factor could duplicate in vitro all the known effects which occur in cartilage following the in vivo administration of growth hormone including increased synthesis of DNA, RNA, collagen and non-collagenous proteins (DAUGHADAY and MARIZ, 1962; DAUGHADAY and REEDER, 1966; SALMON et al., 1968).

An even broader concept of the role of these substances was developed when SALMON and DUVALL (1970b) demonstrated that the "sulfation factor" acted like insulin as a potent stimulator of protein and DNA synthesis in muscle. Subsequently HALL and UTHNE (1972) demonstrated that plasma fractions which stimulated cartilage sulfation contained parallel quantities of non-suppressible insulin-like activity. Somatomedin subsequently was demonstrated to produce the full range of insulin-like effects in fat cells including anti-lipolysis (UNDERWOOD et al., 1972), glucose oxidation (HALL and UTHNE, 1971), and lipid synthesis (CLEMMONS et al., 1974). Recognition of this broad spectrum of actions led to changing the name of this substance to somatomedin, a term implying that it mediates the action of pituitary somatotropin (DAUGHADAY et al., 1972).

VAN WYK et al. (1974) showed that when plasma extracts were purified by isoelectric focusing, somatomedin activity could be separated into at least 2 molecular forms. The fractions containing the greatest amounts of cartilage sulfation activity focused above pH 8, whereas the majority of insulin-like activity focused in more neutral regions[2]. Both kinds of activity were identified in the two zones, however. The biologic activity which isofocuses in the neutral region is probably attributable to IGF-II. Because of its potent stimulatory activity in the cartilage bioassay, the basic material, pI 8.0–8.6 was selected for final purification and given the designation somatomedin-C. All studies pertaining to cell growth from our laboratory have utilized this material.

The isolation of the somatomedins has been handicapped by the lack of any known organ source from which this peptide can be extracted in significant quantity. Most workers have used Cohn Fraction IV of human plasma as the starting material for purification. Although this byproduct of the blood fractionation industry contains about 80% of the somatomedin activity in native plasma, it is normally discarded since it contains no other substances of commercial importance (VAN WYK et al., 1975). The peptide is remarkably stable in plasma and withstands boiling, freezing and thawing, lyophilization and acidification to pH 2 (KOUMANS and DAUGHADAY, 1963; VAN WYK and UNDERWOOD, 1978). It is less stable at highly alkaline pHs.

Although SM-C was initially isolated from acid ethanol extracts of Cohn Fraction IV, as described for the purification of NSILA (JAKOB et al., 1968), the losses with this procedure were excessive and we now simply homogenize the Cohn IV

---

2 This finding has recently been confirmed independently by GINSBERG et al. (1979)

Table 2. Major steps in purification of somatomedin-C[a]

Step	Recovery/liter of original plasma		Specific activity (units/mg protein)	Purification (fold)
	Units[b]	Mg protein[c]		
Native plasma (theoretical)	1,000	75,000	0.013	0
Cohn fraction IV (clarified extract)	630	9,970	0.063	4.8
1st SP Sephadex	549	N.D.	–	–
2nd SP Sephadex	505	98	5.15	4.0×10^2
Sephadex G50 (after 3rd SP Sephadex)	348	19.5	17.8	1.4×10^3
Isofocusing pH 3–10	285	N.D.[d]	–	–
Isofocusing pH 7–10	210	N.D.[d]	–	–
Sephadex G 50 (1st after isofocusing)	172	0.15[e]	1,146	8.8×10^4
Sephadex G 50	159	0.05[e]	3,180	2.4×10^5
After HPLC and final Sephadex G 50	71	0.007	10,142	7.8×10^5

[a] Starting material was 3 kg of Cohn IV paste equivalent to 100 liters of plasma
[b] Recovery determined by SM-C RIA. The standard was a commercial pool of serum from normal adults (Ortho 1777-2)
[c] Proteins determined by Lowry method. Final protein was determined by sum of amino acids in hydrolysate
[d] Proteins could not be measured due to interference by ampholytes
[e] May be spuriously high due to residual ampholytes

paste in 2 M acetic acid containing 0.075 M NaCl and concentrate the supernatant fraction on SP-Sephadex. After a succession of chromatographic steps on cation exchange resins and Sephadex G-50, the basic fraction is separated both from the neutral somatomedins and from the majority of contaminating proteins by isoelectric focusing in a flat bed matrix of Sephadex G-75. Final purification is by high pressure liquid chromatography followed by desalting on Sephadex G-50 in 1.0 M acetic acid. The final product is a basic peptide (pI 8.0–8.6) with a specific activity of 10,142 units/mgm protein. This represents a 780,000-fold purification from native plasma. The recovery at different stages of the purification procedure are shown in Table 2 (SVOBODA et al., 1980).

The best fit for integral amino acids in hydrolysates of pure SM-C is obtained with a structure containing 78 residues. Except for a few unresolved discrepancies, the portions of the amino acid sequence which have been published are identical to those in corresponding regions of IGF-I as described in the next section.

Using a sulfation factor assay to monitor purification, two additional groups have recently reported the isolation of basic somatomedins from Cohn Fraction IV of human plasma (BHAUMICK and BALA, 1979; LIBERTI and MILLER, 1980). The properties of these peptides clearly place them in the same functional grouping with SM-C and IGF-I, although their identity with the latter peptides remains to be established.

2. Insulin-Like Growth Factor I (IGF-I)

Comparisons of insulin-like activity in plasma as measured by bioassay with the concentrations measurable by the specific radioimmunoassay for insulin led to the discovery that only 10% of the bioactive insulin-like material in serum can be accounted for by immunoreactive insulin. The remainder of this biologic activity, which cannot be neutralized with anti-insulin antibody, was initially labeled bound insulin (ANTONIADES, 1961), atypical insulin (SAMAAN et al., 1962), or non-suppressible insulin-like activity (NSILA) (FROESCH et al., 1963). This factor was purified from outdated Cohn B plasma fractions by monitoring the purification steps with a bioassay based on the stimulation of glucose oxidation in rat adipose tissue (BÜRGI et al., 1966). Following the demonstration that somatomedin preparations were intrinsically insulin-like in certain of their properties, a highly purified preparation of NSILA was shown to stimulate mitosis in chick embryo fibroblasts and to stimulate sulfate and thymidine incorporation into chick cartilage (MORRELL and FROESCH, 1973; ZAPF et al., 1978c). Later, KAUFMANN et al. (1978) demonstrated that serum NSILA concentrations are growth hormone dependent, thereby fulfilling all three criteria for a somatomedin.

In 1976a, RINDERKNECHT and HUMBEL reported the final isolation of two closely related peptides which they named NSILA-I and NSILA-II. These peptides were shown to have growth promoting actions in cartilage and to produce insulin-like effects in the fat pad bioassay. Because their growth promoting actions could be demonstrated at concentrations significantly below those required to demonstrate insulin-like effects, they were subsequently renamed insulin-like growth factors I and II. Thus, although the designations NSILA and IGF are synonyms for the same substances, the term NSILA continues to be used in describing earlier studies with incompletely purified preparations. This distinction is not a trivial one, since much of the confusion in the vast literature on the somatomedins and insulin-like peptides has come from attributing properties to one or another of these peptides when, in fact, the purity of the test material was inadequately documented.

In the scheme used for the isolation of IGF, acid ethanol extracts of Cohn Fraction B were extracted with acetic acid (0.5 M) and further purified by repeated chromatography in 1 M acetic acid on Sephadex G-75 and G-50 followed by preparative SDS polyacrylamide gel electrophoresis. IGF-I was separated from IGF-II by ion exchange chromatography using SE-Sephadex. Analytical isoelectric focusing of IGF-I revealed an isoelectric point of 8.4 ± 0.2. IGF-I and II have subsequently undergone sequence analysis and detailed characterization of their molecular properties (RINDERKNECHT and HUMBEL, 1978a). Sequence analysis of IGF-I has revealed remarkable homology with proinsulin (Fig. 2). Fifty percent of the residues in the portions of IGF-I corresponding to the A and B chains of human insulin are identical. The C peptide of IGF-I is 12 residues in length and bears no resemblance to the 30 residue C peptide of human proinsulin. IGF-I also contains an 8 amino acid extension beyond the carboxy terminus of human proinsulin.

BLUNDELL et al. (1978) have utilized computer techniques to construct a hypothetical 3 dimensional model of IGF-I based on its amino acid sequence. They found that those portions of the molecule which are responsible for insulin's antigenic activity (predominantly on the A chain) are, in the case of IGF-I, folded into

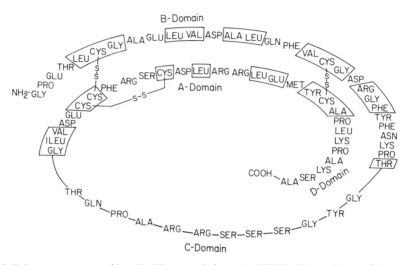

Fig. 2. Primary structure of insulin-like growth factor I (IGF-I). The residues of the A and B domains of IGF-I, which are enclosed in boxes, are those which are identical with the insulin-proinsulin molecule. The C domain, which corresponds to the C peptide of proinsulin, has no homology with the latter. The D domain is an 8 residue extension at the carboxy terminus which does not exist in proinsulin. (Structure determined by Rinderknecht and Humbel, 1978 a.)

the hydrophobic interior of the molecule. The sequences on IGF-I which are homologous to those portions of insulin which react with membrane receptors, however, are more exposed. These findings provide an explanation at the molecular level for the failure of IGF to crossreact with anti-insulin antibodies as well as its ability to crossreact with the insulin receptor.

Hintz et al. (1980) have synthesized the C peptide of IGF-I and raised antibodies to this fragment in rabbits. They found that ^{125}I-IGF-I and ^{125}I-SM-C bind identically to this antibody and that IGF-I, SM-C and the synthetic C peptide of IGF-I all displace the radioiodinated C peptide from binding to the antibody. On the other hand, IGF-II, which contains a different C peptide fragment, failed to react. These data provide confirmatory evidence that IGF-I and SM-C are identical or very similar, whereas IGF-II is a separate substance.

3. Somatomedin in Other Species

Daughaday et al. (1979) have partially purified somatomedin from the serum of rats implanted with growth hormone secreting pituitary tumors. Using a rat cartilage bioassay and a somatomedin receptor assay to monitor purification, they found that nearly all of this serum somatomedin activity focused between pH 8.0 and 8.6. Sequence analysis of this purified material, however, revealed the amino acid sequence of complement C 3 A which apparently had been co-isolated with rat somatomedin. Further isofocusing resolved the two peptides (Jacobs et al., 1978).

Liberti (1975) has purified somatomedin from bovine serum and found that the active substance in cartilage sulfation assays is also a basic peptide with some properties similar to human SM-C/IGF-I.

II. Neutral Somatomedins

1. Insulin-Like Growth Factor II (IGF-II)

During the course of their purification of NSILA from Cohn Fraction B of human plasma, RINDERKNECHT and HUMBEL (1978 b) recognized two molecular forms which could be separated by cation exchange chromatography on SE Sephadex. NSILA-II (IGF-II) proved to be very similar to IGF-I with 70% of the residues in the A and B regions identical. IGF-II has a less positive net charge than IGF-I, and the C peptide of IGF-II is only 8 residues in length vs 12 in IGF-I. The carboxy extension of IGF-II is also shorter than that of IGF-I. Comparison of the biologic activity of these two peptides revealed that IGF-II was several times more active than IGF-I in stimulating glucose oxidation by isolated fat cells, but somewhat less active than IGF-I in promoting sulfate uptake by cartilage explants (ZAPF et al., 1978 c). These differences conform to the relative potencies of the neutral and basic serum fractions in radioreceptor assays for SM-C and insulin (VAN WYK et al., 1975). Furthermore, serum levels of IGF-II were found to be much less growth hormone dependent than those of IGF-I since by specific RIAs for IGF-I and IGF-II, it was found that patients with acromegaly consistently had elevated levels of IGF-I but not of IGF-II (ZAPF et al., 1978 b).

2. Somatomedin-A (SM-A)

Somatomedin-A was initially isolated by HALL (1972) and UTHNE (1973) from acid ethanol extracts of Cohn Fraction IV. Further purification was by ion exchange and gel chromatography and high voltage electrophoresis. FRYKLUND et al. (1974) utilized preparative electrophoresis on cellulose columns to yield the preparation of SM-A which has been used in most of the studies published by these investigators on this material. Two forms of SM-A were described: SM-A$_1$ with 60 residues and SM-A$_2$ with 63 residues. In addition to possessing fewer total residues than the other somatomedins, SM-A differs from other somatomedins in reportedly containing only a single ½ cystine and thus no disulfide bonds. SM-A has a spectrum of biologic activity which is similar to the other somatomedins and it crossreacts with the insulin and somatomedin receptors (RECHLER et al., 1978). Somatomedin-A reacts weakly with an antiserum to somatomedin-C, but an antiserum raised to somatomedin-A in hens reacts better with IGF-I than with somatomedin-A itself (HALL et al., 1979). MSA does not crossreact with this antiserum and IGF-II is less active than somatomedin-A in this assay. Further documentation of the purity and amino acid sequence data are required, however, before the relationship between this substance and the other somatomedins can be established.

3. Multiplication Stimulating Activity (MSA)

After studying the effect of insulin on normal and virus transformed avian embryo fibroblasts, TEMIN (1967) concluded that the concentration of insulin itself is too low to explain the mitogenic effect of serum on these cells; he therefore postulated that this activity might be derived from the insulin-like peptides which had been described by other workers (ANTONIADES et al., 1961; BÜRGI et al., 1966). Using a methodologic scheme similar to that used to isolate insulin-like peptides from human serum, PIERSON and TEMIN (1972) showed that part of the mitogenic activity of calf serum could be attributed to a small molecular weight fraction which also contained insulin-like biologic activity and which stimulated sulfate uptake by cartilage. This substance was not further characterized.

DULAK and TEMIN (1973 a) subsequently reported the presence of multiplication stimulating activity in the serum-free conditioned medium which had been exposed to certain lines of buffalo rat liver cells. This strain of hepatocytes had been used because of their ability to proliferate in serum-free medium. Using a purification scheme similar to that used to purify calf serum, MSA, they purified several peptides with molecular weights estimated to be approximately 10,000 (DULAK and TEMIN, 1973 b). This material was also active in the cartilage sulfation factor assay and in the rat fat pad bioassay for insulin-like activity (SMITH and TEMIN, 1974). The term multiplication stimulating activity (MSA) is now used exclusively to describe these peptides formed by rat hepatocyte cultures.

Two forms of MSA produced by rat liver cells have been purified to apparent homogeneity. The larger peptide, designated MSA II-1 has an estimated molecular weight of 8,700 daltons and an isoelectric point of slightly below neutrality (MOSES et al., 1980a). The other peptide, designated III-2 has an estimated molecular weight of 7,100, but its charge properties have not been characterized. MSA is stable to boiling and acid but is inactivated by reduction and alkylation (NISSLEY and RECHLER, 1977). Two or 3 half-cysteine residues were identified after performic acid oxidation. Sequence analysis of MSA II-1 revealed a blocked N terminus and a carboxy terminal glycine residue.

Comparison between the purified MSA peptides and human somatomedins is complicated both by their different species of origin and their derivation from different sources. The neutral isoelectric point of MSA II-1 and the behavior of both MSA II-1 and MSA III-2 in the radioligand assays, shown in Fig. 1, however, suggest that they are more similar to IGF-II than somatomedin-C and IGF-I.

D. Production of the Somatomedins

I. Somatomedin Production by Organs and Tissue Slices

Perhaps the greatest gap in our understanding of somatomedin physiology is the uncertainty regarding its site of formation and the mechanism of its production. Numerous investigators have used bioassay methods to quantitate somatomedin activity in extracts from various organs. Although early studies revealed that NSILA-like material could be detected in many different organs, no organ containing a concentration greater than serum could be identified (SOLOMON et al., 1967). This finding was later reproduced with cartilage sulfation factor assays. It is important to note that these bioassays often fail to distinguish between the presence of low somatomedin concentrations or the presence of substances which are inhibitory in its assay. These studies have now been repeated using the highly specific radioimmunoassay for somatomedin-C. Extracts of muscle, liver, kidney, pituitary, thymus, brain, etc. from the rat consistently revealed concentrations which, when expressed as units/mg protein, were less than the concentrations in serum

(COPELAND et al., 1980). It may therefore be concluded that either a concentrated organ source of somatomedin does not exist or that somatomedin is stored in a form different from that recognized by antisomatomedin-C antibodies. The conclusion that somatomedin is not stored in significant quantities is supported by the finding that a detectable increase in serum somatomedin concentration following growth hormone administration is observed only after a 6-h lag period.

Lacking a concentrated organ source for extraction, several investigators have used organ perfusion techniques to study the release and control of somatomedin-like peptides. Although most of these studies have supported the liver as one site of somatomedin production, the effects of growth hormone on hepatic somatomedin production have been less conclusive. MCCONAGHEY and SLEDGE (1970) were the first to report increased somatomedin-like activity in liver perfusates following introduction of growth hormone into the perfusion medium. Using a more precise bioassay, however, PHILLIPS et al. (1976) were able to show only minimal increases in the somatomedin activity by perfusing whole rat livers with growth hormone. Likewise, HINTZ et al. (1972 b) showed only minimal increases in bioassayable somatomedin activity using a superfusion technique of liver slices. Utilizing a radioreceptor assay for somatomedin-C and thus controlling for the presence of inhibitors, MARSHALL showed a significant increase in somatomedin-like activity during growth hormone perfusion of rat livers (MARSHALL and VAN WYK, unpublished). However, in each case where growth hormone stimulation has been observed, supraphysiological concentrations of growth hormone (> 50 ng/ml) have been required. In addition, it is unclear whether the increase in somatomedin activity during such experiments is greater than the generalized increase in protein synthesis which occurs in response to growth hormone. Further casting doubt on the specificity of the growth hormone induced somatomedin increase was the report by DAUGHADAY et al. (1976) that insulin and ovine prolactin effected a greater increase in somatomedin production by perfused rat livers than did growth hormone itself. Recently, using improved assay and perfusion techniques, several investigators have been able to demonstrate specific effects of growth hormone, thyroxine, and nutritional factors on hepatic somatomedin production (VASSILOPOULOU-SELLIN et al., 1980; SCHALCH et al., 1979).

Studies of AV differences in both the dog and humans have revealed a somatomedin gradient across the liver (hepatic:portal vein difference). SCHIMPFF et al. (1976) used the Fick principle to calculate somatomedin production rates across the dog liver and calculated a blood production rate of several mg/day. Numerous clinical studies have revealed depressed serum somatomedin concentrations in patients with chronic liver disease (WU et al., 1974; TAKANO et al., 1977; SCHIMPFF et al., 1977). However, since nutritional status is an important component in the maintenance of normal somatomedin concentrations, it is difficult in these studies to segregate the specific role of impaired liver function from the effects of nutritional deficiency in such patients.

Failure to establish a reproducible whole organ perfusion system or tissue slice system for the study of somatomedin secretion has led investigators to experiment with the use of cultured explants of fetal tissues. RECHLER et al. (1979) found that rat fetal liver explants released significant amount of a MSA-like peptide into the

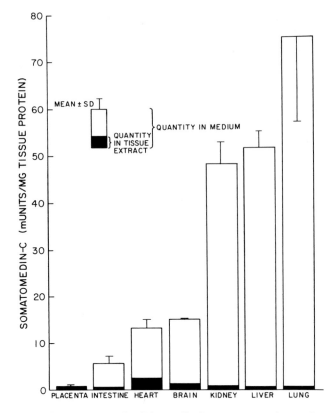

Fig. 3. Immunoreactive somatomedin-C in media from organ explants of multiple 17-day gestation fetal mouse tissues (N = 4). The darkened portion of each bar represents the amount of somatomedin found in extracts of tissues at the beginning of incubation. The entire bar represents the amount found in the medium in which each tissue has incubated over 3 days time. (D'ERCOLE et al., 1980)

medium for at least 4 days under serum-free conditions. Whether the release of this peptide is under the control of growth hormone has not been reported. It is noteworthy that buffalo rat liver MSA production is not stimulated by the exogenous addition of growth hormone. Recently, D'ERCOLE et al. (1980) have shown that fetal mouse liver explants, limb bud explants, and explants from intestine, heart, brain, kidney, and lung produce somatomedin-like peptides which crossreact with human somatomedin-C for binding to its antibody (Fig. 3). These studies have not addressed the hormonal control of this process.

II. Somatomedin Production by Monolayer Cultures

Since the initial reports that monolayer cultures of buffalo rat liver cells, produce MSA (DULAK and TEMIN, 1973a, b) DELARCO and TODARO (1978) have described the production of a MSA-like peptide by SV-40 transformed cells and by a human sarcoma cell line. This material was detected by its ability to crossreact in the MSA radioreceptor assay. POSTEL-VINAY et al. (1979) have described the production of an insulin-like peptide by short term rat hepatocyte cultures and SPENCER (1979) has detected somatomedin-A in similar cultures.

The possibility that human fibroblasts might produce a somatomedin-like growth factor was suggested by reports that human fibroblasts, unlike certain other fibroblast cell lines, are able to proliferate in somatomedin-C deficient serum (MOSES et al., 1978). ATKINSON et al. (1980) have demonstrated that WI-38 fibroblasts (derived from embryonic lung) release a peptide that reacts with somatomedin-C antibodies and that its concentration in media increases after the addition of relatively high concentrations of growth hormone (50 ng/ml). Using tritiated amino acid incorporation studies, they have preliminarily reported that the somatomedin released into the medium is the product of *de novo* synthesis (WEIDMAN et al., 1979).

CLEMMONS et al. (1981 a) recently described the production of somatomedin-like peptides by post-natal human fibroblast cultures. Preincubation of these cells with hGH in concentrations between 10–40 ng/ml (concentrations which are present in normal human plasma) augmented the production of this substance in a dose-dependent manner. Platelet derived growth factor and fibroblast growth factor likewise stimulated somatomedin production and when added with growth hormone, the combined effect was additive. Hydrocortisone, thyroxine, epidermal growth factor and insulin were without effect. Other variables which appeared to control the somatomedin production rate were the age of the fibroblast donor, cell number and the incubation time. These studies indicate that somatomedin production by monolayers may provide a model system for the study of somatomedin biosynthesis and for the study of the control of fibroblast growth (see Sect. G.III.3).

E. Molecular Size and Transport of Somatomedins in Plasma: The Somatomedin Binding Proteins

All of the somatomedin activity in serum is associated with large molecular proteins and little or no activity can be identified in molecular weight fractions corresponding to the free peptides (BÜRGI et al., 1966; JAKOB et al., 1968; ZAPF et al., 1975c; HINTZ and LIU, 1977; MOSES et al., 1976). When chromatographed at neutral pH on Sephadex G-200, 75%–90% of the total somatomedin activity is associated with proteins slightly smaller than IgG (145 K), with minor quantities distributed with smaller molecular weight fractions in a less well defined pattern (Fig. 4a). Usually, no somatomedin activity can be identified in fractions smaller than 30 K (or slightly smaller than serum albumin). It has further been demonstrated that binding proteins for the somatomedins are associated with these large molecular weight fractions. When serum is preincubated with any of the radiolabeled somatomedins prior to gel chromatography, the binding sites for the labeled hormone are found to be distributed somewhat differently than the endogenous somatomedin activity. The major binding site for the labeled somatomedins is associated with the albumin fraction, whereas the minor binding site is associated with the IgG fraction.

The growth hormone status of the donor influences the molecular size distribution of both the binding proteins and the endogenous somatomedin activity (Co-

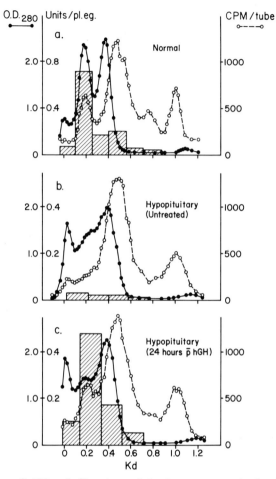

Fig. 4 a–c. Sephadex G-200 gel filtration of 1 ml serum premixed with 30,000 cpm ^{125}I-SM-C at 4 °C for 24 h. (●——●), OD at 280 nm; (o———o), cpm of ^{125}I-SM-C/tube; *hatched bars*, SM-C endogenous activity in units/plasma equivalent. In this system γ-globulin is eluted at Kd of 0.2, albumin at 0.4, and free iodide at 1.0. **a** Normal serum. The concentration of SM-C in whole serum was 1.20 U/ml. **b** Hypopituitary patient, previously untreated with hGH. **c** Same hypopituitary patient, 24 h after a single injection of hGH, 0.8 U/kg IM. (COPELAND et al., 1980)

HEN and NISSLEY, 1976; MOSES et al., 1976; HINTZ and LIU, 1977; KAUFMANN et al., 1978). MOSES et al. (1979 a) demonstrated that the larger of the MSA macromolecular complexes in rat serum disappears after hypophysectomy and reappears after the administration of growth hormone. Since this fraction accounts for a majority of the measurable MSA activity in serum, these workers suggested that growth hormone regulates the production of the binding protein rather than the small molecular weight somatomedin peptide itself (WHITE et al., 1979; MOSES et al., 1976). COPELAND et al. (1980) have recently confirmed the growth hormone dependence of the 145 K binding protein: SM-C complex in human serum by studying the chromatographic profiles of SM-C and ^{125}I-SM-C in hypopituitary dwarfs before and after growth hormone treatment (Figs. 4 b, c).

Although the pattern of binding is similar for all the iodinated somatomedins, ZAPF et al. (1978c) have shown that IGF-II has 3–4 times greater affinity for serum binding sites than does IGF-I. The binding of radiolabeled somatomedins to serum proteins is inhibited in a dose-dependent fashion by co-incubation with graded amounts of unlabeled somatomedin. This observation has been capitalized on by ZAPF et al. (1977) and SCHALCH et al. (1978) to develop competitive protein binding assays for somatomedin peptides (see Section B.III).

The affinity of the serum binding proteins for labeled somatomedins interferes with the detection of somatomedins in whole serum by radioligand techniques and probably also interferes with quantitative detection in bioassays of serum (FUR-LANETTO et al., 1977; MEULI et al., 1978; ZAPF et al., 1979; MEGYESI et al., 1977; HORNER et al., 1978). Since acidification of serum dissociates the small molecular weight peptides from their binding proteins, many investigators routinely chromatograph whole serum on Sephadex G-50 or G-75 in 1 M acetic acid to separate free somatomedin from the binding proteins (MOSES et al., 1980a; GUYDA et al., 1979; HORNER and HINTZ, 1979). After the excluded macromolecular fraction containing the stripped binding proteins is neutralized and rechromatographed on Sephadex G-200, the binding activity is now exclusively in the 38 K region, thus suggesting that the 145 K binding site has been either destroyed by acid or dissociated into subunits (KAUFMANN et al., 1978; MOSES et al., 1979a, 1980). In further studies of the nature of the binding protein complexes in serum, FURLANETTO (1980) was able to identify 3 components by anion exchange chromatography, and concluded that the 145 K complex is composed of 2 subunits, 1 which is acid stable and the other acid labile. By combining these two components, he was able to reconstitute the large binding component found in native serum.

The presence of serum binding proteins is unique for a peptide hormone and has helped to explain why total serum somatomedin concentrations, which average in excess of 200 ng/ml for somatomedin-C, are far in excess of the concentrations required for the stimulation of biologic activity in most in vitro test systems (usually between 1–20 ng/ml). The binding protein also serves to prevent the rapid fluctuations characteristic of most peptide hormones. The serum half-life of somatomedin has been variously estimated to be 2–4h, whereas the half-life in serum of most polypeptide hormones are less than 20 min (COHEN and NISSLEY, 1976; DAUGHADAY et al., 1968; KAUFMANN et al., 1977).

The physiologic role of the somatomedin binding proteins in serum is unclear at present. In the case of the thyroxin binding proteins and the binding proteins for steroid hormones, the binding protein is synthesized in the liver independently from the synthesis of the ligand in endocrine glands. This does not appear to be the case for the somatomedins, however, since the in vitro production of somatomedin by isolated cell systems or organ cultures is usually accompanied by the production of binding proteins (MOSES et al., 1979a). Rat liver cells release into culture medium a binding protein complex of approximately 40,000 molecular weight which has MSA binding properties (MOSES et al., 1979a). Normal rat serum contains both the 145 K (peak II) and 40 K (peak III) complexes. KAUFMANN et al. (1978) have provided data suggesting that following the injection of ^{125}I-

NSILA into normal rats, the labeled material rapidly associates with a complex of approximately 40,000 daltons. This is followed by the appearance of labeled material in the 145,000 molecular weight region. These observations taken together with the data of Furlanetto (1980) (see p. 177) support the hypothesis that the 145 K serum complex contains at least 2 subunits and that at least one of these subunits (40 K) is part of the cellular secretory product. If so, these macromolecular complexes might be analogous to the native forms of nerve growth factor and epidermal growth factor which have been isolated. The gamma subunit of native NGF is an arginine esterase which has the capacity to bind the active beta subunit at the latter's carboxy terminus (Pantazis et al., 1977). A similar situation exists for the native form of epidermal growth factor in which the active 6,000 molecular weight peptide is likewise associated with an arginine esterase which possesses binding capability for the active peptide (Taylor et al., 1974). Before the exact physiologic role of somatomedin and the other growth factors can be fully understood, it will be necessary to determine why these substances are formed as large macromolecular complexes and how their serum complexes serve to regulate the exposure of peripheral cells to the active forms of these growth factors.

F. Control of Somatomedin Concentrations in Blood

Although a vast amount of data has been published concerning the control of somatomedin blood levels in experimental animals and humans, most of these studies were performed using the somatomedin cartilage bioassay. The inherent imprecision of the bioassays as well as their susceptibility to interference by inhibitors has led in certain situations to conclusions which have required revision based on results obtained using radioligand assays.

Radioligand assays while offering advantages of speed, reproducibility, and minimal interference from inhibitors, likewise have significant limitations. Binding proteins may interfere with the ability of antibodies or membrane receptors to recognize somatomedin. Additionally, bioassays measure the effects of stimulators and inhibitors, thus reflecting the net content of active material. Radioligand assays measure only closely related molecules which may be biologically inert.

I. Blood Concentrations in Normal Individuals

1. Effect of Age

The preponderance of evidence suggests that somatomedin concentrations in the fetus and newborn are substantially lower than those found in adult serum and rise during the first years of life (van den Brande and Du Caju, 1974b; D'Ercole et al., 1977; Takano et al., 1976b; Almqvist and Rune, 1961; Parker and Daughaday, 1968; Hall, 1971). Such conclusions are supported by a wealth of data obtained by bioassay and radioreceptor techniques in the human, pig, mouse, and certain other species. Using a radioimmunoassay for MSA, however, Moses et al. (1980b) found that serum of fetal rats near term contains concentrations of cross-reacting substances nearly 20- to 100-fold higher than those found in maternal

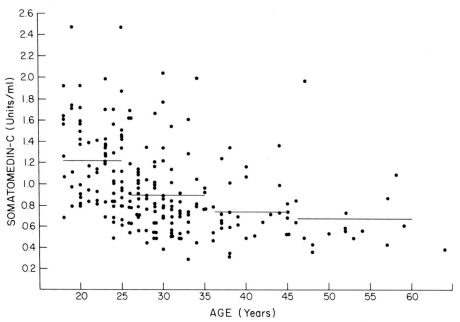

Fig. 5. Somatomedin-C concentration in 220 normal adults. Morning fasting samples were obtained from adult volunteers between the ages of 17 and 82 years. The mean of each decade interval is given (——)

serum. The high fetal levels gradually fell following birth and approached maternal levels by the 25th day of extrauterine life, an age which coincides with the beginning of sexual maturation in this species. These findings are at variance with the findings in humans since by several techniques of measurement, somatomedin levels progressively rise during childhood.

We have recently reviewed this topic using our radioimmunoassay for SM-C/IGF-I and have confirmed the presence of low serum levels in late gestational and cord blood, after which the concentrations rise progressively through the first 13 years of life (FURLANETTO et al., 1977). Peak values are reached at puberty, following which time there is a slow tapering of levels in successive decades (Fig. 5). The reason for the low levels of SM-C in the fetus and neonate are unknown, and this pattern is paradoxical since this is the time of life when the growth rate is maximal. Similar studies with the RIA for IGF-II are critically needed, however, because it is possible that different forms of somatomedin are produced at different stages of life. As indicated in previous sections, MSA has more characteristics in common with IGF-II than with SM-C/IGF-I.

If all somatomedins are indeed low in the human fetus and neonate, this finding might suggest that cartilage is very sensitive to the effects of somatomedin at this time of life and that sensitivity diminishes as this tissue ages. Bioassay data on rabbit (BEATON and SINGH, 1975), rat (HEINS et al., 1970) and porcine cartilage (PHILLIPS et al., 1974) obtained from animals at differing ages indeed indicate decreasing sensitivity of this tissue to somatomedin with increasing age of the donor. The etiology of the rise in somatomedin at puberty is unknown, although the temporal association of this rise with the adolescent growth spurt suggests that sex steroids may play some role.

We have found that plasma somatomedin-C concentrations in 227 normal adults between ages of 20 and 77 followed a log normal distribution with 95% of the values falling between 0.4 and 2.0 units/ml. In any given age group, female values averaged 10%–20% higher than male values. No acute fluctuations have been observed following food intake, and most observers have failed to detect any diurnal variation. Using the constant withdrawal pump technique, however, MIN-UTO et al. (1979) found up to a 30% fall in SM-C concentrations during the early hours of sleep with a slight rise early in the morning. The depression in SM-C during early sleep occured earlier than the growth hormones spikes that occur during sleep and correlated inversely with their magnitude.

2. Effect of Hormonal Status

a) Growth Hormone

When measured by bioassay and radioreceptor techniques, hypopituitary dwarfs consistently have subnormal serum concentrations, but the range of values in these patients frequently overlaps with normal (VAN DEN BRANDE and DU CAJU, 1974 b; DAUGHADAY et al., 1959; ALMQVIST et al., 1961 b; PARKER and DAUGHADAY, 1968; TAKANO et al., 1976 b; MEGYESI et al., 1975; SCHIMPFF and DONNADIEU, 1973; SPENCER and TAYLOR, 1978; HALL, 1970; HEINRICH et al., 1978). FURLANETTO et al. (1977) and COPELAND et al. (1980) have assessed serum levels of SM-C by RIA in approximately 52 hypopituitary children. All sera were found to contain less than 0.4 units/ml and 48 of 52 contained less than 0.2 units/ml (3 standard deviations below the adult mean). Somatomedin values determined by radioreceptor assay were found to correlate with the growth rate in children receiving chronic treatment with growth hormone injections (D'ERCOLE et al., 1977). Induction studies carried out by administration of a single dose of growth hormone to 36 hypopituitary children revealed that the SM-C concentration by RIA increased from a mean of 0.1 units/ml to a mean of 0.63 units/ml (COPELAND et al., 1980). Neither in normal subjects nor in hypopituitary dwarfs was any detectable rise in somatomedin observed before 6 h, and peak values did not occur until about 24 h (Figs. 6 and 7). The variables which control the somatomedin response to this single injection of growth hormone appeared to be nutritional status, previous exposure to growth hormone, and the timing of the previous injection.

Patients with active acromegaly have been found by bioassay and some radioreceptor assays to have 3- to 4-fold elevations of their somatomedin concentration over normal individuals (HALL, 1970; DAUGHADAY et al., 1959; VAN DEN BRANDE and DU CAJU, 1974 b; ALMQVIST et al., 1961 a; SCHIMPFF and DONNADIEU, 1973; SPENCER and TAYLOR, 1978; WEIDEMANN et al., 1979; MEGYESI et al., 1975; TAKANO et al., 1976 b; HEINRICH et al., 1978; D'ERCOLE et al., 1977). However, consistent elevations in this group of patients have not been observed with certain other receptor assays, with protein binding assays, nor with the RIA for IGF-II (ZAPF et al., 1978 b; HEINRICH et al., 1978; MEGYESI et al., 1975; TAKANO et al.,

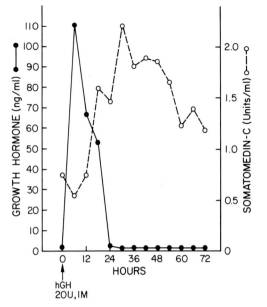

Fig. 6. SM-C (o---o) and growth hormone (●—●) responses to a single intramuscular injection of 20 U hGH to a normal adult male. (COPELAND et al., 1980)

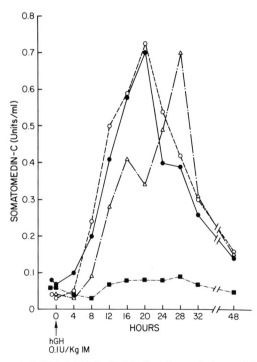

Fig. 7. SM-C responses to hGH, 0.1 U/kg IM, in four hypopituitary children. The single female subject (△-·-·△) was treated with hGH, thyroxine, cortisone acetate, Premarin and Provera. The single "noninducer" (■---■) was on treatment with hGH only. The two remaining male subjects were on treatment with both hGH and thyroxine and patient (●—●), with cortisone acetate additionally. (COPELAND et al., 1980)

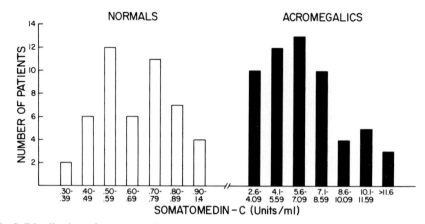

Fig. 8. Distribution of somatomedin-C concentrations in 48 normal subjects and 57 patients with acromegaly. (CLEMMONS et al., 1979)

1976 b; ZAPF et al., 1977; DAUGHADAY et al., 1959). Using our RIA, which has high specificity for SM-C/IGF-I, we have demonstrated that in a population of 57 patients with acromegaly, the mean somatomedin-C concentration was 10 fold greater than the mean concentration in 48 normal control subjects (CLEMMONS et al., 1979). There was no overlap between these 2 groups (Fig. 8). Five patients with active acromegaly who had normal growth hormone levels both in the fasting state and after glucose suppression, had an 8-fold elevation of their mean concentration of somatomedin-C. The individual somatomedin levels in the entire population of 57 patients appeared to correlate better with parameters of disease activity such as heel pad thickness and fasting blood sugar than did growth hormone measurements. These results demonstrated that somatomedin-C radioimmunoassay is a highly useful test in diagnosing acromegaly, correlates better than hGH measurements with disease activity, and may prove to be highly effective in monitoring therapeutic responses to treatment. ZAPF et al. (1978 b) and BALA and BHAUMICK (1979) have reported that IGF-I and "basic somatomedin" levels are elevated in acromegaly by specific RIAs. Since IGF-II (neutral somatomedin) is not consistently elevated in acromegaly, the failure of other assays to detect as striking elevations in acromegaly probably reflects their relative sensitivities to basic and neutral somatomedins.

b) Prolactin and Placental Lactogens

Ovine placental lactogen and ovine growth hormone have been shown to be equally potent in stimulating increases in somatomedin-C levels in hypophysectomized rats (HURLEY et al., 1977). The relationship between the human forms of prolactin, placental lactogen, and growth hormone in inducing somatomedin is less well defined. We and others have observed that patients with small prolactin secreting pituitary adenomas have normal somatomedin levels (BALA and BHAUMICK, 1979; GUYDA et al., 1979). However, patients with large prolactin secreting tumors associated with somatotropin deficiency were also found to have levels within the normal range (1.02 ± 0.44 units/ml) (CLEMMONS et al., 1981 c; GUYDA et al., 1977). A comparable group of patients with non-prolactin secretory pituitary tumors and the destruction of somatotropin production, however, had distinctly subnormal levels (mean 0.22 ± 0.1 units/ml). These results are consistent with the hypothesis that prolactin may be a weak stimulator of somatomedin production in the absence of normal somatotropin secretion. SPENCER (1980), however, was unable to discern by a receptor assay any effect of prolactin secreting tumors on somatomedin concentrations.

c) Thyroid Hormone

Thyroid hormones have been shown to interact with somatomedin in multiple target tissues to promote growth. However, the data describing its effects on somatomedin levels in hypothyroid states are conflicting. Cartilage bioassay studies and data from radioreceptor assays for somatomedin-A and C have shown either minimally depressed or normal levels (BURSTEIN et al., 1978; VAN DEN BRANDE and DU CAJU, 1974 b; TAKANO et al., 1976 b; D'ERCOLE et al., 1977). More consistent reductions of somatomedin concentrations in hypothyroidism have been observed with a protein binding assay for IGF-I, with restoration to normal after thyroxin replacement (DRAZNIN et al., 1978; BURSTEIN et al., 1979). Limited studies with the RIA for SM-C in hypothyroid children are in aggreement with the latter studies (UNDERWOOD, unpublished).

d) Cortisol

The inhibitory effects of excessive amounts of cortisol on bone growth have been well characterized (KILGORE et al., 1979). It is now generally agreed that cortisol directly inhibits cartilage metabolism in vitro and either this or the induction of bioassay inhibitors accounts for the low SM concentrations which have been reported by several investigators using the bioassay in hypercortisolemic states (KERET et al., 1976; TESSLER and SALMON, 1975; PHILLIPS et al., 1975; EISENBARTH and LEBOVITZ, 1974; ELDERS et al., 1975; VAN DEN BRANDE et al., 1975). As measured by radioreceptor assay and radioimmunoassay techniques, somatomedin levels have not been depressed in such patients (FURLANETTO et al., 1977; D'ERCOLE et al., 1977; WEIDEMANN et al., 1979). The effects of physiologic levels of glucocorticoids on somatomedin secretion are unknown.

e) Estrogens

Pharmacologic administration of estradiol to normal or acromegalic patients results in a depression of both bioassayable and radioimmunoassayable somatomedin activity (ALMQVIST et al., 1961 b; SCHWARTZ et al., 1969; WIEDEMANN and SCHWARTZ, 1972; WIEDEMANN et al., 1976; CLEMMONS et al., 1980 a). The changes in radioimmunoassayable levels have been shown to correlate with changes in bone turnover (CLEMMONS et al., 1980 a). Estrogens, unlike cortisol, have no direct effects on cartilage sulfation and the reduction of somatomedin-C concentrations as measured both by bioassay and radioimmunoassay are believed to reflect a decreased blood production rate (WIEDEMANN and SCHWARTZ, 1972). The mechanism by which this might occur has not been studied.

3. Effect of Pregnancy

Since normal human pregnancy is characterized by maternal weight gain and positive nitrogen balance in excess of fetal requirements, several investigators have measured maternal somatomedin concentrations serially during pregnancy. Bioassays and radioreceptor assays of blood and amniotic fluid have yielded conflicting results (DAUGHADAY et al., 1959; CHESLEY, 1962; HINTZ et al., 1977; BALA et al., 1978). In a cross-sectional study of maternal sera during pregnancy, FURLANETTO et al. (1979) observed a progressive rise in SM-C concentrations beginning at the 18th week of gestation with an abrupt fall to normal following delivery. These changes in somatomedin-C concentrations correlated with changes in human placental lactogen concentrations. Two studies support the hypothesis that placenta lactogen induces a rise in maternal somatomedin (Fig. 9). DAUGHADAY and KAPADIA (1978) observed that hypophysectomy in pregnant rats is not followed by a fall in maternal somatomedin concentrations although the levels fall promptly in the post-partum period. Additionally, HURLEY et al. (1977) demonstrated that ovine placental lactogen was a potent inducer of SM-C in hypophysectomized rats. Taken together, these data support the conclusion that placental lactogen induces a rise in SM-C concentrations in maternal serum during late pregnancy and that delivery of the feto-placental unit results in an abrupt fall in these concentrations.

4. Effect of Nutritional Status

Amino acid uptake is essential for normal protein synthesis and growth by cells. Evidence from a variety of sources indicates that normal somatomedin concentrations are severely reduced by nutritional deficiency. PHILLIPS and YOUNG (1976) have demonstrated that a significant fall in somatomedin activity, as measured by bioassay, occurs over the first 72 h of fasting in normal rats. Although in this study an attempt was made to exclude the effects of possible inhibitors on the assay, other investigators have shown that starvation stimulates the appearance in serum of macromolecular factors which inhibit sulfate uptake by cartilage (SALMON, 1975, 1974; VAN DEN BRANDE et al., 1975). PHILLIPS et al. (1979) have recently demonstrated that decreased bioassayable somatomedin activity in starving normal rats results both from lower serum concentrations and the induction of a bioassay inhibitor released by the liver (VASSILOPOULOU-SELLIN et al., 1980).

SHAPIRO and PIMSTONE (1979) have extended these observations using the rat cartilage bioassay to show that decreased protein intake per se accounts for both decreased serum concentrations and the loss of releasable somatomedin activity from liver and that isocaloric diets enriched with carbohydrate or fat cannot compensate for this decrease. In contrast, PHILLIPS et al. (1978) found that refeeding with protein alone resulted in only a transient rise in somatomedin activity and adequate caloric intake was required to maintain normal somatomedin levels.

In humans, somatomedin levels have been shown by both bioassay, radioreceptor assay and SM-C RIA to be depressed in children with kwashiorkor and restored to normal after refeeding (VAN DEN BRANDE and DU CAJU, 1974b; HINTZ et al., 1978; GRANT et al., 1973). This is a unique circumstance where there is complete dissociation between SM-C and growth hormone levels. In untreated kwashiorkor, growth hormone levels are extremely elevated and following the institution of dietary treatment, return to normal before SM-C levels rise. Although

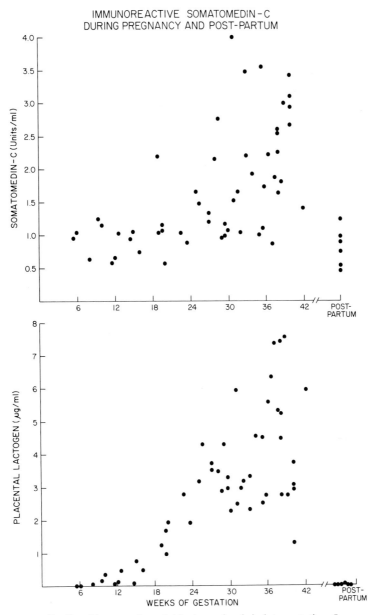

Fig. 9. Somatomedin-C and human placental lactogen levels in late gestation. Immunoreactive somatomedin-C (*top panel*) and placental lactogen (*bottom panel*) are plotted as a function of the duration of gestation. (FURLANETTO et al., 1979)

the mechanism for this disparity has not been worked out, it is noteworthy that the direct anti-insulin actions of growth hormone on carbohydrate and fat metabolism are protein sparing and therefore conducive to survival. Simultaneous reduction in somatomedin production during starvation would also conserve protein for essential biologic functions by limiting expenditures for growth.

G. In Vitro Biological Effects of the Somatomedins

I. Whole Tissue Effects

Studies to determine the mechanisms by which the somatomedins stimulate the growth of target cells were initiated long before extracts were available with sufficient purity to exclude the synergistic effects of other mitogens contained in serum. Most of the early studies were confined to determining the effect on tissue fragments during relatively short incubations and only recently have investigators attempted to assess the effect of pure somatomedins on cells in culture. Nevertheless, much important information has been derived from these studies and will be reviewed briefly here.

1. Cartilage

The original observation of SALMON and DAUGHADAY (1957) that a growth hormone dependent factor in serum stimulates sulfate and thymidine incorporation into cartilage explants has been confirmed using pure peptides (VAN WYK et al., 1978). With the advent of radioligand assays for most of the somatomedins, the current usefulness of these bioassays appears to be in confirming the biologic activity of pure peptide preparations and in discovering their mechanism of action at the cellular level.

In addition to stimulating proteoglycan (SALMON et al., 1967) and DNA synthesis in cartilage (DAUGHADAY and REEDER, 1966), the somatomedins also stimulate the synthesis of RNA (SALMON and DuVALL, 1970a), collagen (DAUGHADAY and MARIZ, 1962) and noncollagenous protein (SALMON and DUVALL, 1970a). Cartilage from pig (VAN DEN BRANDE and DU CAJU, 1974a), rat (DAUGHADAY and MARIZ, 1962), chick (HALL, 1970), rabbit (ASH and FRANCIS, 1975), cow and human (ASHTON and FRANCIS, 1978) are sensitive to the somatomedins and cartilage from young animals is more sensitive than cartilage from older animals (HEINS et al., 1970). Recently, CANALIS and RAISZ (1979) have extended these observations to osteoblast function. They demonstrated that MSA stimulated DNA, collagen, and non-collagen protein synthesis in fetal rat calvaria. Insulin had no effect in this system.

The mechanisms by which the somatomedins elicit the various pleiotypic responses of cartilage generally have been studied by comparing the effect of normal and hypopituitary serum rather than by using purified material. ADAMSON and ANAST (1966) showed that a growth hormone dependent serum factor stimulates K^+ dependent membrane transport of amino acids in cartilage and that sulfate incorporation is dependent on RNA and protein synthesis since it can be inhibited with either puromycin or actinomycin-D. SALMON and his coworkers' (1967) studies using actinomycin-D suggested that sulfate incorporation into proteoglycans was not directly stimulated by somatomedin, but depended on intact synthesis of the protein-polysaccharide complex. Similar conclusions were reached by ELDERS et al. (1977). It is not known whether somatomedin is required simply to maintain the optimal level of RNA and protein synthesis required for growth or whether it is stimulating some more basic process.

The role of low molecular weight nutrients in modulating the cartilage response was illustrated initially by DAUGHADAY and KIPNIS (1966) who showed that optimal concentrations of glutamine and serine are required for maximal sulfate incorporation. Likewise, EISENBARTH et al. (1973) showed that butyrate and octanoate inhibit some of the metabolic effects of serum on cartilage. FROESCH et al. (1976) reported that the activity of NSILA in chick embryo cartilage was greatly potentiated in the presence of thyroid hormones.

High concentrations of partially purified somatomedin have been reported to inhibit hormone stimulated adenylate cyclase activity in cartilage (TELL et al., 1972), although other investigators, using different preparations, have reported no significant changes in cyclic nucleotide levels (RENDALL et al., 1972) or even elevations (LEBOVITZ et al., 1976). These conflicting results may be due to different preparations and dosages, since none of the studies were carried out with pure preparations of somatomedin.

2. Muscle

Early studies with NSILA in rat diaphragm explants showed that this peptide stimulated both ^{14}C glucose incorporation into glycogen and amino acid incorporation into protein (FROESCH et al., 1966). A partially purified preparation of somatomedin also was shown by SALMON and DUVAL (1970 b) to stimulate ^{3}H-leucine incorporation into protein and tritiated thymidine into DNA in the rat diaphragm. UTHNE et al. (1974) demonstrated that partially purified preparations of somatomedin-A stimulated glucose transport, amino acid uptake and protein synthesis by rat diaphragms. Stimulation of membrane transport was not dependent on protein synthesis. More recent studies have shown that highly purified NSILA is capable of stimulating glucose uptake, lactate production, and 3-O methylglycose efflux in the isolated perfused rat heart (MEULI and FROESCH, 1975). Highly purified IGF-I extracts have been shown to increase 2-deoxyglucose transport, glycolysis, and glycogen synthesis in the isolated mouse soleus muscle (POGGI et al., 1979). It is noteworthy that all of the muscle studies reviewed to date have compared somatomedin with native insulin and all the parameters measured are insulin-like in their scope. FLORINI et al. (1977), however, have demonstrated somatomedin effects which are independent of those of insulin on the replication of isolated rat myoblasts. The receptor interactions which mediate these different effects will be addressed in Sect. G.II.

3. Adipose Tissue

NSILA mimics all of the effects of insulin in adipose tissue including the stimulation of ^{14}C glucose oxidation to ^{14}CO$_2$, fatty acids, lipids and glycogen (OELZ et al., 1972). These results were later repeated with partially purified somatomedin extracts on the parameters of glucose oxidation (HALL and UTHNE, 1971; CLEMMONS et al., 1974) and lipid synthesis (CLEMMONS et al., 1974). Somatomedin was also found to inhibit epinephrine stimulated lipolysis (UNDERWOOD et al., 1972). Additionally, CLEMMONS et al. (1974) showed that somatomedin and insulin produced parallel dose response curves, both in stimulating glucose oxidation and lipid synthesis, and that the effects of the two hormones were additive when present together in subsaturating concentrations (Fig. 10). These observations were inter-

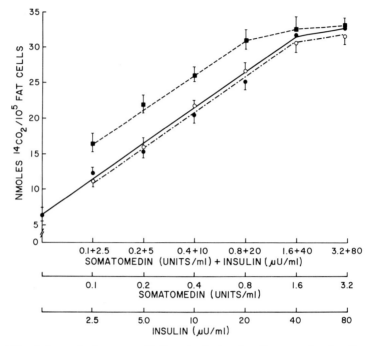

Fig. 10. Stimulation of glucose oxidation in isolated adipocytes by insulin (●—●), somatomedin (○--○) and the simultaneous addition of both hormones (■--■). 1.5×10^5 cells were incubated at 37 °C for 2 hr in 1.0 ml Krebs-Ringer bicarbonate buffer, pH 7.4. Responses to each dose of hormone are expressed as means $\pm$ SE. (CLEMMONS et al., 1974)

preted as evidence that the two hormones act through a common cellular mechanism to stimulate these metabolic processes. NSILA and partially purified somatomedin extracts also have been shown to inhibit epinephrine stimulated adenyl cyclase in adipose tissue membranes (TELL et al., 1972; RENNER et al., 1973).

Recently, ZAPF et al. (1978c) have repeated many of the earlier NSILA studies using purified IGF-I and II preparations. They have confirmed that both peptides provide a potent stimulus for glucose oxidation, 3-O-methylglucose transport, lipogenesis, and antilipolysis in isolated rat adipocytes. IGF-II appears to be slightly more potent in these test systems and both peptides have biologic potencies between 1%–3% that of insulin. These studies also indicate that the majority of the insulin-like effects in fat cells appear to be mediated primarily through the insulin receptor.

II. Correlation Between Biological Responses and Receptor Interactions

Thirty five years have elapsed since F. G. YOUNG (1945) postulated that insulin itself might be the true anabolic hormone which mediates the growth promoting effect of growth hormone. Although Professor YOUNG's hypothesis was based on the anabolic effects of insulin in living animals, TEMIN (1967) (as well as earlier investigators) had pointed out that amongst the known hormones, only insulin could stimulate the proliferation of cultured cells in vitro. Thus, following the observation that partially purified somatomedin preparations could interact with the insulin receptor on cell membranes (HINTZ et al., 1972a), extensive studies were undertaken to explore the possibility that somatomedin might exert its growth promoting effects through interaction with the insulin receptor.

An alternative hypothesis was suggested by the observation of MARSHALL et al. (1974) that an iodinated preparation of somatomedin-C, when incubated with human placental cell membranes, preferentially bound to receptors which are distinct from the insulin receptor. Whereas somatomedin was approximately 2% as effective as insulin in competing with ^{125}I-insulin for binding to the placental insulin receptor, somatomedin-C was 1,700 times more potent than insulin in competing with ^{125}I-somatomedin-C for binding to the somatomedin-C placental receptor (Table 1). Furthermore, the concentrations of somatomedin which were effective in the somatomedin-C receptor assay were of the same order of magnitude as the dosages required to produce a growth-promoting effect in cartilage and in growing cells. Similarly, the concentrations of somatomedin required to compete with insulin for binding to its receptor were similar to the dosages required to evoke an insulin-like response in adipose tissues (CLEMMONS et al., 1974). These observations led to the concept that the growth-promoting actions of somatomedin are mediated through specific somatomedin receptors whereas the insulin-like actions of this hormone are mediated through their crossreactivity with the insulin receptor (VAN WYK et al., 1975).

Similar observations have subsequently been made for each of the somatomedins (MEGYESI et al., 1974; ZAPF et al., 1978c, 1975c; RECHLER et al., 1977b, 1978). MEULI and FROESCH (1976) confirmed the presence of distinct insulin and NSILA receptors in both isolated fat cells and in intact perfused rat heart. These investigators have demonstrated that in fat cells, NSILA (like somatomedin-C) appears to stimulate insulin-like actions through its interactions with insulin receptors, whereas in rat heart, stimulation of 3-O-methyl glucose transport occurs through NSILA receptors (ZAPF et al., 1978c). In addition, these investigators have demonstrated a specific NSILA receptor in chick embryo fibroblasts (ZAPF et al., 1975a).

MSA receptor interactions have been extensively studied in a wide variety of target tissues, including rat liver membranes (RECHLER et al., 1977b), buffalo rat liver 3A2 cells (RECHLER et al., 1978), chick embryo fibroblasts (ZAPF et al., 1975b) and human fibroblasts (RECHLER et al., 1977a). Since the concentrations of MSA capable of displacing ^{125}I-MSA from its receptor were similar to those used to stimulate DNA synthesis in these cell types, it was concluded that the stimulation of DNA synthesis in these target cells was mediated through interaction of MSA with a "growth-promoting receptor." Direct evidence in support of these conclusions has come from receptor and metabolic studies carried out in the presence of the Fab fragment derived from an antiserum containing a high concentration of antibodies directed against the insulin receptor (KING et al., 1980). The Fab fragment inhibited the binding of ^{125}I-insulin to membrane receptors in fat cells and chick fibroblasts but reduced the binding of ^{125}I-MSA by only 14%. Similarly, the Fab fragment inhibited the insulin-like metabolic responses to insulin and MSA in adipose tissue but not the effect of MSA on DNA synthesis in fibroblasts. These observations provide the strongest evidence yet advanced that the growth promoting effects of the somatomedins are effected through a "growth receptor," entirely distinct from the insulin receptor.

Several investigators have compared the distribution of somatomedin and insulin receptors in a wide variety of tissues from fetal and adult rats and pigs (D'ERCOLE et al., 1976b; TAKANO et al., 1976a). These studies demonstrated the ubiquitous distribution of somatomedin receptors in nearly every cell type.

As noted in previous sections and in Fig. 1, all of the somatomedin-like peptides crossreact with the somatomedin-C receptor in human placental preparations. Crossreactivity amongst the various somatomedin peptides in other receptor assays has been reported by RECHLER et al. (1978) using labeled MSA and labeled somatomedin-A. They found that somatomedin-A, somatomedin-C, IGF-I, IGF-II, and MSA II-1 and III-2 were active in the MSA receptor assay in rat liver membranes, buffalo rat liver 3A2 cells, chick embryo fibroblasts, and human fibroblasts (MOSES et al., 1979b). MSA was also found to compete with ^{125}I-IGF-II for binding to a variety of cell lines (RECHLER et al., 1977b). Analysis of these and many similar studies suggests that each peptide has greater specificity in receptor assays when its own peptide is used as the radioligand, but that significant crossreactivity exists amongst the various members of this group.

III. Stimulation of DNA Synthesis and Cell Growth in Tissue Culture

1. Range of Responsive Cell Types

A distinctive feature of the somatomedins is that, like epidermal growth factor and fibroblast growth factor, they are mitogenic for a broad range of cell types. Initial studies characterizing their mitogenicity were performed using tissue slices or explants. However, in recent years their effects on cultured cell types have been the subject of extensive analysis.

Purified somatomedin-C has been demonstrated to stimulate ^{3}H-thymidine incorporation and mitosis in chick embryo fibroblasts in the absence of serum (VAN WYK et al., 1975). Purified somatomedin preparations were subsequently shown to be mitogenic for such diverse cell types as rat myoblasts, human fibroblasts, GH3 cells, rat fibroblasts, ovarian tumor cells, corneal epithelium, chick chondrocytes and fetal rat calvaria (VAN WYK and UNDERWOOD, 1978). Somatomedin-C has been shown to be mitogenic in the 5–10 ng/ml range. These concentrations are well below those that occur in human serum, although the interstitial fluid levels of this peptide are unknown. In each case, however, when somatomedin-C was added alone, it could not substitute completely for the mitogenic effects of 10% whole serum. In similar cell culture systems the concentration of insulin required to produce a mitogenic response is approximately 2 orders of magnitude higher than its physiologic concentration in serum.

Multiplication stimulating activity has been shown to be mitogenic for chick embryo fibroblasts (DULAK and TEMIN, 1973a), rat fibroblasts (FLORINI et al., 1977), human fibroblasts (RECHLER et al., 1977a), 3T3 fibroblasts (NISSLEY et al., 1977a), and rat myoblasts (FLORINI et al., 1977). In addition, the peptide has been shown to stimulate aminoisobutyrate transport in chick embryo fibroblasts (SMITH and TEMIN, 1974) and rat myoblasts (FLORINI et al., 1977), and to stimulate ornithine decarboxylase activity in cultured 3 T 3 fibroblasts (NISSLEY et al., 1977 a). Contrary to expectation, MSA has no measurable effect on the replication of buffalo rat liver cells themselves (NISSLEY et al., 1977 b). HEATON et al. (1980) have recently demonstrated that MSA stimulates tyrosine amino transferase activity in rat hepatoma cells. Somatomedin-A, like the other somatomedins, is mitogenic for human fibroblasts and chick embryo fibroblasts (RECHLER et al., 1978), but the peptide was not found to be active in human glial cell cultures (WASTESON et al., 1975).

Partially purified NSILA was shown to be mitogenic for chick embryo fibroblasts (MORRELL and FROESCH, 1973), and chick chondrocytes (FROESCH et al., 1976). In the latter tissue, the addition of thyroxin produced an additive response. Subsequently, NSILA has been shown to stimulate growth in human fibroblasts (RECHLER et al., 1975) and to stimulate ornithine decarboxylase activity in chick embryo fibroblasts (HASSABACHER and HUMBEL, 1976). Using highly purified IGF-I and II, these investigators have confirmed that at low concentrations (1–25 ng/ml) both peptides are mitogenic for chick embryo fibroblasts (RINDERKNECHT and HUMBEL, 1976 a).

2. Interaction Between Somatomedin and Other Growth Factors in the Cell Cycle

Although the preceding studies have been valuable in establishing the range of responsive cell types and the concentration ranges required, it is only recently that studies have attempted to define the mechanisms by which somatomedin produces its cellular responses. Since neither somatomedin nor any other peptide growth factor is as effective as whole serum in stimulating cell proliferation, many investigators have followed the lead of SATO in evaluating in fully defined media the responses of various cell types to different combinations of peptide growth factors and hormones (HUTCHINGS and SATO, 1978). Studies by HOLLEY and KIERNAN (1974) were amongst the first to point out that when added together fibroblast growth factor and insulin stimulated DNA synthesis in 3 T 3 cells but neither peptide alone had significant mitogenic activity. Other studies by GOSPODAROWICZ and MORAN (1974) illustrated synergism between FGF and dexamethasone in cultured fibroblasts.

Approaching the problem from a different perspective, RUTHERFORD and ROSS (1976) demonstrated that crude extracts of platelets contained a platelet derived growth factor (PDGF) which, when added to heat denatured platelet poor plasma, induces a synergistic increase in DNA synthesis in glial cells and smooth muscle cells. ANTONIADES et al. (1979) subsequently purified this factor to homogeneity and showed that purified PDGF is only mitogenic in the presence of platelet poor plasma.

The observations of PLEDGER et al. (1977) provide a model for explaining the mechanism by which two or more peptide growth factors might interact to stimulate replication. They showed that exposure of quiescent Balb/c 3 T 3 cells to platelet derived growth factor was required before they could respond to growth factors contained in plasma. This initial conditioning of the cells by platelet derived growth factor was termed "competence" and he has applied the term "competence

Table 3. Progression factor activity of somatomedin-C[a]

Addition to confluent Balb/c 3T3 cells

Preincubation (5 h) Incubation (24 h)

Platelet derived growth factor pretreatment for 5 h	Platelet poor plasma	Somatomedin-C ng/ml	% Nuclear labeling at 24 h
+	–	–	2
–	5% from normal donor	–	1
+	5% from normal donor	–	68
+	5% from hypopituitary donor	–	5
+	5% from hypopituitary donor	1	20
+	5% from hypopituitary donor	2.5	35
+	5% from hypopituitary donor	5.0	55
+	5% from hypopituitary donor	7.0	65

[a] This assay was carried out as described by Stiles et al. (1979), by Dr. W. J. Pledger, Department of Pharmacology and Cancer Research Center, University of North Carolina

factors" to those growth factors which are capable of stimulating cells to leave the quiescent state (Pledger et al., 1978). The 12-h lag phase required for "competent" cells to enter DNA synthesis following exposure to platelet poor plasma was termed "progression," and those substances in platelet poor plasma which are necessary for DNA synthesis to occur were termed "progression factors." These findings subsequently have been confirmed by Vogel et al. (1978).

The factors in platelet poor plasma which control progression in Balb/c 3T3 cells subsequently were analysed by Stiles et al. (1979). They found that optimal stimulation of DNA synthesis required the presence of both somatomedin-C and non-growth hormone dependent factors contained in somatomedin-C deficient plasma obtained from hypophysectomized rats or hypopituitary humans. Following the induction of competence by brief exposure to platelet derived growth factor, the addition of platelet poor plasma from hypophysectomized animals resulted in limited progression to a growth arrest point 6 h prior to "S" phase. Subsequent addition of pure somatomedin-C (10 ng/ml) resulted in the initiation of DNA synthesis following a further lag of 6 h. Neither somatomedin-C nor hypopituitary plasma alone were effective in stimulating competent cells to enter the "S" phase of the cell cycle. These relationships are shown in Table 3 and Fig. 11.

Stiles et al. (1979) used these observations to test several growth factors for either competence or progression factor activity. Somatomedin-A and MSA were found to be active as progression factors and insulin was active at much higher concentrations. On the other hand, fibroblast growth factor functioned like PDGF as a "competence factor." The factors contained in hypophysectomized platelet poor plasma which synergize with somatomedin-C to cause progression are not known, although epidermal growth factor, hydrocortisone and transferrin can at least partially substitute for plasma from growth hormone deficient animals (Stiles and Pledger, personal communication). There are multiple other substances con-

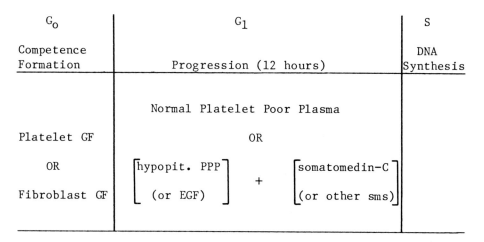

Fig. 11. Sequential action of serum factors in stimulating quiescent Balb/c 3 T 3 cells to enter DNA synthesis. The platelet derived growth factor (or the fibroblast growth factor) renders confluent cells "competent" so that they can respond to "progression factors" in platelet poor plasma (PPP). Hypopituitary PPP (or epidermal growth factor) will advance cells to a point located 6 h prior to the normal onset of DNA synthesis. Somatomedin-C (or other somatomedins) are required to permit further progression and entry into S

tained in hypophysectomized platelet poor plasma, however, that might be necessary for the full expression of serum mitogenic activity.

We have recently demonstrated one mechanism by which PDGF and factor(s) in somatomedin-C deficient PPP interact to condition the cell to respond to somatomedin-C (CLEMMONS et al., 1980b). Exposure of quiescent Balb/c 3 T 3 fibroblasts to PDGF followed by somatomedin-C deficient PPP resulted in a two fold increase in the specific binding of ^{125}I-somatomedin-C to these cells. This increase was accounted for by an increase in receptor number (Fig. 12).

3. Production of Somatomedin-Like Peptides by Cultured Cells and Their Role in Cellular Proliferation

Although these studies have established an important role for somatomedin-C in stimulating "progression" of Balb/c 3 T 3 cells, the mechanism by which this hormone acts in other target cell types and cells from other species is unclear at present. Human fibroblasts for example, do not appear to require the exogenous addition of somatomedin-C (MOSES et al., 1978; AUGUST et al., 1973). These differences in somatomedin dependency of different cell types might be attributable to the species of cell origin, culture conditions, or differences in the method in which serum is prepared. Furthermore, it is possible that certain cell types, such as human fibroblasts, might be exquisitely sensitive to the effects of competence factors, such as PDGF and therefore require only minimal or very low levels of somatomedin to initiate DNA synthesis.

An alternative explanation for reduced growth factor dependency is that these cells might be capable of endogenously producing this growth factor in sufficient

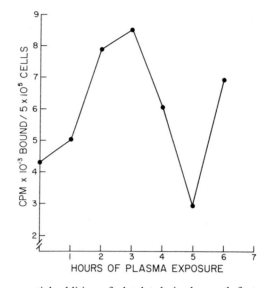

Fig. 12. Effect of sequential addition of platelet derived growth factor then platelet poor plasma on ^{125}I-SM-C binding. Confluent monolayers were washed three times with DME and fresh medium containing approximately 5 μg platelet derived growth factor was added. Following a 5-h incubation, this medium was aspirated and fresh DME containing 5% platelet poor plasma was added. The cells were incubated for the times indicated; subsequently this medium was removed and ^{125}I-SM-C binding was determined. (Clemmons et al., 1980 b)

quantity to proliferate. Evidence in support of the latter model derives from the work of De Larco and Todaro (1978) who demonstrated that SV 40 transformed cells were capable of producing a polypeptide which crossreacted in the MSA radioreceptor assay. They postulated that this peptide enables this cell type to divide under MSA deficient conditions. D'Ercole et al. (1980) also described the production of a somatomedin-like peptide by explants of fetal mouse liver and a number of other tissues. The material released into the medium is capable of supporting DNA synthesis in Balb/c 3 T 3 cells. Rechler et al. (1979) have recently described the production of an MSA-like peptide from fetal rat liver explants. This material has been shown to stimulate cell division in cultures of chick embryo fibroblasts.

The possibility that human fibroblasts might produce somatomedin-like peptides was suggested by several studies which demonstrated production of growth factors by these cells. Human fibroblasts have been found to produce a small peptide which stimulates endothelial cell division (Hakim, 1978) and Millis et al. (1978) have detected in conditioned medium from these cells two factors which stimulate the growth of human fibroblasts themselves. Additionally, Weidman et al. (1979) have demonstrated that human fetal lung fibroblasts produce a somatomedin-C-like peptide which can be quantitated using their specific RIA. This peptide stimulates DNA synthesis in that cell type and the production of this peptide was increased in the presence of high concentrations of growth hormone.

We, therefore, have recently tested three hypotheses related to the role of somatomedin in stimulating fibroblast replication: 1) that somatomedin is in fact rate limiting for human fibroblast replication, b) that growth of these cells in somatomedin deficient medium is possible because of their capacity to produce somatomedins and c) that the capacity of these cells to produce this peptide is under hormonal control. To analyze whether human fibroblasts require somatomedin for proliferation, quiescent cultures were transiently exposed to PDGF and then transferred to either somatomedin-C deficient plasma alone or this medium plus 5 ng/ml somatomedin-C. Following a 48-h incubation during which the media were changed every 2 h to fresh media of the same composition, only the somatomedin-C treated cultures showed significant increase in the percentage of cells synthesizing new DNA. The cells, therefore, appeared to be conditioning the medium with factors required for replication and the exogenous addition of somatomedin-C could substitute for this conditioning process. In additional experiments, it was confirmed that during a 72-h incubation of quiescent postnatal human fibroblasts, the medium became enriched with substances crossreacting in the RIA for SM-C (see Sect. D.II). The concentrations detected were consistent with those necessary to stimulate DNA synthesis in other cell types, and the diluted conditioned medium was capable of supporting DNA synthesis in the somatomedin dependent Balb/c 3 T 3 cell line. This process is under hormonal control since both platelet derived growth factor and human growth hormone stimulate a 3-fold increase in somatomedin production in these cells (CLEMMONS et al., 1981 a).

The importance of this hormonally regulated somatomedin production by human fibroblasts is that it might provide a mechanism by which growth hormone could stimulate generalized cell growth. In contrast, the stimulation of somatomedin production by platelet derived growth factor could represent a non-growth hormone dependent pathway by which fibroblast division could be stimulated in a circumscribed area such as occurs during wound healing. The platelet derived growth factor stimulated pathway might provide an explanation for the normal wound healing observed in hypopituitary patients, a process which could be initiated in vivo by the local release of PDGF.

These studies have also provided a model to address the major question of whether growth hormone stimulates cell replication directly or acts through the induction of somatomedin. We have recently demonstrated that competent, confluent human fibroblasts when treated with hypopituitary platelet poor plasma enter DNA synthesis at a slower rate than when exposed to normal platelet poor plasma. This difference between hypopituitary and normal platelet poor plasma can be abolished by pre-treatment for 6 h of these fibroblast monolayers with either human growth hormone (10 ng/ml) or somatomedin-C (5 ng/ml).

However, if medium containing growth hormone and hypopituitary platelet poor plasma is removed and replenished every two hours with fresh medium containing these substances, replication of fibroblasts is quite abnormal and only 20% of these cells enter DNA synthesis after 48 h of incubation (CLEMMONS and VAN WYK, 1981 b). In contrast, frequent changing of media containing somatomedin-C enriched hypopituitary platelet poor plasma allows replication to proceed at a normal rate. These studies strongly suggest that the conditioning effect of growth hormone on medium is mediated through the production of somatomedin-C, or a somatomedin-like substance. If a similar process can be shown to operate in other cell types, significant progress will have been made towards resolving the issue of which actions of growth hormone are direct and which are mediated through the production of somatomedins.

H. In Vivo Actions of the Somatomedins

Oelz et al. (1970) demonstrated that high dosages of a partially purified NSILA produced prolonged lowering of the blood sugar and stimulation of ^{14}C glucose incorporation into glycogen by muscle. Such acute responses to high dosages may not be relevant to the normal physiologic role of the somatomedins, however, since under normal circumstances the somatomedins circulate as high molecular weight complexes which are largely confined to the vascular spaces. Meuli et al. (1978) demonstrated that whereas perfusion of rat heart with free NSILA produced acute metabolic effects in this tissue, this effect was abolished by simultaneous perfusion with serum binding proteins. Thus the interpretation of in vivo results obtained with high dosages of somatomedin will remain in question until the circumstances under which somatomedins are released from high molecular weight serum complexes into extracellular tissues have been defined.

In a preliminary communication, Uthne (1975) reported that hypophysectomized rats treated with a partially purified preparation of somatomedin-A in a dosage of 14 units/day for 5 days responded with increases in tibial width comparable to those achieved with 20 μg of human growth hormone. Although both increases were statistically greater than control animals, there was no significant increase in body weight. Fryklund et al. (1979) administered somatomedin-A (0.5 mg bid × 4 days) to hypophysectomized rats that had been pretreated for 3 days with thyroxine and cortisone. Although this regimen produced a 20% increase in the incorporation of $^{35}SO_4$ into proteoglycans, human growth hormone induced a 220% increase; furthermore, at a higher somatomedin-A dosage of 2.0 μg bid × 4 days there was a negative response. Somatomedin-A was without effect if the animals had not received pretreatment with thyroxine and cortisol.

More extended studies have been carried out by Van Buul-Offers et al. (1979) who treated Snell dwarf mice for 4 weeks with daily injections of a partially purified somatomedin preparation. One group received human growth hormone 16 units/day, another 1-thyroxine 100 ng/day and the third group, 6 units/day of somatomedin. All three groups experienced approximately 30% increases in their total body weights, changes which were statistically greater than those of control animals. In addition, the body length of somatomedin treated animals increased approximately 5% above controls. Sulfate uptake by cartilage was increased over basal values after three days of treatment. It should be noted that all of these increases, although significantly greater than in control Snell dwarf mice, were not equal to genetically normal mice of the same age. Insulin in a dosage of 0.6 μg/day was ineffective in stimulating growth in these animals.

The first convincing demonstration of in vivo growth promotion by one of the pure somatomedins was an outgrowth of studies by Rothstein and his colleagues

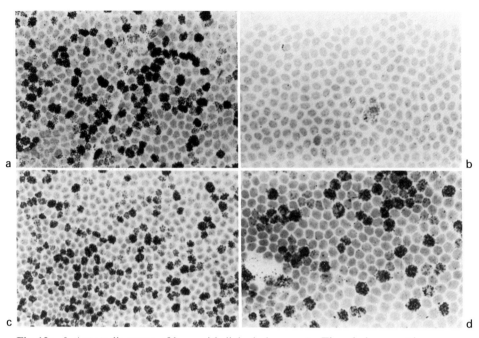

Fig. 13 a–d. Autoradiograms of lens epithelial whole mounts. The whole mount in **a** came from the lens of a normal, intact, postmetamorphic bullfrog. **b** A similar preparation from a frog that had been hypophysectomized. Neither cells in DNA synthesis nor mitosis can be found in the preparation. **c** A whole mount from an animal that was hypophysectomized and later injected with human growth hormone; labeling is evident. **d** Tissue from a hypophysectomized frog that received 700 ng purified somatomedin daily for 10 days. Mitotic figures and incorporation of ^{3}H-thymidine were apparent. (ROTHSTEIN et al., 1980)

(1980) at the Kresge Eye Institute of the hormonal factors which initiate the proliferation of lens epithelium in frogs. Hypophysectomy of bullfrogs is followed 3–4 weeks later by complete cessation of mitosis and DNA synthesis in the epithelium coating the anterior surface of the lens. DNA synthesis and mitosis are resumed following in vivo administration of human, bovine or anuran growth hormone, anuran prolactin or by TSH or T_3. In vitro studies reveal that these cells are arrested in G_0/G_1 and can be stimulated by 20% serum from normal frogs to enter DNA synthesis after a lag of 60 h. Although growth hormone or T_3 have no in vitro effect on these cells, incubation with 20% serum from hypophysectomized frogs advances the cells in G_1 to another arrest point 12 h before the G_1/S boundary. Using this model, ROTHSTEIN et al. (1980) demonstrated that growth hormone administration to hypophysectomized bullfrogs induced a rise in somatomedin-C-like substances to levels 1.7 times greater than those determined in normal animals. Thyroxine administration to hypophysectomized Leopard frogs likewise increased the concentrations of somatomedin-like substances to supra-normal values. When somatomedin-C was administered to the hypophysectomized frogs, abundant DNA synthesis and mitotic activity were detected in the lens epithelium after two weeks of treatment (Fig. 13). Control corneal endothelial cells, which are bathed by the same aqueous humor, showed no such stimulation. This is the first report indicating that the in vivo administration of an essentially pure peptide growth factor can restore DNA synthesis and proliferation in an organ from which both processes had been previously eliminated by experimental manipulation. These studies also provide evidence that the pituitary control of cell proliferation by this tissue is mediated by the somatomedins.

References

Adamson, L.F., Anast, C.S.: Amino acid, potassium, and sulfate transport and incorporation by embryonic chick cartilage: the mechanism of the stimulatory effects of serum. Biochim. Biophys. Acta *121*, 10–20 (1966)

Almqvist, S., Rune, I.: Studies on sulfation factor (SF) activity of human serum: the variation of serum SF with age. Acta Endocrinol. (Kbh.) *36*, 566–576 (1961)

Almqvist, S., Ikkos, D., Luft, R.: Studies on sulfation factor (SF) activity of human serum: the effects of oestrogen and X-ray therapy on serum SF activity in acromegaly. Acta Endocrinol. (Kbh.) *37*, 138–147 (1961 a)

Almqvist, S., Ikkos, D., Luft, R.: Studies on sulfation factor (SF) activity of human serum: serum SF in hypopituitarism and acromegaly. Acta Endocrinol. (Kbh.) *36*, 577–595 (1961 b)

Antoniades, H.N.: Studies on the state of insulin in blood: The state and transport of insulin in blood. Endocrinology *68*, 7–16 (1961)

Antoniades, H.N., Scher, C.D., Stiles, C.D.: Purification of human platelet derived growth factor. Proc. Natl. Acad. Sci. U.S.A. *79*, 1809–1813 (1979)

Ash, P., Francis, M.J.O.: Response of isolated rabbit articular and epiphyseal chondrocytes to rat liver somatomedin. J. Endocrinol. *66*, 71–78 (1975)

Ashton, T.K., Francis, M.J.O.: Response of chondrocytes isolated from human foetal cartilage to plasma somatomedin activity. J. Endocrinol. *76*, 473–477 (1978)

Atkinson, P.R., Weidman, E.R., Bhaumick, B., Bala, R.M.: Release of somatomedin-like activity by cultured WI-38 human fibroblasts. Endocrinology 106, 2006–2012 (1980)

August, G.R., Cheng, R.R., Hung, W., Houck, J.C.: Fibroblast proliferative activity in the sera of growth hormone deficient patients. Horm. Metab. Res. *5*, 340–341 (1973)

Bala, R.M., Bhaumick, B.: Radioimmunoassay of a basic somatomedin: Comparison of various assay techniques and somatomedin levels in various sera. J. Clin. Endocrinol. Metab. *49*, 770–777 (1979)

Bala, R.M., Wright, C., Bardai, A., Smith, G.R.: Somatomedin bioactivity in serum and amniotic fluid during pregnancy. J. Clin. Endocrinol. Metab. *46*, 649–652 (1978)

Beaton, G.R., Singh,V.: Age dependent variation in cartilage response to somatomedin. Pediatr. Res. *9*, 683 (1975)

Bhaumick, B., Bala, R.M.: Purification of a basic somatomedin from human plasma Cohn Fraction IV-1 with physicochemical and radioimmunoassay similarity to somatomedin-C and insulin-like growth factor. Can. J. Biochem. *57*, 1289–1298 (1979)

Blundell, T.L., Bedarkar, S., Rinderknecht, E., Humbel, R.E.: Insulin-like growth factor: A model for tertiary structure accounting for immunoreactivity and receptor binding. Proc. Natl. Acad. Sci. U.S.A. *75*, 180–184 (1978)

Bürgi, H., Müller, W.A., Humbel, R.E., Labhart, A., Froesch, E.R.: Non-suppressible insulin-like activity of human serum. I. Physicochemical properties, extraction and partial purification. Biochim. Biophys. Acta *121*, 349–359 (1966)

Burstein, P.J., Schalch, D.S., Heinrich, U.E., Johnson, C.J.: The effects of hypothyroidism on somatomedin IGF (insulin-like growth factor). Clin. Res. *26*, 411 A (1978)

Burstein, P.J., Draznin, B., Johnson, C.J., Schalch, D.S.: The effect of hypothyroidism on growth, serum growth hormone and growth hormone dependent somatomedin, insulin-like growth factor, and its carrier protein in rats. Endocrinology *104*, 1107–1111 (1979)

Canalis, E., Raisz, L.G.: Effect of multiplication stimulating activity on bone formation in vitro. Presented to the 61st meeting of the Endocrine Society, Anaheim, California, June 18–21 (1979)

Cheek, D.B., Hill, D.E.: The effect of growth hormone on cell and somatic growth. In: Handbook of physiology, Sect. 7: Endocrinology 4, part 2, pp. 159–186. Baltimore, Maryland: Williams and Wilkins Company 1974

Chesley, L.C.: Growth hormone activity in human pregnancy. I. Serum sulfation factor. Am. J. Obstet. Gynecol. *84*, 1075–1080 (1962)

Clemmons, D.R., Van Wyk, J.J.: Somatomedin-C and platelet derived growth factor stimulate human fibroblast replication. J. Cell Physiol. 106, 361–367 (1981 b)

Clemmons, D.R., Hintz, R.L., Underwood, L.E., Van Wyk, J.J.: Common mechanism of action of somatomedin and insulin on fat cells: Further evidence. Isr. J. Med. Sci. *10*, 1254–1262 (1974)

Clemmons, D.R., Van Wyk, J.J., Ridgway, E.C., Kliman, B., Kjellberg, R.N., Underwood, L.E.: Evaluation of acromegaly by radioimmunoassay of somatomedin-C. New Engl. J. Med. *301*, 1138–1142 (1979)

Clemmons, D.R., Underwood, L.E., Ridgway, E.G., Kliman, B., Kjellberg, R.N., Van Wyk, J.J.: Estradiol treatment of acromegaly: Reduction of immunoreactive somatomedin-C and improvement of metabolic status. Am. J. Med. *69*, 571–575 (1980 a)

Clemmons, D.R., Underwood, L.E., Van Wyk, J.J.: Hormonal control of immunoreactive somatomedin production by cultured human fibroblasts. J. Clin. Invest. *67*, 10–17 (1981 a)

Clemmons, D.R., Van Wyk, J.J., Pledger, W.J.: Sequential addition of platelet factor and plasma to Balb/c 3 T 3 fibroblast cultures stimulates somatomedin-C binding early in the cell cycle. Proc. Natl. Acad. Sci. U.S.A. *77*, 6644–6648 (1980 b)

Clemmons, D.R., Underwood, L.E., Ridgway, E.C., Kliman, B., Van Wyk, J.J.: Hyperprolactinemia is associated with increased immunoreative somatomedin-C in hypopituitarism. J. Clin. Endocrinol. Metab. 52, 731–735 (1981 c)

Cohen, K.L., Nissley, S.P.: The serum half-life of somatomedin activity: evidence for growth hormone dependence. Acta Endocrinol. (Kbh.) *83*, 243–258 (1976)

Copeland, K.C., Underwood, L.E., Van Wyk, J.J.: Induction of immunoreactive somatomedin-C in human serum by growth hormone: dose response relationships and effect on chromatographic profiles. J. Clin. Endocrinol. Metab. *50:* 690–697 (1980)

Daughaday, W.H., Kapadia, M.: Maintenance of serum somatomedin activity in hypophysectomized pregnant rats. Endocrinology *102*, 1317–1320 (1978)

Daughaday, W.H., Kipnis, D.M.: The growth promoting and anti-insulin actions of somatotropin. Recent Prog. Horm. Res. *22*, 49–99 (1966)

Daughaday, W.H., Mariz, I.K.: Conversion of proline-C^{14} to labeled hydroxyproline by rat cartilage in vitro: effects of hypophysectomy, growth hormone, and cortisol. J. Lab. Clin. Med. *59*, 741–752 (1962)

Daughaday, W.H., Reeder, C.: Synchronous activation of DNA synthesis in hypophysectomized rat cartilage by growth hormone. J. Lab. Clin. Med. *68*, 357–368 (1966)

Daughaday, W.H., Salmon, W.D. jr., Alexander, F.: Sulfation factor activity of sera from patients with pituitary disorders. J. Clin. Endocrinol. Metab. *19*, 743–758 (1959)

Daughaday, W.H., Heins, J.N., Srivastava, L., Hammer, C.: Sulfation factor: studies of its removal from plasma and metabolic fate in cartilage. J. Lab. Clin. Med. *72*, 803–812 (1968)

Daughaday, W.H., Hall, K., Raben, M.S., Salmon, W.D. jr., Brande, J.L. van den, Van Wyk, J.J.: Somatomedin: Proposed designation for sulphation factor. Nature *235*, 107 (1972)

Daughaday, W.H., Herington, A.C., Phillips, L.S.: The regulation of growth by endocrines. Annu. Rev. Physiol. *37*, 211–244 (1975 a)

Daughaday, W.H., Phillips, L.S., Herington, A.C.: Measurement of somatomedin by cartilage in vitro. Methods Enzymol. *37 B*, 93–109 (1975 b)

Daughaday, W.H., Phillips, L.S., Mueller, M.C.: The effects of insulin and growth hormone on the release of somatomedin by the isolated rat liver. Endocrinology 98, 1214–1219 (1976)

Daughaday, W.H., Mariz, I.K., Daniels, J.S., Jacobs, J.W., Rubin, J.S., Bradshaw, R.A.: Studies on rat somatomedin. In: Somatomedins and growth. Giordano, G., Van Wyk, J.J., Minuto, F. (eds.) New York: Academic Press 1979

D'Ercole, A.J., Foushee, D.B., Underwood, L.E.: Somatomedin-C receptor ontogeny and levels in porcine fetal and human cord serum. J. Clin. Endocrinol. Metab. *43*, 1069–1077 (1976 a)

D'Ercole, A.J., Underwood, L.E., Van Wyk, J.J., Decedue, C.J., Foushee, D.B.: Specificity, topography, and ontogeny of the somatomedin receptor in mammalian tissues. In: Growth Hormone and Related Peptides, pp. 190–201. Amsterdam: Excerpta Medica 1976 b

D'Ercole, A.J., Underwood, L.E., Van Wyk, J.J.: Serum somatomedin-C in hypopituitarism and in other disorders of growth. J. Pediatr. *90*, 375–381 (1977)

D'Ercole, A.J., Applewhite, G.T., Underwood, L.E.: Evidence that somatomedin is synthesized by multiple target tissues in the fetus. Dev. Biol. *75*, 315–328 (1980)

DeLarco, J.E., Todaro, G.J.: A human fibrosarcoma cell line producing multiplication stimulating activity (MSA)-related peptides. Nature *272*, 356–385 (1978)

Draznin, B., Burnstein, D.J., Johnson, C.J., Emler, C.A., Schalch, D.S.: Effect of hypopituitarism and hypothyroidism on insulin-like growth factor and its carrier protein. Presented at the 60th meeting of the Endocrine Society, Miami, Florida, June 14–16 (1978)

Dulak, N.C., Temin, H.M.: A partially purified polypeptide fraction from rat liver cell conditioned medium with multiplication stimulating activity for embryo fibroblasts. J. Cell Physiol. *81*, 153–160 (1973 a)

Dulak, N.C., Temin, H.M.: Multiplication stimulating activity for chicken embryo fibroblasts from rat liver cell conditioned medium: a family of small polypeptides. J. Cell Physiol. *81*, 161–170 (1973 b)

Eisenbarth, G.S., Lebovitz, H.E.: Isolation and characterization of a serum inhibitor of cartilage metabolism. Endocrinology *95*, 1600–1607 (1974)

Eisenbarth, G.S., Beuttel, S.C., Lebovitz, H.E.: Fatty acid inhibition of somatomedin (serum sulfation factor) stimulated protein and RNA synthesis in embryonic chicken cartilage. Biochim. Biophys. Acta *331*, 397–409 (1973)

Elders, M.J., Wingfield, B.S., McNatt, M.L., Clarke, J.S., Hughes, E.R.: Glucocorticoid therapy in children: effect on somatomedin secretion. Am. J. Dis. Child *129*, 1393–1396 (1975)

Elders, M.J., McNatt, M.L., Kilgore, B.S., Hughes, E.R.: Glucocorticoid inhibition of glycosaminoglycan biosynthesis: Decrease of acceptor protein. Biochem. Biophys. Res. Commun. *77*, 557–565 (1977)

Ellis, S., Huble, J., Simpson, M.E.: Influence of hypophysectomy and growth hormone on cartilage sulfate metabolism. Proc. Soc. Exp. Biol. Med. *84*, 603–605 (1953)

Florini, J.R., Nicholson, M.L., Dulak, N.C.: Effects of peptide anabolic hormones on growth of myoblasts in culture. Endocrinology *101*, 32–41 (1977)

Froesch, E.R., Bürgi, H., Ramseier, E.B., Bally, P., Labhart, A.: Antibody suppressible and non-suppressible insulin-like activities in human serum and their physiologic significance. J. Clin. Invest. *42*, 1816–1834 (1963)

Froesch, E.R., Müller, W.A., Bürgi, H., Waldvogel, M., Labhart, A.: Nonsuppressible insulin-like activity of human serum. II. Biological properties of plasma extracts with nonsuppressible insulin-like activity. Biochim. Biophys. Acta *121*, 360–374 (1966)

Froesch, E.R., Zapf, J., Audhya, T.K., Ben-Porath, E., Segen, B.J., Gibson, K.D.: Nonsuppressible insulin-like activity and thyroid hormones: major pituitary-dependent sulfation factors for chick embryo cartilage. Proc. Natl. Acad. Sci. U.S.A. *73*, 2904–2908 (1976)

Fryklund, L., Uthne, K., Sievertsson, H.: Identification of two somatomedin-A active polypeptides and in vivo effects of a somatomedin-A concentrate. Biochem. Biophys. Res. Commun. *61*, 957–962 (1974)

Fryklund, L., Skottner, A., Forsman, A.: Somatomedin A and B: Chemistry and Biology. In: Giordano, G., Van Wyk, J.J., Minuto, F. (eds.), pp. 7–17. Somatomedins and growth. New York: Academic Press 1979

Furlanetto, R.W.: Somatomedin-C binding protein: Evidence for a heterologous subunit structure. J. Clin. Endocrinol. Metab. *51*, 12–19 (1980)

Furlanetto, R.W., Underwood, L.E., Van Wyk, J.J., D'Ercole, A.J.: Estimation of somatomedin-C levels in normals and patients with pituitary disease by radioimmunoassay. J. Clin. Invest. *60*, 648–657 (1977)

Furlanetto, R.W., Underwood, L.E., Van Wyk, J.J., Handwerger, S.: Serum immunoreactive somatomedin-C is elevated late in pregnancy. J. Clin. Endocrinol. Metab. *47*, 695–697 (1979)

Ginsberg, B.H., Kahn, C.R., Roth, J., Megyesi, K., Baumann, G.: Identification and high yield purification of insulin-like growth factors (nonsuppressible insulin-like activities and somatomedins) from human plasma by use of endogenous binding proteins. J. Clin. Endocrinol. Metab. *48*, 43–49 (1979)

Gospodarowicz, D., Moran, J.J.: Stimulation of division of sparse and confluent 3T3 cell populations by a fibroblast growth factor, dexamethasone and insulin. Proc. Natl. Acad. Sci. U.S.A. *71*, 4584–4588 (1974)

Grant, D.B., Hambley, J., Becker, D., Pimstone, B.L.: Reduced sulphation factor in undernourished children. Arch. Dis. Child. *48*, 596–600 (1973)

Guyda, H., Posner, B., Rappaport, R.: Growth hormone (GH) and prolactin (PRL) dependence of somatomedin determined by radioreceptor assay (RRA) and bioassay (SM). Clin. Res. *25*, 681A (1977)

Guyda, H.J., Corvol, M.T., Rappaport, R., Posner, B.I.: Radioreceptor assay of insulin-like peptide in human plasma: Growth hormone dependence and correlation with sulfation activity by two bioassays. J. Clin. Endocrinol. Metab. *79*, 739–747 (1979)

Hakim, E.: Isolation of a growth stimulating agent from human fibroblast cultures. Experimentia *34*, 1515–1517 (1978)

Hall, K.: Quantitative determination of the sulphation factor activity in human serum. Acta Endocrinol. (Kbh.) *63*, 338–350 (1970)

Hall, K.: Effect of intravenous administration of human growth hormone on sulphation factor activity in serum of hypopituitary subjects. Acta Endocrinol. (Kbh.) *66*, 491–497 (1971)

Hall, K.: Human somatomedin determination, occurence, biological activity, and purification. Acta Endocrinol. Suppl. *163*, 1–52 (1972)

Hall, K., Uthne, K.: Some biological properties of purified sulfation factor (SF) from human plasma. Acta Med. Scand. *190*, 137–143 (1971)

Hall, K., Uthne, K.: Human growth hormone and sulfation factor. In: Growth and growth hormone. Pecile, A., Mueller, E. (eds.), pp. 192–198. Amsterdam: Excerpta Medica 1972

Hall, K., Van Wyk, J.J.: Somatomedin. In: Current topics in experimental endocrinology. James, V.H., Martin, L. (eds.), vol. 2, pp. 156–178. New York: Academic Press 1974

Hall, K., Brandt, J., Engerg, G., Fryklund, L.: Immunoreactive somatomedin-A in human serum. J. Clin. Endocrinol. Metab. *48*, 271–278 (1979)

Hassalbacher, G.F., Humbel, R.E.: Stimulation of ornithine decarboxylase activity in chick embryo fibroblasts by nonsuppressible insulin-like activity, insulin and serum. J. Cell. Physiol. *88*, 239–245 (1976)

Heaton, J.H., Schilling, E.E., Gelehiter, T.D., Rechler, M.M., Spencer, C.J., Nissley, S.P.: Induction of tyrosine aminotransferase in HTC rat hepatoma cells by insulin and insulin-like growth factor MSA: Mediation by insulin receptors and MSA receptors. BBA *632*, 198–203 (1980)

Heinrich, W.E., Schalch, D.R., Koch, J.G., Johnson, C.J.: Nonsuppressible insulin-like activity (NSILA) II. Rgulation of serum concentrations by growth hormone and insulin. J. Clin. Endocrinol. Metab. *46*, 672–678 (1978)

Heins, J.N., Garland, J.T., Daughaday, W.H.: Incorporation of ^{35}S-sulfate into rat cartilage explants in vitro: effects of aging on responsiveness to stimulation by sulfation factor. Endocrinology *87*, 688–692 (1970)

Hintz, R.L., Liu, F.: Demonstration of specific plasma protein binding sites for somatomedin. J. Clin. Endocrinol. Metab. *45*, 988–995 (1977)

Hintz, R.L., Clemmons, D.R., Underwood, L.E., Van Wyk, J.J.: Competitive binding of somatomedin to the insulin receptors of adipocytes, chondrocytes, and liver membranes. Proc. Natl. Acad. Sci. U.S.A. *69*, 2351–2353 (1972a)

Hintz, R.L., Clemmons, D.R., Van Wyk, J.J.: Growth hormone induced somatomedin-like activity from liver. Pediatr. Res. *6*, 88 (1972b)

Hintz, R.L., Seeds, J.M., Johnsonbaugh, R.E.: Somatomedin and growth hormone in the newborn. Am. J. Dis. Child. *131*, 1249–1251 (1977)

Hintz, R.L., Suskind, R., Amatayakul, K., Thanangkul, O., Olson, R.: Plasma somatomedin and growth hormone values in children with protein-caloric malnutrition. J. Pediatr. *92*, 153–156 (1978)

Hintz, R.L., Liu, F., Marshall, L.B., Chang, L.: Interaction of somatomedin-C with antibody directed against the synthetic C-peptide region of insulin-like growth factor I. J. Clin. Endocrinol. Metab. *50*, 405–407 (1980)

Holley, R.W., Kiernan, J.A.: Control of the initiation of DNA synthesis in 3 T 3 cells: serum factors. Proc. Natl. Acad. Sci. U.S.A. *71*, 2908–2911 (1974)

Horner, J.M., Hintz, R.L.: Further comparisons of the ^{125}I-somatomedin-A and the ^{125}I-somatomedin-C radioreceptor assays of somatomedin peptide. J. Clin. Endocrinol. Metab. *48*, 959–963 (1979)

Horner, J.M., Liu, F., Hintz, R.L.: Comparison of ^{125}I-somatomedin-A and ^{125}I-somatomedin-C radioreceptor assays for somatomedin peptide content in whole and acid-chromatographed plasma. J. Clin. Endocrinol. Metab. *47*, 1287–1295 (1978)

Hurley, T.W., D'Ercole, A.J., Handwerger, S., Underwood, L.E., Furlanetto, R.W., Fellows, R.E.: Ovine placental lactogen induces somatomedin: a possible role in fetal growth. Endocrinology *101*, 1635–1638 (1977)

Hutchings, S.E., Sato, G.H.: Growth and maintenance of HeLa cells in serum-free medium supplemented with hormones. Proc. Natl. Acad. Sci. U.S.A. *75*, 901–904 (1978)

Jacobs, J.W., Rubin, J.S., Hoagli, T.E., Bogardt, R.A., Mariz, I.K., Daniels, J.J., Daughaday, W.H., Bradshaw, R.A.: Purification, characterization and amino acid sequence of rat anaphylatoxin (C-3 a). Biochemistry *17*, 5031–5038 (1978)

Jakob, A., Hauri, C.H., Froesch, E.R.: Nonsuppressible insulin-like activity in human serum. III. Differentiation of two distinct molecules with nonsuppressible ILA. J. Clin. Invest. *47*, 2678–2688 (1968)

Kaufmann, U., Zapf, J., Torretti, B., Froesch, E.R.: Demonstration of a specific serum carrier protein of nonsuppressible insulin-like activity in vivo. J. Clin. Endocrinol. Metab. *44*, 160–166 (1977)

Kaufmann, U., Zapf, J., Froesch, E.R.: Growth hormone dependence of nonsuppressible insulin-like activity (NSILA) and of NSILA-carrier protein in rats. Acta Endocrinol. (Kbh.) *87*, 716–727 (1978)

Keret, R., Schimpff, R.M., Girard, F.: The inhibitory effect of hydrocortisone on the chicken embryo cartilage somatomedin assay. Horm. Res. *7*, 254–259 (1976)

Kilgore, B.S., McNatt, M.L., Meadors, S., Leex, J.A., Hughes, E.R., Elders, M.J.: Alterations of glycosaminoglycan protein acceptor by somatomedin and inhibition by cortisol. Pediatr. Res. *13*, 96–99 (1979)

King, G.L., Kahn, C.R., Rechler, M.M., Nissley, S.P.: Direct demonstration of separate receptors for growth and metabolic activities of insulin and insulin-like growth factors in rat adipocytes and human fibroblasts. J. Clin. Invest. *66*, 130–140 (1980)

Koumans, J., Daughaday, W.H.: Amino acid requirement for activity of partially purified sulfation factor. Trans. Assoc. Am.Physicians *76*, 152–172 (1963)

Lebovitz, H.E., Drezner, M.K., Neelon, F.A.: Evidence for a role of adenosine 3'5'-monophosphate in growth hormone-dependent serum sulfation factor (somatomedin) action on cartilage. In: Growth hormone and related peptides. Pecile, A., Mueller, E. (eds.), pp. 202–215. Amsterdam: Excerpta Medica 1976

Liberti, J.P.: Purification of bovine somatomedin. Biochem. Biophys. Res. Commun. *67*, 1226–1231 (1975)

Liberti, J.P., Miller, M.S.: The purification and partial characterization of human somatomedin-C. J. Biol. Chem. *255*, 1023–1033 (1980)

Marshall, R.N., Underwood, L.E., Voina, S.J., Foushee, D.B., Van Wyk, J.J.: Characterization of the insulin and somatomedin-C receptors in human placental cell membranes. J. Clin. Endocrinol. Metab. *39*, 283–292 (1974)

McConaghey, P., Sledge, C.B.: Production of "sulphation factor" by the perfused liver. Nature *225*, 1249–1250 (1970)

Megyesi, K., Kahn, C.R., Roth, J. et al.: Insulin and nonsuppressible insulin-like activity (NSILA-s): evidence for separate plasma membrane receptor sites. Biochem. Biophys. Res. Commun. *57*, 307–315 (1974)

Megyesi, K., Kahn, C.R., Roth, J., Gorden, P.: Circulating NSILA-s in man: Preliminary studies of stimuli in vivo and of binding to plasma components. J. Clin. Endocrinol. Metab. *41*, 475–484 (1975)

Megyesi, K., Gordon, P.H., Kahn, C.R.: Lack of a simple relationship between endogenous growth hormone and NSILA-s related peptides. J. Clin. Endocrinol. Metab. *45*, 330–338 (1977)

Meuli, C., Froesch, E.R.: Effects of insulin and of NSILA-s on the perfused rat heart: glucose uptake, lactate production and efflux of 3-O-methyl glucose. Eur. J. Clin. Invest. *5*, 93–99 (1975)

Meuli, C., Froesch, E.R.: Binding of insulin and nonsuppressible insulin-like activity to isolated perfused rat heart muscle: Evidence for two separate binding sites. Arch. Biophys. Biochem. *177*, 31–38 (1976)

Meuli, C., Zapf, J., Froesch, E.R.: NSILA-carrier protein abolishes the action of nonsuppressible insulin-like activity (NSILA-s) on perfused rat heart. Diabetologia *14*, 255–259 (1978)

Millis, A.J., Hoyle, M., Field, B.: Human fibroblast conditioned media contains growth promoting activities for low density cells. J. Cell Physiol. *93*, 17–24 (1978)

Minuto, F., Grimaldi, P., Giusti, M., Baiardi, M., Cocco, R., Giordano, G.: Study of somatomedin-C and growth hormone circadian secretion. In: Somatomedins and growth. Giordano, G., Van Wyk, J.J., Minuto, F. (eds.), pp. 285–288. New York: Academic Press 1979

Morrell, B., Froesch, E.R.: Fibroblasts as an experimental tool in metabolic and hormone studies. II. Effects of insulin and nonsuppressible insulin-like activity (NSILA-s) on fibroblasts in culture. Eur. J. Clin. Invest. *3*, 119–123 (1973)

Moses, A.C., Nissley, S.P., Cohen, K.L., Rechler, M.M.: Specific binding of a somatomedin-like polypeptide in rat serum depends on growth hormone. Nature *263*, 137–140 (1976)

Moses, A.C., Cohen, K.L., Johnsonbaugh, R., Nissley, S.P.: Contribution of human somatomedin activity to the serum growth requirement of human skin fibroblasts and chick embryo fibroblasts in culture. J. Clin. Endocrinol. Metab. *46*, 937–946 (1978)

Moses, A.C., Nissley, S.P., Passamani, J., White, R.M.: Further characterization of growth hormone dependent somatomedin binding proteins in rat serum and demonstration of somatomedin binding proteins produced by rat liver cells in culture. Endocrinology *104*, 536–546 (1979a)

Moses, A.C., Nissley, S.P., Rechler, M.M., Short, P.A., Podskalny, J.M.: The purification and characterization of multiplication stimulating activity (MSA) from media conditioned by a rat liver cell line. In: Somatomedins and growth. Giordano, G., Van Wyk, J.J., Minuto, F. (eds.). London: Academic Press 1979b

Moses, A.C., Nissley, S.P., Short, P.A., Rechler, M.M.: Purification and characterization of multiplication stimulating activity, insulin-like growth factors purified from rat liver cell conditioned medium. Eur. J. Biochem. *103*, 387–400 (1980a)

Moses, A.C., Nissley, S.P., Short, P.A., Rechler, M.M., White, R.M., Knight, A.B., Higa, O.Z.: Elevated levels of insulin-like growth factor, multiplication stimulating activity in fetal rat serum. Proc. Natl. Acad. Sci. U.S.A. *77*, 3649–3653 (1980b)

Nissley, S.P., Rechler, M.M.: Multiplication stimulating activity (MSA): A somatomedin-like polypeptide from cultured rat liver cells. In: Decennial Review Conference in Cell, Tissue and Organ Cultures, NCI Monograph No. 48, pp. 167–178 (1977)

Nissley, S.P., Passamani, J., Short, P.: Stimulation of DNA synthesis, cell multiplication and ornithine decarboxylase in 3 T 3 cells by multiplication stimulating activity (MSA). J. Cell Physiol. *89*, 393–402 (1977a)

Nissley, S.P., Short, P.A., Rechler, M.M., Podskalny, J.M.: Proliferation of buffalo rat liver cells in serum from medium does not depend upon multiplication stimulating activity (MSA). Cell *11*, 441–446 (1977b)

Oelz, O., Jakob, A., Froesch, E.R.: Nonsuppressible insulin-like activity (NSILA) of human serum. V. Hypoglycaemia and preferential metabolic stimulation of muscle by NSILA-s. Eur. J. Clin. Invest. *1*, 48–53 (1970)

Oelz, O., Froesch, E.R., Bunzli, H.F., Humbel, R.E., Ritschard, W.J.: Antibody-suppressible and nonsuppressible insulin-like activities. In: Handbook of physiology. Steiner, D.F., Freinkel, N. (eds.), Sect. 7, vol. 7, pp. 685–702. Baltimore: Williams & Wilkins (1972)

Pantazis, N.J., Blanchard, M.H., Aranson, B.G., Young, M.: Molecular properties of nerve growth factor secreted by L cells. Proc. Natl. Acad. Sci. U.S.A. *74*, 1492–1496 (1977)

Parker, M.L., Daughaday, W.H.: Growth retardation: correlation of plasma GH responses to insulin and arginine with subsequent metabolic and skeletal responses to GH treatment. In: Growth hormone and related peptides. Pecile, A., Muller, E.E. (eds.), pp. 398–407. Amsterdam: Excerpta Medica 1968

Phillips, L.S., Daughaday, W.H.: Bioassay of somatomedin. In: James VHT, ed. Endocrinology: Proceedings of the V International Congress of Endocrinology, vol. 2, pp. 150–161. Amsterdam: Excerpta Medica (1977)

Phillips, L.S., Young, H.S.: Nutrition and somatomedin. I. Effects of fasting and refeeding on serum somatomedin activity and cartilage growth activity in rats. Endocrinology *99*, 304–314 (1976)

Phillips, L.S., Herington, A.c., Daughaday, W.H.: Somatomedin stimulation of sulfate incorporation in porcine costal cartilage discs. Endocrinology *94*, 856–863 (1974)

Phillips, L.S., Herington, A.C., Daughaday, W.H.: Steroid hormone effects on somatomedin. I. Somatomedin action in vitro. Endocrinology *97*, 780–786 (1975)

Phillips, L.S., Herington, A.C., Karl, I.E., Daughaday, W.H.: Comparison of somatomedin activity in perfusates of normal and hypophysectomized rat livers with and without added growth hormone. Endocrinology *98*, 606–614 (1976)

Phillips, L.S., Orawski, A.T., Belosky, D.C.: Somatomedin and nutrition. IV. Regulation of somatomedin activity and growth cartilage activity by quantity and composition of diet in rats. Endocrinology *103*, 121–127 (1978)

Phillips, L.S., Belosky, D.C., Young, H.S., Reichard, L.A.: Nutrition and somatomedin. VI. Somatomedin activity and somatomedin inhibitory activity in serum from normal and diabetic rats. Endocrinology *104*, 1519–1524 (1979)

Pierson, R.W. jr., Temin, H.M.: The partial purification from calf serum of a fraction with multiplication stimulating activity for chicken fibroblasts in cell culture and with nonsuppressible insulin-like activity. J. Cell Physiol. *79*, 319–329 (1972)

Pimstone, B., Shapiro, B.: Somatomedin in human and experimental protein energy malnutrition. In: Somatomedins and growth. Giordano, G., Van Wyk, J.J., Minuto, F. (eds.), pp. 325–328. New York: Academic Press 1979

Pledger, W.J., Stiles, C.D., Antoniades, H.N., Scher, C.D.: Induction of DNA synthesis in Balb/c 3T3 cells by serum components: a reevaluation of the commitment process. Proc. Natl. Acad. Sci. U.S.A. *74*, 4481–4487 (1977)

Pledger, W.J., Stiles, C.D., Antoniades, H.N., Scher, C.D.: An ordered sequence of events is required before Balb/c 3T3 cells become committed to DNA synthesis. Proc. Natl. Acad. Sci. U.S.A. *75*, 2839–2843 (1978)

Poggi, C., LeMarchard-Brustel, Y., Zapf, J., Froesch, E.R., Freychet, P.: Effects and binding of insulin-like growth factor I in isolated soleus muscle of lean and obese mice: Comparison with insulin. Endocrinology *105*, 723–730 (1979)

Posner, B.I., Guyda, H.J., Corvol, M.T., Rappaport, R., Harley, C., Goldstein, S.: Partial purification, characterization, and assay of a slightly acicid insulin-like peptide (ILAs) from human plasma. J. Clin. Endocrinol. Metab. *47*, 1240–1250 (1978)

Postel-Vinay, M.C., Guyda, H., Posner, B., Corvol, M.T.: Production of insulin-like activity (ILAs) by rat hepatocytes in short term culture. In: Somatomedins and growth. Giordano, G., Van Wyk, J.J., Minuto, F. (eds.), pp. 111–118. New York: Academic Press 1979

Rechler, M.M., Podskalny, J.M., Goldfine, I.D.: DNA synthesis in human fibroblasts: stimulation by insulin and nonsuppressible insulin-like activity (NSILAs). J. Clin. Endocrinol. Metab. *39*, 512–521 (1975)

Rechler, M.M., Podskalny, J.M., Nissley, S.P.: Interaction of multiplication stimulating activity with chick embryo fibroblasts demonstrates a growth receptor. Nature *259*, 134–136 (1976)

Rechler, M.M., Nissley, S.P., Podskalny, J.M., Moses, A.C., Fryklund, L.: Identification of a receptor for somatomedin-like polypeptides in human fibroblasts. J. Clin. Endocrinol. Metab. *44*, 820–831 (1977a)

Rechler, M.M., Podskalny, J.M., Nissley, S.P.: Characterization of the binding of multiplication stimulating activity to a receptor for growth polypeptides in chick embryo fibroblasts. J. Biol. Chem. *252*, 3898–3910 (1977b)

Rechler, M.M., Fryklund, L., Nissley, S.P., Hall, K., Podskalny, J.M., Skottner, A., Moses, A.C.: Purified human somatomedin-A and rat multiplication stimulating activity: mitogens for cultured fibroblasts that crossreact with the same growth peptide receptors. Eur. J. Biochem. *82*, 5–12 (1978)

Rechler, M.M., Eisen, H.J., Higa, O.Z., Nissley, S.P., Moses, A.C., Shilling, E.E., Fernoy, I., Bruni, C.B., Phillips, L.S., Baird, K.L.: Characterization of somatomedin (insulin-like growth factor) synthesized by fetal rat organ cultures. J. Biol. Chem. *254*, 7942–7950 (1979)

Rendall, J.L., Dechler, H.K., Lebovitz, H.L.: Cyclic 3'5' Adenosine monophosphate inhibition of sulfation factor activity. Biochem. Biophys. Res. Commun. *46*, 1425–1429 (1972)

Renner, R., Hepp, K.D., Humbel, R.F., Froesch, E.R.: Mechanism of antilipolytic action of NSILA-s: Inhibition of adenylate cyclase in lipocyte ghosts. Hormone Metab. Res. *5*, 56–57 (1973)

Rinderknecht, E., Humbel, R.E.: Polypeptides with non-suppressible insulin-like and cell growth promoting activities in human serum: Isolation, chemical characterization, and some biological properties of forms I and II. Proc. Natl. Acad. Sci. U.S.A. *73*, 2365–2369 (1976a)

Rinderknecht, E., Humbel, R.E.: Amino terminal sequences of two polypeptides from human serum with non-suppressible insulin-like and cell growth-promoting activities: Evidence for structural homology with insulin B chain. Proc. Natl. Acad. Sci. U.S.A. *73*, 4379–4381 (1976b)

Rinderknecht, E., Humbel, R.E.: The amino acid sequence of human insulin-like growth factor I and its structural homology with proinsulin. J. Biol. Chem. *253*, 2769–2776 (1978a)

Rinderknecht, E., Humbel, R.E.: Primary structure of human insulinlike growth factor II. FEBS Letters *89*, 283–286 (1978b)

Rothstein, H., Van Wyk, J.J., Hayden, J.H., Gordon, S.R., Weinseider, A.R.: An in vivo study with somatomedin-C: Restoration of cycle traverse in G_0/G_1 blocked cells of hypophysectomized animals. Science *208*, 410–412 (1980)

Rutherford, R.B., Ross, R.: Platelet factors stimulate fibroblasts and smooth muscle cells quiescent in plasma serum to proliferate. J. Cell Biol. *69*, 196–202 (1976)

Salmon, W.D. jr.: Effects of somatomedin on cartilage metabolism: further observations on an inhibitory serum factor. In: Raiti, S. (ed.), Advances in human growth hormone research. Washington, D.C.: Government Printing Office, pp. 76–97 [DHEW publication no. (NIH) 74-612] 1974

Salmon, W.D. jr.: Interaction of somatomedin and a peptide inhibitor in serum of hypophysectomized and starved, pituitary-intact rats. Adv. Metab. Disord. *8*, 183–199 (1975)

Salmon, W.D. jr., Daughaday, W.H.: A hormonally controlled serum factor which stimulates sulfate incorporation by cartilage in vitro. J. Lab. Clin. Med. *49*, 825–836 (1957)

Salmon, W.D. jr., Du Vall, M.R.: A serum fraction with "sulfation factor activity" stimulates in vitro incorporation of leucine and sulfate into proteinpolysaccharide complexes, uridine into RNA and thymidine into DNA of costal cartilage from hypophysectomized rats. Endocrinology *86*, 721–727 (1970a)

Salmon, W.D. jr., Du Vall, M.R.: In vitro stimulation of leucine incorporation into muscle and cartilage protein by a serum fraction with sulfation factor activity: differentiation of effects from those of growth hormone and insulin. Endocrinology *87*, 1168–1180 (1970b)

Salmon, W.D. jr., Hagen, M.J. von, Thompson, E.Y.: Effects of puromycin and actinomycin in vitro on sulfate incorporation by cartilage of the rat and its stimulation by serum sulfation factor and insulin. Endocrinology *80*, 999–1005 (1967)

Salmon, W.D. jr., Du Vall, M.R., Thompson, E.Y.: Stimulation by insulin in vitro of incorporation of [^{35}S] sulfate and [^{14}C] leucine into protein-polysaccharide complexes, [^{3}H] uridine into RNA, and [^{3}H] thymidine into DNA of costal cartilage from hypophysectomized rats. Endocrinology *82*, 493–499 (1968)

Samaan, N.A., Dempster, W.J., Fraser, R., Please, N.W., Stillman, D.: Further immunological studies on the form of circulating insulin. J. Endocrinol. *24*, 263–277 (1962)

Schalch, D.S., Heinrich, U.E., Koch, J.G., Johnson, C.J., Schlueter, R.J.: Nonsuppressible insulin-like activity (NSILA). I. Development of a new sensitive competitive protein-binding assay for determination of serum levels. J. Clin. Endocrinol. Metab. *46*, 664–671 (1978)

Schalch, D.S., Heinrich, U.E., Draznin, B., Johnson, C.J., Miller, L.L.: Role of the liver in regulating somatomedin activity: Hormonal effects on the synthesis and release of insulin-like growth factor and its carrier protein by the isolated perfused rat liver. Endocrinology *104*, 1143–1151 (1979)

Schimpff, R.M., Donnadieu, M.: Quantitative determination of somatomedin in human serum (^{35}S uptake by embryonic chick cartilage). Biomedicine *19*, 142–147 (1973)

Schimpff, R.M., Donnadieu, M., Gasinovic, J.C., Warnet, J.M., Girard, F.: The liver as source of somatomedin (an in vivo study in the dog). Acta Endocrinol. (Kbh.) *83*, 365–372 (1976)

Schimpff, R.M., Lebrec, D., Donnadieu, M.: Somatomedin production in normal adults and cirrhotic patients. Acta Endocrinol. (Kbh.) *86*, 355–362 (1977)

Schlumpf, U., Heimann, R., Zapf, J., Froesch, E.R.: Nonsuppressible insulin-like activity and sulphation activity in serum extracts of normal subjects, acromegalics and pituitary dwarfs. Acta Endocrinol. (Kbh.) *81*, 28–42 (1976)

Schwartz, E., Echemendia, E., Schiffer, M., Panariello, V.A.: Mechanism of estrogenic action in acromegaly. J. Clin. Invest. *48*, 260–270 (1969)

Shapiro, B., Pimstone, B.: Somatomedin Release from isolated perfused livers of protein malnourished normal rats in response to growth hormone and insulin. In: Somatomedins and growth. Giordano, G., Van Wyk, J.J., Minuto, F. (eds.), pp. 329–334. New York: Academic Press 1979

Smith, G.L., Temin, H.M.: Purified multiplication stimulating activity from rat liver conditioned medium: Comparison of biologic activities with calf serum insulin and somatomedin. J. Cell Physiol. *84*, 181–192 (1974)

Spencer, E.M.: Synthesis by cultured hepatocytes of somatomedin and its binding protein. FEBS Letters *99*, 157–161 (1979)

Spencer, E.M.: Lack of response of serum somatomedin to hyperprolactinemia in humans. J. Clin. Endocrinol. Metab. *50*, 182–185 (1980)

Spencer, G.S.G., Taylor, A.M.: A rapid simplified bioassay for somatomedin. J. Endocrinol. *78*, 83–88 (1978)

Solomon, S.S., Poffenburger, P.L., Hepp, D.K., Fenstin, L.F., Ensinck, J.W., Williams, R.H.: Quantitation and partial characterization of nonsuppressible insulin-like activity in serum and tissue extracts of the rat. Endocrinology *81*, 213–224 (1967)

Stiles, C.D., Capone, G.T., Scher, C.D., Antoniades, H.N., Van Wyk, J.J., Pledger, W.J.: Dual control of cell growth by somatomedin and platelet derived growth factor. Proc. Natl. Acad. Sci. U.S.A. *76*, 1279–1284 (1979)

Svoboda, M.E., Van Wyk, J.J., Klapper, D.G., Fellows, R.E., Grissom, F.E., Schlueter, R.J.: Purification of somatomedin-C from human plasma: chemistry and biologic properties, partial sequence analysis and relationship to other somatomedins. Biochemistry *19*, 790–797 (1980)

Takano, K., Hall, K., Fryklund, L., Sievertsson, H.: Binding of somatomedins to plasma membranes prepared from rat and monkey tissues. Horm. Metab. Res. *8*, 16–24 (1976a)

Takano, K., Hall, K., Ritzen, M., Iselius, L., Sievertsson, H.: Somatomedin-A in human serum, determined by radioreceptor assay. Acta Endocrinol. (Kbh.) *82*, 449–459 (1976b)

Takano, K., Hizuka, N., Shizume, K., Hayashi, N., Motoike, Y., Obata, H.: Serum somatomedin peptides measured by somatomedin-A radioreceptor assay in chronic liver disease. J. Clin. Endocrinol. Metab. *45*, 828–832 (1977)

Taylor, J.M., Mitchell, L., Cohen, S.: Characterization of high molecular weight form of epidermal growth factor. J. Biol. Chem. *249*, 3198–3206 (1974)

Tell, G.P., Cuatrecasas, P., Van Wyk, J.J., Hintz, R.L.: Somatomedin: inhibition of adenylate cyclase activity in subcellular membranes of various tissues. Science *180*, 312–315 (1972)

Termin, H.M.: Studies on carcinogenesis by avian sarcoma viruses. VI. Differential multiplication of uninfected and of converted cells in response to insulin. J. Cell Physiol. *69*, 377–384 (1967)

Tessler, R.H., Salmon, W.D. jr.: Glucocorticoid inhibition of sulfate incorporation by cartilage of normal rats. Endocrinology *96*, 898–902 (1975)

Underwood, L.E., Hintz, R.L., Voina, S.J., Van Wyk, J.J.: Human somatomedin, the growth hormone-dependent sulfation factor, is antilipolytic. J. Clin. Endocrinol. Metab. *35*, 194–198 (1972)

Uthne, K.: Human somatomedins: Purification and some studies on their biologic actions. Acta Endocrinol. (Kbh.) Suppl. *175*, 1–35 (1973)

Uthne, K.: Preliminary studies of somatomedin in vitro and in vivo in rats. Adv. Metab. Disord. *8*, 115–127 (1975)

Uthne, K., Reagan, C.R., Gimpel, L.P., Kostyo, J.L.: Effects of human somatomedin preparations on membrane transport and protein synthesis in the isolated rat diaphragm. J. Clin. Endocrinol. Metab. *39*, 548–554 (1974)

Van Buul-Offers, S., Dumoleijn, L., Hackeng, W., Hoogerbrogge, C.M., Kortel, R.M., Klundert, P. van de, Brande, J.L. van den: The Snell dwarf mouse: interrelationship of growth in length and weight, serum somatomedin activity and sulfate incorporation in costal cartilage during growth hormone, thyroxine and somatomedin treatment. In: Somatomedins and Growth. Giordano, G., Van Wyk, J.J., Minuto, F. (eds.), pp. 281–283. New York: Academic Press 1979

van den Brande, J.L., Du Caju, M.V.L.: An improved technique for measuring somatomedin activity in vitro. Acta Endocrinol. (Kbh.) *75*, 233–242 (1974a)

van den Brande, J.L., Du Caju, M.V.L.: Plasma somatomedin activity in children with growth disturbances. In: Raiti, S. (ed.), pp. 98–115. Advances in human growth hormone research. Washington, D.C.: Government Printing Office,(DHEW publication no. (NIH 74-612) 1974b

van den Brande, J.L., Van Buul, S., Heinrich, U., Van Roon, F., Zurcher, T., Van Steirtegem, A.C.: Further observations on plasma somatomedin activity in children. Adv. Metab. Disord. *8*, 171–181 (1975)

Van Wyk, J.J., Underwood, L.E.: The somatomedins and their actions. In: Biochemical Actions of Hormones. Litwack, G. (ed.), vol. V, pp. 101–148. New York: Academic Press 1978

Van Wyk, J.J., Underwood, L.E., Hintz, R.L., Clemmons, D.R., Voina, S.J., Weaver, R.P.: The somatomedins: a family of insulin-like hormones under growth hormone control. Recent Prog. Horm. Res. *30*, 259–318 (1974)

Van Wyk, J.J., Underwood, L.E., Baseman, J.B., Hintz, R.L., Clemmons, D.R., Marshall, R.N.: Explorations of the insulin-like and growth promoting properties of somatomedin by membrane receptor assays. Adv. Metab. Disord. *8*, 127–150 (1975)

Van Wyk, J.J., Furlanetto, R.W., Plet, A.S., D'Ercole, A.J., Underwood, L.E.: The somatomedin group of peptide growth factors. In: Decennial Review Conference Cell, Tissue and Organ Culture. Natl. Cancer Inst. Monograph No. 48, pp. 141–148. 1978

Van Wyk, J.J., Svoboda, M.E., Underwood, L.E.: Evidence from radioligand assays that somatomedin-C and insulin-like growth factor I are similar to each other and different from other somatomedins. J. Clin. Endocrinol. Metab. *50*, 206–208 (1980)

Vassilopoulou-Sellin, R., Phillips, L.S., Reichard, L.A.: Nutrition and somatomedin. VII. Regulation of somatomedin activity by the perfused rat liver. Endocrinology *106*, 260–267 (1980)

Vogel, A., Raines, E., Kariya, B., Rivest, M.J., Ross, R.: Coordinate control of 3T3 cell proliferation by platelet-derived growth factor and plasma components. Proc. Natl. Acad. Sci. U.S.A. *76*, 2810–2814 (1978)

Wasteson, A., Westermark, B., Uthne, K.: Somatomedin-A and B: demonstration of two different somatomedin-like components in human plasma. Adv. Metab. Disord. 8, 101–114 (1975)

Weidman, E.R., Atkinson, P., Bala, R.M.: Production of somatomedin and mitogenic effects on human fibroblasts. Presented at the 61st meeting of the Endocrine Society, Anaheim, California, June 13–15, 1979

Weidemann, E., Uthne, K., Tang, R.G., Spender, M., Saito, T., Linfoot, J.A.: Serum somatomedin activity by rat cartilage bioassay and human placental membrane radioreceptor assay in acromegaly and Cushings disease. In: Somatomedins and Growth. Giordano, G., Van Wyk, J.J., Minuto, F. (eds.), vol. 23, pp. 289–294. New York: Academic Press 1979

Weidemann, E., Schwartz, E.: Suppression of growth hormone-dependent human serum sulfation factor by estrogen. J. Clin. Endocrinol. Metab. 34, 51–58 (1972)

Weidemann, E., Schwartz, E., Frantz, A.G.: Acute and chronic estrogen effects upon serum somatomedin activity, growth hormone and prolactin in man. J. Clin. Endocrinol. Metab. 42, 942–952 (1976)

Weidemann, E., Schwartz, E., Reddy, N.: Corticosteroid (CS) effects upon serum somatomedin activity (SMA) and upon in vitro cartilage sulfate uptake. Clin. Res. 25, 131 A (1977)

White, R.M., Nissley, S.P., Moses, A.C., Rechler, M.M., Johnsonbaugh, R.E.: Growth hormone dependence of somatomedin binding protein in humans. Presented to the 61st meeting of the Endocrine Society, Anaheim, California, June 13–15, 1979

Wu, A., Grant, D.B., Hambley, J., Levi, A.J.: Reduced serum somatomedin activity in patients with chronic liver disease. Clin. Sci. Mol. Med. 47, 359–366 (1974)

Young, F.G.: Growth and diabetes in normal animals treated with pituitary (anterior lobe) diabetogenic extract. Biochem. J. 39, 515–536 (1945)

Zapf, J., Mader, M., Waldvogel, M., Froesch, E.R.: Nonsuppressible insulin-like activity: biological activity and receptor bindings. Isr. J. Med. Sci. 11, 664–678 (1975a)

Zapf, J., Mader, M., Waldvogel, M., Schalch, D.S., Froesch, E.R.: Specific binding of nonsuppressible insulin-like activity to chicken embryo fibroblasts and to a solubilized fibroblast receptor. Arch. Biochem. Biophys. 168, 630–637 (1975b)

Zapf, J., Waldvogel, M., Froesch, E.R.: Binding of nonsuppressible insulin-like activity to human serum: evidence for a carrier protein. Arch. Biochem. Biophys. 168, 638–645 (1975c)

Zapf, J., Kaufmann, U., Eigenmann, E.G., Froesch, E.R.: Determination of nonsuppressible insulin-like activity in human serum by a sensitive protein binding assay. Clin. Chem. 23, 677–682 (1977)

Zapf, J., Jagars, G., Sand, I., Froesch, E.R.: Evidence for the existence in human serum of large molecular weight nonsuppressible insulin-like activity (NSILA) different from small molecular weight forms. FEBS Letters 90, 136–140 (1978a)

Zapf, J., Rinderknecht, E., Humbel, R.E., Froesch, E.R.: Nonsuppressible insulin-like activity (NSILA) from human serum: recent accomplishments and their physiologic implications. Metabolism 27, 1803–1828 (1978b)

Zapf, J., Schoenle, E., Froesch, E.R.: Insulin-like growth factors I and II: some biological actions and receptor binding characteristics of two purified constituents of nonsuppressible insulin-like activity of human serum. Eur. J. Biochem. 87, 285–296 (1978c)

Zapf, J., Schoenle, E., Jagars, G., Sand, I., Grunwald, J., Froesch, E.R.: Inhibition of the action of nonsuppressible insulin-like activity on isolated fat cells by binding to its carrier protein. J. Clin. Invest. 63, 1077–1084 (1979)

CHAPTER 6

Glucocorticoid Modulation of Cell Proliferation

V. J. CRISTOFALO and B. A. ROSNER

A. Introduction

I. Historical Perspective

Glucocorticoids are known to have profound effects on many biological processes. For example, the important role of adrenal cortical steroids in the regulation of carbohydrate metabolism has been evident since the early 1900's (e.g., see CAHILL, 1971). Their effects on ACTH activity as well as their immunosuppressive and anti-inflammatory properties are also generally well established (FAUCI, 1979). It is now evident that, in addition, glucocorticoids play an important role in the growth and differentiation of mammalian tissues (BALLARD, 1979). At present, many of the effects of glucocorticoids are being studied in cell culture where hormone responsiveness can be monitored under controlled environmental conditions.

One major cellular response to glucocorticoids, in vitro, is altered proliferative activity. This phenomenon has provided investigators with model systems with which to study both the mechanism of glucocorticoid action and factors regulating cellular replication. In recent years, numerous papers describing glucocorticoid effects on cell proliferation in vitro have appeared. Some cells were stimulated, some inhibited, and some were unaffected by glucocorticoids (BERLINER and DOUGHERTY, 1961; GROVE et al., 1977). Our knowledge of the patterns and mechanisms by which glucocorticoids affect cell proliferation is still sketchy. Furthermore, the relevance of the in vitro observations to in vivo physiology remains, for the most part, obscure. However, the evidence available suggests that the effects of glucocorticoids on cell proliferation are varied and represent a wide range of cell-hormone interactions. In this chapter we will review the available literature and consider the details of cell-type specific proliferative responses observed in vitro. We will discuss several alternate mechanisms by which the proliferative activity of glucocorticoid responsive cells might be regulated. These include possible permissive effects (i.e., through the regulation of heterologous hormone activity) or by regulation of the activity or expression of other factors implicated in growth control such as neutral proteases and polyamines. Finally, we will present some recent data from our laboratory that suggest that the glucocorticoid modulation of cell proliferation may, in some cases, be mediated by an induced factor(s) with growth-stimulating activity that is secreted into the growth medium.

II. Glucocorticoids: General Mechanisms of Action

The term "glucocorticoid" has been applied to the class of steroids that has marked effects on carbohydrate metabolism. The structure of hydrocortisone (HC), the ac-

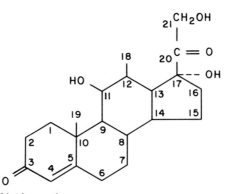

Fig. 1. The structure of hydrocortisone

TARGET CELL

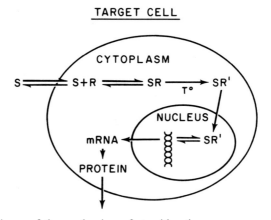

Fig. 2. General scheme of the mechanism of steroid action

tive glucocorticoid in man, is presented in Fig. 1. The key substitutions in the Δ4-pregnene molecule that confer its glucocorticoid activity are the 3,20-keto and the 11β-hydroxyl groups. The structure-activity relationships of glucocorticoids have recently been reviewed by SCHMIT and ROUSSEAU (1979) and WOLFF (1979). In some species, such as the rat, corticosterone, which lacks the 17-hydroxyl substitution of hydrocortisone, is the endogenously active form of the steroid.

A general scheme of the mechanism of steroid action is presented in Fig. 2. Unlike peptide hormones, which mediate their effects through specific receptors at the surface of target cells, steroid hormones must traverse the cell membrane in order to bind to their receptors in the cytosol of target cells. This specific molecular interaction between a steroid hormone and its receptor (JENSEN and JACOBSON, 1962), is the first stage in a multistep mechanism by which steroids can effect changes in gene expression and, consequently, changes in the growth, differentiation, and metabolism of responsive cells.

The mechanism by which steroids enter the cell has been a topic of recurrent debate. Since all steroids are lipophilic to varying degrees, it has been assumed that they entered the cell by passive diffusion. However, there are reports in the literature of facilitated transport of steroids into (MILGROM et al., 1973) and out of (GROSS et al., 1968) steroid-responsive tissues and cells. Once inside the cell, the steroid interacts with specific receptor proteins before it can effect a biological response. These cytosol proteins must fulfill certain biological criteria in order to be classified as steroid receptors: 1) they must demonstrate molecular specificity and high-affinity binding of the steroid; 2) the binding of the steroid to its receptor should be saturable; 3) the steroid-receptor complex (S-R) formed must undergo a temperature-dependent "activation" (S-R*); and 4) the "activated" steroid-receptor complex must be capable of translocating from the cytoplasm into the nucleus enabling it to bind to nuclear acceptor sites.

The nature of the interaction between the steroid-receptor complex with the nuclear acceptor sites and the molecular events that follow (i.e., the induction or repression of specific protein syntheses) are not fully understood. However, the primary mechanism of glucocorticoid and other steroid action in the nucleus is thought to be the regulation of transcription which results in the synthesis of specific messenger ribonucleic acids (mRNA's) which, in turn, results in the synthesis of specific proteins that mediate the hormone's effects. Most of the evidence for this mechanism is derived from experiments involving the timed addition of inhibitors of RNA synthesis such as actinomycin D and cordycepin which block the hormone effects. This aspect of glucocorticoid action has been recently reviewed by MUNCK and LEUNG (1977). Direct demonstration of glucocorticoid-induced mRNA [e.g., the cortisol-induced increase in rat liver tryptophan oxygenase mRNA's (SCHUTZ et al., 1972; SCHUTZ et al., 1973)] has been few in number, but recent development of techniques for the isolation and in vitro translation of mRNA should aid these kinds of investigations. Several comprehensive reviews on the mechanism of steroid action are available for the interested reader (see YAMAMOTO and ALBERTS, 1976; BAULIEU, 1975; GORSKI and GANNON, 1976; MCCARTY and MCCARTY, 1977; CHAN et al., 1978; HIGGINS and GEHRING, 1978; THRALL et al., 1978) and the recently published monograph on glucocorticoid hormone action edited by BAXTER and ROUSSEAU (1979).

B. Hormone Responsiveness by Cell Cultures

I. Various Vertebrate Species

The reported effects of glucocorticoids on the proliferative activity of cells grown in vitro are varied. Early work indicated that fibroblast-like cell proliferation was inhibited by glucocorticoids (RUHMAN and BERLINER, 1965; EPIFANOVA, 1971). It is now apparent that the responses are dependent on such things as donor species, tissue of origin, stage of ontogeny and the cell type which eventually becomes established in culture (GROVE et al., 1977; CRISTOFALO, 1975; CRISTOFALO and ROSNER 1979). Table 1 lists the effect of hydrocortisone on the proliferation of a series of cell lines representing all the vertebrate classes (CRISTOFALO, 1975). In the upper portion are human cells. The results shown for WI-38 and WI-26 cells are typical of those for normal diploid human fetal lung-derived cultures. An increase in both the fraction of cells undergoing DNA synthesis and the harvest density was seen. (However, not all human diploid cultures respond similarly; see Table 2.) The ef-

Table 1. Effect of hydrocortisone on cell proliferation and DNA synthesis in various animal cell lines

Origin			% Change with hydrocortisone	
Animal	Tissue	Line	Cell count	% Labeled cells
Human	Fetal lung	WI-38	+40	+38
		WI-26	+35	+33
	SV40-transformed fetal lung	WI-38 VA13A	−25	0
		WI-26 VA4	−30	0
	Cervical carcinoma	HeLa	−31	−20
African green monkey	Kidney	CV-1	−54	−18
Syrian hamster	Embryo	—	−61	−15
C3H mouse	Arcolar	L	−34	−24
White leghorn chick	Embryo	—	− 9	−16
Lizard	Embryo	GE-1	−20	−35
Frog	Embryo	RP	−20	−40
Fathead minnow	Tail	FHM	−67	−50

fects of HC on the proliferative activity of their SV 40-transformed counterparts, WI-38 VA 13 A and WI-26 VA 4, were remarkably different. These cell lines showed an inhibition of cell growth, but interestingly, HC had no effect on the number of cells synthesizing DNA. HeLa cells (derived from a human cervical carcinoma) showed inhibition of both cell proliferation and DNA synthesis. So, even though both these cell types proliferate indefinitely, their responses to HC are different.

The effects of glucocorticoid treatment on the proliferation of non-human cell lines (CRISTOFALO, 1975; GROVE et al., 1977) were similar to those observed in HeLa cells (Table 1). HC decreased both the fraction of the cells undergoing DNA synthesis and the final cell yields of each cell type tested. It should be noted here that all but one of the cell lines that were inhibited (with chick cells as the exception) are indefinitely proliferating, transformed cell lines. Similarly, other authors have reported inhibition of proliferation for a series of transformed permanently proliferating, cell lines (RUHMANN and BERLINER, 1965: HeLa and L cells; HACKNEY et al., 1970: L929 cells; and SIBLEY et al., 1974: lymphoma cells). These results suggest a possible relationship between transformation and proliferative response to hydrocortisone. On the other hand, mouse 3 T 3 cells, which are indefinitely proliferating but which show many of the growth properties of normal cells, were stimulated by hydrocortisone (TRASH and CUNNINGHAM, 1973; GOSPODAROWICZ and MORAN, 1975 a; GOSPODAROWICZ, 1974; HOLLEY and KIERNAN, 1974; ARMELIN, 1973). In some of these studies the HC potentiated the effect of other factors such as FGF (GOSPODAROWICZ, et al. 1976). In other studies, CLARK et al. (1972) have shown stimulation of ovarian cells by HC and CASTOR and PRINCE (1964) have shown increased proliferation of human fibroblasts by HC.

II. Human Cell Lines

Glucocorticoids affect many different human tissues in vivo (BAXTER and FOR-
SHAM, 1972), and therefore, it is not unexpected that numerous cell lines derived
from normal tissues and those expressing various pathological and/or genetic abnormalities have been studied for hormone responsiveness. Tables 2 and 3 present
the effects of glucocorticoids on the proliferation of various human cell types in
vitro as reported over the last several years.

Table 2 lists the effects of glucocorticoids on the proliferation of normal fibroblastic cell types. As mentioned above, the proliferation of fetal lung-derived fibroblastic cells is stimulated by glucocorticoids. All of the 19 different cell lines tested
were stimulated by one short-term exposure. It is also known that chronic exposure
of these cells to glucocorticoids maintains a larger fraction of cells in the rapidly
proliferating pool resulting in a longer replicative life span of the cell population
(CRISTOFALO, 1970; CRISTOFALO, 1975). Infant lung-derived cells were also stimulated while adult-derived lung cell lines were variable.

Similarly, studies of the effects of glucocorticoids on the proliferation of
neonatal foreskin fibroblastic cells have consistently shown stimulation. However,
recent evidence made available to us by Roger Ladda (personal communication),
who has tested over 25 neonatal foreskin lines has indicated that the proliferation
of the majority of these cell lines was unaffected by the addition of glucocorticoids,
and some showed a slight inhibition. In the one study of an adult foreskin line from
a 78-yr-old donor, proliferation was inhibited by HC. Varied responsiveness also
appears to be the case for normal adult skin-derived fibroblastic cell lines. Though
24 out of 35 of the adult skin lines presented in Table 2 were stimulated by glucocorticoids, it is evident, especially from the studies of GROVE et al. (1977) and RUS-
SELL et al. (1978), in which 6 or more lines were studied under the same conditions,
that several skin cell lines were clearly inhibited, and others were not affected by
the presence of glucocorticoids.

In contrast, the growth inhibitory effect of glucocorticoids on normal human
fetal skin-derived cell lines (Table 2) has been seen in 6 out of 7 different cell lines
tested in our laboratory over the last four years. In this study 3 of the fetal skin
cultures were from the same individual as the fetal lung cultures studied above.
Thus, we are not dealing with genetic differences in responsiveness. ROWE et al.
(1977) reported a 22% decrease in the cell density at confluence of HC-treated fetal
skin cells, however, this result was not significantly different from the control
values when the results of 5 different cell lines were combined. Similarly,
SCHNEIDER and MITSUI (1977) had observed a decrease in the growth rate of HC-
treated fetal skin fibroblastic lines.

It is clear from the above data that cells isolated at different developmental
stages, but which adopt a similar fibroblastic morphology in vitro, express different functional responses to glucocorticoids. These data emphasize the fact that fibroblast-like cells in vitro may display different physiological characteristics.

The response of fetal lung cultures to glucocorticoids may be related to the response of fetal lung in vivo. Lung development involves both cell proliferation as
well as the development of differentiated cell types. The Type II pneumonocytes'
development is enhanced by glucocorticoids (SMITH et al., 1973; SMITH, 1979). Fur-

Table 2. Normal fibroblastic human cell types whose proliferation is modulated by glucocorticoids in vitro[a]

Donor tissue	Glucocorticoid (concentration)	No. tested	No. stimulated	No. inhibited	References
Fetal lung	Cortisone Acetate (6 μM)	2	2	0	MACIEIRA-COEHLO (1966)
	Hydrocortisone (14 μM)	1	1	0	CRISTOFALO (1970)
	Hydrocortisone (0.5 μM)	1	1	0	SMITH et al. (1973)
	Hydrocortisone (0.5 μM)	1	1	0	TAYLOR and POLGAR (1977)
	Hydrocortisone (14 μM)	1	1	0	FULDER (1977)
	Hydrocortisone (14 μM)	7	7	0	GROVE et al. (1977)
	Hydrocortisone (0.14 μM)	6	6	0	ROSNER and CRISTOFALO (unpublished)
Totals		19	19	0	
Infant lung	Hydrocortisone (0.14 μM)	2[b]	2	0	CRISTOFALO and LADDA (unpublished)
Adult lung	Hydrocortisone (14 μM)	3	0	3	GROVE et al. (1977)
	Hydrocortisone (0.14 μM)	3	1	0	ROSNER and CRISTOFALO (unpublished)
Totals		6	1	3	
Neonatal foreskin	Hydrocortisone (14 μM)	4	4	0	GROVE et al. (1977)
	Hydrocortisone (0.14 μM)	2[b]	0[c]	0[c]	CRISTOFALO and LADDA (unpublished)
	Dexamethasone (0.25 μM)	1	1	0	BAKER et al. (1978)
Totals		7	5	0	
Adult foreskin	Hydrocortisone (0.14 μM)	1[b]	0	1	CRISTOFALO and CHARPENTIER (unpubl.)
Fetal skin	Hydrocortisone (14 μM)	2	0	2	GROVE et al. (1977)
Non-genital	Hydrocortisone (0.14 μM)	5	0	4	ROSNER and CRISTOFALO (unpublished)
Totals		7	0	6	
Adult skin	Hydrocortisone (28 μM)	1	0	1	HARVEY et al. (1976)
Non-genital	Hydrocortisone (14 μM)	11	5	2	GROVE et al. (1977)
Fibroblastic	Hydrocortisone (0.3 μM)	9	9	0	ROWE et al. (1977)
	Prednisolone (0.03 μM)	3	3	0	RUNIKIS et al. (1978)
	Hydrocortisone (0.14 μM)	6	4	0	RUSSELL et al. (1978)
	Dexamethasone (0.25 μM)	1	1	0	BAKER et al. (1978)
	Hydrocortisone (0.14 μM)	3	2	0	ROSNER and CRISTOFALO (unpublished)
	Prednisolone (0.22 μM)	1	0	1	HOLLENBERG (1977)
Totals		35	24	4	

[a] Modulation of proliferation as measured by changes in DNA synthesis and/or cell number resulting from a one-time hormone exposure

[b] Modulation of proliferation as measured by direct cell counts at weekly intervals (four or more subcultivations) in the presence or absence of hormones

[c] Of the two lines tested, one demonstrated no growth response while responsiveness of the second was variable

ther support for this view comes from the work of PINSKY et al. (1972) who showed that fibroblastic cell lines derived from neonatal foreskin cultures retain the capacity to metabolize testosterone, whereas cultures derived from non-genital skin do not. In addition, HARPER and GROVE (1979) have shown that the papillary and reticular regions of adult human skin can give rise to cultures with similar fibroblastic morphology but different proliferative capacities. Therefore, the observed differences in glucocorticoid responsiveness of non-fetal skin derived fibroblastic lines may be due, in part, to the region of the dermis from which the cells were derived.

Table 3 presents the recently reported effects of glucocorticoids on the proliferation of other human-derived cell types. STEVENSON et al. (1978), observed stimulated proliferation in mixed bone marrow cell cultures whereas MACA and coworkers (1978a) observed no proliferative effect in treated umbilical vein endothelial cells. However, in another publication, MACA et al. (1978b) demonstrated that glucocorticoids had a marked effect on the cell surface of these endothelial cells which rendered them less susceptible to adhesion by lymphoblastoid cell lines. In a study comparing the growth effects of glucocorticoids in keloid tissue-derived cells, RUSSELL et al. (1978) observed that their proliferative activity was significantly inhibited, whereas the majority of the normal control skin cell lines were stimulated.

The bottom half of Table 3 presents the reported effects of glucocorticoids on the proliferation of cell types representative of human tumors and cancers of epithelial and lymphoid cell origin. The proliferative response of most of these cell types to glucocorticoids is inhibitory, not unlike that described for the cervical carcinoma-derived HeLa cell line. However, there are now clear examples that growth in some human cancer-derived cells in vitro is stimulated upon treatment with glucocorticoids. These cell types include astrocytoma cells (GUNER et al., 1977), mammary fibroadenoma epithelial derived cells (GAFFNEY and PIGOTT, 1978), and also several isolated acute lymphocytic leukemias (CRABTREE et al., 1978). The increasing application of endocrine therapy in the clinical management of human cancers has clearly demonstrated the need for in vitro experimental systems in which hormonal responsiveness can be assessed under defined environmental conditions.

III. Mouse and Rat Cell Lines

Murine cell lines have an indefinite replicative capacity and are also extensively used for the study of hormonal regulation of cell proliferation. Table 4 presents the effects of glucocorticoids on the proliferation of various mouse cell types in culture. With the exception of the 3 T 3 cell lines, in which proliferation is stimulated by glucocorticoids under certain conditions (THRASH and CUNNINGHAM, 1973; RUDLAND et al., 1974; HOLLEY and KIERNAN, 1974), glucocorticoids have consistently inhibited the proliferation of all established mouse cell lines. It is interesting, in this regard, that 3 T 3 cells, although they have an indefinite replicative life span possess some normal growth characteristics such as the density-dependent inhibition of cell proliferation displayed by normal human cell cultures.

The effects of glucocorticoids on the proliferation of rat cell lines (Table 5) present a more complex picture than mouse cell lines in that more variation is observed. Stimulation of cell growth has been observed in myoblast cultures (GUERRIERO and

Table 3. Glucocorticoid effects on the proliferation of other human-derived cell types as measured by changes in DNA synthesis and/or cell number

Cell type (line)	Glucocorticoid (concentration)		Glucocorticoid effect	References
Bone marrow cells (mixed)	Prednisolone	(0.28 μM)	Stimulation	Stevenson et al. (1978)
Umbilical vein endothelial	Dexamethasone	(1 μM)	None	Maca et al. (1978b)
Keloid cells	Hydrocortisone	(1.4 μM)	Inhibition	Russell et al. (1978)
Astrocytoma	Dexamethasone	(31 μM)	Stimulation	Guner et al. (1977)
Breast cancer	Dexamethasone	(0.1 μM)	Inhibition	Osborne et al. (1979)
Mammary fibroadenoma epithelial cells	Hydrocortisone	(14 μM)	Stimulation	Gaffney and Pigott (1978)
Lung alveolar carcinoma	Dexamethasone	(40 μM)	Inhibition	Jones et al. (1978)
Lymphoblastoid	Prednisolone	(1 μM)	Inhibition	Norman et al. (1978)
(CEM-C-7)	Dexamethasone	(1 μM)	Inhibition	Harmon et al. (1979)
Lymphoma	Dexamethasone	(0.1 μM)	Inhibition	Crabtree et al. (1978)
Leukemias	Dexamethasone	(0.1 μM)	Inhibition	Crabtree et al. (1978)
Acute lymphocytic leukemias	Dexamethasone	(0.1 μM)	Inhibition and stimulation	Crabtree et al. (1978)
Acute myelomonocytic leukemia	Dexamethasone	(10 μM)	Inhibition	Iacobelli et al. (1978)

Table 4. Effects of glucocorticoids on mouse cell proliferation[a] in vitro

Cell type (line)	Glucocorticoid (concentration)		Glucocorticoid effect	References
Fibroblastic (L929)	Hydrocortisone	(0.1 μM)	Inhibition	RUHMAN and BERLINER (1965)
Fibroblastic (L929)	Hydrocortisone	(14 μM)	Inhibition	PRATT and ARONOW (1966)
Fibroblastic (Swiss embryo 3T3)	Hydrocortisone	(1.4 μM)	Stimulation	THRASH and CUNNINGHAM (1973)
Fibroblastic (Swiss embryo 3T3)	Hydrocortisone	(1 μM)	Stimulation	RUDLAND et al. (1974)
Fibroblastic (Swiss embryo 3T3)	Dexamethasone	(0.14 μM)	Stimulation	HOLLEY and KIERNAN (1974)
Fibroblastic (ST1, Swiss 3T3 clone)	Hydrocortisone	(3 μM)	Inhibition	ARMELIN and ARMELIN (1977)
Lymphoma (S49)	Hydrocortisone hemisuccinate	(1 μM)	Inhibition	HORIBATA and HARRIS (1970)
Lymphoma (CSP1798)	Hydrocortisone	(1 μM)	Inhibition	STEVENS and STEVENS (1975)
Embryonic Facial Mesenchyme	Dexamethasone	(14.4 μM)	Inhibition	SALOMON and PRATT (1978)
Adrenal Tumor (Y1)	Corticosterone		Inhibition	SAITO et al. (1979)

[a] As measured by changes in DNA synthesis and/or cell number

Table 5. Effects of glucocorticoids on rat cell proliferation[a] in vitro

Cell type (line)	Glucocorticoid (concentration)		Glucocorticoid effect	References
Fibroblastic	Dexamethasone	(0.1 μM)	Inhibition	GUERRIERO and FLORINI (1978)
Hepatocytes	Dexamethasone	(0.2 μM)	Inhibition	RICHMAN et al. (1976)
Myoblasts	Dexamethasone	(0.1 μM)	Stimulation	GUERRIERO and FLORINI (1978)
Pituitary tumor (GH$_3$)	Hydrocortisone	(5 μM)	Inhibition	CLAUSEN et al. (1978)
Fetal calivaria	Dexamethasone	(13 μM)	Inhibition	CHEN and FELDMAN (1979)
Fetal calivaria	Hydrocortisone	(0.03 μM)	Inhibition	DIETRICH et al. (1979)
Mammary tumor (RBA, NMU)	Dexamethasone	(0.1 μM)	Stimulation	VIGNON et al. (1979)
Glioma (C-6)	Hydrocortisone	(>1 μM)	Inhibition	GRASSO and JOHNSON (1977)
Hepatoma	Hydrocortisone	(0.3 μM)	Inhibition	LOEB (1976)

[a] As measured by changes in DNA synthesis and/or cell number

Florini, 1978) and mammary tumor cell cultures (Vignon et al., 1979) whereas inhibition was observed for fibroblast (Guerriero and Florini, 1978), hepatocyte (Richman et al., 1976), pituitary tumor (Clausen et al., 1978), and fetal calivaria (Chen and Feldman, 1979) cultures. Glucocorticoids affected a decrease in the proliferation of rat hepatoma cultures which have been extensively used for the study of glucocorticoid hormone action (Baxter and Tomkins, 1970).

The cell-type specific proliferative responsiveness to glucocorticoids in vitro presented above clearly emphasizes the complexity of hormone cell interactions. Unquestionably, the experimental data are incomplete with respect to concentrations of hormone, specific cell type actually studied, stage of the growth cycle when the hormone was added and a host of other details that one can imagine may be important in the proliferative response.

However, it seems generally clear that with the exception of human fetal lung-derived cultures, and perhaps human neonatal foreskin cultures, cell lines are more often inhibited than stimulated by physiological levels of HC. Human fetal lung responds to HC by the production of surfactant and perhaps the consistent proliferative response seen in fetal lung-derived fibroblasts may reflect the epithelial-mesenchymal interactions involved in this response in vivo.

There is no clear relationship between transformation and response to HC although by and large it seems much more likely that transformed cells in general are inhibited by HC.

To further complicate the picture, all of these observations were made in the presence of serum supplemented medium. When cultures are prepared in chemically defined medium, HC is stimulatory in virtually all cases (Sato and Reid, 1978). Thus, the relationship between serum factors, glucocorticoids and cells requires far more study to sort out the multiple complex aspects in their interaction.

C. Permissive Effects of Glucocorticoids

Permissive effects of glucocorticoids refer to the activity of these hormones at physiological concentrations which results in altered biological responses to other hormones that need not act through the same mechanism as the steroid. For example, Gospodarowicz and Moran (1974, 1975b) have reported the potentiation of FGF proliferative effects by HC in 3T3 cells. In addition, Clark et al. (1972) and Armelin (1973) showed that in the presence of partially purified bovine luteinizing hormone glucocorticoids had a potentiating effect on proliferation. Permissive effects of glucocorticoids are discussed in detail by Thompson and Lippman (1974), Wicks (1974), and Granner (1979). Many of the reported permissive effects of glucocorticoids such as the induction of tyrosine aminotransferase in hepatoma cells (Granner, 1976) involve the action of cyclic nucleotides.

I. Glucocorticoid Modulation of Heterologous Receptors

There is now evidence suggesting that permissive effects may also be mediated through the regulation of the cell's responsiveness to a second hormone through regulation of the receptor concentration for the second hormone. Table 6 lists sev-

Table 6. Glucocorticoid modulation of heterologous receptors

Receptor	Concentration change in response to glucocorticoids	Target cell or tissue	Physiological response affected	References
Thyrotropin releasing hormone	↑	Rat pituitary GH-cells	Growth hormone synthesis, PRL	TASHJIAN et al. (1977)
Epidermal growth factor	↑	Human foreskin fibroblasts	Increased growth	BAKER et al. (1978)
Prolactin	↑	Mouse mammary epithelial cells	Lactation, casein mRNA synthesis	SAKAI and BANERJEE (1979)
Estradiol (nuclear)	↓	Rat uterus	Uterotrophic response to E_2	CAMPBELL (1978)
Prolactin	↓	Rat kidney and adrenal membranes	Salt and water metabolism	MARSHALL et al. (1978)
Crystallizable fragment from antibody	↓	Human promyelocytic cell line HL-60	Phagocytosis	CRABTREE et al. (1979)

eral recent reports in which evidence for glucocorticoid modulation of heterologous hormone receptors has been correlated with altered cellular and tissue responsiveness. The physiological responses affected by glucocorticoid treatment, again, clearly reflect the differentiated state of the target tissue under investigation. Glucocorticoid-induced increases in the concentration of receptors specific for thyrotrophin-releasing hormone (TRH), epidermal growth factor (EGF), and prolactin (PRL), have been correlated with increased growth hormone and PRL synthesis in rat pituitary cells (TASHJIAN et al., 1977), increased cellular proliferation in human fibroblasts (BAKER et al., 1978), and increased casein mRNA synthesis in mouse mammary epithelial cells (SAKAI and BANERJEE, 1979), respectively. Conversely, glucocorticoid-induced decreases in nuclear estradiol (E_2) and in PRL receptors have been correlated with an altered uterorophic response in the rat (CAMPBELL, 1978) and changes in salt and water metabolism in rat kidney (MARSHALL et al., 1978). CRABTREE and co-workers (1979) have also demonstrated that glucocorticoid treatment of human promyelocytic cells results in a decrease in the concentration of Fc receptor.

II. Glucocorticoid Effects on Other Factors Implicated in the Regulation of Cell Growth

The apparent glucocorticoid regulation of heterologous hormone receptor concentration, as in the case of the enhancement of EGF binding and its correlation to increased cell proliferation (BAKER et al., 1978), supports an intriguing model of "permissive" glucocorticoid regulation of cell replication. However, evidence from our laboratory on the HC modulation of WI-38 cell growth suggests that the mechanism may be much more complex. We have observed that after the serum stimulation of quiescent WI-38 cultures, there is a cell cycle-dependent increase in the glucocorticoid specific binding in these cells (CRISTOFALO et al., 1979). The transient increase in receptor concentration during the G_0–G_1 transition corresponds to the period of greatest hormone responsiveness as assayed by saturation density. The fact that under certain conditions serum is a prerequisite for the HC response (CRISTOFALO, 1970) would suggest that some as yet undefined serum components may play a role in the regulation of the metabolism of the receptors or the mechanism of action of the hormone (CRISTOFALO et al., 1979). Although there have been similar reports of cell cycle-dependent regulation of glucocorticoid receptor concentration and responsiveness (MARTIN et al., 1969; CAVENEE and MELNYKOVYCH, 1979; NEILFELD et al., 1977; SMITH et al., 1977), little is known about the underlying mechanism or the serum factors involved.

D. Glucocorticoid Induction of Growth Factors in Normal Human Cells In Vitro

Recent evidence from several laboratories has suggested that at least one more step involving steroid induction of factors with growth stimulating activity (GSA) may be involved in steroid regulation of cell proliferation and differentiation. In previous studies in our laboratory on the mechanism of action of HC, we had shown specific binding of HC and dexamethasone to WI-38 cells (CRISTOFALO and ROSNER, 1979). Quantitatively, the binding was correlated with cell proliferation in that specific binding of the hormone declined as the cells progressed through their replicative life span. Responsiveness to HC or dexamethasone could be assayed, as mentioned above, either by an increased fraction of cells in the rapidly proliferating pool or by an increased saturation density (CRISTOFALO, 1975). This response was cell cycle-specific (CRISTOFALO et al., 1979). The cells were responsive to the hormone only during the period immediately following (0–6 h) stimulation of quiescent cells to enter DNA synthesis. Thus, responsiveness was correlated with increased specific binding of the hormone during this period. A question of major interest was concerned with the mechanism by which an event occurring only during the first 24-h period following stimulation (i.e., hormone binding, presumably followed by changing protein synthesis) could be expressed 9 days later as an increased saturation density. One possibility was that cells were responding to HC by "conditioning" the medium. The modification and conditioning of the growth medium by cells in culture has been recognized for over 30 years (SANFORD et al., 1949). Many "conditioning factors" responsible for enhanced cell growth in vitro have since been isolated and characterized. These include multiplication stimulating activity (MSA) (DULAK and TEMIN, 1973), colony stimulating factor (CSF) (BURGESS et al., 1977), and sarcoma growth factors (SGF's) (DE LARCO and TODARO, 1978). In addition, several cell lines grown in vitro have been shown to synthesize and secrete factors required for cell survival (WYCHE and NOTEBOOM, 1977) and attachment to the growth surface (MILLIS and HOYLE, 1978). Our recent stud-

Table 7. The effect of hydrocortisone-conditioned media on cell proliferation

Seeding material ($t = 0$ h)	Treatment ($t = 24$ h)	Refeeding material ($t = 24$ h)	Relative cell yield ($t = 126$ h)
FM[a]	None	None	100
FM + HC[b]	None	None	139
FM	Remove C-CM[c]; wash	24-h C-CM	100
FM	Remove C-CM; wash	24-h C-CM + HC	104
FM	Remove C-CM; wash	24-h HC-CM[d]	133

[a] FM   – fresh growth medium containing 10% FBS
[b] HC   – hydrocortisone (14 μM)
[c] C-CM   – control cell-conditioned medium
[d] HC-CM – HC-treated cell-conditioned medium

ies have shown that the proliferative response seen with glucocorticoid treatment of WI-38 cells also results in the conditioning of the growth medium with some GSA factor capable of stimulating WI-38 cell growth at a time in the growth cycle when the culture is unresponsive to the hormone itself (CRISTOFALO et al., 1979). Evidence for this interpretation is presented in Table 7, which shows the results of an experiment in which, 24 h after seeding, 24-h conditioned media were removed from replicate cultures incubated in the presence or absence of HC (HC-CM$_{24}$ and CM$_{24}$, respectively). When non-HC-treated cells were refed at 24 h with HC-CM$_{24}$, an increase in cell yield was observed on day 9 that was comparable to the increase obtained for cultures exposed to HC from seeding. These cultures are not responsive to HC alone when it is added at 24 h. As noted above, the responsiveness of WI-38 cells to HC, as measured by increased cell proliferation, is restricted to the $G_0 \rightarrow G_1$ transition period of the cell cycle (CRISTOFALO et al., 1979; ROSNER et al., submitted for publication). In contrast, this HC-induced conditioning factor will stimulate logarithmically growing WI-38 cells at any time during the growth cycle. Preliminary data suggest that the growth stimulating activity is secreted before the onset of DNA synthesis and is heat stable (ROSNER et al., submitted for publication).

Our observation that WI-38 cells condition the growth medium in response to HC is not unique. There are now several reports that demonstrate the production of several classes of bioactive substances in response to steroid treatment both in vivo and in vitro. STEVENSON (1978) has shown that when human blood monocytes are exposed to HC in vitro, they produce a factor that stimulates polymorph migration. SMITH (1979) has shown that lung maturation in the fetal rat could be accelerated after the administration of fibroblast-pneumonocyte factor (FPF). FPF, a heat stable, dialyzable polypeptide, was produced by WI-38 cells in response to glucocorticoids. The activity of FPF may be intrinsically different from the factor(s) we believe are responsible for increased WI-38 cell growth since, as noted by SMITH, the administration of semi-purified FPF affected only lung maturation and not its growth.

Another related series of observations was made by KING et al. (1977) who showed that estrogen-induced protein (IP) from rat uterus was capable of stimulating DNA synthesis in 3 T 6 cells in the absence of serum. Similarly, Sirbasku (1978) has also shown that estrogen could induce the production of growth factors in several rat tissues in vivo. These estrogen-induced growth factors were capable of stimulating the growth of hormone-responsive mammary, pituitary, and kidney tumor cells in vitro (Sirbasku, 1978).

The mechanism by which the HC-induced WI-38 cell conditioned medium factor(s) facilitates enhanced cell proliferation still remains to be elucidated. Based on our preliminary results and those from other laboratories described above, we feel that an additional model of the mechanism of glucocorticoid action on the regulation of cell proliferation should be considered. This new model should include the hormone's induction of "growth factor(s)" that can ultimately regulate the growth of the cell generating this factor and/or act on other cell types. This model of so-called "autocrine" stimulation of cells by factors they themselves produce has been reported for sarcoma cells in culture by De Larco and Todaro (1978). Our studies on WI-38 cells extend this concept to a normal cell's activity in response to a specific hormone.

One possible model which provides a mechanism by which this glucocorticoid-factor acts is based on the known effects of glucocorticoids on prostaglandin metabolism. For example, it has been demonstrated that WI-38 cells synthesize and secrete several classes of prostaglandins including $F_{2\alpha}$ and E_2 (Mets et al., 1979; Taylor and Polgar, 1977), and that these latter two prostaglandins can inhibit WI-38 cell growth (Taylor and Polgar, 1977). In addition, Taylor and Polgar (1977) also demonstrated that inhibitors of prostaglandin synthesis mimic the HC stimulation of human fibroblast cell growth in vitro. The possible correlation between this glucocorticoid-induced factor in WI-38 cells and the regulation of prostaglandin synthesis is further supported by the observation that isolated lungs that are perfused with glucocorticoids induce the synthesis and secretion of a protein that inhibits phospholipase A_2 synthesis (Flower and Blackwell, 1979). Phospholipase A_2 is required for arachidonate release; thus, the inhibition of its activity prevents prostaglandin synthesis. Other models are also possible and elucidation of this mechanism must await further research.

E. Summary and Conclusions

Glucocorticoid hormones affect many biological processes including the processes of growth and proliferation. Advances in cell culture technology have allowed the use of various cell types in vitro as models to study the effects of glucocorticoids under relatively controlled environmental conditions. The focus of this chapter has been on the effects of glucocorticoids on the proliferation of cells in culture derived from various normal and tumor tissues and at different stages of ontogeny.

We have presented in Tables 1–5 the proliferative effects of glucocorticoids on cell cultures derived from normal and tumor tissues of humans, rats, and mice. We also presented data on cell lines derived from members of the other four vertebrate classes. Overall, glucocorticoids inhibit the proliferation of indefinitely proliferating cell lines although there are a few exceptions to this. For example, human astrocytoma and mammary fibroadenoma tumor-derived cell lines as well as several acute lymphotic-leukemias were found to be stimulated by glucocorticoids. Similarly, two rat mammary tumor-derived cell lines were also stimulated. The stimulation seen in mouse embryo 3 T 3 cells presents a more complex situation since these cell lines, though possessing an indefinite replicative capacity, demonstrate certain "normal" growth characteristics such as density-dependent inhibition of growth. Cell lines derived from non-human vertebrate animals also demonstrated growth inhibition in the presence of glucocorticoids. All of these cultures were permanently proliferating except the chick cells.

In contrast to the long held notion that glucocorticoids universally inhibit proliferation of cell cultures, we have presented examples of cell lines derived from fetal lungs and from neonatal foreskins which are stimulated by physiological concentrations of hydrocortisone. Paired cultures derived from lung and skin of the same individual showed opposite responses, thus ruling out genetic differences in responsiveness in these cases at least. Perhaps one of the most important findings to emerge from these studies is the fact that fibroblast-like cells in culture may display differences depending on the tissue of origin and stage of development. They probably retain some differentiated characteristics even after they are established in vitro and have been serially passaged many times. Thus, studies on one fibroblast-like cell line may not be generalizable and for reasons other than the media and culture procedures usually invoked to explain such differences.

The mechanism by which glucocorticoids affect cell proliferation is still not well defined. The presence of specific glucocorticoid receptors has been correlated to growth responsiveness in the majority of the cell lines reported in Tables 1–5 suggesting that these proliferative effects may be the result of altered gene expression. Recent evidence has further suggested that glucocorticoids can regulate the activity of receptors for heterologous hormones which may indirectly or directly play a role in the regulation of cellular replication. Our own recent finding that glucocorticoids induce the appearance of a "factor" with growth stimulating activity into the conditioned medium is consistent with data from several other laboratories suggesting an additional step in the mechanism of steroid hormone action. That is, the production of factors with growth stimulating activity which can regulate the growth of the cell which generates this factor and/or act on other cell types.

Further characterization of these glucocorticoid induced growth factors is indicated. Knowledge of their biochemical properties and mode of action may help further our understanding of the underlying regulatory factors involved in the control of cellular replication.

Acknowledgments. We would like to express our thanks to a number of our laboratory co-workers who worked on unpublished aspects of some of the studies described here. They include Roberta Charpentier, Jane Miller, Marie Skelston, and Jean Wallace. We would also like to express our thanks to Dr. Roger Ladda for making available to us neonatal lung and foreskin derived cultures for our studies as well as his own unpublished data from these cultures.

Studies reported here were supported by NIH grants AG 00378 and training grant CA 09171.

References

Armelin, H.A.: Pituitary extracts and steroid hormones in the control of 3T3 cell growth. Proc. Natl. Acad. Sci. U.S.A. *70*, 2702–2706 (1973)

Armelin, M.C.S., Armelin, H.A.: Serum and hormonal regulation of the "resting-proliferative" transition in a variant of 3T3 mouse cells. Nature *265*, 148–151 (1977)

Baker, J.B., Barsh, G.S., Carney, D.H., Cunningham, D.: Dexamethasone modulates binding and action of epidermal growth factor in serum-free cell cultures. Proc. Natl. Acad. Sci. U.S.A. *75*, 1882–1886 (1978)

Ballard, P.L.: Glucocorticoids and differentiation. In: Glucocorticoid hormone actions. Monographs on endocrinology. Baxter, J.D., Rousseau, G.G. (eds.), vol. 12, pp. 493–515. Berlin, Heidelberg, New York: Springer 1979

Baulieu, E.E.: Some aspects of the mechanism of action of steroid hormones. Mol. Cell. Biochem. *7*, 157–173 (1975)

Baxter, J.D., Forsham, P.H.: Tissue effects of glucocorticoids. Am. J. Med. *53*, 573–589 (1972)

Baxter, J.D., Rousseau, G.G.: Glucocorticoid hormone action: an overview. In: Glucocorticoid hormone action. Monographs on endocrinology. Baxter, J.D., Rousseau, G.G. (eds.), vol. 12, pp. 1–24. Berlin, Heidelberg, New York: Springer 1979

Baxter, J.D., Tomkins, G.M.: The relationship between glucocorticoid binding and tyrosine aminotransferase induction in hepatoma tissue culture cells. Proc. Natl. Acad. Sci. U.S.A. *65*, 709–715 (1970)

Berliner, D.L., Dougherty, T.F.: Hepatic and extrahepatic regulation of corticosteroids. Pharmacol. Rev. *13*, 329–359 (1961)

Burgess, A.W., Camakaris, J., Metcalf, D.: Purification and properties of colony-stimulating factor from mouse lung-conditioned medium. J. Biol. Chem. *252*, 1998–2003 (1977)

Cahill, G.G.: Actions of adrenal cortical steroids on carbohydrate metabolism. In: The human adrenal cortex. Christy, N.P. (ed.), pp. 205–239. New York: Harper & Row (1971)

Campbell, P.S.: The mechanism of the inhibition of uterotrophic responses to acute dexamethasone pretreatment. Endocrinologies *103*, 716–723 (1978)

Castor, C.W., Prince, R.K.: Modulation of the intrinsic viscosity of hyauronic acid formed by human fibroblasts in vitro: the effects of hydrocortisone and colchicine. Biochim. Biophys. Acta *83*, 165–177 (1964)

Cavenee, W., Melnykovych, G.: Elevation of HeLa cell 3-hydroxy-3-methyl glutaryl coenzyme. A reductase activity by glucocorticoids. J. Cell. Physiol. *98*, 199–212 (1979)

Chan, L., Means, A.R., O'Malley, B.W.: Hormone modulation of specific gene expression. Vitam. Horm. *36*, 259–295 (1978)

Chen, T.L., Feldman, D.: Glucocorticoid receptors and actions in subpopulations of cultured rat bone cells. J. Clin. Invest. *63*, 750–758 (1979)

Clark, J.L., Jones, K.L., Gospodarowicz, D., Sato, G.H.: Growth response to hormones by a new rat ovary cell line. Nature New Biol. *236*, 180–181 (1972)

Clausen, P.P.F., Guatvik, K.M., Haugh, E.: Effects of cortisol, 17β-estradiol and thyroliberin on prolactin and growth hormone production, cell growth and cell cycle distribution in cultured rat pituitary tumor cells. J. Cell. Physiol. *94*, 205–214 (1978)

Crabtree, G.R., Munck, A., Smith, K.A.: Glucocorticoids inhibit expression of Fc receptors in the human granulocytic cell line HL-60. Nature *279*, 338–339 (1979)

Crabtree, G.R., Smith, R.A., Munck, A.: Glucocorticoid receptors and sensitivity of isolated human leukemia lymphoma and lymphoma cells. Cancer Res. *38*, 4268–4272 (1978)

Cristofalo, V.J.: Metabolic aspects of aging in diploid human cells. In: Aging in cell and tissue culture. Holeckova, E., Cristofalo, V.J. (eds.), pp. 83–119. New York: Plenum Press 1970

Cristofalo, V.J.: Hydrocortisone as a modulator of cell division and population life span. In: Explorations in aging. Cristofalo, V.J., Roberts, J., Adelman, R.C. (eds.), pp. 57–79. New York: Plenum Press 1975

Cristofalo, V.J., Rosner, B.A.: Modulation of cell proliferation and senescence of WI-38 cells by hydrocortisone. Fed. Proc. *38*, 1851–1856 (1979)

Cristofalo, V.J., Wallace, J.M., Rosner, B.A.: Glucocorticoid enhancement of proliferative activity in WI-38 cells. In: Hormones and cell culture. Sato, G.H., Ross, R. (eds.)., vol. 6, book b, pp. 875–887. Cold Spring Harbor Conferences on Cell Proliferation, Cold Spring Harbor Laboratory Publications (1979)

De Larco, J.E., Todaro, G.J.: Growth factors from murine sarcoma virus-transformed cells. Proc. Natl. Acad. Sci. U.S.A. 75, 4001–4005 (1978)

Dietrich, J.W., Canalis, E.M., Maina, D.M., Raise, L.G.: Effects of glucocorticoids on fetal rate bone collagen synthesis in vitro. Endocrinology 104, 715–721 (1979)

Dulak, N.C., Temin, H.M.: Multiplication-stimulating activity for chicken embryo fibroblasts from rat liver cell conditioned medium: a family of small polypeptides. J. Cell. Physiol. 81, 161–170 (1973)

Epifanova, O.I.: Effects of hormones on the cell cycle. In: The cell cycle and cancer. Baserga, R. (ed.), pp. 145–180. New York: Marcel Dekker 1971

Fauci, A.S.: Immunosuppressive and anti-inflammatory effects of glucocorticoids. In: Glucocorticoid hormone action. Monographs on endocrinology. Baxter, J.D., Rousseau, G.G. (eds.), vol. 12, pp. 449–465. Berlin, Heidelberg, New York: Springer 1979

Flower, R.J., Blackwell, G.J.: Anti-inflammatory steroids induce biosynthesis of A phospholipase A_2 inhibitor which prevents prostaglandin generation. Nature 278, 456–459 (1979)

Fulder, S.T.: The growth of cultured human fibroblasts treated with hydrocortisone and extracts of the medicinal plant Panax ginseng. Exp. Gerontol. 12, 125–131 (1977)

Gaffney, E.V., Pigott, D.: Hydrocortisone stimulation of human mammary epithelial cells. In Vitro 14, 621–624 (1978)

Gorski, J., Gannon, F.: Current models of steroid hormone action: a critique. Ann. Rev. Physiol. 38, 425–450 (1976)

Gospodarowicz, D.: Localization of a fibroblast growth factor and its effect alone and with hydrocortisone in 3T3 cell growth. Nature 249, 123–127 (1974)

Gospodarowicz, D., Moran, J.S.: Stimulation of division of sparse and confluent 3T3 cell populations by a fibroblast growth factor, dexamethasone, and insulin. Proc. Natl. Acad. Sci. U.S.A. 71, 4584–4588 (1974)

Gospodarowicz, D., Moran, J.S.: Mitogenic effect of fibroblast growth factor on early passage cultures of human and murine fibroblasts. J. Cell Biol. 66, 451–457 (1975a)

Gospodarowicz, D., Moran, J.S.: Optimal conditions for the study of growth control in BALB/c 3T3 fibroblasts. Exp. Cell Res. 90, 279–284 (1975b)

Gospodarowicz, D., Weseman, J., Moran, J.S., Linstrom, J.: Effect of fibroblast growth factor on the division and fusion of bovine myoblasts. J. Cell Biol. 70, 395–405 (1976)

Granner, D.K.: Restoration of sensitivity of cultured hepatoma cells to cyclic nucleotides shows the permissive effect of dexamethasone. Nature 259, 572–573 (1976)

Granner, D.K.: The role of glucocorticoid hormones as biological amplifiers. In: Glucocorticoid hormone action. Monographs on endocrinology. Baxter, J.D., Rousseau, G.G. (eds.), vol. 12, pp. 593–661. Berlin, Heidelberg, New York: Springer 1979

Grasso, R.J., Johnson, C.E.: Dose-response relationships between glucocorticoids and growth inhibition in rate glioma monolayer cultures. Proc. Soc. Exp. Biol. Med. 154, 238–248 (1977)

Gross, S.R., Aronow, L., Pratt, W.B.: The active transport of cortisol by mouse fibroblasts growing in vitro. Biochem. Biophys. Res. Commun. 32, 66–72 (1968)

Grove, G.L., Houghton, B.A., Cochran, J.W., Kress, E.D., Cristofalo, V.J.: Hydrocortisone effects on cell proliferation: specificity of response among various cell types. Cell Biol. Internat. Rep. 1, 147–155 (1977)

Guerriero, V. jr., Florini, J.R.: Stimulation by glucocorticoids of myoblast growth at low cell densities. Cell Biol. Int. Rep. 2, 441–446 (1978)

Guner, M., Freshney, R.I., Morgan, D., Freshney, G., Thomas, D.G.T., Graham, D.I.: Effects of dexamethasone and betamethosane on in vitro cultures from human astrocytoma. Brit. J. Cancer 35, 439–447 (1977)

Hackney, J.F., Gross, S.R., Aronson, L., Pratt, W.B.: Specific glucocorticoid binding macromolecular from mouse fibroblasts growing in vitro. A possible steroid receptor for growth inhibition. Mol. Pharmacol. 6, 500–512 (1970)

Harmon, J.M., Norman, M.R., Fowlkes, B.J., Thompson, E.B.: Dexamethasone induces irreversible G_1 arrest and death of a human lymphoid cell line. J. Cell. Physiol. *98*, 267–278 (1979)

Harper, R.A., Grove, G.: Human skin fibroblasts derived from papillary and reticular dermis: differences in growth potential in vitro. Science *204*, 526–527 (1979)

Harvey, W., Grahame, R., Panayi, G.S.: Effect of steroid hormones on human fibroblasts in vitro. Ann. Rheum. Dis. *35*, 148–151 (1976)

Higgins, S.J., Gehring, I.: Molecular mechanisms of steroid hormone action. Adv. Cancer Res. *28*, 313–397 (1978)

Hollenberg, M.: Steroid-stimulated amino acid uptake in cultured human fibroblasts reflects glucocorticoid and anti-inflammatory potency. Mol. Pharmacol. *13*, 150–160 (1977)

Holley, R.W., Kiernan, J.A.: Control of DNA synthesis of 3 T 3 cells: serum factors. Proc. Natl. Acad. Sci. U.S.A. *71*, 2908–2911 (1974)

Horibata, K., Harris, A.W.: Mouse myelomas and lymphomas in culture. Exp. Cell Res. *60*, 61–77 (1970)

Iacobelli, S., Ranelletti, F.O., Longo, P., Riccardi, R., Mastrangelo, R.: Discrepancies between in vivo and in vitro effects of glucocorticoids in myelomonocytic leukemic cells with steroid receptors. Cancer Res. *38*, 4257–4762 (1978)

Jensen, E.V., Jacobson, S.: Basic guides to the mechanism of estrogen action. Recent Prog. Hormone Res. *18*, 387–414 (1962)

Jones, F.L., Anderson, W.S., Addison, J.: Glucocorticoid induced growth inhibition of cells from human lung alveolar cell carcinoma. Cancer Res. *38*, 1688–1693 (1978)

King, R.J.B., Kaye, A.M., Shodell, M.J.: Co-purification of an Oestrogen-induced protein from rat uterus and a factor able to stimulate DNA synthesis in cultured cells. Exp. Cell Res. *109*, 1–8 (1977)

Loeb, J.N.: Corticosteroids and growth. New Engl. J. Med. *295*, 547–552 (1976)

Maca, R.D., Fry, G.L., Hayes, A.D.: Effects of glucocorticoids on the interaction of lymphoblastoid cells with human endothelial cells in vitro. Cancer Res. *38*, 2224–2228 (1978 a)

Maca, R.D., Fry, G.L., Hoak, J.C.: The effects of glucocorticoids on cultured human endothelial cells. Brit. J. Haemotol. *38*, 501–509 (1978 b)

Macieira-Coelho, A.: Action of cortisone on human fibroblasts in vitro. Experientia *22*, 390–391 (1966)

Martin, J.D., Tomkins, G.M., Granner, D.: Synthesis and induction of tyrosine aminotransferase in synchronized hepatoma cells in culture. Proc. Natl. Acad. Sci. U.S.A. *62*, 248–255 (1969)

Marshall, S., Huang, H.H., Kledzik, G.S., Campbell, C.A., Moites, J.: Glucocorticoid regulation of prolactin receptors in kidneys and adrenals of male rats. Endocrinology *102*, 869–875 (1978)

McCarty, K.S. jr., McCarty, K.S. sr.: Steroid hormone receptors in the regulation of differentiation. Am. J. Pathol. *86*, 704–744 (1977)

Mets, T., Korteweg, M., Verdank, G.: Increased prostaglandin F_2 and E_2 production in late passage WI-38 diploid fibroblasts. Cell Biol. Int. Rep. *3*, 691–694 (1979)

Milgrom, E., Atger, M., Baulieu, E.E.: Studies on estrogen entry into uterine cells and on estradiol-receptor complex attachment to the nucleus. Is the entry of estrogen into uterine cells a protein-mediated process? Biochim. Biophys. Acta *320*, 267–283 (1973)

Millis, A.J.T., Hoyle, M.: Fibroblast-conditioned medium contains cell surface proteins required for cell attachment and spreading. Nature *271*, 668–669 (1978)

Munck, A., Leung, K.: Glucocorticoid receptors and mechanisms of action. In: Receptors and mechanism of action of steroid hormones. Pasqualini, J.R. (ed.), chap. 2, pp. 311–397. New York: Marcel Dekker 1977

Neifeld, J.D., Lyman, M.E., Tomey, D.C.: Steroid hormone receptors in normal lymphocytes. J. Biol. Chem. *252*, 2972–2977 (1977)

Norman, M.R., Harmon, J.M., Thompson, E.B.: Use of a human lymphoid cell line to evaluate interactions between prednisolone and other chemotherapeutic agents. Cancer Res. *38*, 4273–4278 (1978)

Osborne, C.K., Huff, K.H., Bronzert, D., Lippman, M.E.: Direct inhibition of growth and antagonism of insulin action by glucocorticoids in human breast cancer cells in culture. Cancer Res. *39*, 2422–2428 (1979)

Pinsky, L., Finkelberg, R., Straisfeld, L., Zilaki, B., Kauffman, M., Hull, G.: Testosterone metabolism by serially subcultured fibroblasts from genital and non-genital skin of individual human donors. Biochem. Biophys. Res. Commun. *46*, 364–369 (1972)

Pratt, W.B., Aronow, L.: The effect of glucocorticoids on protein and nucleic acid synthesis in mouse fibroblasts growing in vitro. J. Biol. Chem. *241*, 5244–5250 (1966)

Richman, R.A., Claus, T.C., Pilpis, S.J., Friedman, D.L.: Hormonal stimulation of synthesis in primary cultures of adult rat hepatocytes. Proc. Natl. Acad. Sci. U.S.A. *73*, 3589–3593 (1976)

Rowe, D.W., Starman, B.D., Fugimoto, W.Y., Williams, R.H.: Differences in growth response to hydrocortisone and ascorbic acid by human diploid fibroblasts. In Vitro *13*, 824–830 (1977)

Rudland, P.S., Seifert, W., Gospodarowicz, D.: Growth control in cultured mouse fibroblasts: induction of the pleiotype and mitogenic responses by purified growth factor. Proc. Natl. Acad. Sci. U.S.A. *71*, 2600–2604 (1974)

Ruhman, A.G., Berliner, D.L.: Effect of steroids on growth of mouse fibroblasts in vitro. Endocrinology *76*, 916–927 (1965)

Runikis, J.O., McLean, D.I., Stewart, W.D.: Growth rate of cultured human fibroblasts increased by glucocorticoids. J. Invest. Dermatol. *70*, 348–351 (1978)

Russell, J.D., Russell, S.B., Trupin, K.M.: Differential effects of hydrocortisone on both growth and collagen metabolism of human fibroblasts from normal and keloid tissue. J. Cell Physiol. *97*, 221–230 (1978)

Saito, E., Mukai, M., Miraki, T., Ichikawa, Y., Humma, M.: Inhibitory effects of corticosterone on cell proliferation and steroidogenesis in the mouse adrenal tumor cell line Y-1. Endocrinology *104*, 487–492 (1979)

Sakai, S., Banerjee, M.: Glucocorticoid modulation of prolactin receptors in mammary cells of lactating mice. Biochim. Biophys. Acta *582*, 79–88 (1979)

Salomon, D.S., Pratt, R.M.: Inhibition of growth in vitro by glucocorticoids in mouse embryonic facial mesenchyme cells. J. Cell. Physiol. *97*, 315–327 (1978)

Sanford, K.K., Earle, W.R., Likely, G.D.: The growth in vitro of single isolated (no line) tissue cells. J. Natl. Cancer Inst. *9*, 229–246 (1949)

Sato, G., Reid, L.: Replacement of serum in cell culture by hormones. In: Biochemistry and mode of action of hormones II. Rickenberg, H.U. (ed.), vol. 20, pp. 219–251. Baltimore: University Park Press 1978

Schmit, J.P., Rousseau, G.G.: Structure and conformation of glucocorticoids. In: Glucocorticoid hormone action. Monographs on endocrinology. Baxter, J.D., Rousseau, G.G. (eds.), vol. 12, pp. 79–95. Berlin, Heidelberg, New York: Springer 1979

Schneider, E.L., Mitsui, Y.O., Au, K.S., Shorr, S.S.: Tissue-specific differences in cultured human diploid fibroblasts. Exp. Cell Res. *108*, 1–6 (1977)

Schutz, G., Beato, M., Feigelson, P.: Messenger RNA for hepatic tryptophan oxygenase: Its partial purification, its translation in a heterologous cell-free system, and its control by glucocorticoid hormones. Proc. Natl. Acad. Sci. U.S.A. *70*, 1218–1221 (1972)

Schutz, G., Killewich, L., Chen, G., Feigelson, P.: Control of the mRNA for hepatic tryptophan oxygenase during hormonal substrate induction. Proc. Natl. Acad. Sci. U.S.A. *72*, 1017–1020 (1973)

Sibley, C.H., Tomkins, G.M.: Isolation of lymphoma cell variants resistant to killing by glucocorticoids. Cell *2*, 213–220 (1974)

Sirbasku, D.: Estrogen induction of growth factors specific for hormone-responsive mammary, pituitary and kidney tumor cells. Proc. Natl. Acad. Sci. U.S.A. *75*, 3786–3790 (1978)

Smith, B., Torday, J.S., Giroud, C.J.P.: The growth promoting effect of cortisol on human fetal lung cells. Steroids *22*, 515–524 (1973)

Smith, K.A., Crabtree, G.R., Kennedy, S.J., Munck, A.U.: Glucocorticoid receptors and glucocorticoid sensitivity of mitogen stimulated and unstimulated human lymphocytes. Nature *267*, 523–326 (1977)

Smith, B.T.: Lung maturation of the fetal rat: acceleration by injection of fibroblast-pneu-monocyte factor. Science *204*, 1094–1095 (1979)

Stevens, J., Stevens, Y.W.: Cortisol-induced lymphocytolysis of P 1798 tumor cells in glu-cose-free, pyruvate-free medium. J. Natl. Cancer Inst. *54*, 1493–1494 (1975)

Stevenson, R.D., Lucie, N.P., Gray, A.C.: Effect of prednisolone on the growth of human bone marrow cells in vitro. Brit. J. Exp. Path. *59*, 467–472 (1978)

Tashjian, A.H., Osborne, R., Maina, D., Kraian, A.: Hydrocortisone increases the number of receptors for thyrotropin-releasing hormone on pituitary cells in culture. Biochem. Biophys. Res. Commun. *79*, 233–340 (1977)

Taylor, L., Polgar, P.: Self regulation of growth by human diploid fibroblasts via prosta-glandin production. FEBS. Letters *79*, 69–72 (1977)

Thompson, E.B., Lippman, M.: Mechanism of action of glucocorticoids. Metabolism *23*, 159–202 (1974)

Thrall, C., Webster, R.A., Spelsberg, T.C.: Receptor interaction with chromatin. In: The cell nucleus. Busch, H. (ed.), vol. 3, pp. 461–529. New York: Academic Press 1978

Thrash, C.R., Cunningham, D.D.: Stimulation of division of density inhibited fibroblasts by glucocorticoids. Nature *242*, 399–401 (1973)

Vignon, F., Chan, P., Rochefort, H.: Hormonal regulation in two rat mammary cancer cell lines: glucocorticoid and androgen receptors. Mol. Cell. Endocrinol. *13*, 191–202 (1979)

Wicks, W.D.: Non-interaction glucocorticoid nucleus activity: molecular events. In: Bio-chemistry of hormones. Rickenberg, H.V. (ed.), vol. 8, p. 811. Boston: Union Park (1974)

Wolff, M.E.: Structure-activity relationship in glucocorticoids. In: Glucocorticoid hormone action. Monographs on endocrinology. Baxter, J.D., Rousseau, G.G. (eds.), vol. 12, pp. 97–107. Berlin, Heidelberg, New York: Springer 1979

Wyche, J.H., Noteboom, W.D.: Requirement of a specific factor for the multiplication of ovarian cells in serum-free medium. Exp. Cell Res. *110*, 135–141 (1977)

Yamamoto, K.R., Alberts, B.M.: Steroid receptors: elements for modulation of eukaryotic transcription. Ann. Rev. Biochem. *45*, 721–746 (1976)

CHAPTER 7

Proteases as Growth Factors

D. D. CUNNINGHAM

A. Introduction

It is now clear that certain proteolytic enzymes play critical roles in the activation of many biological processes including blood coagulation and clot formation (DAVIE and FUJIKAWA, 1975), platelet aggregation and release (DETWILER and WASIEWSKI, 1977), fibrinolysis (MCKEE, 1976), complement activation (MÜLLER-EBERHARD, 1972, 1975), hormone production (STEINER et al., 1975), mammalian reproductive processes (WILLIAMS-ASHMAN, 1975), virus replication (KORANT, 1975) and animal cell division (CARNEY et al., 1978). In these activation processes, the key initiation event involves limited proteolysis... the conversion of inactive species to active forms by selective proteolytic cleavages involving proteases which are usually very specific (NEURATH, 1975; NEURATH and WALSH, 1976a, b). The activation step itself is essentially irreversible since proteolysis is an exergonic reaction and under normal physiological conditions there are no simple biological mechanisms to repair a broken peptide bond. The activation by limited proteolysis can occur in one step; alternatively, it can occur after a series of activation steps in which the product of each step is itself a protease that can activate the next zymogen in the series (cascade). Where biological activations involve cascades, there are opportunities for large amplification effects in which the degree of amplification is largely determined by the number of steps in the cascade. Another important feature of most biological events which are regulated by proteolysis is the possibility for modulation by certain protease inhibitors. By their ability to limit the extent, duration and site of protease action, they provide added specificity to the regulatory systems (NEURATH and WALSH, 1976a, b).

From the above examples of events that are activated by limited proteolysis, it is clear that quite a variety of different biological processes can be regulated by the action of proteases. Although the specific proteolytic activation in each of these different processes is necessarily unique, there are some underlying common features. As a result, the processes that have been studied most thoroughly, particularly clot formation, fibrinolysis and complement activation, can serve as useful model systems for investigations on other biological events known to be regulated by proteolytic enzymes.

B. Stimulation of DNA Synthesis and Cell Divisions by Proteases

I. Cell-Protease Combinations Which Yield Mitogenic Stimulation

The ability of certain proteolytic enzymes to initiate DNA synthesis and cell division was discovered in 1970. That year, BURGER (1970) reported that addition of trypsin, pronase or ficin to cultures of nondividing mouse 3 T 3 cells led to mitosis and cell division with a doubling in cell number. The same year, SEFTON and RUBIN (1970) reported that addition of the same proteases to cultures of nonproliferating secondary chick embryo fibroblasts brought about increased DNA synthesis, mitosis and a twofold increase in cell number. The latter study suggested that the ability of the proteases to stimulate cell division resulted from action on the cells and was not a result of releasing active peptides from protein-containing components of the cell culture medium. In both studies, the use of protease inhibitors strongly indicated that the enzymic activity of the proteases was necessary for initiation of cell division.

The mitogenic action of trypsin on nonproliferating chick embryo fibroblasts has been confirmed in a number of investigations (VAHERI et al., 1974; BLUMBERG and ROBBINS, 1975; CUNNINGHAM and HO, 1975; HALE and WEBER, 1975) but it has been more difficult to repeat the initiation of mouse 3 T 3 cell division by trypsin (GLYNN et al., 1973; HOLLEY and KIERNAN, 1974; CUNNINGHAM and HO, 1975). It now appears that some fresh serum must be added along with the trypsin to observe initiation with 3 T 3 cells (NOONAN, 1976) and that initiation occurs only with certain clones of these cells (NOONAN, personal communication). It has also been reported that trypsin addition stimulates division of cultured baby hamster kidney cells (BROWN and KIEHN, 1977). The stimulation did not occur with nonproliferating cells; trypsin produced an increase in cell number only when the cells were growing in culture medium containing a level of serum that was somewhat rate-limiting for growth (KIEHN and BROWN, 1978). In these studies, it was reported that the growth stimulatory action of trypsin was only partially inhibited by pancreatic or soybean trypsin inhibitor, or by inactivation with diisopropylfluorophosphate. However, these results must be interpreted with caution since the degree of protease inactivation was measured only by spectrophotometric analysis of casein hydrolysis, a relatively insensitive measure of protease activity.

CHEN and BUCHANAN (1975) made the important discovery that addition of highy purified thrombin to nonproliferating cultures of secondary chick embryo fibroblasts stimulated DNA synthesis and cell division. Since thrombin is produced at the site of tissue injury (DAVIE and FUJIKAWA, 1975), this suggested the possibility that it might play an important role in wound healing. Thrombin also stimulates the proliferation of cultured nondividing normal human fibroblasts (POHJAN-PELTO, 1977, 1978; CARNEY et al., 1978), secondary mouse embryo fibroblasts (BUCHANAN et al., 1976; CARNEY et al., 1978) and a line of Chinese hamster lung cells (SIMMER et al., 1980). However, thrombin is not mitogenic for chick embryo fibroblasts transformed by Rous sarcoma virus (ZETTER et al., 1977 a). In addition, as untransformed chick embryo fibroblasts are serially subcultured, they lose their ability to respond to the mitogenic action of thrombin; by the twenty-fifth population doubling, thrombin produces only a small increase in cell number even though all of the cells still respond to the mitogenic action of serum (CARNEY et al., 1978). This indicates that the responsiveness of cells to thrombin is a differentiated function that is lost after repeated passage in cell culture, and might explain why a number of cell lines are unresponsive to the mitogenic action of thrombin (BUCHANAN et al., 1976; CARNEY et al., 1978).

A number of other proteases have been examined to determine if they can stimulate DNA synthesis or cell division in nonproliferating cultured fibroblasts. In general, thrombin is the only protease which reproducibly has been shown to stimulate mammalian fibroblasts, although there is a report that plasmin and trypsin are weakly mitogenic for cultured human fibroblasts (POHJANPELTO, 1977). On the other hand, a number of different proteases have been shown to readily stimulate DNA synthesis or cell division in nonproliferating cultured chick embryo fibroblasts including trypsin (SEFTON and RUBIN, 1970; VAHERI et al., 1974; CUNNINGHAM and HO, 1975; HALE and WEBER, 1975), the γ-subunit of nerve growth factor (an arginyl-esteropeptidase) (GREENE et al., 1971) elastase, bromelin (TENG and CHEN, 1975; ZETTER et al., 1976), collaginase and plasmin (BLUMBERG and ROBBINS, 1975). The mitogenic activity of α-chymotrypsin on chick embryo fibroblasts is variable (BLUMBERG and ROBBINS, 1975; TENG and CHEN, 1975; MARTIN and QUIGLEY, 1978), indicating that the ability of certain preparations to stimulate might be due to contaminants. A number of proteases have been shown to be non-mitogenic or only very weakly mitogenic for chick embryo fibroblasts including α-protease, ficin, subtilisin (TENG and CHEN, 1975), thermolysin and papain (ZETTER et al., 1976). It is noteworthy that the inability of α-protease, ficin, or subtilisin to produce significant DNA synthesis in chick embryo cells was not due to the presence of inhibitors in these preparations or to irreversible damage to the cells during the test period since they did not inhibit the ability of thrombin to stimulate the cells (TENG and CHEN, 1975).

Proteases can also stimulate DNA synthesis, measured by incorporation of ^{3}H-thymidine, in certain cultured lymphocytes. With mouse Balb/c spleen cells, trypsin can stimulate as effectively as lipopolysaccharide (VISHER, 1974). Moreover, over 80% of the transformed cells after stimulation of the Balb/c spleen cells by trypsin had immunoglobulin surface receptors and were B cells. Trypsin can also stimulate DNA synthesis in B lymphocytes from nude (athymic) and AKR mice (KAPLAN and BONA, 1974), human B lymphocytes (GIRARD and FERNANDES, 1976), hamster spleen cultures (HART and STREILEIN, 1976) and in chicken spleen cultures and peripheral lymphocytes (HOVI et al., 1978). Pronase can stimulate DNA synthesis in spleen cell cultures of AKR mice (KAPLAN and BONA, 1974) and human lymphocytes (GIRARD and FERNANDES, 1976), while chymotrypsin has been reported to stimulate hamster spleen cells (HART and STREILEIN, 1976). In addition, an elastase and chymotrypsin-like cathepsin B, purified from human polymorphonuclear leukocytes, reportedly stimulate DNA synthesis in human peripheral lymphocytes and mouse splenocytes (VISCHER et al., 1976). Moreover, there is report that thrombin is mitogenic for cultured mouse spleen cells (CHEN et al., 1976). Thus, quite a variety of different proteases can stimulate DNA synthesis in B lymphocytes; however, T lymphocytes generally are not very responsive to protease stimulation. Studies with trypsin inhibitors have indicated that the proteolytic activity of trypsin is required for stimulation of B lymphocytes (VISCHER, 1974; KAPLAN and BONA, 1974). The stimulation by trypsin of B lymphocytes is thought to be mitogenic rather than immunogenic since it occurs in cultured spleen from nude mice and requires the proteolytic activity of trypsin (KAPLAN and BONA, 1974).

II. Involvement of Proteases in the Action of Other Growth Factors

There are two kinds of experiments which have indicated that proteases might play some role in the action of other growth promoting agents. In a direct approach, it has been shown that proteases can potentiate the action of other growth factors. Indirect studies using protease inhibitors have suggested that endogenous cellular proteases might be involved in the action of certain lymphocyte mitogens.

There are several reports that proteases can augment the action of other growth factors. For example, an arginine esterase that binds epidermal growth factor (EGF) also potentiates the action of EGF (LEMBACH, 1976). In addition, treatment of chick embryo fibroblasts with trypsin increases the ability of insulin to stimulate DNA synthesis in them by about twofold; interestingly, this same treatment increases the ability of the cells to bind ^{125}I-insulin by about twofold, leading to the suggestion that the potentiation by trypsin results from an unmasking of mitogen receptors (RAIZADA and PERDUE, 1976). There is also a report that a chymotrypsin-like protease produced by polymorphonuclear leukocytes enhances the response of mouse thymocytes to phytohemagglutinin (YOSHINAGA et al., 1977); however, this result should be interpreted cautiously since the potentiating protease released by the leukocytes was not highly purified. Studies have shown that thrombin can clearly potentiate the action of some growth factors on certain cells. Large potentiating effects of thrombin were observed with fibroblast growth factor (FGF) and rabbit corneal fibroblasts, serum and mouse 3 T 3 cells, and EGF and prostaglandin F 2 and mouse B 77–3 T 3 cells (ZETTER et al., 1977 b). Thrombin also potentiates the response of human but not bovine endothelial cells to FGF (GOSPODAROWICZ et al., 1978).

The studies suggesting that proteases are involved in stimulation of lymphocytes by certain mitogens are based on the use of protease inhibitors. For example, epsilon amino caproic acid (EACA), tosyl arginine methylester (TAME), tosyl lysine chloromethyl ketone (TLCK), and tosyl phenylalanine chloromethyl ketone (TPCK) inhibit the response of human peripheral blood lymphocytes to phytohemagglutinin (HIRSHHORN, 1971). Also, soybean trypsin inhibitor has been shown to decrease the ability of both concanavalin A and phytohemagglutinin to stimulate DNA synthesis in rat and guinea pig spleen cells; however, it augmented the action of these mitogens in cultures of hamster spleen cells (HART, 1977). There is a recent report that soybean trypsin inhibitor and Trasylol inhibit the response of mouse spleen cells to a number of mitogens (VISCHER, 1979). However, the suggestion has been made from other studies that Trasylol might inhibit lymphocyte triggering by interfering with the "helping" action of certain accessory cells (HIGUCHI et al., 1977). At present, it is difficult to draw firm conclusions from these studies on protease inhibitors. In some of the above experiments, the protease inhibitors affected the rate of ^{3}H-thymidine incorporation in cells that had not been treated with the mitogen. In addition, the protease inhibitors can have pronounced side effects. This is especially true for the alkylating agents TLCK and TPCK. Both have been shown to inhibit protein synthesis in cultured fibroblasts (CHOU et al., 1974).

III. Protease Activity of Other Growth Factors

In view of the ability of certain proteases to stimulate cell proliferation and to potentiate the action of some other growth factors, it is important to ask whether proteolytic activity is a general property of growth factors.

It is presently not clear whether nerve growth factor (NGF) possesses protease activity. This factor is isolated as a complex of α, β, γ, and subunits. The γ-subunit has no known biological activity. The β-subunit is an arginyl-esteropepidase which converts the pro β-subunit to the β-subunit (BERGER and SHOOTER, 1977). The β-subunit possesses the biological activity of NGF (BRADSHAW, 1978). It has been reported that the active NGF is a protease (ORENSTEIN et al., 1978), although it has been questioned whether the β-subunit itself specifically possesses proteolytic activity (CALISSANO and LEVI-MONTALCINI, 1979). The issue remains controversial (YOUNG, 1979).

EGF is also isolated as a high molecular weight complex in which the growth factor peptide is associated with a specific arginine esteropepidase known as the EGF-binding protein (TAYLOR et al., 1970). EGF is synthesized as an Mr 9,000 biosynthetic precursor which is processed to the active Mr 6,000 species; it appears that this processing event is catalyzed by the arginine esteropepidase (FREY et al., 1979). There is a report that the arginine esteropeptidase enhances the mitogenic action of EGF (LEMBACH, 1976), but there have been no reports that EGF itself possesses proteolytic activity.

There have also been no reports of protease activity in other purified peptide growth factor preparations including FGF (GOSPODAROWICZ, 1975), multiplication stimulating activity (SMITH and TEMIN, 1974), insulin-like growth factors (RINDERKNECHT and HUMBEL, 1976a, b) or the platelet-derived growth factor (ANTONIADES et al., 1979; HELDIN et al., 1979).

C. Mechanisms of Protease-Stimulated Cell Proliferation

I. Mitogenic Stimulation by Proteases Under Serum-Free Chemically Defined Conditions

Although the earliest studies on mitogenic stimulation by proteases were conducted with serum-containing cultures (BURGER, 1970; SEFTON and RUBIN, 1970) there is now evidence that serum is not required for protease-stimulated DNA synthesis or cell division. The first report of protease stimulation in serum-free medium was with mouse B lymphocytes (VISCHER, 1974); subsequently, it has been shown that serum is not required for the stimulation of chick fibroblasts by thrombin or trypsin (CHEN and BUCHANAN, 1975; CARNEY et al., 1978) or the stimulation of mammalian fibroblasts by thrombin (POHJANPELTO, 1977; CARNEY et al., 1978).

It should be pointed out that in the above studies the cultures were set up in serum-containing media, and at some time later the cells were changed to serum-free media to arrest cell growth and test the mitogenic action of added proteases. As a result, it is necessary to consider the possibility that some serum components remained associated with the cells and were necessary for the protease-stimulated DNA synthesis or cell division. Although this possibility cannot be ruled out from presently available data, it is noteworthy that the capacity of cultured chick embryo, mouse embryo and human fibroblasts to respond to the mitogenic action of thrombin or trypsin increased with extended culture in serum-free medium up to about 60 h (CARNEY et al., 1978). The most likely explanation of this result is that during the extended culture in serum-free medium the cells "deplete" the medium of residual protease inhibitors (LOW and CUNNINGHAM, unpublished data). All in all, the available data provide evidence that serum is not required for protease-stimulated cell division. This is important, since it indicates that the protease stimulation results from direct action on the cells rather than a protein-containing component of the culture medium. As will be apparent below, the stimulation in serum-free medium has facilitated studies on the mechanism of the mitogenic action of proteases.

It should be pointed out that addition of mitogenic proteases to responsive cells in serum-free medium generally leads to approximately one round of DNA synthesis and cell division. Although an additional round or two can occur under optimal conditions, there have been no reports of sustained cell proliferation in serum-free medium when the only mitogenic agent present is a proteolytic enzyme.

In either the presence or absence of serum, continued presence of the mitogenic protease is required for maximal stimulation. Brief treatments lead to correspondingly small proliferative responses (CUNNINGHAM and HO, 1975; POHJANPELTO, 1978).

II. Requirement of Proteolytic Activity

A fundamental question regarding the mechanism by which proteases stimulate cell proliferation is whether the catalytic activity of the enzyme is required. If it is, studies on the mitogenic mechanism would be facilitated since at least the general nature of the primary biochemical event would be known.

In a number of studies, it has been reported that the stimulation of cell division is inhibited if the proteolytic activity of the enzyme is inhibited by macromolecular inhibitors like soybean trypsin inhibitor or by catalytic-site derivatizing agents like diisopropylfluorophosphate or phenylmethylsulfonyl fluoride (BURGER, 1970; SEFTON and RUBIN, 1970; KAPLAN and BONA, 1974; POHJANPELTO, 1977, GLENN et al., 1980). Thus, these studies indicated that the catalytic activity of the protease is required for mitogenic stimulation. A consideration which must be taken into account in these studies is that the inhibitor might alter properties of the protease in addition to its catalytic activity, and that these other properties might be necessary for the stimulation of cell division. Although it is not possible to entirely rule out this possibility, recent studies have shown that thrombin which has been derivatized at its active-site serine with either diisopropylfluorophosphate or methylsulfonyl fluoride binds to cell surface receptors of Chinese hamster lung cells and mouse embryo cells as effectively as active thrombin; however, the catalytically inactive thrombins are not mitogenic (GLENN et al., 1980). Since the inactivated thrombins bind to the receptors as well as active thrombin, it appears that the conformation of thrombin is not significantly altered by derivatizing its catalytic-site serine. Thus, the inability of the inactive thrombin preparations to stimulate cell proliferation is most likely a result of inhibition of its catalytic activity.

There are two studies which suggest that inhibition of proteolytic activity might not correspondingly reduce the mitogenic activity of a protease (BROWN and KIEHN, 1977; KIEHN and BROWN, 1978). In these studies, the inhibited trypsin lost only part of its mitogenic activity. However, as noted above, a very insensitive spectrophotometric measure of protease activity was used in these experiments, so the "inactive" trypsin might have posessed very substantial protease activity.

On the balance, the present evidence strongly indicates that the catalytic activity of mitogenic proteases is required for stimulation of cell division.

III. Sufficiency of Cell Surface Action

Another important question that is basic to studies on the mechanism by which proteolytic enzymes stimulate cell proliferation is their cellular site of action. In particular, it has been asked whether proteases must be internalized for stimulation, or whether their mitogenic action can result from interaction only with the cell surface. Answers to these questions permit subsequent studies to focus on a particular part of the cell. Experiments showing that trypsin (HODGES et al., 1973) and thrombin (ZETTER et al., 1977 a; MARTIN and QUIGLEY, 1978) are both internalized by cultured cells and that aspects of the stimulation by thrombin are correlated with its internalization indicated that these proteases might act intracellularly to stimulate cell proliferation (ZETTER et al., 1977 a; MARTIN and QUIGLEY, 1978).

An early method to determine if polypeptide hormones or growth factors could bring about their biological effects without entering cells involved linking them to Speharose beads that were excluded from cells (CUATRECASAS, 1969). However, with many of these preparations there was excessive release of the polypeptides from the beads, compromising the conclusions that could be drawn from the experiments (DAVIDSON et al., 1973; GARWIN and GELEHRTER, 1974; KOLB et al., 1975).

To determine if proteases could stimulate cell division by action only at the cell surface, another technique was employed to immobilize the mitogen (CARNEY and CUNNINGHAM, 1977, 1978 a, b). Trypsin or thrombin was attached to carboxylate-derivatized polystyrene beads using a water-soluble carbodiimide reagent so the protease was linked to the beads by a peptide bond. This linkage was sufficiently stable to permit experiments which showed that action of either protease at the cell surface was sufficient to stimulate division of nonproliferating chick embryo cells. Both the trypsin-polystyrene beads and the thrombin-polystyrene beads produced cell number increases that were about one-half as large as the maximal increases brought about by soluble trypsin or thrombin; however, beads with nonmitogenic proteins attached to them did not stimulate cell division. The critical question in these experiments was whether release of protease from the beads into the culture medium or directly into cells could account for the cell number increase produced by the protease beads. To answer this, both trypsin and thrombin were radiolabelled to a high specific activity with ^{125}I so the fate of the proteases could sensitively be followed throughout the experiments. There was a small amount of release from the beads into the culture medium; however, release from the beads directly into cells could not be detected. By adding defined amounts of soluble protease to the culture medium of nonproliferating cells, it was possible to show that the amount of trypsin or thrombin released from the beads into the culture medium was not sufficient to account for any of the cell division caused by the trypsin or thrombin-polystyrene beads. In addition, it was possible to set an upper limit on the amount of protease released from the beads directly into the cytoplasm and then incubate cells with amounts of soluble protease that resulted in the internalization of defined amounts of the protease. These experiments showed that direct release from the beads into the cytoplasm did not account for the cell division brought about by the protease polystyrene beads. Together, these experiments showed that neither trypsin nor thrombin must enter cells to stimulate cell division (CARNEY and CUNNINGHAM, 1977, 1978 a, b). In complimentary studies, QUIGLEY et al. (1979) used a preparation of thrombin linked to Sepharose beads by trichloro-s-triazine, a linkage that is more stable than the one obtained with cyanogen bromide-Sepharose (FINLAY et al., 1978). These experiments also showed that immobilized thrombin could stimulate cell division, although much more immobilized thrombin than soluble thrombin was required to stimulate the cells. As QUIGLEY et al. (1979) emphasized, this could be due to a restricting geometry between the cells and beads, or it could mean that internalization of thrombin must occur to produce maximal cell proliferation. The former alternative seems more likely in view of the very large size of the Sepharose beads compared to the cells. Nevertheless, the latter point is worth considering in future experiments since a demonstration that cell surface action is sufficient to cause cell division does not rule out the possibility that internalized protease might also be able to stimulate cell proliferation.

IV. Cell Surface Receptors for Thrombin

Several studies have revealed cell surface binding sites or receptors for thrombin on cells that are responsive to its mitogenic action. Two studies on the binding of mitogenic concentrations of ^{125}I-thrombin revealed cell surface binding sites for thrombin on cultured chick embryo cells (ZETTER et al., 1977a; MARTIN and QUIGLEY, 1978). To examine the binding of ^{125}I-thrombin that is "specific" (binding that is saturable and competed by nonlabeled thrombin) other investigators used lower concentrations of ^{125}I-thrombin to minimize "nonspecific" binding (binding that is nonsaturable and not competed by nonlabeled thrombin) (CARNEY and CUNNINGHAM, 1978c; PERDUE et al., 1978, 1979; GLENN et al., 1980). In these experiments, nonspecific binding was measured as the amount of radioactivity bound to cultures after incubation in binding medium containing excess nonlabelled thrombin in addition to the ^{125}I-thrombin; specific binding was determined by subtracting nonspecific binding from total radioactivity bound to cultures incubated only with ^{125}I-thrombin. This permitted quantitative experiments on the binding to cell surface receptors.

Studies on the binding of ^{125}I-thrombin to cultured mouse embryo, chick embryo, human foreskin, Chinese hamster lung and mouse epithelial cells have demonstrated high affinity receptors with KDs ranging from 0.2 to 2.0 nM. The number of receptors per cell ranged from about 20 to 200 × 10^3. The thrombin receptors are actually on the cell surface since they can be removed by brief treatments with trypsin (CARNEY and CUNNINGHAM, 1978c). The receptors are specific since ^{125}I-thrombin binding is not competed by prothrombin, EGF or insulin on mouse embryo cells (CARNEY and CUNNINGHAM, 1978a) or EGF, insulin, plasmin, or chymotrypsin on chick embryo, mouse embryo, or mouse epithelial cells (PERDUE et al., 1979). However, ^{125}I-thrombin binding was competed by plasminogen activator (PERDUE et al., 1979).

The relationship of the binding of thrombin to cellular receptors and stimulation of cell division has been studied in mouse embryo cells (CARNEY and CUNNINGHAM, 1978c). Evaluation of the fraction of total thrombin receptors occupied and the amount of cell number increase as a function of thrombin concentration showed that there was a close relationship between receptor occupancy and increase in cell number. Unlike several hormones which appear to have spare receptors on cells, maximal stimulation of cell division by thrombin required occupancy of essentially all of the receptors. This relationship between thrombin receptor occupancy and stimulation of cell division was also shown in other experiments using added serum. Low concentrations of serum that did not stimulate division of non-proliferating mouse embryo cells markedly inhibited the ability of thrombin to bring about cell division in these cells. These low concentrations of serum did not inhibit either the proteolytic activity of thrombin or the nonspecific association of it with the cells. However, specific binding of thrombin to cellular receptors was markedly inhibited by the concentrations of serum that inhibited its ability to stimulate cell division. Analysis of the binding data by Scatchard plots showed that this inhibition by serum resulted from a masking of thrombin receptors on cells and not from binding of thrombin by serum factors (CARNEY and CUNNINGHAM, 1978c).

Recent experiments using modified thrombin preparations have permitted identification of active-site regions of thrombin required for receptor binding and stimulation of cell division (GLENN et al., 1980; PERDUE, personal communication). Nitration or limited proteolysis of thrombin leads to nitro-thrombin or γ-thrombin, respectively. These thrombins are modified at regions necessary for the binding and recognition of fibrinogen that are removed from the catalytic apparatus of thrombin (FENTON et al., 1979). Neither nitro-thrombin nor γ-thrombin bound to the thrombin receptor, indicating that active-site regions distinct from the catalytic apparatus are required for receptor binding (GLENN et al., 1980). The catalytic-site serine of thrombin was derivatized with diisopropylfluorophosphate or methylsulfonyl fluoride, producing thrombins without proteolytic activity. Both of these inactive thrombins bound to Chinese hamster lung and mouse embryo cells, but not to chick embryo or human foreskin cells; they did not stimulate division of any of the cells. From these results, it is possible to conclude that binding of thrombin to the chick and human cells requires the catalytic apparatus of thrombin, whereas binding to the mouse or Chinese hamster lung cells does not. In addition, since the catalytic-site inactivated thrombins bound to the Chinese hamster lung and mouse embryo cells as well as active thrombin but did not stimulate cell division, it can be concluded that the mitogenic activity of thrombin on these cells requires the intact catalytic apparatus of the enzyme to interact with and presumably cleave a protein component of the cellular receptor (GLENN et al., 1980).

It should be pointed out that receptors for thrombin have been carefully studied on platelets (GANGULY, 1974; TOLLEFSON and MAJERUS, 1976; MARTIN et al., 1976) and on human endothelial cells (AWBREY et al., 1979). With these preparations, steady-state binding of ^{125}I-thrombin is achieved much more rapidly than with fibroblast-like cells. It is presently not clear whether the thrombin receptors on human endothelial cells (AWBREY et al., 1979) are involved in the potentiation by thrombin of these cells to the mitogenic action of FGF (GOSPODAROWICZ et al., 1978).

V. Cleavage of Cell Surface Proteins and Stimulation of Cell Division

A number of investigations have been directed at the identification of cell surface proteins whose cleavage is involved in the stimulation of cell division by proteases. The early studies were largely prompted by findings that the proteolytic activity of these enzymes was required for stimulation (BURGER, 1970; SEFTON and RUBIN, 1970), and by studies which showed that a 230–250 K dalton cell surface component which was reduced in many transformed cell lines was quite sensitive to trypsin treatment (HYNES, 1973, 1976; HOGG, 1974). The more recent findings that both trypsin (CARNEY and CUNNINGHAM, 1977) and thrombin (CARNEY and CUNNINGHAM, 1978 a, b) can stimulate cell division by action at the cell surface have contributed to the sustained search for protease-sensitive cell surface components that participate in the stimulation of cell division by these mitogens. Throughout these studies, particular emphasis has been placed on thrombin because of its very high proteolytic specificity (BLOMBÄCK et al., 1977).

In several early studies the possibility was explored that removal of a 230–250 K dalton cell surface component might lead to cell proliferation. This component is readily labeled by ^{125}I and lactoperoxidase and is markedly reduced in some transformed cell lines (HYNES, 1976). It has been called large external transformation sensitive (LETS) protein (HYNES, 1976), cell surface proten (CSP) (YAMADA and WESTON, 1974) and fibronectin (HEDMAN et al., 1978). An early study indicated that it might be involved in the control of cell proliferation since: (1) exponentially growing normal cells have lower levels of the 230–250 K dalton component as detected by iodination than do cells blocked in G_1/G_0 by high density or low serum, and (2) the high levels of this component on the surfaces of cells blocked in G_1/G_0 by low serum fall after addition of serum to stimulate cell proliferation (HYNES and BYE, 1974). However, studies on the ability of various proteases to remove the 230–250 K dalton component (judged by subsequent iodination with ^{125}I and lactoperoxidase) and to stimulate proliferation in chick embryo cells revealed that this component was not removed by thrombin, a mitogenic protease, but was removed by several proteases that were not mitogenic or only weakly mitogenic (TENG and CHEN, 1975; BLUMBERG and ROBBINS, 1975; ZETTER et al., 1976). Thus, simple removal of the 230–250 K dalton component appears to be neither necessary nor sufficient for protease-stimulated cell division. However, it is noteworthy that a recent report shows that thrombin can stimulate both the production and release of this component in cultures of normal human fibroblasts (MOSHER and VAHERI, 1978). Although these studies do not link these thrombin-mediated changes to its ability to stimulate cell division, the possibility should be considered that the events are related in some important way.

Several other cell-surface components have been shown to be cleaved by thrombin, and experiments have been conducted to probe their role in thrombin-stimulated cell division. A thrombin-sensitive surface component of 205 k daltons that is labeled by ^{131}I and lactoperoxidase was identified on cultured chick embryo cells (TENG and CHEN, 1976) but subsequent studies showed that is was also removed by proteases that were not mitogenic for these cells (ZETTER et al., 1976). It was also reported that on mouse splenocytes there is a 45 k dalton surface component that is removed by thrombin. The relationship of its removal to the mitogenic action of thrombin on these cells was not examined (CHEN et al., 1976). However, this component might be similar to a 43 k dalton cell surface component on chick embryo cells that appears to be involved in the thrombin-stimulated division of these cells (GLENN and CUNNINGHAM, 1979). In the latter investigation the relationship between specific cell surface cleavages and stimulation of cell division by thrombin was examined in studies on cultured chick embryo cells that were responsive to the mitogenic action of thrombin, and also on chick embryo cells that did not divide after thrombin treatment but which remained responsive to serum. Both responsive and unresponsive cells possessed a cell surface component of 43 k daltons that was labeled with ^{125}I and lactoperoxidase. With the responsive cells, this component was removed by treatments with thrombin that led to cell division. However, the 43 k dalton component was not removed by thrombin from the chick embryo cells that were unresponsive to its mitogenic action. Also, the component was not removed after serum stimulation of chick embryo cells, showing that its removal was not simply a consequence of initiating cell proliferation. These studies strongly indicated that removal of the 43 k dalton surface component was necessary for thrombin-stimulated division of chick embryo cells (GLENN and CUNNINGHAM, 1979). The likelihood that this component was the cell surface receptor for thrombin was suggested by experiments using a photoreactive crosslinking derivative of thrombin. These studies indicated that the molecular weight of the thrombin receptor is also about 43,000. Although it should be emphasized that the correspondence in the approximate molecular weights of these components could be coincidental, these studies indicated that the cell surface component whose cleavage is required for thrombin-stimulated cell division is likely the thrombin receptor (GLENN and CUNNINGHAM, 1979).

The above studies on cleavage of cell surface components by proteases employed one-dimensional polyacrylamide gels to analyze cell surface components labeled by lactoperoxidase-catalyzed iodination. Although these experiments have yielded valuable information, it will be critical in future studies to apply procedures to maximize the resolution of thrombin-sensitive cell surface components. In particular, it will be important to analyze these components on two-dimensional gels after they have been labeled by additional procedures including metabolic labeling. This should permit thorough analyses of thrombin-sensitive surface components that have already been identified as well as a search for other components whose cleavage might be neccessary for thrombin-stimulated cell division.

VI. Perspectives and Future Studies

At this early stage in our understanding of how thrombin and other proteases bring about cell division, there are several findings which will require additional study before it will be possible to determine their relationship to mitogenic stimulation. For example, about 50% of the thrombin that specifically binds to the surface of cultured human fibroblasts becomes covalently linked to a component that at first appeared to be the thrombin receptor (BAKER et al., 1979). Additional studies (BAKER et al., 1980; Low et al., 1981) have revealed that this component is present not only on the cell surface but also in the cytoplasm; in addition, it is released into the culture medium. It becomes linked to thrombin by an ester linkage involving the catalytic-site serine of thrombin; thus this linkage inactivates thrombin. Because this component can also become linked to other serine proteases, it has been named protease-nexin (PN). Protease-PN complexes formed in the culture medium bind to cells via a heparin-sensitive site, and these complexes are then internalized. It appears that formation of the thrombin-PN complex is not required for thrombin-stimulated cell division since it is possible to markedly inhibit complex formation without diminishing the capacity of thrombin to stimulate the cells. However, PN might function in other ways. For example, it might regulate levels of serine proteases at or near the cell surface membrane by internalizing them so they can be degraded (BAKER et al., 1980; LOW et al., 1981). Recent studies have shown that there is also a binding site or receptor for thrombin that is distinct from PN. This was demonstrated in experiments using ^{125}I-thrombin inactivated at its catalytic site serine with diisopropylfluorophosphate. This inactivated thrombin will not form complexes with PN yet it will bind to a thrombin receptor of Chinese hamster lung cells. Since PN does not compete for binding of the inactivated thrombin to these cells, there must be a receptor for thrombin on the cells distinct from PN (LOW and CUNNINGHAM, unpublished data). These studies have revealed that the binding of thrombin to fibroblasts is complex and that several cellular components are involved in the process. At present, it seems that the component that thrombin must cleave to stimulate cell division (GLENN and CUNNINGHAM, 1979) is probably the thrombin receptor that is distinct from PN. However, before this thrombin-sensitive component can be unequivocally equated with the thrombin receptor, it will be necessary to identify the thrombin receptor on two dimensional gels as an isolated component so its cleavage by thrombin can be studied. At present, it seems that a promising way to do this will involve a photoreactive crosslinking derivative of catalytic-site inactivated thrombin.

The requirement for cell surface proteolysis by thrombin to stimulate cell division (CARNEY and CUNNINGHAM, 1978 a, b; GLENN and CUNNINGHAM, 1979; GLENN et al., 1980) suggests two alternative models for the stimulation (CUNNINGHAM et al., 1979). In one, the required cleavage could be in a cell surface component that is a negative effector which prevents cell division; in its intact state the component would participate in the development of a negative signal to prevent cell division. In the other alternative model, the required cleavage by thrombin would produce a specific fragment which is a positive effector and signals the cells to divide. If the required cleavage by thrombin turns out to be in a surface component that nonmitogenic proteases can also cleave, then the second model would be strongly implicated. If so, it would then be fruitful to carefully follow the fate of cell surface peptide fragments from components that thrombin must cleave to stimulate cell division. Such fragments could initiate the signal to proliferate by some action within the plasma membrane. On the other hand, even though thrombin itself does not have to be internalized to stimulate cell division, it is possible that internalization of a fragment of some thrombin-sensitive cell surface component might be necessary to produce the response.

D. Possible Role of Protease-Stimulated Cell Proliferation in Physiological Processes

Although trypsin will readily stimulate proliferation of B lymphocytes and of chick embryo fibroblasts in cell culture, it is unlikely that it is a physiologically important mitogen for these cells since it is not a circulating enzyme and is not found in most tissues. On the other hand, it is much more likely that the stimulation by thrombin represents a biologically important event. Prothrombin is present at high levels in plasma, and at the site of tissue damage it is converted to thrombin. Although it is difficult to accurately estimate the thrombin concentration at such a site over a period of time that is necessary for it to bring about maximal cell division, the concentration of prothrombin in plasma is at least several times higher than the concentration of thrombin required to fully stimulate cells in culture. Thus, it has been reasonable to predict that stimulation of cell division by thrombin plays an important role in tissue repair following injury (CHEN and BUCHANAN, 1975). The demonstration of specific receptors for thrombin on a variety of cells (CARNEY and CUNNINGHAM, 1978c; PERDUE et al., 1978, 1979; GLENN et al., 1980) strongly indicates that its interactions with these cells is physiologically important. The significance of the ability of elastase, bromelin (TENG and CHEN, 1975; ZETTER et al., 1976) or collaginase (BLUMBERG and ROBBIN,s 1975) to stimulate cell division is less clear. Studies on the possibility that plasmin might be at least partly responsible for the uncontrolled proliferation of certain transformed cells is discussed below.

Studies on levels of extracellular or cell surface protease activity as a function of growth rate and position in the cell cycle have indicated correlations which suggest that in cell culture levels of endogenous cell surface proteases might be involved in the control of cell cycle traverse. For example, one study showed that cell surface or extracellular protease activity, measured by ^{3}H-acetyl casein hydrolysis, was significantly higher in cells with short doubling times than in cells with relatively long doubling times (HATCHER et al., 1976). In addition, studies on several synchronized cell lines showed a marked increase in cell surface protease activity during or before mitosis (HATCHER et al., 1976). Subsequent experiments revealed that when normal smooth muscle cells or fibroblasts stopped proliferating at high cell density a decrease in cell surface protease activity occurred. In contrast, no decrease in extracellular or surface protease activity or the rate of cell proliferation was observed with transformed melanoma or epidermal cells (HATCHER et al., 1977). Similar results have been obtained on the release of plasminogen activator by cultured cells (CHOU et al., 1977a). In these studies, plasminogen activator release was evaluated by measuring the fibrinolytic activity of the culture medium. As nonconfluent mouse 3T3 cells grew to confluency and the cells stopped proliferating, there was a decrease in the secretion of plasminogen activator even though the intracellular level remained relatively high. In contrast, dense cultures of 3T3 cells transformed by SV40 virus continued to secrete high levels of plasminogen activator (CHOU et al., 1977a).

In another study, (CHOU et al., 1977b) addition of Ca^{++} to 3T3 cells increased both secreted and cell-associated plasminogen activator activity. A similar effect on plasminogen activator was produced by adding the ionophore A23,187 in the presence of normal levels of Ca^{++}. It was suggested that the Ca^{++} stimulation of plasminogen activator production and secretion might be related to the mitogenic effect of Ca^{++} on these cells (CHOU et al., 1977b).

It is noteworthy that direct measurements of cell surface protease activity have implicated this activity in the stimulation of lymphocytes by mitogens (TÖKES et al., 1978). The stimulation of DNA synthesis in lymphocytes by mitogens or periodate requires the presence of glass-adherant leukocytes. After mitogen or periodate treatment, there is a marked increase in surface protease activity of the adherent cells; contact between the treated adherent cells and untreated T lymphocytes then leads to DNA synthesis in the latter cells (TÖKES et al., 1978).

Although the above studies do not rule out the possibility that enhanced cell surface or released protease activity might be a consequence of stimulating cell proliferation, it will be important in future studies to investigate the alternative important possibility that endogenous cell surface protease activity might participate in the control of movement of cells through the cell cycle.

There is now much evidence that many transformed or malignant cells contain and release high levels of certain proteases compared to normal cells (SYLVEN and BOIS-SVENSSEN, 1965; OSSOWSKI et al., 1973 a; BOSMANN et al., 1974; BOSMANN and HALL, 1974; QUIGLEY, 1979). Many investigators have asked whether this increased proteolytic activity is responsible for some of the aberrant properties of these cells (for a review of this topic, see QUIGLEY, 1979). Of interest here is whether the uncontrolled growth characteristic of most transformed cells might result from the high levels of extracellular or cell surface protease activity. Although an early study indicated that the proliferation of several transformed mouse 3 T 3 cell lines was much more sensitive than untransformed 3 T 3 cells to inhibition by several protease inhibitors (SCHNEBLI and BURGER, 1972), a subsequent investigation showed that the effects with TPCK and TLCK were not selective for transformed cells and that these agents probably decreased cell proliferation because they inhibited protein synthesis (CHOU et al., 1974). In addition, the high fibrinolytic activity characteristic of many transformed cells appears not to be responsible for their abnormal growth properties. This has been concluded from studies using plasminogen-free culture media (OSSOWSKI et al., 1973 b) and by adding EACA to inhibit plasmin activity (ROBLIN et al., 1975). Thus, there is no convincing evidence to date which links the increased protease activity of many transformed cells to their abnormal growth properties. On the other hand, it appears that plasmin is the only protease for which there is adequate information to suggest it is not involved. Transformed and malignant cells undoubtedly release quite a spectrum of proteases, so the basic question remains largely unanswered.

E. Conclusions

Many studies have now shown that several different proteases can stimulate proliferation of cultured animal cells. Although these experiments have not yet been extended to some cell types or proteases that will be important to investigate in future studies, some general trends emerge from the results to date. For example, cultured early passage chick embryo fibroblasts divide in response to several proteases including thrombin, trypsin, elastase, collaginase, and plasmin. In contrast, mammalian fibroblast-like cells appear not to respond to such a wide spectrum of proteases; thrombin is the only one that has been shown to reproducibly stimulate the division of these cells. Several proteases can stimulate DNA synthesis in B lymphocytes from a number of different species. In general, cell strains are more responsive to proteolytic stimulation than cell lines, but there are exceptions.

Studies on the mechanism by which proteases bring about cell division have been carried out mostly using thrombin. Its proteolytic activity is required for cell activation; because it is a very specific protease, it has been the one of choice to look for cleavages that are required for protease-stimulated cell division. In addition, thrombin is mitogenic in serum-free medium, permitting experiments to be conducted under chemically defined conditions. Thrombin does not have to be internalized to produce cell division, so studies on its mechanism of action have focused on the cell surface. Binding experiments using ^{125}I-thrombin have revealed binding sites or receptors that are specific for thrombin and there is evidence that thrombin must bind to these receptors to stimulate cell division. Studies using cells that are responsive or unresponsive to the mitogenic action of thrombin have shown that it must cleave a cell surface component of 43,000 daltons to activate chick embryo cells. The molecular weight of the thrombin receptor on these cells is also about 43,000 daltons determined in experiments using a photoreactive cross-linking derivative of ^{125}I-thrombin. Together, these studies have indicated that the component that thrombin must cleave to stimulate the cells is probably its cell surface receptor. However, the specific binding of thrombin by fibroblasts is complex and involves several components; there are many questions that remain to be answered concerning the nature of these components, their cleavage by thrombin, and their role in thrombin-stimulated cell division. The requirement for the proteolytic activity of thrombin for cell activation indicates that it should be fruitful to examine the fate of peptide fragments from cell surface components that are cleaved by thrombin. Even though thrombin itself does not have to be internalized to stimulate cell division, these peptide fragments might have an intracellular site of action.

Although there is little concrete information regarding the biological significance of protease-stimulated cell division, it has been suggested that the stimulation by thrombin might be involved in tissue repair following injury. Thrombin is produced at the site of a wound, and might function not only in clot formation but also to signal the cell division necessary for wound healing. It has also been suggested that increased levels of proteases secreted by many transformed cells might be responsible for the uncontrolled growth characteristic of them. Although there is experimental evidence that the increased fibrinolytic activity of some transformed cells is not responsible for their growth to high densities, the other proteolytic activities that are elevated in certain transformed cells have not yet been adequately studied to reach meaningful conclusions.

References

Antoniades, H., Scher, C., Stiles, C.: Purification of human platelet-derived growth factor. Proc. Natl. Acad. Sci. U.S.A. *76*, 1809–1813 (1979)

Awbrey, B., Hoak, J., Owen, W.: Binding of human thrombin to cultured human endothelial cells. J. Biol. Chem. *254*, 4092–4095 (1979)

Baker, J., Simmer, R., Glenn, K., Cunningham, D.: Thrombin and epidermal growth factor become linked to cell surface receptors during mitogenic stimulation. Nature *278*, 743–745 (1979)

Baker, J., Low, D., Simmer, R., Cunningham, D.: Protease-Nexin: A cellular component that links thrombin and plasminogen activator and mediates their binding to cells. Cell *21*, 37–45 (1980)

Berger, E.A., Shooter, E.M.: Evidence for pro-β-nerve growth factor, a biosynthetic precursor of β-nerve growth factor. Proc. Natl. Acad. Sci. U.S.A. *74*, 3647–3651 (1977)

Blombäck, B., Hessel, B., Hogg, D., Claesson, G.: Substrate specificity of thrombin on proteins and synthetic substrates. In: Chemistry and biology of thrombin. Lundblad, r., Fenton, J. II, Mann, K. (eds.), p. 275. Ann Arbor: Ann Arbor Sci. Publ. 1977

Blumberg, P.M., Robbins, P.W.: Effect of proteases on activation of resting chick embryo fibroblasts and on cell surface proteins. Cell *6*, 137–147 (1975)

Bosmann, H.B., Hall, T.C.: Enzyme activity in invasive tumors of human breast and colon. Proc. Natl. Acad. Sci. U.S.A. *71*, 1833–1837 (1974)

Bosmann, H.B., Lockwood, T., Morgan, H.: Surface biochemical changes accompanying primary infection with Rous sarcoma virus. II. Proteolytic and glycosidase activity and sublethal autolysis. Exp. Cell Res. *83*, 25–30 (1974)

Bradshaw, R.A.: Nerve growth factor. Ann. Rev. Biochem. *47*, 191–216 (1978)

Brown, M., Kiehn, D.: Protease effects on specific growth properties of normal and transformed baby hamster kidney cells. Proc. Natl. Acad. Sci. U.S.A. *74*, 2874–2878 (1977)

Buchanan, J.M., Chen, L.B., Zetter, B.R.: Protease-related effects in normal and transformed cells. In: Cancer enzymology. Schultz, J., Ahmad, F. (eds.), p. 1. New York: Academic Press 1976

Burger, M.M.: Proteolytic enzymes initiating cell division and escape from contact inhibition of growth. Nature *227*, 170–171 (1970)

Calissano, P., Levi-Montalcini, R.: Is NGF an enzyme? Nature, News, Views *280*, 359 (1979)

Carney, D., Cunningham, D.: Initiation of chick cell division by trypsin action at the cell surface. Nature *268*, 602–606 (1977)

Carney, D., Cunningham, D.: Cell surface action of thrombin is sufficient to initiate division of chick cells. Cell *14*, 811–823 (1978a)

Carney, D., Cunningham, D.: Transmembrane action of thrombin initiates chick cell division. J. Supramol. Struct. *9*, 337–350 (1978b)

Carney, D., Cunningham, D.: Role of specific cell surface receptors in thrombin-stimulated cell division. Cell *15*, 1341–1349 (1978c)

Carney, D., Glenn, K.C., Cunningham, D.: Conditions which affect initiation of animal cell division by trypsin and thrombin. J. Cell. Physiol. *95*, 13–22 (1978)

Chen, L.B., Buchanan, J.M.: Mitogenic activity of blood components. I. Thrombin and prothrombin. Proc. Natl. Acad. Sci. U.S.A. *72*, 131–135 (1975)

Chen, L.B., Teng, N.N.H., Buchanan, J.M.: Mitogenicity of thrombin and surface alterations on mouse splenocytes. Exp. Cell Res. *101*, 41–46 (1976)

Chou, I.-N., Black, P.H., Roblin, R.O.: Non-selective inhibition of transformed cell growth by a protease inhibitor. Proc. Natl. Acad. Sci. U.S.A. *71*, 1748–1752 (1974)

Chou, I.N., ODonnell, S., Black, P., Roblin, R.: Cell density-dependent secretion of plasminogen activator by 3T3 cells. J. Cell. Physiol. *91*, 31–38 (1977a)

Chou, I.N., Roblin, R., Black, P.: Calcium stimulation of plasminogen activator secretion/production by Swiss 3T3 cells. J. Biol. Chem. *252*, 6256–6259 (1977b)

Cuatrecasas, P.: Interaction of insulin with the cell membrane: the primary action of insulin. Proc. Natl. Acad. Sci. U.S.A. *63*, 450–457 (1969)

Cunningham, D.D., Ho, T.S.: Effects of added proteases on concanavalin A-specific agglutinability and proliferation of quiescent fibroblasts. In: Proteases and biological control. Cold Spring Harbor Conferences on Cell Proliferation. Reich, E., Rifkin, D.B., Shaw, E. (eds.), vol. 2, p. 795. New York: Cold spring Harbor Laboratory 1975

Cunningham, D., Glenn, K., Baker, J., Simmer, R., Low, D.: Mechanisms of thrombin-stimulated cell division. J. Supramol. Struct. (1979)

Davidson, M., Van Herle, A., Gerschenson, L.: Insulin and Sepharoseinsulin effects on tyrosine transaminase levels in cultured rat liver cells. Endocrinology *92*, 1442–1446 (1973)

Davie, E.W., Fujikawa, K.: Basic mechanisms in blood coagulation. Ann. Rev. Biochem. *44*, 799–829 (1975)

Detwiler, T.C., Wasiewski, W.W.: The equilibrium reaction of thrombin with platelets. In: Chemistry and biology of thrombin. Lundblad, R.L., Fenton, J.W. II, Mann, K.G. (eds.), p. 465. Ann Arbor: Ann Arbor Science Publishers, Inc. 1977

Fenton, J. II, Landis, B., Walz, D., Bing, D., Feinman, R., Zabinski, M., Sonder, S., Berliner, L., Finlayson, J.: Human thrombin: preparative evaluation, structural properties, and enzymic specificity. In: The Chemistry and physiology of the human plasma proteins. Bing, D. (ed.), p. 151. New York: Pergamon Press 1979

Finlay, T., Troll, W., Levy, M., Johnson, A., Hodgins, L.: New methods for the preparation of biospecific adsorbents and immobilized enzymes utilizing trichloro-triazene. Anal. Biochem. *87*, 77–90 (1978)

Frey, P., Forand, R., Maciag, T., Shooter, E.M.: The biosynthetic precursor of epidermal growth factor and the mechanism of its processing. Proc. Natl. Acad. Sci. U.S.A. *76*, 6294–6298 (1979)

Ganguly, P.: Binding of thrombin to human platelets. Nature *247*, 306–307 (1974)

Garwin, J., Gelehrter, T.: Induction of tyrosine aminotransferase by Sepharose-insulin. Arch. Biochem. Biophys. *164*, 52–59 (1974)

Girard, J.P., Fernandes, B.: Studies on the mitogenic activity of trypsin, pronase and neuraminidase on human peripheral blood lymphocytes. Eur. J. Clin. Invest. *6*, 347–353 (1976)

Glenn, K., Cunningham, D.: Thrombin-stimulated cell division involves proteolysis of its cell surface receptor. Nature *278*, 711–714 (1979)

Glenn, K., Carney, D., Fenton, J. II, Cunningham, D.: Thrombin active-site regions required for fibroblast receptor binding and initiation of cell division. J. Biol. Chem. *255*, 6609–6616 (1980)

Glynn, R.D., Thrash, C.R., Cunningham, D.D.: Maximal concanvalin A-specific agglutinability without loss of density-dependent growth control. Proc. Natl. Acad. Sci. U.S.A. *70*, 2676–2677 (1973)

Gospodarowicz, D.: Purification of a fibroblast growth factor from bovine pituitary. J. Biol. Chem. *250*, 2515–2520 (1975)

Gospodarowicz, D., Brown, K., Birdwell, C., Zetter, B.: Control of proliferation of human vascular endothelial cells. Characterization of the response of human umbilical vein endothelial cells to fibroblast growth factor, epidermal growth factor and thrombin. J. Cell Biol. *77*, 774–788 (1978)

Greene, L.A., Tomita, J.T., Varon, S.: Growth-stimulating activities of mouse submaxillary esteropeptidases on chick embryo fibroblasts in vitro. Exp. Cell Res. *64*, 387–395 (1971)

Hale, A.H., Weber, M.J.: Hydrolase and serum treatment of normal chick embryo cells: effects on hexose transport. Cell *5*, 245–252 (1975)

Hart, D.A.: The effect of soybean trypsin inhibitor on concanavalin A and phytohemagglutinin stimulation of hamster, guinea pig, rat, and mouse lymphoid cells. Cell. Immunol. *32*, 146–159 (1977)

Hart, D.A., Streilein, J.S.: Stimulation of a subpopulation of hamster lymphoid cells by trypsin and chymotrypsin. Exp. Cell Res. *102*, 246–252 (1976)

Hatcher, V., Wertheim, M., Rhee, C., Tsien, G., Burk, P.: Relationship between cell surface protease activity and doubling time in various normal and transformed cells. Biochim. Biophys. Acta *451*, 499–510 (1976)

Hatcher, V., Oberman, M., Wertheim, M., Rhee, C., Tsien, G., Burk, P.: The relationship between surface protease activity and the rate of cell proliferation in normal and transformed cells. Biochem. Biophys. Res. Commun. *76*, 602–608 (1977)

Hedman, K., Vaheri, A., Wartiovaara, J.: External fibronection of cultured human fibroblasts is predominantly a matrix protein. J. Cell Biol. *76*, 748–760 (1978)

Heldin, C.H., Westermark, B., Wasteson, A.: Platelet-derived growth factor: purification and partial characterization. Proc. Natl. Acad. Sci. U.S.A. *76*, 3722–3726 (1979)

Higuchi, S., Ohkawara, S., Nakamura, S., Yoshinaga, M., The polyvalent protease inhibitor, trasylol, inhibits DNA synthesis of mouse lymphocytes by an indirect mechanism. Cell. Immunol. *34*, 395–405 (1977)

Hirshhorn, R., Grossman, J., Troll, W., Weissman, G.: The effect of epsilon amino caproic acid and other inhibitors of proteolysis upon the response of human peripheral blood lymphocytes to phytohemagglutinin. J. Clin. Invest. *50*, 1206–1217 (1971)

Hodges, G., Livingston, D., Franks, L.: The localization of trypsin in cultured mammalian cells. J. Cell Sci. *12*, 887–902 (1973)

Hogg, N.: A comparison of membrane proteins of normal and transformed cells by lacto-peroxidase labeling. Proc. Natl. Acad. Sci. U.S.A. *71*, 489–492 (1974)

Holley, R.W., Kiernan, J.A.: Control of the initiation of DNA synthesis in 3 T 3 cells: serum factors. Proc. Natl. Acad. Sci. U.S.A. *71*, 2908–2911 (1974)

Hovi, T., Suni, J., Hortlins, L., Vaheri, A.: Stimulation of chicken lymphocytes by T and B cell mitogens. Cell. Immunol. *39*, 70–78 (1978)

Hynes, R.: Alteration of cell surface proteins by viral transformation and proteolytis. Proc. Natl. Acad. Sci. U.S.A. *70*, 3170–3174 (1973)

Hynes, R.: Cell surface proteins and malignant transformation. Biochim. Biophys. Acta *458*, 73–107 (1976)

Hynes, R.O., Bye, J.M.: Density and cell cycle dependence of cell surface proteins in hamster fibroblasts. Cell *3*, 113–120 (1974)

Kaplan, J.G., Bona, C.: Proteases as mitogens. Exp. Cell Res. *88*, 388–394 (1974)

Kiehn, D., Brown, M.: Studies on the nature of protease-induced growth stimulation in normal and transformed BHK cells. J. Cell. Physiol. *97*, 169–176 (1978)

Kolb, J., Renner, R., Hepp, K., Weiss, L., Wieland, O.: Re-evaluation of Sepharose-insulin as a tool for the study of insulin action. Proc. Natl. Acad. Sci. U.S.A. *74*, 248–252 (1975)

Korant, B.D.: Regulation of animal virus replication by protein cleavage. In: Proteases and Biological Control. Cold Spring Harbor Conferences on Cell Proliferation. Reich, E., Rifkin, D.B., Shaw, E. (eds.), vol. 2, p. 621. New York: Cold Spring Harbor Laboratory 1975

Lembach, K.: Induction of human fibroblast proliferation by epidermal growth factor (EGF): enhancement by an EGF-binding arginine esterase and by ascorbate. Proc. Natl. Acad. Sci. U.S.A. *73*, 183–187 (1976)

Low, D., Baker, J., Koonce, W., Cunningham, D.: Released protease-nexin regulates cellular binding, internalization and degradation of serine proteases. Proc. Natl. Acad. Sci. U.S.A. *78*, 2340–2344 (1981)

Martin, B.M., Quigley, J.P.: Binding and internalization of ^{125}I-thrombin in chick embryo fibroblasts: possible role in mitogenesis. J. Cell. Physiol. *96*, 155–164 (1978)

Martin, B., Wasiewski, W., Fenton, J. II, Detwiler, T.: Equilibrium binding of thrombin to platelets. Biochemistry *15*, 4886–4893 (1976)

McKee, P.A.: Fibrin formation and dissolution. In: Proteolysis and physiological regulation. Miami Winter Symposium. Ribbons, D.W., Brew, K. (eds.), vol. II, p. 239. New York: Academic Press 1976

Mosher, D., Vaheri, D.: Thrombin stimulates the production and release of a major surface-associated glycoporatein (fibronectin) in cultures of human fibroblasts. Exp. Cell Res. *112*, 323–334 (1978)

Müller-Eberhard, H.J.: The molecular basis of the biological activities of complement. In: The Harvey lectures, ser. 66, p. 75. New York: Academic Press 1972

Müller-Eberhard, H.J.: Initiation of membrane attack by complement: assembly and control of C 3 and C 5 convertase. In: Proteases and biological control. Cold Spring Harbor Conferences on Cell Proliferation. Reich, E., Rifkin, D.B., Shaw, E. (eds.), vol. 2, p. 229. New York: Cold Spring Harbor Laboratory 1975

Neurath, H.: Limited proteolysis and zymogen activation. In: Proteases and biological control. Cold Spring Harbor Conferences on Cell Proliferation. Reich, E., Rifkin, D.B., Shaw, E. (eds.), vol. 2, p. 51. New York: Cold Spring Harbor Laboratory 1975

Neurath, H., Walsh, K.A.: The role of proteases in biological regulation. In: Proteolytic and physiological regulation. Miami Winter Symposium. Ribbons, D.W., Brew, K. (eds.), vol. II, p. 29. New York: Academic Press 1976 a

Neurath, H., Walsh, K.A.: role of proteolytic enzymes in biological regulation (a review). Proc. Natl. Acad. Sci. U.S.A. *73*, 3825–3832 (1976 b)

Noonan, K.D.: Role of serum in protease-induced stimulation of 3 T 3 cell division past the monolayer stage. Nature *259*, 573–576 (1976)

Orenstein, N.S., Dvorak, H.F., Blanchard, M.H., Young, M.: Nerve growth factor: a protease that can activate plasminogen. Proc. Natl. Acad. Sci. U.S.A. *75*, 5497–5500 (1978)

Ossowski, L., Unkeless, J., Tobia, A., Quigley, J., Rifkin, D., Reich, E.: An enzymatic function associated with transformation of fibroblasts by oncogenic viruses II. Mammalian fibroblast cultures transformed by DNA and RNA tumor viruses. J. Exp. Med. *137*, 11–126 (1973 a)

Ossowski, L., Quigley, J., Kellerman, G., Riech, E.: Fibrinolysis associated with oncogenic transformation. Requirement of plasminogen for correlated changes in cellular morphology, colony formation in agar and cell migration. J. Exp. Med. *138*, 1056–1064 (1973 b)

Perdue, J., Lubenskyi, W., Kivity, E., Susanto, I.: Identification and characterization of thrombin receptors on Avian cells. J. Cell Biol. *79*, 50 a (1978)

Perdue, J., Kivity, E., Lubenskyi, W.: Characterization of thrombin binding to cultured avian and mouse fibroblasts and mouse epithelial cells. J. Cell Biol. *83*, 255 a (1979)

Pohjanpelto, P.: Proteases stimulate proliferation of human fibroblasts. J. Cell. Physiol. *91*, 387–392 (1977)

Pohjanpelto, P.: Stimulation of DNA synthesis in human fibroblasts by thrombin. J. Cell. Physiol. *95*, 189–194 (1978)

Quigley, J.: Proteolytic enzymes of normal and malignant cells. In: Surface of normal and malignant cells. Hynes, R. (ed.), p. 247. New York: John Wiley and Sons 1979

Quigley, J., Martin, B., Goldfarb, R., Scheiner, C., Muller, W.: Involvement of serine proteases in growth control and malignant transformation. In: Hormones and cell culture. Cold Spring Harbor Conferences on Cell Proliferation. Sato, G., Ross, R. (eds.), vol. 6, p. 219. New York: Cold Spring Harbor Laboratory 1979

Raizada, M.K., Perdue, J.F.: Mitogen receptors in chick embryo fibroblasts. Kinetics, specificity, unmasking, and synthesis of ^{125}I-insulin binding sites. J. Biol. Chem. *251*, 6445–6455 (1976)

Rinderknecht, E., Humbel, R.: Polypeptides with nonsupressible insulin-like and cell-growth promoting activities in human serum: isolation, chemical characterization and some biological properties of forms I and II. Proc. Natl. Acad. Sci. U.S.A. *73*, 2365–2369 (1976 a)

Rinderknecht, E., Humbel, R.: Amino-terminal sequence of two polypeptides from human serum with nonsuppressible insulin-like and cell-growth-promoting activities: evidence for structural homology with insulin B chain. Proc. Natl. Acad. Sci. U.S.A. *73*, 4379–4381 (1976 b)

Roblin, R., Chou, I.H., Black, P.: Role of fibrinolysin T activity in properties of 3 T 3 and SV 3 T 3 cells. In: Proteases and biological control. Cold Spring Harbor Conferences on Cell Proliferation. Reich, E., Rifkin, D.B., Shaw, E. (eds.), vol. 2, p. 869. New York: Cold Spring Harbor Laboratory 1975

Schnebli, H., Burger, M.: Selective inhibition of growth of transformed cells by protease inhibitors. Proc. Natl. Acad. Sci. U.S.A. *69*, 3825–3827 (1972)

Sefton, B.M., Rubin, H.: Release from density dependent growth inhibition by proteolytic enzymes. Nature *227*, 843–845 (1970)

Simmer, R.L., Baker, J.B., Cunningham, D.D.: Direct linkage of thrombin to its cell surface receptors in different cell types. J. Supramol. Struct. *12*, 245–257 (1979)

Smith, G.L., Temin, H.: Purified multiplication stimulating activity from rat liver cell conditioned medium: comparison of biological activities with calf serum, insulin and somatomedin. J. Cell. Physiol. *8ö4*, ö181–192 (1974)

Steiner, D.F., Kemmler, W., Tager, H.S., Rubenstein, A.H., Lernmark, A., Zuhlki, H.: Proteolytic mechanisms in the biosynthesis of polypeptide hormones. In: Proteases and biological control. Cold Spring Harbor Conferences on Cell Proliferation. Reich, E., Rifkin, D.B., Shaw, E. (eds.), vol. 2, p. 531. New York: Cold Spring Harbor Laboratory 1975

Sylven, B., Bois-Svenssen, I.: On the chemical pathology of interstitial fluid. I. Proteolytic activities in transplanted mouse tumors. Cancer Res. *25*, 458–468 (1965)

Taylor, J.M., Cohen, S., Mitchell, W.M.: Epidermal growth factor: high and low molecular weight forms. Proc. Natl. Acad. Sci. U.S.A. *67*, 164–171 (1970)

Teng, N.N.H., Chen, L.B.: The role of surface proteins in cell proliferation as studied with thrombin and other proteases. Proc. Natl. Acad. Sci. U.S.A. *72*, 413–417 (1975)

Teng, N., Chen, L.B.: Thrombin-sensitive surface protein of cultured chick embryo cells. Nature *259*, 578–580 (1976)

Tökes, Z., Bruszewski, W., Obrien, R.: Surface proteolytic activity in cell interactions: activation of human peripheral lymphocytes in contact with mitogen-treated glass-adherant leukocytes. Birth defects: Original Article Series *14*, 195–202 (1978)

Tollefson, D., Majerus, P.: Evidence for a single class of thrombin binding sites on human platelets. Biochemistry *15*, 2144–2149 (1976)

Vaheri, A., Ruoslahti, E., Hovi, T.: Cell surface and growth control of chick embryo fibroblasts in culture. In: Control of proliferation in animal cells. Cold Spring Harbor Conferences on Cell Proliferation. Clarkson, B., Baserga, R. (eds.), vol. 1, p. 305. New York: Cold Spring Harbor Laboratory 1974

Vischer, T.L.: Stimulation of mouse B lymphocytes by trypsin. J. Immunol. *113*, 58–62 (1974)

Vischer, T.L.: Protease inhibitors reduce mitogen induced lymphocyte stimulation. Immunology *36*, 811–813 (1979)

Vischer, T.L., Bretz, U., Baggiolini, M.: In vitro stimulation of lymphocytes by neutral proteinases from human polymorphonucleur leukocyte granules. J. Exp. Med. *144*, 863–872 (1976)

Williams-Ashman, H.B.: Introductory overview of the participation of proteinases and their regulators in mammalian reproductive physiology. In: Proteases and biological control. Cold Spring Harbor Conference on Cell Proliferation. Reich, E., Rifkin, D.B., Shaw, E. (eds.), vol. 2, p. 677. New York: Cold Spring Harbor Laboratory 1975

Yamada, K., Weston, J.: Isolation of a major cell surface glycoprotein from fibroblasts. Proc. Natl. Acad. Sci. U.S.A. *71*, 3493–3496 (1974)

Yoshinaga, M., Nakamura, S., Higuchi, S.: Enhancement of phytohemagglutinin-induced DNA synthesis of mouse thymocytes and T-lymphocytes by a neutral protease of polymorphonuclear leukocytes. Acta Pathol. Jpn. *27*, 857–868 (1977)

Young, M.: Yes, NGF is an enzyme. Nature, News Views *281*, 15 (1979)

Zetter, B.R., Chen, L.B., Buchanan, J.M.: Effects of protease treatment on growth, morphology, adhesion and cell surface proteins of secondary chick embryo fibroblasts. Cell *7*, 407–412 (1976)

Zetter, B.R., Chen, L.B., Buchanan, J.M.: Binding and internalization of thrombin by normal and transformed chick cells. Proc. Natl. Acad. Sci. U.S.A. *74*, 596–600 (1977a)

Zetter, B.R., Sun, T.T., Chen, L.B., Buchanan, J.M.: Thrombin potentiates the mitogenic response of cultured fibroblasts to serum and other growth promoting agents. J. Cell. Physiol. *92*, 233–240 (1977b)

CHAPTER 8

Nerve Growth Factor*

Mary J. Koroly and M. Young

Nerve growth factor (NGF) is a protein, discovered many years ago, which has the remarkable ability to stimulate nerve outgrowth from certain sympathetic and embryonic sensory ganglia. In what follows, it will be seen that NGF arises from a variety of sources in nature – including some which at first glance might appear to be unrelated to the nervous system and nerve growth. Furthermore, NGF has been shown to exist in several different molecular forms, some of which display certain chemical and biological effects other than stimulation of neurite outgrowth.

In the review presented here, we have tried to outline some of the problems presented by this interesting protein. No attempt has been made to be exhaustive in this treatment since several thorough reviews on the subject have recently appeared (Hogue-Angeletti et al., 1975; Bradshaw and Young, 1976; Young et al., 1976; Mobley et al., 1977; Bradshaw, 1978; Server and Shooter, 1978). It is our purpose here to try to place in perspective current knowledge of NGF which will eventually form the basis for understanding its physiologic function.

A. Discovery of NGF and its Early History

Before proceeding to more recent studies, it is important to consider the pioneering work of Levi-Montalcini and her colleagues which led to the discovery of NGF as well as to much of our current information in the field. (See Levi-Montalcini, 1952, 1965, 1968; Levi-Montalcini and Angeletti, 1961 a, 1964, 1968 for detailed reviews and references.)

I. NGF was first discovered as a soluble diffusible factor produced by two spontaneously arising mouse sarcomas termed 180 and 37. When these tumors were implanted into mouse embryos, they soon became invaded by large numbers of sensory and sympathetic fibers. The factor responsible has never been isolated from these tumors, and the reason why they contain it is not known.

II. The preceeding observation soon led to development of a sensitive in vitro bioassay for the factor. Fragments of these sarcomas were cultured on semisolid medium together with senory or sympathetic ganglia from 7–9 day old chick embryos. After about 10–24 h, the ganglia displayed a dense halo of nerve fibers. This test for nerve-growth promoting activity remains today as one of the most reliable measures of the biological activity of NGF.

* Preparation of this review was supported by grants CA-28110 and NS-16321 from the National Institutes of Health

III. Following demonstration of nerve-growth promoting activity in sarcomas 180 and 37, studies were undertaken to purify the responsible factor. For this purpose, snake venom phosphodiesterase was used to inactivate nucleic acids – with the startling result that snake venom itself is a potent source of NGF. [See LEVI-MONTALCINI and ANGELETTI (1968) and TU (1977) and references therein.] Snake venom NGF has now been purified and its properties will be discussed later. The reason why snake venom contains the factor is not known.

IV. In 1958, mouse submandibular glands were found to contain large amounts of NGF (LEVI-MONTALCINI and ANGELETTI, 1964). Levels in male glands greatly exceed those of female glands and this property stems from the fact that these organs are responsive to androgens such as testosterone. Prepubescent male glands contain much less NGF than they do after puberty; administration of testosterone to female animals increases NGF levels; and castration of males reduces glandular NGF (LEVI-MONTALCINI and ANGELETTI, 1964). The reason why mouse salivary glands contain NGF is not known.

V. In 1960, COHEN reported the purification of NGF from submandibular glands and in that same year LEVI-MONTALCINI and BOOKER used the purified protein to demonstrate two more remarkable phenomena. The first was that administration of NGF to animals produced excessive growth of sympathetic ganglia (LEVI-MONTALCINI and BOOKER, 1960a) and the second was that treatment of animals with an antiserum to NGF selectively destroyed the sympathetic chain ganglia (LEVI-MONTALCINI and BOOKER, 1960b). The term immunosympathectomy was used to describe this effect. Subsequently, immunosympathectomy has been used as a pharmacologic tool and that aspect is treated fully in the volume by ZAIMIS and KNIGHT (1972).

The foregoing list presents some of the most central features of NGF biology and they have given rise to the widely held view that NGF is somehow involved in development and maintenance of the sensory and sympathetic nervous system. Against this background, we turn now to consideration of the multiple molecular forms of NGF and their chemical properties.

B. Multiple Molecular Forms of NGF

One of the problems associated with efforts to unravel the physiologic function of NGF is the fact that several different molecular forms of the protein have been isolated. These different forms arise depending upon the purification scheme employed, and we shall discuss each of them in turn.

I. Cohen's NGF

In 1960, COHEN first reported purification of NGF from mouse submandibular glands. This protein displayed potent nerve growth promoting activity when tested in the sensory ganglia assay system and it possessed a sedimentation coefficient of 4.3S. From this number, COHEN estimated the molecular weight of NGF to be about 44,000. This form of NGF was used to prepare antiserum for use with the immunosympathectomy studies referred to earlier. To our knowledge, the molecular properties of this protein have not received further study since Cohen's original description of it. However, as we shall see later on, this species does in fact exist in crude submandibular gland extracts and it represents a biologically active degradation product of a much larger form of NGF. The degradation reaction which gives rise to Cohen's form of the protein occurs during the process of purification.

II. 7S NGF

In 1967, Varon et al. reported isolation of a high molecular weight form of salivary gland NGF. This protein displayed a sedimentation coefficient of 7S and from this value a molecular weight of 140,000 was estimated. Further studies revealed that 7S NGF is stable only between pH 5–8 (Varon et al., 1968). Outside this pH range, the molecule dissociates into 3 kinds of subunits termed α, β, and γ. This dissociation reaction is reversible. Isolation of the β subunit reveals that it is the only one which exhibits nerve growth promoting activity (Varon et al., 1968).

Subsequently, it was shown that 7S NGF possesses esterolytic activity upon certain N-substituted lysine and arginine esters (e.g., α-N-benzoyl-L-arginine ethyl ester (Greene et al., 1968, 1969). Moreover, upon dissociation of the 7S-complex into its subunits, the γ-subunit is the one which displays the esterolytic activity. The function of the α-subunit is not known.

Kinetic studies on the hydrolysis of these synthetic esters revealed a curious phenomenon. When 7S NGF was diluted from high to low protein concentrations in the enzyme assay solution, the maximal hydrolytic rate was achieved only after an initial lag phase of lower velocity (Greene et al., 1969). The isolated γ subunit did not exhibit this lag phase, and it was therefore proposed that the three subunits of 7S NGF are in equilibrium with one another as well as with the parent 7S-complex. According to this model, 7S NGF exhibits little or no enzymic activity. Upon dilution of the protein, dissociation occurs with subsequent appearance of activity – hence the lag phase (Greene et al., 1969). As will be seen later, this model is unlikely to be correct, simply on kinetic grounds. Further work revealed that 7S NGF and its subunits exist in multiple molecular forms (Smith et al., 1968). For example, upon electrophoresis, both the α and γ subunits exist in at least 3 different recognizable forms. These findings indicated that 7S NGF is heterogeneous in the sense that it can contain different types of subunits (Smith et al., 1968). Each of the three types of subunits (α, β, and γ) has a molecular weight on the order of 26,000. The subunit structure of 7S NGF has not been determined, although it has been proposed that it contains 2α, 2γ, and 1β subunits (Server and Shooter, 1978). It should be pointed out that we have been unable to find any data in the literature which directly support such a model.

Several studies have demonstrated that 7S NGF is appreciably unstable even at relatively high protein concentrations. For example, in their original paper on isolation of the protein, Varon et al. (1967) observed that, during DEAE-cellulose chromatography, extremely high flow rates (on the order of 350 ml/h) were required to achieve good recovery. At more normal flow rates, recovery was poor and NGF activity eluted over many fractions. The authors proposed that this behavior was related to slow dissociation of the 7S-complex during chromatography.

Other studies from the same laboratory have supported the concept that 7S NGF is a relatively unstable protein and that it exists in equilibrium with its subunits (VARON et al., 1968; GREENE et al., 1969; SMITH et al., 1968, 1969). Direct ultracentrifugal measurements of 7S NGF stability by BAKER (1975) revealed considerable dissociation at pH 7.4 and protein concentrations below 1 mg/ml. These findings have been confirmed (PANTAZIS et al., 1977b). Possible explanations for this dissociation reaction will be presented when we discuss later the properties of NGF in crude mouse salivary gland extracts.

In 1975, PATTISON and DUNN made the important observation that 7S NGF contains 1–2 g of tightly bound Zn(II) per mole. Values ranged from 1.7 to 2.0. In contrast, the isolated subunits of 7S NGF do not contain stoichiometrically significant amounts of Zn(II). Treatment of 7S NGF with metal ion chelating reagents significantly enhanced the esterolytic activity of the protein. Consequently, these authors proposed that zinc ion is somehow critically involved in the structure of 7S NGF and that the enzyme activation reaction results from chelation of this metal (PATTISON and DUNN, 1975). In later studies, they concluded that the chelator-induced activation is a freely reversible process (PATTISON and DUNN, 1976a). (See also PATTISON and DUNN, 1976b; BOTHWELL and SHOOTER, 1978.) The function of Zn(II) in the structure of 7S NGF will be discussed again when we consider the NGF zymogen.

III. 2.5S NGF

In 1969, BOCCHINI and ANGELETTI presented a purification scheme for salivary gland NGF that was distinctly different from that used for the 7S form. This scheme employed a carboxymethylcellulose chromatography step at pH 5.0 and yielded a pure protein of molecular weight 26,000. This protein was later given the name 2.5S NGF based upon its sedimentation coefficient.

The 2.5S protein is highly biologically active in producing neurite outgrowth in vitro and it is chemically closely related to the β-subunit of 7S NGF as follows. ANGELETTI et al (1973) have determined the complete primary structure of 2.5S NGF and it consists of 2 identical non-covalently linked polypeptide chains (ANGELETTI, R.H., et al., 1971). Each chain contains 118 amino acids and 3 disulfide bonds. 2.5S NGF differs from the β component of 7S NGF in that during isolation of the 2.5S protein by the procedure of BOCCHINI and ANGELETTI (1969), limited proteolysis occurs at both the N- and C-termini (ANGELETTI et al., 1973). The amino terminal modification of 2.5S NGF results in removal of the first eight residues, while the carboxyl terminal cleavage results in loss of the C-terminal arginine. If the β subunit of 7S NGF is isolated directly from the 7S-complex, these proteolytic cleavages do not occur (MOORE et al., 1974). It is the chromatography step at pH 5.0 (or below) which gives rise to 2.5S NGF since at that pH, the β subunit of 7S NGF dissociates. β is then converted to the 2.5S form by the above-mentioned limited proteolysis which does not affect biological activity, at least when tested in the sensory ganglion assay system (MOORE et al., 1974).

Sequence studies have shown that 2.5S NGF and proinsulin are structurally related (FRAZIER et al., 1972). For example, the regions of proinsulin which yield the

B and A chains of insulin display 30% and 52% indentities with the corresponding regions of 2.5S NGF. Also there is some similarity in disulfide pairing between the two proteins (see BRADSHAW, 1978, for a detailed account). Thus, it has been suggested that insulin and NGF may share a common evolutionary origin and that they arise secondary to gene duplication.

Several years ago it was proposed that β NGF might be synthesized initially as a larger precursor (a pro-β-NGF) (ANGELETTI and BRADSHAW, 1971; MOORE et al., 1974). The arguments for this go as follows. The C-terminal residue of both chains of β NGF is arginine (ANGELETTI and BRADSHAW, 1971) and removal of this residue by carboxypeptidase B prevents reassociation of des-arginine β NGF with the other subunits to form 7S NGF (MOORE et al., 1974). Moreover, the arginine esterase activity of the γ subunit is largely inhibited in the complex. Thus, it has been suggested that the C-terminal arginine of β NGF is situated at, and thereby inhibits, the active site of the γ subunit in the 7S-complex. According to this model, the γ subunit is involved in post-translationally processing a pro-β by cleavage at arginine, and then the γ subunit active site binds to this arginine, thereby rendering γ enzymically inactive. Enzyme activity is thus proposed to appear when γ dissociates from that specific arginine. For several reasons, to be discussed later, this mechanism appears to be an unlikely one. On the other hand, BERGER and SHOOTER (1977) have presented evidence that the approximately 13,000 molecular weight chains of β NGF are initially synthesized as a 22,000 molecular weight protein which can be cleaved by the γ subunit to yield β NGF.

IV. Snake Venom NGF

The venoms of many poisonous snakes contain NGF, including the elapid, crotalid and viper families – and there is no experimental information on why this should be so. The NGF from cobra *(Naja naja)* venom has been purified and studied extensively. This protein has a molecular weight of 28,000 and it is composed of two similar and possibly identical noncovalently-linked chains (HOGUE-ANGELETTI et al., 1976). Sequence analyses reveal that these chains contain 116 residues and that of 73 of them, 61% are identical to mouse 2.5S NGF. Thus, the primary structure of this form of NGF is appreciably conserved over a wide range of the evolutionary scale. It should be pointed out here that, like the purification procedure for 2.5S NGF, carboxymethyl cellulose chromatography at acid pH has been used for the snake protein (HOGUE-ANGELETTI et al., 1976). Thus, it is possible that a larger molecule, such as 7S NGF, might be present in venom, and have undergone dissociation during the purification scheme employed. It has been shown, however, that cobra NGF does not interact with the α and γ subunits of mouse 7S NGF (HOGUE-ANGELETTI et al., 1976).

PEARCE and coworkers (1972) have studied yet another form of NGF from *Vipera russelli*. This protein, unlike those of mouse and cobra, is a complex glycoprotein which contains N-acetylglucosamine, mannose, galactose, fucose, and N-acetylneuraminic acid. The primary structure of this NGF is not yet known. (For a comprehensive review of snake NGF, see HOGUE-ANGELETTI and BRADSHAW, 1977.)

V. What is the Naturally Occurring Form
of NGF in the Mouse Salivary Gland?

The observation that NGF can be isolated from mouse salivary glands in several different molecular forms raises the question of which one is the predominant, or naturally synthesized, species. This question has recently been approached by asking what are the molecular properties of NGF in crude submandibular gland extracts at physiologic pH values. For this purpose, a radioimmunoassay (YOUNG et al., 1979) specific for 2.5S NGF was used to detect all forms of NGF which are immunochemically recognizable by antibody to 2.5S NGF.

Initial studies, utilizing gel filtration chromatography, revealed that crude gland extracts contain multiple species of NGF (YOUNG et al., 1978). The predominant form is a high molecular weight species (see below) and it is closely similar to the 7S form of NGF first discovered by VARON et al. (1967). Other species of the protein which can be detected include the form originally found by COHEN 20 years ago (approximately 40,000 molecular weight) (COHEN, 1960) as well as a protein which, on the basis of size, is indistinguishable from 2.5S (or β) NGF itself (molecular weight 26,000). Using a variety of chromatographic and electrophoretic procedures, at least six different forms of NGF can be identified (YOUNG et al., 1978).

Several lines of evidence indicate that the lower molecular weight forms arise from the largest species by dissociation, probably secondary to proteolytic damage during the process of isolation. First, when gland extracts are examined by gel filtration, increasing amounts of the smaller forms appear as a function of higher dilutions of the extracts. This finding is indicative of a concentration dependent dissociation reaction. Second, this process is irreversible in that the smaller forms cannot recombine to form the largest species, suggesting that some form of covalent modification has occurred during the purification procedure. Finally, as discussed earlier, 2.5S NGF has been shown to differ from the β subunit of 7S NGF in that it has suffered proteolytic modification during isolation (ANGELETTI et al., 1973). Thus it would appear that the smaller forms are degradation products of the single and largest NGF protein found in gland extracts (YOUNG et al., 1978).

This large form of NGF has now been purified to homogeneity (YOUNG et al., 1978). Equilibrium sedimentation studies reveal that it has a molecular weight close to 116,000 and a partial specific volume of 0.688 ml/g. At this point, it is useful to compare similarities and differences in the properties of this high molecular form of NGF with those of 7S NGF.

First is the matter of molecular weight. As noted earlier, VARON et al. (1967) estimated the mass of 7S NGF to be about 140,000 based solely upon the sedimentation coefficient. Subsequently, BAKER (1975) measured its molecular weight by sedimentation equilibrium. Although BAKER detected some dissociation of the protein, he found a weight of 137,000 and assumed a partial volume of 0.73 which was calculated from the amino acid composition. It should be pointed out that measured and calculated values of partial volumes of proteins sometimes differ significantly.

Second, it will be recalled from the section on 7S NGF that very high ion-exchange column flow rates were necessary for purification, and the necessity for these high flow rates was attributed to dissociation of the oligomer even at relatively high protein concentrations. This behavior is not characteristic of the 116,000 molecular weight protein under discussion here. This form is highly stable at μg/ml protein concentrations and very slow ion-exchange flow rates were used to purify it (YOUNG et al., 1978). Thus, we infer that some chemical process has occurred during isolation of 7S NGF and, as discussed above, the most likely candidate is proteolysis. Taken together, all available evidence suggests that the 116,000 molecular weight NGF and 7S NGF are one and the same protein. The major exception to this is that one is unstable in dilute solution and the other retains its high molecular weight structure. Presumably, this property stems from the different procedures used to isolate the two proteins (VARON et al., 1967; YOUNG et al., 1978).

C. The NGF-Zymogen and Its Enzymic Properties

I. Autocatalytic Self-activation

The important observation of GREENE et al. (1969) that 7S NGF can hydrolyze certain synthetic arginine and lysine esters led us to ask whether the 116,000 molecular weight NGF shows this property. It does. When the 116,000 molecular weight protein is diluted from high to low protein concentrations, TAME can be hydrolyzed and, like 7S NGF, the maximum velocity is achieved only after a lag phase of lower velocity (YOUNG, 1979). In this regard, it will be recalled that GREENE et al. (1969) attributed this lag phase to slow dissociation of 7S NGF, with concomitant gradual appearance of enzyme activity.

The 116,000 molecular weight protein exhibits the same characteristic lag phase of 7S NGF, but for several reasons, dissociation cannot be used to explain it. First, this protein has been shown to be stable at the low concentrations used for the enzyme assays (YOUNG et al., 1978). Second, when the lag phase is examined as a function of protein concentration, the length of the lag phase increases greatly as the enzyme concentration is reduced (YOUNG, 1979). On chemical kinetic grounds, such behavior is inconsistent with a dissociation reaction. For example, in the reaction $AB \rightleftharpoons A + B$, where AB is inactive and A is active, the half-life of this first-order reaction (and hence duration of the lag phase) is independent of the initial protein concentration. However, with the 116,000 molecular weight enzyme, *the rate of appearance of enzyme activity actually increases with time*. Third, a dissociation reaction of the kind postulated by GREENE et al. (1969) should be reversible. That is, once the lag phase has occurred (following dilution of the protein with concomitant dissociation), it should be restorable after reconcentration of the protein. With the 116,000 molecular weight NGF, it is not (YOUNG, 1979).

Taken together, all available evidence indicates that the 116,000 molecular weight protein, and most likely 7S NGF as well, represents an autocatalytic self-activating enzyme system and thus, as isolated from the mouse salivary gland, the protein exists as a catalytically inactive zymogen (YOUNG, 1979). For a reaction of the kind shown by Eq. (1)

$$P \rightarrow P', \tag{1}$$

where P is zymogen and P' is active enzyme, and where the reaction is catalyzed by P', then the rate of autoactivation is given by

$$\frac{d[P']}{dt}=k[P][P'].\tag{2}$$

Here k is the second-order rate constant of activation. If A is the total protein concentration and X is the concentration of active enzyme at time t, then

$$\frac{dx}{x(A-x)}=kdt.\tag{3}$$

Integrating this equation yields

$$kAt=-\ln\left(\frac{A-x_t}{x_t}\right)+\ln\left(\frac{A-x_0}{x_0}\right),\tag{4}$$

where x_t = activity at time t, and x_0 = activity at time 0. This equation has been used successfully to analyze the kinetics of activation of the NGF zymogen. As expected, values of k (about 14×10^6 l/mole/min) do not change over a wide range of protein concentrations, whereas values for the calculated half-life increase considerably as A decreases.

Several predictions can be made about how an autocatalytic system should behave under different conditions. One is that it should be an irreversible reaction and it is (YOUNG, 1979). A second is that any reagent which inhibits enzyme activity should prolong the lag phase. Conversely, any reagent which stimulates activity should shorten the lag phase. From other studies it has been shown that diisopropylphosphofluoridate (DFP) inhibits the activated zymogen according to second-order kinetics and thus this enzyme is a member of the general class of serine proteases (ORENSTEIN et al., 1978). When activated NGF is treated with DFP to partially inhibit its activity, the lag phase is significantly prolonged as expected. For reasons to be discussed later, EDTA enhances enzyme activity, and it also shortens the lag phase (YOUNG, 1979).

In light of the above findings, it is pertinent here to examine the post-translational processing hypothesis of a pro-β-NGF by the γ subunit of 7S NGF (ANGELETTI and BRADSHAW, 1971; MOORE et al., 1974). According to this model, a pro-β molecule is processed by γ, which then is inhibited because the C-terminal arginine of the processed β binds to the active site of γ. Based upon the evidence that autocatalytic activation, not dissociation, is responsible for the lag phase, this mechanism appears to be unlikely for two reasons. First, the NGF zymogen does not dissociate. Second, it is difficult to see how an inactive zymogen, whose full enzyme activity remains to be expressed following self-activation, could be responsible for converting a precursor to β NGF, since the postulated post-translational cleavage event is finished *before* the enzyme has been activated (YOUNG, 1979). Again, we do not mean to imply that a precursor form of β does not exist (see BERGER and SHOOTER, 1977), only that the zymogen seems an unlikely candidate for processing it.

II. Regulation of the Autoactivation Reaction

We have already noted that 7S NGF has been shown to contain 1–2 g-atoms of tightly bound Zn(II) per mole (PATTISON and DUNN, 1975). To see whether the NGF-zymogen contains this metal, solutions of the protein were dialyzed exhaustively against metal free buffers and analyzed by atomic absorption spectroscopy. Results show that the zymogen (based upon a molecular weight of 116,000) contains 1.05 ± 0.15 (SD) g-atom Zn(II) per mole protein (YOUNG and KOROLY, 1980). Parallel studies with ^{3}H-DFP also indicate that after the zymogen has been activated, it reacts stoichiometrically with 1 mole DFP per mole protein. This result implies that the activated enzyme contains but one active site serine (YOUNG and KOROLY, 1980).

Recent studies have now shown that autocatalytic activation of the NGF-zymogen is under the strict control of Zn(II) (YOUNG and KOROLY, 1980). As long as this metal ion is bound to the protein, autocatalysis is completely prevented. Removal of the metal, either by high dilution of the zymogen, or by chelation, results in autoactivation. This process is irreversible. Once activation is complete, it cannot be reversed by addition of Zn(II). To our knowledge, this type of built-in control mechanism is unique in enzymology. We know of no enzyme system in which an autocatalytic activation reaction is regulated by a metal ion. Surely this must be important in the physiologic mechanism of action of NGF – yet we do not know what that may be.

III. Activation of Plasminogen by NGF

The observation that the NGF zymogen, once activated, is a member of the serine protease family of enzymes led directly to the question: what other naturally-occurring substrates might NGF hydrolyze. One such substrate is plasminogen which is also activated by a variety of highly specific serine proteases, including urokinase and the tissue activator which has been shown to be secreted by cells in culture (ROBBINS et al., 1967; UNKELESS et al., 1974). Both of these enzymes, like NGF, also display a high degree of specificity for an arginine residue on the C-terminal side of the bond to be split. Recent studies have now shown that autocatalytically activated NGF-zymogen can in fact activate plasminogen with subsequent lysis (by plasmin) of a fibrin clot. This reaction is strictly plasminogen dependent in that no hydrolysis of fibrin is observed in the absence of plasminogen. Although NGF is not nearly as active as human urokinase in converting plasminogen to plasmin, chromatographic as well as electrophoretic studies have demonstrated that the plasminogen-activating activity of NGF is not due to a contaminating protease (ORENSTEIN et al., 1978). The high degree of specificity of the enzyme is further shown by the observation that the activated zymogen does not hydrolyze casein at a detectable rate.

The biologic significance of plasminogen activation by activated NGF is not clear and it could well be that plasminogen is not a naturally-occurring substrate for NGF at all. Nevertheless, activation of plasminogen by NGF is an activity, displayed by this protease, upon a naturally-occurring non-neural substrate.

D. Role of the Submandibular Gland in Secretion of NGF

In 1973, HENDRY and IVERSEN reported that bilateral removal of the submandibular glands of adult mice caused a gradual reduction in the plasma concentrations
of β NGF as measured by radioimmunoassay. In male animals, the concentration
decay half-life was 9 days, and concentrations reached a minimum 33 days after
operation. These authors also reported that normal male mice have higher circulating levels of NGF than do females – a finding which would be consistent with
the known differences in gland content of the protein between the two sexes. Thus
they concluded that the normal circulating level of NGF is dependent upon intact
submandibular glands.

The observation that the decrease in plasma concentration of NGF is such a
slow process following gland removal is puzzling, particularly in light of the work
of ISHII and SHOOTER (1975). These authors observed that when ^{125}I-NGF is injected intravascularly, the half-life of disappearance from blood is far more rapid
(10–30 min).

To study this problem further we have measured serum levels of 2.5 S NGF by
radioimmunoassay over the time period 30 min to 35 days following bilateral
sialoadenectomy (MURPHY et al., 1977a). Results showed no significant changes
in serum NGF levels at any time point over this interval. Furthermore, there were
no significant differences in serum NGF concentrations between adult male and
female mice. In contrast, measurements of NGF levels in mouse saliva yielded an
entirely different story. Male mouse submandibular saliva contains an extraordinarily high concentration of NGF (150 µg/ml) and female saliva contains considerably less (about 1 µg/ml) (MURPHY et al., 1977a). Therefore, the sexual dimor-
phic characteristic of mouse submandibular glands is not reflected by serum levels
of NGF, but by the concentration in saliva. WALLACE and PARTLOW (1976) have
also found high levels of NGF in saliva following stimulation by α-adrenergic
agents. Thus, present available evidence indicates that the salivary glands are not
endocrine glands with respect to NGF secretion. Rather, they are exocrine. The
role of NGF in saliva is not known, but it may play some unsuspected role some-
where in the alimentary tract.

Gel filtration studies have shown that, like salivary gland extracts, mouse saliva
contains at least two biologically active forms of NGF (MURPHY et al., 1977b).
One of these has an apparent molecular weight of 114,000 and is probably the same
as the NGF zymogen which has been purified from salivary glands (MURPHY et
al., 1977b, 1979). The other species is similar in size to 2.5 S NGF. Moreover, the
larger NGF appears to be continuously degraded in saliva, probably enzymically,
to yield the smaller protein. The physiological reason for two biologically active
(sensory ganglion assay system) forms of NGF in saliva is not clear. In studies on
saliva from an individual mouse, samples can be obtained which contain only the
larger form, and it could be that slow degradation simply stems from the presence
of proteases in saliva (MURPHY et al., 1977b).

E. Secretion of NGF by Cells in Culture

In 1974, it was shown that mouse L cells and 3 T 3 cells synthesize and secrete a
nerve growth factor that is biologically and immunologically similar to the salivary

Table 1. Cells which secrete NGF in culture[a]

Cell type	Source
L cells	Mouse
3T3 cells	Mouse
SV40 3T3 cells	Mouse
Primary fibroblasts	Chick
Neuroblastoma	Mouse
Melanoma	Mouse
Myoblasts	Rat
Glioma	Rat
Glioblastoma	Human
Primary skin fibroblasts	Human
Primary synovial fibroblasts	Human

[a] For references, see Table 1 in Bradshaw and Young (1976)

gland factor (OGER et al., 1974). Secretion of NGF by L cells is pertinent to the well-established observation that these cells are a source of so-called conditioned medium in that they produce factors which can stimulate proliferation of other unrelated cells.

Several studies have now shown that many different kinds of cells in culture secrete NGF. These include a variety of both transformed as well as primary cells. (See Table 1 and references in BRADSHAW and YOUNG, 1976.) One cell type which is particularly interesting is mouse and human neuroblastoma (MURPHY et al., 1975) since several pieces of information indicate that these cells in culture can respond to 2.5S NGF (GOLDSTEIN and BRODEUR, 1973; KOLBER et al., 1974). Moreover, mouse neuroblastoma cells display receptors for 2.5S NGF and this property is a function of the interphase cell cycle, with receptors appearing largely in late G_1 (REVOLTELLA et al., 1974, 1975; BOSMAN et al., 1975). Human melanoma cells have also been shown to have NGF receptors (FABRICANT et al., 1977) and to respond biologically to the factor (SHERWIN et al., 1979). These observations raise several interesting possibilities. One is that certain cells within a clone secrete NGF and other cells respond to it. Or perhaps a given cell may secrete NGF at one stage of the cell cycle, and respond to it, in a functional sense, at another stage. If so, this could represent an autoregulatory growth mechanism (MURPHY et al., 1975). Yet the basic reason why cells secrete NGF in culture is not known.

So far, cell secreted NGF has not been purified. However, from gel filtration studies, a few things can be said about the properties of the factor secreted by mouse L cells (PANTAZIS et al., 1977a) and by rat muscle cells (MURPHY et al., 1977c). Both of these macromolecules are highly stable in dilute solution and they have molecular weights in the range 140,000–160,000. Both also contain, as part of their structures, a molecule which is closely similar in size and immunological properties to 2.5S (β) NGF. Until cell secreted NGF is purified it will not be possible to see whether it displays biologic functions that are different from those of salivary gland NGF.

F. Biological Effects of NGF Related to the Nervous System and Neural Crest Derivatives

I. Sympathetic and Sensory Neurons

An extensive literature exists to support a physiologic role for NGF at a specific stage in the development of avian and mammalian sensory ganglia and in the growth and long-term maintenance of certain sympathetic neurons in vivo (LEVI-MONTALCINI and ANGELETTI, 1968; LEVI-MONTALCINI, 1976; HENDRY, 1976; MOBLEY et al., 1977). The bulk of the evidence which implicates NGF as an essential component for normal growth and development of the peripheral nerve system is as follows.

1. Systemic injections of relatively high concentrations of 2.5 S NGF (repeated injections of ∼ 10 µg NGF/g body weight) into newborn chicks (LEVI-MONTALCINI and BOOKER, 1960 a), rodents (LEVI-MONTALCINI, 1964; BANKS et al., 1975), and kittens (EDWARDS et al., 1966) result in hypertrophy and possibly hyperplasia (LEVI-MONTALCINI and BOOKER, 1960 a; LEVI-MONTALCINI, 1976) of adrenergic cells in the pre- and paravertebral sympathetic ganglia, and hypertrophy of these neurons in adults (LEVI-MONTALCINI and BOOKER, 1960 a; LEVI-MONTALCINI, 1965; ANGELETTI et al., 1971). There is a marked stimulation of fiber outgrowth in the ganglia (LEVI-MONTALCINI and ANGELETTI, 1968) and a sizeable increase in the density of adrenergic terminal plexuses in several peripheral organs after 7 S or 2.5 S NGF treatment (LEVI-MONTALCINI and ANGELETTI, 1968; BJERRE et al., 1975) which may be reversible with time (BJERRE et al., 1975). These gross morphological changes are accompanied by elevated levels of the adrenergic neurotransmitter norepinephrine (CRAIN and WIEGAND, 1961; BJERRE et al., 1975) and of two specific enzymes in its anabolic sequence, tyrosine hydroxylase (TOH) (THOENEN et al., 1971; HENDRY and IVERSEN, 1971) and dopamine-β-hydroxylase (DOH) (THOENEN et al., 1971). The dorsal root ganglia also increase in volume and cell numbers in response to exogenous NGF during a restricted period of development (LEVI-MONTALCINI and ANGELETTI, 1968). However, the growth factor elicits no apparent structural or functional alterations in mature animals (LEVI-MONTALCINI and ANGELETTI, 1968; STOECKEL et al., 1975 a).

2. Administration of NGF antiserum by direct injection (LEVI-MONTALCINI and ANGELETTI, 1966; BANKS et al., 1979) or via mother's milk (LEVI-MONTALCINI and ANGELETTI, 1961 b; GORIN and JOHNSON, 1979) to newborn rodents leads to destruction of much of the peripheral adrenergic nervous system. This immunosympathectomy succeeds only in the early stages in the animal's maturation (LEVI-MONTALCINI and ANGELETTI, 1966) and preferentially destroys different ganglia at specific times (ZAIMIS et al., 1965; KLINGMAN and KLINGMAN, 1967). For reasons as yet unclear, antibodies to 2.5 S NGF do not affect the short adrenergic neurons that innervate the urogenital tract and brown adipose tissue (LEVI-MONTALCINI, 1965).

Anti-NGF given to animals over 2 weeks of age causes marked, but transient, morphological and biochemical alterations in sympathetic ganglia (LEVI-MONTALCINI and ANGELETTI, 1961 b; ANGELETTI, P.U., et al., 1971; HENDRY and IVERSEN, 1971). Even maintenance of high antibody titres in adult rats for a period of months results in changes that are reversed with decreasing antibody levels (OTTEN et al., 1979 b). Sensory neurons of the dorsal root ganglia are not sensitive to NGF antiserum injected any time after birth. However, since sensory ganglia in the offspring of autoimmune female rats do not fully develop, these neurons may be responsive to NGF prior to birth (GORIN and JOHNSON, 1979).

Although neutralization of endogenous NGF by NGF antiserum is the most likely mode of action of the NGF antibody (ENNIS et al., 1979), the mechanism of interaction between NGF antiserum and the native form(s) of NGF is not completely understood (HENDRY, 1976; ENNIS et al., 1979). Thus, caution must be exercised in interpretation of results. It may be well to note that antiserum to snake venom NGF does not immunosympathectomize rodent neonates, perhaps due to differences in antigenic determinants of mouse and snake venom factors (BANKS et al., 1979).

3. Chemical sympathectomy with agents such as 6-hydroxydopamine and vinblastine or surgical disruption of the growing adrenergic nerve fibers in young animals result in degeneration of ganglia (for review, see TANZIER et al., 1969; HENDRY, 1976; LEVI-MONTALCINI, 1979). NGF administered shortly after trauma offsets destruction of the neuronal cell bodies. Once the animal has matured to the stage of end-organ innervation, no irreversible damage of the ganglia occurs secondary to trauma, but NGF antiserum prevents NGF-stimulated recovery (BJERRE et al., 1973 a, 1974 a; HENDRY, 1975; LEVI-MONTALCINI et al., 1975; BANKS and WALTER, 1977).

4. In vitro studies of responsive sympathetic and sensory ganglia in organ or dissociated cell cultures provide a description of a multitude of events that occur as a consequence of NGF in the medium. Marked changes in cellular morphology epitomized by neurite extension and vesicle formation occur in tandem with a pronounced increase in the uptake (LEVI-MONTALCINI and ANGELETTI, 1968; HORII and VARON, 1977; SKAPER and VARON, 1979) and incorporation of precursors into protein, RNA and lipids, and in the utilization of carbohydrates (LEVI-MONTALCINI and ANGELETTI, 1968; LARRABEE, 1972). An increased production of epinephrine in adrenergic cells [see IKENO and GUROFF (1979) and references therein] and of the putative neurotransmitter substance P in the sensory neurons (SCHWARTZ and COSTA, 1979; KESSLER and BLACK, 1980) takes place. However, absence of NGF from the medium results in the death of these neurons (LEVI-MONTALCINI and ANGELETTI, 1968). This fact, together with the disruption of nerve fibers (axotomy) that may occur during preparation of ganglia for culture (HENDRY, 1976) necessitates caution in equating normal biologic responses with those in vitro. Nevertheless, as will be reported below, certain of these responses to NGF are mimicked by a cell line that does not require added NGF for survival.

Excellent coverage of the above research prior to the mid 1970's is available (LEVI-MONTALCINI and ANGELETTI, 1968; TANZIER, 1969; LARRABEE, 1972; LEVI-MONTALCINI, 1976). During the 1970's, more precise definitions of the cellular responses to NGF, especially with regard to neurite extension and norepinephrine production, became available. The following briefly summarizes this information. It must be noted, however, that many questions still remain unanswered, especially about the molecular form(s) of NGF operative in vivo and the basic mechanism(s) of NGF action.

Neurite extension, the earliest and most spectacular structural change elicited by NGF and the basis of the bioasay which defines a substance as having nerve growth activity (LEVI-MONTALCINI and ANGELETTI, 1968), depends on the assembly and organization of neurotubules and neurofilaments. RNA synthesis is not required for fiber outgrowth (PARTLOW and LARRABEE, 1971; BURNHAM and VARON, 1974), at least initially (LEVI-MONTALCINI et al., 1974). Apparently, sufficient tubulin (MIZEL and BAMBURG, 1975), actin and other precursors of polymer assembly are available in the cytoplasm to start the elongation process. The observation that NGF interacts in vitro with both tubulin (CALISSANO and COZZARI, 1974) and actin and activates actomyosin ATPase (CALISSANO et al., 1978) supports a possible direct role in assembly and/or stabilization of one or both of these protein polymers. Vinblastine and cytochalasin B, inhibitors of microtubule and microfilament polymerization, respectively, destroy immature sympathetic neurons of newborn mice (MENESINI-CHEN et al., 1977), rats and hamsters (JOHNSON, 1978). This effect is ameliorated by 2.5 S NGF, perhaps because it protects against dissociation of preformed microtubules (MENESINI-CHEN et al., 1977). Vinblastine, even after chronic administration, has no notable influence on neurons of older animals (JOHNSON, 1978), and, in amounts sufficient to destroy most sympathetic neurons, this drug does not harm sensory nerves in the dorsal root or nodose ganglia in 2-day old animals (JOHNSON et al., 1978). Anti-NGF has no effect at this time either. The nature of any interplay in vivo among NGF, microtubules and microfilaments is not known. NGF may effect initiation or regulation of the polymerization process, nucleation sites for microtubule organization or the dynamics of the reversible dimer-polymer equilibrium. Further, no explanation can be given as to why such low amounts of NGF protect against vinblastine in vivo, whereas much higher amounts are required in vitro (MENESINI-CHEN et al., 1977).

The most pronounced metabolic event within the same time-frame as enlargement of the sympathetic ganglia and neurite outgrowth in vivo is an increase in protein, including ornithine decarboxylase (THOENEN et al., 1971; MACDONNELL et al., 1977a), RNA polymerase (HUFF et al., 1978) and the rate-limiting enzymes catalyzing norepinephrine synthesis (THOENEN et al., 1971). 7S and 2.5S NGF evoke an elevation in the specific activities of TOH and DOH in rodent ganglia (THOENEN et al., 1971; HENDRY and IVERSEN, 1971; YU et al., 1977; MACDONNELL et al., 1977a, b), but oxidation of the tryptophan residues of 2.5 S NGF renders the molecule inactive. In vivo, NGF appears to induce de novo synthesis of these latter enzymes (STÖCKEL et al., 1974b; MACDONNELL et al., 1977b; OTTEN et al., 1977). A specific increase in TOH mRNA (MACDONNELL et al., 1977b) and in the phosphorylation of a specific nuclear protein has been reported (YU et al., 1978). Whether or not added NGF can stimulate TOH synthesis in sympathetic ganglia in vitro without the aid of other modulatory factors is not resolved (OTTEN and THOENEN, 1976; NAGAIAH et al., 1977; MAX et al., 1978; ROHRER et al., 1978). MAX et al. (1978) report that NGF alone elicits maximum TOH activity in organ cultures of superior cervical ganglia. On the other hand, perhaps due to different culture conditions, GUROFF and colleagues (YU et al., 1977; MACDONNELL et al., 1977b) claim only a relatively small stimulation of TOH synthesis over that of other proteins. However, these authors find that organ cultures of ganglia from rats treated with NGF prior to sacrifice respond to NGF by an Actinomycin D insensitive increase in TOH activity comparable to that induced by NGF in vivo. If animals treated with Actinomycin D or anti-NGF were the donors of the ganglia, NGF could not markedly elevate TOH levels in culture (MacDonnell et al., 1977 b). Although these observations have not been confirmed, this work suggests that NGF alone may not have a direct influence at the level of transcription, but rather may work in concert with other agents. Glucocorticoids enhance NGF-mediated enzyme induction in organ cultures (Otten and Thoenen, 1976; Nagaiah et al., 1977; Rohrer et al., 1978). In vivo, biologically active exogenous NGF in concentrations which induce TOH stimulates adrenocorticotropin (ACTH) secretion and, within minutes, leads to prolonged high plasma levels of glucocorticoids (Nagaiah et al., 1978; Otten et al., 1979 a). This may suggest a physiological interplay between NGF and the steroid hormones.

A number of studies have focused on the role of cAMP as a potential intracellular mediator of at least some of the actions of NGF. However, there is as yet no convincing evidence to prove, or disprove, an intermediary role. A summation of the evidence pro and con direct cAMP involvement has been presented (Ikeno and Guroff, 1979). More recently, support for cyclic nucleotide participation in the NGF response, at least in PC-12 cells, has been offered by one lab (Garrels and Schubert, 1979), while another (Landreth et al., 1980) has opposed the premise that NGF action is mediated through cAMP-dependent calcium mobilization.

II. Adrenal Cells

Adrenal medullary cells derive from the neural crest and synthesize and store epinephrine as do synaptic neurons. However, adrenal chromaffin cells, unlike neurons, are endocrine in nature and release their catecholamine products into the circulation. The morphology of fully differentiated chromaffin cells is not sensitive to NGF (Levi-Montalcini, 1965; Angeletti et al., 1972), but TOH and DOH levels do increase, albeit to a less dramatic extent than in neurons (Otten et al., 1977). Adrenal cells do not require exogenous NGF to survive in culture, perhaps because they are able to produce their own (Harper et al., 1976), and NGF antiserum has no major impact on the mature adrenal medulla (Levi-Montalcini, 1965; Angeletti et al., 1972). However, immature chromaffin cells do take on the biochemical and structural characteristics of sympathetic nerve cells in response to NGF both in vivo and in vitro (Otten et al., 1977; Unsicker et al., 1978; Aloe and Levi-Montalcini, 1979). NGF injections into fetal and infant rats stimulate development of neuron-like cells at the expense of adrenal medulla cells (Aloe and Levi-Montalcini, 1979). Chromaffin cells isolated from young rats and incubated with NGF exhibit fiber outgrowth, and the extension of these processes can be abolished by the concentration of glucocorticoids normally found in the adrenal environment (Unsicker et al., 1978). These same amounts of steroids enhance NGF-mediated induction of TOH (Unsicker et al., 1978). Although a tempting hypothesis, the implication of steroid-NGF interaction is still subject to question (Aloe and Levi-Montalcini, 1979).

III. Pheochromocytoma Cells

Recent work, primarily from Greene's laboratory (Tischler and Greene, 1975; Greene and Tischler, 1976; Dichter et al., 1977; Connolly et al., 1979; Lucken-

BILL-EDDS et al., 1979) has centered on the definition of NGF action on a clonal noradrenergic line of rat adrenal medullary pheochromocytoma cells (PC-12). These cells are sensitive to NGF, but not dependent upon it for survival. Without NGF, this cell line shares the differentiated properties of adrenal chromaffin cells, but shortly after supplementation of the medium with nanogram amounts of NGF, there is a reorganization of the cell surface which is coupled temporally to NGF binding (BANERJEE et al., 1973; YANKNER and SHOOTER, 1979) and internalization. There seems to be a concomitant alteration in the adhesive properties of the cells (SCHUBERT and WHITLOCK, 1977; SCHUBERT et al., 1978), a rapid and selective uptake of amino acids (MCGUIRE and GREENE, 1979), an induction of ornithine decarboxylase (GREENE and MCGUIRE, 1978; HATANAKA et al., 1978) although not of TOH (EDGAR and THOENEN, 1978) and an increase in cell volume and protein content. Within several days after NGF addition, the cell line acquires the phenotypic properties of normal sympathetic neurons – i.e., appearance of synaptic vesicles, cessation of cell division, development of electrical excitability and a stimulation of neurite outgrowth which is segmented into a de novo RNA dependent and independent pathway (BURSTEIN and GREENE, 1978).

IV. Central Nervous System

NGF may have some influence on the tissues of the central nervous system (CNS), although any biologic significance has yet to be established. NGF receptors are present in the brain (SZUTOWICZ et al., 1976a, b). Injection of 2.5 S or 7 S NGF into adult rat brain accelerates axosomal outgrowth from transected central noradrenaline, dopamine, and indolamine neuron sytems (BJÖRKLUND and STENEVI, 1972; BJERRE et al., 1973 b) and, even in the newt, NGF enhances regeneration of axotomized optic nerves (TURNER and DELANEY, 1979). Anti-NGF blocks regrowth of the adrenergic fibers (BJERRE et al., 1974 b; TURNER and DELANEY, 1979). However, noradrenaline synthesis is not stimulated in the CNS by NGF and specific retrograde transport of NGF does not occur (SCHWAB et al., 1979). NGF injected intraventricularly, but not systemically, elevates ornithine decarboxylase (ODC) levels in brains of mature rats (LEWIS et al., 1978) to a greater extent than occurs in young rats. Further, NGF in the brain increases ODC levels in liver, kidney, adrenals, perhaps mediated via the increased production and/or release of adrenocorticotropic hormone and the resultant rise in plasma steroids observed within relatively brief periods of time (NAGAIAH et al., 1978). Intracranial administration of picomole amounts of NGF evokes a dose-dependent thirst and intense appetite for sodium in rats for unknown reasons (LEWIS et al., 1979).

The extent of dependence of NGF on other neural and non-neural tissue is not known since most cell types examined secrete NGF into the environment, perhaps as a means of inter- or intracellular regulation. Indeed, some of the events observed in tissue culture and/or in vivo may represent an interplay between NGF, nerve cells, and other cell types and/or their products. The molecular structure of NGF per se may also be a critical determinant in the nature and extent of response, and even in the type of cell which exhibits any response.

G. Retrograde Transport and Tropic Effects

The accumulated literature on the mode of action of NGF supports the hypothesis that NGF is internalized during the life span of sympathetic and sensory neurons via two routes: through the membrane of the cell body and through retrograde axonal transport (HENDRY et al., 1974 a, b; IVERSEN et al., 1975; STOECKEL et al., 1975 a; STÖCKEL et al., 1975 b; JOHNSON et al., 1978; BRUNSO-BECHTOLD and HAMBURGER, 1979 and for review, see LEVI-MONTALCINI, 1976; HENDRY, 1976; BRADSHAW, 1978; SERVER and SHOOTER, 1978). No selective retrograde transport of NGF appears to occur in the central nervous system (SCHWAB et al., 1979) or in motor fibers (STÖCKEL et al., 1975 b), THOENEN and his colleagues (STÖCKEL et al., 1974 a; PARAVICINI et al., 1975; STÖCKEL et al., 1975 b; STOECKEL and THOENEN, 1975; STOECKEL et al., 1976), using a variety of experimental systems, have shown that retrograde movement is relatively selective for the biologically active 2.5 S or β forms of NGF, although specificity may reside in receptor site(s) at the nerve terminals rather than within the transport system per se. The transport process is saturable and can be abolished by axon transection, by drugs and, in neonates only, by anti-NGF.

A recent study using physiologic concentrations of high specific activity [125]I-2.5 S NGF essentially confirms the existence of the retrograde mode of transport, and offers a precise quantitative description of NGF internalization, direction and rate of transport, and turnover (JOHNSON et al., 1978). Once accumulated, a substantial portion of intracellular NGF appears associated with the nucleus.

The combined observations of retrograde transport, NGF secretion by fibroblasts and other cell types in vitro, the presence of NGF activity in serum and in a variety of tissues, and the apparent growth of responsive neurons towards a concentrated source of NGF in vivo and in vitro (CHAMLEY and DOWELL, 1975; CAMPENOT, 1977; EBENDAL and JACOBSON, 1977; LETOURNEAU, 1978; GUNDERSEN and BARRETT, 1979) suggest that NGF may act as a tropic factor. This aspect of possible NGF function is fully discussed in several recent reviews (HENDRY, 1976; LEVI-MONTALCINI, 1976; BLACK, 1978; VARON and BUNGE, 1978; LEVI-MONTALCINI, 1979).

H. Cellular NGF Receptors

The cellular site(s) of action of NGF is not clear. In what follows, we shall discuss studies on the interaction of 2.5 S (or β) NGF with certain cell structures.

In 1973, FRAZIER et al. observed that covalent conjugates of NGF with Sepharose were able to elicit neurite outgrowth in the sensory ganglion assay system. Control experiments indicated that this was probably not due to release of NGF from the Sepharose matrix. Further work also demonstrated that little, if any, NGF is solubilized upon interaction of NGF-Sepharose with ganglia. These experiments represented the first direct evidence for cell surface receptors.

Subsequent studies with ^{125}I-NGF have revealed high affinity neuronal plasma membrane binding sites for the radiolabelled protein which could be displaced by unlabelled NGF ($K_a \sim 10^{10}$ l/mole) (HERRUP and SHOOTER, 1973; BANERJEE et al.,

1973; FRAZIER et al., 1974a). Several kinds of peripheral tissues including heart and brain have also been shown to bind ^{125}I-NGF (FRAZIER et al., 1974b, c). Chemical destruction of the sympathetic nervous supply of the heart appears not to change specific binding and it has thus been suggested that the binding sites are part of the tissues themselves and not of sympathetic nerve endings (FRAZIER et al., 1974b). In brain, NGF receptors appear to reside in synaptosomes (SZUTOWICZ et al., 1976a, b). NGF receptors have also been detected using both human and hamster melanoma cells in culture (FABRICANT et al., 1977; SHERWIN et al., 1979). The growth factor has been shown to increase survival of melanoma cells which have been deprived of serum growth factors. Thus, it may be that since melanoma cells have been shown to secrete NGF (MURPHY et al., 1975; SHERWIN et al., 1979), they may also respond functionally to it, perhaps as part of an autoregulatory mechanism. Utilizing embryonic chick sensory ganglia, HERRUP and SHOOTER (HERRUP and SHOOTER, 1975) have shown that binding of ^{125}I-NGF increases at day 8 and then decreases after day 14. This finding can be correlated with the time course of biological responsiveness of these ganglia during development and suggests that ultimate loss of responsiveness may be due to loss of functional receptors (HERRUP and SHOOTER, 1975).

A plasma membrane receptor for NGF has been solubilized by treatment of both sympathetic (BANERJEE et al., 1976) and sensory (ANDRES et al., 1977) ganglia with nonionic detergents. Studies on the receptor solubilized from rabbit superior cervical ganglionic neurons indicate that it is a high affinity binding protein, with molecular weight near 135,000 (COSTRINI and BRADSHAW, 1979; COSTRINI et al., 1979). The equilibrium dissociation constant for the NGF-receptor interaction has been found to be $2-8 \times 10^{-10}$ M, and this value is in good agreement with that observed with intact cellular receptors (COSTRINI and BRADSHAW, 1979).

Two distinct classes of NGF receptors have now been recognized. One of these is the high affinity class associated with the plasma membrane as discussed above. The other is located in the nucleus and appears to be bound to chromatin (ANDRES et al., 1977). Consequently, it has been suggested that NGF may have a bimodal mechanism of action whereby the growth factor first interacts with the plasma membrane, followed by internalization and subsequent interaction with the nucleus. Other studies utilizing rat PC-12 pheochromocytoma cells in culture have also suggested the existence of a site of action at the nucleus (YANKNER and SHOOTER, 1979; MARCHISIO et al., 1980). All of these studies are pertinent to the finding discussed earlier, that NGF is retrogradely transported to the cell bodies of responsive neurons following interaction with the plasma membrane and cellular internalization. On the other hand, it is not entirely clear that NGF does not operate primarily at the level of the plasma membrane, particularly in light of the studies with NGF-Sepharose conjugates (FRAZIER et al., 1973).

I. Other Biological Actions of the 116,000 Molecular Weight NGF

Despite the high concentrations of NGF produced by the submandibular gland, very few studies have focused on possible biologic action(s) of NGF not associated with the neural crest derivatives.

I. Effect of Saliva and NGF on Wounds

HUTSON et al. (1979) demonstrated that mouse submandibular glands have a marked influence on the rate of wound contraction. The rate of healing of experimentally-induced wounds in mice with glands intact was significantly accelerated over that in sialoadenectomized animals. Further, the substance(s) which enhances wound contraction was contained in the submandibular and sublingual gland saliva and applied to the wound area by the licking process. These results were confirmed by LI et al. (1980 b).

Saliva contains high concentrations of the 116,000 molecular weight NGF zymogen (MURPHY et al., 1977 a, 1979). Since granulation tissue has NGF activity (LEVI-MONTALCINI and ANGELETTI, 1961 a) and since fibroblasts, a major cell component of this tissue, secrete NGF in culture (YOUNG et al., 1975), it could be that NGF is a factor that could stimulate wound healing (YOUNG et al., 1975). To test this hypothesis, LI et al. (1980 b) applied physiological concentrations of pure NGF zymogen (YOUNG et al., 1978; YOUNG, 1979) topically to superficial skin wounds of sialoadenectomized mice. The rate of wound contraction was markedly accelerated over that in buffer-treated controls and attained the same rate of decrease in wound area as control animals permitted communal licking. This effect is specific to the enzymically active, high molecular weight form of NGF, and is not mimicked by 2.5 S NGF, DFP-inactivated NGF, trypsin or urokinase. Thus, promotion of wound healing may be one of the physiological functions of NGF in saliva.

II. Effect of Saliva and NGF on Stress Ulcers

Since 116,000 molecular weight NGF (YOUNG et al., 1978) accelerated the rate of healing of external wounds, it was asked whether NGF might play a similar role in the digestive tract (LI et al., 1980 a). Gastric ulcers in mice were induced in 1½ h by physical restraint at 4°. Contrary to expectation, mice stimulated by isoproterenol to secrete more saliva (GOSS, 1978) displayed a significant increase in the number of gastric ulcers produced. Comparable amounts of pure salivary gland NGF administered by stomach tube to sialoadenectomized animals had the same effect. On the other hand, 2.5 S NGF, DFP-inactivated NGF, urokinase, and bovine serum albumin elicited only the control number of gastric lesions. The biological meaning, if any, of this unexpected action of high-molecular weight NGF is unknown.

K. Summary and Perspectives

It has long been believed that NGF is intimately involved in the maintenance and development of certain elements of the sensory and autonomic nervous systems. Certainly, the most compelling reason for this view came from the pioneering work of LEVI-MONTALCINI and coworkers on NGF induced hypertrophy of the sympathetic ganglia as well as the phenomenon of immunosympathectomy. While there is nothing in this review which is intended to suggest that NGF may not function primarily in the nervous system, several features of the problem do not neatly fit together. For example, why is NGF present in such high concentrations in saliva and the venoms of many poisonous snakes? Why do so many cells secrete the factor? Is it part of a mechanism which allows cells to communicate with the nervous system or is it somehow related to a more general regulatory function for cell-cell communication and growth? Why does salivary gland NGF exist as a zymogen? One interesting possibility here is that the molecule is designed to remain inactive until it recognizes its naturally-occurring substrate(s) which can remove Zn(II) and thereby initiate autocatalytic activation.

Finally, it should be emphasized that the majority of biologic studies on NGF have utilized the 26,000 molecular weight 2.5 S (or β) NGF. Yet this protein is but a part of a much larger salivary gland NGF zymogen and of a much larger cell-secreted factor. The predominant form of the naturally-occurring and naturally-secreted salivary gland NGF is not the 2.5 S protein and it may be that this will have to be taken into account before the true physiologic function of NGF can be clearly appreciated.

References

Aloe, L., Levi-Montalcini, R.: Nerve growth factor-induced transformation of immature chromaffin cells in vivo into sympathetic neurons: effect of antiserum to nerve growth factor. Proc. Natl. Acad. Sci. U.S.A. 76, 1246–1250 (1979)

Andres, R.Y., Jeng, I., Bradshaw, R.A.: Nerve growth factor receptors: identification of distinct classes in plasma membranes and nuclei of embryonic dorsal root neurons. Proc. Natl. Acad. Sci. U.S.A. 74, 2785–2789 (1977)

Angeletti, P.U., Levi-Montalcini, R., Caramia, F.: Analysis of the effects of the antiserum to nerve growth factor in adult mice. Brain Res. 27, 343–355 (1971)

Angeletti, P.U., Levi-Montalcini, R., Kettler, R., Theonen, H.: Comparative studies on the effect of the nerve growth factor in sympathetic ganglia and adrenal medulla in newborn rats. Brain Res. 44, 197–206 (1972)

Angeletti, R.H., Bradshaw, R.A.: Nerve growth factor from mouse submaxillary gland; amino acid sequence. Proc. Natl. Acad. Sci. U.S.A. 68, 2417–2420 (1971)

Angeletti, R.H., Bradshaw, R.A., Wade, R.D.: Subunit structure and amino acid composition of mouse submaxillary gland nerve growth factor. Biochemistry 10, 463–469 (1971)

Angeletti, R.H., Hermodson, M.A., Bradshaw, R.A.: Amino acid sequences of mouse 2.5 S nerve growth factor. II. Isolation and characterization of the thermolytic and peptic peptides and the complete covalent structure. Biochemistry 12, 100–115 (1973)

Baker, M.E.: Molecular weight and structure of 7 S nerve growth factor protein. J. Biol. Chem. 250, 1714–1717 (1975)

Banerjee, S.P., Snyder, S.H., Cuatrecasas, P., Greene, L.A.: Binding of nerve growth factor receptor in sympathetic ganglia. Proc. Natl. Acad. Sci. U.S.A. 70, 2519–2523 (1973)

Banerjee, S.P., Cuatrecasas, P., Snyder, S.H.: Solubilization of nerve growth factor receptors of rabbit superior cervical ganglia. J. Biol. Chem. 251, 5680–5685 (1976)

Banks, B.E.C., Walter, S.J.: The effects of postganglionic axotomy and nerve growth factor in the superior cervical ganglion of developing mice. J. Neurocytol. 6, 287–297 (1977)

Banks, B.E.C., Charlwood, K.A., Edwards, D.C., Vernon, C.A., Walter, S.J.: Effects of nerve growth factors from mouse salivary glands and snake venom in the sympathetic ganglia of neonatal and developing mice. J. Physiol. (Lond.) 247, 289–298 (1975)

Banks, B.E.C., Carstairs, J.R., Vernon, C.A.: Differing effects of superior cervical ganglia in neonatal mice produced by antisera to nerve growth factors from mice and snakes. Neuroscience 4, 1145–1155 (1979)

Berger, E.A., Shooter, E.M.: Evidence for pro-β-nerve growth factor, a biosynthetic precursor to β-nerve growth factor. Proc. Natl. Acad. Sci. U.S.A. 74, 3647–3651 (1977)

Bjerre, B., Björklund, A., Mobley, W.: A stimulatory effect by nerve growth factor in the regrowth of adrenergic nerve fibers in the mouse peripheral tissues after chemical sympathectomy with 6-hydroxydopamine. Cell Tissue Res. 146, 15–44 (1973a)

Bjerre, B., Björklund, A., Stenevi, U.: Stimulation of growth of new axonal sprouts from lesioned monoamine neurones in adult rat brain by nerve growth factor. Brain Res. *60*, 161–176 (1973b)

Bjerre, B., Björklund, A., Edwards, D.C.: Axonal regeneration of peripheral adrenergic neurons: effect of antiserum to nerve growth factor in mouse. Cell Tissue Res. *148*, 441–476 (1974a)

Bjerre, B., Björklund, A., Stenevi, U.: Inhibition of the regenerative growth of central noradrenergic neurons by intracerebrally administered anti-NGF serum. Brain Res. *74*, 1–18 (1974b)

Bjerre, B., Björklund, A., Mobley, W., Rosengren, E.: Short- and long-term effect of nerve growth factor on the sympathetic nervous system in the adult mouse. Brain Res. *94*, 263–277 (1975)

Björklund, A., Stenevi, U.: Nerve growth factor: stimulation of regenerative growth of central noradrenergic neurons. Science *175*, 1251–1253 (1972)

Black, I.B.: Regulation of autonomic development. Ann. Rev. Neurosci. *1*, 183–214 (1978)

Bocchini, V., Angeletti, P.U.: The nerve growth factor: purification as a 30,000 molecular weight protein. Proc. Natl. Acad. Sci. U.S.A. *64*, 787–794 (1969)

Bosman, C., Revoltella, R., Bertolini, L.: Phagocytosis of nerve growth factor-coated erythrocytes in neuroblastoma rosette-forming cells. Cancer Res. *35*, 896–905 (1975)

Bothwell, M.A., Shooter, E.M.: Thermodynamics of interaction of the subunits of 7 S nerve growth factor. The mechanism of activation of the esteropeptidase activity by chelators. J. Biol. Chem. *253*, 8458–8464 (1978)

Bradshaw, R.A.: Nerve growth factor. Annu. Rev. Biochem. *47*, 191–216 (1978)

Bradshaw, R.A., Young, M.: Nerve growth factor: recent developments and perspectives. Biochem. Pharmacol. *25*, 1445–1449 (1976)

Brunso-Bechtold, J.K., Hamburger, V.: Retrograde transport of nerve growth factor in chicken embryo. Proc. Natl. Acad. Sci. U.S.A. *76*, 1494–1496 (1979)

Burnham, P.A., Varon, S.: Biosynthetic activities of dorsal root ganglia in vitro and the influence of nerve growth factor. Neurobiology *4*, 57–70 (1974)

Burstein, D.E., Greene, L.A.: Evidence for RNA synthesis-dependent and -independent pathways in stimulation of neurite outgrowth by nerve growth factor. Proc. Natl. Acad. Sci. U.S.A. *75*, 6059–6063 (1978)

Calissano, P., Cozzari, C.: Interaction of nerve growth factor with the mousebrain neurotubule protein(s). Proc. Natl. Acad. Sci. U.S.A. *71*, 2131–2135 (1974)

Calissano, P., Monaco, G., Castellani, L., Mercanti, D., Levi, A.: Nerve growth factor potentiates actomyosin adenosinetriphosphatase. Proc. Natl. Acad. Sci. U.S.A. *75*, 2210–2214 (1978)

Campenot, R.B.: Local control of neurite development by nerve growth factor. Proc. Natl. Acad. Sci. U.S.A. *74*, 4516–4519 (1977)

Chamley, J.H., Dowell, J.J.: Specificity of nerve fibre "attraction" to autonomic effector organs in tissue cultures. Exp. Cell Res. *90*, 1–7 (1975)

Cohen, S.: Purification of a nerve-growth promoting protein from the mouse salivary gland and its neuro-cytotoxic antiserum. Proc. Natl. Acad. Sci. U.S.A. *46*, 302–311 (1960)

Connolly, J.L., Greene, L.A., Viscarello, R.R., Riley, W.D.: Rapid, sequential changes in surface morphology of PC12 pheochromocytoma cells in response to nerve growth factor. J. Cell Biol. *82*, 820–827 (1979)

Costrini, N.V., Bradshaw, R.A.: Binding characteristics and apparent molecular size of detergent-solubilized nerve growth factor receptor of sympathetic ganglia. Proc. Natl. Acad. Sci. U.S.A. *76*, 3242–3245 (1979)

Costrini, N.V., Kogan, M., Kukreja, K., Bradshaw, R.A.: Physical properties of the detergent-extracted NGF receptor of sympathetic ganglia. J. Biol. Chem. *254*, 11242–11246 (1979)

Crain, S.M., Wiegand, R.G.: Catecholamine levels of mouse sympathetic ganglia following hypertrophy produced by salivary nerve-growth factor. Proc. Soc. Exp. Biol. Med. *107*, 663–665 (1961)

Dichter, M.A., Tischler, A.S., Greene, L.A.: Nerve growth factor-induced increase in electrical excitability and acetylcholine sensitivity of a rat pheochromocytoma cell line. Nature *268*, 501–504 (1977)

Ebendal, T., Jacobson, C.-O.: Tissue explants affecting extension and orientation of axons in cultured chick embryo ganglia. Exp. Cell Res. *105*, 379–387 (1977)

Edgar, D.H., Thoenen, H.: Selective enzyme induction in a nerve growth factor-responsive pheochromocytoma cell line (PC12). Brain Res. *154*, 186–190 (1978)

Edwards, D.C., Fenton, E.L., Kakeri, S., Large, B.J., Papadaki, L., Zaimis, E.: Effects of nerve growth factor in new-born mice, rats, and kittens. J. Physiol. (Lond.) *186*, 10 P– 12 P (1966)

Ennis, M., Pearce, F.L., Vernon, C.A.: Some studies on the mechanism of action of antibodies to nerve growth factor. Neuroscience *4*, 1391–1398 (1979)

Fabricant, R.N., Delarco, J.E., Todaro, G.J.: Nerve growth factor receptors on human melanoma cells in culture. Proc. Natl. Acad. Sci. U.S.A. *74*, 565–569 (1977)

Frazier, W.A., Angeletti, R.H., Bradshaw, R.A.: Nerve growth factor and insulin. Structure similarities indicate an evolutionary relationship reflected by physiological action. Science *176*, 482–488 (1972)

Frazier, W.A., Boyd, L.F., Bradshaw, R.A.: Interaction of nerve growth factor with surface membranes: biological competence of insolubilized nerve growth factor. Proc. Natl. Acad. Sci. U.S.A. *70*, 2931–2935 (1973)

Frazier, W.A., Boyd, L.F., Bradshaw, R.A.: Properties of the specific binding of ^{125}I-nerve growth factor to responsive peripheral neurons. J. Biol. Chem. *249*, 5513–5519 (1974a)

Frazier, W.A., Boyd, L.F., Szutowicz, A., Pulliam, M.W., Bradshaw, R.A.: Specific binding sites for ^{125}I-nerve growth factor in peripheral tissues and brain. Biochem. Biophys. Res. Commun. *57*, 1096–1103 (1974b)

Frazier, W.A., Boyd, L.F., Pulliam, M.W., Szutowicz, A., Bradshaw, R.A.: Properties and specificity of binding sites for ^{125}I-nerve growth factor in embryonic heart and brain. J. Biol. Chem. *249*, 5918–5923 (1974c)

Garrels, J.I., Schubert, D.: Modulation of protein synthesis by nerve growth factor. J. Biol. Chem. *254*, 7978–7985 (1979)

Goldstein, M.N., Brodeur, G.M.: The effect of nerve growth factor and dibutyryl cyclic AMP on acetylcholinesterase in human and mouse neuroblastomas. Anat. Rec. *175*, 330 (1973)

Gorin, P.D., Johnson, E.M.: Experimental autoimmune model of nerve growth factor deprivation: Effects on developing peripheral sympathetic and sensory neurons. Proc. Natl. Acad. Sci. U.S.A. *76*, 5382–5386 (1979)

Goss, R.J.: The physiology of growth, p. 240. New York: Academic Press Inc. 1978

Greene, L.A., McGuire, J.C.: Induction of ornithine decarboxylase by nerve growth factor dissociated from effects on survival and neurite outgrowth. Nature *276*, 191–194 (1978)

Greene, L.A., Tischler, A.S.: Establishment of a noradrenergic clonal line of rat adrenal pheochromocytoma cells which respond to nerve growth factor. Proc. Natl. Acad. Sci. U.S.A. *73*, 2424–2428 (1976)

Greene, L.A., Shooter, E.M., Varon, S.: Enzymatic activities of mouse nerve growth factor and its subunits. Proc. Natl. Acad. Sci. U.S.A. *60*, 1383–1388 (1968)

Greene, L.A., Shooter, E.M., Varon, S.: Subunit interaction and enzymatic activity of mouse 7S nerve growth factor. Biochemistry *8*, 3735–3741 (1969)

Gundersen, R.W., Barrett, J.N.: Neuronal chemotaxis: Chick dorsal-root axons turn toward high concentrations of nerve growth factor. Science *206*, 1079–1080 (1979)

Harper, G.P., Pearce, F.L., Vernon, C.A.: Production of nerve growth factor by the mouse adrenal medulla. Nature *261*, 251–253 (1976)

Hatanaka, H., Otten, U., Thoenen, H.: Nerve growth factor-mediated selective induction of ornithine decarboxylase in rat pheochromocytoma; a cyclic AMP-independent process. FEBS Letters *92*, 313–316 (1978)

Hendry, I.A.: The response of adrenergic neurones to axotomy and nerve growth factor. Brain Res. *94*, 87–97 (1975)

Hendry, I.A.: Control in the development of the vertebrate sympathetic nervous system. Rev. Neurosci. *2*, 149–194 (1976)

Hendry, I.A., Iversen, L.L.: Effect of nerve growth factor and its antiserum on tyrosine hydroxylase activity in mouse superior cervical sympathetic ganglion. Brain Res. *29*, 159– 162 (1971)

Hendry, I.A., Iversen, L.L.: Reduction in the concentration of nerve growth factor in mice after sialectomy and castration. Nature *243*, 500–504 (1973)

Hendry, I.A., Stach, R., Herrup, K.: Characteristics of the retrograde axonal transport system for nerve growth factor in the sympathetic nervous system. Brain Res. *82*, 117–128 (1974a)

Hendry, I.A., Stöckel, K., Thoenen, H., Iversen, L.L.: The retrograde axonal transport of nerve growth factor. Brain Res. *68*, 103–121 (1974b)

Herrup, K., Shooter, E.M.: Properties of the β nerve growth factor receptor of avian dorsal root ganglia. Proc. Natl. Acad. Sci. U.S.A. *70*, 3884–3888 (1973)

Herrup, K., Shooter, E.M.: Properties of the β-nerve growth factor receptor in development. J. Cell Biol. *67*, 118–125 (1975)

Hogue-Angeletti, R.A., Bradshaw, R.A.: Snake Venoms. In: Handbook exp. pharmacology. Lee, C.Y. (ed.), pp. 361–371. Berlin, Heidelberg, New York: Springer 1977

Hogue-Angeletti, R.A., Bradshaw, R.A., Frazier, W.A.: Nerve growth factor: structure and mechanism of action. Adv. Metab. Disord. *8*, 285–299 (1975)

Hogue-Angeletti, R.A., Frazier, W.A., Jacobs, J.W., Niall, H.D., Bradshaw, R.A.: Purification, characterization, and partial amino acid sequence of nerve growth factor from cobra venom. Biochemistry *15*, 26–34 (1976)

Horii, Z.-I., Varon, S.: Nerve growth factor action on membrane permeation to exogenous substrates in dorsal root ganglionic dissociates from the chick embryo. Brain Res. *124*, 121–133 (1977)

Huff, K., Lakshmanan, J., Guroff, G.: RNA polymerase activity in the superior cervical ganglion of the neonatal rat: the effect of nerve growth factor. J. Neurochem. *31*, 599–606 (1978)

Hutson, J.M., Niall, M., Evans, D., Fowler, R.: Effect of salivary glands on wound contraction in mice. Nature *279*, 793–795 (1979)

Ikeno, T., Guroff, G.: Growth regulation by nerve growth factor. Mol. Cell. Biochem. *28*, 67–91 (1979)

Ishii, D.N., Shooter, E.M.: Regulation of nerve growth factor synthesis in mouse submaxillary glands by testosterone. J. Neurochem. *25*, 843–851 (1975)

Iversen, L.L., Stöckel, K., Thoenen, H.: Autoradiographic studies of the retrograde axonal transport of nerve growth factor in mouse sympathetic neurones. Brain Res. *88*, 37–43 (1975)

Johnson, E.M. jr.: Destruction of the sympathetic nervous system in neonatal rats and hamsters by vinblastine: Prevention by concomitant administration of nerve growth factor. Brain Res. *141*, 105–118 (1978)

Johnson, E.M. jr., Andres, R.Y., Bradshaw, R.A.: Characterization of the retrograde transport of nerve growth factor (NGF) using high specific activity [^{125}I]NGF. Brain Res. *150*, 319–331 (1978)

Kessler, J.A., Black, I.B.: Nerve growth factor stimulates the development of substance P in sensory ganglia. Proc. Natl. Acad. Sci. U.S.A. *77*, 649–652 (1980)

Klingman, G., Klingman, J.: Catecholamines in peripheral tissues of mice and cell counts of sympathetic ganglion after the prenatal and postnatal administration of nerve growth factor antiserum. Int. J. Neuropharmacol. *6*, 501–508 (1967)

Kolber, A.R., Goldstein, M.N., Moore, B.W.: Effect of nerve growth factor on the expression of colchicine-binding activity and 14-3-2 protein in an established line of human neuroblastoma. Proc. Natl. Acad. Sci. U.S.A. *71*, 4203–4207 (1974)

Landreth, G., Cohen, P., Shooter, E.M.: Ca^{2+} transmembrane fluxes and nerve growth factor action on a clonal cell line of rat phaeochromocytoma. Nature *283*, 202–204 (1980)

Larrabee, M.G.: Metabolism during development in sympathetic ganglia of chickens: effects of age, nerve growth factor and metabolic inhibitors. In: Nerve growth factor and its antiserum. Zaimis, E. (ed.), pp. 71–87. London: Athlone Press 1972

Letourneau, P.C.: Chemotactic response of nerve fiber elongation to nerve growth factor. Dev. Biol. *66*, 183–196 (1978)

Levi-Montalcini, R.: Effects of mouse tumor transplantation on the nervous system. Ann. N.Y. Acad. Sci. *55*, 330–343 (1952)

Levi-Montalcini, R.: The nerve growth factor. Ann. N.Y. Acad. Sci. *118*, 149–170 (1964)

Levi-Montalcini, R.: The nerve growth factor: its mode of action on sensory and sympathetic nerve cells. Harvey Lect. *60*, 217–259 (1965)

Levi-Montalcini, R.: Growth control of the nervous system. Ciba Foundation Symposium. Wolstenholme, G.E.W., O'Connor, M. (eds.), pp. 126–142. Boston: Little, Brown & Co. 1968

Levi-Montalcini, R.: The nerve growth factor: its role in growth, differentiation and function of the sympathetic adrenergic neuron. Prog. Brain Res. *45*, 235–258 (1976)

Levi-Montalcini, R.: Recent studies on the NGF-target cells interaction. Differentiation *13*, 51–53 (1979)

Levi-Montalcini, R., Angeletti, P.U.: Biological properties of a nerve growth promoting protein and its antiserum. In: Proc. 4th Intern. Neurochemical Symp. on Regional Neurochemistry. Kety, S.S., Elkes, J. (eds.), pp. 362–376. New York: Pergamon Press 1961 a

Levi-Montalcini, R., Angeletti, P.U.: Growth control of the sympathetic system by a specific protein factor. Q. Rev. Biol. *36*, 99–108 (1961 b)

Levi-Montalcini, R., Angeletti, P.U.: Hormonal control of the NGF content in the submaxillary salivary glands of mouse. In: Salivary glands and their secretions. Scheebny, L.M., Meyer, J. (eds.), pp. 129–141. Oxford: Pergamon Press 1964

Levi-Montalcini, R., Angeletti, P.U.: Immunosympathectomy. Pharmacol. Rev. *18*, 619–628 (1966)

Levi-Montalcini, R., Angeletti, P.U.: Nerve growth factor. Physiol. Rev. *48*, 534–569 (1968)

Levi-Montalcini, R., Booker, B.: Excessive growth of the sympathetic ganglia evoked by a protein isolated from mouse salivary gland. Proc. Natl. Acad. Sci. U.S.A. *46*, 373–384 (1960 a)

Levi-Montalcini, R., Booker, B.: Destruction of the sympathetic ganglia in mammals by an antiserum to a nerve-growth protein. Proc. Natl. Acad. Sci. U.S.A. *46*, 384–391 (1960 b)

Levi-Montalcini, R., Revoltella, R., Calissano, P.: Microtubule proteins in the nerve growth factor mediated response. Interaction between the nerve growth factor and its target cells. Recent Prog. Horm. Res. *30*, 635–669 (1974)

Levi-Montalcini, R., Aloe, L., Mugnaini, E., Oesch, F., Thoenen, H.: Nerve growth factor induces volume increases and enhances tyrosine hydroxylase synthesis in chemically axotomized sympathetic ganglia of newborn rats. Proc. Natl. Acad. Sci. U.S.A. *72*, 595–599 (1975)

Lewis, M.E., Lakshmanan, J., Nagaiah, K., MacDonnell, P.C., Guroff, G.: Nerve growth factor increases activity of ornithine decarboxylase in rat brain. Proc. Natl. Acad. Sci. U.S.A. *75*, 1021–1023 (1978)

Lewis, M.E., Avrith, D.B., Fitzsimons, J.T.: Short latency drinking and increased Na appetite after intracerebral microinjections of NGF in rats. Nature *279*, 440–442 (1979)

Li, A.K.C., Koroly, M.J., Jeppsson, B.W., Schattenkerk, M.E., Young, M., Malt, R.A.: Nerve growth factor potentiates stress-induced gastric ulceration. Submitted for publication, 1980 a

Li, A.K.C., Koroly, M.J., Schattenkerk, M.E., Malt, R.A., Young, M.: Nerve growth factor: acceleration of the rate of wound healing in mice. Proc. Natl. Acad. Sci. U.S.A. *77*, 4379–4381 (1980 b)

Luckenbill-Edds, L., Van Horn, C., Greene, L.A.: Fine structure of initial outgrowth of processes induced in a pheochromocytoma cell line (PC 12) by nerve growth factor. J. Neurocytol. *8*, 493–511 (1979)

MacDonnell, P.C., Nagaiah, K., Lakshmanan, J., Guroff, G.: Nerve growth factor increases activity of ornithine decarboxylase in superior cervical ganglia of young rats. Proc. Natl. Acad. Sci. *74*, 4681–4684 (1977 a)

MacDonnell, P.C., Tolson, N., Guroff, G.: Selective de novo synthesis of tyrosine hydroxylase in organ culture of rat superior cervical ganglia after in vivo administration of nerve growth factor. J. Biol. Chem. *252*, 5859–5863 (1977 b)

Marchisio, P.C., Naldini, L., Calissano, P.: Intracellular distribution of nerve growth factor in rat pheochromocytoma PC 12 cells: evidence for a perinuclear and intranuclear location. Proc. Natl. Acad. Sci. U.S.A. *77*, 1656–1660 (1980)

Max, S.R., Rohrer, H., Otten, U., Thoenen, H.: Nerve growth-mediated induction of tyrosine hydroxylase in rat superior cervical ganglia in vitro. J. Biol. Chem. *253*, 8013–8015 (1978)

McGuire, J.C., Greene, L.A.: Rapid stimulation by nerve growth factor of amino acid uptake by clonal PC 12 Pheochromocytoma cells. J. Biol. Chem. *254*, 3362–3367 (1979)

Menesini-Chen, M.G., Chen, J.S., Calissano, P., Levi-Montalcini, R.: Nerve growth factor prevents vinblastine destructive effects on sympathetic ganglia in newborn mice. Proc. Natl. Acad. Sci. U.S.A. *74*, 5559–5563 (1977)

Mizel, S.B., Bamburg, J.R.: Studies on the action of nerve growth factor. II. Neurotubule protein levels during neurite outgrowth. Neurobiology *5*, 283–290 (1975)

Mobley, W.C., Server, A.C., Ishii, D.N., Riopelle, R.J., Shooter, E.M.: Nerve growth factor. New Engl. J. Med. *297*, Three parts: 1096–1104, 1149–1158, 1211–1218 (1977)

Moore, J.B. jr., Mobley, W.C., Shooter, E.M.: Proteolytic modification of the β Nerve growth factor protein. Biochemistry *13*, 833–840 (1974)

Murphy, R.A., Pantazis, N.J., Arnason, B.G.W., Young, M.: Secretion of a nerve growth factor by mouse neuroblastoma cells in culture. Proc. Natl. Acad. Sci. U.S.A. *72*, 1895–1898 (1975)

Murphy, R.A., Saide, J.D., Blanchard, M.H., Young, M.: Nerve growth factor in mouse serum and saliva; role of the submandibular gland. Proc. Natl. Acad. Sci. U.S.A. *74*, 2330–2333 (1977a)

Murphy, R.A., Saide, J.D., Blanchard, M.H., Young, M.: Molecular properties of the nerve growth factor secreted in mouse saliva. Proc. Natl. Acad. Sci. U.S.A. *74*, 2672–2676 (1977b)

Murphy, R.A., Singer, R.H., Saide, J.D., Pantazis, N.J., Blanchard, M.H., Byron, K.S., Arnason, B.G.W., Young, M.: Synthesis and secretion of a high molecular weight form of nerve growth factor by skeletal muscle cells in culture. Proc. Natl. Acad. Sci. U.S.A. *74*, 4496–4500 (1977c)

Murphy, R.A., Pantazis, N.J., Papastavros, M.: Epidermal growth factor and nerve growth factor in mouse saliva: a comparative study. Dev. Biol. *71*, 356–370 (1979)

Nagaiah, K., MacDonnell, P., Guroff, G.: Induction of tyrosine hydroxylase synthesis in rat superior cervical ganglia in vitro by nerve growth factor and dexamethasone. Biochem. Biophys. Res. Commun. *75*, 832–837 (1977)

Nagaiah, K., Ikeno, T., Lakshmanan, J., MacDonnell, P., Guroff, G.: Intraventricular administration of nerve growth factor induces ornithine decarboxylase in peripheral tissues of the rat. Proc. Natl. Acad. Sci. U.S.A. *75*, 2512–2515 (1978)

Oger, J., Arnason, B.G.W., Pantazis, N.J., Lehrich, J.R., Young, M.: Synthesis of nerve growth factor by L and 3T3 cells in culture. Proc. Natl. Acad. Sci. U.S.A. *71*, 1554–1558 (1974)

Orenstein, N.S., Dvorak, H.F., Blanchard, M.H., Young, M.: Nerve growth factor: a protease that can activate plasminogen. Proc. Natl. Acad. Sci. U.S.A. *75*, 5497–5500 (1978)

Otten, U., Thoenen, H.: Modulatory role of glucocorticoids on NGF-mediated enzyme induction in organ cultures of sympathetic ganglia. Brain Res. *111*, 438–441 (1976)

Otten, U., Schwab, M., Gagnon, C., Thoenen, H.: Selective induction of tyrosine hydroxylase and dopamine β-hydroxylase by nerve growth factor: comparison between adrenal medulla and sympathetic ganglia of adult and newborn rats. Brain Res. *133*, 291–303 (1977)

Otten, U., Baumann, J.B., Girard, J.: Stimulation of the pituitary-adrenocortical axis by NGF. Nature *282*, 413–414 (1979a)

Otten, U., Goedert, M., Schwab, M., Thibault, J.: Immunization of adult rats against 2.5 S NGF: effects on the peripheral nervous system. Brain Res. *176*, 79–90 (1979b)

Pantazis, N.J., Blanchard, M.H., Arnason, B.G.W., Young, M.: Molecular properties of the nerve growth factor secreted by L cells. Proc. Natl. Acad. Sci. U.S.A. *74*, 1492–1496 (1977a)

Pantazis, N.J., Murphy, R.A., Saide, J.D., Blanchard, M.H., Young, M.: Dissociation of the 7 S-nerve growth factor complex in solution. Biochemistry *16*, 1525–1530 (1977b)

Paravicini, U., Stoeckel, K., Thoenen, H.: Biological importance of retrograde axonal transport of nerve growth factor in adrenergic neurons. Brain Res. *84*, 279–291 (1975)

Partlow, L.M., Larrabee, M.G.: Effects of a nerve-growth factor, embryo age and metabolic inhibitors on growth of fibres and on synthesis of ribonucleic acid and protein in embryonic sympathetic ganglia. J. Neurochem. *18*, 2101–2118 (1971)

Pattison, S.E., Dunn, M.F.: On the relationship of zinc ion to the structure and function of the 7 S nerve growth factor protein. Biochemistry *14*, 2733–2739 (1975)

Pattison, S.E., Dunn, M.F.: On the mechanism of divalent metal ion chelator induced activation of the 7S nerve growth factor esteropeptidase. Activation by 2,2′,2″-Terpyridine and by 8-Hydroxyquinoline-5-sulfonic acid. Biochemistry *15*, 3691–3696 (1976a)

Pattison, S.E., Dunn, M.F.: On the mechanism of divalent metal ion chelator induced activation of the 7 S nerve growth factor esteropeptidase. Thermodynamics and kinetics of activation. Biochemistry *15*, 3696–3703 (1976b)

Pearce, F.L., Banks, B.E., Banthorpe, D.V., Berry, A.R., Davies, H.S., Vernon, C.A.: The isolation and characterization of nerve growth factor from the venom of *Vipera russelli*. Eur. J. Biochem. *29*, 417–425 (1972)

Revoltella, R., Bertolini, L., Pediconi, M., Vigneti, E.: Specific binding of nerve growth factor (NGF) by murine C 1300 neuroblastoma cells. J. Exp. Med. *140*, 437–451 (1974)

Revoltella, R., Bosman, C., Bertolini, L.: Detection of nerve growth factor sites on neuroblastoma cells by rosette formation. Cancer Res. *35*, 890–895 (1975)

Robbins, K.C., Summaria, L., Hsieh, B., Shah, R.J.: The peptide chains of human plasma. J. Biol. Chem. *242*, 2333–2342 (1967)

Rohrer, H., Otten, U., Thoenen, H.: On the role of RNA synthesis in the selective induction of tyrosine hydroxylase by nerve growth factor. Brain Res. *159*, 436–439 (1978)

Schubert, D., Whitlock, C.: Alteration of cellular adhesion by nerve growth factor. Proc. Natl. Acad. Sci. U.S.A. *74*, 4055–4058 (1977)

Schubert, D., Lacorbiere, M., Whitlock, C., Stallcup, W.: Alterations in the surface properties of cells responsive to nerve growth factor. Nature *273*, 718–723 (1978)

Schwab, M.E., Otten, U., Agid, Y., Thoenen, H.: Nerve growth factor in the rat CNS: absence of specific retrograde axonal transport and tyrosine hydroxylase induction in locus coeruleus and substantia nigra. Brain Res. *168*, 473–483 (1979)

Schwartz, J.P., Costa, E.: Nerve growth factor-mediated increase of the substance P content of chick embryo dorsal root ganglia. Brain Res. *170*, 198–202 (1979)

Server, A.C., Shooter, E.M.: Nerve growth factor. Adv. Prot. Chem. *31*, 339–409 (1978)

Sherwin, S.A., Sliski, A.H., Todaro, G.J.: Human melanoma cells have both nerve growth factor and nerve growth factor-specific receptors on their cell surfaces. Proc. Natl. Acad. Sci. U.S.A. *76*, 1288–1292 (1979)

Skaper, S.D., Varon, S.: Sodium dependence of the nerve growth factor-regulated hexose uptake in chick embryo ganglionic cells. Brain Res. *172*, 303–313 (1979)

Smith, A.P., Varon, S., Shooter, E.M.: Multiple forms of the nerve growth factor protein and its subunits. Biochemistry *7*, 3259–3268 (1968)

Smith, A.P., Greene, L.A., Fisk, H.R., Varon, S., Shooter, E.M.: Subunit equilibria of the 7 S nerve growth factor protein. Biochemistry *8*, 4918–4926 (1969)

Stöckel, K., Paravicini, U., Thoenen, H.: Specificity of the retrograde axonal transport of nerve growth factor. Brain Res. *76*, 413–421 (1974a)

Stöckel, K., Solomon, F., Paravicini, U., Thoenen, H.: Dissociation between effects of nerve growth factor on tyrosine hydroxylase and tubulin synthesis in sympathetic ganglia. Nature *250*, 150–151 (1974b)

Stöckel, K., Thoenen, H.: Retrograde axonal transport of nerve growth factor: specificity and biological importance. Brain Res. *85*, 337–341 (1975)

Stöckel, K., Schwab, M., Thoenen, H.: Specificity of retrograde transport of nerve growth factor (NGF) in sensory neurons: a biochemical and morphological study. Brain Res. *89*, 1–14 (1975a)

Stöckel, K., Schwab, M., Thoenen, H.: Comparison between the retrograde axonal transport of nerve growth factor and tetanus toxin in motor, sensory and adrenergic neurons. Brain Res. *99*, 1–16 (1975b)

Stöckel, K., Guroff, G., Schwab, M., Thoenen, H.: The significance of retrograde transport for the accumulation of systemically administered nerve growth factor in the rat superior cervical ganglion. Brain Res. *109*, 271–284 (1976)

Szutowicz, A., Frazier, W.A., Bradshaw, R.A.: Subcellular localization of nerve growth factor receptors. J. Biol. Chem. *251*, 1516–1523 (1976a)

Szutowicz, A., Frazier, W.A., Bradshaw, R.A.: Subcellular localization of nerve growth factor receptors. J. Biol. Chem. *251*, 1524–1528 (1976b)

Tanzier, J.P., Thoenen, H., Snipes, R.L., Richards, J.G.: Recent developments in the ultrastructure of adrenergic nerve endings in various experimental conditions. Prog. Brain Res. *31*, 33–46 (1969)

Thoenen, H., Angeletti, P.U., Levi-Montalcini, R., Kettler, R.: Selective induction by nerve growth factor of tyrosine hydroxylase and dopamine-β-hydroxylase in the rat superior cervical ganglia. Proc. Natl. Acad. Sci. U.S.A. *68*, 1598–1602 (1971)

Tischler, A.S., Greene, L.A.: Nerve growth factor-induced process formation by cultured rat phaeochromocytoma cells. Nature *258*, 341–342 (1975)

Tu, A.T.: Venoms: chemistry and molecular biology, pp. 361–371. New York: John Wiley and Sons Inc. 1977

Turner, J.E., Delaney, R.K.: Retinal ganglion cell response to axotomy and nerve growth factor in the regenerating visual system of the newt *(Notophthalmus viridescens)*: an ultrastructural morphometric analysis. Brain Res. *171*, 197–212 (1979)

Unkeless, J., Dan, K., Kellerman, G.M., Reich, E.: Fibrinolysis associated with oncogenic transformation. Partial purification and characterization of the cell factor, a plasminogen activator. J. Biol. Chem. *249*, 4295–4305 (1974)

Unsicker, K., Krisch, B., Otten, U., Thoenen, H.: Nerve growth factor-induced fiber outgrowth from isolated rat adrenal chromaffin cells: impairment by glucocorticoids. Proc. Natl. Acad. Sci. U.S.A. *75*, 3498–3502 (1978)

Varon, S., Bunge, R.P.: Trophic mechanisms in the peripheral nervous system. Ann. Rev. Neurosci. *1*, 327–361 (1978)

Varon, S., Nomura, J., Shooter, E.M.: The isolation of the mouse nerve growth factor protein in a high molecular weight form. Biochemistry *6*, 2202–2209 (1967)

Varon, S., Nomura, J., Shooter, E.M.: Reversible dissociation of the mouse nerve growth factor protein into different subunits. Biochemistry *7*, 1296–1303 (1968)

Wallace, L.J., Partlow, L.M.: α-Adrenergic regulation of secretion of mouse saliva rich in nerve growth factor. Proc. Natl. Acad. Sci. U.S.A. *73*, 4210–4214 (1976)

Yankner, B.A., Shooter, E.M.: Nerve growth factor in the nucleus: interaction with receptors on the nuclear membrane. Proc. Natl. Acad. Sci. U.S.A. *76*, 1269–1273 (1979)

Young, M.: Proteolytic activity of nerve growth factor: a case of autocatalytic activation. Biochemistry *18*, 3050–3055 (1979)

Young, M., Koroly, M.J.: The nerve growth factor zymogen. Stoichiometry of the active site serine and the role of Zn(II) in controlling autocatalytic selfactivation. Biochemistry *19*, 5316–5321 (1980)

Young, M., Oger, J., Blanchard, M.H., Asdourian, H., Amos, H., Arnason, B.G.W.: Secretion of a nerve growth factor by primary chick fibroblast cultures. Science *187*, 361–362 (1975)

Young, M., Murphy, R.A., Saide, J.D., Pantazis, N.J., Blanchard, M.H., Arnason, B.G.W.: Studies on the molecular properties of nerve growth factor and its cellular biosynthesis and secretion. In: Surface membrane receptors. Bradshaw, R.A., Frazier, W.A., Merrell, R.C., Gottlieb, D.I., Hogue-Angeletti, R.A. (eds.), pp. 247–267. New York: Plenum Press 1976

Young, M., Saide, J.D., Murphy, R.A., Blanchard, M.H.: Nerve growth factor: multiple dissociation products in homogenates of the mouse submandibular gland. Purification and molecular properties of the intact undissociated form of the protein. Biochemistry *17*, 1490–1498 (1978)

Young, M., Blanchard, M.H., Saide, J.D.: NGF: radioimmunoassay and bacteriophage immunoassay. In: Methods of hormone radioimmunoassay. Jaffe, B.M., Behrman, H.R. (eds.), pp. 941–958. New York: Academic Press 1979

Yu, M.W., Nikodijevic, B., Lakshmanan, J., Rowe, V., MacDonnell, P., Guroff, G.: Nerve growth factor and the activity of tyrosine hydroxylase in organ cultures of rat superior cervical ganglia. J. Neurochem. *28*, 835–842 (1977)

Yu, M.W., Horii, S., Tolson, N., Huff, K., Guroff, G.: Increased phosphorylation of a specific nuclear protein in rat superior cervical ganglia in response to nerve growth factor. Biochem. Biophys. Res. Commun. *81*, 941–946 (1978)

Zaimis, E., Knight, J., editors: Nerve growth factor and its antiserum. London: Athlone Press of the University of London 1972

Zaimis, E., Birk, L., Callingham, B.A.: Morphological, biochemical and functional changes in the sympathetic nervous system of rats treated with nerve growth factor-antiserum. Nature *206*, 1220–1222 (1965)

CHAPTER 9

The Role of Cold Insoluble Globulin (Plasma Fibronectin) in Cell Adhesion In Vitro

F. GRINNELL

A. Introduction

It is appropriate that a volume on cell growth factors include a chapter on factors in cell adhesion. Throughout the early phases during which cell culture techniques were developed, cells were almost always grown attached to extracellular substrata and rarely grown in suspension. The role of adhesion in growth control became of particular interest following the observation that transformed cell variants could be selected from cell populations by growth in soft agar (MACPHERSON and MONTAGNIER, 1964). Subsequently, it was demonstrated that normal cells required attachment to a substratum in order to grow, a property that was called "anchorage dependence" (STOKER et al., 1968). To this day, the best in vitro correlate of malignant transformation is still loss of "anchorage dependence" (e.g. KAHN and SHIN, 1979). Lymphocytes are the only normal cells whose growth is apparently anchorage independent and adhesion of these cells to culture vessels is normally very tenuous.

Experiments have been carried out to determine if anchorage dependence can be satisfied by cell attachment to a substratum per se or if cell spreading is required for growth. When cells were grown on glass beads or fibers of different dimensions, it was found that cell spreading was the critical parameter. That is, cells were not able to grow unless attached to particles that were sufficiently large to permit cell spreading (MAROUDAS, 1972, 1973a, b). The requirement for cell spreading was studied thoroughly in experiments employing modified tissue culture dishes on which cells spread to different degrees (FOLKMAN and MOSCONA, 1978). There was a marked correlation between the degree of cell spreading and the ability of the cells to enter S phase.

The molecular basis for the cell spreading requirement in cell growth is unknown. It is possible that spread cells bind or respond better than suspended cells to hormone-like serum factors critical for cell growth (e.g. HOLLY, 1975). In support of this idea is the observation that some cells that are normally anchorage dependent will grow in suspension if the serum concentration is very high (PAUL et al., 1971; O'NEILL et al., 1979). However, it should be pointed out that cell spreading has pleotropic effects on cell surface related properties and receptors, more than one of which may be important in growth control. Compared to round cells, cells which are spread demonstrate increases in transport rates (FOSTER and PARDEE, 1969; OTSUKA and MOSKOWITZ, 1975), disappearance of cell surface anionic sites (GRINNELL et al., 1975, 1976), alteration in mobility of con A binding sites (WRIGHT and KARNOVSKY, 1979), and loss of adhesiveness of the upper cell surface (DI PASQUALE and BELL, 1974; VASILIEV and GELFAND, 1976).

Because cell adhesion is required for cell growth, there has been considerable interest in understanding the factors which play a role in adhesion (for a recent comprehensive review see Grinnell, 1978). During the past several years, it has become clear that cold insoluble globulin (CIG), a 440,000 dalton glycoprotein present in plasma at a concentration of about 0.25 gms/l (Mosesson and Umfleet, 1970; Mosesson et al., 1975), is an important adhesion factor for fibroblasts as well as other cell types. The purpose of this review is to describe the role of CIG in cell adhesion to tissue culture dishes and to other surfaces such as bacteriological dishes, collagen, and fibrin. Some experimental and technical data relevant to the activity and purification of CIG are presented. Also, there is a section on the adhesion and spreading of cells which do not require CIG but secrete their own adhesion factors. Finally, a discussion of the mechanism of action of CIG is included.

B. Cold Insoluble Globulin and Fibronectin: Terminology

In this review, CIG refers to the plasma protein and fibronectin refers to the cell surface protein which is closely related to CIG [for reviews on fibronectin see Vaheri and Mosher (1978); Yamada and Olden (1978)]. The distinction between CIG and fibronectin is made because the precise relationship between CIG and fibronectin and the exact roles which each play in situ are unresolved problems. Current evidence suggests that CIG and fibronectin exhibit somewhat distinct biological activities (Yamada and Kennedy, 1979) despite their immunological cross reactivity and many other similar physical properties (Alexander et al., 1978, 1979). The CIG/fibronectin antigen is located predominantly in the connective tissue and basement membranes of adult tissues (Stenman and Vaheri, 1978; Schachner et al., 1978; Fryand, 1979); however, since the antisera which have been used do not distinguish CIG from fibronectin, it is not clear which molecule is being detected in these studies. Finally, fibronectin is synthesized by a variety of different cell types including fibroblasts (Ruoslahti et al., 1973), endothelial cells (Jaffe and Mosher, 1978; Macarak et al., 1978; Birdwell et al., 1978), and epithelial cells (Smith et al., 1979), but the source of CIG in plasma has not been established.

C. The Role of CIG in Cell Adhesion to Tissue Culture Dishes

I. Historical Overview

The question of whether serum proteins are required for cell adhesion was controversial. In a 1960 symposium, results from one laboratory were presented demonstrating that serum promoted adhesion (Weiss, 1961) and from another demonstrating that serum inhibited adhesion (Taylor, 1961). Around the same time, a report appeared suggesting that the fetal serum component called fetuin was an adhesion promoting substance (Fisher et al., 1958); however, ion exchange chromatography of fetuin revealed that the active factor was a contaminating glycoprotein (Lieberman et al., 1959). In retrospect, the properties reported for the partially purified adhesion glycoprotein indicate that it was probably CIG.

The problem in understanding the precise role of serum in adhesion developed because of the failure to recognize that initial cell attachment can occur either actively or passively depending upon whether or not serum is present in the incubations [see GRINNELL (1978) for a review]. Briefly, in the presence of serum, protein components adsorb to the material surface, e.g. plastic or glass, and form the substratum sites for adhesion. Attachment of cells in the presence of serum is active. That is, it does not occur at 4 °C or if cells are treated with inhibitors of glycolysis and oxidative phosphorylation; fixatives; sulfhydryl binding reagents; or high concentrations of trypsin. Cell attachment in the absence of serum is passive; it is unaffected by any of the above treatments. Cell spreading is active since it can be inhibited by the same conditions that prevent active cell attachment.

II. Identification of CIG as the Serum Factor in Cell Attachment and Spreading

A systematic analysis of serum, to determine what factors were required for fibroblast adhesion to tissue culture dishes, was initiated several years ago using an assay for fibroblast cell spreading (GRINNELL, 1976 a, b). By utilizing a spreading assay the problem of distinguishing active and passive initial cell attachment was avoided. The results of these studies indicated that there was a component in fetal calf serum required for spreading of baby hamster kidney (BHK) cells. The factor responsible for the activity was also demonstrated to be present in calf, porcine, human, rabbit, and chicken sera. It was able to promote the spreading of chinese hamster ovary cells, HeLa cells, and L cells as well as BHK cells. Cell spreading occurred only if the factor was adsorbed on culture dishes. If other proteins were first adsorbed to the culture dishes, e.g. bovine albumin (BSA) or immunoglobulin, the factor no longer promoted cell spreading, even when it was added to the incubations in large excess. In a subsequent study (GRINNELL et al., 1977) the fetal calf factor was purified and two active glycoprotein components of 12.5 S and 9 S were isolated. A distinguishing characteristic of the factor was its low isoelectric point of 4.0.

In order to determine whether the cell spreading factor was required for cell attachment per se, it was necessary to devise a method of preventing passive cell attachment. This was accomplished by covering adhesion sites on the culture dishes with bovine serum albumin. Technically, this was accomplished by treatment of the dishes with a 10 mg/ml solution of BSA for 10 min. at 22 °C. Subsequently, BHK cells were not able to attach to the dishes. Experiments were then carried out in which BSA and the isolated spreading factor were allowed to compete for the critical sites on the culture dishes. When this was done, it was found that the extent of cell attachment to the dishes depended upon the amount of spreading factor to which the dishes were exposed. Therefore, the serum factor that promoted cell spreading was also a cell attachment factor (GRINNELL et al., 1977).

Around the same time as these experiments were being carried out, studies were reported (PEARLSTEIN, 1976) indicating a close relationship between a serum factor active in cell adhesion to collagen (KLEBE, 1974), and the fibroblast surface protein, fibronectin, that had been shown to be similar to CIG (RUOSLAHTI and VAHERI, 1975). Experiments were subsequently performed to determine if CIG was the

same as the factor required for cells to spread on tissue culture substrata (GRIN-
NELL and HAYS, 1978 a). These experiments were carried out with human plasma
and it was found that CIG and the spreading factor present in serum were identical
according to chromatographic, electrophoretic and immunological criteria.

Recently, CIG has been shown to play a role in the adhesion of a variety of cell
types other than fibroblasts (BENSUSAN et al., 1978; ORLY and SATO, 1979; CHIQUET
et al., 1979), and as an opsonin in the phagocytic activity of some macrophages
(SABA et al., 1978). Nevertheless, not all cell types utilize CIG. For instance, chondrocyte adhesion requires a different serum factor (HEWITT et al., 1980) and the
presence of CIG can alter chondrocyte differentiation under some conditions
(WEST et al., 1979).

III. Analysis of CIG Adsorption to Tissue Culture Dishes

Some of the concepts discussed in the preceeding section are illustrated in Fig. 1–4.
The pattern of adsorption of CIG on tissue culture dishes is shown in Fig. 1. Dishes
were treated with increasing concentrations of CIG and then part of the CIG was
removed by gentle scraping with a flat teflon rod. Subsequently, the dishes were
treated with BSA to cover the material surface sites from which CIG was removed.
Then the dishes with adsorbed proteins were stained with the immunoglubulin
fraction of a specific anti-CIG antiserum. This antiserum was produced by adsorbing crude anti-CIG antiserum against CIG-depleted serum (GRINNELL and FELD,
1979). The uniform deposition of CIG on the dishes was shown by the even pattern
of fluorescence and there was a relatively sharp line (arrows) where the CIG was
removed by the scraping technique. When cells were plated out on dishes prepared
in this manner it was observed that cells attached on the regions where CIG was
located but not on the regions where the material surface was coated with BSA
(Fig. 2). Similarly, in Fig. 3 A, B, cells were attached and spread in the region of
CIG on the material surface (arrows) but not in the regions from which CIG was
removed. Also, when the material surface was first covered with BSA and then part
of the BSA removed and then the dishes treated with CIG (Fig. 3 C), the cells were
attached to the region from which BSA had been removed (arrows) but not elsewhere.

The concentration dependence of CIG in cell adhesion is shown in Fig. 4. In this
experiment, culture dishes were treated with increasing CIG concentratiorfs and
then with BSA to cover remaining adsorption sites on the dishes. Then cells were
added and attachment and spreading measured. It is worthwhile noting that a dramatic increase in cell attachment and spreading occurred over a relatively small
change in CIG concentration. This observation has recently been emphasized and
the possibility has been proposed that attachment is a threshold phenomenon

$\longrightarrow$

Fig. 1 a–e. Adsorption of CIG to tissue culture dishes. Falcon Nr. 3001 tissue culture dishes
were treated with 1 ml adhesion medium (GRINNELL and HAYS, 1978 a) containing human
CIG (267 units/mg) as follows: **a** 0.3 units; **b** 0.6 units; **c** 1.5 units; **d** 3 units; **e** 6 units. After
10 min at 22 °C the dishes were rinsed thoroughly and portions of the dishes were scraped
with a Teflon rod (GRINNELL and FELD, 1979). Subsequently, the dishes were processed for
indirect immunofluorescence with specific anti-CIG antiserum as described previously
(GRINNELL and FELD, 1979). X 850

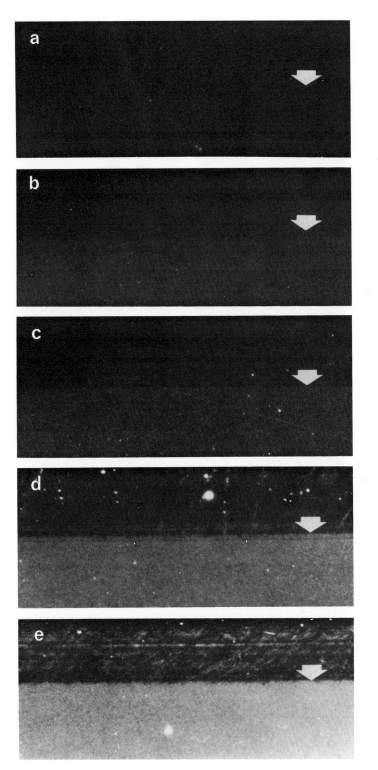

Fig. 1

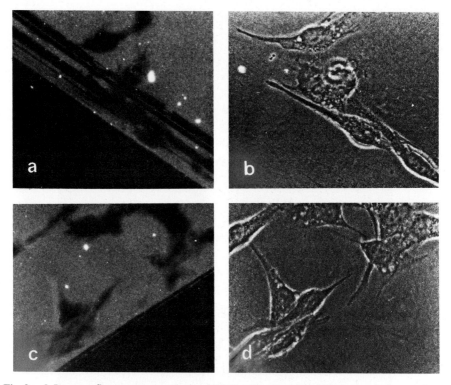

Fig. 2 a–d. Immunofluorescence analysis of BHK cells attached to CIG coated tissue culture dishes. Tissue culture dishes were treated with 4 units/ml CIG and scraped with a Teflon rod. The dishes were then treated with 1 ml adhesion medium containing 10 mg/ml bovine albumin for 10 min at 22 °C. Next, the dishes were incubated at 37 °C for 45 min with 7.5×10^5 BHK-21-13s cells in 1 ml adhesion medium. Finally, the dishes were processed for indirect immunofluorescence with specific anti-CIG antiserum. See the legend to Fig. 1 for other details. **b, d** phase contrast; **a, c** fluorescence. X 715

requiring a certain number of cell-CIG interactions. Based upon critical measurements, it was concluded that 45,000 CIG molecules are required beneath each BHK cell for complete cell spreading to occur (HUGHES et al., 1979).

IV. Technical Comments on CIG Purification

One unit of CIG activity is defined as the amount of CIG required to promote complete spreading of 0.75×10^6 BHK cells in 1 ml of medium in a 45 min incubation at 37 °C (GRINNELL, 1976 b; cf. Fig. 4). The specific activity of CIG preparations is generally about 400–500 using the technique for purification involving salt pre-

————————————————————————————▶

Fig. 3 a–c. Adhesion of BHK cells to CIG and BSA coated tissue culture dishes. Tissue culture dishes were **a, b** treated with CIG (4 units/ml), scraped, and then treated with BSA (10 mg/ml), or **c** treated with BSA, scraped, and then treated with CIG. Subsequently, the dishes were incubated with BHK cells. See the legends to Figs. 1 and 2 for other details. **a, c** X 100; **b** X 270

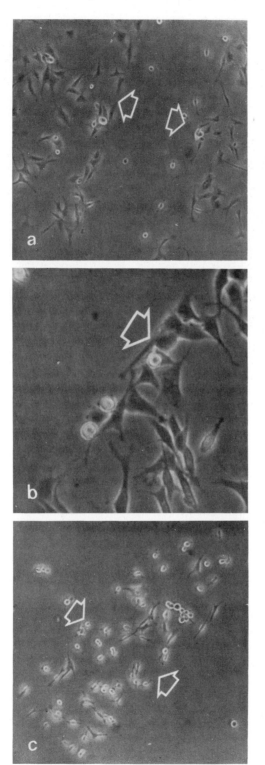

Fig. 3

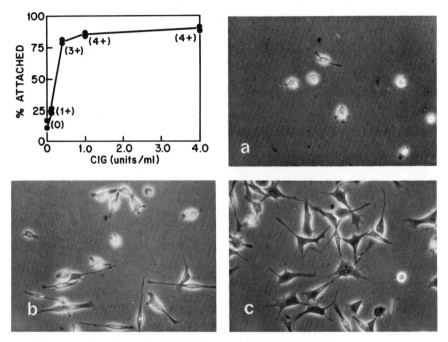

Fig. 4 a–c. CIG concentration dependence of BHK cell adhesion to tissue culture dishes. Tissue culture dishes were treated with CIG at the concentrations indicated followed by BSA (10 mg/ml). Subsequently, the dishes were incubated with BHK cells. See the legends to Figs. 1 and 2 for other details. **a** 0.1 unit/ml CIG; **b** 0.5 units/ml CIG; **c** 1 unit/ml CIG. X 175

cepitation and ion-exchange chromatography (GRINNELL and HAYS, 1978 a). In the original method described, the recovery of total activity was low. However, by omitting the 56 °C heat treatment, which was part of the procedure, it has been found that recovery of total activity can be increased to greater than 50% (GRINNELL et al., 1980). The purified protein can be stored without loss of activity or solubility for several months at 20 °C if the concentration is ≤ 1 mg/ml. A technique for purifying CIG has been described using affinity chromatography on gelatin-sepharose (ENGVALL and RUOSLAHTI, 1977). Eluting CIG from the gelatin-sepharose column with 8 M urea results in a substantial loss of spreading activity (50%–80%) and the eluted protein often precipitates upon storage. Eluting the gelatin sepharose column with 1 M NaBr, 20 mM Na acetate, at pH 5 (MOSHER, personal communication) permits full recovery of activity.

V. Other Components Which Promote Cell Spreading

Two components that promote cell spreading have been observed in fetal calf serum (GRINNELL et al., 1977) chicken serum (THOM et al., 1979) and human serum (KNOX and GRIFFITHS, 1979). With chicken serum, the smaller component was shown to be antigenically related to the larger (THOM et al., 1979) and with fetal calf serum, an anti-CIG antiserum was found to inhibit the activity of the mixed factor (GRINNELL et al., 1977). Figure 5 shows an experiment in which substrata were treated with human plasma and then with specific anti-CIG antiserum (see

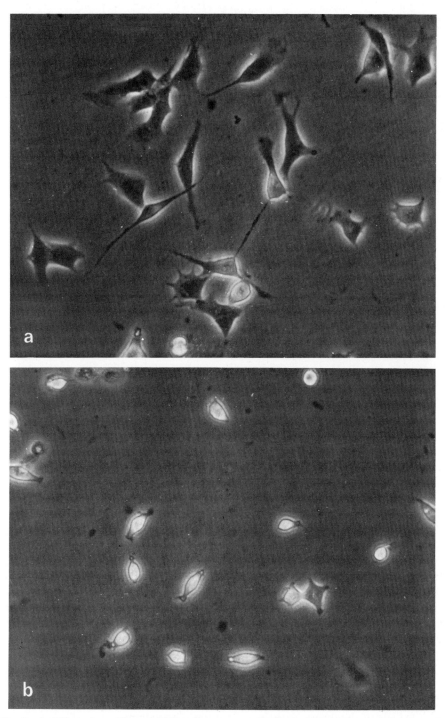

Fig. 5 a, b. Effect of specific anti-CIG antiserum on cell adhesion to plasma coated tissue culture dishes. Tissue culture dishes were treated with 1 ml adhesion medium containing 1% human plasma for 10 min at 22 °C. The dishes were then (**b**) treated with specific anti-CIG antiserum (1.5 mg/ml) for 30 min at 37 °C. Subsequently, the dishes were incubated with BHK cells. A similar concentration of pre-immune serum (**a**) had no effect. See the legends to Figs. 1 and 2 for other details. X 350

above). Subsequently, adhesion was tested with BHK cells. The antibody partially inhibited cell attachment and almost completely inhibited cell spreading. In addition, spreading activity was removed from plasma when it was clotted at 4 °C (GRINNELL and HAYS, 1978 a) or when plasma was passed through a gelatin-sepharose column. These results indicate that CIG is the major plasma component involved in cell spreading. Smaller molecules are likely to be proteolytic fragments of CIG. CIG is very susceptive to proteolytic digestion by a variety of different enzymes and several fragments have been isolated which retain cell spreading promoting activity (RUOSLAHTI and HAYMANN, 1979; HAHN and YAMADA, 1979 a, b).

Notwithstanding the singular role of CIG emphasized above, plasma and serum components other than CIG may influence aspects of cell adhesion other than cell spreading. Following initial cell attachment and spreading, many types of fibroblasts [but not all, see (GRINNELL, 1978) for a review] form "adhesion plaques" (BRUNK et al., 1971; ABERCROMBIE et al., 1971; BRAGINA et al., 1976). Although it has been inferred that plaques are involved in cell motility, one recent study indicates that plaques may be associated with more firmly anchored, less motile cells (COUCHMAN and REES, 1979). In any event, it has been reported that spreading of some cells on CIG substrate does not result in the appearance of plaques unless serum components in addition to CIG are present in the medium (BADLEY et al., 1978). This is not the case, however, with human fibroblasts (GRINNELL, 1980 a).

D. The Role of CIG in Cell Adhesion to Surfaces Other than Tissue Culture Dishes

I. Bacteriological Dishes

It is known that cells attach and grow poorly on bacteriological dishes and other non-wettable surfaces (reviewed in GRINNELL, 1978). Transformed cells will grow preferentially compared to normal cells on such surfaces because of the anchorage dependence requirement of normal cells. The method used by many commercial firms for preparing tissue culture dishes is to treat the non-wettable bacteriological dishes by the glow-discharge technique (MAROUDAS, 1973 c; GRINNELL, 1978). This disrupts the plastic polymer and results in an increase in the wettability of the plastic surface. Preliminary results (GRINNELL and FELD, unpublished) indicate that CIG adsorbs to bacteriological dishes with a lower affinity than to wettable tissue culture dishes. However, following treatment of bacteriological dishes with CIG at concentrations of 4–8 units/ml, complete cell attachment and spreading occurred on the dishes. Even at lower concentrations of CIG, there were some areas of the dishes where adhesion occurred. These findings indicate that bacteriological dishes might be suitable for growing normal cells if the dishes are coated with CIG before adding cells to the dishes in their normal growth medium.

II. CIG and the Adhesion of Cells to Collagen

The role of a serum protein in cell adhesion to collagen substrata was established in studies using chinese hamster ovary (CHO) cells (KLEBE, 1974). The substrata were prepared by air drying collagen gels which had been polymerized at high pH and this technique results in denaturation of the collagen as indicated by sensitivity of the substrata to trypsin digestion (GRINNELL and MINTER, 1978). Serum was found to be an absolute requirement for adhesion to occur to these denatured collagen substrata. Unlike material surfaces, no passive adhesion process was observed on the collagen substrata. Subsequently, CIG was determined to be the serum component active in promoting adhesion (PEARLSTEIN, 1976) and the region of collagen to which the CIG bound was identified (KLEINMAN et al., 1976). It was found that adsorption of CIG to the collagen was necessary for cell adhesion (PEARLSTEIN, 1978), similar to material surfaces.

Although CIG appears to be absolutely required for cell adhesion to denatured collagen, several reports have indicated that cells can attach to native collagen substrata in the absence of CIG (LINSENMAYER et al., 1978; GRINNELL and MINTER, 1978; SCHOR and COURT, 1979). The possibility has been raised that there might be a non-specific adhesion mechanism dependent upon the presence of calcium phosphate deposits in the native collagen gels (KLEINMAN et al., 1979b). However, the evidence favors a specific interaction. With BHK cells, adhesion occurred on native collagen gels even when the gels were prepared in the absence of phosphate and the adhesion incubation reactions were carried out in the absence of phosphate. Moreover, metabolic inhibitors prevented adhesion to native collagen which suggests it is an active process and not passive. Finally, BHK cells not only attached to native collagen in the absence of CIG but were able to spread. Another possibility is that collagen stimulates BHK cells to secrete their own fibronectin, but anti-CIG antiserum had no effect on BHK cell adhesion to native collagen (GRINNELL and MINTER, 1979).

Figure 6 shows the attachment and spreading of BHK cells on dried collagen gels in the presence of CIG. The surface of the dried gels was relatively flat and the pronounced three dimension fibrillar network typical of native collagen was absent (cf. Fig. 7). The cells were spread in triangular shapes typical of their behavior on two dimensional material surfaces and there were many lamellapodial extensions. In marked contrast, Fig. 7 shows the attachment and spreading of BHK cells on native collagen gels in which a three dimensional lattice of collagen fibrils was preserved. The cells did not spread as well as on denatured substrata; however, those cells that were spread were bipolar with filipodial extensions often penetrating into the collagen. It is interesting that the attachment and spreading of BHK cells on native collagen substrata in the presence of CIG was intermediate between cells on denatured collagen and cells on native collagen without CIG. In the presence of CIG, some lamellapodia were apparent and these stayed on top of the collagen; they did not penetrate the three dimensional matrix. Thus, even though CIG is not absolutely required for cell adhesion to collagen, it may modify the interaction. (GRINNELL and BENNETT, 1981) The ability of fibroblasts to interact with collagen in the absence of CIG is consistent with the ligand-receptor hypothesis (see Sect. F.I, below) since fibroblasts have been reported to have a collagen receptor (GOLDBERG, 1979).

Studies have been carried out to determine the mechanism by which CIG binds to collagen. These studies have been predominantly concerned with measuring the non-covalent binding of CIG to denatured collagen; binding to native collagen is generally much weaker (ENGVALL et al., 1978; JILEK and HORMANN, 1978). It

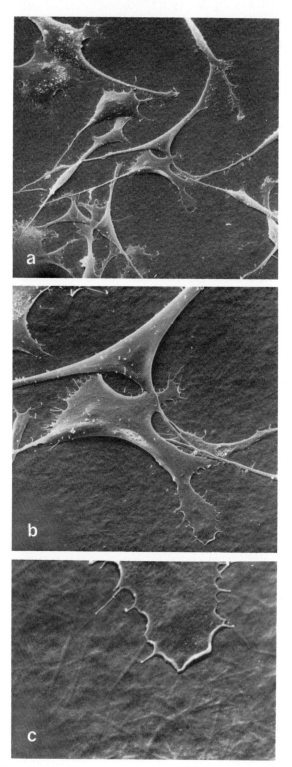

Fig. 6 a–c. Adhesion of BHK cells to dried collagen substrata. Dried collagen substrata were prepared in tissue culture dishes as described previously (GRINNELL and MINTER, 1978). Subsequently, the dishes were incubated for 2 h at 37 °C with BHK cells in 1.0 ml adhesion medium containing 30 units/ml of CIG (231 units/mg). The dishes were then fixed and processed for scanning electron microscopy (GRINNELL and FELD, 1979). See the legends to Figs. 1 and 2 for other details. **a** X 450; **b** X 1100; **c** X 3250

Fig. 7a–c. Adhesion of BHK cells to native, hydrated collagen gels. Native, hydrated collagen gels were prepared in tissue culture dishes as described previously (Grinnell and Minter, 1978). Subsequently, the dishes were incubated with BHK cells for 2 h at 37 °C and then processed for scanning electron microscopy. See the legends to Figs. 1, 2, and 6 for other details. **a** X 850; **b** X 2550; **c** X 8450

should be mentioned that the binding of CIG to native collagen can be enhanced by glycosaminoglycans, a property which may be functionally important in situ (JILEK and HORMANN, 1979). In any event, several laboratories have isolated fragments of the CIG and fibronectin that contain denatured collagen binding activity, the smallest of which is 30,000 daltons (BALIAN et al., 1979; WAGNER and HYNES, 1979; RUOSLAHTI et al., 1979; HAHN and YAMADA, 1979 a; GOLD et al., 1979). There is agreement that this region of the molecule is rich in sulfhydryl groups and it has been demonstrated that the sulfhydryl groups are required for binding to denatured collagen (BALIAN et al., 1979; WAGNER and HYNES, 1979). In addition to non-covalent binding of CIG to collagen, it has been shown that CIG can be covalently crosslinked to collagen through the action of plasma transglutaminase (Factor XIII), fibrin stabilizing factor (MOSHER et al., 1979).

III. CIG and the Adhesion of Cells to Fibrin and Fibrinogen

The adhesion of BHK cells to fibrin and fibrinogen is also mediated by CIG (GRINNELL et al., 1980). As with collagen, there was no passive attachment of cells to fibrinogen or fibrin. Pretreatment of fibrinogen or fibrin substrate with CIG promoted subsequent cell attachment and spreading. However, very high levels of CIG were required unless the CIG was covalently linked to the fibrin or fibrinogen through the action of plasma transglutaminase (Factor XIII). The ability of CIG to be linked to fibrin by plasma transglutaminase has been studied in detail and it has been shown that a glutamine residue of CIG is involved in cross-linking (MOSHER et al., 1980). Quantitative binding studies with radiolabeled CIG demonstrated that the enhancement of adhesion activity caused by covalent crosslinking of CIG to the fibrin or fibrinogen could not be accounted for simply by the absolute amount of CIG bound (GRINNELL et al., 1980). Therefore, the possibility must be considered that the orientation of CIG on the fibrin or fibrinogen is critical for activity.

The requirement for CIG in fibroblast adhesion to fibrin and fibrinogen has interesting implications with regard to the growth of fibroblasts in fibrin clots. It was shown that fibroblasts grow poorly in clots prepared with Factor XIII deficient plasma (BECK et al., 1961; UEYAMA and URAYAMA, 1978). This can be understood as a failure of fibroblasts to attach properly to the clot because CIG is not covalently incorporated.

E. Cell Spreading in the Absence of Serum or CIG

In previous sections, the focus of attention was on cells that require the addition of CIG in order to spread. These are all permanent cell lines. However, there are a number of cell types, notably diploid fibroblasts, which are able to spread in serum free medium (TAYLOR, 1961; WITKOWSKI and BRIGHTON, 1971; RAJARAMAN et al., 1974). Since diploid cells generally have higher levels of fibronectin and secrete more of this material into the medium than permanent cell lines (MOSHER, 1977; YAMADA et al., 1977) it seemed likely that the diploid fibroblasts might secrete their own CIG-like protein, i.e. fibronectin, onto the culture dish, and interact with this material (GRINNELL, 1978).

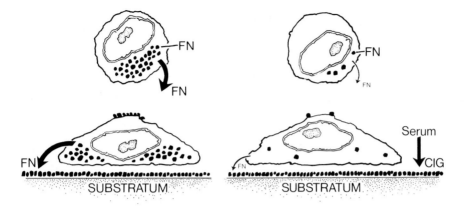

Fig. 8. Activities of CIG and fibronectin in cell adhesion. See text for explanation. (Modified from GRINNELL, 1978)

In recent studies (GRINNELL and FELD, 1979), it was found that human skin fibroblasts deposit fibronectin on tissue culture dishes during initial adhesion in serum free medium as determined by indirect immunofluorescence with specific anti-CIG antiserum. Initial attachment and subsequent spreading of the cells could be interferred with by covering protein adsorption sites on the dishes with BSA. The time course of fibronectin deposition on the dish surface was measured using a BHK cell adhesion assay. It was found that anti-CIG antiserum inhibited the ability of the human fibroblast secreted material to promote cell spreading. The results confirmed the hypothesis which is shown in Fig. 8. Cells able to produce and secrete sufficient fibronectin do not require serum for cell spreading. Cells that do not produce fibronectin or do not produce enough of this material require serum as a source for CIG.

The concept presented in Fig. 8 is important for understanding cell growth in serum-free medium, which is generally accomplished by adding specific hormones or hormone related substances to the defined medium (e.g. WU and SATO, 1978). Often, corticosteroids such as hydrocortisone or dexamethasone are included which promote cell adhesion (BALLARD and TOMKINS, 1970) and the deposition of fibronectin on tissue culture dishes (FURCHT et al., 1979). Thus, the ability of cells to grow in serum free medium probably depends in part upon using conditions that promote secretion of endogenous cell adhesion factors.

F. Mechanism of Action of CIG

I. Ligand Receptor Hypothesis

In previous sections, phenomenological and technical aspects concerning CIG and cell adhesion have been emphasized. It is now appropriate to consider the mechanism by which CIG promotes cell adhesion and the identity of the cell surface receptor. It has been proposed that CIG promotes attachment and spreading of cells by a ligand-receptor like interaction between CIG molecules adsorbed on the ma-

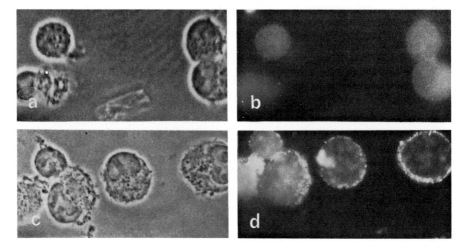

Fig. 9 a–d. Binding of CIG to BHK cell surfaces. The experiments contained 10^6 BHK cells in 1 ml adhesion medium with 10 mg/ml BSA. In **c** and **d**, CIG (231 units/mg) was added at a concentration of 35 units/ml. The experiments were incubated for 45 min at 37 °C at which time they were diluted to 5 ml with adhesion medium and centrifuged. Subsequently, the pellets were fixed and processed for indirect immunofluorescence with specific anti-CIG antiserum. See the legends to Figs. 1 and 2 for other details. X 650

terial surface and an appropriate receptor on the cell surface (Grinnell, 1978). This hypothesis is based in part upon the results of model system studies in which known ligands directed against the cell surface have been shown to promote cell attachment and spreading. Thus, treatment of material surfaces with antibodies against cell surface antigens (Giaever and Ward, 1978; Grinnell and Hays, 1978 b; Stocker and Heusser, 1979), plant lectins (Rutishauser et al., 1974; Grinnell and Hays, 1978 b; Seglen and Fossa, 1978; Hughes et al., 1979), polycationic proteins (Lieberman and Ove, 1958; Macieira-Coelho et al., 1974; McKeehan and Ham, 1976; Grinnell and Hays, 1978 b), and other proteins for which there are specific cell surface receptors (North, 1968; Alexander and Henkart, 1976; Hook et al., 1977) promotes cell attachment and spreading on the surfaces. In this context, cell attachment and spreading can be considered to be a general response of cells to the presence of a ligand on a material surface. In some respects, this is analogous to the capping of cell surface receptors by multivalent ligands in solution (De Petris and Raff, 1973).

II. Cooperativity and CIG Binding

That CIG binds to cell surfaces can be shown by indirect immunofluorescence as shown in Fig. 9. BHK cells in suspension demonstrated little staining with the specific anti-human CIG antiserum. However, when cells were treated with high concentrations of CIG and then washed and stained with anti-CIG antiserum, a fluorescent ring of staining was observed (Fig. 9 D). Even though binding of CIG to BHK cells was demonstrated by immunofluorescence, attempts at studying the binding properties of the cell surface receptor with radiolabeled CIG have been unsuccessful. Also, CIG does not aggregate BHK cells in suspension and soluble CIG does not compete with CIG adsorbed on culture dishes to prevent cell spreading. For these reasons, we and others have proposed that binding of CIG to the cell surface is a weak interaction. However, when CIG is adsorbed on a material surface, the cooperative effect of multiple simultaneous interactions is sufficient to stabilize adhesion (GRINNELL, 1978; THOM et al., 1979).

III. Interaction of Multimeric CIG with BHK Cells

In order to study the cell surface receptors for CIG a multimeric form of CIG was produced by adsorbing CIG on the surfaces of 0.76 μ latex polystyrene beads (GRINNELL, 1979 b). Latex beads treated with bovine serum albumin interacted poorly with BHK cells. On the other hand, beads coated with CIG were found to bind to the surfaces of BHK cells. If the beads were coated with BSA, they were then unable to bind to the cells even in the presence of excess CIG. As would be expected of a multimeric form of CIG, incubation with the CIG-beads caused BHK cells to aggregate.

The binding of CIG beads to the cells occurred all over the cell surfaces indicating an initially random distribution of the receptors. However, CIG beads did not bind to the upper surfaces of BHK cells that had spread previously on CIG coated culture dishes. This can be interpreted to indicate that during spreading, the initially random CIG receptors translocate to the portion of the cell surface adjacent to the CIG substratum and are no longer present on the upper cell surface. A similar translocation occurs with the F_C receptors of macrophages spread on immune complex coated substrata (MICHL et al., 1979). Alternatively, CIG receptors still may be present on the upper surface of spread cells, but in an arrangement or condition unfavorable to cell-bead interactions. For instance, if multiple CIG-receptor interactions are required for stable binding, then receptor clustering may be required and receptor mobility may be decreased on spread cells compared to round cells (WRIGHT and KARNOVSKY, 1979). A receptor-clustering model would be consistent with the idea that adhesion is a threshold phenomenon (HUGHES et al., 1979; WEIGEL et al., 1979) and would explain why cell-bead binding was slower at 4 °C than at 22 °C or 37 °C.

Binding of CIG beads to the cells at 37 °C was followed by endocytosis and the beads ended up inside cytoplasmic structures which appeared to be secondary lysosomes. This observation suggests that the CIG receptor is capable of causing cell endocytosis or cell spreading depending upon the size of the CIG coated surface and indicates that cell spreading on culture dishes may be an attempt by the cell to endocytose a particle of infinite diameter as was suggested some time ago for macrophages (NORTH, 1968, 1970).

The cell receptor for CIG probably contains a protein component since treatment of cells with 0.01 mg/ml trypsin for 10 min at 37 °C inhibited the ability of cells to interact with the beads. Also, the binding of CIG beads to the cells was found to be independent of divalent cations indicating that these are not required for activity or stability of the receptor (GRINNELL, 1979).

IV. Inhibition of Cell Adhesion with Gangliosides

Studies on the identity of the cell surface receptors for CIG have been carried out
in the cell-collagen adhesion system. It has been demonstrated that gangliosides
compete with cells for binding to the CIG molecules adsorbed to collagen substrata
(Kleinman et al., 1979a). The gangliosides did not bind directly to cells or to collagen. In particular, GT_1 and GD_{1a} were the most effective inhibitors. In other experiments it was found that the ceramide portions of the gangliosides and the
monosaccharides had no effect but that the oligosaccharides were somewhat inhibitory. Also, oxidation of the ganglioside sialic acid residues blocked the inhibitory activity. Other molecules tested including polysaccarides and glycosaminoglycans had no effect on cell adhesion except hyaluronate which was slightly inhibitory. These findings suggest that a cell surface ganglioside may be part of the CIG
binding site and that the carbohydrate portion is involved. Thus, the CIG-receptor
may be similar to the cell receptor for thyrotropic stimulating hormone and some
other glycoprotein hormones that have both glycoprotein and glycolipid components (Critchley et al., 1979).

V. Isolation of Cell-Substratum Adhesion Sites

Another approach to studying the cell surface receptor for CIG has been to identify
the cell surface components that are left associated with the CIG coated material
surface after cells are removed. The idea is to remove attached cells so as to leave
behind putative adhesion sites (Culp, 1978). A number of cell cytoplasmic components are also left behind, therefore, the system is extremely complex. In any
event, when deposition of cell surface materials on the CIG coated material surface
has been studied temporarily during initial cell adhesion, the results have suggested
a role for cell surface glycosaminoglycans and in particular heparan sulfate (Culp,
1978; Culp et al., 1978). There is also evidence from chemical crosslinking studies
that cell surface fibronectin is closely associated with glycosaminoglycans (Perkins
et al., 1979).

VI. Studies on the Active Site(s) of CIG

Finally, an approach to characterizing the active site of CIG has been to modify
adsorbed CIG with various reagents to determine which groups on the CIG molecule are required for biological activity (Grinnell and Minter, 1979). The modifications were carried out after adsorption of CIG on culture dishes since they
could have altered the ability of CIG to adsorb and thereby, modified the expression of CIG activity indirectly. Also, some experiments were carried out with
radiolabeled CIG to control for the possibility that CIG was removed from the
substrata by the modifications. However, because of the small amounts of CIG adsorbed on the substrata it was not possible to reisolate the treated CIG to prove
that the expected modifications had actually taken place. In any event, the results
were that inhibition of CIG biological activity occurred following treatment of adsorbed CIG with proteases, with lactoperoxidase or chloramine T under conditions
of tyrosine iodination, with N-bromosuccinimide under conditions of tryptophan

oxidation, or with carbodiimide under conditions whereby carboxyl groups would be blocked. However, a variety of treatments that modify carbohydrate components, amino groups, or sulfhydryl groups had no effect. These studies suggest that the cell binding activity of CIG depends upon protein portions of the molecule and that carboxyl groups, tyrosine residues, and tryptophan residues might be involved at the active site or in maintaining the active conformation of CIG.

G. Summary

The purpose of this chapter has been to discuss some aspects of cell adhesion and the role of cold insoluble globulin. Since attachment and spreading are growth requirements of normal cells, CIG can be considered to be a growth factor for the many cell types which require CIG for adhesion. It should be clear that CIG plays a central role in adhesion: it is present in all the plasma and sera that have been tested; it works with a large number of different cell types although it has been studied mostly with fibroblasts; and it is involved in cell adhesion to material surfaces, collagen, and fibrin.

Acknowledgments. MARIAN FELD assisted with immunofluorescence studies and MARYLYN BENNETT and KATHY HEWITT assisted with scanning electron microscopy. The author's research has been supported predominantly by a grant from the National Institutes of Health, CA 14609.

References

Abercrombie, M., Heaysman, J.E.M., Pegrum, S.H.: The locomotion of fibroblasts in culture. Exp. Cell Res. *67*, 359–367 (1971)
Alexander, E., Henkart, P.: The adherence of human F_C receptor-bearing lymphocytes to antigen-antibody complexes. II. Morphological alterations induced by the substrate. J. Exp. Med. *143*, 329–347 (1976)
Alexander, S.S., Colonna, G., Edelhoch, H.: Structure and stability of human plasma cold insoluble globulin. J. Biol. Chem. *254*, 1501–1505 (1979)
Alexander, S.S., Colonna, G., Yamada, K.M., Pastan, I., Edelhoch, H.: Molecular properties of a major cell surface protein from chick embryo fibroblasts. J. Biol. Chem. *253*, 5820–5824 (1978)
Badley, R.A., Lloyd, C.W., Woods, A., Carruthers, L., Allcock, C., Rees, D.A.: Mechanisms of cellular adhesion. III. Preparation and preliminary characterization of adhesions. Exp. Cell Res. *117*, 231–244 (1978)
Balian, G., Click, E.M., Crouch, E., Davidson, J.M., Bornstein, P.: Isolation of a collagen-binding fragment from fibronectin and cold-insoluble globulin. J. Biol. Chem. *254*, 1429–1432 (1979)
Ballard, P.L., Tomkins, G.M.: Glucocorticoid-induced alterations of the surface membrane of culture hepatoma cells. J. Cell Biol. *47*, 222–234 (1970)
Beck, E., Duckert, F., Ernst, M.: The influence of fibrin stabilizing factor on the growth of fibroblasts in vitro and wound healing. Thromb. Diathes. Haemorr. *6*, 485–491 (1961)
Bensusan, H.B., Koh, T.L., Kim, H.G., Murray, B.A., Culp, L.A.: Evidence that fibronectin is the collagen receptor on platelet membranes. Proc. Natl. Acad. Sci. U.S.A. *75*, 5864–5868 (1978)
Birdwell, C.R., Gospadarowicz, D., Nicolson, G.L.: Identification, localization and role of fibronectin on cultured bovine endothelial cells. Proc. Natl. Acad. Sci. U.S.A. *75*, 3273–3277 (1978)

Bragina, E.E., Vasiliev, J.M., Gelfand, I.M.: Formation of bundles of microfilaments during spreading of fibroblasts on the substrate. Exp. Cell. Res. *97*, 241–248 (1976)

Brunk, U., Ericsson, J.L.E., Ponten, J., Westermark, B.: Specialization of cell surfaces in contact-inhibited human glial-like cells in vitro. Exp. Cell Res. *67*, 407–415 (1971)

Chiquet, M., Puri, E.C., Turner, D.C.: Fibronectin mediates attachment of chicken myoblasts to gelatin-coated substratum. J. Biol. Chem. *254*, 5475–5482 (1979)

Couchman, J.R., Rees, D.A.: Actomyosin organization for adhesion, spreading, growth and movement in chick fibroblasts. Cell Biol. Int. Reports *3*, 431–439 (1979)

Critchley, D.R., Ansell, S., Dilks, S.: Glycolipids: A class of membrane receptors. Biochem. Soc. Transact. *7*, 314–319 (1979)

Culp, L.: Biochemical determinants of cell adhesion. Curr. Top. Mem. Trans. *11*, 327–396 (1978)

Culp, L.A., Rollins, B.J., Buniel, J., Hitri, S.: Two functionally distinct pools of glycosaminoglycans in the substratum adhesion site of murine cells. J. Cell Biol. *79*, 788–801 (1978)

De Petris, S., Raff, M.C.: Fluidity of the plasma membrane and its implications for cell movement. In: Locomotion of tissue cells. Ciba Foundation Symposium *14*, 27–41 (1973)

Di Pasquale, A., Bell, P.B., jr.: The upper cell surface: its inability to support active cell movement in culture. J. Cell Bio. *62*, 198–214 (1974)

Engvall, E., Ruoslahti, E.: Binding of soluble form of fibroblast surface protein, fibronectin, to collagen. Int. J. Cancer *20*, 1–5 (1977)

Engvall, E., Ruoslahti, E., Miller, E.J.: Affinity of fibronectin to collagens of different genetic types and to fibrinogen. J. Exp. Med. *147*, 1584–1595 (1978)

Fisher, H.W., Puck, T.T., Sato, G.: Molecular growth requirements of single mammalian cells: The action of fetuin in promoting cell attachment to glass. Proc. Natl. Acad. Sci. U.S.A. *44*, 4–10 (1958)

Folkman, J., Moscona, A.: Role of cell shape in growth control. Nature *273*, 345–349 (1978)

Foster, D.O., Pardee, A.B.: Transport of amino acids by confluent and non-confluent 3T3 and polyoma virus-transformed 3T3 cells growing on glass coverslips. J. Biol. Chem. *244*, 2675–2681 (1969)

Fryand, O.: Studies on fibronectin in skin. I. Indirect immunofluorescence studies in normal human skin. Br. J. Dermatol. *101*, 263–270 (1979)

Furcht, L.T., Mosher, D.F., Wendelschafer-Crabb, G., Woodbridge, P.A., Foidart, J.M.: Dexamethasone-induced accumulation of a fibronectin and collagen extracellular maitrix in transformed human cells. Nature *277*, 393–395 (1979)

Giaever, I., Ward, E.: Cell adhesion to substrates containing adsorbed or attached IgG. Proc. Natl. Acad. Sci. U.S.A. *75*, 1366–1368 (1978)

Gold, L.I., Garcia-Pardo, A., Frangione, B., Franklin, E.C., Pearlstein, E.: Subtilisin and cyanogen bromide cleavage products of fibronectin that retain gelatin-binding activity. Proc. Natl. Acad. Sci. U.S.A. *76*, 4803–4807 (1979)

Goldberg, B.: Binding of soluble type I collagen molecules to the fibroblast plasma membrane. Cell *16*, 265–275 (1979)

Grinnell, F.: The serum dependence of baby hamster kidney cell attachment to a substratum. Exp. Cell Res. *97*, 265–276 (1976a)

Grinnell, F.: Cell spreading factor: Occurrence and specificity of action. Exp. Cell Res. *102*, 51–62 (1976b)

Grinnell, F.: Cellular adhesiveness and extracellular substrata. Int. Rev. Cytol. *53*, 65–144 (1978)

Grinnell, F.: Visualization of cell-substratum adhesion plaques by antibody exclusion. Cell Biol. Int. Reports (in press) (1980a)

Grinnell, F.: Interactions between plasma fibronectin coated latex beads and baby hamster kidney cells. J. Cell Biol. *86*, 104–112 (1980b)

Grinnell, F., Bennett, M.: Fibroblast adhesion on collagen substrata in the presence and absence of plasma fibronectin. J. Cell Sci. (in press) (1981)

Grinnell, F., Feld, M.: Initial adhesion of human fibroblasts in serum-free medium. Possible role of secreted fibronectin. Cell *17*, 117–129 (1979)

Grinnell, F., Hays, D.G.: Cell adhesion and spreading factor: Similarity to cold insoluble globulin in human serum. Exp. Cell Res. *115*, 221–229 (1978a)

Grinnell, F., Hays, D.G.: Induction of cell spreading by substratum adsorbed ligands directed against the cell surface. Exp. Cell Res. *116*, 275–284 (1978b)

Grinnell, F., Minter, D.: Attachment and spreading of baby hamster kidney cells to collagen substrata: Effects of cold-insoluble globulin. Proc. Natl. Acad. Sci. U.S.A. *75*, 4408–4412 (1978)

Grinnell, F., Minter, D.: Cell adhesion and spreading factor: Chemical modification studies. Biochim. Biophys. Acta *550*, 92–99 (1979)

Grinnel F., Tobleman, M.Q., Hackenbrock, C.R.: The distribution and mobility of anionic sites on the surfaces of baby hamster kidney cells. J. Cell Biol. *66*, 470–479 (1975)

Grinnell, F., Tobleman, M.Q., Hackenbrock, C.R.: Initial attachment of baby hamster kidney cells to a substrata. Ultrastructural analysis. J. Cell Biol. *70*, 707 (1976)

Grinnell, F., Hays, D., Minter, D.: Cell adhesion and spreading factor: Partial purification and properties. Exp. Cell Res. *110*, 175–190 (1977)

Grinnell, F., Feld, M., Minter, D.: Cell adhesion to fibrinogen and fibrin substrata: Role of cold insoluble globulin (plasma fibronectin). Cell *19*, 517–525 (1980)

Hahn, L.-H.E., Yamada, K.M.: Identity and isolation of a collagen-binding fragment of the adhesive glycoprotein fibronectin. Proc. Natl. Acad. Sci. U.S.A. *76*, 1160–1163 (1979a)

Hahn, L.-H.E., Yamada, K.M.: Isolation and characterization of active fragments of the adhesive glycoprotein fibronectin. Cell *18*, 1043–1051 (1979b)

Hewitt, A.T., Kleinman, H.K., Pennypacker, J.P., Martin, G.R.: Identification of an adhesion factor for chondrocytes. Proc. Natl. Acad. Sci. U.S.A. *77*, 385–388 (1980)

Holley, R.W.: Control of growth of mammalian cells in cell culture. Nature *258*, 487–490 (1975)

Hook, M., Rubin, K., Oldberg, A., Obrink, B., Vaheri, A.: Cold insoluble globulin mediates the adhesion of rat liver cells to plastic petri dishes. Biochem. Biophys. Res. Commun. *79*, 726–733 (1977)

Hughes, R.C., Pena, S.D.J., Clark, J., Dourmashkin, R.R.: Molecular requirements for the adhesion and spreading of hamster fibroblasts. Exp. Cell Res. *121*, 307–314 (1979)

Jaffe, E.A., Mosher, D.F.: Synthesis of fibronectin by cultured human endothelial cells. J. Exp. Med. *147*, 1779–1791 (1978)

Jilek, F., Hormann, H.: Cold insoluble globulin (fibronectin) IV. Affinity to soluble collagen of various types. Hoppe-Seylers Z. Physiol. Chem. *359*, 247–250 (1978)

Jilek, F., Hormann, H.: Fibronectin (cold insoluble globulin) VI. Influence of heparin and hyaluronic acid on the binding of native collagen. Hoppe-Seylers Z. Physiol. Chem. *360*, 597–603 (1979)

Kahn, P., Shin, S.-I.: Cellular tumorigenicity in nude mice. J. Cell Biol. *82*, 1–16 (1979)

Klebe, R.J.: Isolation of a collagen dependent cell attachment factor. Nature *250*, 248–251 (1974)

Kleinman, H.K., McGoodwin, E.B., Klebe, R.J.: Localization of the cell attachment region in types I and II collagens. Biochem. Biophys. Res. Commun. *72*, 426–432 (1976)

Kleinman, H.K., Martin, G.R., Fishman, P.H.: Ganglioside inhibition of fibronectin-mediated cell adhesion to collagen. Proc. Natl. Acad. Sci. U.S.A. *76*, 3369 (1979a)

Kleinman, H.K., McGoodwin, E.B., Rennard, S.I., Martin, G.R.: Preparation of collagen substrates for cell attachment: Effect of collagen concentration and phosphate buffer. Anal. Biochem. *94*, 308–312 (1979b)

Knox, P., Griffiths, S.: A cell spreading factor in human serum that is not cold insoluble globulin. Exp. Cell Res. *123*, 421–423 (1979)

Lieberman, I., Ove, P.: A protein growth factor for mammalian cells in culture. J. Biol. Chem. *233*, 637–642 (1958)

Lieberman, I., Lamy, F., Ove, P.: Nonidentity of fetuin and protein growth (flattening) factor. Science *129*, 43–44 (1959)

Linsenmayer, T.F., Gibney, E., Toole, B.P., Gross, J.: Cellular adhesion to collagen. Exp. Cell Res. *116*, 470–474 (1978)

Macarak, E.J., Kirby, E., Kirt, T., Kefalides, N.A.: Synthesis of cold-insoluble globulin by cultured calf endothelial cells. Proc. Natl. Acad. Sci. U.S.A. *75*, 2621–2625 (1978)

Macieiri-Coelho, A., Berumen, L., Avrameas, S.: Properties of protein polymers as substratum for cell growth in vitro. J. Cell. Physiol. *83*, 379–388 (1974)

MacPherson, I., Montagnier, L.: Agar suspension culture for the selective assay of cells transformed by polyoma virus. Virology *23*, 291–294 (1964)

Maroudas, N.G.: Anchorage dependence: Correlation between amount of growth and diameter of bead, for single cells grown on individual glass beads. Exp. Cell Res. *74*, 337–342 (1972)

Maroudas, N.G.: Growth of fibroblasts on linear and planar anchorages of limiting dimensions. Exp. Cell Res. *81*, 104–110 (1973a)

Maroudas, N.G.: Chemical and mechanical requirements for fibroblast adhesion. Nature *244*, 353–355 (1973b)

Maroudas, N.G.: New methods for large-scale culture of anchorage dependent cells. New Tech. Biophys. Cell Biol. *1*, 67–86 (1973c)

McKeehan, W.L., Ham, R.G.: Stimulation of clonal growth of normal fibroblasts with substrata coated with basic polymers. J. Cell Biol. *71*, 727–734 (1976)

Michl, J., Pieczonka, M.M., Unkless, J.C., Silverstien, S.C.: Effects of immobilized immune complexes of F_C and complement-receptor function in resident and thioglycollate-elicited mouse peritoneal macrophages. J. Exp. Med. *150*, 607–621 (1979)

Mosesson, M.W., Umfleet, R.A.: The cold insoluble globulin of human plasma. I. Purification, primary characterization, and relationship to fibrinogen and other cold-insoluble fraction components. J. Biol. Chem. *245*, 5728–5736 (1970)

Mosesson, M.W., Chen, A.B., Huseby, R.M.: The cold-insoluble globulin of human plasma: Studies of its essential structural features. Biochim. Biophys. Acta *386*, 509–524 (1975)

Mosher, D.F.: Distribution of a major surface-associated glycoprotein, fibronectin, in cultures of adherent cells. J. Supramol. Struct. *6*, 551–557 (1977)

Mosher, D.F., Schad, P.E., Kleinman, H.K.: Cross-linking of fibronectin to collagen by blood coagulation Factor XIIIa. J. Clin. Invest. *64*, 781–787 (1979)

Mosher, D.F., Schad, P.E., Vann, J.M.: Cross-linking of collagen and fibronectin by Factor XIIIa: Localization of participating glutaminyl residues to a tryptic fragment of fibronectin. J. Biol. Chem. *255*, 1181–1188 (1980)

North, R.J.: The uptake of particulate antigens. J. Reticuloendothel. Soc. *5*, 203–229 (1968)

North, R.J.: Endocytosis. Semin. Hematol. *7*, 161–171 (1970)

O'Neill, C.H., Riddle, P.N., Jordan, P.W.: The relation between surface area and anchorage dependence of growth in hamster and mouse fibroblasts. Cell *16*, 909–918 (1979)

Orly, J., Sato, G.: Fibronectin mediates cytokinesis and growth of rat follicular cells in serum-free medium. Cell *17*, 295–305 (1979)

Otsuka, H., Moskowitz, M.: Difference in transport of leucine in attached and suspended 3T3 cells. J. Cell. Physiol. *85*, 665–674 (1975)

Paul, D., Lipton, A., Klinger, I.: Serum factor requirements of normal and simian virus 40-transformed 3T3 mouse fibroblasts. Proc. Natl. Acad. Sci. U.S.A. *68*, 645–652 (1971)

Pearlstein, E.: Plasma membrane glycoprotein which mediates adhesion of fibroblasts to collagen. Nature *262*, 497–499 (1976)

Pearlstein, E.: Substrate activation of cell adhesion as a prerequisite for cell attachment. Int. J. Cancer *22*, 32–35 (1978)

Perkins, M.E., Ji, T.H., Hynes, R.O.: Cross-linking of fibronectin to sulfated proteoglycans at the cell surface. Cell *16*, 941–952 (1979)

Rajaraman, R., Rounds, D.E., Yen, S.P.S., Rembaum, A.: A scanning electron microscope study of cell adhesion and spreading in vitro. Exp. Cell Res. *88*, 327–339 (1974)

Ruoslahti, E., Hayman, E.G.: Two active sites with different characteristics in fibronectin. FEBS Letters *97*, 221–224 (1979)

Ruoslahti, E., Vaheri, A.: Interaction of soluble fibroblast surface antigen with fibrinogen and fibrin. Identity with cold insoluble globulin of human plasma. J. Exp. Med. *141*, 497–501 (1975)

Ruoslahti, E., Vaheri, A., Kuusela, P., Linder, E.: Fibroblast surface antigen: A new serum protein. Biochim. Biophys. Acta *322*, 352–358 (1973)

Ruoslahti, E., Hayman, E.G., Kuusela, P., Shively, J.E., Engvall, E.: Isolation of a tryptic fragment containing the collagen-binding site of plasma fibronectin. J. Biol. Chem. *254*, 6054–6059 (1979)

Rutishauser, U., Yahara, I., Edelman, G.M.: Morphology, motility and surface behavior of lymphocytes bound to nylon fibers. Proc. Natl. Acad. Sci. *71*, 1149–1153 (1974)

Saba, T.M., Blumenstock, F.A., Weber, P., Kaplan, J.E.: Physiologic role of cold insoluble globulin in systemic host defense: Implications of its characterization as the opsonic α_2 surface binding glycoprotein. Ann. N.Y. Acad. Sci. *312*, 43–55 (1978)

Schachner, M., Schoonmaker, G., Hynes, R.O.: Cellular and subcellular localization of LETS protein in the nervous system. Brain Res. *158*, 149–158 (1978)

Schor, S.L., Court, J.: Different mechanisms in the attachment of cells to native and denatured collagen. J. Cell Sci. *38*, 267–281 (1979)

Seglen, P.O., Fossa, J.: Attachment of rat hepatocytes in vitro to substrata of serum protein, collagen or concanavalin A. Exp. Cell Res. *116*, 199–206 (1978)

Smith, H.S., Riggs, J.L., Mosesson, M.W.: Production of fibronectin by human epithelial cells in culture. Cancer Res. *39*, 4138–4144 (1979)

Stenman, S., Vaheri, A.: Distribution of a major connective tissue protein, fibronectin, in normal human tissues. J. Exp. Med. *147*, 1054–1064 (1978)

Stocker, J.W., Heusser, C.H.: Methods for binding cells to plastic: Application to a solid-phase radioimmunoassay for cell-surface antigens. J. Immunol. Methods *26*, 87–95 (1979)

Stoker, M., O'Neill, C., Berryman, S., Waxman, V.: Anchorage and growth regulation in normal and virus-transformed cells. Int. J. Cancer *3*, 683–693 (1968)

Taylor, A.C.: Attachment and spreading of cells in culture. Exp. Cell Res. Suppl. *8*, 154–173 (1961)

Thom, D., Powell, A.J., Rees, D.A.: Mechanisms of cellular adhesion. J. Cell Sci. *35*, 281–305 (1979)

Ueyama, M., Urayama, T.: The role of Factor XIII in fibroblast proliferation. Jpn. J. Exp. Med. *48*, 135–142 (1978)

Vaheri, A., Mosher, D.F.: High molecular weight, cell surface associated glycoprotein (fibronectin) lost in malignant transformation. Biochim. Biophys. Acta *516*, 1–25 (1978)

Vasiliev, J.M., Gelfand, J.M.: Morphogenetic reactions and locomotory behavior of transformed cells in culture. In: Fundamental aspects of metastasis, Weiss, L. (ed.), pp. 71–98. New York: American Elsevier 1976

Wagner, D.D., Hynes, R.O.: Domain structure of fibronectin and its relation to function. Disulfides and sulfhydryl groups. J. Biol. Chem. *254*, 6746–6754 (1979)

Weigel, P.H., Schnarr, R.L., Kuhlenschmidt, M.S., Schmell, E., Lee, R.T., Lee, Y.C., Roseman, S.: Adhesion of hepatocytes to immobilized sugars: A threshold phenomenon. J. Biol. Chem. *254*, 10830–10838 (1979)

Weiss, L.: The measurement of cell adhesion. Exp. Cell Res. Suppl. *8*, 141–153 (1961)

West, C.M., Lanza, R., Rosenbloom, T., Lowe, M., Holtzer, H., Advalovic, N.: Fibronectin alters the phenotypic properties of cultured chick embryo fibroblasts. Cell *17*, 491–501 (1979)

Witkowski, J.A., Brighton, W.D.: Stages of spreading of human diploid cells on glass surfaces. Exp. Cell Res. *68*, 372–380 (1971)

Wright, T.C., jr., Karnovsky, M.J.: Relationships between cell-substratum interactions and the distribution of concanvalin A receptors of mouse embryo fibroblasts. Exp. Cell Res. *123*, 377–382 (1979)

Wu, R., Sato, G.H.: Replacement of serum in cell culture by hormones: A study of hormonal regulation of cell growth and specific gene expression. J. Toxicol. Environ. Health *4*, 427–448 (1978)

Yamada, K.M., Kennedy, D.W.: Fibroblast cellular and plasma fibronectins are similar but not identical. J. Cell Biol. *80*, 492–498 (1979)

Yamada, K.M., Olden, K.: Fibronectins: Adhesive glycoproteins of cell surface and blood. Nature *275*, 179–184 (1978)

Yamada, K.M., Yamada, S.S., Pastran, I.: Quantitation of a transformation sensitive, adhesive cell surface glycoprotein. J. Cell Biol. *74*, 649–654 (1977)

CHAPTER 10

Membrane-Derived Inhibitory Factors

P. Datta

A. Introduction and Perspective

Cells of animal origin require a complex medium containing small molecular weight nutrients and serum for growth. Serum is a complex mixture of substances which include growth factors, polypeptide hormones, enzymes as well as various small molecules such as ions, cyclic nucleotides and steroids. Usually, the culture density of cells is proportional to the concentration of serum in the medium. When grown to high saturation density, anchorage-dependent cells become nonproliferative exhibiting density-dependent inhibition of growth. Cells seeded in low serum concentration also become quiescent at sparse culture density. The quiescent cells, whether they are confluent or sparse, remain viable for some time and can be stimulated to grow again (i.e., progress to the proliferative state) by a variety of agents that act as "mitogens." Highly transformed cells, on the other hand, exist only in proliferative state, lose their density-dependent inhibition of growth and continue to grow until the medium is depleted at which time cell death occurs. For up to date and critical coverage of the subject matter, the reader is referred to the reviews by Holley (1975), Baserga (1976), Pardee et al. (1978), and Rudland and Jiminez De Asua (1979).

One of the intriguing questions in cell biology is how do proliferating cells achieve quiescence? An understanding of the factors or mechanisms involved in this physiologic transition would be an important step forward toward uncovering the programmed events in the growth control of mammalian cells (cf. Baserga, 1978).

In the past decade, evidence has been gathered to indicate that, depending on the cell line, density-dependent regulation of growth may be influenced by the limitation of growth factors available to cells by a diffusion-limited process, destruction of factors in crowded cultures, a decrease in the number of available "receptors" for growth factors, and accumulation of inhibitory substances in the medium. Semiquantitative estimates of the relative contribution of each of these factors to density-inhibited growth control have led to the suggestion by Holley et al. (1978) that the loss of "receptor" sites on cell surface and release of inhibitors into the culture medium appear to participate equally in arresting growth in the kidney epithelial cell line, BSC-1. On the other hand, inactivation of serum factors in crowded cultures apparently plays a major part in the proliferation of 3 T 3 mouse fibroblasts. Other lines of evidence (for example see Holley, 1972; Pardee, 1975; Whittenberger and Glaser, 1977; Natraj and Datta, 1978a, b) suggest that components of the cell surface may participate directly by providing a regulatory

Table 1. Some examples of "mitogens" that stimulate quiescent cells

Nature of the substance	Name	Cell types used	References
Polypeptide growth factors	Epidermal growth factor	3T3	Armelin (1973)
	Fibroblast growth factor	3T3	Gospodarowicz (1974)
	Platelet-derived growth factor	Smooth muscle cells	Ross et al. (1974)
Lectin	Concanavalin A	Lymphocytes	Powell and Leon (1970)
Hormones	Insulin	3T3	Holley and Kiernan (1974)
	Prostaglandin $F_{2\alpha}$	3T3	Jiminez De Asua et al. (1975)
Enzymes	Trypsin	3T3	Burger (1970)
	Thrombin	Chick embryo fibroblasts	Chen and Buchanan (1975)
	α-Mannosidase	Lymphocytes	Paus and Steen (1978)
	Neuraminidase	Chick embryo fibroblasts	Vaheri et al. (1972)
Divalent cations	Ca^{++}	3T3	Dulbecco and Elkington (1975)
	Zn^{++}	Chick embryo fibroblasts	Rubin (1975)

signal for the onset of quiescence both in density-inhibited and sparse cell populations. What follows is a brief review of the control of DNA synthesis and cell division by membrane derived factors in mouse fibroblasts. The reader should consult the preceding chapter for a timely discussion of "Chalones."

Before beginning this review, it is appropriate to consider the rationale that led to the search for molecules on the cell surface that may influence cell growth in some specific way. It is known for some time that many substances such as proteases, cyclic nucleotides, lectins and certain cations can interact with surface membrane and trigger the progression of quiescent (G_0) cells in conditioned medium through at least one round of cell division (Table 1). Although a general mechanism to explain how these diverse agents stimulate cell growth has not yet been formulated, it is likely that some alterations in the cell membrane must occur in order to allow the G_0 cells to obtain components from the conditioned medium that are neccessary for renewed DNA synthesis and cell division. To put this concept in another way, the conditioned medium may not be completely depleted of growth-requiring substances, but it may contain them in low concentrations not sufficient to sustain the growth of cells which have not been exposed to surface-acting agents. A logical extension of this notion is that under restrictive culture conditions cells are able to sense the impending depletion of factors required for normal growth, and respond by elaborating some cell surface component(s) that acts as a positive regulatory signal to achieve quiescence. Under favorable nutritional status, the cells may not synthesize the regulatory factor or produce it in an inactive form.

These predictions led to the discovery of a fibroblast growth regulatory factor in sparse quiescent cells, and a plasma membrane-derived component in density-inhibited mouse fibroblasts which prevent DNA synthesis and cell division by arresting cells in G_1.

B. Some Properties of Membrane-Derived Inhibitory Factors

The initial indication that the plasma membrane of quiescent cells contains an inhibitory component comes from the experiments of WHITTENBERGER and GLASER (1977): a membrane preparation from density-inhibited Swiss 3 T 3 cells when added to a culture of growing 3 T 3 cells reduced the rate of ^{3}H-thymidine incorporation into acid-precipitable material. The extent of inhibition was dependent on membrane concentration (as expressed by units of phosphodiesterase activity, a membrane marker enzyme), and increased with incubation time. The inhibitory activity is heat-labile, and the effect is reversible in that the plating efficiency of membrane-treated cells after trypsinization is similar to that seen with control cells. That the inhibition of DNA synthesis is not the result of removal by membranes of some necessary growth factors in the bulk medium was shown as follows: when growing 3 T 3 cells were exposed to membrane preparations in complete medium for 24 h, a significant inhibition of DNA synthesis was observed. The "used" medium recovered from these dishes, when added to fresh cultures of 3 T 3 cells after removal of membranes by centrifugation, did not influence DNA synthesis.

Two recent observations by WHITTENBERGER et al. (1978, 1979) indicate that membrane preparations from sparse growing cells have less of the inhibitory factor, and that the inhibitory activity is associated with an intrinsic membrane component which cannot be extracted by high concentrations of sodium pyrophosphate; only when the membranes are incubated with the nonionic detergent octylglucoside, did a significant fraction of the inhibitory material become "solubilized" (that is the active component is not sedimented at $100,000 \times g$ for 30 min).

In an independent study, NATRAJ and DATTA (1978b) detected the presence of an active component in quiescent BALB/C 3 T 3 cells (designated fibroblast growth regulatory factor, FGRF) which inhibits DNA synthesis and cell division in growing 3 T 3 cells. The factor is loosely attached to the cell surface and is easily extracted from intact cells by incubating in serum-free medium containing a low concentration of urea; it is nondialyzable and appears to be protein as judged by trypsin sensitivity and heat lability. After gel filtration through Sephadex G-200, partially purified FGRF at a concentration of 4 µg/ml reduced both the rate of ^{3}H-thymidine incorporation into DNA and the nuclei labeling index in 3 T 3 cells by 50%–60%. A transient exposure of cells to the factor for 1 h followed by incubation in its absence for 20 h was sufficient to elicit its inhibitory effect. Extracts obtained in an identical manner from quiescent cells that had been preincubated in situ with uridine diphosphate N-acetyl-D-glucosamine (UDP-GlcNAc) did not inhibit DNA synthesis (see below). Also, extracts from actively growing cells were inactive. As of this writing, the inhibitory factor has not been purified sufficiently to examine its physical and chemical characteristics. Thus, it is too early to decide whether FGRF extracted from sparse quiescent BALB/C 3 T 3 cells is similar to, or different from, the active component solubilized from the plasma membrane-enriched fraction of density-inhibited Swiss 3 T 3 cells. It is noteworthy in this context that a high molecular weight polypeptide, which also inhibits DNA synthesis, has recently been detected in the culture medium of density-inhibited BSC-1 cells (HOLLEY et al., 1978); interestingly, the inhibitor does not act on 3 T 3 cells indicating some cell specificity.

C. Cell Cycle-Dependent Inhibition of DNA Synthesis and Cell Division

Before assigning a physiological role to a substance that shows inhibitory effect on cell growth, it is important to rule out the trivial effects such as general toxicity, loss of anchorage dependence, and inactivation or sequestration by the added material of some needed growth factors in the culture medium (cf. BASERGA, 1976). Experiments with FGRF and the plasma membrane-enriched fraction (or octylglucoside extracts of membrane preparations) show that the added inhibitory factors do not preferentially detach cells from the culture dish, and the inhibitory effect is still observed when treated cells are incubated in fresh serum-containing medium (after removal of excess unbound material); the inhibition is at least partially reversible by either exhaustive washing or trypsin treatment (WHITTENBERGER and GLASER, 1977; WHITTENBERGER et al., 1978; NATRAJ and DATTA, 1978 b; NATRAJ and DATTA, unpublished work).

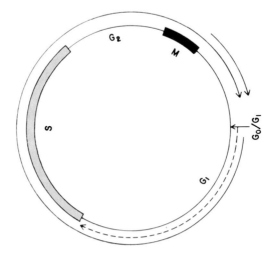

Fig. 1. Schematic representation of the action of FGRF in the cell cycle. Addition of FGRF at any point during late G_1, S, or G_2 arrests cells in early G_1 of the next cycle. FGRF added to cells in early G_1 blocks progression of cells in the same cycle. When added to serum-arrested Go/G_1 cells, FGRF delays their entry into the S phase (----) but eventually they are able to traverse G_1. FGRF-mediated growth arrest point may or may not be identical with the Go/G_1 arrest point. (Adapted from DATTA and NATRAJ, 1980)

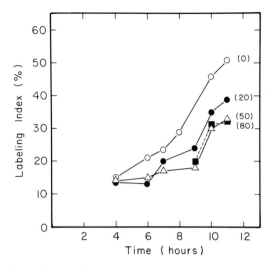

Fig. 2. Kinetics of entry into S phase of G_1-arrested cells in the presence of varying levels of FGRF added at the time of serum stimulation. Concentration of FGRF, expressed as µg protein per ml culture, is indicated next to each curve. Percent nuclei labeled was determined by autoradiography. (Adapted from DATTA and NATRAJ, 1980)

The most compelling evidence for a possible physiological role of FGRF comes from a detailed analysis of the cell cycle-dependent inhibition of DNA synthesis in synchronized populations of 3 T 3 cells (DATTA and NATRAJ, 1980). Addition of FGRF to cells in late G_1 or early S phase does not inhibit DNA synthesis in the immediate S phase, but only a small fraction of the post-mitotic cells enters the next S phase; the time needed for the cells to enter mitosis is the same in the presence or absence of FGRF. Cells exposed to FGRF in late S/early G_2 show reduced DNA synthesis in the following S, and late S/early G_2 cells appear to be more sensitive to inhibition than the cells in G_1. One plausible explanation for this observation is that alterations in the membrane environment as the cells begin to round up for mitosis may allow a more efficient interaction of FGRF with the cell surface. These results suggest that the physiological effect of FGRF is to arrest cells in G_1 thus preventing their entry into a new round of cell cycle (Fig. 1).

By comparing the ability of mitotic cells to synthesize DNA in the next S phase with and without FGRF added at various times after plating out in complete medium, it became apparent that the cells are arrested in early G_1. That is, FGRF added to mitotic cells approximately 3 h after they were plated out had no effect on DNA synthesis; the time needed for these cells to enter S phase was about 9 h. This notion is also supported by the observation that FGRF added to serum-arrested cells at the time of serum stimulation delayed their entry into S phase (see Fig. 2).

Membrane preparations from density-inhibited Swiss 3 T 3 cells also appear to arrest cells in G_1 (WHITTENBERGER et al., 1979). The kinetics of entry of membrane-inhibited cells into S phase are identical to those seen with serum-deprived cells after stimulation with serum, or when density-inhibited cells are trypsinized and replated in complete medium with serum. Further, the flow microfluorometric analysis of membrane-inhibited cells shows an intermediate value of their DNA content that lies between those seen with sparse growing and density-inhibited G_0 cells.

Because complete inhibition of DNA synthesis by FGRF or the membrane fraction isolated from density-inhibited cells has not been achieved, an important question to be answered is whether or not cells exhibit differential sensitivity to the action of the inhibitory factors. Based on the evidence that the apparent rate of DNA synthesis in nonsynchronous cells (pulsed with ^{3}H-thymidine) decreases logarithmically with time, and that the inhibition is increased both by increasing membrane concentration or, at a given membrane concentration, by decreasing the amount of serum present, Whittenberger et al. (1979) proposed that a constant fraction of the cell population is arrested in G_1 during a given cell cycle. A logical extension of this proposal is that each cell which escapes inhibition has an equal probability of being arrested at a later time. It is too early to decide whether this notion is correct especially when little information is available regarding the site in the cell cycle where the membranes act to arrest cells in G_1, the minimum threshold concentration of serum factors that can overcome the inhibition, and the presence of contaminants in the plasma membrane-enriched fractions which, under some experimental situations, lead to detachment of cells from the petri dishes (see Whittenberger et al., 1979).

Nevertheless, when taken together, these results clearly ascribe a physiological role of the membrane-derived factors. It is noteworthy that of the four distinct phases of the cell cycle (G_1, S, G_2, and M) the G_1 period is a major site for control of animal cell proliferation. Beside serum starvation nutritional manipulations, for example amino acid deprivation (Tobey, 1973), and the inhibitory effect of interferon (Balkwill and Taylor-Papadimitriou, 1978) have been correlated with cells' arrest in G_1. Based on a series of kinetic experiments, Pardee (1974) has proposed the existence of a restriction-point (R-point) control which is a switching point in G_1 that regulates the reentry of normal cells into a new round of cell cycle. Whether or not the FGRF-mediated growth-arrest point is identical to the R-point remains to be established.

D. The Role of Glycosylation in the Control of DNA Synthesis

It was mentioned earlier that extract from growing cells does not inhibit DNA synthesis, and that incubation of quiescent 3T3 cells with UDP-GlcNAc also yields inactive FGRF. That the effect is specific for UDP-GlcNAc is noted by the observation that quiescent cells preincubated with guanosine diphosphate mannose, uridine diphosphate glucose or uridine diphosphate galactose produce active FGRF (Natraj and Datta, 1978b). Because UDP-GlcNAc serves only as a substrate for glycosyltransferases that transfer N-acetylglucosaminyl (GlcNAc) residues onto oligosaccharide chains (Roseman, 1970), the most likely interpretation of this finding is that glycosylation of some cell surface acceptors with GlcNAc residues leads to the conversion of active FGRF to an inactive form. Several experimental observations (Natraj and Datta, 1978a; Natraj and Datta, 1978b) are consistent with this proposal. Incubation of serum-deprived quiescent 3T3

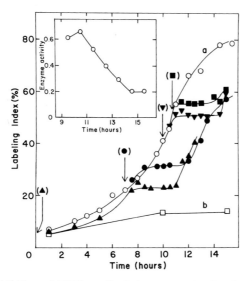

Fig. 3. Transient inhibition of DNA synthesis in serum-stimulated quiescent 3 T 3 cells by the action of N-acetyl-β-D-glucosaminidase. Aliquots of enzyme in phosphate-buffered saline were added to separate dishes at the times designated by the arrows (enzyme concentration, 0.45 unit/ml of culture medium). Percent nuclei labeled was determined by autoradiography. The *inset* shows disappearance of N-acetylglucosaminidase activity (added at 9.5 h after serum stimulation) from the culture medium as a function of time. (NATRAJ et al., in preparation)

cells with UDP-^{3}H-GlcNAc results in a time- and concentration-dependent incorporation of ^{3}H-GlcNAc onto cell surface acceptors. Cells stimulated with serum for 7 h prior to incubation with labeled UDP-GlcNAc incorporate less than half the amount of GlcNAc, suggesting that cell surface acceptors of quiescent cells are underglycosylated, at least in terms of GlcNAc residues, and that serum-stimulated cells already have a large fraction of their acceptors charged with this amino sugar. When inactive FGRF extracted from UDP-GlcNAc-pretreated quiescent cells is exposed to purified beef kidney N-acetyl-β-D-glucosaminidase in vitro, the inactive form of the factor is converted to an active form as judged by inhibition of DNA synthesis in growing 3 T 3 cells. Although it appears highly likely, no evidence exists as yet to conclude that *direct* glycosylation or deglycosylation of FGRF molecules is involved in the interconversion between active and inactive forms.

The findings described above have led to the proposal (NATRAJ and DATTA, 1978 b) that the onset of quiescence by serum deprivation and the appearance of active FGRF on the surface of 3 T 3 cells are causally related, and that conversion of the active FGRF to an inactive form under favorable nutritional status may be viewed as a switch that turns off the regulatory factor and allows DNA synthesis to resume. According to this model, inactive glycosylated FGRF will be found on the surface of serum-stimulated quiescent cells during their subsequent transit through G_1. Some experimental data which support this prediction are displayed in Fig. 3 (NATRAJ et al., in preparation). When serum-stimulated 3 T 3 cells are exposed to purified N-acetyl-β-D-glucosaminidase, there is a transient inhibition in

DNA synthesis as measured by the nuclei labeling technique. Regardless of the time of addition of the enzyme, a small but constant number of cells continues to enter S phase before DNA synthesis is shut off; further, after the "plateau period" during which no new nuclei are labeled, the enzyme-treated cells recover and begin to enter the S phase at almost the same rate as that seen with the control culture. Upon continued incubation, cells complete their entry into S phase, indicating the transient reversible nature of the inhibition phenomenon. Several control experiments (NATRAJ et al., in preparation) show that (a) active FGRF can be extracted from the enzyme-treated growing cells, (b) incubation of enzyme-treated cells with UDP-GlcNAc which glycosylates surface acceptors yields inactive FGRF, and (c) cells prelabeled metabolically with ^{14}C-glucosamine release labeled free GlcNAc after treatment with N-acetyl-β-D-glucosaminidase. One general conclusion from these results is that serum-stimulated quiescent cells contain nonfunctional FGRF which is converted to an active form by removal of GlcNAc residues from some cell surface acceptors by the action of N-acetylglucosaminidase.

The above data do not suggest a mechanism as to how the unglycosylated acceptors present on the surface of quiescent cells become glycosylated after serum stimulation. It is possible to envisage that direct glycosylation of surface components with GlcNAc residues from UDP-GlcNAc may occur in vivo (and this is achieved by incubating in situ intact quiescent cells with UDP-GlcNAc, see NATRAJ and DATTA, 1978a, b). However, a more likely physiologically relevant mechanism may be to replace deglycosylated components with newly synthesized glycosylated molecules on the cell surface. Whatever the exact mechanism may be, it is clear that glycosylation-deglycosylation reactions play an important part in the reversible interconversion of active and inactive forms of the regulatory factor which is intimately involved in the transition of mouse fibroblasts between proliferative and quiescent states, and vice versa (Fig. 4). Future experiments should reveal whether or not this model is correct.

E. Inhibition of DNA Synthesis in Virus-Transformed Cells

Do membrane-derived inhibitory factors influence DNA synthesis in virus-transformed cells? Experiments from two laboratories (WHITTENBERGER et al., 1978; DATTA and NATRAJ, 1980) indicate that extracts from quiescent 3T3 cells, which arrest growth of untransformed 3T3 cells, have no effect on DNA synthesis in 3T3 cells transformed by Simian Virus 40 (SV-3T3). This is perhaps expected because transformed cells exist in proliferative state and, in most cases, appear to have lost the normal growth controls (PARDEE, 1974; HOLLEY, 1975; SCHER et al., 1978). Surprisingly however, DATTA and NATRAJ (1980) found that the transformed cells contain an active regulatory factor which, at comparable protein concentrations, inhibits DNA synthesis in untransformed cells to an extent similar to that seen with FGRF from quiescent 3T3 cells. In contrast, WHITTENBERGER and GLASER (1977) found only a marginal inhibition of DNA synthesis when untransformed cells were exposed to a membrane preparation from SV-3T3 cells. This apparent discrepancy may be due to different growth conditions of the transformed cell lines. DATTA and NATRAJ (1980), for example, grew SV-3T3 in 4% serum for 72 h without medium change followed by incubation for 12 h in serum-free medium which significantly reduced the growth of cells, a condition that led to a state of "quiescence." Whereas, WHITTENBERGER and GLASER (1977) used SV-3T3 cells grown to multilayer density in 10% serum for their membrane preparations; presumably these cells were still growing at a significant rate, and as mentioned above, extracts from growing cells do not inhibit DNA synthesis.

The reason why active FGRF isolated from untransformed or transformed cells does not inhibit DNA synthesis in SV-3T3 cells is not known. There are several possibilities including (a) SV-3T3 cells lack the restriction-point (R-point) control (PARDEE, 1974) and a functional restriction point is necessary for the action of FGRF; (b) at least one other component is required for the expression of the inhibitory effect of FGRF and SV-3T3 cells do not have the second component or have it in an altered form; and (c) SV-3T3 cells produce their own unique growth factor(s) (cf. DE LARCO and TODARO, 1978) which compensates for the negative regulatory effect of FGRF. Further research should clarify this situation.

F. Surface Membranes and Nutrient Uptake

One distinguishing feature of quiescent cells is their decreased rate of uptake of various metabolites; when stimulated to grow by a variety of mitogens, these cells exhibit a sharp increase in transport activity within minutes. It has been proposed by HOLLEY (1972) that the transport process itself may act as a primary determinant for the transition between quiescent and proliferative states by regulating the supply of essential nutrients inside of cells, although a cause and effect relationship has not been demonstrated clearly as yet. Whatever the exact mechanism may be, it is reasonable to assume that some changes in the surface membrane of quiescent cells might influence the uptake of nutrients from external medium.

LIEBERMAN et al. (1979) have shown that the rates of uptake of α-aminoisobutyric acid and uridine are decreased in growing 3T3 cells after the addition of a plasma membrane-enriched fraction from density-inhibited cells, whereas, the rates of uptake of 2-deoxyglucose and phosphate are not. These data are interpreted to mean that reduced uptake of α-aminoisobutyric acid and uridine, but not 2-deoxyglucose or phosphate, are related to "contact inhibition" of growth.

Recently, NATRAJ and DATTA (1978a) reported that preincubation of sparse quiescent 3T3 cells with UDP-GlcNAc in conditioned medium results in the glycosylation of cell surface acceptor with GlcNAc residues and a concomitant stimulation of uptake of 2-deoxyglucose, uridine, and α-aminoisobutyric acid over that seen in conditioned medium alone. The UDP-GlcNAc-treated cells exhibited higher V_{max} without a change in the K_m values, indicating increased uptake potential for these metabolites. Thus, it is plausible that restoration of uncharged surface acceptors by glycosylation with GlcNAc residues might be a regulatory signal for the increased uptake of small molecules by the quiescent cells. It may be recalled in this context that quiescent 3T3 cells contain an active FGRF which may be converted to an inactive form by glycosylation of surface acceptors with GlcNAc residues; on the other hand, deglycosylation of cell surface acceptors of growing cells by N-acetyl-β-D-glucosaminidase produces active FGRF as detected by the inhibi-

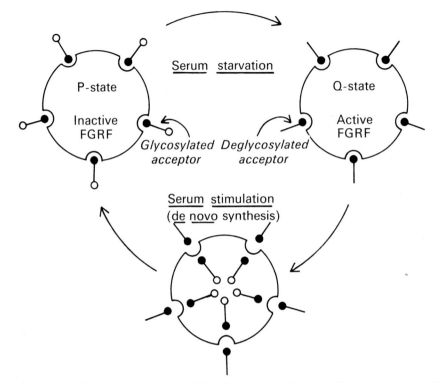

Fig. 4. A model for the reversible transition between proliferative (*P*) and quiescent (*Q*) states of mouse fibroblasts

tion of DNA synthesis. This series of findings raises the intriguing possibility that the reversible transition from quiescent to proliferative states may be regulated by a glycosylation-deglycosylation reaction of surface acceptors with GlcNAc residues: when acceptors are uncharged, active FGRF arrests cells in G_1 and the quiescent cells have low metabolic activity including uptake of nutrients; replacement of uncharged acceptors with glycosylated molecules allows the cells' progression through G_1 by neutralizing the inhibitory potential of FGRF and stimulating transport activity for the essential metabolites. These relationships are depicted schematically in Fig. 4.

G. Prospects and Conclusions

Proliferation of animal cells occurs through a series of precisely programmed events and, under normal physiological conditions, the temporal sequence of the cell cycle is rigidly maintained. From the past several years of research a general picture is emerging to establish the existence of inhibitory factors on cell membranes that regulate DNA synthesis and cell division in mouse fibroblasts in culture. The specific physiological effect of these inhibitory components is to arrest cells in G_1, thus preventing their entry into a new round of the cell cycle. One of these inhibitors appears to be an intrinsic membrane component which is found in greater amounts in density-inhibited cells than in sparsely growing cells; growth-arrest by surface membranes of sparsely growing cells presumably occurs by the same mechanism as that seen in density-dependent inhibition of growth in confluent cultures where increased cell density increases the concentration of the inhibitory factor at or near the cells' surface resulting in loss of proliferative ability.

Fibroblast growth regulatory factor (FGRF), on the other hand, is a loosely attached macromolecular component which appears to accumulate on surface of cells during their transit from proliferative to quiescent state under restrictive nutritional conditions. When exposed to balanced nutritional status, the inhibitory factor is turned off and the actively proliferating cells now contain inactive FGRF. It is reasonable to propose that both the plasma membrane-associated inhibitor and the cell surface FGRF are normally synthesized during active cell growth without exhibiting any apparent deleterious effects; only when the cells reach confluency or exhaust the required growth factors, do these factors exert their regulatory influences to arrest the normal proliferative cycle.

The purification and a detailed characterization of these membrane derived inhibitory factors have not yet been achieved. Judging from the experience of other laboratories with polypeptide growth factors and hormones, a major time commitment is anticipated to achieve these goals. Nevertheless, availability of purified material is essential for future studies on their modes of action. Specifically, one would like to know where in the cell cycle do they act in order to arrest cells' growth in G_1? What is the chemical nature of the inhibitory signals? How are the signals transmitted through the membrane and how do they interact with the cellular machinery? It is hoped that an analysis of the structure and function of these factors will in the near future provide an insight into the mechanisms of growth control in mammalian cells.

References

Armelin, H.A.: Pituitary extracts and steroid hormones in control of 3T3 cell growth. Proc. Natl. Acad. Sci. U.S.A. 70, 2702–2706 (1973)

Balkwill, F., Taylor-Papadimitriou, J.: Interferon affects both G_1 and $S + G_2$ in cells stimulated from quiescence to growth. Nature 274, 798–800 (1978)

Baserga, R.: Multiplication and division of mammalian cells. New York: Marcel Dekker 1976

Baserga, R.: Resting cells and the G_1 phase of the cell cycle. J. Cell. Physiol. 95, 377–386 (1978)

Burger, M.: Proteolytic enzymes initiating cell division and escape from contact inhibition of growth. Nature 227, 170–171 (1970)

Chen, L.B., Buchanan, J.M.: Mitogenic activity of blood components. 1. Thrombin and prothrombin. Proc. Natl. Acad. Sci. U.S.A. 72, 131–135 (1975)

Datta, P., Natraj, C.V.: Fibroblast growth regulatory factor inhibits DNA synthesis in BALB/C 3T3 cells by arresting in G_1. Exp. Cell Res. 125, 431–439 (1980)

De Larco, J.E., Todaro, G.J.: A human fibrosarcoma cell line producing multiplication stimulating activity (MSA)-related peptides. Nature 272, 356–358 (1978)

Dulbecco, R., Elkington, J.: Induction of growth in resting fibroblastic cell cultures by Ca^{++}. Proc. Natl. Acad. Sci. U.S.A. 72, 1584–1588 (1975)

Gospodarowicz, D.: Localization of a fibroblast growth factor and its effect alone and with hydrocortisone on 3T3 cell growth. Nature 249, 123–127 (1974)

Holley, R.W.: A unifying hypothesis concerning the nature of malignant growth. Prac. Natl. Acad. Sci. U.S.A. *69*, 2840–2841 (1972)

Holley, R.W.: Control of growth of mammalian cells in cell culture. Nature *258*, 487–490 (1975)

Holley, R.W., Kiernan, J.A.: Control of initiation of DNA synthesis in 3T3 cells: Serum factors. Proc. Natl. Acad. Sci. U.S.A. *71*, 2908–2911 (1974)

Holley, R.W., Armour, R., Baldwin, J.H.: Density-dependent regulation of growth of BSC-1 cells in cell culture: Growth inhibitors formed by the cells. Proc. Natl. Acad. Sci. U.S.A. *75*, 1864–1866 (1978)

Jiminez De Asua, L., Clingan, D., Rudland, P.S.: Initiation of cell proliferation in cultured mouse fibroblasts by prostaglandin $F_{2\alpha}$. Proc. Natl. Acad. Sci. U.S.A. *72*, 2724–2728 (1975)

Lieberman, M.A., Raben, D.M., Whittenberger, B., Glaser, L.: Effect of plasma membranes on solute transport in 3T3 cells. J. Biol. Chem. *254*, 6357–6361 (1979)

Natraj, C.V., Datta, P.: Uridine diphosphate N-acetylglucosamine stimulates uptake of nutrients by quiescent BALB/C 3T3 cells. Proc. Natl. Acad. Sci. U.S.A. *75*, 3859–3862 (1978a)

Natraj, C.V., Datta, P.: Control of DNA synthesis in growing BALB/C 3T3 mouse cells by a fibroblast growth regulatory factor. Proc. Natl. Acad. Sci. U.S.A. *75*, 6115–6119 (1978b)

Natraj, C.V., Hilfinger, J.M., Datta, P.: Deglycosylation of surface acceptors prevents progression of 3T3 cells from G_1 to S. (in preparation)

Pardee, A.B.: A restriction point for control of normal animal cell proliferation. Proc. Natl. Acad. Sci. U.S.A. *71*, 1286–1290 (1974)

Pardee, A.B.: The cell surface and fibroblast proliferation. Some current research trends. Biochim. Biophys. Acta *417*, 153–172 (1975)

Pardee, A.B., Dubrow, R., Hamlin, J.L., Kletzien, R.F.: Animal cell cycle. Ann. Rev. Biochem. *47*, 715–750 (1978)

Paus, E., Steen, H.B.: Mitogenic effect of α-mannosidase on lymphocytes. Nature *272*, 452–454 (1978)

Powell, A.E., Leon, M.A.: Reversible interaction of human lymphocytes with the mitogen concanavalin A. Exp. Cell Res. *62*, 315–325 (1970)

Roseman, S.: The synthesis of complex carbohydrates by multiglycosyl-transferase systems and their potential function in intercellular adhesion. Chem. Phys. Lipids *5*, 270–297 (1970)

Ross, R., Glomset, J., Kariya, B., Harker, L.: A platelet-dependent serum factor that stimulates the proliferation of arterial smooth muscle cells in vitro. Proc. Natl. Acad. Sci. U.S.A. *71*, 1207–1210 (1974)

Rubin, H.: Nonspecific nature of the stimulus to DNA synthesis in cultures of chick embryo cells. Proc. Natl. Acad. Sci. U.S.A. *72*, 1676–1680 (1975)

Rudland, P.S., Jiminez de Asua, L.: Action of growth factors in the cell cycle. Biochim. Biophys. Acta *560*, 91–133 (1979)

Scher, C.D., Pledger, W.J., Martin, P., Antoniades, H.N., Stiles, C.D.: Transforming viruses directly reduce the cellular growth requirement for a platelet derived growth factor. J. Cell. Physiol. *97*, 371–380 (1978)

Tobey, R.A.: Production and characterization of mammalian cells reversibly arrested in G_1 in isoleucine-deficient medium. In: Methods in cell biology. Prescott, D.M. (ed.), vol. VI, pp. 67–112. New York: Academic Press 1973

Vaheri, A., Rouslahti, E., Nordling, S.: Neuraminidase stimulates division and sugar uptake in density-inhibited cell cultures. Nature New Biol. *238*, 211–212 (1972)

Whittenberger, B., Glaser, L.: Inhibition of DNA synthesis in cultures of 3T3 cells by isolated surface membranes. Proc. Natl. Acad. Sci. U.S.A. *74*, 2251–2255 (1977)

Whittenberger, B., Raben, D., Glaser, L.: Regulation of the cell cycle of 3T3 cells in culture by a surface membrane-enriched cell fraction. J. Supramol. Struct. *10*, 307–327 (1979)

Whittenberger, B., Raben, D., Lieberman, M.A., Glaser, L.: Inhibition of growth of 3T3 cells by extract of surface membranes. Proc. Natl. Acad. Sci. U.S.A. *75*, 5457–5461 (1978)

CHAPTER 11

Diffusible Factors in Tissue Cultures

Louise Harel

A. Introduction

It is well known that cells of higher, multicellular organisms interact in a complex way compared with populations of unicellular organisms such as bacteria and protozoa.

Culture of cells in vitro has been used as a simple model to study these complex interactions. One important aspect of this research is related to the observation of a change in the characteristics of cell-cell interactions when the cells are transformed (i.e., rendered malignant). Thus, many studies have been conducted to compare normal and malignant cells with the hope of improving knowledge of the mechanism of cell transformation. For example, it has often been observed since the work of Earle et al. (1951) that the rate of proliferation of a normal cell culture is dependent on the cell density of seeding. There is an optimum cell density, the "feeder density" as it has been called by Stoker (1967a), for which the rate of cell proliferation is maximum. It decreases at lower or higher cell densities.

Normal cells in particular show density-dependent regulation of growth. They stop growing at a certain saturation density and at this density the culture remains healthy but quiescent for some time (the cells enter a Go phase).

In contrast, highly transformed cells have lost density-dependent regulation of growth and they multiply continuously until they exhaust the nutrients of the medium. The culture does not become quiescent and the cells die if they are not refed with fresh medium.

Cell-cell interaction has been studied both between cells of the same type and between cells of different types or even species, particularly between normal and transformed cells. We will not discuss all the different types of interactions between cells but only those that take place by diffusion of a molecule from one cell to another. This diffusion may require close contact between cells and takes place across gap junctions as in metabolic cooperation. Interactions can also occur between cells not in close contact; the molecule is released from a cell, is diffused in the medium, and reaches another cell at long or short range. Concerning this last type of interaction, we will examine:

1. The problem of feeder layers and the diffusion in the medium of stimulating factors such as the multiplication-stimulating activity (MSA) released by Buffalo strain of rat cells.

2. The problem of density-dependent inhibition and the diffusion in the medium of inhibitory factors.

3. Diffusion of factors which may enhance cell transformation.

We will not discuss the different factors produced by cells of the hemopoietic system. They are the subject of other chapters of this book.

B. Metabolic Cooperation

SUBAK-SHARPE et al. (1966), BURK et al. (1968), and SUBAK-SHARPE et al. (1969) have shown that mutant cells incapable of incorporating exogenous hypoxanthine into their nucleic acids become capable of doing so when cocultivated with wild type cells. Mixed cultures of mutant and wild type cells were labeled with ^{3}H-hypoxanthine, and autoradiography of the fixed cultures showed that every mutant cell, like the wild type, incorporated the labeled base. The cells had to be in contact and cytoplasmic bridges between the cooperating cells were described.

SUBAK-SHARPE (1965) has shown that the deficiency of the mutant cells was not due to their impermeability to hypoxanthine but to a deficiency in an enzyme, the hypoxanthine-phosphoribosyl transferase (HPRT). Therefore, the effect of contact between cells is not due to a change in the permeability to hypoxanthine.

It was demonstrated later that metabolic cooperation between normal and HPRT-deficient cells (HPRT$^-$) was the result of transfer from normal to mutant cells of a product of the enzyme, a nucleotide or nucleotide derivative, rather than of transfer of enzyme or informational macromolecules leading to the synthesis of enzyme (Cox et al., 1970, 1972, 1974). Generally in metabolic cooperation, only low-molecular-weight compounds are transferred from one cell to another. This transfer can be observed within 1 h if mutant cells and wild type cells are in contact (Cox et al., 1974). The contact-dependent metabolic cooperation is characterized by the normal cells being the donor and the mutant cells being the recipient, the former transferring to the latter a lacking metabolite. But ZOREF et al. (1977) have shown also that transfer may occur from a mutant donor cell to normal cell. The mutant cells transfer the resistance to feedback inhibition by purine derivatives.

Metabolic cooperation does not require energy synthesis; since treatment of cells with either DNP or a combination of azide and fluoride, which markedly inhibit ATP synthesis, permits a proportion of labeled nucleotides to be transferred from the wild type to mutant cells (HPRT$^-$) (Cox et al., 1972). It neither requires protein synthesis nor is it under the immediate control of the cell nucleus. Enucleated normal cells efficiently communicate with HPRT$^-$ mutant cells. Like the nontreated myoblasts, cytoplasms, prepared from myoblasts, are able to cooperate even 20 h after enucleation (BOLS and RINGERTZ, 1979). By contrast karyoplasts are unable to communicate with intact cells (Cox et al., 1976). Hence it has been suggested that the transfer proceeds via junctions between cells. Supporting this hypothesis are observations made by use of freeze etching techniques and electron microscopy, which show that both metabolic coupling and ionic coupling in fibroblasts (Chinese hamster line) are associated with the appearance of "gap" junctions (GILULA et al., 1972). The structure of gap junctions was described: They appear as arrays of closely packed subunits 80–90 Å in diameter which traverse the two closely apposed membrane (BENEDETTI et al., 1976; CASPAR et al., 1977). The junctional proteins can be separated into several components, the major having a molecular weight of 26,000 (EHRHART and CHAUVEAU, 1977; HENDERSON et al., 1979).

Correlated electrophysiology and metabolic cooperation studies show that cell types incapable of junctional communication are also incapable of metabolic cooperation (GILULA et al., 1972; AZARNIA et al., 1972).

Cytochalasin B, which dissociates mirofilaments and binds to cell membranes, reduced metabolic cooperation, while Colcemid, which dissociates microtubules, had little effect (COX et al., 1974). Communication may take place among cells that apparently lack gap junctions, but this requires prolonged contact.

Metabolic cooperation is selective and specific. Not all cells are able to undergo metabolic cooperation. Specificity is also exhibited in the nature of substances transferred so that, according to COX, the phenotype ($G6PD^-$) of mutant cells is not corrected by contact with normal cells ($G6PD^+$) (COX et al., 1972). However, PITTS and FINBOW (1977) have reported that not only nucleotides but also sugar phosphate, choline phosphate, and the cofactor tetrahydrofolate are exchanged freely between junction-forming cells in culture.

FUJIMOTO et al. (1971) introduced another method to demonstrate metabolic cooperation between cells. They used toxic purine analogues such as 6-thioguanine. This toxic purine was converted into nucleotide by the normal cells but not by cells deficient in hypoxanthine guanine phosphoribosyl transferase activity (fibroblasts established from skin biopsies from patients with Lesch-Nyhan syndrome). In conditions where de novo synthesis of purine is abolished by aminopterin in mixed cultures, the toxic nucleotide synthetized by nondeficient cells is transferred to mutant cells which die. In this case the metabolic cooperation can be described as the "kiss of death." By contrast, in the presence of azaserine, a strong selection of phenotypically normal cells takes place. In mixed cultures, azaserine fails to destroy the mutant cells. Here the ($HPRT^-$) cells with a blocked ability to utilize hypoxanthine can, through metabolic contact, receive the "kiss of life" from contact with ($HPRT^+$) cells that are capable of utilizing hypoxanthine as their sole purine source.

PITTS and SIMMS (1977) used another method which does not require the prior isolation of mutant cells for the study of metabolic cooperation. The donor cells are first labeled with a radioactive precursor such as ^{3}H-uridine and then washed with unlabeled medium. During this wash, ^{3}H-uridine is lost, but the nucleotides are not, because the cytoplasmic membrane is impermeable to nucleotides. After 3 h of labeling and washing, ^{3}H counts are found to be half in the nucleotide pool and half in the nucleic acids. When these prelabeled cells are cocultured with unlabeled recipient cells, radioactive materials are transferred from donor cells to recipient cells in contact, as observed by autoradiography. Radioactive material is not transferred when cells are very close to, but not in contact with, donor cells. The formation of gap junctions between cells is certainly required. Between cells which do not form junctions, like L 921, there was no transfer of nucleotides. PITTS and SIMMS have shown that macromolecules are not transferred by intercellular junctions.

Recently LEDBETTER and LUBIN (1979) have introduced a new method for measurement of cell-cell coupling. Using $^{86}Rb^+$ as a measure of intracellular K^+, they found K^+ transferred from ouabain-resistant cells to ouabain-sensitive cells. They define an "index of cooperation" which can be used to measure cell interaction in different cell combinations. An index of cooperation greater than 0 requires cell contact, since no enhancement of intracellular $^{86}Rb^+$ in the culture occurs when contact between two cell types is prevented. The index of cooperation they found for different cell combinations agrees with other measures such as electrical coupling. Coculture of ouabain-sensitive and ouabain-resistant cells in the presence of ouabain also leads to restoration of the capacity of sensitive cells for protein synthesis and growth.

Metabolic cooperation has been and continues to be used to approach different problems.

STOKER (1967 b) has used metabolic cooperation to study the growth regulation in cocultures of normal and transformed cells. ^{3}H-thymidine and ^{3}H-hypoxanthine incorporation were investigated by autoradiography in mixed cultures of polyoma-transformed BHK_{21} cells and freshly isolated mouse fibroblasts. Thymidine incorporation was inhibited in transformed cells when they were in contact with stationary layers of normal cells which themselves showed a low proportion of thymidine incorporation. Furthermore, when a variant of polyoma-transformed cells deficient in inosinic pyrophosphorylase was cultured in contact with normal mouse embryo cells, ^{3}H-hypoxanthine was incorporated into nucleic acid of transformed deficient cells. STOKER concluded from these results that junctions exist between normal and transformed cells and that the growth regulation factors might also be transferred between transformed and normal cells.

CORSARO and MIGEON (1977) detected differences in metabolic cooperation between normal and transformed human cells in culture. The communication is significantly reduced when various types of transformed cells are used as either donors or recipients in the contact feeding assay.

Metabolic cooperation was also used by STOKER (1975) to determine if contacts between cells are important in the density-dependent inhibition of cell growth in culture. Cytochalasin B, which abolished most preexisting cooperation between cells both in the layer and in the wound edges of cultures, did not change the topoinhibition of growth. He concluded that alteration in capacity to form stable intercellular junctions is not a necessary feature of the topoinhibition phenomenon. Therefore, density-dependent inhibition is not likely to be due to the contacts between cells.

Metabolic cooperation has been used in demonstration of junctional communication between cells, but it is less direct than the techniques of fluorescence tracer injection or electrical measurement. However, although a positive finding is not sufficient to demonstrate junctional transfer, it does demonstrate cell-cell interaction. It is of interest to note that the basis for the apparent loss of metabolic cooperation in variant (mec$^-$) cells derived by selection from a (mec$^+$) line is a deficiency of intercellular junctions (WRIGHT et al., 1976a). The deficiency can be reversed by treatment of the cells with db-cAMP and theophylline, which also restore the metabolic cooperation. The difference between mec$^-$ cells and the mec$^+$ parental line seems to lie in an altered membrane component involved in cell morphology or in gap junction formation (WRIGHT, 1976b).

Using a method we described previously, PITTS and SIMMS (1977) characterized cell lines according to their ability or inability to form intercellular junctions. They distinguished different cell types, such as BHK, 3T3 Py, and SV 3T3, which are able to transfer low-molecular-weight compounds rapidly, HeLa, BSC_1, and

CHO, which are moderately able to transfer the small molecules, and L 929, HTC, and H_{35}, which are unable to transfer the molecules.

It was of interest to know if differentiated cells were able to interact with non-differentiated cells in mixed cultures and if, through metabolic cooperation, they might be affected by the presence of multipotential cells. NICOLAS et al. (1978) introduced a simple method to study metabolic cooperation between mouse embryonal carcinoma cells and their differentiated derivatives.

In HAT (hypoxanthine/aminopterin/thymidine) medium, cells depleted in HPRT die and lyse. If such cells are labeled with ^{14}C-thymidine, the amount of lysis can be measured by the fraction of radioactivity released into the medium. Metabolic cooperation between these cells and wild type variants can thus be determined by measuring the decrease of radioactivity released from mixed populations.

NICOLAS et al. (1978) used this method to determine an index of intercellular communication between mouse embryonal carcinoma (EC) cells and other cell types. EC cells did not cooperate with differentiated cells of various properties or origins. In contrast, they cooperated with cells of all other EC lines tested, including a human teratocarcinoma line. This would imply that the surface structures involved have been conserved during evolution. The fact that the EC line did not cooperate with differentiated cells indicates that a group of cells can remain isolated from differentiated cells as differentiation proceeds. The isolation from HGPRT⁻ embryonal carcinoma cell line PC 13 TG 8 of a variant R 5/3 defective in metabolic cooperation (SLACK et al., 1978) should give valuable information about the mechanism of metabolic cooperation and its role in embryonic development.

Metabolic cooperation has also been studied in research on the mechanism of promotion in carcinogenesis and particularly that by 12-O-tetradecanoyl phorbol-13-acetate (10^{-7}–10^{-9} M), which greatly enhances tumorigenesis in cells previously initiated by exposure to a low dose of carcinogen.

MURRAY and FITZGERALD (1979) labeled mouse epidermal cells (HEL/37) by incubation with ^{3}H-uridine. The cells were then washed and cocultured with unlabeled 3 T 3 cells. Metabolic cooperation was determined by transfer of label from HEL/37 cells to adjoining 3 T 3 cells. In the presence of phorbol ester, metabolic cooperation was inhibited. Nonpromoting derivatives of these esters did not inhibit transfer of label. These results support the hypothesis that phorbol esters act on the cell membrane. Furthermore, it is possible to predict that cells treated with phorbol esters might acquire independence from their environment.

Metabolic cooperation was also studied in human embryo fibroblasts (IMR-90) as a function of serial subcultivation of cells with the aim of demonstrating changes in cell surface in aging cells. The data of KELLEY et al. (1979) reveal a significant reduction in metabolic cooperation between human diploid fibroblasts as the cells lose proliferative capabilities. The decrease in transfer of nucleotides between aged cultures is affected by altered cellular capabilities for the assembly of gap junctions, even though the available surface area for contact exists, particularly when cells are seeded at high densities. However, it is not clear how the reduction of metabolic cooperation is related to the loss of proliferative capability in late passage cultures.

One interesting feature in metabolic cooperation is that the control of metabolism of different cells may be coupled. In coculture metabolic variations in one cell type affect the other cell type. In mixed cultures of HGPRT and wild type mammalian cells coupled by permeable junctions, the purine nucleotide pools equilibrate between the two cell types prior to incorporation into nucleic acids, even when the two cell types are mixed in different proportions. In the presence of high concentrations of exogenous hypoxanthine, an inhibition of the de novo purine biosynthesis pathway may be observed in the mutant cells, probably as a result of the purine nucleotide pools expended in the wild type cells and consequently by metabolic cooperation also in mutant cells (Sheridan et al., 1975, 1979).

These results clearly show that the mixed cultures acquire different properties than the two cell types examined separately. It becomes a "tissue phenotype." It is possible that in vivo some characteristic properties of tissues are a consequence of cell interactions, since gap junctions are a common feature of most animal cells. We have to keep in mind that the disruption of the tissue will change the properties of the separated cell types.

C. Diffusion of Stimulating Factors

I. The Feeder Effect

Earle et al. (1951) were the first to study the rate of cell multiplication as a function of the seeding density. They determined the minimum number of L. Strain fibroblasts required in a given volume of medium in order to assure growth. [This density was called "the feeder density" by Stoker (1967 a).] Their investigations emphasized the necessity of conditioning the nutrient media by the living cells. As a result of these observations, they were able to produce clones from a variety of cells by growing individual cells in capillary tubes where diffusible factors essential for multiplication remain in association with the cells (Sanford et al., 1948; Likely et al., 1952; Perry et al., 1957).

Puck and Marcus (1955) had the idea of conditioning the medium by a feeder layer of irradiated cells and so developed simple methods for cloning single cells. The cells to be cloned are seeded at very low cell density on the feeder layer. In this condition, a high proportion of cells grow into colonies. Since that report, this cloning technique has often been used and continues to be used, particularly to grow cells freshly isolated from organs and to establish cell lines (Epstein and Kaplan, 1979); the feeder cells used are either the syngeneic cells previously irradiated or cells from another species, either irradiated (Grogan et al., 1970; Malmquist and Brown, 1974; Bird and James, 1975; Graves, 1978) or mitomycin-treated (Winger et al., 1977; Taylor-Papadimitriou et al., 1978). Kirkland et al. (1979) used a feeder layer of mitomycin-treated human fibroblasts to increase the plating efficiency of both normal and malignant cells.

Pope et al. (1974) and Schneider and Zur Hausen (1975) observed an enhancement of transformation of human lymphocytes of B origin by coculture with macrophages. Pope et al. postulated a specific cooperation between macrophages and lymphocytes in the transformation of the latter. However, the results of Schneider and Zur Hausen suggest that macrophages exert a feeder layer effect which is responsible for the prolonged survival of the lymphocyte and this would increase the percent transformation by Epstein-Barr Virus.

To explain the mode of action of the feeder layer, several hypothesis have been proposed. It was first suggested that the cells must condition the surrounding medium with metabolic products. The minimum essential nutritional requirement was defined by EAGLE (1965) for the growth of cells, but this medium did not allow the growth of sparse cultures. Therefore, it was assumed that some metabolites synthetized by the cells leak from the cells into the medium. In sparse culture the accumulation of these molecules in the medium is too low compared to the required intracellular concentration of this substance. For example, the accumulation of serine by HeLa cells is not sufficient when the cells are seeded at concentrations lower than 10^2 cells/ml (EAGLE and PIEZ, 1962).

FISHER and PUCK (1956), using HeLa cells as feeder layers for the growth of sparse cultures of HeLa cells, showed that feeder cells are able to release inositol into the medium, which allows the growth of sparse culture in inositol-free medium. However, it appears that enrichment of the medium with a metabolite is not sufficient to explain the feeder effect, because the medium previously conditioned by the growth of nonirradiated cells is not effective in promoting colony formation (PUCK and MARCUS, 1955). Hence PUCK and MARCUS suggested that the feeder cells supplied a short-lived diffusible factor necessary to sustain colony formation.

For WEISS et al. (1975), growth of low-density populations of mammalian cells may be achieved as effectively by seeding cells on cellular microexudates as by use of the feeder layers.

STOKER and SUSSMAN (1965) used X-irradiated mouse cells as feeder layers and $BHK_{21}C_{13}$ cells as dependent cells. They show that the feeder effect depends on the continuous integrity of the feeder cells. By using marked cells, they demonstrated that the feeder effect is not due to more effective sticking of the dependent cells in the neighborhood of feeder cells. Attachment occurs as readily on the baze areas of a surface as in the proximity of feeder cells. The feeder effect does not require contact between the feeder cells and dependent cells, because the enhancement will diffuse through agar to dependent cells suspended 0.5 mm above the feeder layer. However, it depends on the distance between cells being maximum within 0.2 mm of the feeder cells and is detectable with diminishing effect up to 1.5 mm. Furthermore, the active principle is not dialyzable but passes through millipore filter. From these results STOKER and SUSSMAN assumed that large unstable molecules are involved in the feeder effect.

REIN and RUBIN (1971) have shown that when chick cells are seeded at a low density in fresh medium they not only fail to grow but within a day also lose their ability to grow. Addition of pyruvate to the culture medium increases the survival of cells. The protective effect of sodium pyruvate suggests that cells at low concentration in fresh medium lose pyruvate or metabolically related compounds to the medium and this depletion is lethal. Other studies are compatible with this conclusion (DANES and PAUL, 1961; KIELER, 1960; WHITFIELD and RIXON, 1961). It must be noted that pyruvate protects the cells but does not enhance cell multiplication. The addition of conditioned medium (medium conditioned by incubation with a dense culture) or macromolecules but not small molecules from conditioned medium would sustain growth of low-density culture. REIN and RUBIN suggested that macromolecules that enhance cell growth are relatively stable. If the feeder effect of feeder layer is larger at short range it is because the enhancing factor diffuses away from the feeder cells so slowly that if the cells are close together the concentration of conditioning factor in the medium near the cell becomes high enough to support cell growth.

We also have to note that mouse EC cells which are dependent on feeder layers for growth and differentiation may be grown and differentiated in the absence of feeder layers only by adding mercaptoethanol to the medium (OSHIMA, 1978). The meaning of this result is not clear.

Perhaps the mechanism of fedder effect depends on the cells used. In some cases, the enhancing factors in conditioned medium were found stable and have been isolated. This is the case especially of tumor growth factors, which will be described in one chapter of this book and of MSA factor, found in conditioned medium of rat liver cells, the properties of which will be described briefly in the next section. A review of this subject was recently published (NISSLEY and RECHLER, 1978).

II. Multiplication-Stimulating Activity from Rat Cells

Studying the control of multiplication of uninfected rat cells and rat cells converted by murine Sarcoma virus, TEMIN (1970) observed that some rat cells are able to multiply in vitro in the absence of serum.

1. Purification of MSA

From conditioned medium of a line (BRL 3 A) of rat liver cells (Buffalo strain), DULAK and TEMIN (1973 a) purified a polypeptide fraction with MSA for chicken and rat embryo fibroblasts. The MSA activity of conditioned medium is defined by its ability to stimulate the incorporation of ^{3}H-thymidine into acid insoluble material in stationary chicken embryo fibroblasts incubated in the presence of ^{3}H-thymidine.

The method used by DULAK and TEMIN to isolate MSA from conditioned medium was similar to that used by PIERSON and TEMIN (1972) to purify a MSA from calf serum. After chromatography on Dowex 50 and gel electrophoresis, the specific activity was increased from 700 to 27,000 units/mg protein. One unit of MSA activity is defined as the amount of activity equivalent to that contained in 1 mg calf serum protein. MSA was inactivated by mercaptoethanol and dithiothreitol. The rat liver MSA resembled both the MSA from calf serum and somatomedin. It stimulated sulfate incorporation in cartilage and had nonsuppressible insulinlike activity (NSILA-S). It differs from the calf serum MSA fraction in that it does not have antitrypsin activity. DULAK and TEMIN (1973 b) have shown that MSA from rat liver cells conditioned medium resides in a family of at least four polypeptide components which have nearly identical apparent molecular weights (about 10,000 daltons) but which differ in their net electrical charges.

It is of interest to note that the assumption which led to the isolation of MSA is wrong. In effect TEMIN looked for MSA activity in conditioned medium of Buffalo rat liver cells because these cells were able to multiply in serum-free medium, whereas the ability of the BRL cells to multiply in serum-free medium is independent of the level of MSA in the medium (NISSLEY et al., 1977). This is one more case of a wrong assumption leading to an interesting discovery.

The purification of MSA from conditioned medium of Buffalo rat liver cells was continued by NISSLEY et al. (1976). After absorption of conditioned medium on Dowex at neutral pH and elution at pH 11 as in the DULAK and TEMIN method, the enriched fraction was filtered on Sephadex G 75 in 1 M acetic acid and further purified by disk preparative acrylamide gel electrophoresis. A single protein band bearing the activity was obtained and was used for biochemical characterization (NISSLEY and RECHLER, 1978).

The molecular weight of MSA was 8,700 instead of 10,000 daltons. It was stable in prolonged storage in 1 M acetic acid and remained active after boiling at pH 5.5. Like DULAK and TEMIN, NISSLEY and RECHLER found that MSA posseses a single disulfide bond which is important for biological activity. Recently MOSES et al. (1980a) identified at least seven distinct MSA polypeptides in conditioned rat cell medium. The spectrum of biological activities and chemical characteristics suggested that all of these polypeptides were closely related. This conclusion was supported by the study of immunoreactivity of the multiple species of MSA which demonstrated extensive but not complete immunological cross reactivity of the various MSA polypeptides (MOSES et al., 1980b).

2. Metabolic Effects of MSA

The MSA factors enhance glucose and aminoisobutyric acid (AIB) transport in chick fibroblasts (SMITH and TEMIN, 1974) and induce ornithine decarboxylase activity in 3 T 3 cells (NISSLEY et al., 1976). For MERRIL et al. (1977) MSA increased the maximal velocity of AIB uptake in myoblasts as well as in myotubes but had no effect on K_m. Both the basal and MSA stimulated rate of AIB uptake are sodium dependent (MERRIL et al., 1977; DERR and SMITH, 1980).

Addition of MSA to quiescent chicken embryo fibroblasts rapidly stimulates ouabain sensitive Na$^+$, K$^+$-ATPase activity as measured by the rate of ^{86}Rb uptake. So an early event in the stimulation of cell proliferation via MSA as well as serum is an activation of membrane Na$^+$-K$^+$ATPase with an accumulation of K$^+$ inside the cells (SMITH, 1977; MERRIL et al., 1977). However, MSA stimulation of AIB transport is independent of K$^+$ accumulation in myoblast. Under conditions in which the MSA-stimulated accumulation of cell K$^+$ was prevented by ouabain a substantial stimulation of AIB uptake was still observed (MERRIL et al., 1979). The addition of MSA to intact chick embryo fibroblasts does not decrease the basal content of cAMP in the cells. However, it decreases the accumulation of cAMP in cells treated with prostaglandin E$_1$ and causes an inhibition of the AMP cyclase activity (ANDERSON et al., 1979).

It is of interest to note thate the effect of MSA on DNA synthesis was found in various cell lines but with consistent differences in the percent stimulation observed (NISSLEY and RECHLER, 1978).

As known, in cells infected by some temperature-sensitive mutant of Rous Sarcoma virus (RSV) the transforming gene is not expressed at restrictive temperature and expressed at permissive temperature. The expression of the transforming gene, the serum growth factors, and the MSA induce an increase of DNA synthesis. KNAUER and SMITH (1979) compared the early biochemical events initiated by the addition of MSA or serum or the switch on of the transforming genes. They concluded from their experiments that the presence of serum in the medium enhances the proliferative response of quiescent (RSV) infected cells shifted to the permissive temperature. In contrast, the presence of MSA has no additional effect on the response of cells shifted to the permissive temperature in serum-free medium. The results were consistent with the hypothesis that both MSA and the expression of a transforming genes turn on the same activities.

3. Cell Surface Receptors

Using ^{125}I MSA cell surface receptors for MSA have been identified in different systems: chick embryo fibroblast, human skin fibroblast, the BRL 3 A_2 cultured rat liver cell lines, and also purified rat liver plasma membranes.

The competition for the membrane binding between ^{125}I MSA and nonlabeled different related peptides gave the following results:

a) When MSA was reduced and carboxymethylated it was inactive in the radioreceptor assay.

b) Only somatomedin A and NSILAS (RECHLER et al., 1977 a, b; MEGYESI et al., 1976) purified from human plasma consistently and potently inhibited ^{125}I MSA binding. Somatomedin B, growth hormone, glucagon, epidermal, and nerve and fibroblast growth factors did not inhibit the binding of ^{125}I MSA.

c) Competition with insulin produces different results according to the cells used (RECHLER et al., 1977 a, b; RECHLER and PODSKALNY, 1976). In rat liver membranes and cultures of BRL 3 A_2 cells, insulin and proinsulin did not inhibit ^{125}I MSA binding. In contrast, in chick embryo and human skin fibroblasts insulin inhibited ^{125}I MSA binding completely. However, ^{125}I insulin binding to human fibroblast was slightly inhibited by MSA. Furthermore, the analysis of data according to Scatchard is consistent with a single class of noninteracting binding sites for MSA and with heterogeneity of binding sites and/or interactions among binding sites for insulin. Finally, some cell types, such as the IH 9 human lymphoblastoid cell line and turkey erythrocytes, do not bind MSA but have a well-characterized insulin receptor.

Since all the biochemical and biological properties of MSA and somatomedin A and C are very similar, MSA has been termed somatomedinlike polypeptides, used as an analog of rat somatomedin, and often studied with somatomedin peptides.

D. Diffusion of Inhibitory Factors

I. Problem of Density-Dependent Inhibition

As was noted previously, the rate of cell multiplication depends on the cell density of the culture. Even when the medium is changed daily, the rate of proliferation decreases when the density of the culture increases over an optimum density (the feeder density). This density-dependent inhibition (DDI) of growth, often termed "contact inhibition," is particularly interesting, not only because it may be connected with the mechanism that controls the growth of cells in the developing animal but also because the rate of proliferation of tumor cells in vitro is less dependent than that of normal cells on the density of the culture and this may be related to the observation that tumor cells seem to grow unrestrictedly in vivo. Different cell lines, under the same conditions of medium change, show different critical saturation densities. Mouse 3 T 3 cells that were isolated under conditions selecting culture having low saturation density cease to grow when they are just contiguous (TODARO and GREEN, 1963). This is why mouse 3 T 3 cells have often been used to study the mechanism of DDI. In contrast, the BHK$_{21}$ line of hamster fibroblasts has a high saturation density (STOKER, 1967 a). However, BHK$_{21}$ cells are smaller than 3 T 3 cells and this had to be considered (HAREL, unpublished results). In most cases cells transformed spontaneously (after many different passages) or by tumor virus grow into multilayers much thicker than the cells from which they were derived (GOLDE, 1962; VOGT and DULBECCO, 1960; STOKER and MACPHERSON, 1961; RAPP and WESTMORELAND, 1976).

The effect of cocultivation of normal cells and virus-transformed cells, which are less subject to growth inhibition has also been investigated. STOKER et al. (1966) studied the growth of polyoma-transformed BHK$_{21}$ (Py) cells in mixed cultures with normal mouse fibroblasts. Py cells were inhibited only after contact with fibroblasts that were no longer dividing, and the results supported, without demonstrating, the concept that the inhibitory effect was not due to a general change in the medium. On the contrary, MACINTYRE and PONTEN (1967) concluded from the study of mixed cultures of embryonic lung fibroblasts and Rous sarcoma virus-transformed fibroblasts that the normal cells exerted no inhibition on the transformed cells. Recently VAN DER BOSH and MAIER (1979) found an inhibition of SV 40 3 T 3 cell growth in a density-dependent manner when they were cocultivated with 3 T 3 cells. SAKIYAMA et al. (1978) examined the effect of confluent monolayers on the proliferation of cells inoculated on this monolayer. The growth of density-inhibited cells was reduced when they were inoculated on a monolayer of homologous cells. In contrast, the growth pattern of transformed cells was not inhibited when seeded on homologous cell sheets. In mixed culture of different types of density-inhibited cells or density-inhibited cells and transformed cells, the superinoculated cells were or were not inhibited, depending on the strain of cells used. It is of interest to note that growth inhibition, even between homologous cells, disappeared once the monolayer cells were fixed by glutaraldehyde or osmic acid. This could suggest that some kind of interaction is taking place to inhibit the growth of the top layer of cells.

1. Mechanism of DDI

Different hypotheses have been proposed to explain the DDI of proliferation in cultures of normal cells and the decrease of the DDI in transformed cells.

The most simple hypothesis assumes that dense cultures modify the culture medium more rapidly than sparse cultures. For example, in the presence of dense cultures, the medium is acidified rapidly and this would decrease the rate of cell growth since the rate of cell proliferation is pH dependent (CECCARINI and EAGLE, 1971; FROEHLICH and ANASTASSIADES, 1974). CECCARINI and EAGLE suggest that the optimum pH varies from 6.9 to 7.8 with the different cell strains and that the saturation density of some cultures is increased if the pH is stabilized by the addition of nonvolatile organic buffers. However, FODGE and RUBIN (1975) concluded from their own results that neither the intracellular nor extracellular accumulation of lactic acid nor the accompanying reduction in pH is sufficient to explain DDI of the rate of multiplication of chick cells.

It may also be assumed that growth factors of the medium are depleted more rapidly in the presence of dense cultures than in the presence of sparse cultures. If cultures of transformed cells are not so inhibited as cultures of normal cells at large cell densities, it is because transformed cells have a decreased requirement for serum growth factor (HOLLEY, 1975). However, this hypothesis, at least in the form outlined here, has been rejected for the following reasons:

1. Medium removed from nonproliferating cultures is sufficient to support growth of actively growing cultures.

2. If, in a quiescent culture of normal cells at saturation density, a wound is made by removing a narrow strip of cells, cells migrate from the edge of the wound and incorporate thymidine into DNA, while other cells still in the confluent layer, only a few micrometers away in the same medium, remain in a quiescent state (TO-DARO et al., 1965). However, in wounded cultures of transformed cells the rate of DNA synthesis is the same in the cells near the wound or in the layer (DULBECCO and STOKER, 1970).

The differential ability of the cells in the wound to initiate DNA synthesis, in comparison to those in the layer, has been defined as "topoinhibition" by DULBEC-CO (1970). In normal cell cultures the topoinhibition has given support to the view that contact between cells might be responsible for inhibition of normal cell proliferation, an assumption which was first put forward on the bases of the well-known observations of ABERCROMBIE (1962, 1979) on the contact inhibition of movement. This hypothesis, which gave its name to DDI of cell proliferation (often called "contact inhibition"), still has its proponents (WHITTENBERGER and GLASER, 1977; NATRAJ and DATTA, 1978) and will be defended in another chapter of this book. To this hypothesis can be added that of FOLKMAN and MOSCONA (1978). They cultured untransformed cells on modified substrata such that the extent of cell spreading was accurately controlled. They found that cell shape is closely related to DNA synthesis and growth. They concluded that DDI may be related to the change in cell spreading when the culture becomes crowded.

However, different results do not agree with the assumption that physical contact between cells is responsible for the DDI of growth.

1. CANAGARATNA and RILEY (1975), using a technique which allowed areas of cells of different densities to be seeded on the same slide and to grow in the same medium, have reported that the rate of DNA synthesis (followed by the incorporation of ^{3}H-thymidine) in the 3 T 3 cells was dependent on local cell densities but seemed to be independent of the number of contacts between cells. MARTZ and STEINBERG (1972) observed also that visible cell-cell contacts are not the direct cause of postconfluence inhibition of growth. CANAGARATNA and RILEY found that, contrary to a number of previous studies, DDI is exhibited in relatively sparse cultures, beginning at 0.5×10^4 cells/cm^2. At this density, contacts between cells are very few. The experiments of CANAGARATNA and RILEY were of particular interest because they clearly showed that depletion of medium is not the only limiting factor of cell growth. From the results published by CANAGARATNA and RILEY (1975) and also from those published by NORDENSKJOLD et al. (1970) and by CLARKE et

al. (1970), HAREL and JULLIEN (1976 a, b) deduced that, in conditions where the exhaustion of medium does not interfere, the rate of DNA synthesis is inversely related to the square root of cell density. This relationship shows that the number of contacts between cells is not involved in the decrease of DNA synthesis, which in this case would be proportional to the square of the cell density.

2. In cells in culture (LEVINE et al., 1965), the rate of macromolecules synthesis is correlated to the cell density. There is also a correlation between cell density and the rate of uptake of different nutrients: phosphate (BLADE et al., 1966; CUNNINGHAM and PARDEE, 1969), uridine (CUNNINGHAM and PARDEE, 1969), deoxyglucose (BOSE and ZLOTNICK, 1973), and 2-aminoisobutyric acid (DUBROW et al., 1978). In untransformed cells, the rate of uptake of these different nutrients is inhibited when the cell density increases. However, in cells transformed by oncogenic virus, the rate of uptake of these nutrients is less inhibited by cell density (PARDEE, 1975). Furthermore, the addition of serum to saturation density culture of normal cells induces an increase of the uptake of these nutrients within a few minutes (CUNNINGHAM and PARDEE, 1969; JULLIEN and HAREL, 1976), and only about 12 h later does it induce an increase of DNA synthesis (MARTY DE MORALES et al., 1974). Therefore, it was assumed that the rate of uptake of the nutrients might be a growth-limiting factor (HOLLEY, 1972). It is interesting to note that in 3 T 3 cells, phosphate uptake (HAREL and JULLIEN, 1976 a) and deoxyglucose uptake (BOSE and ZLOTNICK, 1973) begin to decrease in cultures of low cell density (5×10^3 cells/cm^2). At this density, the surface available on the substratum or contacts between cells do not limit cell metabolism.

3. DULBECCO and ELKINGTON (1973) tried to clarify the role of cell-to-cell contacts in DDI under conditions in which the availability of medium is constant and only the surface of the dishes changes. With 3 T 3 cells, they found that when the cell inoculum and the serum concentration were high, the final cell density was equal on dishes of all sizes and depended only on the volume of medium. It increased when the medium was changed daily. With epithelial cells, which have a high topoinhibition but a low serum requirement (DULBECCO, 1970), growth was limited primarily by the restriction of dish surface when 10% serum was used. However, when the dish surfaces differed by a factor of eight, the cell number differed only by a factor of five. DULBECCO and ELKINGTON concluded that contacts between cells do not appear to limit the multiplication of 3 T 3 cells, but that the availability of medium factors would limit proliferation. However, we believe that their results did not eliminate the possibility that the decrease of cell growth at saturation density was also due to an accumulation of inhibitory factors in the medium.

4. In the wounding experiment, loss of contact between the migrating cells at the border of a confluent layer was not essential for growth stimulation. In the presence of cytochalasin B, which prevents locomotion of cells and maintains the original topography, the thymidine labeling index was still increased along the wound boundary (STOKER and PIGGOTT, 1974).

5. Confluent quiescent cultures of 3 T 3 cells can be stimulated to grow (with or without a medium change) simply by increasing the velocity of the medium flow over the cell sheet (STOKER, 1973) or by shaking the cultures (STOKER and PIGGOTT, 1974). From these results and those with cytochalasin B, STOKER concluded that the short-range effect on cell growth cannot be due only to cell contact, that growth

in confluent 3 T 3 cells is limited by a diffusion barrier in the medium in the unstirred region close to the cell surface, and that this barrier limits uptake of essential nutrients. This limitation would be released by increased fluid velocity. The existence of such a diffusion barrier remains a subject of controversy (WHITTENBERGER and GLASER, 1978; HOLLEY and BALDWIN, 1979).

6. For HOLLEY (1975), factors that control growth are probably polypeptide hormones or hormonelike material. He assumes that if crowded cultures of normal cells are inhibited it is because they need a higher concentration of the macromolecular growth factors due to a diffusion boundary layer or to a decrease of cell surface and surface receptors for growth factors. HOLLEY et al. (1977), studying the density-dependent regulation of BSC_1 cell growth, found that serum can be replaced to some extent by epidermal growth factor (EGF) and that cells from low density culture bind much more ^{125}I-EGF than do cells from high density culture. HOLLEY et al. suggested that a decrease in the number of available EGF receptor sites per cell and the accompanying decrease in sensitivity of the cells to EGF contribute to density-dependent regulation of growth of these cells. It is obvious, and we agree with HOLLEY, that cells need different types of nutrients and growth factors in order to proliferate. However, the following questions can be raised: Why, in dense cultures, do the cells have a decrease in the number of available EGF receptor sites? Is this decrease due to an effect of cell interaction?

For OKUDA and KIMURA (1978) the amount per cell of factor(s) present in serum plays a major role in the stimulation of DNA synthesis in the resting cultures when the cell density is low. At high cell density factors or conditions other than the factor of serum are additionally involved.

As mentioned by STOKER and PIGGOTT (1974), the results of experiments on shaking of saturation density 3 T 3 cell cultures might also be interpreted in an entirely different manner: the diffusion barrier would not limit the uptake of serum factors but would limit the loss of inhibitory factors produced by the cells. Shaking dense cultures would increase cell metabolism by increasing the probability that inhibitor molecules, released by the cells, would be diluted in the medium instead of being trapped by the cell membrane.

Actually, among the earliest hypotheses stated was the one proposing that DDI is due to diffusion, in the medium, of inhibitory factors produced by the cells (SWANN, 1957; PARDEE, 1964; STOKER, 1967a). As we will see, this hypothesis allows the interpretation of numerous results. Furthermore, a growing number of published studies reveal the presence of metabolism inhibitors and cell multiplication inhibitors in the culture medium.

Although it may be premature to affirm that these diffusible factors are truly responsible for DDI, the published data, taken together, show this hypothesis to be probable. Recently HOLLEY (1980), discussing the different assumptions on DDI, arrives at the same conclusion.

II. Inhibitory Factors Released by Cells

1. Inhibitors from 3 T 3 Cells

YEH and FISHER (1969) have observed a rapid increase in RNA synthesis in 3 T 3 cells when medium of saturation density cultures was replaced by fresh medium (as was also reported by TODARO et al., 1965). They tried to demonstrate an inhibitor of RNA synthesis in post-3-day medium from contact-inhibited culture (CI medium). With such a medium, RNA synthesis in sparse cultures was greatly decreased compared to that with fresh medium. When CI medium was diluted either with fresh medium or saline medium, RNA synthesis was increased, showing that the depletion of growth factors is not the only cause of the decrease of RNA synthesis in CI medium. When the CI medium was dialyzed against the same volume of fresh medium (without serum), the inhibition was found with the dialysate and the dialysand. YEH and FISHER concluded that the inhibitor was a low-molecular-weight compound. This makes the interpretation of their results difficult. They do not formally demonstrate that the inhibition observed with used medium, dialyzed or not, diluted with fresh medium or not, is not due to a depletion of low-molecular-weight nutrients in the medium.

However, YEH and FISHER found that in the presence of CI medium from CI 3T3 cells, the RNA synthesis of HeLa cells of SV61 cells was less inhibited than the RNA synthesis of 3T3 cells. The used medium from HeLa cells or SV61 cultures had no inhibitory effect on 3T3 cells or on HeLa and SV61 cells. Today, the results obtained by YEH and FISHER with HeLa cells or SV61 cells may be explained by the release of growth factors by tumor cells. It is also of interest that after exhaustive dialysis of CI medium the dialysate was still 20% more inhibitory than fresh medium toward RNA synthesis. Perhaps YEH and FISHER have incorrectly interpreted their results; the inhibitor molecule may be of high molecular weight.

PARISER and CUNNINGHAM (1971) tried to show that inhibitors of uridine and phosphate transport are released into the medium by confluent cultures of 3T3 cells. The medium of confluent 3T3 cultures was inhibitory compared to fresh medium and remained inhibitory even when 10% fresh serum was added to this used medium. These results indicated that the depletion of transport-stimulating serum factors could not alone account for the reduced activity of used medium. Dialysis of 3T3 used medium against fresh medium, without serum, significantly increased its transport-stimulating activity, but the activity was also restored by dialysis against buffered isotonic saline as well as against fresh defined medium. Phosphate determination showed that the cells did not release phosphate in used medium and that the specific phosphate radioactivity was the same in fresh and used medium. In addition, uridine uptake by Py 3T3 cells was much less affected by 3T3 used medium than was uptake by 3T3 cells. From these results, PARISER and CUNNINGHAM concluded that low-molecular-weight inhibitor of phosphate and uridine transport were released into the medium by 3T3.

Like the results of YEH and FISHER, those of PARISER and CUNNINGHAM suggested (but did not demonstrate) that inhibitors are released into the medium: In fact, it is difficult to rule out definitely the possibility that the increase of phosphate and uridine transport observed after the dialysis of used medium was not due to replacement of some defined medium component exhausted by cells. It is true that the activity of the used medium was restored by dialysis against saline medium, but the authors did not determine the effect on phosphate uptake of dialysis of fresh medium against a saline; hence the interpretation is not clear. We note, however, that after exhaustive dialysis against fresh medium, the used medium remained 10% inhibitory compared to fresh medium; therefore, the inhibitor may be a high-molecular-weight compound.

By different approaches, HAREL et al. (1975, 1978) also tried to demonstrate the release of inhibitors of phosphate metabolism and cell growth by dense cultures of 3 T 3 cells in experimental conditions where exhaustion of the medium was minimized.

1. 3 T 3 cell cultures grown to stationary phase in monolayer were trypsinized, washed, and seeded in fresh medium at different concentrations; phosphate uptake was determined immediately in cells still in suspension. HAREL et al. (1975) observed that the phosphate uptake was maximum when the cell concentration was 10^5 cells/ml. Dilution of cells from 10^5 to $2.5 \; 10^4$ cells/ml led to a 17% decrease of phosphate metabolism. When the cell concentration increased from 10^5 to 10^6 cells/ml, the phosphate uptake and the incorporation of phosphate in organic acid soluble compound was decreased by 35%. These results are reminiscent of those obtained with monolayer cultures of 3 T 3 cells of different densities, which also showed that the metabolism of cells is maximum for an optimum cell density and decreases for smaller and larger cell densities. It must be noted that a change in phosphate uptake as a function of cell concentration was not observed when 3 T 3 cells transformed by SV_{40} oncogenic virus were trypsinized and seeded in the same conditions. The "concentration-dependent" inhibition of phosphate uptake in 3 T 3 cells was not due to exhaustion of the medium, since the effect was observed within 20 min following the incubation of cells in fresh medium + 10% serum. Furthermore, experiments with fresh medium + 5%, 10%, or 20% serum showed the same concentration-dependent inhibition. It was not due to variation of pH of the medium, which was buffered with bicarbonate and N-tris(hydroxymethyl)methyl-2-aminoethane sulfonic acid (TES) and did not change during the incubation period. It was also verified that aeration of the suspension was sufficient and that agglutination of cells was not responsible for the decrease observed. From these results the authors concluded that some interaction between normal cells exists, and the diffusion of an inhibitor continuously released into the medium is the most plausible explanation.

2. HAREL et al. (1978) studied the effect on phosphate metabolism of shaking dense cultures of 3 T 3 cells incubated in fresh medium with or without 10% serum. They observed that incorporation of phosphate into cells and particularly the phosphorylation of small organic compounds (phosphorylation linked to ATP synthesis) were increased by shaking the 3 T 3 cultures. This response was not obtained when SV_{40} transformed cell cultures were shaken in the same conditions. Thus shaking 3 T 3 denses cultures immediately increases phosphate metabolism and, later, DNA synthesis (STOKER and PIGGOTT, 1974). The oxygenation of the medium did not seem to be involved (STOKER and PIGGOTT, 1974). From these results it appeared that the metabolism of dense cultures of 3 T 3 cells was not limited by a diffusion barrier affecting the uptake of serum component as suggested by STOKER and PIGGOTT (1974), since the increase of phosphate metabolism was observed when 3 T 3 cells were shaken in fresh medium in the presence or absence of serum. More likely, the increase of phosphate metabolism was due to a change in the diffusion barrier affecting the concentration of an inhibitor released by the cells. This assumption was demonstrated in another way.

3. 3 T 3 cells were seeded on glass slides at high or low cell densities. After 2 days of culture, slides from sparse cultures were taken at random and placed in new

Petri dishes side by side either with slides bearing sparse cultures or with slides bearing dense cultures. HAREL et al. (1978) observed an inhibition of phosphate metabolism in sparse 3 T 3 cell cultures which were shaken in the presence of dense cultures. The depletion of growth factors did not seem to be involved, since the inhibition of phosphate metabolism was observed as early as 20 min after the beginning of coincubation of sparse cultures in the presence of dense cultures. Furthermore, after 24 h cocultivation of sparse with dense cultures as described above, the growth of sparse cultures was clearly inhibited. The inhibition of phosphate metabolism and cell growth was much less when the sparse and dense cultures were cocultivated but not shaken. These findings suggest that an inhibitor is released by dense cultures of 3 T 3 cells. Even if this inhibitor is labile, the condition of cocultivation allowed the inhibitor to reach the sparse cultures and affect their growth rate. By ^{3}H-thymidine labeling of nonhomogeneously seeded monolayers cultures of 3 T 3 cells, CANAGARATNA et al. (1977) and CHAPMAN et al. (1978) showed that a correlation exists between the local labeling index and the overall cell density in the culture, suggesting long-range effects due to diffusible factor.

STECK et al. (1979) confirmed the findings of HAREL et al. (1978). They found a decrease of DNA synthesis in sparse cultures cocultivated in the presence of dense cultures instead of in the presence of sparse culture. Furthermore, used medium of confluent 3 T 3 cultures inhibited DNA synthesis of sparse cultures of 3 T 3 cells. The inhibition was lost after dialysis of used medium. However, when the protein was precipitated with ammonium sulfate and the concentrated proteins were filtered on Sephadex G_{15}, the inhibition was found with the protein fraction.

2. Inhibitors from WI$_{38}$ Cells

Experiments by GARCIA GIRALT et al. (1970) suggested that exhaustion of the division potential of WI$_{38}$ human fibroblasts was due to the progressive accumulation of an inhibitor. Evidence of a specific inhibitor of WI$_{38}$ human fibroblast growth has been presented by HOUK et al. (1972). This inhibitor was found in extracts of WI$_{38}$ cells from cultures at saturation density or in used medium from monolayer cultures of these cells after 2, 3, or 4 days incubation. The used medium was dialyzed against 200 volume of distilled water and then centrifuged and lyophilized. The ^{3}H-thymidine incorporation in DNA of sparse culture of WI$_{38}$ cells was determined in the presence or absence of lyophilized protein. The inhibition was concentration dependent. It is of interest that dialyzed used medium did not inhibit ^{3}H-thymidine incorporation in human mitogen-stimulated lymphocytes or in HeLa cells. The inhibition of DNA synthesis may be observed within 3 h after addition of inhibitor. The molecular weight of the inhibitor estimated by the use of Diaflo membrane was between 3 and 5×10^4, and its apparent isoelectric point was approximately pH 3.5–4.3.

3. Inhibitors from Chinese Hamster Fibroblasts

FROESE (1971) observed a delay of division in log phase Chinese hamster cells (V_{79-1}) when they were adjacent to a confluent layer of the same cells across a gap of about 30 µm. He assumed that a diffusible growth inhibitor substance was present in the vicinity of confluent cells and looked for this molecule in the used medium of confluent cultures of Chinese hamster fibroblasts. This medium was concentrated by filtration through a dialyzing membrane and the filtrate was found to inhibit the growth of sparse cultures of cells.

Once more, it was noted that this inhibitor exhibited some specificity, since it did not inhibit the growth of HeLa cells. The decrease in cell number was observed after 6–8 h, which agrees with the supposition that the cells are temporarily held up in the G 1 or G 2 phase of the cell cycle.

4. Inhibitors from a Melanocytic Line

From a highly malignant hamster melanoma cell line (RPMI No 1846), LIPKIN and KNECHT (1974) obtained a contact-inhibited (CI), flat, very cohesive aneuploid hamster amelanotic melanoma cell line with markedly reduced proliferative rate and saturation density. Conditioned mediums from both cell lines were compared in an attempt to demonstrate the release, by the contact-inhibited cell lines (FF), of a macromolecule that would be responsible for the decrease of the growth rate and of the saturation density. The conditioned medium was prepared by incubating saturated cultures both of the highly malignant (RPMI) and the CI cell line for 48 h in fresh medium without serum. From lyophilisate of conditioned medium of the CI cell line, the authors used G_{200} gel filtration to separate a fraction able to restore "contact inhibition" of the cell division in the highly malignant cell line. Cultures treated by this fraction were flat, oriented, and fibroblastlike, and they showed a 55% decrease in saturation density with no loss of viability. By contrast, the same fraction separated from conditioned medium of the highly malignant cell line had no effect.

Both morphological and growth inhibitory effects of the melanocyte inhibitor factor (MCIF) transcend the species barrier, extending equally to malignant melanocytes of man and mouse (LIPKIN and KNECHT, 1975, 1976). An analogous fraction to MCIF was obtained by Sephadex G_{200} chromatography of conditioned culture media from human epidermal cells (LIPKIN et al., 1978). This fraction was functionally identical to MCIF in both morphological and growth inhibitory effects on malignant melanocytes. The major protein present in MCIF is a glycoprotein with a molecular weight of 160,000. Since plant glycoproteins can act as mitotic inhibitors on transformed cells it was assumed by the authors that MCIF acts in the same manner. Both morphological changes and DDI of growth induced by MCIF are preceded by a fall in cyclic GMP and a rise in cyclic AMP (KNECHT and LIPKIN, 1977).

5. Inhibitors from BSC$_1$ Cells

HOLLEY et al. (1978) studied the mechanism of density-dependent regulation of growth of BSC$_1$ cells (an epithelial cell line of African green monkey kidney origin) and tried to verify that inhibitors formed by the cells and released in the medium play an important role in limiting growth at high cell densities. In conditioned medium of dense cultures of BSC$_1$ cells at least three inhibitors were found by HOLLEY et al. (1978): lactic acid, ammonia, and an "unidentified" inhibitor. It was assumed that the cells produced an unidentified inhibitor because it was possible to restore most of the growth activity of the conditioned medium just by shaking this medium in a glass bottle before replacing it into the culture. It was later shown that the unidentified inhibitor was destroyed not only by shaking but also by heating or storing the medium in the absence of cells. For HOLLEY et al. the inhibitory effect on cell growth was largely due to the unidentified inhibitor; lactate and ammonium ions were probably less important.

E. Diffusion of Transforming Factors

This section will consist of two parts: the first concerns the diffusion of protein factors released by cells and able to induce or increase malignant transformation; the second concerns metabolic interaction between two cell types which increases or allows chemical carcinogenesis.

I. Diffusion of Protein Factors Which Enhance Malignant Transformation

1. Transformed cells release into the culture medium various macromolecules which are not released or are released in much lower amounts by untransformed cells. We are referring in particular to the tumor growth factors and the plasminogen activators. Recently, GOTTESMAN (1978) showed that a polypeptide of 35,000 molecular weight is released into the medium in relatively large amounts by transformed cells and in much smaller amounts by nontransformed fibroblasts. SENGER et al. (1979) have examined the proteins secreted into the growth medium by normal and transformed cells and found that transformed cell lines from several mammalian species secrete proteins of about 58,000 molecular weight. Secretion of this protein occurs with spontaneously transformed cells and cells transformed by either DNA or RNA virus. It is probably the surface antigen described by DE LEO et al. (1979), by KRESS et al. (1979), and by CHANG et al. (1979).

The biological role of these 55–58 K protein is unknown. However, there are some publications showing the diffusion of transforming factors from transformed cells; these factors not only increase DNA synthesis in target cells but also change the morphology and the anchorage dependency of the cells.

In 1973, BURK indicated that the medium of SV_{28} cells (baby hamster kidney cells transformed by SV_{40} virus) contained a factor which enhances the migration of 3 T 3 cells. This factor was found neither in calf serum nor in the medium of cultures of untransformed $BHK_{21}C_{13}$ cells. Furthermore, the most highly purified preparation of the migration factor also promoted overgrowth of 3 T 3 cells to higher density and prolonged cell survival in the absence of serum. KRYCEVE et al. (1976) report evidence for the production by chick cells and hamster cells (BHK_{21}) transformed by SCHMIDT RUPPIN Rous sarcoma virus (SR-RSV) of a factor which is released into the medium of cultures and increases focus formation by the BRYAN Rous sarcoma virus (presumably partially defective for cell transformation). The number of foci produced by infection of chick embryo fibroblasts with Bryan sarcoma virus was enhanced 2- to 7-fold when the infected chick embryo fibroblasts were incubated in the medium from culture of RSV-transformed chick embryo fi-

broblasts or transformed BHK cells instead of in culture medium from nontransformed cells.

Chicken cultures infected with a temperature-sensitive mutant (Fu 19) of SR-
RSV have transformed morphology and properties at permissive temperature
(37 °C) and normal phenotype at nonpermissive temperature (41 °C). When these
cells are shifted from 37 °C to 41 °C, they acquire normal phenotype. But in the
presence of conditioned medium from RSV-transformed cells, this reversion was
inhibited (KRYCEVE et al., 1976). The authors have shown that RS $^2/_3$ cells (BHK$_{21}$
cells transformed by the SR-RSV) produce factor(s) which enhance focus formation and this factor(s) was operationally termed "transformation-enhancing factor" (TEF). This factor is nondialyzable and thermolabile and is presumably a protein. Its molecular weight is between 10^5 and 2.10^5 daltons.

The factor is also produced by BHK$_{21}$ cells (C$_{13}$MSV and C$_{13}$Py) transformed
by other tumor viruses: murine sarcoma virus or polyoma virus. The medium of
normal secondary hamster fibroblasts was devoid of enhancing activity, whereas
the medium of spontaneously transformed BHK$_{21}$ cells was active. The mechanism of enhancement of formation of BRSV foci remains unknown. It is clear,
however, that the enhancing activity is not associated with virus particle, since
transformed BHK cells did not produce virus. It may be the product of the viral
transforming gene or some activated cell coded product(s) (KRYCEVE et al., 1976).

Murine sarcoma virus-transformed mouse fibroblasts also produce transforming factors and release them into the medium (DE LARCO and TODARO, 1978).
Three major peaks of activity were separated from the medium with apparent molecular weights of 25,000, 12,000, and 7,000. These factors (termed SGF) not only
stimulate cells to divide but also permit rat fibroblasts to grow progressively in
agar. Neither the untransformed parental cell nor the DNA virus-transformed cells
released similar products. The compounds of all three molecular weights are also
capable of competing for membrane epidermal growth factor (EGF) when tested
with ^{125}I-labeled EGF. However, they differ from mouse EGF in their molecular
weights, their inability to react with anti-EGF antibodies, and also in their ability
to convert cells to anchorage-independent growth. If cell morphology and anchorage-independent growth in agar are taken as indices of the transformed phenotype
(which may not be precisely the case: GAUDRAY et al., 1978), the SGF may be direct
effectors of cell transformation. Recently, SGF was purified by binding and elution
from cells rich in receptor for epidermal growth factor (DE LARCO et al., 1980). This
method yielded a peptide that was essentially homogeneous as determined by
isoelectric focusing. The p$_I$ for ^{125}I SGF was 6.8 compared to 4.4 for ^{125}I EGF and
the biological activity comigrated with the radioactivity. When subjected to
sodium dodecyl-sulfate polyacrylamide gel electrophoresis, the purified ^{125}I SGF
gave a discrete band and migrated as a peptide of approximately 10,000 molecular
weight.

II. Metabolic Interaction
Which Enhances the Chemical Transformation of Cells

As is well known, a number of chemical carcinogens need to be activated before
producing malignant transformation. Aflatoxin B$_1$, 2-acetylamine fluorene

(AAF), and dimethylnitrosamine (DMN) are metabolized by the S_9 fraction of rat liver homogenates (AMES et al., 1973). They are also activated by primary culture of rat liver epithelial cells. HUBERMAN and SACHS (1976) have studied mutagenesis and carcinogenesis with different polycyclic hydrocarbons in Chinese hamster cells. They developed a cell-mediated mutagenesis assay in which cells with appropriate marker for mutagenese are cocultivated with lethally irradiated cells able to activate the different carcinogens. During cocultivation, the reactive metabolites are transferred to the target cells and induce mutagenesis and carcinogenesis. LANGENBACH et al. (1978) studied mutagenesis of V_{79} Chinese hamster cells by using primary cultures of rat liver cells in coculture. The carcinogens N-nitrosodimethylamine, N-nitrosodiethylamine, and aflatoxin B_1 induced ouabain-resistant mutants of V_{79} cells only when the liver cells were present in the culture. The mutation frequency increased with increasing numbers of liver cells seeded.

MONDAL et al. (1979) have developed epithelial cell lines from regenerating mouse liver which metabolize 3-methylcholanthrene, aflatoxin B_1, AAF, and DMN and produce cytotoxic metabolites. A monolayer of these cells, X-irradiated (5,600 rads), was used in coculture with $C_3H/10$ $T_1/2$ cells, and 20 h later the cells were treated with aflatoxin (0.2–0.1 µg/ml). The dishes with $C_3H/10$ $T_1/2$ cells alone were treated at the same time. In the coculture, aflatoxin B_1 which was activated by X-irradiated liver cells killed 25%–40% target cells and produced transformed type III foci. In contrast, in dishes in which only $C_3H/10$ $T_1/2$ cells were seeded, aflatoxin B_1 at this concentration had no effect. This method offers new possibilities for chemical transformation in vitro and will certainly be used in current research.

F. Concluding Remarks

The cell culture in vitro is a simplified model for the study of metabolic interactions which occur in tissue and organs in vivo, but already the metabolic interactions between cells in culture are multiple and complex.

Metabolic cooperation, which requires contact between cells (since gap junction must occur), is certainly important in the life of cells forming a tissue in vivo. Via the gap junction different ions and low-molecular-weight compounds may be transferred from one cell to another and the control of metabolic activity of different cells may be coupled. As a result a tissue will present some unity and express its phenotypic characteristic properties. Metabolic cooperation requires specific surface structure. It may occur between normal and transformed cells, it did not occur between mouse embryonal carcinoma and differentiated cells. So as soon as the differentiation proceeds cells are no more affected by their cellular precursors.

Metabolic interaction may occur also at distance by diffusion of factors reacting with other cells at long or short ranges. The cells release into the culture medium different types of small and large molecular weight compounds.

Some of these compounds enhance cell multiplication and are absolutely required to "condition" the medium when the cells in culture are seeded below "the feeder density" (the density for which the cells are able to "condition" their own medium and grow). The feeder compounds are also produced by X-irradiated cells.

In a cloning method a monolayer of these cells is used to "feed" the cells to be cloned, seeded in low density. In some cases in which the mechanism of feeder effect was studied it was found that the enhancing factors are macromolecules which must be more or less labile according to the cells studied. They diffused through agar to dependent cells suspended 0.5 mm above the feeder layer, but they did not really accumulate in the medium. However, diffusion of stable enhancing factors from cells was also demonstrated, especially in the case of numerous tumor growth factors and multiplication stimulating factor (MSA) from Buffalo rat cells. MSA is a family of polypeptide with properties similar to somatomedin A and C.

As was shown, several results suggest that density-dependent inhibition of growth is due to the diffusion into the medium of inhibitory molecules. However, only a few published data reveal the presence of inhibitors of metabolism and cell growth in the medium of dense cultures.

Several explanations may be offered for the difficulty of demonstrating the presence of inhibitory factors in used medium of dense cultures. First, it had to be shown that the inhibition found with the used medium is due not only to depletion of medium in nutrients and growth factors. Also, in conditioned medium, enhancing and inhibitory factors are present together and the inhibitory factors must be first separated from enhancing factors and then purified in order to demonstrate their existence. Such a purification will be possible if the inhibitory factors are not labile or if they could be stabilized during the different steps of purification. Though difficult, such work has to be undertaken.

If the density-dependent inhibition of growth is due, in nontransformed cells, to the release of inhibitory factors it would be of great interest to establish whether transformed cells release also inhibitory factors and whether they have receptors for these factors.

Transformed cells release into the medium macromolecules different from those diffusing from nontransformed cells. We do not yet know the function of these molecules. However, recently it has been shown that macromolecule factors diffusing from transformed cells may be able to change the morphology and metabolic properties of surrounding nontransformed cells. These factors seem to be coded by the cells and not by the viruses. This result may open a new field in the study of carcinogenesis.

Acknowledgements. I greatly acknowledge the cleaver assistance of J. BANNELIER and the Institut National da la Santé et de la Recherche Médicale for the grant ATP 82.79.114.

References

Abercrombie, M.: Contact-dependent behavior of normal cells and the possible significance of surface changes in virus-induced transformation. Cold Spring Harbor Symp. Quant. Biol. *27*, 427–431 (1962)

Abercrombie, M.: Contact inhibition and malignancy. Nature *281*, 259–262 (1979)

Ames, B.N., Durston, W.E., Yamasaki, E., Lee, F.D.: Carcinogens are mutagens: a simple test system combining liver homogenates for activation and bacteria for detection. Proc. Natl. Acad. Sci. U.S.A. *70*, 2281–2285 (1973)

Anderson, W.B., Wilson, J., Rechler, M.M., Nissley, S.P.: Effect of multiplication stimulating activity (MSA) on intracellular cAMP levels and adenylate cyclase activity in chick embryo fibroblasts. Exp. Cell Res. *120*, 47–53 (1979)

Azarnia, R., Michalke, W., Loewenstein, W.R.: Intercellular communication and tissue growth. VI – Failure of exchange of endogenous molecules between cancer cells with defective functions and uncancerous cells. J. Membr. Biol. *10*, 247–258 (1972)

Benedetti, E.L., Dunia, I., Bentzel, C.J., Vermorken, A.J., Kibbelaar, M., Bloemendal, H.: A portrait of plasma membrane specializations in eye lens epithelium and fibers. Biochim. Biophys. Acta *457*, 353–384 (1976)

Bird, M.M., James, D.W.: The culture of previously dissociated embryonic chick spinal cord cells on feeder layers of liver and kidney and the development of paraformaldehyde induced fluorescence upon the former. J. Neurocytl. *4*, 633–646 (1975)

Blade, E., Harel, L., Hanania, N.: Variation du taux d'incorporation du phosphore dans les cellules en fonction de leurs concentrations et inhibition de contact. Exp. Cell Res. *41*, 473–482 (1966)

Bols, N.C., Ringertz, N.R.: A study of metabolic cooperation with established myoblast cell lines. Exp. Cell Res. *120*, 15–23 (1979)

Bose, S.K., Zlotnick, B.: Growth and density dependent inhibition of deoxyglucose transport in Balb/3T3 cells and its absence in cells transformed by murine sarcoma virus. Proc. Natl. Acad. Sci. U.S.A. *70*, 2374–2378 (1973)

Burk, R.R.: A factor from a transformed cell line that affects cell migration. Proc. Natl. Acad. Sci. U.S.A. *70*, 369–372 (1973)

Burk, R.R., Pitts, J.D., Subak-Sharpe, J.H.: Exchange between hamster cells in culture. Exp. Cell Res. *53*, 297–301 (1968)

Canagaratna, M.C., Riley, P.A.: The pattern of density dependant growth inhibition in murine fibroblasts. J. Cell. Physiol. *85*, 271–281 (1975)

Canagaratna, M.C.P., Chapman, R., Ehrlich, E., Sutton, P.M., Riley, P.A.: Evidence for long-range effects in density-dependent inhibition of proliferation. Differentiation *9*, 157–160 (1977)

Caspar, D.L.D., Goodenough, D.A., Makowski, L., Phillips, W.C.: Gap junction structures. Correlated electron microscopy and X-ray diffraction. J. Cell. Biol. *74*, 605–628 (1977)

Ceccarini, C., Eagle, H.: pH as a determinant of cellular growth and contact inhibition. Proc. Natl. Acad. Sci. U.S.A. *68*, 229–233 (1971)

Chang, C., Simmons, D.T., Martin, M.A., Mora, P.T.: Identification and partial characterization of new antigens from Simian virus 40-transformed mouse cells. J. Virol. *31*, 463–471 (1979)

Chapman, R.E., Marsh, E., Sutton, P.M., Riley, P.A.: Studies of long range density effects in the proliferation of 3T3 and RLCW cells in recirculated medium. Differentiation *10*, 159–164 (1978)

Clarke, G.D., Stoker, M.G.P., Ludlow, A., Thornton, M.: Requirement of serum for DNA synthesis in BHK_{21} cells: effects of density, suspension and virus transformation. Nature *227*, 798–801 (1970)

Corsaro, C.M., Migeon, R.: Comparison of contact mediated communication in normal and transformed human cells in culture. Proc. Natl. Acad. Sci. U.S.A. *74*, 4476–4480 (1977)

Cox, R.P., Krauss, M.R., Balis, M.E., Dancis, J.: Evidence for transfer of enzyme product as the basis of metabolic cooperation between tissue culture fibroblasts of Lesch-Nylan disease and normal cells. Proc. Natl. Acad. Sci. U.S.A. *67*, 1573–1579 (1970)

Cox, R.P., Krauss, M.R., Balis, M.E., Dancis, J.: Communication between normal and enzyme deficient cells in tissue culture. Exp. Cell Res. *74*, 251–268 (1972)

Cox, R.P., Krauss, M.R., Balis, M.E., Dancis, J.: Metabolic cooperation in cell culture: studies of the mechanism of cell interaction. J. Cell. Physiol. *84*, 237–252 (1974)

Cox, R.P., Krauss, M.R., Balis, M.E., Dancis, J.: Studies on cell communication with enucleated human fibroblasts. J. Cell. Biol. *71*, 693–703 (1976)

Cunningham, D.D., Pardee, A.B.: Transport changes rapidly initiated by serum addition to "contact inhibited" 3T3 cells. Proc. Natl. Acad. Sci. U.S.A. *64*, 1049–1056 (1969)

Danes, B.S., Paul, J.: Environmental factors influencing respiration of L. cells strain. Exp. Cell Res. *24*, 344–355 (1961)

De Larco, J.E., Todaro, G.J.: Growth factors from murine sarcoma virus transformed cells. Proc. Natl. Acad. Sci. U.S.A. *75*, 4001–4005 (1978)

De Larco, J., Reynolds, R., Carlberg, K., Engle, C., Todaro, G.J.: Sarcoma growth factor (SGF) from mouse Sarcoma virus (MSV) transformed cells: purification by binding and elution from epidermal growth factor (EGF) receptor-rich cells. J. Biol. Chem. *255*, 3685–3690 (1980)

De Leo, A.B., Jay, G., Appella, E., Dubois, G., Law, L.W., Old, L.J.: Detection of a transformation-related antigen in chemically induced sarcomas and other transformed cells of the mouse. Proc. Natl. Acad. Sci. U.S.A. *76*, 2420–2424 (1979)

Derr, J.T., Smith, G.L.: Regulation of amino acid transport in chicken embryo fibroblasts by purified Multiplication-stimulating activity (MSA). J. Cell. Physiol. *102*, 55–62 (1980)

Dubrow, R., Pardee, A.B., Pollack, R.: 2-Amino-isobutyric acid and 3-O-methyl-D-glucose transport in 3 T 3, SV 40-transformed 3 T 3 and revertant cell lines. J. Cell. Physiol. *95*, 203–211 (1978)

Dulak, N., Temin, H.M.: A partially purified polypeptide fraction from rat liver cell conditioned medium with multiplication-stimulating activity for embryo fibroblasts. J. Cell. Physiol. *81*, 153–160 (1973 a)

Dulak, H., Temin, H.M.: Multiplication-stimulating activity for chicken embryo fibroblasts from rat liver cell conditioned medium: a family of small polypeptides. J. Cell. Physiol. *81*, 161–170 (1973 b)

Dulbecco, R.: Topoinhibition and serum requirement of transformed and untransformed cells. Nature *227*, 802–806 (1970)

Dulbecco, R., Elkington, J.: Conditions limiting multiplication of fibroblasts and epithelial cells in dense cultures. Nature *246*, 197–199 (1973)

Dulbecco, R., Stoker, M.G.: Conditions determining initiation of DNA synthesis in 3 T 3 cells. Proc. Natl. Acad. Sci. U.S.A. *66*, 204–210 (1970)

Eagle, H.: Metabolic controls in cultured mammalian cells. Science *148*, 42–51 (1965)

Eagle, H., Piez, K.: The population-dependent requirement by cultured mammalian cells for metabolites which they can synthesize. J. Exp. Med. *116*, 29–43 (1962)

Earle, W.R., Sanford, K.K., Evans, V.J., Waltz, H.K., Shannon, J.E.: The influence of inoculum size on proliferation in tissue cultures. J. Natl. Cancer Inst. *12*, 133–153 (1951)

Ehrhart, J.C., Chauveau, J.: The protein component of mouse-hepatocyte gap junctions. FEBS Letters *78*, 295–299 (1977)

Epstein, A.L., Kaplan, H.S.: Feeder layer and nutritional requirements for the establishment and cloning of human malignant lymphoma cell lines. Cancer Res. *39*, 1748–1759 (1979)

Fisher, H.W., Puck, T.T.: On the functions of X-irradiated "feeder" cells in supporting growth of single mammalian cells. Proc. Natl. Acad. Sci. U.S.A. *42*, 900–906 (1956)

Fodge, D.W., Rubin, H.: Glucose utilization, pH reduction and density dependent inhibition in cultures of chick embryo fibroblasts. J. Cell. Physiol. *85*, 635–642 (1975)

Folkman, J., Moscona, A.: Role of cell shape in growth control. Nature *273*, 345–349 (1978)

Froehlich, J.E., Anastassiades, T.P.: Role of pH in fibroblast proliferation. J. Cell. Physiol. *84*, 253–260 (1974)

Froehlich, J.E., Anastassiades, T.P.: Possible limitation of growth in human fibroblast cultures by diffusion. J. Cell. Physiol. *86*, 567–580 (1975)

Froese, G.: Regulation of growth in chinese hamster cells by a local inhibitor. Exp. Cell. Res. *65*, 297–306 (1971)

Fujimoto, W.Y., Subak-Sharpe, J.H., Seegmiller, J.E.: Hypoxanthine-guanine Phosphoribosyl-transferase deficiency chemical agents selective for mutant or normal cultures fibroblasts in mixed and heterozygote cultures. Proc. Natl. Acad. Sci. U.S.A. *68*, 1516–1519 (1971)

Garcia Giralt, E., Berumen, L., Macieira-Coelho, A.: Growth inhibitory activity in the supernatants of nondividing WI 38 cells. J. Natl. Cancer Inst. *45*, 649–655 (1970)

Gaudray, P., Rassoulzadegan, M., Cuzin, F.: Expression of simian virus early genes in transformed rat cells is correlated with maintenance of the transformed phenotype. Proc. Natl. Acad. Sci. U.S.A. 75, 4987–4991 (1978)

Gilula, N.B., Reeves, O.R., Steinbach, A.: Metabolic coupling ionic coupling and cell contacts. Nature 235, 262–265 (1972)

Golde, A.: Chemical changes in chick embryo cells infected with Rous Sarcoma virus. Virology 16, 9–20 (1962)

Gottesman, M.M.: Transformation-dependent secretion of a low molecular weight protein by murine fibroblasts. Proc. Natl. Acad. Sci. U.S.A. 75, 2767–2771 (1978)

Graves, J.A.: Use of enzyme – deficient cell lines as feeder layers. In Vitro 14, 506–509 (1978)

Grogan, E.A., Enders, J.F., Miller, G.: Trypsinized placental cell cultures for the propagation of viruses and as "feeder-layers". J. Virol. 5, 406–409 (1970)

Harel, L., Jullien, M.: Relation between the rate of phosphate uptake or DNA synthesis and the mean intercellular distance in mouse and hamster cell cultures. J. Microsc. Biol. Cell. 26, 75–78 (1976a)

Harel, L., Jullien, M.: Evaluation of proximity inhibition of DNA synthesis in 3T3 cells. J. Cell. Physiol. 88, 253–254 (1976b)

Harel, L., Jullien, M., Blat, C.: Control by cell interaction of phosphate uptake in 3T3 cells. Exp. Cell. Res. 90, 201–210 (1975)

Harel, L., Jullien, M., De Monti, M.: Diffusible factor(s) controlling density inhibition of 3T3 cells growth: a new approach. J. Cell. Physiol. 96, 327–332 (1978)

Henderson, D., Eibl, M., Weber, K.: Structure and Biochemistry of mouse hepatic gap. junctions. J. Mol. Biol. 132, 193–218 (1979)

Holley, R.W.: A unifying hypothesis concerning the nature of malignant growth. Proc. Natl. Acad. Sci. U.S.A. 69, 2840–2841 (1972)

Holley, R.W.: Control of growth of mammalian cells in cell culture. Nature 258, 487–490 (1975)

Holley, R.W.: Control of growth of kidney epithelial cells in: Control mechanisms in cultured animal cells. Ed. by: Shields, R., Levi-Montalcini, R., Iacobelli, S., Jimenez Dearua, L.: Sous Presse 1980

Holley, R.W., Baldwin, J.H.: Cell density is determined by a diffusion-limited process. Nature 278, 283–284 (1979)

Holley, R.W., Armour, R., Baldwin, J.H., Brown, K.D., Yeh, Y.C.: Density-dependent regulation of growth of BSC_1 cells in cell culture. Control of growth by serum factors. Proc. Natl. Acad. Sci. U.S.A. 74, 5046–5050 (1977)

Holley, R.W., Armour, R., Baldwin, J.H.: Density-dependent regulation of growth of BSC-1 cells in cell culture: growth inhibitors formed by the cells. Proc. Natl. Acad. Sci. U.S.A. 75, 1864–1866 (1978)

Houck, J.C., Weil, R.L., Sharma, V.K.: Evidence for a fibroblast chalone. Nature New Biol. 240, 210–211 (1972)

Huberman, E., Sachs, L.: Mutability of different genetic loci in mammalian cells by metabolically activated carcinogenic polycyclic hydrocarbons. Proc. Natl. Acad. Sci. U.S.A. 73, 188–192 (1976)

Jullien, M., Harel, L.: Stimulation by serum of the phosphorylation reactions in density-inhibited 3T3 cells. Exp. Cell. Res. 97, 23–30 (1976)

Kelley, R.O., Vogel, K.G., Crissman, H.A., Lujan, C.J., Skipper, B.E.: Development of the aging cell surface. Exp. Cell. Res. 119, 127–143 (1979)

Kieler, J.: Influence of CO_2 tension on the respiration of Yoshida ascites tumor cells. J. Natl. Cancer Inst. 25, 161–176 (1960)

Kirkland, W., Yang, N.S., Jorgensen, T., Longley, C., Furmanski, P.: Growth of normal and malignant human mammary epithelial cells in culture. J. Natl. Cancer Inst. 63, 29–39 (1979)

Knauer, D.J., Smith, G.L.: Regulation of the proliferative response in Rous Sarcoma virus transformed chickens embryo fibroblasts by serum and multiplication-stimulating activity (MSA). J. Cell. Physiol. 100, 311–322 (1979)

Knecht, M.E., Lipkin, G.: Biochemical studies of a protein which restores contact inhibition of growth to malignant melanocytes. Exp. Cell. Res. 108, 15–22 (1977)

Kress, M., May, E., Cassingena, R., May, P.: Simian virus 40-transformed cells express new species of protein precipitable by anti-simian virus 40 tumor serum. J. Virol. *31*, 472–483 (1979)

Kryceve, C., Vigier, P., Barlati, S.: Transformation-enhancing factor(s) produced by virus-transformed and established cells. Int. J. Cancer *17*, 370–379 (1976)

Langenbach, R., Freed, H.J., Huberman, E.: Liver cell-mediated mutagenesis of mammalian cells by liver carcinogens. Proc. Natl. Acad. Sci. U.S.A. *75*, 2564–2867 (1978)

Ledbetter, M.L., Lubin, M.: Transfert of potassium. A new measure of cell-cell coupling. J. Cell Biol. *80*, 150–165 (1979)

Levine, E.M., Becker, Y., Brooke, C.B., Eagle, H.: Contact inhibition, macromolecular synthesis and polyribosomes in cultured human diploid fibroblasts. Proc. Natl. Acad. Sci. U.S.A. *53*, 350–356 (1965)

Likely, G.D., Sanford, K.K., Earle, W.R.: Further studies on the proliferation in vitro of single isolated tissue cells. J. Natl. Cancer Inst. *13*, 177–187 (1952)

Lipkin, G., Knecht, M.: A diffusible factor restoring contact inhibition of growth to malignant melanocytes. Proc. Natl. Acad. Sci. U.S.A. *71*, 849–853 (1974)

Lipkin, G., Knecht, M.E.: Restoring contact inhibition of growth to malignant melanocytes of man, mouse, and hamster. Schweiz. Med. Wochenschr. *105*, 1360–1364 (1975)

Lipkin, G., Knecht, M.E.: Contact inhibition of growth is restored to malignant melanocytes of man and mouse by a hamster protein. Exp. Cell Res. *102*, 341–348 (1976)

Lipkin, G., Knecht, M.E., Rosenberg, M.: A potent inhibitor or normal and transformed cell growth derived from contact-inhibited cells. Cancer Res. *38*, 635–643 (1978)

MacIntyre, E., Ponten, J.: Interaction between normal and transformed bovine fibroblast in culture I – Cells transformed by Rous Sarcoma virus. J. Cell Sci. *2*, 309–322 (1967)

Malmquist, W.A., Brown, C.G.: Establishment of Theileria parva infected lymphoblastoid cell lines using homologous feeder layers. Res. Vet. Sci. *16*, 134–135 (1974)

Marty de Morales, M., Blat, C., Harel, L.: Changes in the phosphorylation of non-histone chromosomal proteins in relationship to DNA and RNA synthesis in $BHK_{21}C_{13}$ cells. Exp. Cell Res. *86*, 111–119 (1974)

Martz, E., Steinberg, H.S.: The role of cell-cell contact in "contact" inhibition of cell division: a review and new evidence. J. Cell. Physiol. *79*, 189–210 (1972)

Megyesi, K., Kahn, C.R., Roth, J.: The NSILA-s receptor in liver plasma membranes: characterization and comparison with the insulin receptor. J. Biol. Chem. *250*, 8990–8996 (1976)

Merrill, G.F., Florini, J.R., Dulak, N.C.: Effects of multiplication stimulating activity (MSA) on AIB transport into myoblast and myotube cultures. J. Cell. Physiol. *93*, 173–182 (1977)

Merrill, G.F., Dulak, N.C., Florini, J.R.: MSA stimulation of AIB transport is dependent of K^+ accumulation in myoblasts. J. Cell. Physiol. *100*, 343–350 (1979)

Mondal, S., Lillehaug, J.R., Heidelberger, C.: Cell mediated activation of aflatoxin-B1 to transform C3H/10 T ½ cells. Proc. A. A. C.R. Abstract *20*, 62 (1979)

Moses, A.C., Nissley, S.P., Short, P.A., Rechler, M.M., Podskalny, J.M.: Purification and characterization of multiplication stimulating activity (MSA): insulin-like growth factors purified from rat liver cell conditioned media. Eur. J. Biochem. *103*, 387–400 (1980a)

Moses, A.C., Nissley, S.P., Short, P.A., Rechler, M.M.: Immunological cross-reactivity of multiplication stimulating activity (MSA) polypeptides. Eur. J. Biochem. *103*, 401–408 (1980b)

Murray, A., Fitzgerald, D.J.: Tumor promoters inhibit metabolic cooperation in cocultures of epidermal and 3T3 cells. Biochem. Biophys. Res. Commun. *91*, 395–401 (1979)

Natraj, C.V., Datta, P.: Control of DNA synthesis in growing BALB/c 3T3 mouse cells by a fibroblast growth regulatory factor. Proc. Natl. Acad. Sci. U.S.A. *75*, 6115–6119 (1978)

Nicolas, J.F., Jacob, M., Jacob, F.: Metabolic cooperation between mouse embryonal carcinoma cells and their differentiated derivatives. Proc. Natl. Acad. Sci. U.S.A. *75*, 3293–3296 (1978)

Nissley, S.P., Rechler, M.M.: Multiplication-stimulating activity (MSA): a somatomedin-like polypeptide from cultured rat liver cells. Natl. Cancer Inst. Monogr. *48*, 167–177 (1978)

Nissley, S.P., Passamani, J., Short, P.: Stimulation of DNA synthesis, cell multiplication and ornithine decarboxylase in 3 T 3 cells by multiplication-stimulating activity (MSA). J. Cell. Physiol. *89*, 393–402 (1976)

Nissley, S.P., Short, P.A., Rechler, M.M., Podskalny, J.N., Coon, H.G.: Proliferation of Buffalo rat liver cells in serumfree medium does not depend upon multiplication-stimulating activity (MSA). Cell *11*, 441–446 (1977)

Nordenskjold, B.A., Skoog, L., Brown, M.C., Reichard, P.: Deoxyribonucleotide pools and deoxyribonucleic acid synthesis in cultured mouse embryo cells. J. Biol. Chem. *245*, 5360–5368 (1970)

Okuda, A., Kimura, G.: Serum stimulation of DNA synthesis in rat 3 Y 1 cells. Dependence on cell density, serum concentration and ratio of cell number to medium volume. Exp. Cell. Res. *111*, 55–62 (1978)

Oshima, R.: Stimulation of the clonal growth and differentiation of feeder layer dependent mouse embryonal carcinoma cells by β-mercaptoethanol. Differentiation *11*, 149–155 (1978)

Pardee, A.: Cell division and a hypothesis of cancer. Natl. Cancer Inst. Monogr. *14*, 7–18 (1964)

Pardee, A.B.: The cell surface and fibroblast proliferation. Some current research trends. Biochim. Biophys. Acta *417*, 153–172 (1975)

Pariser, R.P., Cunningham, D.D.: Transport inhibitors released by 3 T 3 mouse cells and their relation to growth control. J. Cell Biol. *49*, 525–529 (1971)

Perry, W.P., Sanford, K.K., Evans, V.J., Hyatt, G.W., Earle, W.R.: Establishment of clones of epithelial cells from human skin. J. Natl. Cancer Inst. *18*, 709–713 (1957)

Pierson, R.W., Temin, H.M.: The partial purification from calf serum of a fraction with multiplication-stimulating activity for chicken fibroblasts in cell culture and with non-suppressible insulin-like activity. J. Cell. Physiol. *79*, 319–330 (1972)

Pitts, J.D., Finbow, M.E.: Functional permeability and its consequences. In: Intercellular communication. Mello, W.C. de (ed.), pp. 61–86. New York: Plenum 1977

Pitts, J.D., Simms, J.W.: Permeability of junctions between animal cells. Exp. Cell Res. *104*, 153–163 (1977)

Pope, J.H., Scott, W., Moss, D.J.: Cell relationships in transformation of human leukocytes by Epstein-Barr virus. Int. J. Cancer *14*, 122–129 (1974)

Puck, T.T., Marcus, P.I.: A rapid method for viable cell titration and clone production with HeLa cells in tissue culture: the use of X-irradiated cells to supply conditioning factors. Proc. Natl. Acad. Sci. U.S.A. *41*, 432–437 (1955)

Rapp, F., Westmoreland, D.: Cell transformation by DNA-containing viruses. Biochim. Biophys. Acta *458*, 167–211 (1976)

Rechler, M.M., Podskalny, J.M.: Insulin receptors in cultured human fibroblasts. Diabetes *25*, 250–255 (1976)

Rechler, M.M., Podskalny, J.M., Nissley, S.P.: Characterization of the binding of multiplication stimulating activity to a receptor for growth polypeptides in chick embryo fibroblasts. J. Biol. Chem. *252*, 3898–3910 (1977 a)

Rechler, M.M., Nissley, S.P., Podskalny, J.M. et al.: Identification of a receptor for somatomedin-like polypeptides in human fibroblasts. J. Clin. Endocrinol. Metab. *44*, 820–831 (1977 b)

Rein, A., Rubin, H.: On the survival of chick embryo cells at low concentrations in culture. Exp. Cell. Res. *65*, 209–214 (1971)

Sakiyama, H., Terasima, T., Sato, K.: Effects of confluent monolayers of density-inhibited and transformed cells on the growth of superinoculated cells. Cancer Res. *38*, 2854–2858 (1978)

Sanford, K.K., Earle, W.R., Likely, G.D.: The growth in vitro of single isolated tissue cells. J. Natl. Cancer Inst. *9*, 229–246 (1948)

Schneider, U., Zur Hausen, H.: Epstein-Barr virus-induced transformation of human leukocytes after cell fractionation. Int. J. Cancer *15*, 59–66 (1975)

Senger, D.R., Wirth, D.F., Hynes, R.O.: Transformed mammalian cells secrete specific proteins and phosphoproteins. Cell *16*, 885–893 (1979)

Sheridan, J.D., Finbow, M.E., Pitts, J.D.: Metabolic cooperation in culture: possible involvement of junctional transfer in regulation of enzyme activities. J. Cell Biol. *67*, 396a (1975)

Sheridan, J., Finbow, M.E., Pitts, J.D.: Metabolic interactions between animal cells through permeable intercellular junctions. Exp. Cell Res. *123*, 111–117 (1979)

Slack, C., Morgan, R.H.M., Hooper, M.L.: Isolation of metabolic cooperation-defective variants from mouse embryonal carcinoma cells. Exp. Cell Res. *117*, 195–205 (1978)

Smith, G.L.: Increased ouabain-sensitive [86]Rubidium uptake after mitogenic stimulation of quiescent chicken embryo fibroblasts with purified multiplication-stimulating activity. J. Cell Biol. *73*, 761–767 (1977)

Smith, G.L., Temin, H.M.: Purified multiplication-stimulating activity from rat liver conditioned medium: comparison of biological activities with calf serum. J. Cell. Physiol. *84*, 181–192 (1974)

Steck, P.A., Voss, P.G., Wang, J.L.: Growth control in cultured 3T3 fibroblasts. Assays of cell proliferation and demonstration of a growth inhibiting activity. J. Cell Biol. *83*, 562–575 (1979)

Stoker, M.: Contact and short-range interactions affecting growth of animal cells in culture. Curr. Top. Dev. Biol. *2*, 107–128 (1967a)

Stoker, M.G.P.: Transfer of growth inhibition between normal and virus-transformed cells: autoradiographic studies using marked cells. J. Cell Sci. *2*, 293–304 (1967b)

Stoker, M.G.P.: Role of diffusion boundary layer in contact inhibition of growth. Nature *246*, 200–203 (1973)

Stoker, M.G.P.: The effects of topoinhibition and cytochalasin B on etabolic cooperation. Cell *6*, 253–257 (1975)

Stoker, M.G.P., Mac Pherson, I.: Studies on transformation of hamster cells by polyoma virus in vitro. Virology *14*, 359–370 (1961)

Stoker, M.G.P., Piggott, D.: Shaking 3T3 cells: further studies on diffusion boundary effects. Cell *3*, 207–215 (1974)

Stoker, M.G.P., Sussman, M.: Studies on the action of feeder layers in cell culture. Exp. Cell. Res. *38*, 645–653 (1965)

Stoker, M.G.P., Shearer, M., O'Neill, C.: Growth inhibition of polyoma-transformed cells by contact with static normal fibroblasts. J. Cell Sci. *1*, 297–310 (1966)

Subak-Sharpe, J.H.: Biochemically marked variants of the syrian hamster fibroblast cell line BHK$_{21}$ and its derivatives. Exp. Cell Res. *38*, 106–119 (1965)

Subak-Sharpe, J.H., Burk, R.R., Pitts, J.D., May, J.: Metabolic cooperation by cell to cell transfer between genetically different mammalian cells in tissue culture. Heredity *21*, 342–343 (1966)

Subak-Sharpe, H., Burk, R.R., Pitts, J.D.: Metabolic cooperation between biochemically marked mammalian cells in tissue culture. J. Cell Sci. *4*, 353–367 (1969)

Swann, M.M.: The control of cell division: a review I – General mechanisms. Cancer Res. *17*, 727–758 (1957)

Taylor-Papadimitriou, J., Shearer, M., Walting, D.: Growth requirements of calf lens epithelium in culture. J. Cell. Physiol. *95*, 95–103 (1978)

Temin, H.M.: Control of multiplication of uninfected rat cells and rat cells converted by murine Sarcoma virus. J. Cell. Physiol. *75*, 107–120 (1970)

Todaro, G.J., Green, H.: Quantitative studies of the growth of mouse embryo cells in culture and their development into established lines. J. Cell Biol. *17*, 299–313 (1963)

Todaro, G.J., Lazar, G.K., Green, H.: The initiation of cell division in a contact-inhibited mammalian cell line. J. Cell. Physiol. *66*, 325–334 (1965)

Van der Bosch, J., Maier, H.: Density-dependent growth inhibiting interactions between 3T3 and SV40-3T3 cells in mixed cultures. Z. Naturforsch. *34*, 272–278 (1979)

Vogt, M., Dulbecco, R.: Virus-cell interaction with a tumor-producing virus. Proc. Natl. Acad. Sci. U.S.A. *46*, 365–370 (1960)

Weiss, L., Poste, G., Mac Kearnin, A., Willett, K.: Growth of mammalian cells on substrates coated with cellular micro exudates. I. Effect on cell growth at low population densities. J. Cell Biol. *64*, 135–145 (1975)

Whitfield, J.G., Rixon, R.H.: The effect of dilution and carbon dioxide on the metabolic properties of suspension cultures of strain L mouse cells. Exp. Cell Res. *24*, 177–180 (1961)

Whittenberger, B., Glaser, L.: Inhibition of DNA synthesis in cultures of 3 T 3 cells by isolated surface membranes. Proc. Natl. Acad. Sci. U.S.A. *74*, 2251–2255 (1977)

Whittenberger, B., Glaser, L.: Cell saturation density is not determined by a diffusion-limited process. Nature *272*, 821–823 (1978)

Winger, L.A., Nowel, P.C., Daniele, R.P.: Sequential proliferation induced in human peripheral blood lymphocytes by mitogen – I. growth of 1000 lymphocytes in Feeder layer cultures. J. Immunol. *118*, 1763–1767 (1977)

Wright, E.D., Slack, C., Goldfarb, P.S.G., Subak-Sharpe, J.H.: Investigation of the basis of reduced metabolic cooperation in MEC-cells. Exp. Cell Res. *103*, 79–91 (1976 a)

Wright, E.D., Goldfarb, P.S.G., Subak-Sharpe, J.H.: Isolation of variant cells with defective metabolic cooperation (MEC) from polyome virus transformed syrian hamster cells. Exp. Cell Res. *103*, 63–77 (1976 b)

Yeh, J., Fisher, H.W.: A diffusible factor which sustains contact inhibition of replication. J. Cell Biol. *40*, 382–388 (1969)

Zoref, E., De Vries, A., Sperling, O.: Transfer of resistance to selective conditions from fibroblasts with mutant feedback-resistant phosphoribosyl pyrophosphate synthetase to normal cells. A form of metabolic cooperation. Adv. Exp. Med. Biol. *76 A*, 80–84 (1977)

CHAPTER 12

Hemopoietic Colony Stimulating Factors

D. METCALF

A. Introduction and Terminology

The concept that the continuous formation of blood cells might be under the control of specific humoral regulatory factors received early support from the demonstration that a humoral factor, termed erythropoietin, was involved in the regulation of erythropoiesis (CARNOT and DEFLANDRE, 1906). However it was not until the introduction in 1966 of methods for growing colonies of normal hemopoietic cells in semisolid culture, that progress was made in identifying and characterizing corresponding specific regulators for other hemopoietic populations.

The first hemopoietic colonies to be grown in such cultures were composed of neutrophilic granulocytes and/or macrophages and the proliferation of cells to form colonies was stimulated by using a variety of cell underlayers or media conditioned by various cells (BRADLEY and METCALF, 1966; ICHIKAWA et al., 1966). With the demonstration that certain sera were also able to stimulate colony formation, the operational term "colony stimulating factor" "CSF" was coined to denote the active factor present in such materials (METCALF, 1970). An alternative term used to describe this active factor was macrophage-granulocyte inducer (MGI) (LANDAU and SACHS, 1971).

Following the introduction of the granulocyte-macrophage colony-forming technique, comparable culture systems were developed that could support the proliferation of eosinophil, erythroid, megakaryocyte and mixed hemopoietic colonies. In each case, evidence was produced that the proliferation in vitro of these various hemopoietic cells is dependent on specific factors. The generic term "colony stimulating factor" has been extended to include these more recently discovered factors but it is now necessary to identify the specific factor by addition of a prefix indicating the target cells responding, e.g. eosinophil colony stimulating factor, EO-CSF; megakaryocyte colony stimulating factor, MEG-CSF, etc.

Since most studies are still performed using crude or incompletely characterized stimulating material, many workers prefer to refer to their active preparations as having, or containing, "colony stimulating activity" (CSA), the term colony stimulating factor (CSF) being reserved either for purified preparations of colony stimulating factor or for crude materials that contain a previously characterized factor. In the interests of simplicity, the term "colony stimulating factor" (CSF) will be used throughout this review but the reader consulting original papers should bear in mind the alternative terminology used in many of these papers.

Analytical studies on the cells that proliferate in vitro to generate colonies of differentiating hemopoietic cells have shown that such cells are the specific precursors of the relevant hemopoietic populations. For example, the cells forming granulocyte-macrophage (GM) colonies in vitro lie intermediate in differentiation between the multipotential hemopoietic stem cells, able to self-generate and form all classes of hemopoietic cells, and the morphologically-identifiable early members of the granulocyte-macrophage population, the myeloblasts and myelocytes. This sequence for granulocytic cells can be represented as: multipotential stem cell → GM-progenitor (colony-forming) cell → myeloblast → myelocyte → metamyelocyte → polymorph.

Multipotential stem cells were first identified and enumerated by their capacity to form hemopoietic colonies in the spleen of irradiated syngeneic recipients (TILL and McCULLOCH, 1961) and are termed "Colony-Forming Units, Spleen" (CFU-S). A less frequently used term is hemopoietic stem cell (HSC). Since the GM progenitor cells are normally identified and enumerated by their ability to form colonies in semisolid culture, these cells have been termed granulocyte-macrophage colony-forming cells (GM-CFC). Many authors prefer to use the term Colony-Forming Unit, Culture (CFU-C) for these cells. While this latter term served well enough when GM colonies were the only hemopoietic colonies able to be grown in vitro, the term is no longer adequate now that all hemopoietic cells can be cloned in comparable cultures. A variety of rather unsatisfactory terms has been used to describe these other precursor cells and it seems more logical to use a similar terminology to that used for the stimulating factors, e.g. eosinophil colonyforming cell (EO-CFC), etc.

Although both T- and B-lymphoid cells can now be cloned in semisolid medium (see review METCALF, 1977), the situation differs from that existing with hemopoietic cells in two major respects: (a) most cells forming lymphoid colonies appear to be mature lymphoid cells rather than genuine progenitor cells, and (b) colony formation requires stimulation by non-specific mitogens, e.g. the agar mitogen for B-lymphocytes or PHA for T-lymphocytes. Although there is some evidence for candidate regulatory factors produced by macrophages and possibly other cells, it is at present unclear whether these correspond in nature to the regulatory factors for hemopoietic cells and the present review will be restricted to factors controlling non-lymphoid hemopoietic populations.

B. Culture of Hemopoietic Colonies in Semisolid Medium

Semisolid cultures are usually prepared in 35 or 50 mm petri dishes and a full description of the techniques used and problems encountered with semisolid cultures has been published elsewhere (METCALF, 1977). In the techniques, a dispersed cell suspension, e.g. of marrow cells, is added to tissue culture medium (e.g., Modified Eagle's Medium or McCoy's Medium) containing a gelling agent. The most commonly used agent for gelling is agar, but methylcellulose or plasma gel clots have been used extensively, particularly in studies on erythroid colony formation.

Although it is possible to grow good granulocyte-macrophage colonies in serum-free, fully-characterized, culture medium (GUILBERT and ISCOVE, 1976), the growth of hemopoietic colonies is usually performed using media with a high serum concentration (final concentration 15%–30%). Hemopoietic colony formation is strongly influenced by the nature of this serum for reasons that have not been fully clarified. Many fetal calf, horse or human sera contain non-specific inhibitors of hemopoietic colony formation. Even when inhibitory lipoproteins are removed, different batches of sera vary widely in their capacity to support the proliferation of various types of hemopoietic cell and sera need to be pretested before use in any of the hemopoietic culture techniques.

Hemopoietic cultures are routinely incubated at 37 °C for 7–14 days without media change or refeeding in an atmosphere kept humidified to prevent drying of the cultures. Most workers use an atmosphere of 5%–10% CO_2 in air although it has been shown that more sustained colony growth can be achieved by using 5% oxygen in the atmosphere rather than the higher percentage present in air (BRADLEY et al., 1978).

C. Detection and Assay of Colony-Stimulating Factors

The in vitro proliferation of hemopoietic cells to form colonies is dependent on the continuous presence in the cultures of adequate concentrations of the appropriate colony stimulating factor. These stimulating factors can be provided by adding 0.1 or 0.2 ml of a liquid preparation containing CSF to the culture dish before addition of the 1 ml of agar-medium containing the target cell suspension. The CSF is mixed with the culture medium before gelling occurs. Alternatively, the stimulating factor or cells producing the factor can be included in an agar underlayer in the culture dish.

A dose-response relationship exists between the concentration of CSF in the medium and the number and size of colonies developing and the cultures therefore can be used to bioassay CSF concentrations. The usual parameter used to assess CSF concentrations is the sigmoid dose-response relationship between CSF concentration and colony number. Calculation of CSF concentration is based on the linear portion of the dose-response curve. Since the number of colonies developing also depends on the number of available colony-forming cells, CSF concentrations need to be related to the number of target cells used. For example, if a culture contained 75,000 bone marrow cells and 0.1 ml of a 1 : 16 dilution of the CSF preparation stimulated 60 colonies to develop, the CSF concentration can be expressed as 9,600 Units per ml of the CSF preparation (Fig. 1 a). If 150,000 marrow cells had been used, the same CSF concentration would stimulate 120 colonies to develop and be expressed as 19,200 Units per ml. Obviously therefore, with this system of recording CSF concentrations, the total number of cells cultured must be stated clearly.

One method for avoiding this difficulty is to express CSF Units as the reciprocal of the highest dilution of the preparation able to stimulate detectable colony formation (VAN DEN ENGH, 1974). Thus, in the example in Fig. 1 b, the CSF concentration could be recorded as 640 Units/ml.

A number of culture systems support the simultaneous formation of more than one type of hemopoietic colony. Such cultures can be used to simultaneously estimate the concentrations of the different regulators involved provided that colony formation by one cell type does not interfere with the formation of colonies by other cells. Where more than one type of colony develops, it is necessary to perform

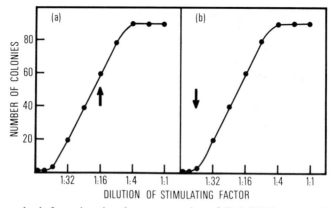

Fig. 1. Two methods for estimating the concentration of GM-CSF in assay cultures of bone marrow cells: **a** based on the linear portion of the colony number dose-response curve, and **b** based on the lowest dilution of test material stimulating detectable colony formation. *Arrows* indicate reference points for calculations

differential colony counts and this usually requires sampling of the colonies and the use of an appropriate differential stain, e.g. Luxol fast blue, for eosinophil colonies; benzidine, for erythroid colonies or acetylcholinesterase for megakaryocyte colonies.

One of the most common problems encountered in attempting to detect or assay colony stimulating factors is the coexistence of inhibitory materials in the crude preparations being tested. Assays on such crude material often produces gross underestimates of the CSF levels present or may even fail to detect CSF. The minimal requirements to avoid this error are to assay serial dilutions of the test material, looking for evidence of high dose inhibition and to observe the consequences of mixing the test material with an active preparation of CSF. In no instance has a semi-purified CSF been found to exhibit genuine high dose inhibition of colony formation and all examples of apparent high dose inhibition have subsequently been shown to be due to the presence of inhibitors in the preparation.

Calculations have shown that, with purified preparations of GM-CSF or erythropoietin and with good culture conditions, the semisolid cultures are extremely sensitive bioassay systems capable of detecting 10^{-11}–10^{-12} M concentrations of these regulators. It has been suggested that the use of radioimmunoassays might circumvent the necessity for bioassaying CSF levels (STANLEY, 1979). There are however certain problems involved in the use of radioimmunoassays. It is known that antigenically-distinct forms of GM-CSF exist even within one inbred animal. Furthermore it has proved extremely difficult to stimulate the formation of high titers of specific antibody to CSF. Even when such antisera have been generated there are suspicions that the antibodies may not discriminate adequately between biologically-active and biologically-inactive molecules. For these various reasons almost all work to date on the assay of CSF's has employed bioassays. While radioimmunoassays could theoretically shorten assay procedures from 7 days to perhaps 1–2 days, the level of technology required is no less complex than for tissue culture assays and a useful role for radioimmunoassays has yet to be demonstrated.

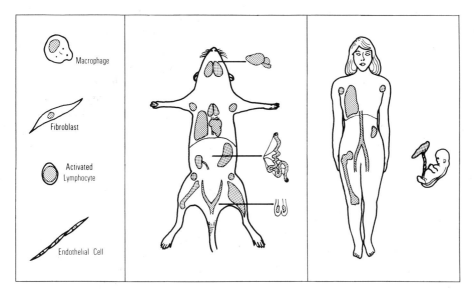

Fig. 2. Cells and organs known to contain and produce GM-CSF in the mouse and man. Many other organs may have a similar capacity but have yet to be tested

D. Granulocyte-Macrophage Colony Stimulating Factor (GM-CSF)

Initial studies on granulocyte-macrophage colony formation demonstrated that underlayers of various mouse tissues could stimulate colony formation, as could medium harvested from cultures of such tissues (conditioned medium) (BRADLEY and METCALF, 1966; ICHIKAWA et al., 1966). It was subsequently shown that serum from normal or leukemic mice could stimulate colony formation (ROBINSON et al., 1967) and similar activity was also demonstrable in human urine (ROBINSON et al., 1969; METCALF and STANLEY, 1969). The biological activity of these crude materials appeared to reside in a definable population of macromolecules and the term GM-CSF is the current name for the active molecule in such material.

I. Sites of Production of GM-CSF

It was recognized early in these studies that the situation with GM-CSF must differ significantly from that with classical hormones in that there was clearly not a single organ source of GM-CSF (Fig. 2). GM-CSF could be extracted from many organs of normal or endotoxin-injected mice in concentrations higher than exist in the serum (Fig. 2) and the GM-CSF's extracted from these different tissues appeared to differ in physical nature even though they stimulated the formation of apparently similar GM colonies (BRADLEY et al., 1971; SHERIDAN and STANLEY, 1971; SHERIDAN and METCALF, 1972). Parallel studies showed that certain continuous cell lines of normal or neoplastic cells could also produce GM-CSF, the most frequently used system being the production of GM-CSF by L-cell cultures (AUSTIN et al., 1971).

A wide variety of adult mouse organs are also capable, on incubation, of releasing GM-CSF into the medium (BRADLEY and SUMNER, 1968; NICOLA et al., 1979 a). It was initially thought possible that the GM-CSF present in, and released by, tissues might possibly be GM-CSF stored by the tissues after absorption of circulating GM-CSF. However, this possibility was eliminated by the demonstration that GM-CSF release into conditioned medium was blocked by puromycin or chlorheximide, indicating that the process depended on active protein synthesis (SHERIDAN and METCALF, 1973 a; METCALF, unpublished data). In the case of these organ-generated GM-CSF's, the process did not appear to depend on cell division in vitro since the process was radioresistant and was not blocked by metabolic inhibitors of cell division.

GM-CSF is also extractable from, or produced by, a variety of fetal tissues (BRADLEY and METCALF, 1966; PLUZNIK and SACHS, 1966; BRADLEY and SUMNER, 1968; BRADLEY et al., 1971; JOHNSON and METCALF, 1978 a).

Although less extensive studies have been performed on human tissues, the situation appears to be essentially similar to that for the mouse. GM-CSF is produced in vitro by peripheral blood cells (PIKE and ROBINSON, 1970) and in this system the monocytes appear to be the major source (MOORE and WILLIAMS, 1972; CHERVENICK and LO BUGLIO, 1972; GOLDE and CLINE, 1972; MOORE et al., 1973). However, after mitogen stimulation, T-lymphocytes can also produce GM-CSF (CLINE and GOLDE, 1974; PRIVAL et al., 1974; SHAH et al., 1977). The placenta has been shown to actively produce GM-CSF in vitro (BURGESS et al., 1977 a) as have the spleen (PARAN et al., 1970) lung (FOJO et al., 1977) and vascular endothelial (KNUDTZON and MORTENSEN, 1975) and marrow adherent cells (GREENBERG et al., 1978).

It was demonstrated that the injection into mice of endotoxin or a variety of bacterial antigens caused a dramatic elevation of serum-GM-CSF levels commencing within minutes of injection and reaching peak levels 3–6 h after injection (METCALF, 1971; QUESENBERRY et al., 1972). Tissue GM-CSF levels also rose following such injections (SHERIDAN and METCALF, 1972).

The source of the GM-CSF appearing in the serum after the injection of endotoxin has not been determined. The radioresistance of the response (METCALF, 1971) appears to eliminate lymphoid cells as playing a significant role, but potentially any of the various tissues known to produce GM-CSF could contribute. For example, human endothelial cell monolayers produce greatly increased levels of GM-CSF in vitro following the addition of endotoxin (QUESENBERRY et al., 1978). Similarly, macrophages release increased amounts of GM-CSF into culture medium following the addition of endotoxin (HAYES et al., 1972; SHERIDAN and METCALF, 1973 a; CHERVENICK et al., 1973; EAVES and BRUCE, 1974; STABER et al., 1978).

In order to determine that a particular cell type can synthesize GM-CSF it is necessary to: (a) obtain relatively large numbers of such cells, (b) free them of contaminating cells, (c) demonstrate formally that the GM-CSF detected is actually synthesized and not simply the release of stored GM-CSF, (d) perform such tests on cells recently removed from the body and (e) culture the cells in a manner adequate for the survival of healthy cells. In this context, information from continuously propagable cell lines or neoplastic cell lines is potentially unreliable since a capacity to produce GM-CSF may be acquired by derepression of the genome or be suppressed following some secondary change occurring in vitro. Assays on fluid harvested from cultures of various cells often require preliminary chemical fractionation of the culture medium either to concentrate low levels of GM-CSF or to remove inhibitory materials that suppress colony formation and mask the detection of GM-CSF.

From these considerations it becomes evident that the apparently innocent question "Which cells make GM-CSF?" involves complex technical difficulties. As a consequence, firm information is available for only a very limited number of cells. To date, only four normal cell types seem to have clearly been shown to be able to synthesize GM-CSF – fibroblasts, mitogen-stimulated lymphocytes, endothelial cells and monocyte-macrophages (Fig. 2). It is not known whether all cells of these types can synthesize GM-CSF or whether only certain subpopulations of these cells possess this functional activity. Polymorphs appear unable to produce GM-CSF but no studies have been reported on carefully fractionated polymorph conditioned media and since polymorphs release a variety of inhibitors, the failure of these cells to produce GM-CSF may be more apparent than real.

At the present time therefore it is not possible to interpret the significance of the observations that all tissues contain and synthesize GM-CSF. It may be that *all* cells in the body have some capacity to synthesize this factor or, alternatively, the capacity of all tissues to produce GM-CSF may simply be due to the fact that all tissues contain sufficient endothelial cells, macrophages, lymphocytes or fibroblasts to produce detectable levels of GM-CSF.

It has been observed that many neoplastic cells produce GM-CSF in culture, e.g. lung tumor cells, melanoma cells, myelomonocytic and lymphoid leukemic cells. This may reflect the capacity of all normal cells to produce GM-CSF or may be due simply to derepression of the genome of cultured neoplastic cells.

While the recent recognition of the close similarity of many mouse GM-CSF's (see below) has removed much of the apparent complexity involving the biology of GM-CSF production and has strengthened the view that GM-CSF is likely to be a genuine regulator of granulocyte and monocyte production in vivo, it has not really resolved the question of which cells produce GM-CSF.

An alternative approach to this question would be to use fluorescein-conjugated anti-CSF sera to determine which tissue cells contain unusual amounts of GM-CSF. To date only one antiserum has been employed in such studies – an anti-L cell GM-CSF. This appeared to show that tissue macrophages were the major cells containing GM-CSF (SHADDUCK and CAVALLO, 1973). However, the interpretation of the data is difficult since this antiserum is not capable of inactivating several major types of mouse GM-CSF, e.g. that synthesized by mouse lung tissue or by activated mouse lymphocytes (SHADDUCK and METCALF, 1975) and thus the antiserum could not be expected to detect the cells producing these GM-CSF's. Furthermore, a cell actively synthesizing GM-CSF need not necessarily store significant concentrations of the molecule and might well be undetected by this procedure.

Little is known of where GM-CSF is produced or stored within cells but it has been reported that membrane fractions of human peripheral blood leukocytes, presumably containing some peripheral cytoplasmic material, contain most of the

GM-CSF extractable from L-cells (Price et al., 1975 a). Parallel studies on L-cells in active cell cycle suggested that GM-CSF may be transfered to the cell surface during the S phase of the cell cycle and be shed following metaphase (Cifone and Defendi, 1974).

The only conclusions possible at present on the sites of production of GM-CSF in the body are that all organs produce GM-CSF and that GM-CSF is present in the serum and urine. Several cell types are known to synthesize GM-CSF – fibroblasts, lymphocytes, macrophages, and epithelial cells – and it may be that many other or all cells can synthesize varying amounts of this molecule.

II. Purification and Chemical Nature of GM-CSF

Early attempts were made to purify GM-CSF using as starting materials human urine and mouse embryo cell conditioned medium. A relatively pure preparation of GM-CSF was obtained from human urine and the molecule was shown to be a glycoprotein with an apparent molecular weight of 45,000 (Stanley and Metcalf, 1969; Stanley et al., 1975). Parallel fractionation studies on mouse embryo cell conditioned medium indicated that the active factor was a protein with an apparent molecular weight of 65,000–70,000 (Landau and Sachs, 1971).

In subsequent studies, GM-CSF was obtained in chemically pure form from medium conditioned by lung tissue from mice injected with endotoxin. This GM-CSF was shown to be a neuraminic acid-containing glycoprotein of molecular weight 23,000 (Sheridan and Metcalf, 1973 b; Sheridan et al., 1974; Burgess et al., 1977 b). Carbohydrate appeared to comprise only 10%–20% of the molecule and removal of carbohydrate by a variety of carbohydrases did not affect its capacity to stimulate granulocyte-macrophage colony formation. The molecule has been radiolabeled with ^{125}I without apparent loss of biological activity (Burgess and Metcalf, 1977 a). Attempts to obtain a biologically-active fragment of the molecule by digestion with peptidases have been unsuccessful. This form of GM-CSF is capable of stimulating colony formation at molar concentrations as low as 10^{-11}. At high concentrations, a high proportion of colonies formed by mouse marrow cells is composed of granulocytes or mixed granulocytes and macrophages whereas at low concentrations, only macrophage colonies develop (Burgess and Metcalf, 1977 b).

In parallel work, GM-CSF was also purified from medium conditioned by mouse L-cells (fibroblasts). This form of GM-CSF was also found to by a glycoprotein but had an apparent molecular weight of 60,000 and, even at high concentrations, tends to stimulate the formation of a relatively high proportion of macrophage colonies (Stanley et al., 1976; Stanley and Heard, 1977; Shadduck et al., 1979).

Antisera raised against relatively pure L-cell-derived GM-CSF were not able to suppress colony formation stimulated by mouse lung-derived GM-CSF but did inhibit colony formation stimulated by a variety of other mouse GM-CSF's and human urine GM-CSF (Shadduck and Metcalf, 1975). Use of an antiserum to L-cell GM-CSF in a radioimmunoassay gave similar evidence of cross-reactivity between various GM-CSF's (Stanley, 1979).

Initial studies on GM-CSF's extracted from, or synthesized in vitro by, a variety of mouse organs, suggested that these GM-CSF's might differ widely in physical properties (STANLEY et al., 1971; SHERIDAN and STANLEY, 1971; SHERIDAN and METCALF, 1972; NICOLA et al., 1979a). However, a more detailed analysis of 17 different GM-CSF's synthesized in vitro by adult mouse tissues revealed that the major apparent charge differences could be eliminated by pretreatment with neuraminidase. Furthermore, although initial size estimates by gel filtration suggested that some GM-CSF's had molecular weights as high as 200,000, when semi-purified preparations of these GM-CSF's were pretreated with neuraminidase then subjected to gel filtration under dissociating conditions (using 6 M guanidine hydrochloride), all exhibited an apparent molecular weight of 23,000. Thus, despite some variability in carbohydrate content and some antigenic differences, all adult organ-derived GM-CSF's appear to be closely similar glycoproteins of molecular weight 23,000 and all share the characteristics of stimulating both granulocyte and macrophage colony formation, granulocyte colony formation being stimulated preferentially at high concentrations and macrophage colony formation at low concentrations (NICOLA et al., 1979a).

Studies on the GM-CSF synthesized in vitro by pokeweed mitogen stimulated spleen cells have also shown this GM-CSF to be a glycoprotein of molecular weight 24,000 although two subsets of GM-CSF were separable by hydrophobic chromatography – one preferentially stimulating macrophage colony formation (BURGESS et al., 1978a, b, 1979).

Evidence suggesting that two distinct forms of GM-CSF might exist had previously been obtained from an analysis of the differential sensitivity to trypsin of the GM-CSF in medium conditioned by mouse peritoneal tissues (HORIUCHI and ICHIKAWA, 1977). A larger, trypsin-resistant, molecule was observed to preferentially stimulate macrophage colony formation and a smaller trypsin-sensitive molecule to stimulate granulocyte colony formation.

Studies on the GM-CSF produced by mouse yolk sacs indicated that this molecule preferentially stimulated macrophage colony formation and had an apparent molecular weight of 70,000 (JOHNSON and BURGESS, 1978). Subsequent analysis of this material under reducing conditions has revealed that the material is somewhat heterogeneous with molecular weights in the range of 23,000–36,000. Re-analysis under dissociating conditions of the GM-CSF present in L-cell conditioned medium and pregnant mouse uterus extract (also preferentially stimulating macrophage colony formation) has indicated a similar heterogeneity and molecular weights in the range 26,000–54,000 (NICOLA, BURGESS, and METCALF, unpublished data).

Thus the present situation with respect to mouse GM-CSF's is that there appear to be two types: (a) a 23,000 molecular weight type, stimulating both granulocyte and macrophage colony formation, produced by most adult tissues and present in the serum, and (b) a type with a molecular weight range of 23,000–54,000 preferentially (but not exclusively) stimulating macrophage colony formation, present in L-cell and yolk sac conditioned medium and extractable from pregnant mouse uterus. These two groups of GM-CSF share antigenic cross-reactivity (SHADDUCK and METCALF, 1975; STANLEY, 1979), although as mentioned above, at least two antigenically-distinct forms of the 23,000 molecule appear to exist.

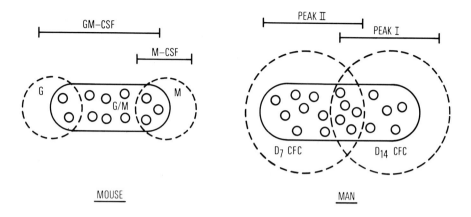

Fig. 3. Schematic illustration of the various subpopulations of granulocyte-macrophage progenitor cells in mouse and human marrow. In the mouse, GM-CSF can stimulate cells forming granulocyte *G* mixed *GM* or macrophage *M* colonies but M-CSF has a more restricted action. Two subsets of granulocyte colony-forming cells have been identified in human marrow by scoring cultures at 7 and 14 days of incubation. These subsets are stimulated preferentially by corresponding subsets of GM-CSF molecules produced by placental tissue

The situation with human GM-CSF's is less clear. Re-analysis of human urine GM-CSF under dissociating conditions has indicated that this molecule has an apparent molecular weight of 36,000 (Nicola, Burgess, and Metcalf, unpublished data). However this molecule is capable of stimulating only a minor degree of proliferation by human marrow cells and is unable to stimulate genuine colony formation (Metcalf, 1974a).

The GM-CSF extractable from and produced by human white cells has been reported to be heterogeneous and to have a variety of apparent molecular weights ranging from 150,000 to 14,700. However, these analyses were not performed using dissociating conditions and much of the heterogeneity observed may have been due to the anomalous behavior of GM-CSF on gel filtration refered to above (Prival et al., 1974; Price et al., 1975b; Shah et al., 1977). Medium conditioned by leukemic cells was reported to contain only one molecular form of GM-CSF (Price et al., 1975b).

Studies on GM-CSF produced in vitro by human placental tissue indicated that the molecule has an apparent molecular weight of 30,000 (Nicola et al., 1978) but no evidence has been obtained that the active material is a glycoprotein. Again, separation using hydrophobic chromatography revealed the presence of two types of GM-CSF in this material – one preferentially stimulating granulocyte colony formation (seen as 7 day colonies) and the other macrophage colonies (seen as more slowly developing 14 day colonies). The two forms of GM-CSF were indistinguishable on the basis either of molecular weight or charge properties (Nicola et al., 1979b). Separative studies on human marrow cells have shown that the two forms of GM-CSF act on two partially segregatable subsets of GM progenitor cells (Fig. 3).

Similar results have been reported by Fojo et al. (1977) working with GM-CSF derived from human lung tissue. Two forms of GM-CSF were detected by electrophoresis and again evidence was obtained that these acted selectively on distinct subsets of target progenitor cells.

GM-CSF with colony stimulating activity for human marrow cells has also been shown to be synthesized by the various tumor cells (Asano et al., 1977; Ohno et al., 1978; Di Persio et al., 1978) and preliminary chemical characterization indicated that the active materials had apparent molecular weights in the range 22,000–50,000.

III. Mechanisms of Action of GM-CSF

GM-CSF is essential for the in vitro proliferation of all cells in the granulocyte-macrophage pathway from the earliest committed progenitor cell (GM-colony-forming cell, GM-CFC) to the last cells capable of proliferation – the myelocytes and promonocytes. In cultures in which no endogenous GM-CSF is being generated, no cell divisions occur unless adequate concentrations (approximately 10^{-11} M) of GM-CSF are present continuously. Transfer of dividing colony cells to cultures lacking GM-CSF is followed rapidly by cessation of cell division (Metcalf and Foster, 1967a; Paran and Sachs, 1968).

Since the action of GM-CSF can be observed using chemically pure GM-CSF in serum-free medium (Guilbert and Iscove, 1976; Iscove, Burgess, and Metcalf, unpublished data) and in cultures containing only a single cell (Metcalf et al., 1980a) this action of GM-CSF appears to be a direct one on GM target cells and does not require the intervention of any other cell type or any interaction with uncharacterized serum proteins. Serum proteins can however significantly modify the responsiveness of cells to GM-CSF (see below).

The stimulating effects of GM-CSF appear to be specific for cells of the GM pathway. GM-CSF is unable to stimulate the formation of colonies of eosinophils, megakaryocytes, erythroid cells or T- and B-lymphoid cells, and has no effect on the proliferation in vitro of fibroblasts, or a wide variety of tumor cell lines (Metcalf, 1977). When injected in vivo, no proliferative effects have been observed on any cell population other than the GM population in the marrow and spleen (Metcalf and Stanley, 1971a).

An extensive literature accumulated in the first half of this century on the apparent ability of a wide variety of extracts or foreign materials to increase granulocyte and monocyte levels when injected into recipient animals. This led to the view that many types of non-specific material could directly stimulate granulopoiesis. It is important in the context of discussing the specificity of action of GM-CSF to emphasize two points: (a) in all instances in which non-specific agents induce changes in granulopoiesis in vivo, major elevations in GM-CSF levels occur (see for example, analysis of the responses to endotoxin and bacterial antigens (McNeill, 1970; Metcalf, 1971) or casein (Watt et al., 1979), and (b) all crude materials containing GM-CSF that stimulate GM colony formation in vitro *fail* to have any residual effects in vitro when the GM-CSF is removed by fractionation. The conclusion from these observations is that GM-CSF is the only agent so far identified that can stimulate the proliferation of GM cells.

If GM-CFC are not in cycle when cultures are initiated, GM-CSF has been shown to force such cells into the S phase of the cell cycle within 3 h (MOORE and WILLIAMS, 1973). GM-CSF also has a concentration-dependent influence on the total cell cycle time of dividing GM colony cells during early colony development, the cycle time being shortened progressively as GM-CSF concentrations are increased (METCALF and MOORE, 1973).

If GM-CFC are cultured in the absence of GM-CSF, most either die or lose their capacity for proliferation within 24–48 h (METCALF, 1970). This phenomenon of in vitro death in the absence of a stimulating factor has been observed with other regulatory factors, e.g. nerve growth factor, epidermal and fibroblast growth factors and therefore appears not to be an unusual property. However the phenomenon has led to the criticism that GM-CSF may merely be a "survival factor" permitting target cells to remain alive and allowing them to exhibit an intrinsic capacity for proliferation that requires no real stimulation. If this were so then, by implication, GM-CSF would be unlikely to function in vivo as a genuine growth stimulating factor. The "survival" hypothesis has been excluded in the case of the action of GM-CSF on macrophage progenitor cells (M-CFC) since these cells survive well in vitro for some days in the absence of GM-CSF, but promptly begin proliferation following the addition of GM-CSF (LIN and STEWART, 1974). Furthermore, in view of the action of GM-CSF on cell cycle status of colony-forming cells and the concentration-dependent action on colony cell cycle times, it seems reasonable to regard this molecule as a genuine proliferative stimulus.

One of the most notable features of the proliferation of normal GM colonies when stimulated by GM-CSF is the progressive differentiation of all or most of the colony cells to mature, post-mitotic, polymorphs, and macrophages. By morphological, functional and membrane marker criteria this maturation appears to duplicate the processes observed in vivo. This has led many workers to conclude that GM-CSF induces, or activates, differentiation of cells in the GM population (hence the alternative term, macrophage-granulocyte inducer, MGI). GM-CSF may indeed induce differentiation but, equally, it may be that differentiation is a secondary event triggered by the number of cell divisions which undifferentiated GM precursor cells have undergone. On this latter hypothesis, GM-CSF need do no more than stimulate cell division, the associated differentiation events being preprogrammed in the genome of the responding cells.

However, certain observations suggest that GM-CSF positively influences differentiation: (a) high concentrations of GM-CSF stimulate the formation of large colonies in which cellular differentiation is delayed relative to that seen in cultures containing low concentrations of GM-CSF (METCALF and MOORE, 1973). This is in fact likely to be the major mechanism responsible for the increased colony growth rates observed with high GM-CSF concentrations (ROBINSON et al., 1967). If differentiation was programmed to occur after a fixed number of cell divisions, the reverse association would be expected, the larger the colony, the greater the content of differentiated cells, (b) GM-CSF activates RNA and protein synthesis in post-mitotic polymorphs (BURGESS and METCALF, 1977 b, c), increases the phagocytic and cytocidal activities of macrophages (HANDMAN and BURGESS, 1979) and increases the synthesis of plasminogen activator by macrophages (LIN and GORDON, 1979). To the degree that such functions can be regarded as differentia-

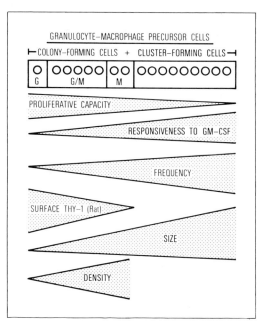

Fig. 4. Granulocyte-macrophage precursor cells in the mouse or rat exhibit a range of sizes, density and expression of surface antigens that permit partial segregation of subsets of precursor cells. These subsets also express varying proliferative capacity and responsiveness to GM-CSF. Colony-forming cells are the ancestors of cluster-forming cells but *G*, *G/M*, and *M* colony-forming cells do not necessarily constitute an ancestral sequence

tive events, clearly GM-CSF has a rapid and readily measurable influence on the expression of specialised cell functions, (c) in cultures of myeloid or myelomonocytic leukemic cells, purified GM-CSF stimulates the development of differentiated polymorphs and macrophages while having minimal effects on colony growth rates (LOTEM and SACHS, 1978; METCALF, 1979a). Furthermore, GM-CSF can suppress the self-generation of leukemic colony-forming cells even in rapidly growing colonies (METCALF, 1979a).

The GM system is of unusual interest in that single progenitor (colony-forming cells) can be shown to generate both polymorphs and macrophages, both arms of these alternative pathways being dependent on stimulation by GM-CSF (MOORE et al., 1972; METCALF et al., 1980a). It has been observed that in cultures containing high concentrations of purified GM-CSF, the formation of polymorph colonies is favored whilst in the presence of low GM-CSF concentrations, most colonies are composed wholly of macrophages (BURGESS and METCALF, 1977b). This appeared to be convincing evidence of the influence of GM-CSF on differentiation. However the situation is more complex than previously supposed in that GM-CFC are quite heterogeneous and subpopulations exist that have the capacity only to form granulocytes and corresponding subpopulations only to form macrophages (Fig. 4) (METCALF and MACDONALD, 1975; WILLIAMS and VAN DEN ENGH, 1975; BYRNE et al., 1977; BOL and WILLIAMS, 1979). Furthermore, granulocyte colony-forming cells require high concentrations of GM-CSF to be stimulated to proliferate whilst macrophage colony-forming cells are highly responsive to low GM-CSF concentrations (METCALF and MACDONALD, 1975). Thus, in cultures containing low concentrations of GM-CSF, only the highly responsive macrophage-forming cells would be stimulated to proliferate and most of the colonies developing would be composed of macrophages. While this heterogeneity of GM-CFC can explain much of the observed association between GM-CSF concentration and colony morphology, it does not account entirely for the phenomenon since, in the mouse, most GM-CFC are bipotential and the relationship between GM-CSF concentration and colony morphology must therefore be based mainly on the varying differentiation within bipotential colonies.

On balance, the available evidence suggests that GM-CSF probably does have a direct influence on the differentiation of GM cells, distinct from its action as a proliferative stimulus. Since the biochemical events in differentiation are likely to differ in many respects from those involved in cell division, GM-CSF may have multiple sites of action within the target cell. A very similar situation appears to exist in erythropoiesis, where the analogous glycoprotein regulator, erythropoietin, not only stimulates cell division in erythroid precursor cells but also induces hemoglobin synthesis in maturing erythroid cells.

The effects of GM-CSF on the two end cells of the GM population (the polymorph and macrophage) are of some interest. Under the action of GM-CSF, the functional activity of these cells rapidly increases, suggesting that when an animal is placed at risk by an infection, GM-CSF plays two roles: (a) it causes an immediate increase in the functional activity of pre-existing cells, and (b) in a more slowly evolving response, it stimulates the formation of new cells.

As discussed earlier, two broad groups of mouse GM-CSF's exist: (a) a group exemplified by the prototype purified from mouse lung conditioned medium, all of molecular weight 23,000 and all able to stimulate the preferential formation of granulocytic or macrophage colonies depending on the concentration used, and (b) a second more heterogeneous group of higher molecular weights (23,000–54,000) that stimulate mainly macrophage colony formation, regardless of the concentration used. With the reservation that this second group of CSF's can stimulate the formation of some granulocytic and mixed granulocytic-macrophage colonies, this group of CSF's could be conveniently labeled M-CSF's. Members of the M-CSF group include the CSF's present in pregnant mouse uterus extracts, L-cell and yolk sac conditioned media.

From what was discussed above concerning the heterogeneity of GM progenitor cells, it might be anticipated that the M-CSF's might not be able to stimulate the proliferation of all GM progenitor cells but have an action restricted mainly to those progenitors preprogrammed to a macrophage pathway of differentiation. It has in fact been observed that M-CSF's appear to stimulate only a limited subset of GM progenitor cells in adult mouse marrow (WILLIAMS and VAN DEN ENGH, 1975; BOL and WILLIAMS, 1979). In view of the derivation of two of these M-CSF's from pregnant uterus and yolk sac, it is of interest that most of the GM progenitors in the early fetus appear to be preprogrammed to macrophage formation (JOHNSON and METCALF, 1978 a, b, c). Since little granulopoiesis occurs in early fetal life, it is possible that the bias towards macrophage formation in early fetal life is the consequence of the almost exclusive presence of GM progenitors programed to macrophage formation and the availability only of M-CSF's to stimulate cell proliferation.

One of the striking features of GM colonies grown in vitro is the large variation in size between individual colonies. Separation of subpopulations of GM precursor cells by velocity sedimentation, density or surface antigens has shown that this heterogeneity is based on the existence of subsets of precursor cells with a varying capacity for proliferation, ranging from cells able to form very large colonies of 10,000–20,000 cells to cells able only to form clusters of 3 or 4 cells (Fig. 4). Because these subsets can be partially segregated by physical separation methods, varying colony or cluster size cannot be ascribed simply to stochastic events aborting clonal expansion by causing terminal differentiation of colony cells to non-dividing end cells.

In cultures stimulated by different concentrations of GM-CSF, a broad spectrum of colony sizes is always observed. Thus, in a culture containing low GM-CSF concentrations a clone size range of 10–200 cells may be observed, whereas in cultures containing high GM-CSF concentrations, clone size may range from 100–2,000 cells. The largest observed GM colonies approximate 20,000 by day 14 (BRADLEY et al., 1978). Because of the known heterogeneity of GM progenitor cells, these observations are difficult to interpret in terms of GM-CSF action. It is possible that, for a given increment of GM-CSF concentration, *all* colonies may increase in size by a constant increment. If so, this may be based on a combination of effects including shorter cell cycle times, delay in maturation time to post-mitotic cells and increase in the proportion of parental type progeny. On the other hand, an inverse correlation has been observed between the responsiveness of a cell to stimulation and the size of the clone generated by that cell. The cells generating the largest colonies seem also to be the cells with the highest threshhold for stimulation. Because of these considerations, the observed association between increasing GM-CSF concentrations and increasing colony size might simply be based on the progressive recruitment of colony-forming cells with a higher and higher threshhold for stimulation and a greater and greater capacity for extended proliferation. A resolution of these two broad alternatives could probably be achieved by separating the initial two daughter cells of a developing colony, placing each cell in a separate culture dish, then adding different GM-CSF concentrations to stimulate subsequent proliferation and completion of colony formation. A size difference between such daughter colonies would indicate that GM-CSF was able to directly alter the total number of progeny generated by individual precursor cells.

Recent experiments using purified GM-CSF have shown that this regulator is also able to stimulate the initial proliferation of some multipotential hemopoietic cells and of cells able eventually to generate colonies of erythroid, eosinophil or megakaryocytic cells when the appropriate stimulus (spleen conditioned medium) is added subsequently (METCALF et al., 1980a). Culture of single cells of this type with purified GM-CSF was shown to be able to stimulate up to five cell divisions without influencing the ultimate differentiation pathway of the cells. For example, erythroid progenitor cells subsequently formed colonies wholly composed of hemoglobin-containing erythroid cells even though GM-CSF had stimulated the first five cell divisions. The reason why GM-CSF fails to stimulate the formation of ery-

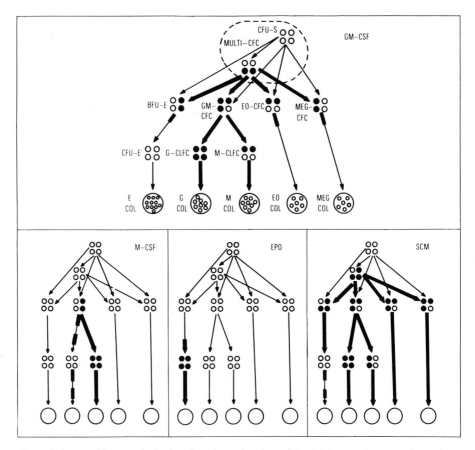

Fig. 5. Scheme of hemopoiesis showing sites of action of GM-CSF, M-CSF, erythropoietin *EPO* and pokeweed mitogen-stimulated spleen conditioned medium *SCM*. Purified GM-CSF is able to stimulate the initial cell divisions of some multipotential, erythroid, eosinophil and megakaryocyte precursor cells *(heavy lines)* but not the terminal cell division leading to the formation of colonies of maturing cells. (METCALF et al., 1980 b)

throid, eosinophil or megakaryocytic colonies appears to be that it is unable to stimulate the *terminal* proliferation and differentiation of cells in these subpopulations. Parallel studies have shown that CSF preferentially stimulating macrophage colony formation (M-CSF) is unable to stimulate the initial proliferation of multipotential or erythroid precursors (METCALF et al., 1980 b) (Fig. 5).

The phenomenon is of some general biological interest as an example of stimulation of cell division by an "inappropriate" regulator that is not associated with any obvious commitment or differentiation in the "appropriate" pathway for that regulator – a clear dissociation between a proliferative effect and the differentiating effects of the same molecule.

Preliminary evidence using radiolabeled GM-CSF has suggested the presence on target GM cells of a receptor for the molecule (STANLEY, personal communication) but nothing is known of subsequent intracellular events following binding of the regulator to the cell membrane. In cultures stimulated by GM-CSF there was only a marginal depletion of GM-CSF from the medium even when low molar concentrations were used (METCALF, 1970). A particularly intriguing situation exists with the myelomonocytic leukemia, WEHI-3 since the growth and differentiation of these cells is responsive to GM-CSF (METCALF, 1979 a). WEHI-3 cells are highly unusual in not only synthesizing GM-CSF but also EO-CSF, MEG-CSF, and E-CSF (METCALF et al., 1969, 1974; WILLIAMS and JACKSON, 1978; METCALF, unpublished data). Is there an internal traffic of GM-CSF molecules within these cells? For example, does GM-CSF synthesized in cytoplasm directly enter the cell nucleus or must it first be secreted, attached to a receptor and be reinternalised in order to act on WEHI-3 cells as a target cell?

IV. Factors Modifying Responsiveness to GM-CSF

Certain factors influence the responsiveness of target GM cells to stimulation by GM-CSF. There is considerable variation in responsiveness of GM populations from different mouse strains. C 57 BL cells are notably more responsive than comparable cells from CBA or BALB/c mice (McNEILL and FLEMING, 1973; METCALF and RUSSELL, 1976).

Studies on the serum of various mouse strains have shown the presence of lipoproteins that in high concentration suppress colony formation by GM cells and at lower concentrations selectively suppress granulocyte colony formation in favor of macrophage colony formation (CHAN, 1971; METCALF and RUSSELL, 1976). It is intriguing that levels of these serum lipoproteins are particularly high in CBA and BALB/c mice – mice whose marrow cells characteristically generate unusually high proportions of macrophage colonies in culture. Furthermore, preincubation of C 57 BL marrow cells with such lipoproteins not only reduces their responsiveness to stimulation by GM-CSF but also leads to the selective formation of macrophage colonies (CHAN, 1971; METCALF and RUSSELL, 1976). It is unclear whether these phenomena are based on selective suppression of granulocyte-forming subpopulations or on modification of the maturation pathway of bipotential cells.

Although serum lipoproteins do not appear to bind or inactivate GM-CSF (METCALF and RUSSELL, 1976), inhibition of colony formation by these lipoproteins can be partially overcome by increasing the concentration of GM-CSF. It is not clear therefore whether the lipoproteins merely make target cells relatively unresponsive or also have a more indirect interaction with GM-CSF.

Apparently similar lipoprotein inhibitors have been demonstrated in human sera (CHAN et al., 1971; GRANSTROM, 1973; GRANSTROM et al., 1972), and potentially could represent modulators of GM-CSF action in vivo. However these very light density lipoproteins appear non-specific in their inhibitory effects since colony formation by many types of normal hemopoietic cells and by various neoplastic cells is also inhibited (METCALF and RUSSELL, 1976; METCALF, unpublished data). Further, what appear to be the same lipoproteins have also been reported to be inhibitors of liver cell proliferation in vitro (LEFFERT et al., 1978). Despite the intriguing differential effects on granulocytic versus macrophage colony formation, it is doubtful whether these lipoproteins represent specific inhibitors or modulators of GM-CSF action.

It has been shown that addition of certain normal sera, e.g. human, mouse, horse, to cultures of mouse bone marrow cells can strongly potentiate GM colony formation stimulated by GM-CSF. The potentiating effects of such sera are seen most clearly when submaximal concentrations of GM-CSF are used (METCALF et al., 1975a; VAN DEN ENGH and BOL, 1975; MABRY et al., 1975). The ability of a serum to potentiate GM colony formation may require prior removal of inhibitory lipoproteins before being demonstrable. The potentiating activity of a serum is strongly increased by the injection of endotoxin 3–15 h prior to serum collection (METCALF et al., 1975a; VAN DEN ENGH and BOL, 1975). At present nothing is known of the physical properties of this potentiating factor nor has its mechanism of action been characterized.

A number of normal and neoplastic cells contain and release materials able to suppress or diminish GM colony formation (see review by VOGLER and WINTON, 1975). Since some of these materials have been derived from relatively pure populations of granulocytes (PAUKOVITS, 1971; AARDAL et al., 1977; LORD et al., 1977), the possibility has been raised that these inhibitors may represent "chalones" – a proposed class of tissue-specific inhibitors that could potentially balance stimulating factors such as GM-CSF. Analysis has indicated that one group of these candidate chalones are polypeptides of low molecular weight (300–5,000) (MAUER et al., 1976; AARDAL et al., 1977; PAUKOVITS and HINTERBERGER, 1978). Further work is necessary to determine their mechanism of action, whether polymorphs are the sole cellular source and whether the inhibitors are absolutely specific for GM cells.

One general problem with inhibitors of colony formation is to establish whether their action results from interference with GM-CSF or from cytotoxic or inhibitory effects on GM cells. On general grounds, cytotoxic inhibitors are unlikely to function in vivo as genuine modulators of granulopoiesis. Toxic effects could be demonstrable by preincubation of target cells with the inhibitor, followed by thorough washing, prior to culture. By this criterion, granulocyte extracts (chalones) appear to be nontoxic for colony-forming cells (AARDAL et al., 1977). Alternatively, interference with GM-CSF action may be detectable by the demonstration of altered dose-response curves in the presence of the inhibitor. This appears to be the situation with cortisone-mediated inhibition of GM colony formation (METCALF, 1969) since an altered responsiveness has been documented (McNEILL and FLEMING, 1973). Since excess concentrations of GM-CSF overcome the inhibition of granulocyte extracts, altered responsiveness to GM-CSF may also be involved in this system (AARDAL et al., 1977).

It was reported that prostaglandin E was a potent inhibitor of GM colony formation. However in the initial studies this effect appeared to be non-specific in that colony formation by B-lymphocytes was also inhibited (KURLAND et al., 1978). More recent studies have indicated that, in very low concentrations, prostaglandin E has a highly selective inhibitory action on macrophage colony formation (WILLIAMS, 1979) and this agent is a promising candidate for an inhibitory modulator of some aspects of proliferation in the GM system.

V. Factors Influencing GM-CSF Production and Levels

At least three types of GM-CSF production appear to occur in vivo:
1) a "steady state" continuous production, influenced by the developmental stage of the tissue and the variable capacity of different cells to synthesize GM-CSF,

2) accelerated GM-CSF production in response to infections and bacterial products involving radio-resistant macrophage and endothelial cells, and
3) accelerated GM-CSF production following lymphocyte activation.

There are in addition a number of less well defined situations in which elevated GM-CSF levels are observed, some of which may involve alterations in degradation and/or excretion of GM-CSF.

1. Steady State Production of GM-CSF

A number of continuous cell lines exhibit a capacity to synthesize GM-CSF. The different levels of activity of these cell lines may indicate genuine differences in the intrinsic capacity of different cells to synthesize GM-CSF under basal "steady state" conditions, although it is not necessarily valid to equate functional activity in vitro with a similar capacity in vivo. In some situations, inhibition of cell proliferation in vitro has been observed to lead to an increased capacity of the cells to synthesize GM-CSF (RALPH and NAKOINZ, 1977; RALPH et al., 1977). Again the significance of this observation is unclear since in a number of examples the cell lines studied were undifferentiated populations of leukemic cells and it seems likely that the inhibitors may have first caused macrophage differentiation and only subsequently increased GM-CSF production (HONMA et al., 1978). In these examples, arrest of proliferation may not have been the direct cause of GM-CSF production and the data may simply reflect the greater capacity of one cell type (macrophages) to produce GM-CSF than another cell (undifferentiated blast cells).

Studies on fetal mice have shown a very high per cell capacity of fetal cells to produce GM-CSF in vitro and very high circulating GM-CSF levels in the blood (JOHNSON and METCALF, 1978 a). This raises the possibility that basal levels of CSF production may fluctuate in different tissues during ontogeny.

In man and gray collie dogs with the disease cyclic neutropenia, cyclic fluctuations in serum and urine GM-CSF levels have been observed (DALE et al., 1972; MANGALIK and ROBINSON, 1973; MOORE et al., 1974; GUERRY et al., 1974). As recurrent infections complicate cyclic neutropenia, the situation is complex (GREENBERG et al., 1976) but the evidence suggests strongly that cyclic fluctuations in basal production of GM-CSF may occur and be responsible, in part at least, for the disease state. A less pronounced cyclic rhythm in polymorph levels is seen in normal humans and some fluctuation in the production rate of GM-CSF may occur in normal humans.

2. Increased GM-CSF Production in Response to Infections and Bacterial Products

Infections of all types – bacterial, viral, and protozoal – have been observed to cause major elevations in serum and urine GM-CSF levels in humans and mice during the febrile stage of the infection (METCALF and FOSTER, 1967 b; FOSTER et al., 1968 a, b; METCALF and WAHREN, 1968; METCALF and BRADLEY, 1970; WAHREN and ECSENYI, 1971; MCNEILL and KILLEN, 1971; TRUDGETT et al., 1973; BROJORGENSON and KNUDTZON, 1977). Conversely, serum GM-CSF levels are abnormally low in germ-free mice (METCALF et al., 1967). From these observations, it seems likely that recent clinical or subclinical infections or exposure to microbial products are major determinants of the somewhat variable GM-CSF levels observed in man and conventionally-reared animals.

The mechanisms operating have been best characterized for endotoxin and the purified bacterial cell-wall components – Lipid A, lipoprotein, and murein. Injection of endotoxin into mice produces a rapid rise in serum and urine levels of GM-CSF up to 50–100-fold above normal (METCALF, 1971; CHERVENICK, 1972; QUESENBERRY et al., 1972). The response reaches maximal levels within 3 h and elevated GM-CSF levels persist for 24 h. Paralleling the rise in serum GM-CSF levels are rises in the levels of GM-CSF extractable from all tissues (SHERIDAN and METCALF, 1972) and in the levels of GM-CSF synthesized in vitro by various tissues from endotoxin-injected mice (SHERIDAN and METCALF, 1973 a; BERAN, 1975; NICOLA et al., 1979 a; METCALF, unpublished data).

The apparent rises in serum GM-CSF levels far exceed those observed in tissue extracts or conditioned media, the latter usually being only 2- to 5-fold. Two factors may contribute to an over-estimate of GM-CSF levels in post-endotoxin (or postinfection) serum: (a) a fall in the levels of serum lipoproteins that normally partially inhibit GM-colony formation in vitro (GRANSTROM, 1973), and (b) a rise in the levels of serum potentiating factor which amplifies colony stimulation by GM-CSF (METCALF et al., 1975 a; VAN DEN ENGH and BOL, 1975). Although both these changes do result in some over-estimation of GM-CSF levels in post-endotoxin serum, studies on chemically fractionated serum clearly revealed that GM-CSF levels in the serum are elevated to a higher level than are the levels of GM-CSF seen in tissue extracts and conditioned media made from endotoxin-injected animals (NICOLA et al., 1979 a; BURGESS and METCALF, unpublished data).

The response to endotoxin is not inhibited by prior irradiation and is not T-cell dependent since nude (congenitally athymic) mice produced typical responses (METCALF, 1971; METCALF, unpublished data). A number of studies have shown that the addition of endotoxin or purified bacterial cell-wall components to cultures of macrophages leads to an increase in GM-CSF levels in the medium within 3–6 h (HAYES et al., 1972; SHERIDAN and METCALF, 1973 a; CHERVENICK et al., 1973; RUSCETTI and CHERVENICK, 1974; EAVES and BRUCE, 1974; STABER et al., 1978). This acute rise may be due to accelerated release of preformed GM-CSF since the phenomenon is not blocked by puromycin (SHERIDAN and METCALF, 1973 a). A comparable release of GM-CSF from blood cells has been reported following exposure to gram-positive bacteria (BOLIN and ROBINSON, 1977). However the more slowly developing increased GM-CSF levels observed in media conditioned by organs from endotoxin-injected mice are dependent on protein synthesis and appear to reflect a genuine stimulation of GM-CSF synthesis following the injection of endotoxin (SHERIDAN and METCALF, 1973 a; METCALF and BURGESS, unpublished data).

A similar response to the addition of endotoxin has been observed in cultures of endothelial cells (QUESENBERRY et al., 1978) and, as discussed earlier, present data do not permit any firm conclusions on how many cell types in the body respond to endotoxin or similar material by the increased synthesis and release of GM-CSF.

Rechallenge of mice with endotoxin or bacterial antigens evokes little or no rise in GM-CSF levels (METCALF, 1974 b; QUESENBERRY et al., 1975) (Fig. 6). Such mice

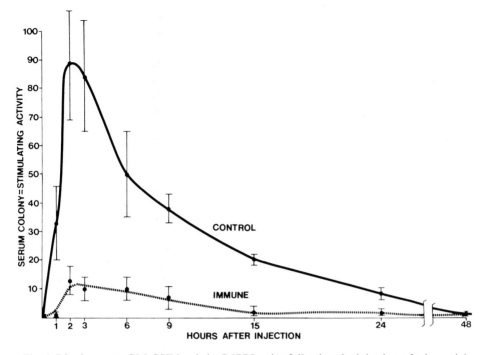

Fig. 6. Rise in serum GM-CSF levels in C 57 BL mice following the injection of a bacterial antigen (Salmonella polymerised flagellin). Note the depressed GM-CSF response in mice previously injected with flagellin (immune) versus the response in mice preinjected with saline (control). (METCALF, 1977)

respond normally to other non-cross-reacting bacterial preparations. In some cases the depressed responsiveness may be due to antibody formation of the original material injected since transfer of serum from refractory mice to normal mice depresses their responsiveness to challenge injection (METCALF, 1974 b). However this does not appear to be the situation in the depressed responsiveness observed after a large priming injection of lipid A (STABER, 1979) and several mechanisms may operate to confer depressed responsiveness on pre-injected animals.

It has been observed that C 3 H/HeJ mice fail to exhibit elevated serum GM-CSF levels following the injection of endotoxin (APTE and PLUZNIK, 1976). Genetic analysis of crosses and backcrosses between responsive (C 3 HeB) and unresponsive (C 3 H/HeJ) mice showed that the capacity to develop elevated GM-CSF responses is determined by a single dominant gene. Endotoxin also causes comparable GM-CSF rises in man (GOLDE and CLINE, 1975) but no genetic anomalies have yet been reported to be associated with an inability to respond to endotoxin, comparable with the situation seen in C 3 H/HeJ mice.

With the use of bacterial cell-wall products, GM-CSF levels rise rapidly but are sustained for no more than 24 h after which time depressed responsiveness to rechallenge becomes quite marked. This raises some problems in explaining how a human or mouse manages to sustain the high serum and urine GM-CSF levels observed for periods as long as weeks during persistent infections (e.g. in leukemic

patients). Certain agents such as BCG or C. parvum appear to be able to evoke
more sustained rises in GM-CSF levels (LADISCH et al., 1979; FOSTER et al., 1977),
suggesting that microbial products with adjuvant-like action may extend the acute
short-lived response seen with simple bacterial products. Alternatively, in natural
infections, a succession of antigenically distinct microbial products may be generated and in sequence sustain a continuous stimulation. Whatever the mechanisms
involved, it needs to be remembered that in natural infections, elevated GM-CSF
levels are maintained until the fever resolves and that the acute responses observed
in the laboratory may not reveal the complete mechanisms involved.

3. Increased GM-CSF Production Following Lymphocyte Stimulation

Lymphoid organs contain relatively low levels of extractable GM-CSF and when
GM-CSF appears in media conditioned by solid fragments of lymph node or
spleen tissue, the production does not appear to be lymphocyte-mediated since the
process is radioresistant (SHERIDAN and STANLEY, 1971; NICOLA et al., 1979 a; MET-
CALF, unpublished data). Similarly, suspensions of lymphoid organ cells usually are
capable of producing only low levels of detectable GM-CSF.

The situation differs radically when the medium is altered to favor lymphocyte
proliferation by the addition of 2-mercapto-ethanol and lymphocytes are stimulated to proliferate by co-culture with allogeneic cells or the addition of a lymphocyte
mitogen such as phytohemagglutinin, pokeweed mitogen or concanavalin A. This
subject has recently been reviewed in detail (see METCALF, 1979 b) but the essential
findings can be summarised here. When lymphoid populations are stimulated by
phytohemagglutinin, the T-lymphocytes synthesize greatly increased amounts of
GM-CSF in a response that appears not to be dependent on interaction with macrophages (MCNEILL, 1973; PARKER and METCALF, 1974 a, b; CLINE and GOLDE,
1974; RUSCETTI and CHERVENICK, 1975; PRIVAL et al., 1974; AYE et al., 1974, 1975).

In cultures containing pokeweed mitogen or concanavalin A, a more complex
response occurs involving an interaction between adherent cells (presumably macrophages) and T-lymphocytes, as a consequence of which not only are increased
amounts of GM-CSF synthesized but also the corresponding regulators EO-CSF,
MEG-CSF, and E-CSF stimulating the formation of eosinophil, megakaryocyte
and pure and mixed erythroid colonies (PARKER and METCALF, 1974a; METCALF
et al., 1974, 1975b; RUSCETTI and CHERVENICK, 1975; NAKEFF and DANIELS-
MCQUEEN, 1976; METCALF, 1978; METCALF and JOHNSON, 1978). This production
system is extremely radiosensitive (D_0 120–238 rads) and although the evidence is
incomplete, the data suggest that the T-lymphocytes are the cells synthesizing the
various CSF's after interaction with macrophages. It is not clear whether each lymphocyte synthesizes all four CSF's or whether lymphocyte subsets exist, each synthesizing only one type of CSF. Experiments using cloned hybridomas produced
by fusing pokeweed mitogen-primed spleen cells to lymphoma cells have produced
some evidence for the occurrence of bi-functional hybridomas secreting both GM-
CSF and EO-CSF (HOWARD et al., 1979) and further experiments of this type may
resolve this question.

In mixed leukocyte cultures of mouse spleen or lymph node cells, the same type
of process occurs as in pokeweed mitogen-stimulated cultures, with the synthesis

of all four types of CSF if culture conditions are carefully controlled (METCALF et al., 1975b).

Somewhat similar events have been observed in cultures of sensitized lymphoid cells following addition of the sensitizing antigen. Thus RUSCETTI et al. (1976) observed the production of GM-CSF and EO-CSF following the addition of Trichinella spiralis or BCG to sensitized populations and we have observed in spleen cell suspensions from mice infested with Mesocestoides corti the synthesis of all four types of CSF following addition of M. corti larvae. Spleen cells from infected nude mice or from uninfected mice did not exhibit this response (METCALF and MITCHELL, unpublished data).

These in vitro observations suggest that, if lymphocyte populations are stimulated to proliferate in vivo either during immune responses or graft-versus-host reactions, similar elevated production of GM-CSF and other types of CSF could be expected. Because these situations usually also involve exposure to bacterial antigens it is not easy to document such situations in vivo as distinct from endotoxin-type responses not specifically involving lymphoid cells. However several observations represent possible examples of lymphocyte-generated GM-CSF production. Thus HARA et al. (1974) and SINGER et al. (1977) have observed elevated GM-CSF levels in mice and humans with graft-versus-host disease and serum GM-CSF levels are elevated in mice bearing antigenic tumors even in germfree mice (METCALF et al., 1967; HIBBERD and METCALF, 1971).

4. Other Situations Modifying GM-CSF Levels

Serum GM-CSF levels have been shown to be elevated in most mice and many humans with spontaneously developing tumors, or in mice with transplanted tumors or tumors induced by irradiation or viruses (ROBINSON et al., 1967; METCALF and FOSTER, 1967b; METCALF et al., 1967; FOSTER et al., 1968b; METCALF et al., 1971; HIBBERD and METCALF, 1971; ASANO et al., 1977).

In each case, the possibility exists that the observed rises may have been due in part to secondary exposure to microbial products because of the debilitated state of the animal or patient. Furthermore, since many mouse tumors exhibit antigens foreign to the host, some element of lymphocyte stimulation may exist and evoke increased GM-CSF production by lymphocytes. It is also possible that tumor tissue per se can activate the GM system in a host response involving neither microbial nor antigenic stimulation, since in mice substantial changes in GM populations and their proliferative activity are consistent features of the tumor-bearing state (HIBBERD and METCALF, 1971; CLAESSON and JOHNSON, 1978; McCARTHY, 1979).

At least some tumor populations are exceptionally good producers of GM-CSF. The WEHI-3 myelomonocytic leukemia is an outstanding example in which the cloned cells have been shown to actively synthesize GM-CSF, EO-CSF, MEG-CSF, and E-CSF (METCALF et al., 1969; METCALF and MOORE, 1971; WILLIAMS et al., 1978; 1979; METCALF, unpublished data).

Similar studies have been reported using human tumor cells. Studies of ASANO et al. (1977) have been particularly revealing in this context. Lung tumor grafts from a patient exhibiting a granulocytosis were established in nude mice and shown to continue to actively secrete GM-CSF, leading to elevated serum GM-CSF levels and a marked granulocytosis in the host animal. A survey of patients with lung tumors showed a close correlation between granulocyte levels in the patient and the capacity of the tumor tissue to secrete GM-CSF in vitro (ASANO, personal communication).

It may be therefore that in some patients or animals with cancer, much of the elevated GM-CSF levels observed are actually derived from the tumor cells themselves.

In studies on the effect of irradiation on GM-CSF levels, it was observed that irradiation could induce two types of dose-related elevations of serum GM-CSF levels – an acute response evident within 6 h (HALL, 1969) followed by a fall then a more slowly developing rise in serum levels over a period of some days (MORLEY et al., 1972). The latter type of response appears to be due to secondary endotoxemia presumably following gut damage. Germ-free mice do not exhibit the slowly developing rise in serum GM-CSF levels following irradiation (MORLEY et al., 1972) but will do so if mono contaminated with Gram negative but not Gram positive organisms (CHANG and POLLARD, 1973).

In the regenerative period following irradiation damage to hemopoietic populations, a pronounced rise occurs in the capacity of bone stromal cells to synthesize GM-CSF immediately before regeneration is observed of granulopoietic populations (CHAN and METCALF, 1972, 1973). This is preceded by a fall in serum levels of the lipoproteins that act in vitro as somewhat non-specific inhibitors of GM-CSF-stimulated colony formation (CHAN and METCALF, 1973) and these lipoproteins have been shown to depress bone stromal production of GM-CSF in vitro (BERAN, 1975). It is uncertain whether this sequence of changes represents cause and effect but it may be that serum lipoproteins somehow modulate stromal production of GM-CSF and must first fall before local GM-CSF production can rise and thus stimulate regeneration of granulopoietic tissue. It is equally possible however that there is no significant interrelationship between these various changes that occur near the onset of regeneration.

Few studies have been reported on metabolic factors that might influence GM-CSF levels. Studies on the half-life in the serum GM-CSF of injected antigenically-distinct GM-CSF or radiolabeled purified GM-CSF indicated a short half-life of 3–7 h (METCALF and STANLEY, 1971 b; BURGESS and METCALF, 1977a). Injection of large doses of cortisone were observed to cause a rapid fall in serum GM-CSF levels (METCALF, 1969). Addition of lithium carbonate to cultures of organ or cell suspensions can increase GM-CSF production (HARKER et al., 1977; SPITZER et al., 1979) and serum and urine GM-CSF have been observed to increase in patients under therapy with lithium carbonate (TISMAN et al., 1973; GUPTA et al., 1976).

Conversely, it has been reported that very low concentrations of lactoferrin can suppress GM-CSF production by human marrow adherent cells (BROXMEYER et al., 1978). Previous studies had shown that polymorphs can suppress GM-CSF production by adherent human marrow cells (HEIT et al., 1974; BROXMEYER et al., 1977) and endotoxin-stimulated monocytes (MAHMOOD and ROBINSON, 1978), serum GM-CSF levels in mice and the production of GM-CSF by various mouse tissues (BROXMEYER, 1978). These various phenomena may be mediated by the lactoferrin present in polymorphs.

High serum colony stimulating activity has been observed in patients with advanced renal disease. This was coupled with a low urinary excretion of GM-CSF (CHAN, 1972). It is unclear how much of the GM-CSF observed in the serum is cleared to the urine by the kidney. Urinary GM-CSF in man differs in molecular weight and biological activity from other GM-CSF's derived from human tissue, suggesting some metabolic modification of the molecule before it enters the urine. Futhermore, since renal tissue actively produces GM-CSF (NICOLA et al., 1979a) some urinary GM-CSF may actually be locally produced in the kidney. Nephrectomy or ligation of the ureters caused moderate elevations in serum GM-CSF levels (CHAN, 1970; FOSTER and MIRAND, 1970) but did not significantly alter the decline in serum GM-CSF levels following an acute response to endotoxin (SHERIDAN and METCALF, 1973c). Following the injection of radiolabeled GM-CSF, much of the radiolabel appearing in the urine was no longer associated with macromolecules able to bind to concanavalin A, although some apparently intact molecules were observed in the first hour after intravenous injection (BURGESS and METCALF, 1977a; BURGESS and METCALF, unpublished data). It may be therefore that elimination in the urine is one metabolic fate of GM-CSF although this may not be the major pathway of GM-CSF removal.

VI. Role of GM-CSF In Vivo

Although a stimulating factor may exert a highly specific and reproducible action on target cells in vitro, certain basic requirements are needed to establish that the factor functions in a comparable manner in vivo: (a) the factor must be demonstrable in adequate concentrations in the serum or tissues, (b) levels of the factors must fluctuate in vivo in accord with expectations during variations in the proliferative activity of the target cell in vivo, and (c) injections of the factor must elicit effects in vivo comparable with those observed in vitro.

While it is possible to advance hypotheses why the above criteria may not be able to be met in specific instances, e.g. short-range factors not reaching serum, secretion or transport in a metabolically inactive form or complex networks of interacting regulators in vivo, the onus remains with the investigator to meet most or all of the above criteria if a factor detected during in vitro studies is to be acceptable as a significant regulatory factor in vivo.

The basis weakness in evidence of types (a) and (b) is that such evidence is indirect and does not necessarily prove the functional activity of the factor in vivo since more than one factor may coexist or fluctuate in parallel and the genuine in vivo regulator may not be detectable in vitro.

With respect to GM-CSF, the evidence for the in vivo functional activity of the molecule is extremely strong by criteria (a) and (b) but currently incomplete and inadequate by criterion (c). GM-CSF is demonstrable in the serum, urine and tissues in concentrations more than adequate to stimulate the proliferation of GM target cells in vitro. Levels of GM-CSF fluctuate to a marked degree in situations involving perturbations of granulocyte and monocytemacrophage formation and such fluctuations precede changes in proliferative activity of the candidate target cells by an interval compatible with the known proliferation and maturation times of the target cell population. The only major deficiency in this convincing body of indirect evidence is the lack of an experimental system for suppressing GM-CSF levels in vivo and thus an inability to determine whether GM-CSF suppression leads to a depressed production of granulocytes and macrophages.

Measurement of the effects of injected GM-CSF is complicated by certain logistic problems and the complexity of the target cell population itself: (a) GM-CSF has a short serum half-life of three to seven hours. If purified GM-CSF is to be injected into a mouse, the amounts needed are in excess of those available at present. The material needs to be of syngeneic origin to the test animal to avoid complications resulting from antigenic stimulation of the recipient and the material must of course also be endotoxin-free. If impure material is used, interpretation of the observed effects is open to the criticism that the changes observed may be due simply to contaminating agents in the preparation; (b) While it might be considered logical to monitor GM-CFC levels in recipient animals, this presupposes that GM-CSF stimulates self-replication of GM-CFC (for which there is no in vitro evidence) or stimulates the formation of GM-CFC's from more ancestral cells. While there is some evidence for this latter process (METCALF et al., 1980a), the generation of GM-CFC from stem cells is obviously a complex process involving other possibly more significant control systems. Unless selfreplication of GM-CFC occurs or a significantly increased generation of new GM-CFC's is provoked, then GM-CSF would in fact be expected to *deplete* pre-existing GM-CFC's by stimulating them to form more differentiated progeny. If both processes occur simultaneously, the effects would be mutually antagonistic and would potentially balance one another; (c) The only logical method of assessing GM-CSF action in vivo is to monitor the production rate of polymorphs and monocytes. The difficulty here is that peripheral blood levels of polymorphs and monocytes are known not necessarily to reflect production rates due to various complex processes including variable release and margination of mature cells and the exponential life span of such cells in the circulation (ATHENS, 1969). For this reason, some investigators have chosen to use "captive" target cell populations in vivo by enclosing marrow cells in millipore diffusion chambers in the injected animal.

The data so far published have indicated that impure preparations of GM-CSF will elicit modest rises in marrow and spleen levels of GM-CFC and their immediate progeny, the GM cluster-forming cells (BRADLEY et al., 1969; METCALF and STANLEY, 1971a) and, as assessed in animals labeled with tritiated thymidine, an increased formation of blood polymorphs and monocytes but not major elevations in the levels of these cells. Injection of antiserum to GM-CSF reduced granulopoiesis in marrow populations enclosed in diffusion chambers (SHADDUCK et al., 1978).

While these data are consistent with a genuine in vivo function of GM-CSF it must be said that the effects are disappointingly small and somewhat inconsistent. Certainly there is a clear discrepancy between the difficulty in demonstrating the in vivo effects of GM-CSF and the great ease of showing in vivo effects of the corresponding regulator for erythropoiesis, erythropoietin, even accepting that the life history of mature red cells is much simpler and easier to monitor than that of polymorphs or monocytes.

A classical method for observing the in vivo effects of a hormone or regulator is to implant into recipient animals a slowly growing tumor that secretes the agent under study. This provides a continuous production source of the agent within the animal over long periods. Recent studies on nude mice implanted with human tumors secreting GM-CSF have produced spectacular evidence of stimulation of granulopoiesis, reversible by resection of the tumor (ASANO et al., 1977). While there is a risk that such tumors may also be secreting another stimulating factor that is the real in vivo agent, this argument seems a little extreme. It may be therefore that, in the immediate future, such model systems may provide the best insight into the type of in vivo responses to be expected when sufficient purified GM-CSF becomes available for extensive in vivo studies.

E. Eosinophil Colony Stimulating Factor (EO-CSF)

In vivo studies in mice have demonstrated that the ability to develop increased levels of eosinophils following sensitization appears to depend on a humoral factor derived from T-lymphocytes (BASTEN and BEESON, 1970; McGARRY et al., 1971; MILLER and McGARRY, 1976).

In this context, it is of interest that when mouse spleen or lymph node cells are stimulated to proliferate in vitro by the addition of pokeweed mitogen or concanavalin A, a complex interaction occurs between adherent cells (presumably macrophages) and T-lymphocytes leading to the synthesis by the T-lymphocytes of a factor able to stimulate the formation by mouse marrow cells of two types of eosinophil colony – loose dispersed colonies of incompletely granulated cells (PARKER and METCALF, 1974a; METCALF et al., 1974, 1975b; METCALF and JOHNSON, 1978) and compact colonies of highly granulated cells (JOHNSON and METCALF, 1980a). Spleen cells from nude (congenitally athymic) mice lack the capacity to produce this eosinophil CSF (EO-CSF) (METCALF and JOHNSON, 1978) but are able to do so following grafting of thymus tissue to the animal (METCALF, 1978).

RUSCETTI et al. (1976) demonstrated the production of a similar eosinophil stimulating factor by incubating spleen cells from mice sensitized to Trichinella spiralis or BCG with the sensitizing antigen. We have demonstrated a similar phenomenon by adding *M. corti* larvae to cultures of spleen cells from mice infested with *M. corti* (METCALF and MITCHELL, unpublished data).

The general behavior and properties of EO-CSF in stimulating eosinophil colony formation in agar cultures of marrow cells are similar to those described in detail for GM-CSF. Similar dose-response relationships can be observed between EO-CSF concentration and the number and growth rate of colonies developing (METCALF et al., 1974, 1978), EO-CFC do not survive in vitro in the absence of added EO-CSF (METCALF et al., 1974) and again heterogeneity has been demonstrated in eosinophil progenitor cells (JOHNSON and METCALF, 1980a).

As discussed earlier, purified GM-CSF appears able to stimulate the initial division of at least some eosinophil colony-forming cells but is unable to sustain the terminal proliferation and differentiation involved in eosinophil colony formation (METCALF and JOHNSON, 1979; METCALF et al., 1980a).

Studies on the molecular nature of EO-CSF in pokeweed mitogen stimulated mouse spleen conditioned medium have indicated that it is a glycoprotein of molecular weight 24,000. Although EO-CSF is partially separable form other CSF's in spleen conditioned medium using either isoelectric focusing or hydrophobic chromatography, EO-CSF has not yet been obtained in pure form (METCALF et al., 1978; BURGESS et al., 1978 b, 1979).

In the mouse, one other source has been demonstrated for EO-CSF. The myelomonocytic leukemia WEHI-3 B is capable of synthesizing EO-CSF, a property which so far appears to be unique for this tumor (METCALF et al., 1969, 1974).

In the human, a different situation appears to exist since EO-CSF able to stimulate eosinophil colony formation by human marrow cells is produced by unfractionated peripheral blood cells and by the placenta (CHERVENICK and BOGGS, 1971; SHOHAM et al., 1974; ZUCKER-FRANKLIN and GRUSKY, 1974; PARMLEY et al., 1976; DAO et al., 1977; DRESCH et al., 1977). This EO-CSF has an apparent molecular weight of 32,000 but again has not been produced in pure form (BURGESS et al., 1978 b; NICOLA et al., 1978, 1979 b). Production of EO-CSF by activated human lymphocytes has not yet been reported. Human EO-CSF appears to be unable to stimulate eosinophil colony formation by mouse bone marrow cells.

EO-CSF so far has not been able to be detected in mouse or human serum and despite the experimental evidence for the involvement of a lymphocyte-derived humoral factor in the production of eosinophilia, it is uncertain what role EO-CSF plays in regulating eosinophil production in vivo. For example, nude mice fail to develop an eosinophilia in response to *M. corti* infestation (JOHNSON et al., 1979) and nude spleen cells are unable to produce EO-CSF in vitro (METCALF, 1978). However, nude mice are capable of producing basal numbers of eosinophils which suggests either that a second source of EO-CSF exists in the mouse or that basal eosinophil production is controlled by some other mechanism.

F. Megakaryocyte Colony Stimulating Factor (MEG-CSF)

The situation with the control of megakaryocyte proliferation in vitro has a number of parallels to that just discussed for eosinophil production. Mouse megakaryocyte colonies can be stimulated to proliferate in vitro by use of pokeweed mitogen- or concanavalin A-stimulated spleen conditioned medium (METCALF et al., 1975 b; METCALF and JOHNSON, 1978) or by medium conditioned by WEHI-3 B cells (WILLIAMS and JACKSON, 1978; WILLIAMS et al., 1978). Megakaryocyte colony formation has also been reported following stimulation by fibroblasts (NAKEFF et al., 1975) or by medium conditioned by PHA-stimulated lymphoid cells or L-cells (NAKEFF and DANIELS-MCQUEEN, 1976). Again there is evidence for the existence of two distinct subpopulations of precursor cells – the first forming small colonies of large mature cells and the second forming large colonies containing all developmental stages of megakaryocyte formation (METCALF et al., 1975 b; WILLIAMS and JACKSON, 1978).

As was true for eosinophil colony formation, purified GM-CSF appears able to directly stimulate the initial proliferation of at least some megakaryocyte precursor cells but is unable to stimulate the terminal proliferative divisions and endomitoses involved in the formation of mature megakaryocytes (METCALF and JOHNSON, 1979; METCALF et al., 1980 a). It is uncertain whether this collaboration between GM-CSF and MEG-CSF plays a significant role in the control of megakaryocyte formation in vivo as no factor with a capacity to directly stimulate mega-karyocyte proliferation in vitro has so far been detected in the serum or in tissue extracts.

Fractionation studies on spleen conditioned medium have indicated that the active factor, MEG-CSF, is a glycoprotein of molecular weight 24,000 (METCALF et al., 1978; BURGESS et al., 1978 b, 1979) and both types of megakaryocyte pre-cursor cell appear to respond to the same factor.

A curious situation exists with respect to the stimulation of megakaryocyte col-ony formation in plasma gel cultures by mouse or human marrow cells in that ad-dition of erythropoietin has been reported to be an effective stimulus for colony formation (McLEOD et al., 1976). The only published data have involved the use of relatively crude preparations of erythropoietin (from human urine or anemic sheep plasma) and it may be that a genuine megakaryocyte-specific factor is pres-ent as a contaminant in such preparations. Purified human erythropoietin appears to have no capacity to stimulate megakaryocyte colony formation in agar by mouse bone marrow cells or to support the survival of megakaryocyte colony-forming cells (METCALF and JOHNSON, 1979; METCALF et al., 1980a) although hu-man erythropoietin is an effective stimulus for mouse erythropoietic cells.

The in vivo role of MEG-CSF as assayed in vitro is unclear. An extensive lit-erature exists based on in vivo studies of a humoral factor termed "thrombo-poietin" that stimulates an increased production of platelets and possibly an in-creased production of megakaryocytes (DE GABRIELE and PENINGTON, 1967; EBBE, 1970; COOPER, 1970). Supernatants from human embryo kidney cell lines contain-ing thrombopoietic stimulating factor or serum containing thrombopoietin have been reported to amplify megakaryocyte colony formation in vitro (NAKEFF, 1977; WILLIAMS et al., 1979) but are not themselves capable of stimulating megakaryo-cyte colony formation. However in this laboratory, serum preparations containing thrombopoietin were found neither to stimulate colony formation nor to modify MEG-CSF-stimulated colony formation (LEVIN and METCALF, unpublished data).

G. Erythropoietin and Erythroid Colony Stimulating Factor

The existence of a humoral regulator controlling erythropoiesis was detected by in vivo studies (CARNOT and DEFLANDRE, 1906). This factor was termed erythropoi-etin and extensive in vivo studies since then have established that erythropoietin is able to stimulate red cell formation and hemoglobin synthesis by erythropoietic cells (see review PESCHELE et al., 1978). Initially it was believed that the sole source of erythropoietin was the kidney but more recent studies have indicated that other cell types probably also have a capacity to synthesize erythropoietin or an erythro-poietic precursor, e.g. fetal liver Kupffer cells (ZUCALI and MIRAND, 1978; NAUGHTON et al., 1978).

STEPHENSON et al. (1971) reported that addition of preparations containing hu-man erythropoietin to cultures of mouse fetal liver cells was able to stimulate the formation of small colonies of hemoglobin-synthesizing erythroid cells and sub-sequent work extended these observations to cultures of human erythropoietic cells (TEPPERMAN et al., 1974). The cells forming these small erythroid colonies were termed "CFU-E" (colony-forming unit, erythroid) and analysis showed that these were relatively mature precursor cells corresponding in differentiation to the cluster-forming cells of the granulocyte-macrophage system.

The purification of erythropoietin from human urine has been achieved and it has been shown to be a neuraminic acid-containing glycoprotein of molecular weight 39,000 (MIYAKE et al., 1977). Addition of purified erythropoietin to methylcellulose cultures or agar cultures of mouse fetal liver or bone marrow cells will stimulate the formation of small erythroid colonies of the type derived from CFU-E (VAN ZANT and GOLDWASSER, 1978; JOHNSON and METCALF, 1977). As was discussed for GM-CSF, erythropoietin has two distinct actions on target erythropoietic cells – stimulation of proliferation and induction of hemoglobin synthesis. This implies that erythropoietin may influence several distinct metabolic pathways in target erythroid cells.

Addition of higher concentrations of erythropoietin to semisolid cultures of mouse or human hemopoietic cells leads to formation of single or multicentric erythroid colonies of much larger size. The cell initiating such colonies is termed the "burst-forming unit" erythroid (BFU-E) and this cell appears analogous with the hemopoietic progenitor cells forming granulocyte-macrophage colonies in vitro (AXELRAD et al., 1973; ISCOVE and SIEBER, 1975; HEATH et al., 1976; OGAWA et al., 1977).

Initially, it was assumed from this finding that erythropoietin was the regulator of erythropoietic proliferation from the earliest committed erythroid progenitor cell onwards with an in vitro role analogous to that of GM-CSF. However more recent data have cast some doubt on the conclusion that erythropoietin regulates the early proliferative events in erythropoiesis. Data from certain kinetic experiments in vivo raised the possibility that some other regulator must control the early proliferative steps (LAJTHA et al., 1971) and this was supported by the observation that the number and proliferative status of BFU-E were not influenced by anemia or hypoxia, procedures elevating erythropoietin levels (ISCOVE, 1977; HARA and OGAWA, 1977; WAGEMAKER et al., 1977). Furthermore, culture of BFU-E in the presence only of erythropoietin did not permit the survival or the proliferation of these cells (ISCOVE, 1978; METCALF and JOHNSON, 1979).

Although burst erythroid colony formation can be stimulated by the addition of high concentrations of preparations containing erythropoietin, these preparations have usually been impure. Furthermore, burst formation is markedly increased by addition of white cell-, kidney- or T-lymphocyte-conditioned medium (AYE, 1977; WAGEMAKER, 1978; AXELRAD et al., 1978; ISCOVE and GUILBERT, 1978; NATHAN et al., 1978) or embryo fibroblast underlayers (GHIO et al., 1979). The active component in such preparations has been termed "burst promoting activity" (BPA) or "burst feeder activity" (BFA) and this agent has been proposed as the major regulator of the early, apparently erythropoietin-independent, proliferation of cells in the erythroid pathway. As discussed above, purified GM-CSF also stimulates the survival and early proliferation of erythroid cells and since all BPA preparations used so far contain GM-CSF, this latter regulator could itself be the so-called "burst promoting activity."

It is uncertain whether BPA plays a significant role in the intact animal but T-lymphocytes or theta-positive cells have been observed to influence the proliferative activity of erythroid populations in certain experimental situations and a BPA-like factor may be involved (GOODMAN et al., 1978; WIKTOR-JEDRZEJCZAK et al., 1977).

Complicating the above analysis of factors controlling erythropoiesis in vitro has been a series of observations indicating an alternative method for stimulating erythropoietic proliferation and differentiation in vitro. Medium conditioned by pokeweed mitogen or concanavalin A-stimulated mouse spleen cells has the capacity to stimulate the formation of erythroid colonies by mouse fetal liver or adult marrow hemopoietic cells. These colonies include the small colonies derived from CFU-E but also typical burst erythroid colonies, half of which are mixed erythroid colonies containing other hemopoietic cells (JOHNSON and METCALF, 1977, 1978a, 1980b; METCALF and JOHNSON, 1978; METCALF et al., 1979). This conditioned medium contains no detectable erythropoietin as assayed in polycythemic mice (JOHNSON and METCALF, 1977) or by heme synthesis in suspension cultures of rabbit marrow cells (FIRKIN and JOHNSON, unpublished data).

Medium conditioned by WEHI-3B leukemic cells also is capable of stimulating pure mixed erythroid colony formation by mouse fetal liver cells (METCALF, WALKER, and BURGESS, unpublished data). Fetal hemopoietic cells appear more responsive to spleen conditioned medium (SCM) than adult cells in that colonies achieve a larger size and colony cells exhibit much more complete differentiation and hemoglobin synthesis. Addition of even high concentrations (up to 10 units/ml) of erythropoietin to cultures of adult cells does not increase colony numbers or hemoglobin content (JOHNSON and METCALF, 1980b) but in cultures of fetal liver cells, addition of purified erythropoietin consistently increased colony numbers and hemoglobin content (METCALF and JOHNSON, 1979).

Spleen conditioned medium can stimulate erythroid colony formation in cultures of single colony-forming cells (JOHNSON and METCALF, 1977) and the action of E-CSF is therefore a direct one on the hemopoietic cells. However since all cultures of this type contain human or fetal calf serum, low levels of erythropoietin are present and may conceivably be acting in collaboration with E-CSF to stimulate the formation of maturing erythroid colonies.

As indicated above, spleen conditioned medium of this type also contains GM-CSF, EO-CSF, and MEG-CSF and the erythroid stimulating factor has not yet been separated successfully from these other factors. The E-CSF in spleen conditioned medium is a glycoprotein of molecular weight 24,000 (METCALF et al., 1978; BURGESS et al., 1978b, 1979). Because mouse erythropoietin has not been characterized, it is not known whether the E-CSF in spleen conditioned medium has any close relationship with mouse erythropoietin. The fact that E-CSF differs in physical properties from purified human urinary erythropoietin is not necessarily relevant in view of the known differences between mouse and human GM-CSF.

Because of the current uncertainty regarding the interrelationship between erythropoietin and E-CSF, it is not possible to discuss fully the in vivo role of the erythropoietic regulators detectable in vitro. There is an extensive body of evidence implicating erythropoietin in the control of erythropoiesis in vivo but E-CSF has not been positively identified in vivo and the same possibilities and reservations exist with respect to an in vivo role of E-CSF that were discussed previously for EO-CSF and MEG-CSF.

H. Final Comments

Work in vitro on hemopoietic colony stimulating factors has strongly supported the general concept that cellular proliferation and differentiation in all tissues is under the control of target cell-specific humoral regulatory factors. If the data on hemopoietic stimulating factors have general relevance it becomes necessary to envisage a new class of humoral regulators characterized by a number of common features: (a) their glycoprotein nature, (b) their production by many different cell types, (c) their highly selective target cell specificity, (d) their dual action on target cells – growth stimulation and induction of differentiation, and (e) the heterogeneity and responsiveness of target cells based on target cell subpopulation differences.

Much less is known of tissue-specific growth inhibitory factors but these must be presumed to exist to ensure controlled regulation of cellular proliferative activity.

While the reasons for presupposing the existence of humoral regulators of hemopoietic tissue have always been compelling because of the highly dispersed nature of these populations, it is not unreasonable to expect that similar control systems exist for other cell populations such as the liver or kidney.

The highly specialised nature of mature hemopoietic cells and the availability of the present efficient cloning systems and purified regulatory molecules make these ideal model populations for solving many basic biological questions concerning the nature and control of cell division and differentiation. The greatest ultimate value of this work on hemopoiesis may be however to point the way for comparable studies on cells from all other tissues.

References

Aardal, N.P., Laerum, O.D., Paukovits, W.R., Mauer, H.R.: Inhibition of agar colony formation by partially purified granulocyte extracts (chalone). Virchows Archiv [Cell Pathol.] 24, 27–39 (1977)

Apte, R.N., Pluznik, D.H.: Genetic control of lipopolysaccharide-induced generation of serum colony stimulating factor and proliferation of splenic granulocyte/macrophage precursor cells. J. Cell. Physiol. 89, 313–324 (1976)

Asano, S., Urabe, A., Okabe, T., Sato, N., Kondo, Y., Ueyama, Y., Chiba, S., Ohsawa, N., Kosaka, K.: Demonstration of granulopoietic factor(s) in the plasma of nude mice transplanted with a human lung cancer and in the tumor tissue. Blood 49, 845–852 (1977)

Athens, J.W.: Granulocyte kinetics in health and disease. Natl. Cancer Inst. Monogr. 30, 135–155 (1969)

Austin, P.E., McCulloch, E.A., Till, J.E.: Characterisation of the factor in L-cell conditioned medium capable of stimulating colony formation by mouse marrow cells in culture. J. Cell. Physiol. 77, 121–133 (1971)

Axelrad, A.A., McLeod, D.L., Shreeve, M.M., Heath, D.S.: Properties of cells that produce erythrocyte colonies in vitro. In: Hemopoiesis in culture. Robinson, W.A. (ed.), pp. 226–234. DHEW Publication (NIH) 74-205 Washington 1973

Axelrad, A.A., McLeod, D.L., Suzuki, S., Shreeve, M.M.: Regulation of population size of erythropoietic progenitor cells. In: Differentiation of normal and neoplastic hematopoietic cells. Clarkson, B., Marks, P.A., Till, J.E. (eds.), pp. 155–163. New York: Cold Spring Harbor Laboratory 1978

Aye, M.T.: Erythroid colony formation in cultures of human marrow. Effect of leukocyte conditioned medium. J. Cell. Physiol. 91, 69–78 (1977)

Aye, M.T., Niko, Y., Till, J.E., McCulloch, E.A.: Studies of leukemic cell populations in culture. Blood *44*, 205–219 (1974)

Aye, M.T., Till, J.E., McCulloch, E.A.: Interacting populations affecting proliferation of leukemic cells in culture. Blood *45*, 485–493 (1975)

Basten, A., Beeson, P.B.: Mechanism of eosinophilia. II. Role of the lymphocyte. J. Exp. Med. *131*, 1288–1305 (1970)

Beran, M.: The influence of mouse sera on colony formation and on the production of colony stimulating factor in vivo. Exp. Hematol. *3*, 309–318 (1975)

Bol, S., Williams, N.: The maturation state of three types of granulocyte-macrophage progenitor cells from mouse bone marrow. J. Cell. Physiol. *102*, 233–244 (1980)

Bolin, R.W., Robinson, W.A.: Bacterial, serum and cellular modulation of granulopoietic activity. J. Cell. Physiol. *92*, 145–154 (1977)

Bradley, T.R., Metcalf, D.: The growth of mouse bone marrow cells in vitro. Aust. J. Exp. Biol. Med. Sci. *44*, 287–300 (1966)

Bradley, T.R., Sumner, M.A.: Stimulation of mouse bone marrow colony growth in vitro by conditioned medium. Aust. J. Exp. Biol. Med. Sci. *46*, 607–618 (1968)

Bradley, T.R., Metcalf, D., Sumner, M., Stanley, R.: Characteristics of in vitro colony formation by cells from haemopoietic tissues. In: Hemic cells in vitro. In Vitro. Farnes, P. (ed.), vol. 4, pp. 22–35. Philadelphia: Williams and Wilkins 1969

Bradley, T.R., Stanley, E.R., Sumner, M.A.: Factors from mouse tissues stimulating colony growth of mouse bone marrow cells in vitro. Aust. J. Exp. Biol. Med. Sci. *49*, 595–603 (1971)

Bradley, T.R., Hodgson, G.S., Rosendaal, M.: The effect of oxygen tension on hemopoietic and fibroblast cell proliferation in vitro. J. Cell. Physiol. *97*, 517–522 (1978)

Bro-Jorgensen, K., Knudtzon, S.: Changes in hemopoiesis during the course of acute LCM virus infection in mice. Blood *49*, 47–57 (1977)

Broxmeyer, H.E.: Inhibition in vivo of mouse granulopoiesis by cell-free activity derived from human polymorphonuclear neutrophils. Blood *51*, 889–901 (1978)

Broxmeyer, H.E., Moore, M.A.S., Ralph, P.: Cell-free granulocyte colony inhibiting activity derived from human polymorphonuclear neutrophils. Exp. Hematol. *5*, 87–102 (1977)

Broxmeyer, H.E., Smithyman, A., Eger, R.R., Meyers, P.A., Sousa, M. de: Identification of lactoferrin as the granulocyte-derived inhibitor of colony stimulating activity production. J. Exp. Med. *148*, 1052–1067 (1978)

Burgess, A.W., Metcalf, D.: Serum half-life and organ distribution of radiolabeled colony stimulating factor in mice. Exp. Hematol. *5*, 456–464 (1977a)

Burgess, A.W., Metcalf, D.: Colony-stimulating factor and the differentiation of granulocytes and macrophages. In: Experimental hematology today. Baum, S.J., Ledney, G.D. (eds.), pp. 135–146. Berlin, Heidelberg, New York: Springer 1977b

Burgess, A.W., Metcalf, D.: The effect of colony stimulating factor on the synthesis of ribonucleic acid by mouse bone marrow cells in vitro. J. Cell. Physiol. *90*, 471–484 (1977c)

Burgess, A.W., Wilson, E.M.A., Metcalf, D.: Stimulation by human placental conditioned medium of hemopoietic colony formation by human marrow cells. Blood *49*, 573–583 (1977a)

Burgess, A.W., Camakaris, J., Metcalf, D.: Purification and properties of colony-stimulating factor from mouse lung-conditioned medium. J. Biol. Chem. *252*, 1998–2003 (1977b)

Burgess, A.W., Metcalf, D., Russell, S.: Regulation of hematopoietic differentiation and proliferation by colony stimulating factors. In: Differentiation of normal and neoplastic hematopoietic cells. Clarkson, B., Marks, P.A., Till, J.E. (eds.), pp. 339–357. New York: Cold Spring Harbor Laboratory 1978a

Burgess, A.W., Metcalf, D., Nicola, N.A., Russell, S.H.M.: Purification and characterization of cell specific colony stimulating factors. In: Hematopoietic cell differentiation. Golde, D.W., Cline, M.J., Metcalf, D., Fox, C.F. (eds.), pp. 399–416. New York: Academic Press 1978b

Burgess, A.W., Metcalf, D., Russell, S.H.M., Nicola, N.A.: Comparison of granulocyte-macrophage, megakaryocyte, eosinophil and erythroid colony stimulating factors produced by mouse spleen cells. Biochem. J. *185*, 301–314 (1980)

Byrne, P., Heit, W., Kubanek, B.: The in vitro differentiation of density subpopulations of colony-forming cells under the influence of different types of colony-stimulating factor. Cell Tissue Kinet. *10*, 341–351 (1977)

Carnot, P., Deflandre, G.: Sur l'activite hemopoietique du serum au cours de la regeneration du sang. Compt. Rend. Acad. Sci. *143*, 384–386 (1906)

Chan, S.H.: Studies on colony stimulating factor (CSF). Role of the kidney in clearing serum CSF. Proc. Soc. Exp. Biol. *134*, 733–737 (1970)

Chan, S.H.: Influence of serum inhibitors on colony development in vitro by bone marrow cells. Aust. J. Exp. Biol. Med. Sci. *49*, 553–564 (1971)

Chan, S.H.: Bone marrow colony stimulating factor (CSF) and inhibitor levels in renal disease. Rev. Europ. Etudes Clin. Biol. *17*, 686–690 (1972)

Chan, S.H., Metcalf, D.: Local production of colony-stimulating factor within the bone marrow: Role of non-hematopoietic cells. Blood *40*, 646–653 (1972)

Chan, S.H., Metcalf, D.: Local and systemic control of granulocytic and macrophage progenitor cell regeneration after irradiation. Cell Tissue Kinet. *6*, 185–197 (1973)

Chan, S.H., Metcalf, D., Stanley, E.R.: Stimulation and inhibition by normal human serum of colony formation in vitro by bone marrow cells. Brit. J. Haematol. *20*, 329–341 (1971)

Chang, C.F., Pollard, M.: Effects of microbial flora on levels of colony stimulating factor in serums of irradiated CFW mice. Proc. Soc. Exp. Biol. Med. *144*, 177–180 (1973)

Chervenick, P.A.: Effect of endotoxin and post-endotoxin plasma on in vitro granulopoiesis. J. Lab. Clin. Med. *79*, 1014–1020 (1972)

Chervenick, P., Boggs, D.R.: In vitro growth of granulocytic and mononuclear cell colonies from blood of normal individuals. Blood *37*, 131–135 (1971)

Chervenick, P.A., Lo Buglio, A.F.: Human blood monocytes: Stimulation of granulocyte and mononuclear formation in vitro. Science *178*, 164–166 (1972)

Chervenick, P.A., Ruscetti, F.W., Lo Buglio, A.F.: Monocyte-macrophage stimulation of granulocyte and mononuclear colonies in vitro. In: Hemopoiesis in culture. Robinson, W.A. (ed.), p. 117. Washington: DHEW Publication No. (NIH) 74-205 1973

Cifone, M., Defendi, V.: Cyclic expression of a growth conditioning factor (MGF) on the cell surface. Nature *252*, 151–153 (1974)

Claesson, M.H., Johnson, G.R.: The effect of syngeneic lymphoid tumours upon mouse B-lymphocyte and granulocyte-macrophage colony forming cells. Eur. J. Cancer *14*, 515–524 (1978)

Cline, M.J., Golde, D.W.: Production of colony-stimulating activity by human lymphocytes. Nature *248*, 703–704 (1974)

Cooper, G.W.: The regulation of thrombopoiesis. In: Regulation of hematopoiesis. Gordon, A.S. (ed.), pp. 1611–1629. New York: Appleton-Century-Crofts 1970

Dale, D.C., Alling, D.W., Wolff, S.M.: Cyclic hematopoiesis. The mechanism of cyclic neutropenia in grey collie dogs. J. Clin. Invest. *51*, 2197–2204 (1972)

Dao, C., Metcalf, D., Bilski-Pasquier, G.: Eosinophil and neutrophil colony-forming cells in culture. Blood *50*, 833–839 (1977)

De Gabriele, G., Penington, D.G.: Regulation of platelet production. Thrombopoietin. Brit. J. Haematol. *13*, 210–215 (1967)

Di Persio, J., Brennan, J.K., Lichtman, M.A., Speiser, B.L.: Human cell lines that elaborate colony-stimulating activity for the marrow cells of man and other species. Blood *51*, 507–519 (1978)

Dresch, C., Johnson, G.R., Metcalf, D.: Eosinophil colony formation in semisolid cultures of human bone marrow cells. Blood *49*, 835–844 (1977)

Eaves, A.C., Bruce, W.R.: In vitro production of colony-stimulating activity. I. Exposure of mouse peritoneal cells to endotoxin. Cell Tissue Kinet. *7*, 19–30 (1974)

Ebbe, S.: Megakaryocytopoiesis. In: Regulation of hematopoiesis. Gordon, A.S. (ed.), pp. 1587–1610. New York: Appleton-Century-Crofts 1970

Fojo, S.S., Wu, M.-C., Gross, M.A., Yunis, A.A.: The isolation and characterization of a colony-stimulating factor from human lung. Biochim. Biophys. Acta *494*, 92–99 (1977)

Foster, R.F., Mirand, E.A.: Bone marrow colony stimulating factor following ureteral ligation in germfree mice. Proc. Soc. Exp. Biol. *133*, 1223–1227 (1970)

Foster, R., Metcalf, D., Kirchmyer, R.: Induction of bone marrow colony stimulating activity by a filterable agent in leukemic and normal mouse serum. J. Exp. Med. *127*, 853–866 (1968a)

Foster, R., Metcalf, D., Robinson, W.A., Bradley, T.R.: Bone marrow colony stimulating activity in human sera. Brit. J. Haematol. *15*, 147–159 (1968b)

Foster, R.F., MacPherson, B.R., Browdie, D.A.: Effect of Corynebacterium parvum on colony-stimulating factor and granulocyte-macrophage colony formation. Cancer Res. *37*, 1349–1355 (1977)

Ghio, R., Bianchi, G., Lowenberg, B., Dicke, K.A., Ajmar, J.: Effects of fibroblasts on the growth of erythroid progenitor cells in vitro. Exp. Hemat. *5*, 341–347 (1979)

Golde, D.W., Cline, M.J.: Identification of the colony-stimulating cell in human peripheral blood. J. Clin. Invest. *52*, 2981–2983 (1972)

Golde, D.W., Cline, M.J.: Endotoxin-induced release of colony-stimulating activity in man. Proc. Soc. Exp. Biol. Med. *149*, 845–848 (1975)

Goodman, J.W., Basford, N.L., Shinpock, S.G., Chambers, Z.E.: An amplifier cell in hemopoiesis. Exp. Hematol. *6*, 151–160 (1978)

Granstrom, M.: Studies on inhibitors of bone marrow colony formation in human sera and during a viral infection. Exp. Cell. Res. *82*, 426–432 (1973)

Granstrom, M., Wahren, B., Gahrton, G., Killander, D., Foley, G.E.: Inhibitors of the bone marrow colony formation in sera of patients with leukemia. Int. J. Cancer *10*, 482–488 (1972)

Greenberg, P.L., Bax, I., Levin, J., Andrews, T.M.: Alteration of colony stimulating factor output, endotoxemia and granulopoiesis in cyclic neutropenia. Am. J. Hematol. *1*, 375–385 (1976)

Greenberg, P.L., Mara, B., Heller, P.: Marrow adherent cell colony-stimulating activity production in acute myeloid leukemia. Blood *52*, 362–378 (1978)

Guerry, D., Adamson, J.W., Dale, D.C., Wolff, S.M.: Human cyclic neutropenia. Urinary colony stimulating factor and erythropoietin levels. Blood *44*, 257–262 (1974)

Guilbert, L.J., Iscove, N.N.: Partial replacement of serum by selenite, transferrin, albumin and lecithin in haemopoietic cell cultures. Nature *263*, 594–595 (1976)

Gupta, R.C., Robinson, W.A., Kurnick, J.E.: Felty's syndrome. Effect of lithium on granulopoiesis. Am. J. Med. *61*, 29–32 (1976)

Hall, B.M.: The effects of whole-body irradiation on serum colony stimulating factor and in vitro colony-forming cells in the bone marrow. Brit. J. Haematol. *17*, 553–561 (1969)

Handman, E., Burgess, A.W.: Stimulation by granulocyte-macrophage colony stimulating factor of Leishmania tropica killing by macrophages. J. Immunol. *122*, 1134–1137 (1979)

Hara, H., Ogawa, M.: Erythropoietic precursors in murine blood. Exp. Hematol. *5*, 161–165 (1977)

Hara, H., Kitamura, Y., Kawata, T., Kanamura, A., Nagai, K.: Synergism between lymph node and bone marrow cells for production of granulocytes. II. Enhanced colony-stimulating activity of sera of mice with graft-versus-host reaction. Exp. Hematol. *2*, 43–49 (1974)

Harker, G.W., Rothstein, G., Clarkson, D.W., Athens, J.W., McFarlane, T.L.: Enhancement of colony-stimulating activity production by lithium. Blood *49*, 263–267 (1977)

Hayes, E.G., Forsen, N.R., Rodensky, D., Craddock, C.G.: The effect of endotoxin on colony-stimulating factor (CSF) release from granulocytes and monocytes in vitro. Blood *40*, 949 (1972)

Heath, D.S., Axelrad, A.A., McLeod, D.L., Shreeve, M.M.: Separation of the erythropoietin responsive progenitors BFU-E and CFU-E in mouse bone marrow by unit gravity sedimentation. Blood *47*, 777–792 (1976)

Heit, W., Kern, P., Kubanek, B., Heimpel, H.: Some factors influencing granulocyte colony formation in vitro by human white blood cells. Blood *44*, 511–515 (1974)

Hibberd, A.D., Metcalf, D.: Proliferation of macrophage-granulocyte precursors in response to primary and transplanted tumors. Isr. J. Med. Sci. *7*, 202–210 (1971)

Honma, Y., Kasukabe, T., Hozumi, M.: Production of differentiation-stimulating factor in cultured mouse myeloid leukemia cells treated by glucocorticoids. Exp. Cell Res. *111*, 261–267 (1978)

Horiuchi, M., Ichikawa, Y.: Control of macrophage and granulocyte colony formation by two different factors. Exp. Cell Res. *110*, 79–85 (1977)

Howard, M., Burgess, A., McPhee, D., Metcalf, D.: Production of a T-cell hybridoma secreting hemopoietic regulatory molecules: Granulocyte-macrophage and eosinophil colony-stimulating factors. Cell *18*, 993–999 (1979)

Ichikawa, Y., Pluznik, D.H., Sachs, L.: In vitro control of the development of macrophage and granulocyte colonies. Proc. Natl. Acad. Sci. U.S.A. *56*, 488–495 (1966)

Iscove, N.N.: The role of erythropoietin in regulation of population size and cell cycling of early and late erythroid precursors in mouse bone marrow. Cell Tissue Kinet. *10*, 323–334 (1977)

Iscove, N.N.: Erythropoietin-independent stimulation of early erythropoiesis in adult marrow cultures by conditioned media from lectin-stimulated mouse spleen cells. In: Hematopoietic cell differentiation. Golde, D.W., Cline, M.J., Metcalf, D., Fox, C.F. (eds.), pp. 37–52. New York: Academic Press 1978

Iscove, N.N., Guilbert, L.J.: Erythropoietin-independence of early erythropoiesis and a 2-regulator model of proliferative control in the hemopoietic system. In: In vitro aspects of erythropoiesis. Murphy, M.J. (ed.), pp. 3–7. Berlin, Heidelberg, New York: Springer 1978

Iscove, N.N., Sieber, F.: Erythroid progenitors in mouse bone marrow detected by macroscopic colony formation in culture. Exp. Hematol. *3*, 32–43 (1975)

Johnson, G.R., Burgess, A.W.: Molecular and biological properties of a macrophage colony-stimulating factor from mouse yolk sacs. J. Cell Biol. *77*, 35–47 (1978)

Johnson, G.R., Metcalf, D.: Pure and mixed erythroid colony formation in vitro stimulated by spleen conditioned medium with no detectable erythropoietin. Proc. Natl. Acad. Sci. U.S.A. *74*, 3879–3882 (1977)

Johnson, G.R., Metcalf, D.: Sources and nature of granulocyte-macrophage colony stimulating factor in fetal mice. Exp. Hematol. *6*, 327–335 (1978a)

Johnson, G.R., Metcalf, D.: Characterization of mouse fetal liver granulocyte-macrophage colony-forming cells using velocity sedimentation. Exp. Hematol. *6*, 246–256 (1978b)

Johnson, G.R., Metcalf, D.: Clonal analysis in vitro of fetal hepatic hematopoiesis. In: Differentiation of normal and neoplastic hematopoietic cells. Clarkson, B., Baserga, R. (eds.), pp. 49–62. Cold Spring Harbor: Cold Spring Harbor Laboratory 1978c

Johnson, G.R., Metcalf, D.: Nature of cells forming erythroid colonies in agar after stimulation by spleen conditioned medium. J. Cell. Physiol. *94*, 243–252 (1978d)

Johnson, G.R., Metcalf, D.: Detection of a new type of mouse eosinophil colony by Luxol Fast Blue staining. Exp. Hematol. *8*, 549–561 (1980a)

Johnson, G.R., Metcalf, D.: Multipotential hemopoietic colony formation in agar cultures stimulated by spleen conditioned medium. In: Experimental hematology. Baum, S.J. (ed.). Berlin, Heidelberg, New York: Springer 1980b (in press)

Johnson, G.R., Nicholas, W.L., Metcalf, D., McKenzie, I.F.C., Mitchell, G.F.: The peritoneal population of mice infected with Mesocestoides corti as a source of eosinophils. Int. Arch. Allergy Appl. Immunol. *59*, 315–322 (1979)

Knudtzon, S., Mortensen, B.T.: Growth stimulation of human bone marrow cells in agar culture by vascular cells. Blood *46*, 937–943 (1975)

Kurland, H.I., Broxmeyer, H.E., Pelus, L.M., Bockman, R.S., Moore, M.A.S.: Role for monocyte-macrophage derived colony stimulating factor and prostaglandin E in the positive and negative feedback control of myeloid stem cell proliferation. Blood *52*, 388–407 (1978)

Ladisch, S., Reaman, G.H., Poplack, D.G.: Bacillus Calmette-Guerin enhancement of colony stimulating activity and myeloid colony formation following administration of cyclophosphamide. Cancer Res. *39*, 2544–2546 (1979)

Lajtha, L.G., Gilbert, C.W., Guzman, E.: Kinetics of haemopoietic colony growth. Brit. J. Haematol. *20*, 343–354 (1971)

Landau, T., Sachs, L.: Characterization of the inducer required for the development of macrophage and granulocytic colonies. Proc. Natl. Acad. Sci. U.S.A. *68*, 2540–2544 (1971)

Leffert, H.L., Koch, K.S., Rubalcava, B., Sell, S., Moran, T., Boorstein, R.: Hepatocyte growth control. In vitro approaches to problems of liver regeneration and function. In: Third Decennial Review Conference. Cell, tissue and organ culture. Sanford, K.S. (ed.), pp. 87–101. NCI Monograph 48 (1978)

Lin, H.-S., Gordon, S.: Secretion of plasminogen activator by bone marrow-derived mononuclear phagocytes and its enhancement by colony-stimulating factor. J. Exp. Med. *150*, 231–245 (1979)

Lin, H.-S., Stewart, C.C.: Peritoneal exudate cells. I. Growth requirements of cells capable of forming colonies in soft agar. J. Cell. Physiol. *83*, 369–378 (1974)

Lord, B.I., Testa, N.G., Wright, E.G., Banerjee, R.K.: Lack of effect of a granulocyte proliferation inhibitor on their committed progenitor cell. Biomedicine *26*, 163–169 (1977)

Lotem, J., Sachs, L.: In vivo induction of normal differentiation in myeloid leukemia cells. Proc. Natl. Acad. Sci. U.S.A. *75*, 3781–3785 (1978)

Mabry, J., Carbone, P.P., Bull, J.M.: Amplification of colony stimulating activity in human serum by interaction with CSF from other sources. Exp. Hematol. *3*, 354–361 (1975)

Mahmood, T., Robinson, W.A.: Granulocyte modulation of endotoxin-stimulated colony stimulating activity (CSA) production. Blood *51*, 879–887 (1978)

Mangalik, A., Robinson, W.A.: Cyclic neutropenia. The relationship between urine granulocyte colony stimulating activity and neutrophil count. Blood *41*, 79–84 (1973)

Mauer, H.R., Weiss, G., Laerum, O.D.: Starting procedures for the isolation and purification of granulocyte chalone activities. Blut *33*, 161–170 (1976)

McCarthy, J.H.: Haemopoietic progenitor cell responses in mice with the transplanted lymphoid leukaemia ABE-8. Brit. J. Cancer *40*, 144–151 (1979)

McGarry, M.P., Speirs, R.S., Jenkins, V.K., Trentin, J.J.: Lymphoid cell dependence of eosinophil response to antigen. J. Exp. Med. *134*, 801–814 (1971)

McLeod, D.L., Shreeve, M.M., Axelrad, A.A.: Induction of megakaryocyte colonies with platelet formation in vitro. Nature *261*, 492–494 (1976)

McNeill, T.A.: Antigenic stimulation of bone marrow colony forming cells. III. Effect in vivo. Immunology *18*, 61–72 (1970)

McNeill, T.A.: Release of bone marrow colony stimulating activity during immunological reactions in vitro. Nature *244*, 175–176 (1973)

McNeill, T.A., Fleming, W.A.: Cellular responsiveness to stimulation in vitro. Strain differences in hemopoietic colony forming cell responsiveness to stimulating factor and suppression of responsiveness by glucocorticoids. J. Cell. Physiol. *82*, 49–58 (1973)

McNeill, T.A., Killen, M.: Hemopoietic colony-forming cell responses in mice infected with ectromelia virus. Infect. Immun. *4*, 323–330 (1971)

Metcalf, D.: Cortisone action on serum colony-stimulating factor and bone marrow in vitro colony-forming cells. Proc. Soc. Exp. Biol. Med. *132*, 391–394 (1969)

Metcalf, D.: Studies on colony formation in vitro by mouse bone marrow cells. II. Action of colony stimulating factor. J. Cell. Physiol. *76*, 89–100 (1970)

Metcalf, D.: Acute antigen-induced elevation of serum colony stimulating factor (CSF) levels. Immunology *21*, 427–436 (1971)

Metcalf, D.: Stimulation by human urine or plasma of granulopoiesis by human marrow cells in agar. Exp. Hematol. *2*, 157–173 (1974a)

Metcalf, D.: Depressed responses of the granulocyte-macrophage system to bacterial antigens following preimmunization. Immunology *26*, 1115–1125 (1974b)

Metcalf, D.: Hemopoietic colonies. In vitro cloning of normal and leukemic cells. Berlin, Heidelberg, New York: Springer 1977

Metcalf, D.: The control of neutrophil and macrophage production at the progenitor cell level. In: Experimental hematology today 1978. Baum, S.J., Ledney, G.D. (eds.), pp. 35–46. Berlin, Heidelberg, New York: Springer 1978

Metcalf, D.: Clonal analysis of the action of GM-CSF on the proliferation and differentiation of myelomonocytic leukemic cells. Int. J. Cancer *24*, 616–623 (1979a)

Metcalf, D.: Production of stimulating factors by lymphoid tissues. In: Biology of the lymphokines. Cohen, S., Pick, E., Oppenheim, J. (eds.), pp. 515–540. New York: Academic Press 1979b

Metcalf, D., Bradley, T.R.: Factors regulating in vitro colony formation by hematopoietic cells. In: Regulation of hematopoiesis. Gordon, A.S. (ed.), pp. 187–215. New York: Appleton-Century-Crofts 1970

Metcalf, D., Foster, R.: Behavior on transfer of serum stimulated bone marrow colonies. Proc. Soc. Exp. Biol. Med. *126*, 758–762 (1967a)

Metcalf, D., Foster, R.: Bone marrow colony-stimulating activity of serum from mice with viral-induced leukemia. J. Natl. Cancer Inst. *39*, 1235–1245 (1967b)

Metcalf, D., Johnson, G.R.: Production by spleen and lymph node cells of conditioned medium with erythroid and other hemopoietic colony stimulating activity. J. Cell. Physiol. *96*, 31–42 (1978)

Metcalf, D., Johnson, G.R.: Interactions between purified GM-CSF, purified erythropoietin and spleen conditioned medium on hemopoietic colony formation in vitro. J. Cell. Physiol. *99*, 159–174 (1979)

Metcalf, D., MacDonald, H.R.: Heterogeneity of in vitro colony- and cluster-forming cells in the mouse marrow. Segregation by velocity sedimentation. J. Cell. Physiol. *85*, 643–654 (1975)

Metcalf, D., Moore, M.A.S.: Haemopoietic cells. Amsterdam: North-Holland 1971

Metcalf, D., Moore, M.A.S.: Regulation of growth and differentiation in haemopoietic colonies growing in agar. In: Haemopoietic stem cells. Wolstenholme, G.E.W., O'Connor, M. (eds.), pp. 157–175. Amsterdam, North-Holland: Elsevier-Excerpta Medica 1973

Metcalf, D., Russell, S.: Inhibition by mouse serum of hemopoietic colony formation in vitro. Exp. Hematol. *4*, 339–353 (1976)

Metcalf, D., Stanley, E.R.: Quantitative studies on the stimulation of mouse bone marrow colony growth in vitro by normal human urine. Aust. J. Exp. Biol. Med. Sci. *47*, 453–466 (1969)

Metcalf, D., Stanley, E.R.: Haematological effects in mice of partially purified colony stimulating factor (CSF) prepared from human urine. Brit. J. Haematol. *21*, 481–492 (1971a)

Metcalf, D., Stanley, E.R.: Serum half-life in mice of colony stimulating factor prepared from human urine. Brit. J. Haematol. *20*, 549–556 (1971b)

Metcalf, D., Wahren, B.: Bone marrow colony stimulating activity of sera in infectious mononucleosis. Brit. Med. J. *1968 III*, 99–101

Metcalf, D., Foster, R., Pollard, M.: Colony stimulating activity of serum from germfree, normal and leukemic mice. J. Cell. Physiol. *70*, 131–132 (1967)

Metcalf, D., Moore, M.A.S., Warner, N.L.: Colony formation in vitro by myelomonocytic leukemic cells. J. Natl. Cancer Instit. *43*, 983–1001 (1969)

Metcalf, D., Chan, S.H., Gunz, F.W., Vincent, P., Ravich, R.B.M.: Colony-stimulating factor and inhibitor levels in acute granulocytic leukemia. Blood *38*, 143–152 (1971)

Metcalf, D., Parker, J., Chester, H.M., Kincade, P.W.: Formation of eosinophilic-like granulocytic colonies by mouse bone marrow cells in vitro. J. Cell. Physiol. *84*, 275–290 (1974)

Metcalf, D., MacDonald, H.R., Chester, H.M.: Serum potentiation of granulocyte and macrophage colony formation in vitro. Exp. Hematol. *3*, 261–273 (1975a)

Metcalf, D., MacDonald, H.R., Odartchenko, N., Sordat, B.: Growth of mouse megakaryocyte colonies in vitro. Proc. Natl. Acad. Sci. U.S.A. *72*, 1744–1748 (1975b)

Metcalf, D., Russell, S., Burgess, A.W.: Production of hemopoietic stimulating factors by pokeweed mitogen-stimulated spleen cells. Transplant. Proc. *10*, 91–94 (1978)

Metcalf, D., Johnson, G.R., Mandel, T.E.: Colony formation in agar by multipotential hemopoietic cells. J. Cell. Physiol. *98*, 401–420 (1979)

Metcalf, D., Johnson, G.R., Burgess, A.W.: Direct stimulation by purified GM-CSF of the proliferation of multipotential and erythroid precursor cells. Blood *55*, 138–147 (1980a)

Metcalf, D., Burgess, A.W., Johnson, G.R.: Stimulation of multipotential and erythroid precursor cells by GM-CSF. In: Experimental hematology today 1979. Baum, S.J. (ed.). Berlin, Heidelberg, New York: Springer 1980b (in press)

Miller, A.M., McGarry, M.P.: A diffusible stimulator of eosinophilopoiesis produced by lymphoid cells as demonstrated with diffusion chambers. Blood *48*, 293–300 (1976)

Miyake, T., Kung, C.K.-H., Goldwasser, E.: Purification of human erythropoietin. J. Biol. Chem. *252*, 5558–5564 (1977)

Moore, M.A.S., Williams, N.: Physical separation of colony stimulating cells from in vitro colony forming cells in hemopoietic tissue. J. Cell. Physiol. *80*, 195–206 (1972)

Moore, M.A.S., Williams, N.: Functional, morphologic and kinetic analysis of the granulo-cyte-macrophage progenitor cell. In: Hemopoiesis in culture. Robinson, W.A. (ed.), pp. 16–26. DHEW Publication No. (NIH) 74-205 Washington (1973)

Moore, M.A.S., Williams, N., Metcalf, D.: Purification and characterisation of the in vitro colony-forming cell in monkey hemopoietic tissue. J. Cell. Physiol. *79*, 283–292 (1972)

Moore, M.A.S., Williams, N., Metcalf, D.: In vitro colony formation by normal and leu-kemic hematopoietic cells. Interaction between colony-forming and colony-stimulating cells. J. Natl. Cancer Inst. *50*, 591–602 (1973)

Moore, M.A.S., Spitzer, G., Metcalf, D., Penington, D.G.: Monocyte production of colony stimulating factor in familial cyclic neutropenia. Brit. J. Haematol. *27*, 47–55 (1974)

Morley, A., Quesenberry, P., Bealmear, P., Stohlman, F., Wilson, R.: Serum colony stimu-lating factor levels in irradiated germfree and conventional CFW mice. Proc. Soc. Exp. Biol. Med. *140*, 478–480 (1972)

Nakeff, A.: Colony-forming unit, megakaryocyte (CFU-M). Its use in elucidating the kinet-ics and humoral control of the megakaryocyte committed progenitor cell compartment. In: Experimental hematology today. Baum, S.J., Ledney, G.D. (eds.), pp. 111–123. Ber-lin, Heidelberg, New York: Springer 1977

Nakeff, A., Daniels-McQueen, S.: In vitro colony assay for a new class of megakaryocyte precursor. Colony-forming unit megakaryocyte (CFU-M). Proc. Soc. Exp. Biol. Med. *151*, 587–590 (1976)

Nakeff, A., Dicke, K.A., Noord, J. van: Megakaryocytes in agar cultures of mouse bone marrow. Ser. Hematol. *8*, 1–21 (1975)

Nathan, D.G., Chess, L., Hillman, D.G., Clarke, B., Breard, J., Merler, E., Housman, D.E.: Human erythroid burst-forming unit: T-cell requirement for proliferation in vitro. J. Exp. Med. *147*, 324–339 (1978)

Naughton, B.A., Gordon, A.S., Piliero, S.J., Liu, P.: Extrarenal erythropoietin. In: In vitro aspects of erythropoiesis. Murphy, M.J. (ed.), pp. 194–217. Berlin, Heidelberg, New York: Springer 1978

Nicola, N.A., Metcalf, D., Johnson, G.R., Burgess, A.W.: Preparation of colony stimulat-ing factors from human placental conditioned medium. Leukemia Res. *2*, 313–322 (1978)

Nicola, N.A., Burgess, A.W., Metcalf, D.: Similar molecular properties of granulocyte-mac-rophage colony-stimulating factors produced by different organs in vitro and in vivo. J. Biol. Chem. *254*, 5290–5299 (1979 a)

Nicola, N.A., Metcalf, D., Johnson, G.R., Burgess, A.W.: Separation of functionally dis-tinct human granulocyte-macrophage colony-stimulating factors. Blood *54*, 614–627 (1979 b)

Ogawa, M., MacEachern, M.D., Avila, L.: Human marrow erythropoiesis in culture. II. Heterogeneity in the morphology, time course of colony formation and sedimentation velocities of the colony forming cells. Am. J. Hemat. *3*, 29–36 (1977)

Ohno, T., Seki, M., Shikita, M.: Colony-stimulating factors active on human bone marrow cells from a Yoshida sarcoma cell line. Blood *51*, 911–918 (1978)

Paran, M., Sachs, L.: The continued requirement for inducer for the development of mac-rophage and granulocyte colonies. J. Cell. Physiol. *72*, 247–250 (1968)

Paran, M., Sachs, L., Barak, Y., Resnitzky, P.: In vitro induction of granulocyte differenti-ation in hematopoietic cells from leukemic and non-leukemic patients. Proc. Natl. Acad. Sci. U.S.A. *67*, 1542–1549 (1970)

Parker, J.W., Metcalf, D.: Production of colony-stimulating factor in mitogen-stimulated lymphocyte cultures. J. Immunol. *112*, 502–510 (1974 a)

Parker, J.W., Metcalf, D.: Production of colony-stimulating factor in mixed leucocyte cul-tures. Immunology *26*, 1039–1049 (1974 b)

Parmley, R.T., Ogawa, M., Spicer, S.S., Wright, N.J.: Ultrastructure and cytochemistry of bone marrow granulocytes in culture. Exp. Hematol. *4*, 75–89 (1976)

Paukovits, W.R.: Control of granulocyte production. Separation and chemical identifica-tion of a specific inhibitor (chalone). Cell Tissue Kinet. *4*, 539–549 (1971)

Paukovits, W.R., Hinterberger, W.: Molecular weight and some chemical properties of the granulocytic chalone. Blut *37*, 7–18 (1978)

Peschele, C., Magli, M.C., Cillo, C., Lettieri, F., Pizzella, F., Migliaccio, G., Soricelli, A., Scala, G., Mastroberardino, G., Sasso, G.F.: Recent advances in erythropoietin physiology. In: In vitro aspects of erythropoiesis. Murphy, M.J. (ed.), pp. 227–239. Berlin, Heidelberg, New York: Springer 1978

Pike, B., Robinson, W.A.: Human bone marrow colony growth in agar gel. J. Cell. Physiol. *76*, 77–84 (1970)

Pluznik, D., Sachs, L.: The induction of clones of normal "mast" cells by a substance from conditioned medium. Exp. Cell. Res. *43*, 553–563 (1966)

Price, G.B., McCulloch, E.A., Till, J.E.: Cell membranes as sources of granulocyte colony stimulating activities. Exp. Hematol. *3*, 227–233 (1975a)

Price, G.B., Senn, J.S., McCulloch, E.A., Till, J.E.: The isolation and properties of granulocytic colony stimulating activities from medium conditioned by human peripheral leucocytes. Biochem. J. *148*, 209–217 (1975b)

Prival, J.T., Paran, M., Gallo, R.C., Wu, A.M.: Colony-stimulating factors in cultures of human peripheral blood cells. J. Natl. Cancer Inst. *53*, 1583–1588 (1974)

Quesenberry, P., Morley, A., Stohlman, F., Rickard, K., Howard, D., Smith, M.: Effect of endotoxin on granulopoiesis and colony stimulating factor. New Engl. J. Med. *286*, 227–232 (1972)

Quesenberry, P., Halperin, J., Ryan, M., Stohlman, F.: Tolerance to the granulocyte releasing and colony stimulating factor elevating effects of endotoxin. Blood *45*, 789–800 (1975)

Quesenberry, P.J., Gimbrone, M.A., McDonald, M.J.: Endothelial-derived colony stimulating actitivy. Exp. Hematol. *6*, Suppl. 3, 4 (1978)

Ralph, P., Nakoniz, I.: Direct toxic effects of immunopotentiators on monocytic myelomonocytic and histiocytic or macrophage tumor cells in culture. Cancer Res. *37*, 546–550 (1977)

Ralph, P., Broxmeyer, H.E., Nakoinz, I.: Immunostimulators induce granulocyte-macrophage colony stimulating activity and block proliferation in a monocyte tumor cell line. J. Exp. Med. *146*, 611–616 (1977)

Robinson, W., Metcalf, D., Bradley, T.R.: Stimulation by normal and leukemic mouse sera of colony formation in vitro by mouse bone marrow cells. J. Cell. Physiol. *69*, 83–92 (1967)

Robinson, W.A., Stanley, E.R., Metcalf, D.: Stimulation of bone marrow colony growth in vitro by human urine. Blood *33*, 396–399 (1969)

Ruscetti, F.W., Chervenick, P.A.: Release of colony-stimulating factor from monocytes by endotoxin and polyinosinic-polycytidylic acid. J. Lab. Clin. Med. *83*, 64–72 (1974)

Ruscetti, F.W., Chervenick, P.A.: Release of colony-stimulating activity from thymus-derived lymphocytes. J. Clin. Invest. *55*, 520–527 (1975)

Ruscetti, F.W., Cypess, R.H., Chervenick, P.A.: Specific release of neutrophilic- and eosinophilic-stimulating factors from sensitized lymphocytes. Blood *47*, 757–765 (1976)

Shadduck, R.K., Cavallo, T.: Immunofluorescent localization of granulocyte colony stimulating factor (CSF). Clin. Res. *21*, 567 (1973)

Shadduck, R.K., Metcalf, D.: Preparation and neutralization characteristics of an anti-CSF antibody. J. Cell. Physiol. *86*, 247–252 (1975)

Shadduck, R.K., Carsten, A.L., Chikkappa, G., Cronkite, E.: Inhibition of diffusion chamber (DC) granulopoiesis by anti-CSF serum. Proc. Soc. Exp. Biol. Med. *158*, 542–549 (1978)

Shadduck, R.K., Waheed, A., Pigoli, C., Boegel, F., Higgins, L.: Fractionation of antibodies to L-cell colony stimulating factor by affinity chromatography. Blood *53*, 1182–1190 (1979)

Shah, R.G., Caporale, L.H., Moore, M.A.S.: Characterization of colony-stimulating activity produced by human monocytes and phytohemagglutinin-stimulated lymphocytes. Blood *50*, 811–821 (1977)

Sheridan, J.W., Metcalf, D.: Studies on the bone marrow colony stimulating factor (CSF): Relation of tissue CSF to serum CSF. J. Cell. Physiol. *80*, 129–140 (1972)

Sheridan, J.W., Metcalf, D.: CSF production and release following endotoxin. In: Hemopoiesis in culture. Robinson, W.A. (ed.), p. 135. Washington: DHEW Publication No. (NIH) 74-205 1973 a

Sheridan, J.W., Metcalf, D.: A low molecular weight factor in lung-conditioned medium stimulating granulocyte and monocyte colony formation in vitro. J. Cell. Physiol. *81*, 11–24 (1973 b)

Sheridan, J.W., Metcalf, D.: The bone marrow colony stimulating factor (CSF). Relation of serum CSF to urine CSF. Proc. Soc. Exp. Biol. *144*, 785–788 (1973 c)

Sheridan, J.W., Stanley, E.R.: Tissue sources of bone marrow colony stimulating factor. J. Cell. Physiol. *78*, 451–460 (1971)

Sheridan, J.W., Metcalf, D., Stanley, E.R.: Further studies on the factor in lung-conditioned medium stimulating granulocyte and monocyte colony formation in vitro. J. Cell. Physiol. *84*, 147–158 (1974)

Shoham, D., Ben David, E., Rozenszajn, L.A.: Cytochemical and morphologic identification of macrophages and eosinophils in tissue cultures of normal human bone marrow. Blood *44*, 221–233 (1974)

Singer, J.W., James, M.C., Thomas, E.D.: Serum colony stimulating factor. A marker for graft-versus-host disease in humans. In: Experimental hematology today. Baum, S.J., Ledney, G.D. (eds.), pp. 221–231. Berlin, Heidelberg, New York: Springer 1977

Spitzer, G., Verma, D.S., Barlogie, B., Beran, M.A., Dicke, K.A.: Possible mechanisms of action of lithium on augmentation of in vitro spontaneous myeloid colony formation. Cancer Res. *39*, 3215–3219 (1979)

Staber, F.G.: Diminished response of granulocyte-macrophage colony-stimulating factor (GM-CSF) in mice after presensitization with bacterial cell-wall components. Exp. Hematol. *8*, 120–133 (1980)

Staber, F.G., Gisler, R.H., Schumann, G., Tarcsay, L., Schlafli, E., Dukor, P.: Modulation of myelopoiesis by different bacterial cell wall components. Induction of colony stimulating activity (by pure preparations, low molecular weight degradation products and a synthetic low molecular analog of bacterial cell-wall components) in vitro. Cell Immunol. *37*, 174–187 (1978)

Stanley, E.R.: Colony stimulating factor (CSF) radioimmunoassay: Detection of a CSF subclass stimulating macrophage production. Proc. Natl. Acad. Sci. U.S.A. *76*, 2969–2973 (1979)

Stanley, E.R., Heard, P.M.: Factors regulating macrophage production and growth. J. Biol. Chem. *252*, 4305–4312 (1977)

Stanley, E.R., Metcalf, D.: Partial purification and some properties of the factor in normal and leukaemic human urine stimulating mouse bone marrow colony growth in vitro. Aust. J. Exp. Biol. Med. Sci. *47*, 467–483 (1969)

Stanley, E.R., Bradley, T.R., Sumner, M.A.: Properties of mouse embryo conditioned medium factor(s) stimulating colony formation by mouse bone marrow cells grown in vitro. J. Cell. Physiol. *78*, 301–318 (1971)

Stanley, E.R., Hansen, G., Woodcock, J., Metcalf, D.: Colony stimulating factor and the regulation of granulopoiesis and macrophage production. Fed. Proc. *34*, 2272–2278 (1975)

Stanley, E.R., Cifone, M., Heard, P.M., Defendi, V.: Factors regulating macrophage production and growth. Identity of colony-stimulating factor and macrophage growth factor. J. Exp. Med. *143*, 631–647 (1976)

Stephenson, J.R., Axelrad, A.A., McLeod, D.L., Shreeve, M.M.: Induction of colonies of hemoglobin-synthesizing cells by erythropoietin in vitro. Proc. Natl. Acad. Sci. U.S.A. *68*, 1542–1546 (1971)

Tepperman, A.D., Curtis, J.E., McCulloch, E.A.: Erythropoietic colonies in cultures of human marrow. Blood *44*, 659–669 (1974)

Till, J.E., McCulloch, E.A.: A direct measurement of the radiation sensitivity of normal mouse bone marrow cells. Radiat. Res. *19*, 213–222 (1961)

Tisman, G., Herbert, V., Rosenblatt, S.: Evidence that lithium induces granulocyte proliferation. Elevated serum B_{12} binding capacity in vivo and granulocyte colony proliferation in vitro. Brit. J. Haematol. *24*, 767–771 (1973)

Trudgett, A., McNeill, T.A., Killen, M.: Granulocyte-macrophage precursor cell and colony-stimulating factor responses of mice infected with Salmonella typhimurium. Infect. Immun. *8*, 450–455 (1973)

Engh, G.J. van den: Quantitative in vitro studies on stimulation of murine haemopoietic cells by colony stimulating factor. Cell Tissue Kinet. *7*, 537–548 (1974)

Engh, G. van den, Bol, S.: The presence of a CSF enhancing activity in the serum of endotoxin-treated mice. Cell Tissue Kinet. *8*, 579–587 (1975)

Zant, G. van, Goldwasser, E.: Competitive effects of erythropoietin and colony-stimulating factor. In: Differentiation of normal and neoplastic hematopoietic cells. Clarkson, B., Marks, P.A., Till, J.E. (eds.), pp. 165–177. New York: Cold Spring Harbor Laboratory 1978

Vogler, W.R., Winton, E.F.: Humoral granulopoietic inhibitors: A review. Exp. Hematol. *3*, 337–353 (1975)

Wagemaker, G.: Induction of erythropoietin responsiveness in vitro. In: Hematopoietic Cell Differentiation. Golde, D.W., Cline, M.J., Metcalf, D., Fox, C.F. (eds.), pp. 109–118. New York: Academic Press 1978

Wagemaker, G., Ober-Kieftenburg, V.E., Brouwer, A., Peters-Slough, M.F.: Some characteristics of in vitro erythroid colony and burst-forming units. In: Experimental hematology today. Baum, S.J., Ledney, G.D. (eds.), pp. 103–110. Berlin, Heidelberg, New York: Springer 1977

Wahren, B., Ecsenyi, M.: Cell growth regulating factors in viral infections and other diseases. Exp. Cell Res. *66*, 396–400 (1971)

Watt, S.M., Burgess, A.W., Metcalf, D.: Isolation and surface labeling of murine polymorphonuclear neutrophils. J. Cell. Physiol. *100*, 1–21 (1979)

Wiktor-Jedrzejczak, W., Sharkis, S., Ahmed, A., Snell, K.W.: Theta-sensitive cell and erythropoiesis. Identification of a defect in W/W^v anemic mice. Science *196*, 313–315 (1977)

Williams, N.: Preferential inhibition of murine macrophage colony formation by prostaglandin E. Blood *53*, 1089–1094 (1979)

Williams, N., Jackson, H.: Regulation of the proliferation of murine megakaryocyte progenitor cells by cell cycle. Blood *52*, 163–170 (1978)

Williams, N., Engh, G.J. van den: Separation of subpopulations of in vitro colony-forming cells form mouse marrow by equilibrium density centrifugation. J. Cell. Physiol. *86*, 237–245 (1975)

Williams, N., Jackson, H., Sheridan, A.P.L., Murphy, M.J., Elste, A., Moore, M.A.S.: Regulation of megakaryopoiesis in long-term murine bone marrow cultures. Blood *51*, 245–255 (1978)

Williams, N., McDonald, T.P., Rabellino, E.M.: Maturation and regulation of megakaryocytopoiesis. Blood Cells *5*, 43–55 (1979)

Zucali, J.R., Mirand, E.A.: In vitro aspects of erythropoietin production. In: In vitro aspects of erythropoiesis. Murphy, M.J. (ed.), pp. 218–224. Berlin, Heidelberg, New York: Springer 1978

Zucker-Franklin, D., Grusky, G.: Ultrastructural analysis of hematopoietic colonies derived from human peripheral blood. J. Cell Biol. *63*, 855–863 (1974)

CHAPTER 13

Inhibition of Hematopoietic Cell Proliferation

J. H. FITCHEN AND M. J. CLINE *

A. Introduction

In healthy post-natal human beings, hematopoiesis is restricted to the bone marrow. As mature blood cells are utilized or die in the periphery, they are continuously replenished by the bone marrow. For years, it has been recognized that blood cell renewal is accomplished by maturation of morphologically identifiable precursors (blast cells) in the bone marrow. More recently, with the development of hematopoietic cell culture techniques, the concept has evolved that these morphologically recognizable precursor cells are in turn replenished from hematopoietic stem cells. In this hierarchical scheme of hematopoiesis, hematopoietic stem cells have the dual capacities of self-renewal and differentiation; that is, they are capable both of maintaining their own numbers and of giving rise to progeny committed to a single line of differentiation. A model of hematopoiesis based on this scheme is depicted in Fig. 1.

Despite the massive numbers of cells produced and turned out by the bone marrow (e.g., approximately 10^{11} granulocytes/day in a healthy adult), peripheral blood counts are maintained between remarkably narrow limits under normal circumstances. In addition, the bone marrow has the ability to expand production dramatically in response to stress. This increased production is largely restricted to the specific cell type appropriate to a particular stress (e.g., erythroid hyperplasia in response to hemolysis, granulocytic hyperplasia in response to bacterial infection).

The remarkable precision and flexibility of the bone marrow, both in meeting ongoing blood cell requirements and in responding to specific stresses, implies that proliferation of this tissue is modulated by sensitive biological control systems. These controls, positive or negative, may be applied at several different levels of hematopoiesis. For example, in the case of granulopoiesis, one can envisage many points at which control mechanisms might act: they may control a) the rate of movement of cells from the pluripotent stem cell compartment into the compartment of stem cells committed to granulopoiesis, b) the proliferative state of the committed stem cell compartment, c) the number of divisions or time of transit in the mitotic pool, d) the rate of entry of cells from the marrow storage compartment into the peripheral blood, e) the distribution of granulocytes between marginal and circulating pools, f) the rate of egress of cells from the blood into the tissues, and g) the life-span of mature cells. Of the possibilities, influences on the committed

* From the Department of Medicine, Center for the Health Sciences, University of California at Los Angeles, and the Wadsworth Veterans Administration Hospital, Los Angeles

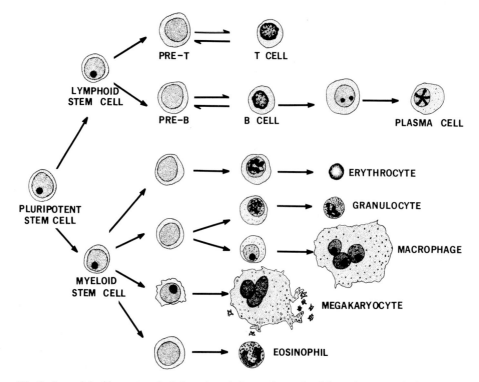

Fig. 1. A model of hematopoiesis based on information gained from hematopoietic culture techniques. Reproduced with permission from Churchill-Livingston

stem cell pool, their dividing progeny, and release from the marrow storage compartment are probably most critical in controlling the numbers of neutrophils in the circulation.

Such a model is depicted schematically in Fig. 2. Because of the pyramidal nature of hematopoiesis with few stem cells and progressively increasing numbers of more differentiated cells, controlling influences applied to proximal events (e.g., expansion of the committed stem cell compartment) would be expected to have a delayed but more profound effect on granulopoiesis. Release control would be more suited to rapid and evanescent responses.

It seems likely that the control of hematopoiesis is accomplished by a combination of positive and negative influences. By adjustments in the balance between these opposing forces, blood cell production might be increased, decreased, or maintained at a constant rate.

This chapter focuses on the inhibitory influences that may affect hematopoiesis. It should be remembered that inhibitors of hematopoiesis probably do not act unopposed. Observed phenomena may reflect the net effect of several influences, particularly when living systems are employed. With the use of the vitro systems, one can attempt to isolate a discrete control mechanism. Although such studies may suggest the existence of a particular inhibitory or stimulatory effect, the role of such effects in vivo may be over-emphasized or over-simplified by in vitro studies.

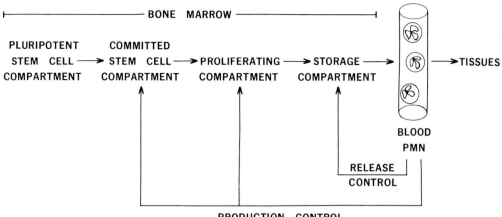

Fig. 2. Most likely controls on the level of circulating PMNs, as indicated by present evidence, are control at the level of production by the stem cell and proliferating compartments, and control of release from the bone marrow storage compartment. Reproduced with permission from CHURCHILL-LIVINGSTON

These potential problems notwithstanding, the bulk of research pertinent to inhibitors of hematopoietic cell proliferation has relied heavily on in vitro culture techniques. Therefore, in what follows, we will first summarize the technology and nomenclature of hematopoietic culture techniques. After presenting this background information, we will review the evidence for the existence and importance of inhibitors of hematopoiesis. Hematopoietic stimulatory factors are discussed elsewhere in this volume (see METCALF chapter 12).

B. Hematopoietic Techniques and Nomenclature

A glossary of terms used in hematopoietic research is given in Table 1. The origin of these terms and the experimental work from which they were derived is discussed below. A review of the topic of hematopoietic stem cells and the techniques used to study them has been recently published (QUESENBERRY and LEVITT, 1979).

I. Pluripotent Stem Cells

In 1961, TILL and McCULLOCH (1961) demonstrated that murine bone marrow cells injected into lethally irradiated syngeneic mice formed discrete nodules in the spleens of recipient animals. On histologic section, some of these nodules were found to consist of erythroid, granulocytic and megakaryocytic elements. The cell that gave rise to these nodules was designated the "colony-forming unit, spleen" or CFU-S. That each colony was derived from a single cell was proven by the demonstration that all of the cells in a given colony were of a uniform karyotype when mixtures of bone marrow cells with unique chromosomal markers were injected into recipient mice (CHEN and SCHOOLEY, 1968). Cell suspensions prepared from a single splenic nodule and reinjected into a second lethally irradiated mouse formed new colonies, demonstrating the capacity of CFU-S for self-renewal. In ad-

Table 1. Glossary of terms used in hematopoietic research

Clonogenic assay	An assay for clones or colonies arising from a single stem cell
Stem Cell	A primitive cell capable of self-renewal *and* differentiation
Commitment	The decision of a stem cell to differentiate along a single pathway, e.g., erythroid
CFU – S	Colony-forming unit-spleen; a measure of the pluripotent stem cell for erythroid, granulocyte, megakaryocyte, and eosinophil development
CFU – C (CFU – G, M)	Colony-forming unit-culture; also called colony-forming unit-granulocyte/macrophage. Forms colonies of granulocytes and/or macrophages in agar when stimulated by CSA
BFU – E	Burst-forming unit-erythroid; a primitive RBC stem cell, relatively erythropoietin insensitive, which forms large colonies or "bursts" of hemoglobinized cells in vitro
CFU – E	Colony-forming unit-erythroid. The immediate antecedent of the pronormoblast, this cell gives rise to small clusters of hemoglobinized cells in the presence of low concentrations of erythropoietin in vitro
CFU – eos; M, TL, etc.	Colony-forming unit-eosinophil, -megakaryocyte, -T lymphocyte, etc.
CSA	Colony-stimulating activity; a family of glycoproteins stimulating CFU – C

dition, the progeny of a single CFU-S (i.e., one colony) were able to repopulate the entire hematopoietic organ in lethally irradiated recipients (METCALF and MOORE, 1971).

The existence of pluripotent hematopoietic stem cells in man is inferred from studies of hematologic disorders that may represent neoplastic transformation of stem cells and from experience with human bone marrow transplantation. The Philadelphia chromosome, a specific marker for chronic myelogenous leukemia (CML), is found in cells of the erythroid, granulocytic, monocytic and megakaryocytic series in patients with CML (WHANG et al., 1963; GOLDE et al., 1977). This finding suggests that the neoplastic transformation in CML takes place at the level of a stem cell common to all three lines of marrow differentiation. Further, studies in women with CML who are heterozygous for G 6 PD have shown that red cells, granulocytes, platelets, and non-T lymphocytes contain only one type of G 6 PD isoenzyme (FIALKOW et al., 1977). Similar findings have been reported in polycythemia vera (ADAMSON et al., 1976) and myeloid metaplasia (JACOBSON et al., 1978). Taken together, these observations suggest that a hematopoietic stem cell exists in man that is common to erythroid, granulocytic, and megakaryocytic cells.

Another line of evidence for the existence of pluripotent hematopoietic stem cells in man is experience with bone marrow transplantation in the treatment of aplastic anemia. Simple infusion of bone marrow from identical twins is effective in restoring normal hematopoiesis in the majority of cases (GALE, 1979). Bone marrow transplantation from allogeneic, HLA-identical siblings, preceded by potent immunosuppressive conditioning, produces hematologic restoration in 60%–70% of patients (CAMITTA and THOMAS, 1978). Sections of bone marrow biopsies performed in the early days and weeks following transplantation reveal discrete "colonies" of hematopoiesis comprised of elements from the erythroid, granulocytic and megakaryocytic series (CLINE et al., 1977). The unproven assumption with both syngeneic and allogeneic transplantation is that stem cells contained in the infused marrow are responsible for hematopoietic reconstitution. Since stem cells represent a tiny fraction of the cells in the infused marrow, it is also possible, however, that some other critical cellular element or environmental influence is provided by the transplanted marrow.

II. Unipotent (Committed) Stem Cells

In 1966, BRADLEY and METCALF (1966) in Australia and PLUZNIK and SACHS (1965) in Israel described a system for growing granulocyte-macrophage colonies from bone marrow suspended in soft agar. The in vitro culture system had three requirements: responsive stem cells from bone marrow or other hematopoietic tissue, a hematopoietic hormone to serve as stimulus, and a semisolid matrix that would support growth and keep the progeny of a single cell together as a "colony". The number of colonies formed was shown to depend on the number of progenitor cells and the concentration of the stimulus in the culture dish.

Over the ensuing years, techniques were developed for growing many classes of hematopoietic precursor cells in vitro, using human cells as well as murine cells. PIKE and ROBINSON (1970) reported a double-layer agar system that employed a "feeder" layer of peripheral blood leukocytes to stimulate the formation of granulocyte-macrophage colonies from human bone marrow. Subsequently, assays for the growth of erythroid, eosinophilic, megakaryocytic, and B- and T-lymphocytic colonies were described (STEPHENSON et al., 1971; ISCOVE et al., 1974; DRESCH et al., 1977; NAKEFF and DANIELS-MCQUEEN, 1976; METCALF et al., 1975; ROSENSZAJN et al., 1975). Recently, a technique has been described by FAUSER and MESSNER (1979) for the in vitro assay of multipotential stem cells in man. If confirmed, this technology will prove to be an important new tool in the armamentarium of hematopoietic research.

Another new technology, still in evolution, is the use of liquid culture techniques that will preserve stem cells in vitro for long periods of time. In the most common type of long-term culture system, adherent cells from mouse bone marrow are used to form a "stroma" on the surface of plastic culture flasks. The stroma consists of epithelioid cells, macrophages and giant fat-containing cells and probably other cell types. After several weeks in vitro, the adherent cells are "recharged" with a fresh inoculum of bone marrow cells which generate pluripotent and committed stem cells for weeks to months (DEXTER et al., 1977a, 1977b). It appears that this system will also be applicable to studies of human hematopoiesis (HOCKING and GOLDE, 1980).

With information gained from the various clonogenic assays described above, a scheme of the early phases of hematopoietic differentiation can be constructed. Figure 3 is a representation of our current knowledge of the interrelationships between stem cells showing the corresponding assays with which they are measured.

C. Inhibitors of Hematopoiesis

A variety of inhibitors of hematopoiesis have been described in experimental animals and in man. Some of these inhibitors may be involved in the control of he-

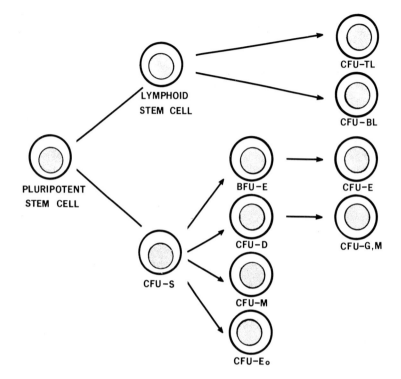

Fig. 3. Interrelationships between hematopoietic stem cells as determined by in vitro and in vivo assays. For definitions of the various assay abbreviations, see the Glossary (Table 1)

matopoiesis under normal circumstances. Others have been identified in hematologic disease states and may be of pathogenic importance. However, the relevance and, in some cases, the very existence of these inhibitors is controversial.

I. "Physiologic" Inhibitors of Hematopoiesis

Much of the work on putative physiologic inhibitors of hematopoiesis has focused on the control of granulopoiesis. This focus reflects in part the fact that in vitro assays for granulocyte-macrophage stem cells are relatively easy to perform and have been available the longest, and in part the widely held conviction that red cell production is primarily regulated by erythropoietin through positive feedback control.

Two classes of "physiologic" inhibitors of granulopoiesis have been described. One class is thought to act directly on granulocyte-macrophage stem cells (CFU-C) or their immediate progeny; the other class is thought to exert its effect by modulating the production of colony-stimulating activity (CSA) by mononuclear phagocytes. Table 2 lists the "physiologic" inhibitors that may have physiologic significance and which have been described since the introduction of culture systems for growth of hematopoietic cells in semi-solid suspension systems.

Table 2. Reported "physiologic" inhibitors of granulopoiesis

Inhibitor	Site of Action	References
Prostaglandins (E series)	CFU−C	KURLAND and MOORE (1977); KURLAND et al. (1978)
Colony inhibitory factor	CSA production	BROXMEYER (1978), BROXMEYER et al. (1977a, 1978)
Granulocyte chalone	CFU−C and progeny	PAUKOVITS (1971), RYTOMAA (1973), BATEMAN (1974), LORD et al. (1974), LAERUM and MAURER (1973)
Serum "lipoproteins"	CFU−C	BERAN (1974), CHAN et al. (1970, 1971)

1. Prostaglandins

Prostaglandins of the E series (PGE_1 and PGE_2) have been reported to produce potent inhibition of the proliferation of CFU-C (KURLAND and MOORE, 1977; KURLAND et al., 1978). KURLAND and MOORE (1977) showed that the effect of colony stimulating activity (CSA) on CFU-C could be effectively limited by PGE at concentrations as low as $10^{-10} M$. This inhibitory effect could be overcome by increasing concentrations of CSA.

In subsequent experiments, KURLAND and associates (1978) presented evidence which suggested that monocytes (light-density, adherent bone marrow cells) served both stimulatory and inhibitory roles in the modulation of committed stem cell proliferation. They proposed that these opposing forces were mediated by monocyte production of both CSA and PGE. They reported a progressive increase in CFU-C colony numbers when increasing numbers of monocytes were added to feeder layers in a two-layer agar culture system. Above a critical number of monocytes (2.3×10^5/culture dish) no further increase in colony formation was observed. If indomethacin (a PGE synthetase inhibitor) was added to the system, PGE production by monocytes was suppressed and a linear increase in CFU-C colony numbers occurred at monocyte concentrations above 2.3×10^5/culture dish. Further, by exploiting the relative species specificity of CSA action (murine CSA does not stimulate human CFU-C), KURLAND et al. (1978) adduced evidence that PGE production by monocytes is regulated by the concentrations of CSA to which monocytes (macrophages) are exposed. In complex experiments, they incubated murine peritoneal macrophages in the presence of high concentrations of murine CSA. The amount of PGE contained in the resultant conditioned medium varied directly with the concentration of murine CSA in the mixture. In addition, medium conditioned by macrophages exposed to high concentrations of murine CSA produced potent inhibition of colony formation by human CFU-C.

The rather elaborate scheme of the control of granulopoiesis based on the observations outlined above remains tentative. It relies on several assumptions: 1) that indomethacin specifically blocks macrophage production of prostaglandins as opposed to other cytotoxic cell products, 2) that the relative numbers of macrophages and CFU-C used in vitro accurately reflect the ratio of these cellular elements in vivo, and 3) that the concentrations of CSA used in vitro are comparable to those that exist in the marrow microenvironment. At present, the latter two assumptions are impossible to test in the intact animal.

2. Neutrophil-Derived Colony Inhibitory Factor

BROXMEYER et al. (1977b) reported that crude extracts of normal neutrophils inhibit the production of CSA by monocytes and macrophages. They termed this material "colony inhibitory factor" or CIF. CIF had no direct effect on CFU-C stimulated in vitro by an exogenous source of CSA. Studies of patients with chronic myelogenous leukemia (CML) revealed two apparent CIF-related defects: CML-neutrophils were defective in producing CIF, and CML-CSA-producing cells were less responsive to CIF than normal CSA-producing cells (BROXMEYER, 1978).

The possible relevance of CIF to in vivo granulopoietic regulation was suggested by BROXMEYER et al. (1977a) based on the effects of human CIF in cyclophosphamide-treated mice. Administration of human CIF significantly blunted the recovery of marrow CFU-C and peripheral blood neutrophils that ordinarily follows cyclophosphamide treatment, but was without effect on colony formation by splenic B lymphocytes. Depending on the timing of CIF injection in relation to cyclophosphamide treatment, granulopoiesis appeared to be affected by suppressed CSA production alone or by suppressed CSA production together with direct inhibition of CFU-C.

Recently, BROXMEYER et al. (1978) have reported biochemical and immunological evidence that CIF is identical to lactoferrin. They found that purified lactoferrin, when fully iron saturated, inhibited CSA production by adherent mononuclear blood cells at concentrations as low as $10^{-17}M$. Despite this impressive potency, the role of lactoferrin (CIF) as a bonafide physiologic regulator of granulopoiesis remains to be proven. In the rare genetic disorder of lactoferrin deficiency, major abnormalities of granulopoiesis are not observed. The apparent direct affect of CIF on CFU-C in the cyclophosphamide-treated mouse, the lack of convincing clinical correlates, and the non-specific growth-inhibitory effects of lactoferrin (e.g., on bacteria) suggest that the data should be interpreted with caution.

3. Chalones

Chalones have been defined as tissue-specific, species non-specific inhibitors of cellular proliferation. They were initially proposed in studies of skin wound healing (BULLOUGH and LAWRENCE, 1960) and epidermal carcinogenesis (IVERSON, 1960). Subsequently, numerous investigators have studied putative granulocyte chalones (RYTOMAA, 1973; PAUKOVITZ, 1971; BATEMAN, 1974; LORD et al., 1974; LAERUM and MAURER, 1973). Despite the attractiveness of the concept that products of mature end cells produce negative feedback inhibition of their immature progenitors, the existence of granulocyte chalone remains uncertain. The uncertainty reflects the use of non-homogeneous tissues for preparation of inhibitor, reliance on indirect assays for measuring inhibition of cell proliferation, heterogeneous target cell populations containing multiple proliferating cell types, and failure to purify chalone and exclude the non-specific action of potentially cytotoxic materials such as neutrophil hydrolytic enzymes.

We re-examined the subject of granulocyte chalones using purified populations of human neutrophils as starting material and a variety of target tissues and assays of cell proliferation (HERMAN et al., 1978). We failed to find neutrophil-derived inhibitory substances with the cellular specificities required of granulocyte chalone. These results notwithstanding, the chalone theory remains attractive for its relative simplicity and is deserving of further study.

A detailed discussion of the general topic of chalones is presented elsewhere in this volume (see IVERSON chapter 17).

4. Serum Inhibitors

Serum from mice (BERAN, 1974), normal humans (CHAN et al., 1971), and patients with leukemia (CHAN et al., 1970) has been reported to inhibit CFU-C colony growth in agar culture. CHAN et al. (1971) suggested that serum inhibitors were lipoprotein in nature based on the findings that serum inhibitory activity was removed by heating or ether treatment and precipitated by dialysis. BERAN (1974) showed that the inhibitory effect of serum on in vitro granulopoiesis disappeared 7–10 days after whole-body irradiation in mice, but that inhibitory activity persisted in the serum of irradiated mice transplanted with syngeneic bone marrow cells. This observation prompted the suggestion that serum "lipoprotein" inhibitors were maintained either directly or indirectly by cells of bone marrow origin.

Since serum contains a myriad of biologically active substances and starting materials have often been obtained from patients with rapidly changing clinical disease, the role of serum inhibitors in the control of granulopoiesis in vivo is unknown.

II. Inhibitors of Hematopoiesis in Disease

The possibility that inhibitors of hematopoiesis might be of importance in the mediation of hematologic disease has been most extensively explored in leukemia and bone marrow failure syndromes. In leukemia, the hypothesis has been advanced that normal hematopoietic cells are suppressed by leukemic cells rather than simply being replaced by them. In the bone marrow failure syndromes (aplastic anemia or isolated failure of a single line of hematopoietic differentiation), recent attention has been focused on the possibility that autoimmune suppression of hematopoietic stem cells might sometimes mediate marrow failure.

1. Leukemic Inhibitors

Several investigators have explored the possibility that leukemic cells may suppress the proliferation of normal hematopoietic clones during leukemic relapse. In theory, hematologic recovery following the induction of remission by chemotherapy could be explained by the release of normal hematopoietic stem cells from inhibitory constraints. Despite the logic of this hypothesis, in vitro culture studies have yielded conflicting results. MORRIS et al. (1975) found moderate to marked reductions of granulocyte-macrophage colony formation by normal human bone marrow grown in agar together with peripheral blood or bone marrow cells from 7 of 9 patients with acute myelomonocytic leukemia. CHIYODA et al. (1975) reported modest inhibition of colony formation when normal marrow was plated in agar with cells from 3 patients with acute leukemia. In co-culture studies with normal marrow and cells from 12 patients with acute leukemia, KNUDTZON and MORTEN-SEN (1976) found several different effects: cells from 5 of the patients were stimulatory (presumably as the result of CSA elaboration), cells from 3 patients were inhibitory, and cells from 4 patients had neither effect. Other workers have found no inhibition of colony growth by normal marrow co-cultured with leukemic cells (GREENBERG et al., 1971; ROBINSON et al., 1971). The disparate results may relate to technical differences such as the source of CSA utilized in the various studies, or to differences in the distribution of sub-types of acute leukemia in the different studies. Even when inhibition has been demonstrated, no attempt has been made to determine if the inhibitory effect is restricted to hematopoietic cells or to characterize the nature or mechanism of action of the inhibitory activity. The relevance of leukemic inhibitors to clinical disease is therefore largely speculative.

We examined a cell line derived from the bone marrow of a patient with blast crisis of chronic myelogenous leukemia for the presence of hematopoietic inhibitory activity (OLOFSSON and CLINE, 1978). Medium conditioned by this cell line (designated K 562) inhibited the proliferation of normal CFU-C and mitogen stimulated lymphocytes, and had weak activity against human CFU-E and murine CFU-C. No activity was demonstrable against fibroblasts, epithelial cell lines or B lymphocytes. Activity against CFU-C and proliferating lymphocytes was dependent on the concentration of the inhibitor and the duration of exposure. Inhibitory activity was not demonstrable against cells already in S phase or in mitosis. Entry of cells into S phase is a possible point of attack. K 562 inhibitor was stable on heating at 56 °C for 30 min and had an apparent molecular weight of $>300,000$ Daltons. Because the inhibitor was produced by a cell line in continuous culture and was tested using in vitro systems, its relevance to possible in vivo leukemic suppressors of hematopoiesis is uncertain.

2. Immune Suppression of Hematopoiesis

Immunologic destruction of mature blood cells, in such disorders as autoimmune hemolytic anemia and idiopathic thrombocytopenic purpura, has been recognized for years. Only recently has it been appreciated that immune phenomena may play a role in the injury of immature hematopoietic cells. In the past five years, an extensive literature devoted to this subject has accumulated, and the topic has been recently reviewed (CLINE and GOLDE, 1978). Interest in the possibility that hematopoietic cell proliferation might be suppressed by humoral or cell-mediated immune mechanisms has been prompted by two observations: 1) the demonstration that pure red cell aplasia is probably mediated by antibodies cytotoxic to red cell precursors (see below), and 2) autologous recovery of hematopoiesis in patients with aplastic anemia following immunosuppressive conditioning and attempted allogeneic bone marrow transplantation (JEANNET et al., 1974; THOMAS et al., 1976; SPECK et al., 1976; TERRITO, 1977; BARAN et al., 1976; SPECK et al., 1977).

a) Humoral Inhibitors

Antibody inhibitors of hematopoiesis which have been demonstrated using in vitro bone marrow culture techniques are summarized in Table 3 (KRANTZ, 1974; CLINE et al., 1976; FITCHEN et al., 1979; FITCHEN and CLINE 1980; HOFFMAN et al., 1979). In a series of elegant experiments, KRANTZ (1974) clearly demonstrated that pure red cell aplasia is often mediated by IgG antibodies cytotoxic to normoblasts. His studies were particularly convincing because he showed that IgG antibodies from patients with pure red cell aplasia were cytotoxic not only to normoblasts from healthy volunteers, but also produced lysis of immature red cells in autologous studies. Further, treatment of patients with cyclophosphamide resulted in disappearance of cytotoxic antibody and clinical remission. The experience with pure red cell aplasia underlines two important criteria that should be applied to the interpretation of studies purporting to demonstrate immune suppression of hematopoiesis: 1) autologous inhibition must be demonstrated before immune mediation of bone marrow failure can be accepted, and 2) therapeutic efforts which successfully remove immune inhibitors should be associated with clinical improvement. The demonstration of autologous inhibition is particularly important because most patients with bone marrow failure syndromes have been exposed to alloantigens as the result of repeated transfusion. Alloantibody to these antigens could produce misleading results since normal hematopoietic progenitor cells are known to express histocompatibility antigens such as HLA (FITCHEN and CLINE, 1979) and I a-like (CLINE and BILLING, 1977) antigens.

The insights gained from using in vitro hematopoietic culture techniques to study possible immune suppression of hematopoiesis, and the many pitfalls that are encountered along the way, are exemplified by a series of experiments that we performed in studies of patients with bone marrow failure syndromes. We first studied a woman with dramatic episodes of panleukopenia and recurrent infections (CLINE et al., 1976). She had complement-independent cytotoxic activity against CFU-C in both the IgG and IgM components of serum. Inhibitor titer, obtained over a 15-month period, varied with clinical status. Serum obtained during relapse was inhibitory for her own remission phase cells. Inhibitory activity was undetectable and remission occurred following treatment with cyclophosphamide. We next studied a woman with serologic evidence of systemic lupus erythematosus and aplastic anemia whose serum contained complement-dependent IgG antibody that

Table 3. Humoral immune inhibitors of hematopoiesis

Disease	Antibody	Target	Reference
Pure red cell aplasia	IgG	Normoblasts	KRANTZ (1974)[a]
Episodic panleukopenia	IgG, IgM	CFU−C	CLINE et al. (1976)[a]
Aplastic anemia	IgG	CFU−C, CFU−E	FITCHEN et al. (1979), FITCHEN and CLINE (1980)
Eosinophilic fasciitis and aplastic anemia	IgG	CFU−C, CFU−E, BFU−E	HOFFMAN et al. (1979)[a]

[a] Indicates autologous inhibition documented

Table 4. Serum CFU−C inhibitors in 61 patients with aplastic anemia

Etiology	No. of patients	No. with inhibitors
Drug-induced	9	0
Post-hepatitic	5	1
Idiopathic	47	14

inhibited both granulocytic and erythroid colony formation by normal human bone marrow in vitro (Fitchen et al., 1979). Her inhibitor titer fell and her hematologic status returned to normal coincident with a course of intensive plasmapheresis. However, relapse phase serum failed to inhibit colony formation by her own remission phase bone marrow. Despite the temporal relationship of recovery to plasmapheresis, our failure to demonstrate autologous inhibition of hematopoiesis made the pathogenetic importance of the antibody uncertain.

Based on experience with these two patients, we undertook a retrospective analysis of sera from 20 patients who had undergone bone marrow transplantation for aplastic anemia (Fitchen et al., 1980). We reasoned that inhibitory activity of recipient serum against hematopoietic progenitor cells might identify patients at risk for graft rejection, and might shed light on the pathogenesis of both graft rejection and primary bone marrow failure. 5 of the 20 sera had potent, complement-dependent inhibitory activity against normal human CFU-C. Graft rejection occurred in 4 of these 5 patients. 15 of the 20 sera had no demonstrable inhibitory activity against normal CFU-C, and 14 of these patients achieved sustained engraftment. Encouraged by these results, we performed additional experiments on 24 sera provided by the Seattle Bone Marrow Transplant Team. No correlation between the presence of CFU-C inhibitors and bone marrow graft outcome was noted in this group. Barrett et al. (1978) performed similar studies with sera from 28 patients transplanted for aplastic anemia. They detected inhibitory activity in the serum of 8 of 13 patients who failed to engraft or underwent graft rejection, and in 2 of 9 patients who achieved sustained engraftment. They were unable to demonstrate inhibition of autologous bone marrow by serum from patients with inhibitory activity against allogeneic targets.

Finally, we surveyed sera from a large number of patients with various forms of bone marrow failure (Fitchen and Cline, 1980). The results of these experiments are summarized in Table 4. Fifteen of the sera produced potent inhibition of CFU-C colony formation by bone marrow from normal volunteers. Inhibitors were not detected in patients with drug induced aplastic anemia, but were common in patients with idiopathic aplasia (28%). The characteristics of the inhibitors are presented in Table 5. Their antibody nature was suggested by the finding that inhibitory activity was complement-dependent in 14 out of 15 cases, and was contained in immunoglobulin fractions in all five sera where sufficient quantities were available to carry out fractionation procedures. Nearly all inhibitory sera also contained cytotoxic activity against allogeneic lymphocytes (lympho-

Table 5. Characteristics of CFU – C inhibitors in aplastic anemia

Feature	No. studied	No. positive
C′-dependent	15	14
Ig Fraction	5	5
Associated with lymphocytotoxins	15	14
Activity absorbed by platelets	9	1
Activity absorbed by B cells	4	0

Table 6. CFU – C inhibitors in patients receiving multiple transfusions

Reason for transfusion	No. of patients	No. with inhibitors
Open heart surgery	30	3
Chronic hemodialysis	5	0
Acute leukemia	7	1
Refractory anemia	4	2

cytotoxins), suggesting that cytotoxicity to CFU-C might result from reaction with HLA or Ia-like antigens. However, inhibitory activity against CFU-C was only rarely removed by absorption with random-donor platelets or B lymphocytes. Despite the potent activity of inhibitory sera against allogeneic targets, we were unable to demonstrate convincingly that these sera were inhibitory to autologous marrow. These findings suggested that antibodies reacted with alloantigens and were perhaps induced by repeated transfusions. To examine this possibility, we studied sera from patients who had received multiple transfusions for reasons other than aplastic anemia. As shown in Table 6, inhibitors were relatively infrequent (though clearly demonstrable) in patients who had received multiple transfusions during open heart surgery. These patients had their transfusion exposure on a single occasion, and may therefore not represent an appropriate control group for patients with bone marrow failure who receive transfusions over a prolonged time period. Despite chronic transfusion exposure, inhibitors were also infrequent (8%) in patients with chronic renal failure or acute leukemia, but the fact that these patients are immunosuppressed may affect their ability to develop antibody inhibitors. In a small group of patients with refractory anemia, inhibitors were frequent.

The studies outlined above underscore the importance of documenting autologous inhibition in studies of immune suppression of hematopoiesis. Also important are convincing clinical correlates. In the index case, we were able to demonstrate autologous inhibition and clinical improvement following immunosuppressive therapy. With the second patient, clinical improvement occurred coincident with removal of inhibitory antibody by plasmapheresis, but we were unable to demonstrate autologous inhibition. Further complexities were encountered in larger scale studies employing allogeneic target cells. Other published reports suffer from similar problems.

FREEDMAN et al. (1979) studied 6 patients with aplastic anemia and found serum inhibitory activity against allogeneic CFU-C and CFU-E in one. Importantly, IgG from this patient's serum also produced moderate to marked inhibition of autologous CFU-C and CFU-E. However, this study, like ours, must be interpreted with caution for several reasons. First, unlike most patients with aplastic anemia (KURNICK et al., 1971), this patient's marrow formed normal numbers of erythroid and granulocytic colonies in vitro. Second, co-culture of the patient's peripheral blood lymphocytes with autologous marrow also produced inhibition of colony formation. Cellular fractions enriched for B lymphocytes had the most activity, and the authors postulated that this was due to antibody secretion during co-culture. Finally, the patient died before therapeutic intervention directed at removing putative autoantibody could be accomplished.

HOFFMAN et al. (1979) were also able to demonstrate autologous inhibition of in vitro colony formation in a patient with eosinophilic fasciitis and aplastic anemia, but their study, like that of FREEDMAN et al. (1979) suffers from the lack of clinical improvement in association with immunosuppressive therapy.

Taken in sum, our studies and those reported by other laboratories create the tantalizing impression that some cases of aplastic anemia and other forms of bone marrow failure may result from antibody-mediated suppression of hematopoiesis. However, conclusive proof of this impression is lacking, and at present we would conclude that antibody-mediated suppression of hematopoiesis is probably a rare event.

b) Cell-Mediated Immune Suppression

The importance of documenting activity against autologous marrow cells is also apparent from analysis of a series of published reports on the possibility that cellular immune mechanisms might mediate bone marrow failure. Initial evidence suggesting lymphocyte-mediated suppression of hematopoiesis in patients with Diamond-Blackfan syndrome (HOFFMAN et al., 1976) and acquired aplastic anemia (HOFFMAN et al., 1977) were followed by several contradictory reports (NATHAN et al., 1978; FREEDMAN and SAUNDERS, 1977; SINGER et al., 1978). In studies of 16 patients with aplastic anemia, SINGER et al. (1978) found suppression of CFU-C colony formation in 3 patients when HLA-unrelated marrow cells were used as targets. However, no suppression was noted if HLA-matched marrow targets were employed.

BAGBY and GABOUREL (1979) have reported 3 patients (2 of them untransfused) with rheumatologic disease and granulopoietic failure in whom T lymphocytes suppressed colony formation by normal marrows. Importantly, removal of T lymphocytes by sheep erythrocyte rosetting techniques or cortisol treatment produced a marked increase in colony formation by patients' marrows in vitro. Colony formation by normal control marrows was not enhanced by T-cell depletion or cortisol treatment. When patients' T cells were added back to autologous marrow cell suspensions, colony formation was again reduced. The effect of cortisol treatment in vitro correlated with clinical responses to prednisone. Similar studies were performed by HAAK et al. (1977) in 10 patients with aplastic anemia. They found a 3- to 14-fold increase in colony formation by aplastic marrows after depletion of T cells by E-rosetting. However, in their studies, colony formation by normal marrows was increased as much as four-fold after T depletion, making the results in aplastic marrows difficult to interpret.

In vitro treatment of bone marrow with anti-thymocyte globulin (ATG) has been reported to increase CFU-C colony formation by marrow cells from patients with aplastic anemia (AMARE et al., 1978; FAILLE et al., 1979). In two studies, the in vitro findings were predictive of favorable responses to therapeutic trials of ATG, although the responses were often modest and usually transient. In one of the studies (FAILLE et al., 1979), the ATG employed also stimulated colony formation by normal bone marrow cells. As pointed out by the authors, this finding weakens the argument that the effect of ATG, in vitro and in vivo, is the removal of pathological auto-suppressive T cells.

TOROK-STORB et al. (1979) studied erythroid colony formation by peripheral blood cells from 35 patients treated with bone marrow transplantation for aplastic anemia. Depletion of T cells by E-rosetting resulted in increased BFU-E-derived colony formation in 4 of the 35 patients. Only 1 of these 4 patients achieved sustained marrow engraftment. Co-culture of patients' cells with donor (HLA-matched) cells produced inhibition of colony formation in 7 patients. In two instances, inhibition was marked, and both patients rejected their marrow grafts. Although these studies suggest that "suppressor" lymphocytes may play a role in bone marrow graft rejection, they do not establish that primary bone marrow failure was mediated by cellular autoimmunity.

ABDOU et al. (1978) demonstrated inhibition of normal colony formation by T cells from three neutropenic patients with Felty's syndrome. Cells from non-neutropenic Felty's patients, from patients with rheumatoid arthritis without neutropenia, and from patients with drug-induced neutropenia were not inhibitory. Detailed co-culture studies in one patient demonstrated that the suppressor efficacy of cells from various compartments was spleen > bone marrow > peripheral blood. Splenectomy in this patient transiently corrected the neutropenia and eliminated the inhibitory effect of peripheral blood mononuclear cells. Some patients in this study had been transfused, and except for the transient response to splenectomy in one patient, no clinical correlates were reported.

Perhaps the most intriguing observations bearing on the possibility of immune mediation of marrow failure have come from experience with bone marrow transplantation for aplastic anemia in identical twins. Two recent reports document initial failure of bone marrow transplantation to cure aplastic anemia in identical twins (GALE, 1979; APPELBAUM et al., 1979). Successful engraftment occurred after immunosuppressive conditioning and a second infusion of bone marrow. This experience, though far from conclusive, suggests that the last chapter in the story of immune suppression of hematopoiesis has yet to be written.

D. Summary and Conclusions

The concentration of blood cells within the circulation is normally maintained within narrow limits. In response to changes in demands for blood cells, as might occur with hemorrhage or infection, the bone marrow demonstrates flexibility in its rate of cell production. The normal steady state condition and the adaptive response to stress are probably achieved by a complex balance of stimulatory and inhibitory factors. Despite much research, however, only one stimulatory factor has been identified with certainty. This factor is *erythropoietin*, the glycoprotein hormone involved in the stimulation of erythropoiesis. A family of glycoprotein hormones designated *colony stimulating activity* are thought to play a corresponding role in the positive regulation of granulocyte and monocyte production; however, definitive proof of such a role is still lacking.

The possible involvement of a variety of negative regulators in the control of normal granulopoiesis is theoretically attractive and has been the subject of many scholarly papers; unfortunately, we still lack definitive evidence that such inhibitors play a role in the day-to-day control of granulocyte production. The reason for the discrepancy between an abundant literature on physiologic "regulators" and a lack of definitive proof of their relevance in vivo probably lies in the relative simplicity of adding putative inhibitor or stimulator in vitro and measuring an effect, and in the difficulties of dissecting out a role for a particular regulator in the complex system of checks and balances that operates in the intact animal. The same limitations exist in regard to interpretation of the role of inhibitors derived from leukemic cells. These inhibitors can often be demonstrated to suppress normal granulopoiesis in cell culture, but their role in producing the failure of hematopoiesis in patients afflicted with acute leukemia is still uncertain.

The strongest argument for the existence of inhibitors of hematopoiesis can be made for "pathologic" inhibitors demonstrable in a few well-studied cases of bone marrow failure. The potential pitfalls in the analysis of such inhibitors are multiple. Nevertheless, autoaggressive suppressors of hematopoiesis do occur in rare patients with selective or global failure of the hematopoietic cell lines of the bone marrow. These inhibitors may respond to appropriate therapeutic measures with the consequent relief of the suppression and restoration of hematopoiesis.

References

Abdou, N.I., Napombejara, C., Balentine, L., Abdou, N.L.: Suppressor cell-mediated neutropenia in Felty's syndrome. J. Clin. Invest. *61*, 738–743 (1978)

Adamson, J.W., Fialkow, P.J., Murphy, S., Prchal, J.F., Steinmann, L.: Polycythemia vera: stem-cell and probable clonal origin of disease. New Engl. J. Med. *295*, 913–916 (1976)

Amare, M., Abdou, N.L., Robinson, M.G., Abdou, N.I.: Aplastic anemia associated with bone marrow suppressor T-cell hyperactivity: Successful treatment with antithymocyte globulin. Am. J. Hemtol. *5*, 25–32 (1978)

Appelbaum, F.R., Fefer, A., Cheever, M.A., Hansen, J., Greenberg, P.D., Thomas, E.D.: Treatment of aplastic anemia with identical twin bone marrow transplantation. Blood (Suppl. 1) *54*, 227a (1979)

Bagby, G.C., Gabourel, J.D.: Neutropenia in three patients with rheumatic disorders: Suppression of granulopoiesis by cortisol sensitive thymus-dependent lymphocytes. J. Clin. Invest. *64*, 72–82 (1979)

Baran, D.T., Griner, P.F., Klemperer, M.R.: Recovery from aplastic anemia after treatment with cyclophosphamide. New Engl. J. Med. *295*, 1522–1523 (1976)

Barrett, A.J., Faille, A., Saal, F., Balitrand, N., Gluckman, E.: Marrow graft rejection and inhibition of growth in culture by serum in aplastic anemia. J. Clin. Pathol. *31*, 1244–1248 (1978)

Bateman, A.E.: Cell specificity of chalone-type inhibitors of DNA synthesis released by blood leukocytes and erythrocytes. Cell Tissue Kinet. *7*, 451–461 (1974)

Beran, M.: Serum inhibitors of in vitro colony formation: Relation to haematopoietic tissue in vivo. Exp. Hematol. *2*, 58–64 (1974)

Bradley, E.R., Metcalf, D.: The growth of mouse bone marrow cells in vitro. Aust. J. Exp. Biol. Med. Sci. *44*, 287–300 (1966)

Broxmeyer, H.E.: Inhibition in vivo of mouse granulopoiesis by cell-free activity derived from human polymorphonuclear neutrophils. Blood *51*, 889–901 (1978)

Broxmeyer, H.E., Mendelson, N., Moore, M.A.S.: Abnormal granulocyte feedback regulation of colony forming and colony stimulating activity-producing cells from patients with chronic myelogenous leukemia. Leukemia Res. *1*, 3–12 (1977a)

Broxmeyer, H.E., Moore, M.A.S., Ralph, P.: Cell-free granulocyte colony inhibiting activity derived from human polymorphonuclear neutrophils. Exp. Hematol. *5*, 87–102 (1977b)

Broxmeyer, H.E., Smithyman, A., Eger, R.R., Meyers, P.A., Sousa, M. de.: Identification of lactoferrin as the granulocyte-derived inhibitor of colony-stimulating activity production. J. Exp. Med. *148*, 1052–1067 (1978)

Bullough, W.S., Laurence, E.B.: The control of epidermal mitotic activity in the mouse. Proc. R. Soc. Med. B *151*, 517–536 (1960)

Camitta, B.M., Thomas, E.D.: Severe aplastic anemia: A prospective study of the effect of androgens in transplantation on haematological recovery and survival. Clin. Haematol. *7*, 587–595 (1978)

Chan, S.H., Metcalf, D.: Inhibition of bone marrow colony formation by normal and leukaemic human serum. Nature *227*, 845–846 (1970)

Chan, S.H., Metcalf, D., Stanley, E.R.: Stimulation and inhibition by normal human serum of colony formation in vitro by bone marrow cells. Br. J. Haematol. *20*, 329–341 (1971)

Chen, M.G., Schooley, J.C.: A study of the clonal nature of spleen colonies using chromosome markers. Transplantation *6*, 121–126 (1968)

Chiyoda, S., Mizoguchi, H., Kosaka, K., Takaku, F., Miura, Y.: Influence of leukaemic cells on the colony formation of human bone marrow cells in vitro. Br. J. Cancer *31*, 355–358 (1975)

Cline, M.J., Billing, R.: Antigens expressed by B lymphocytes and myeloid stem cells. J. Exp. Med. *146*, 1143–1145 (1977)

Cline, M.J., Golde, D.W.: Immune suppression of hematopoiesis. Am. J. Med. *64*, 301–310 (1978)

Cline, M.J., Golde, D.W.: Controlling the production of blood cells. Blood *53*, 157–165 (1979)

Cline, M.J., Opelz, G., Saxon, A., Golde, D.W.: Autoimmune panleukopenia. New Engl. J. Med. *295*, 1489–1493 (1976)

Cline, M.J., Gale, R.P., Golde, D.W.: Discrete clusters of hematopoietic cells in the marrow cavity of man after bone marrow transplantation. Blood *50*, 709–712 (1977)

Dexter, T.M., Allen, T.D., Lajtha, G.: Conditions controlling the proliferation of hematopoietic stem cells in vitro. J. Cell. Physiol. *91*, 335–344 (1977a)

Dexter, T.M., Moore, M.A.S., Sheridan, A.P.C.: Maintenance of hemopoietic stem cells and production of differentiated progeny in allogeneic and semiallogeneic bone marrow chimeras in vitro. J. Exp. Med. *145*, 1612–1616 (1977b)

Dresch, C., Johnson, G.R., Metcalf, D.: Eosinophil colony formation in semisolid cultures of human bone marrow cells. Blood *49*, 835–844 (1977)

Faille, A., Barrett, A.J., Balitrand, N., Ketels, F., Gluckman, E., Najean, Y.: Effect of an-
    tilymphocyte globulin on granulocyte precursors in aplastic anemia. Br. J. Haematol.
    42, 371–380 (1979)
Fauser, A.A., Messner, H.A.: Identification of megakaryocytes, macrophages, and eosino-
    phils in colonies of human bone marrow containing neutrophilic granulocytes and ery-
    throblasts. Blood *53*, 1023–1027 (1979)
Fialkow, P.J., Jacobson, R.J., Papayannopoulou,T.: Chronic myelocytic leukemia: clonal
    origin in a stem cell common to the granulocyte, erythrocyte, platelet and monocyte/
    macrophage. Am. J. Med. *63*, 125–130 (1977)
Fitchen, J.H., Cline, M.J.: Human myeloid progenitor cells express HLA antigens. Blood
    53, 794–797 (1979)
Fitchen, J.H., Cline, M.J.: Serum inhibitors of myelopoiesis. Br. J. Haematol. *44*, 1–16
    (1980)
Fitchen, J.H., Gale, R.P., Cline, M.J.: Serum inhibitors of in vitro hematopoiesis and graft
    outcome in bone marrow transplantation for aplastic anemia. Transplantation *29*, 223–
    226 (1980)
Fitchen, J.H., Cline, M.J., Saxon, A., Golde, D.W.: Serum inhibitors of hematopoiesis in
    a patient with aplastic anemia and systemic lupus erythematosus: Recovery after ex-
    change plasmapheresis: Am. J. Med. *66*, 537–542 (1979)
Freedman, M.H., Saunders, E.F.: Diamond-Blackfan syndrome. Clin. Res. *25*, 339 A (1977)
Freedman, M.H., Gelfand, E.W., Saunders, E.F.: Acquired aplastic anemia: Antibody-me-
    diated hematopoietic failure. Am. J. Hematol. *6*, 135–141 (1979)
Gale, R.P.: Bone marrow transplantation in identical twins with aplastic anemia. Blood
    (Suppl.) *54*, 229 a (1979)
Golde, D.W., Burgaleta, C., Sparkes, R.S., Cline, M.J.: The Philadelphia chromosome in
    human macrophages. Blood *49*, 367–370 (1977)
Greenberg, P.L., Nichols, W.C., Schrier, S.L.: Granulopoiesis in acute myeloid leukemia
    and preleukemia. New Engl. J. Med. *284*, 1225–1232 (1971)
Haak, H.L., Goselink, H.M., Veenhof, W., Pellinkhof-Stadelmann, S., Kleiverda, J.K.,
    Tevelde, J.: Acquired aplastic anemia in adults. IV. Histological and CFU studies in
    transplanted and non-transplanted patients. Scand. J. Hematol. *19*, 159–171 (1977)
Herman, S.P., Golde, D.W., Cline, M.J.: Neutrophil products that inhibit cell proliferation:
    Relation to granulocytic "chalone". Blood *51*, 207–219 (1978)
Hocking, W.G., Golde, D.W.: Long-term human bone marrow cultures. Blood *56*, 118–124
    (1980)
Hoffman, R., Zanjani, E.D., Vila, J., Zalusky, R., Lutton, J.D., Wasserman, L.R.: Dia-
    mond-Blackfan syndrome: Lymphocyte-mediated suppression of erythropoiesis. Sci-
    ence *193*, 899–900 (1976)
Hoffman, R., Zanjani, E.D., Lutton, J.D., Zalusky, R., Wasserman, L.R.: Suppression of
    erythroid-colony formation by lymphocytes from patients with aplastic anemia. New
    Engl. J. Med. *296*, 10–13 (1977)
Hoffman, R., Dainiak, N., Sibrack, L., Pober, J.S., Waldron, J.A.: Antibody-mediated
    aplastic anemia and diffuse fasciitis. New Engl. J. Med. *300*, 718–721 (1979)
Iscove, N.N., Sieber, F., Winterhalter, K.H.: Erythroid colony formation in cultures of
    mouse and human bone marrow: analysis of the requirements for erythropoietin by gel
    formation and affinity chromatography on agarose-concanavalin A. J. Cell. Physiol.
    83, 309–320 (1974)
Iverson, O.H.: Cell metabolism in experimental skin carcinogenesis. Acta Pathol. Micro-
    biol. Scand. *50*, 17–24 (1960)
Jacobson, R.T., Salo, A., Fialkow, P.J.: Agnogenic myeloid metaplasia: a clonal prolifer-
    ation of hematopoietic stem cells with secondary myelofibrosis. Blood *51*, 189–194
    (1978)
Jeannet, M., Rubinstein, A., Pelet, B., Kummer, H.: Prolonged remission of severe aplastic
    anemia after ALG treatment and HL-A-semi-incompatible bone-marrow cell transfu-
    sion. Transplant. Proc. *6*, 359–363 (1974)
Knudtzon, S., Mortensen, B.T.: Interaction between normal and leukaemic human cells in
    agar culture. Br. J. Haematol. *17*, 369–378 (1976)

Krantz, S.B.: Pure red-cell aplasia. New Engl. J. Med. *291*, 345–350 (1974)

Kurland, J., Moore, M.A.S.: Modulation of hemopoiesis by prostaglandins. Exp. Hematol. *5*, 357–373 (1977)

Kurland, J.I., Bockman, R.S., Broxmeyer, H.E., Moore, M.A.S.: Limitation of excessive myelopoiesis by the intrinsic modulation of macrophage-derived prostaglandin E. Science *199*, 552–555 (1978)

Kurnick, J.E., Robinson, W.A., Dickey, C.A.: In vitro granulocytic colony-forming potential of bone marrow from patients with granulocytopenia and aplastic anemia. Proc. Soc. Exp. Biol. Med. *137*, 917–920 (1971)

Laerum, O.D., Maurer, H.R.: Proliferation kinetics of myelopoietic cells and macrophages in diffusion chambers after treatment with granulocyte extracts (chalone). Virchows Arch. (Cell Pathol.) *14*, 293–305 (1973)

Lord, B.I., Cercek, L., Cercek, B., Shah, G.P., Dexter, T.M., Lajtha, L.G.: Inhibitors of haematopoietic cell proliferation? Specificity of action within the haematopoietic system. Br. J. Cancer *29*, 168–175 (1974)

Metcalf, D., Moore, M.A.S.: Hematopoietic cells. In: Frontiers in biology. Neuberger, A., Tatum, E.L. (eds.), p. 109. Amsterdam, London: North-Holland 1971

Metcalf, D., Nossal, G.J.V., Warner, N.L., Miller, J.F.A.P., Mandel, T.E., Layton, J.E., Gutman, G.A.: Growth of B-lymphocyte colonies in vitro. J. Exp. Med. *142*, 1534–1549 (1975)

Morris, T.C.M., McNeill, T.A., Bridges, J.M.: Inhibition of normal human in vitro colony forming cells by cells from leukemic patients. Br. J. Cancer *31*, 641–648 (1975)

Nakeff, A., Daniels-McQueen, S.: In vitro colony assay for a new class of megakaryocyte precursor: colony-forming unit megakaryocyte (CFU-M): Proc. Soc. Exp. Biol. Med. *151*, 587–590 (1976)

Nathan, D.G., Hillman, D.G., Chess, L., Alter, B.P., Clarke, B.J., Breard, J., Housman, D.E.: Normal erythropoietic helper cells in congenital hypoplastic (Diamond-Blackfan) anemia. New Engl. J. Med. *298*, 1049–1051 (1978)

Olofsson, T., Cline, M.J.: Inhibitor of hematopoietic cell proliferation derived from a human leukemic cell line. Blood *52*, 143–152 (1978)

Paukovits, W.R.: Control of granulocyte production: separation and chemical identification of a specific inhibitor. Cell Tissue Kinet. *4*, 539–547 (1971)

Pike, B.L., Robinson, W.A.: Human bone marrow colony growth in agar-gel. J. Cell. Physiol. *76*, 77–84 (1970)

Pluznik, D.H., Sachs, L.: The cloning of normal "mast" cells in tissue culture. J. Cell. Physiol. *66*, 319–324 (1965)

Quesenberry, P., Levitt, L.: Hematopoietic stem cells. New Engl. J. Med. *301*, 755–760 (1979)

Robinson, W.A., Kurnick, J.E., Pike, B.L.: Colony growth of human leukemic peripheral blood cells in vitro. Blood *38*, 500–508 (1971)

Rosenszajn, L.A., Shoham, D., Kalechman, I.: Clonal proliferation of PHA-stimulated human lymphocytes in soft agar culture. Immunology *21*, 1041–1055 (1975)

Rytomaa, T.: Role of chalone in granulopoiesis. Br. J. Haematol. *24*, 141–146 (1973)

Singer, J.W., Brown, J.E., James, M.C., Doney, K., Warren, R.P., Storb, R., Thomas, E.D.: Effect of peripheral blood lymphocytes from patients with aplastic anemia or granulocyte colony growth from HLA-matched and -mismatched marrow: Effect of transfusion sensitization. Blood *52*, 37–46 (1978)

Speck, B., Cornu, P., Jeannet, M., Nissen, C., Burri, H.P., Groff, P., Nagel, G.A., Buckner, C.D. Autologous marrow recovery following allogeneic marrow transplantation in a patient with severe aplastic anemia. Exp. Hematol. *4*, 131–137 (1976)

Speck, B., Gluckman, E., Haak, H.L., Rood, J.J. van: Treatment of aplastic anemia by antilymphocyte globulin with and without allogeneic bone marrow infusions. Lancet *1977 II*, 1145–1148

Stephenson, J.R., Axelrod, A.A., McLeod, D.L., Shreeve, M.M.: Induction of colonies of hemoglobin-synthesizing cells by erythropoietin in vitro. Proc. Natl. Acad. Sci. U.S.A. *68*, 1542–1546 (1971)

Territo, M.C.: Autologous bone marrow repopulation following cyclophosphamide and allogeneic marrow transplantation in aplastic anemia. Br. J. Haematol. *36*, 305–312 (1977)

Thomas, E.D., Storb, R., Giblett, E.R., Longpre, B., Weiden, P.L., Fefer, A., Witherspoon, R., Clift, R.A., Buckner, C.D.: Recovery from aplastic anemia following attempted marrow transplantation. Exp. Hematol. *4*, 97–102 (1976)

Till, J.E., McCulloch, E.A.: A direct measurement of the radiation sensitivity of normal mouse bone marrow cells. Radiat. Res. *14*, 213–222 (1961)

Torok-Storb, B., Storb, R., Thomas, E.D.: In vitro studies of 35 patients with aplastic anemia: Correlation with marrow graft failure or graft rejection. Blood (Suppl. 1) *54*, 231 a (1979)

Whang, J., Frei, E., Ill, Tjio, J.H., Carbone, P.P., Brecher, G.: The distribution of the Philadelphia chromosome in patients with chronic myelocytic leukemia. Blood *22*, 664–673 (1963)

CHAPTER 14

Inducers and Inhibitors of Leukemic Cell Differentiation in Culture

JANET ABRAHM and G. ROVERA

A. Introduction

Bone marrow cells differentiate into well-defined lineages, producing erythroid, myeloid, monocytic, megakaryocytic, and lymphoid cells (see reviews by METCALF and MOORE, 1971; QUESENBERRY and LEVITT, 1979a, b, c). The stem cells in the bone marrow have the potential of giving rise to cells of different lineages. The stem cells have been only partially identified in man, but a cell clearly capable of reconstituting bone marrow has been identified in the mouse (METCALF, 1977). The stem cell committment to differentiation is best described by a stochastic model (TILL et al., 1964). A stem cell, at random, either becomes committed to a particular lineage and gives rise to morphologically recognizable progenitor cells of that lineage, or it remains uncommitted and retains the ability to generate other stem cells. Asymmetric cell division, in which a stem cell gives rise both to differentiated progenitor and undifferentiated cells is not thought to contribute significantly to maintaining the stem cell pool. The progenitor cells then divide further to form the terminally differentiated cells characteristic of each lineage. The end cells are incapable of further proliferation. They express differentiated functions and are capable of exerting feedback regulations on the bone marrow precursor cells (MOORE, 1977; MOORE et al., 1978).

In leukemias, however, a block of differentiation at the level of the stem cells or progenitor cells is thought to exist (see CLINE and GOLDE, 1979). Myeloblastic leukemias, for example, may develop because of the inability of myeloblast progenitor cells to undergo further differentiation, and promyelocytic leukemias may be the result of a differentiation block at the level of the promyelocytes. It is not presently clear if the differentiation block in acute leukemias at the level of progenitor cells reflects the fact that the lesion responsible for the leukemic transformation is present only in the progenitor cells or is present but not expressed in earlier stem cells. Evidence for the second possibility has been reported in the case of chronic myeloid leukemias in which phenotypic alterations are seen in the myeloid lineage but genetic alterations involve the platelets, red cell precursors and non-T lymphocytes as well as the myeloid lineage (WHANG et al., 1963; FIALKOW et al., 1978).

The differentiation block in the leukemias is probably not always complete, since in some cases a fraction of the cells can differentiate normally. The differentiative block typical of the leukemias is sometimes reversible. Studies by SACHS and co-workers (for review, see SACHS, 1978) and by FRIEND and co-workers (for review, see FRIEND, 1980) in murine systems have shown that, by using certain inducers, the block can be overcome and normal differentiated cells can be produced.

Some of these inducers are physiological regulatory polypeptides present in the organism (SACHS, 1978), whereas others are chemical compounds not normally present in the organism. The latter compounds are the subject of this review.

B. Leukemic Cell Lines Utilized for Differentiation Studies in Culture

In the past 15 years, a number of cell lines have been established in culture from lymphoblastic, myeloid, myelomonocytic, and monocytic leukemias. However, only a few lines have been used extensively in differentiation studies since only a few are sensitive to the inducers of differentiation (COLLINS et al., 1978). These sensitive cell lines are generally characterized by a predominance of well-differentiated progenitor cells committed to either an erythroid or a myeloid/monocytic differentiation pathway. The cell lines derived from a spontaneous mouse myeloid leukemia and those derived from virus-induced mouse erythroleukemias have been the ones most widely used. More recently, human leukemic cell lines have been established that are sensitive to inducers of differentiation.

A variety of clones of mouse myeloid leukemic cells have been used by SACHS and co-workers (for review, see SACHS, 1978). These undifferentiated myeloid cells derive from a line established by ICHIKAWA (1969). These cells, called Ml, were originally obtained from a spontaneous myeloid leukemia in an SL mouse. From these cell lines, LOTEM and SACHS (1974) were able to isolate clones which were sensitive (D+) or insensitive (D−) to inducers of differentiation (SACHS, 1978). The undifferentiated myeloid cells have characteristics of myeloblasts without expressing the markers for complement receptors and γ-Fc receptors, lysozyme or phagocytosis. SACHS and other groups (for example, OKABE et al., 1977) have also isolated resistant clones unresponsive to different inducers tested.

The mouse erythroleukemia cell lines have usually been derived from Friend virus-transformed mouse erythroleukemia cells (PATULEIA and FRIEND, 1967; DUBE et al., 1974), or in some cases Rauscher virus erythroleukemia cells (MIAO and FIELDSTEEL, 1979). These cell lines consist mainly of proerythroblast-like cells. However, a fraction of the cells consists of more mature elements of the erythroid lineage including orthochromatic erythroblasts which contain hemoglobin (FRIEND et al., 1971). Lines with a higher frequency of spontaneous differentiation derived from the Friend erythroleukemia lines have been isolated by ROVERA and SURREY (1977). Also, resistant clones unable to differentiate have been developed (PAUL and HICKEY, 1974; ROVERA and SURREY, 1978; OHTA et al., 1976; IKAWA et al., 1976a, b) through different manipulations in culture.

Two human cell lines derived from patients with leukemia have recently been shown to be sensitive to inducers of differentiation. COLLINS et al. (1977, 1978) developed a promyelocytic leukemia cell line (HL 60) obtained from a female patient with acute promyelocytic leukemia. This line has been extensively characterized (GALLAGHER et al., 1979). The majority of the cells in culture are promyelocytes with immature characteristics as shown by electron microscopy (GALLAGHER et al., 1979; ROVERA et al., 1979b). However a small fraction of the population consists of myeloblasts, myelocytes, metamyelocytes, and a few granulocytes.

KOEFFLER et al. (1979) have developed another cell line, KGl, which also consists mainly of myeloblasts and promyelocytes and is responsive to inducers of differentiation. A human cell line derived from a patient with chronic myeloid leukemia in blast crisis was established by LOZZIO and LOZZIO in 1975 and was believed to be originally a myeloid stem cell. This line consists essentially of highly immature blast cells. However, ANDERSSON and co-workers (1979a, b) and RUTHERFORD and co-workers (1979) have shown that this line can be induced to differentiate into cells containing hemoglobin, and both non-induced and induced cells contain surface proteins such as glycophorin and spectrin which are characteristic of the erythroid lineage.

C. Inducers of Differentiation in Culture

A large amount of data on inducers of mouse cell lines have accumulated since the pioneering work of SACHS (for review see SACHS, 1978) and FRIEND (for review, see FRIEND, 1980) but only in the last two years have inducers of differentiation of human leukemias been identified. The most detailed analyses of inducers of leukemic differentiation have been done in the mouse erythroleukemia system. FRIEND et al. (1971) reported that mouse erythroleukemia cells in culture could be induced to differentiate into orthochromatic erythroblasts and synthesize hemoglobin by four days' treatment with 2% dimethylsulfoxide (DMSO). That report was the first to clearly indicate that a simple chemical compound could induce differentiation of leukemic cells into terminally differentiated end cells, expressing differentiation markers. The same work showed that DMSO-treated leukemic cells had a decreased level of leukemogenicity when injected into DBA/2 mice.

A number of other chemical compounds have since been found to induce differentiation of proerythroblast murine erythroleukemia cells. These different compounds have been reviewed by FRIEND (1980), TERADA et al. (1978b) and HARRISON (1977). Table 1 lists most of the inducers of mouse erythroleukemic cell differentiation reported to date. There is a large variety of compounds including polar-planar compounds, fatty acids, purines, and purine derivatives, inhibitors of DNA and RNA synthesis, physical agents such as X-ray and UV, phorbol diesters with known tumor-promoting activity, glycosides, and hemin. It is important to note that not all the agents act on every clone or cell line of mouse erthyroleukemia cells, and not all of the agents are equally effective.

One general characteristic of the inducers of differentiation in Friend erythroleukemia cells is that none can recruit 100% of the cells to terminally differentiate. The efficiency varies from a small percentage of cells, just above the spontaneous background level, to values as high as 99% with the most active inducers of differentiation such as hexamethylene bisacetamide (HMBA). The active concentration of the different compounds able to elicit terminal differentiation also has a very wide range, from nano- to millimolar. Since only the ability to induce hemoglobin has been measured in most of these agents, it is not clear whether all the inducers can also elicit the other markers of terminal erythroid differentiation. Hemin, for example, clearly does not induce several of these markers (HOUSMAN et al., 1978). Another interesting feature of some of these compounds is that they

Table 1. Inducers of mouse erythroleukemia cells

Inducer	Concentration	% Benzidine pos. cells	Cell clone	Reference
Acetamide	250 mM	56	745	TANAKA et al. (1975)
Acetic acid	40 mM	19	3TCl-1	TAKAHASHI et al. (1975)
Actinomycin D	5–10 ng/ml	NR	3TCl-2	EBERT et al. (1976)
	1–2 ng/ml	90	745	TERADA et al. (1978a)
Adipic acid	50 mM	10	745	REUBEN et al. (1978)
Adriamycin	5–20 ng/ml	NR	3TCl-2	EBERT et al. (1976)
Benzamide	10 mM	70	745	REUBEN et al. (1978)
	25 mM	85	745	TERADA et al. (1979)
Bispropionamide	5 mM	90	745	REUBEN et al. (1978)
Bleomycin	3 µg/ml	60	3TCl-1,2	SUGANO et al. (1973)
	50 µg/ml	60	745	SCHER and FRIEND (1978)
Bromodeoxyuridine	5 µg/ml	50	F4Cl-2	ADESNIK and SNITKIN (1978)
Bromouracil	9 mM	60	745	TERADA et al. (1979)
Butyric acid	0.5–1.5 mM	45	3TCl-2	LEDER and LEDER (1975)
	0.5–0.75 mM	38	3TCl-1	TAKAHASHI et al. (1975)
	1.5 mM	70	745	REUBEN et al. (1978)
Butyrl choline	1.8 mM	NR	3TCl-2	SCHER and FRIEND (1978)
Cyclic AMP	10^{-4} M	NR	FSD1/F4	DUBE et al. (1974)
Cycloheximide	100 ng/ml	NR	3TCl-2	EBERT et al. (1976)
Cytosine arabinoside	1×10^{-6} M	NR	3TCl-1	SUGANO et al. (1973)
	0.005–1 µg/ml	NR	3TCl-2	EBERT et al. (1976)
1,5 Diaminopentane	5 mM	25	745	REUBEN et al. (1978)
N,N'-Dimethyl formamide	100 mM	25	745	SCHER et al. (1973)
	130 mM	40	745	PREISLER and LYMAN (1975)
	150 mM	62	745	TANAKA et al. (1975)
N-Dimethyl rifampicin	100 µg/ml	NR	3TCl-1	SUGANO et al. (1973)
Dimethyl sulfoxide	280 mM	85	745	FRIEND et al. (1971)
L-Ethionine	6.7 mM	20–30	745	CHRISTMAN et al. (1977)
Etiocholanolone	10^{-7} M	NR	FSD1/F4	DUBE et al. (1974)
Glutaric acid	50 mM	6	745	REUBEN et al. (1978)
Glycerol	5%	NR	3TCl-1	SUGANO et al. (1973)
	5%	NR	3TCl-2	EBERT et al. (1976)
Hemin	10^{-4} M	NR	745, 3TCl-2	ROSS and SAUTNER (1976)
Hexamethylene bisacetamide	5 mM	99	745	REUBEN et al. (1976)
Hexanoic acid	5 mM	30	745	REUBEN et al. (1978)
Hydroxyurea	8 µg/ml	NR	3TCl-2	EBERT et al. (1976)
Hypoxanthine	3.67 mM	89	745	GUSELLA and HOUSMAN (1976)

Table 1 (continued)

Inducer	Concentration	% Benzidine pos. cells	Cell clone	Reference
	3.67 mM	51	FSD	GUSELLA and HOUSMAN (1976)
Isobutyric acid	6 mM	47	3TCl-1	TAKAHASHI et al. (1975)
N-Methyl acetamide	34 mM	50	745	TANAKA et al. (1975)
	50 mM	70	745	PREISLER et al. (1976b)
N-Methyl formamide	250 mM	49	745	TANAKA et al. (1975)
N-Methyl nicotinamide	8 mM	99	745	TERADA et al. (1979)
1-Methyl-2-piperidone	10 mM	64	745	TANAKA et al. (1975)
N-Methyl pyrrolidinone	30 mM	97	745	TANAKA et al. (1975)
Mithramycin	0.2 µg/ml	NR	3TCl-2	EBERT et al. (1976)
Mitomycin C	1 µg/ml	NR	3TCl-1	SUGANO et al. (1973)
	5 µg/ml	NR	3TCl-2	EBERT et al. (1976)
Nicotinamide	40 mM	60	745	TERADA et al. (1979)
Ouabain	0.15 mM	25	745	BERNSTEIN et al. (1976)
Piperidone	100 mM	28	745	TANAKA et al. (1975)
Potassium	90 mM	35	745	MAGER et al. (1979)
Propionamide	120 mM	36	745	TANAKA et al. (1975)
Propionic acid	2 mM	34	3TCl-1	TAKAHASHI et al. (1975)
	5 mM	50	745	REUBEN et al. (1978)
Prostaglandin E_1	1×10^{-4} M	39	F4N	TABUSE et al. (1977)
	5×10^{-5} M	3	745	TABUSE et al. (1977)
Pyrazinamide	45 mM	30	745	TERADA et al. (1979)
Pyridine-N-oxide	100–110 mM	41–53	745	TANAKA et al. (1975)
	130 mM	56	745	PREISLER and LYMAN (1975) PREISLER et al. (1976b)
2-Pyrrolidinone	90 mM	39	745	TANAKA et al. (1975)
Selenium dioxide	0.05–1 mM	NR	745, 3TCl-1,2	EBERT and MALININ (1979)
Selenium trioxide	0.05 mM	NR	745, 3TCl-1,2	EBERT and MALININ (1979)
Suberic acid	50 mM	19	745	REUBEN et al. (1978)
Succinic acid	30 mM	4	745	REUBEN et al. (1978)
TPA	10^{-8}–10^{-4} M	25–60	RV-133, RV-187	MIAO et al. (1978)
Tetramethylurea	8.6 mM	83	745	PREISLER et al. (1976b)
Thymine	18 mM	85	745	TERADA et al. (1979)
Thymine riboside	7.5 mM	20	745	TERADA et al. (1979)
Triethylene glycol	250 mM	51	745	TANAKA et al. (1975)
Ultraviolet light	NR	NR	745	SCHER and FRIEND (1978)
Valeric acid	5 mM	30	745	REUBEN et al. (1978)
X-ray	1,400 rad	60	745	SCHER and FRIEND (1978)

Table 2. Inducers of mouse myeloid leukemia cells[a]

Inducer	Concentration	Reference
Peptide MGI	NR	FIBACH et al. (1972)
Poly (I)	100–200 µg/ml	TOMIDA et al. (1978)
Lectins		
Concanavalin A	NR	SACHS (1978)
Phytohemagglutinin	NR	SACHS (1978)
Pokeweed mitogen	NR	SACHS (1978)
Polycyclic hydrocarbons		
Benzo(a)pyrene	NR	SACHS (1978)
Dimethylbenz(a)-anthracene	NR	SACHS (1978)
Steroids		
Aldosterone	$10^{-5}\,M$	HONMA et al. (1977b)
Corticosterone	$10^{-5}\,M$	HONMA et al. (1977b)
Dexamethasone	$10^{-7}\text{–}10^{-5}\,M$	HONMA et al. (1977b)
Dexamethasone	$10^{-7}\,M$	LOTEM and SACHS (1975a)
Estradiol	$1\,\mu M$	LOTEM and SACHS (1975a)
Hydrocortisone	$10^{-5}\,M$	HONMA et al. (1977a, b)
Hydrocortisone	NR	SACHS (1978)
Prednisolone	$10^{-7}\,M$	LOTEM and SACHS (1975a)
Other		
Actinomycin D	0.01–0.03 µg/ml	LOTEM and SACHS (1975b)
Adriamycin	NR	SACHS (1978)
5-Bromodeoxyuridine	0.5–1 µg/ml	LOTEM and SACHS (1975b)
Cytosine Arabinoside	0.1 µg/ml	LOTEM and SACHS (1975b)
Daunomycin	NR	SACHS (1978)
Dimethyl sulfoxide	1%	KRYSTOSEK and SACHS (1976)
5-Fluorodeoxyuridine	0.001–0.003 µg/ml	LOTEM and SACHS (1975b)
Hydroxyurea	5.5 µg/ml	LOTEM and SACHS (1975a, b)
Lipid A	10 µg/ml	WEISS and SACHS (1978)
Lipopolysaccharide	10 µg/ml	WEISS and SACHS (1978)
Mitomycin C	0.5 µg/ml	LOTEM and SACHS (1975b)
Nitrosoguanidine	NR	SACHS (1978)
TPA	1 µg/ml	LOTEM and SACHS (1979)
Thymidine	100 µg/ml	LOTEM and SACHS (1975b)
Tunicamycin	0.1–1 µg/ml	NAKAYASU et al. (1980)

[a] Various M-1 derived clones assessed for maturation in various ways

can occasionally function as inducers in some erythroleukemia cell lines and inhibitors in other erythroleukemia cell lines. This is true of 12-O-tetradecanoyl-phorbol-13-acetate (TPA), an inducer of differentiation of Rauscher-transformed murine erythroleukemia cells (MIAO et al., 1978) and an inhibitor of differentiation of Friend virus-transformed erythroleukemia cells (ROVERA et al., 1977; YAMASAKI et al., 1977). BuDR, which is able to induce some clones (ADESNIK and SNITKIN, 1978) but not others of Friend virus-transformed erythroleukemia cell lines, inhibits differentiation in most of them (SCHER et al., 1973; BICK and CULLEN, 1978).

Inducers of differentiation of mouse myeloid leukemic cells have been discussed and reviewed by SACHS (1978). The inducers and their active dose range are reported in Table 2. Terminal differentiation of the mouse leukemia cells is represented

Table 3. Inducers of human myeloid leukemia cells

Inducer	Concentration	Target cell	Mature cell	Reference
Acetamide	150 mM	HL60	Metamyelocytes, bands, polys	COLLINS et al. (1978)
Actinomycin D	1 ng/ml		Metamyelocytes, bands, polys	LOTEM and SACHS (1979)
Butyric Acid	0.6 mM	HL60	Metamyelocytes, bands	COLLINS et al. (1978)
Dimethyl-formamide	60 mM	HL60	Metamyelocytes, bands, polys	COLLINS et al. (1978)
Dimethyl sulfoxide	1.25%	HL60	Metamyelocytes, bands, polys	COLLINS et al. (1978)
N-methyl-acetamide	20 mM	HL60	Metamyelocytes, bands, polys	COLLINS et al. (1978)
N-methyl-formamide	150 mM	HL60	Metamyelocytes, bands, polys	COLLINS et al. (1978)
1-Methyl-2-piperidone	4 mM	HL60	Metamyelocytes, bands, polys	COLLINS et al. (1978)
Piperidone	37.5 mM	HL60	Metamyelocytes, bands	COLLINS et al. (1978)
Retinoic Acid	1 μM	HL60	Granulocytes	BREITMAN et al. (1980)
TPA	8×10^{-10} M 1.6×10^{-11} M	HL60	Myelocytes, metamyelocytes	HUBERMAN and CALLAHAN (1979)
TPA	1.6×10^{-8} M	HL60	Macrophages	ROVERA et al. (1979a, b)
TPA	1 μg/ml	7-M18	Macrophages	LOTEM and SACHS (1979)
TPA	10^{-7} M	KG 1	Macrophages	KOEFFLER et al. (1979)
TPA	0.8 nM	HL60	Macrophages	GILBERT et al. (1979)
6-Thioguanine	5×10^{-7} M	HL60	Myeloid	PAPAC et al. (1979)
Triethylene glycol	100 mM	HL60	Metamyelocytes, bands, polys	COLLINS et al. (1978)
Tunicamycin	0.1–1 μg/ml	HL60	Myelocytes	NAKAYASU et al. (1980)

by formation of populations of macrophages and granulocytes. Markers of differentiation that are expressed during induction of myeloid cells, apart from granulocytic or macrophagic morphology, include the presence of C 3 receptors, of γ-Fc receptors, immunophagocytosis, and lysozyme. Again, as in the Friend erythroleukemia system, there is a wide range of inducer compounds including steroids, peptides, lectins, polycyclic hydrocarbons, inhibitors of RNA synthesis and DNA replication, some polar-planar compounds and phorbol diesters, and again, not all the clones of myeloid leukemia respond to the inducers with development of all the markers. SACHS and co-workers (1978) have shown that the program of terminal differentiation in the myeloid series involves several independent steps which can be independently expressed in some of the clones treated with particular inducers.

Table 3 lists the inducers of myeloid differentiation of human leukemic cells in culture. These inducers have been tested predominantly in the human promyelocytic leukemia cell line HL 60 developed by COLLINS et al. (1978). Terminal differentiation of HL 60 cells results in the formation of metamyelocytes and granulocytes. The most active inducers are DMSO, dimethylformamide, butyric acid, HMBA, and hypoxanthine (GALLAGHER et al., 1979). Retinoic acid has also been recently reported to induce granulocytes (SELONICK et al., 1980). NAKAYASU et al. (1980) have reported that tunicamycin can induce HL 60 to become myelocytes. However, in these studies no good markers of the myelocytes were used and the conclusion of myelocytic differentiation was based purely on morphological grounds. The phorbol diester TPA, on the other hand, characteristically induces differentiation of human HL 60 and human cell line KGl into macrophages without production of myeloid elements (ROVERA et al., 1979 a, b; KOEFFLER et al., 1979; LOTEM and SACHS, 1979; GILBERT et al., 1979). However, HUBERMAN and CALLAHAN (1979) reported that TPA at a concentration of about 3 orders of magnitude lower (10^{-12}) than the one reported by the other investigators may induce granulocytic differentiation. These data have not been confirmed at the present time. As in the case of the Friend erythroleukemia system and that of the mouse myeloid leukemia system, all of the inducers of myeloid differentiation in human cells, with the possible exception of TPA, are unable to recruit 100% of the treated population to differentiate. ROVERA et al. (1979 b, 1980) have reported that contact between TPA and the target HL 60 cells for even a short period of time will induce these cells to differentiate into nonproliferating macrophages, with no proliferating cells remaining. The different markers of differentiation, however, were not expressed in 100% of the TPA-treated cells: even when all of the cells were rendered unable to proliferate, a small fraction (2%) failed to express the characteristic macrophagic marker enzyme, alpha naphthyl acetate esterase, and less than 50% of the cells developed phagocytosis and complement receptors.

Differentiation of the human erythroid leukemic cell line, K 562, can be induced by butyric acid and hemin (ANDERSSON et al., 1979 a; RUTHERFORD et al., 1979), whereas this line has been found to be unresponsive to DMSO, hypoxanthine, and HMBA, which are efficient inducers of erythroid differentiation in the Friend erythroleukemia cells. The induced differentiation of these cells is highly atypical, and the treated cells do not bear a morphological resemblance to orthochromatic erythroblasts (ANDERSSON et al., 1979 a, b). It is not yet clear whether hemin induces the complete program of terminal differentiation in these cells, or as has been found in mouse erythroleukemia cells, only the synthesis of hemoglobin with no arrest of cellular proliferation. The effect of TPA on the K 562 cells is presently the subject of debate. HUBERMAN and CALLAHAN (1979) reported that TPA induces K 562 cells to differentiate along the myeloid lineage, develop phagocytosis and show morphology of mature myeloid cells. However, KOEFFLER and coworkers (1979) have reported that K 562 cells are unresponsive to the effect of TPA. We have observed (unpublished data) that the K 562 cells treated with TPA do develop phagocytic activity but not the morphological markers of myeloid differentiation. Spontaneous erythroid differentiation, present in low levels in these cells, completely disappears after TPA treatment. Associated with this phenomenon is the development in a minority of cells of the enzyme alpha naphthyl acetate esterase but not that of lysozyme (CIOE et al., unpublished data). It is possible that variation in the effects of TPA on these cells is due to the great heterogeneity of these cell lines which have been passed through several laboratories in the world (for discussion, see HARRISON, 1979).

The data available on inducers of differentiation of human leukemic cells are still relatively scanty, and it is likely that further studies will show that other inducers of differentiation, already known to be active either in the Friend erythroleukemia cell system or in the mouse myeloid leukemia cell system will also be active in the human myeloid leukemia cells. Some of the steroids that are known to induce differentiation of the mouse myeloid leukemia cells, however, have been reported by GALLAGHER et al. (1979) to be ineffective in HL 60 cells. The critical question of which of any of these inducers are active on leukemic cells freshly obtained from experimental animals or patients remains unanswered at present. PEGORARO and co-workers (1980) have reported the effect of one inducer, TPA, on human leukemic cells freshly obtained from peripheral blood or bone marrow. In the great majority of myeloid and myelomonocytic leukemias, TPA does induce differentiation of the cells into macrophages, while it is inactive in lymphoid leukemias. There are no convincing studies of the activity of the inducers on normal human or mouse myeloid stem cells or progenitor cells. A few recent papers suggest that in the mouse, TPA acts on myeloid stem cell precursors selectively, inducing the formation of macrophage colonies rather than the usual mixed macrophage granulocyte colonies (STUART and HAMILTON, 1979; GREENBERGER, 1979; LOTEM and SACHS, 1979).

D. Mechanism of Action of Inducers of Differentiation

The mechanism of action of all the inducers of differentiation, both in myeloid and erythroid leukemia, is not well understood, though several hypotheses have been advanced. One theory suggests that inducers of differentiation cause a production of physiological inducers by the cells which in turn cause the cells to differentiate. The second theory proposes that the inducers of differentiation, possibly by acting at the level of the membranes, increase the sensitivity of the cells to physiological inducers already present among the serum factors. The third theory hypothesizes that some of these inducers act on DNA and interfere specifically with the expression of groups of genes.

Some evidence supporting the first theory is given by SACHS and co-workers (for review, see SACHS, 1978). WEISS and SACHS (1978) have reported that the lipid A portion of the lipopolysaccharide inducer indirectly induces differentiation of clones of mouse myeloid leukemia cells by inducing them to produce their differentiation-inducing polypeptide MGI. The same mechanism has been invoked in the induced differentiation by TPA of human and mouse myeloid leukemia cells (LOTEM and SACHS, 1979). Again in this case, MGI would be the real inducer of differentiation. This interpretation does not, however, explain why only macrophages rather than mixed colonies are produced.

The second theory is supported by studies in the mouse erythroid and myeloid leukemia systems. In the erythroleukemia system the continued presence of the inducer is often required in order to achieve differentiation. It is therefore difficult to determine sensitization by pretreatment with and then removal of the inducer.

Induction of differentiation of Friend erythroleukemia cells by ouabain, an inhibitor of the membrane located sodium and potassium dependent ATPase, and by high potassium/low sodium medium alone (Mager et al., 1979) supports the theory of inducer activity directed at the membrane or affecting membrane permeability. The physical and chemical characteristics of several of the inducers which contain highly hydrophobic groups also supports the theory. Further, in the mouse and human myeloid leukemia systems, Nakayasu et al. (1980) showed that tunicamycin, an inhibitor of protein glycosylation, induced differentiation.

Lotem and Sachs (1979) have also suggested that TPA could increase the sensitivity of the cells to exogenous MGI present in the serum. Okabe et al. (1979) showed that inducers of differentiation such as actinomycin D and DMSO sensitize cells to the physiological polypeptide inducers. Actinomycin D alone would not induce differentiation in the treated mouse myeloid leukemia cells but these previously resistant cells would become sensitive after actinomycin D treatment to inducers such as the protein factor present in ascitic fluid.

The third theory, a direct action on DNA, is supported by several authors. Scher et al. (1978) and Terada et al. (1978c) have suggested that DNA can be selectively damaged by treatment with some inducers of differentiation and this damage, as well as the damage induced by UV and X-radiation, could be responsible for the process of differentiation. Inhibitors of DNA replication and RNA synthesis are also possible inducers of differentiation. The process would then probably involve an alteration in the proliferative rate of the cells, rather than the expression of specific genes involved in the process of differentiation. There is presently no evidence that specific gene groups are activated, even though very early studies on inducers like DMSO and other polar-planar compounds suggested this. Genetic studies have been done in the attempt to isolate clones of Friend erythroleukemia cells that show patterns of cross-sensitivity or cross-resistance to the various inducers of differentiation (Paul and Hickey, 1974; Rovera and Bonaiuto, 1976; Harrison et al., 1978; Ohta et al., 1976; Gusella and Housman, 1976; Ikawa et al., 1976a, b; Preisler et al., 1976c). However, several of these variant clones show very high reversion frequencies. While Harrison et al. (1978) were unable to show characteristic patterns of cross-sensitivity or cross-resistance, Rovera and Surrey (1978) reported that most of the clones resistant to DMSO induction were also resistant to a large number of other chemical compounds but not to butyric acid or hemin. On this basis, they proposed three different classes of inducers: Class A contained the most common inducers, Class B was represented by butyric acid, and Class C, by hemin.

Gazitt and Friend (1980) support these sub-divisions in the sense that the inducers grouped in Class A on a genetic basis were all found to stimulate ornithinedecarboxylase (ODC) activity and were inhibited by DMSO and TPA. Inducers in Class B, such as sodium butyrate, actinomycin D and the aminonucleoside puromycin, had little or no stimulatory effect on ODC and were inhibited only by BuDR. Rovera and Bonaiuto (1976) and Leder and Leder (1975) also recognized that butyric acid and DMSO act in different ways. Though this information still does not explain why and how different inducers of differentiation act on the target cells, it does suggest that different inducers may utilize common metabolic pathways.

E. Clonal Analysis of the Induction of Differentiation

The characteristics of the process of induction of differentiation have been studied at the clonal level in great detail in the Friend erythroleukemia system, but only a few inducers of differentiation have been examined (DMSO, HMBA, and butyric acid). In the first hours of the induction process, increased levels of spectrin synthesis (EISEN et al., 1977) and the appearance of a nuclear protein IP 25 (KEPPEL et al., 1977) are seen. These changes are reversible if the inducer is removed shortly thereafter. After longer exposure to the inducer, heme synthesis (PREISLER and GILADI, 1975), globin synthesis (BOYER et al., 1972) and the arrest of cellular proliferation occur. These changes are irreversible. One of the inducing agents, hemin, has a rather unique effect in inducing differentiation in the Friend erythroleukemia cells. It has been reported that treatment of the cells with hemin does not cause an arrest of cellular proliferation (HOUSMAN et al., 1978; DABNEY and BEAUDET, 1977) despite the fact that hemin induces synthesis of hemoglobin (ROSS and SAUTNER, 1976). Therefore these data suggest that hemoglobin synthesis and arrest of cellular proliferation can be uncoupled.

Studies of cell committment to differentiation have been done by cloning the Friend cells after treatment with the inducer for different lengths of time to determine whether or not differentiation occurred. GUSELLA et al. (1976) have shown that cells from a mass culture treated with the inducer for an adequate length of time give rise to colonies of differentiated progeny when subsequently cloned in the absence of the inducer. There was a latent period of 12–18 h following addition of the inducer before terminally differentiated (committed) cells appeared. However the proportion of committed cells then rose rapidly to almost 90% if the inducer was kept in the cultures for more than 48 h. It appears that the kinetics of committment are most consistent with a stochastic model. ROVERA and BONAIUTO (1976) have shown at the clonal level that there is always a fraction of leukemic cells that, despite the continuous presence of the inducer, will not differentiate but continue to proliferate. This fraction has been calculated to be about 1 in 10,000 cells if DMSO is continuously present over a period of at least 12 days. Essentially the same data have been reported for HMBA treatment of cells. HMBA is a more efficient inducer of differentiation of Friend erythroleukemia cells than DMSO, and about 99% of cells can be induced to differentiate within 4 days after treatment (OHTA et al., 1976). Again, the kinetics of commitment are compatible with the stochastic model (FIBACH et al., 1977).

The induction of differentiation of human leukemic cells (HL 60) in the presence of TPA has been reported by ROVERA et al. (1979 b, 1980) to have quite different characteristics. In contrast with other inducers of differentiation of Friend erythroleukemia cells, TPA does not need an extensive period of contact with the cells in order to induce differentiation in the total population. A contact period of a short as 5 min with a relative high concentration of the inducer (10^{-6} M) is adequate. The cells treated with the inducers either become adherent macrophage-like cells or they lyse in suspension, but survivors among 10^8 cells cannot be isolated.

Since it has been shown by ROVERA et al. (1980) that 50% of the cellular population treated with TPA does not undergo DNA synthesis before differentiating, a clonal colony assay of the type described to study commitment for the Friend ery-throleukemia cells cannot be done, nor is it necessary. One could, of course, also interpret the kinetics of induction by TPA as a probabilistic system in which the probability of the leukemic cells differentiating in the presence of the drug is about 100% after a few minutes of contact with the drug. The chance of not differentiat-ing could be theoretically explained by the need of a stem cell population to survive even in the presence of the most intensive stimuli. As discussed by HOUSMAN et al. (1978) in fact, commitment steps in a probabilistic manner are not restricted to leu-kemic cells but have also been shown in a variety of normal cellular systems.

F. Inducers of Leukemic Cell Differentiation and the Cell Cycle

Treatment of undifferentiated cycling cells with inducers results in a withdrawal of the cells into a G_0–G_1 resting phase. GUSELLA and HOUSMAN (1976) have shown by clonal analysis that the colonies formed after treatment of Friend erythroleu-kemia cells with DMSO have a limited size, compatible with a maximum of 4–5 cell divisions which occur after addition of the inducer. At that point, the cells are incapable of further proliferation. Essentially the same information was given by FIBACH et al. (1977). ROVERA et al. (1980) have shown that after TPA induction of HL60 cells into macrophages, the cells predominantly arrest in the G_1 phase of the cell cycle, as shown by quantitating cellular DNA with a scanning microden-sitometer.

   Whether or not DNA replication is required in the differentiation of leukemia cells treated with an inducer has been the subject of several investigations using the murine erythroleukemia cell system. The studies of McCLINTOCK and PAPACON-STANTINOU (1974), LEVY et al. (1975) and WIENS et al. (1976) all suggested that DNA synthesis was required for induced differentiation, whereas LEDER et al. (1975) suggested that these cells could be induced to differentiate with butyric acid in the absence of DNA synthesis. More recently, LEVENSON et al. (1980) have de-termined the relationship between DNA synthesis and differentiation of murine erythroleukemia cells induced by DMSO by blocking the cells with fluoro-deoxyuridine and thymidine. They concluded that DNA synthesis was not required for murine erythroleukemia cells to be committed to differentiate, and suggested that cellular cytotoxicity of the cell cycle inhibitors may have accounted for the re-sults of earlier studies. ROVERA et al. (1980) have shown that DNA synthesis is not required for TPA-treated HL60 cells to differentiate into macrophages. In this case, no cell cycle inhibitors were used. In both the murine erythroleukemia system and in the TPA-induced HL60 system, preliminary evidence has been given that commitment to differentiation takes place in the late G_1 phase of the cell cycle (GELLER et al., 1978; GAMBARI et al., 1978; ROVERA et al., 1980).

G. Inhibitors of Differentiation

Inhibitors of the process of differentiation of leukemic cells have been identified and studied with the hope of utilizing them as probes to dissect the program of dif-ferentiation and possibly to understand more about the mechanism of the differ-

Table 4. Inhibitors of spontaneous or induced erythroid differentiation

Inhibitor	Concentration	Percent inhibition	Reference
Aldosterone	$10 \ \mu M$	62	Scher et al. (1978)
Bromodeoxyuridine	$10^{-5} \ M$	70	Scher et al. (1973)
	$2-6 \times 10^{-6} \ M$	50–90	Bick and Cullen (1978)
Corticosterone	$10^{-8} \ M$	50	Lo et al. (1978)
	$1-10 \ \mu M$	68–75	Scher et al. (1978)
Cortisol	$10^{-8} \ M$	50	Lo et al. (1978)
Deoxycorticosterone	$10^{-7} \ M$	40	Lo et al. (1978)
	$1 \ \mu M$	40	Scher et al. (1978)
Dexamethasone	$10^{-8} \ M$	90	Lo et al. (1978)
	$0.1-1 \ \mu M$	97	Scher et al. (1978)
Etiocholanolone	$10 \ \mu M$	29	Scher et al. (1978)
Hydrocortisone	$10^{-5} \ M$	30	Lo et al. (1978)
	$1 \ \mu M$	91	Scher et al. (1978)
	$10^{-6}-10^{-7} \ M$	98	Santoro et al. (1978)
Hypoxanthine	$20 \ \mu M$	NR	Lacour et al. (1980)
Indomethacin	$10 \ \mu M$	18	Scher et al. (1978)
Ingenol dibenzoate	$3-10$ ng/ml	50	Yamasaki et al. (1977)
Interferon	$500-2,000$ U	55–80	Rossi et al. (1977a, b)
Mezerein	$10-30$ ng/ml	50	Yamasaki et al. (1977)
Phorbol 12, 13 Dibenzoate	$1.6 \times 10^{-7} \ M$	70–80	Rovera et al. (1977)
Phorbol 12, 13 Didecanoate	$1.6 \times 10^{-7} \ M$	70–90	Rovera et al. (1977)
	$3-10$ ng/ml	50	Yamasaki et al. (1977)
Progesterone	$10^{-5} \ M$	90	Lo et al. (1978)
	$1 \ \mu M$	10	Scher et al. (1978)
12-0-Tetradecanoyl-phorbol-13-acetate	$1.6 \times 10^{-7} \ M$	70–90	Rovera et al. (1977)
	12 ng/ml	50	Yamasaki et al. (1977)

entiation block typical of leukemic cells. The inhibitors of differentiation of Friend erythroleukemia cells, in which most of the studies have been done, are listed in Table 4.

Lo et al. (1978) and Scher et al. (1978) have reported that corticosteroids are effective in blocking the differentiation of murine erythroleukemia cells, inhibiting both heme and globin synthesis. Lo et al. (1978) have shown that dexamethasone at a concentration of $10^{-9} \ M$ interferes with the binding of the globin mRNA to the polysomes, whereas at a concentration of $10^{-8} \ M$, it causes a decrease in the rate of globin mRNA accumulation in the cytoplasm. Tsiftsoglou et al. (1979) have shown that the inhibition of erythroid differentiation of murine erythroleukemia cells by dexamethasone is due to inhibition of induced commitment to the differentiative program. The results presented suggest that dexamethasone exerts an inhibitory effect by blocking commitment to terminal erythroid differentiation. Thus, it appears that the final effects of both phorbol diesters and steroids on differentiating cells are similar.

Rossi and co-workers (1977 a, b) have reported that interferon inhibits DMSO-induced erythroid differentiation of Friend erythroleukemia cells. This inhibition of differentiation could be fully reversed by washing away the interferon. Along with the inhibition of differentiation was an inhibition of growth (MATARESE and ROSSI, 1977). In both of these studies, interferon was used in the amount of several hundred units/ml.

TPA inhibits both induced and spontaneous differentiation of Friend erythroleukemia cells (YAMASAKI et al., 1977; ROVERA et al., 1977). As shown by FIBACH et al. (1979), TPA inhibits the expression of all the inducible erythroid characteristics, including commitment to terminal cell division, accumulation of globin mRNA and synthesis of globins, spectrin, heme synthetic enzymes, and heme. FIBACH et al. (1979) have also postulated that TPA produces its inhibitory effect at the level of the plasma membranes. More details about the effects of TPA on the process of terminal differentiation are given in the review by ABRAHM and ROVERA (1980).

BuDR affects differentiation of both malignant (SILAGI and BRUCE, 1970) and normal (STOCKDALE et al., 1964) cells, and inhibits differentiation of Friend erythroleukemia cells when applied at a concentration of approximately 10^{-5} M. The inhibitory effect on these cells extends to protein and globin mRNA synthesis without drastically altering cell viability. Inhibition of induced differentiation by BuDR can be reversed with thymidine or deoxycytidine but only if they are added within the first 6–10 h of cell growth in the presence of BuDR (BICK and CULLEN, 1978). The mechanism of BuDR's inhibition of differentiation, however, is not known. Results of PREISLER and co-workers (1973) and BICK and CULLEN, 1978) are consistent with a DNA-mediated mode of action.

LACOUR et al. (1980) have reported that hypoxanthine at a very low concentration (20 mM) acts as an inhibitor of DMSO-induced differentiation.

It is interesting that among the inhibitors of differentiation reported in Table 4, TPA, BuDR, and hypoxanthine also function as inducers of differentiation. Usually the concentrations for the inducing effect and the inhibitory effect are different. For example, TPA can function as an inducer (MIAO et al., 1978) at a concentration of 1×10^{-5} M, but as an inhibitor at a concentration of 1×10^{-7} M in Friend erythroleukemia cells. However, TPA is an inducer at 1×10^{-8} M in human myeloid leukemia cells and mouse myeloid leukemia cells (ROVERA et al., 1979 a, b; LOTEM and SACHS, 1979). BuDR can induce some clones of Friend erythroleukemia cells to differentiate and can inhibit most of the other clones. Hypoxanthine, is an inducer of differentiation at a concentration of 1–6 mM (GUSELLA and HOUSMAN, 1976; LACOUR et al., 1980) and an inhibitor at a concentration of 20 μM.

H. Inducers of Differentiation and Treatment of Leukemias In Vivo

Inducers of differentiation of leukemic cells have been used in the attempt to treat experimental leukemias by two different approaches. The first consisted of pretreating the cells in vitro with the drugs inducing differentiation and then introducing into the animal the differentiating leukemic cells and determining whether the differentiated population could still produce transplantable tumors. An example of such an approach was reported by FRIEND et al. (1971). The cells were treated for four days with DMSO and more than 80% of the cell population was differentiated. However, since DMSO does not induce complete differentiation of murine erythroleukemia cells it was not surprising to find that the treated population was still tumorigenic though tumor onset was delayed. The second approach was to directly treat the experimental animals carrying a transplantable leukemia with different inducers of differentiation. PREISLER et al. (1976a) have reported that DBA/2 mice carrying Friend murine erythroleukemia were treated with either *N*-methylacetamide, dimethylacetamide or tetramethylurea. Survival was only occasionally prolonged, but the agents significantly inhibit leukemic cell proliferation in the spleen and to a lesser extent in the marrow. The same results were reported by PREISLER et al. (1976a) when instead of murine erythroleukemia cells, mouse myeloid leukemia cells were used. In this study it was also reported that the livers of treated mice were frequently enlarged and yellow, and preliminary pathological studies suggested hepatic necrosis. Inability to treat murine erythroleukemia cells in vivo with inducers of differentiation was also discussed by MARKS and RIFKIND (1978) and could be explained by the fact that it is difficult to achieve an efficient dose in vivo and that induction of differentiation of the leukemic cells is never 100%. It is noteworthy, however, that in the murine erythroleukemia cell system, the Friend erythroleukemia virus could superinfect the animals despite the treatment with inducers of differentiation. Therefore, this system does not appear to be ideal for in vivo studies.

In addition to the studies of PREISLER et al. (1976a) using mouse myeloid leukemia cells, HONMA et al. (1979) reported that treatment of mice injected with sensitive cells of clone Ml of myeloid leukemia with inducers of differentiation was only successful in prolonging the survival time of the tumor-bearing animals. HONMA et al. (1979) also showed that there was an increased differentiation of leukemic cells in millipore chambers implanted in those animals. However, it was apparent from their data that a fraction of cells always escaped the inducers of differentiation and that this fraction was probably responsible for the progress of the disease.

Further studies to delineate the role of in vivo use of chemical inducers of differentiation, or in vitro treatment of cells that will then be infused back into the animal or man face several problems. The inducers' activities in in vitro model cell lines may not be a reliable predictor of activity in various individual patients' leukemic cells. TPA's effects have been mentioned, and more information on inducers with similar non-selective abilities, but with less toxicity, would be useful. The ideal inducer would act at a relatively low concentration without great toxicity, and should be able to induce more than 99% of the cell population, even with a very short contact period. It should be specific for particular lineages, and selective enough to spare early normal stem cells that would recolonize the marrow. It is interesting that some of the chemicals, e.g. 6-thioguanine and actinomycin D, and some of the physical agents, e.g. radiation, which can induce some differentiation in clones of leukemic cells (PAPAC et al., 1979; TERADA et al., 1978c; SCHER et al., 1978) have already been used in leukemic chemotherapy, albeit as inhibitors of cellular proliferation.

An inducer that could increase the sensitivity of leukemic cells to physiologic stimuli, or that could itself induce differentiation would be a welcome addition to the leukemia therapies presently available.

Acknowledgement. The research quoted, which was done in our laboratories, was supported by grants CA 24273, CA 10815, CA 25875 and HL 07439 from the NIH. We thank Marina Hoffman for editorial assistance.

References

Abrahm, J., Rovera, G.: The effect of tumor promoting phorbol diesters on terminal differentiation of cells in culture. Mol. Cell. Biochem. *31*, 165–175 (1980)

Adesnik, M., Snitkin, H.: Induction of erythroid differentiation in Friend leukemia cells by bromodeoxyuridine. J. Cell. Physiol. *95*, 307–311 (1978)

Andersson, L.C., Jokinen, M., Gahmberg, C.G.: Induction of erythroid differentiation in the human leukemia cell line K 562. Nature *278*, 364–365 (1979 a)

Andersson, L.C., Nilsson, K., Gahmberg, C.G.: K 562 a human erythroleukemic cell line. Int. J. Cancer *23*, 143–147 (1979 b)

Bernstein, A., Hunt, D.M., Crichley, V., Mak, T.W.: Induction by ouabain of hemoglobin synthesis in cultured Friend erythroleukemia cells. Cell *9*, 375–381 (1976)

Bick, M.D., Cullen, B.R.: Bromodeoxyuridine inhibition of Friend leukemia cell induction by butyric acid: time course of inhibition, reversal and effect of other base analogs. Somatic Cell Gen. *2*, 545–558 (1978)

Boyer, S.H., Woo, K.D., Noyes, A.N., Young, R., Scher, W., Friend, C., Preisler, H.D., Bank, A.: Hemoglobin biosynthesis in murine virus induced leukemic cells in vitro: structure and amounts of globin chains produced. Blood *40*, 823–835 (1972)

Breitman, T.R., Selonick, S.E., Collins, S.I.: Induction of differentiation of the human promyelocytic cell line (HL 60) by retinoic acid. Proc. Natl. Acad. Sci. U.S.A. *77*, 2936–2940 (1980)

Christman, J.K., Price, P., Pedrinan, L., Acs, G.: Correlation between hypomethylation of dNA and expression of globin genes in Friend erythroleukemia cells. Eur. J. Biochem. *81*, 53–61 (1977)

Cline, M.J., Golde, D.W.: Controlling the production of blood cells. Blood *53*, 157–165 (1979)

Collins, S.J., Gallo, R.C., Gallagher, R.E.: Continuous growth and differentiation of human myeloid leukemic cells in suspension culture. Nature *270*, 347–349 (1977)

Collins, S.J., Ruscetti, F.W., Gallagher, R.E., Gallo, R.C.: Terminal differentiation of human promyelocytic leukemia cells induced by dimethyl sulfoxide and other polar compounds. Proc. Natl. Acad. Sci. U.S.A. *75*, 2458–2462 (1978)

Dabney, B.J., Beaudet, A.L.: Increase in globin chains and globin mRNA in erythroleukemia cells in response to hemin. Arch. Biochem. Biophys. *179*, 106–112 (1977)

Dube, S.K., Gaedick, G., Kluge, N., Weismann, B.J., Melderis, H., Steinheider, G., Crozier, T., Beckmann, H., Ostertag, W.: Hemoglobin synthesizing mouse and human erythroleukemic cell lines as model systems for the study of differentiation and control of gene expression. Proc. 4th Int. Symp. of the Princess Takamatsu Cancer Research Fund, Tokyo, University Park Press, 99–132 (1974)

Ebert, P.S., Malinin, G.I.: Induction of erythroid differentiation in Friend murine erythroleukemic cells by inorganic selenium compounds. Biochem. Biophys. Res. Commun. *86*, 340–349 (1979)

Ebert, P.S., Wars, I., Buell, D.N.: Erythroid differentiation in cultured Friend leukemia cells treated with metabolic inhibitors. Cancer Res. *36*, 1809–1813 (1976)

Eisen, H., Bach, R., Emery, R.: Induction of spectrin in erythroleukemia cells transformed by Friend virus. Proc. Natl. Acad. Sci. U.S.A. *74*, 3898–3902 (1977)

Fialkow, P.J., Denman, A.M., Jacobson, R.J., Lowenthal, M.N.: Chronic myeloid leukemia. Origin of some lymphocytes from leukemic stem cells. J. Clin. Invest. *62*, 815–823 (1978)

Fibach, E., Landau, T., Sachs, L.: Normal differentiation of myeloid leukemic cells induced by a differentiation inducing protein. Nature New Biol. *23*, 276–278 (1972)

Fibach, E., Reuben, R.C., Rifkind, R.A., Marks, P.A.: Effect of hexamethylene bisaceta-mide on the commitment to differentiation of murine erythroleukemia cells. Cancer Res. *37*, 440–444 (1977)

Fibach, E., Gambari, R., Shaw, P.A., Maniatis, G., Reuben, R.C., Sassa, S., Rifkind, R.A., Marks, P.A.: Tumor promoter-mediated inhibition of cell differentiation: suppression of expression of erythroid functions in murine erythroleukemia cells. Proc. Natl. Acad. Sci. U.S.A. *76*, 1906–1910 (1979)

Friend, C.: Differentiation and neoplasia. Berlin, Heidelberg, New York: Springer 1980 (in press)

Friend, C., Scher, W., Holland, J.G., Sato, T.: Hemoglobin synthesis in murine virus in-duced leukemia cells in vitro: stimulation of erythroid differentiation by DMSO. Proc. Natl. Acad. Sci. U.S.A. *68*, 378–382 (1971)

Gallagher, R., Collins, S., Trujillo, J., McCredie, K., Ahearn, M., Tsai, S., Metzgar, R., Au-lakh, G., Ting, R., Ruscetti, F., Gallo, R.: Characterization of the continuous differ-entiating myeloid cell line (HL-60) from a patient with acute promyelocytic leukemia. Blood *54*, 713–733 (1979)

Gambari, R., Terada, M., Bank, A., Rifkind, R.A., Marks, P.A.: Synthesis of globin mRNA in relation to the cell cycle during induced murine erythroleukemia differentiation. Proc. Natl. Acad. Sci. U.S.A. *75*, 3801–3804 (1978)

Gazitt, Y., Friend, C.: Polyamine biosynthesis enzymes in the induction and inhibition of FL cell differentiation. Cancer Res. (1980)

Geller, R., Levenson, R., Housman, D.: Significance of the cell cycle in commitment of mu-rine erythroleukemia cells to erythroid differentiation. J. Cell. Physiol. *95*, 213–222 (1978)

Gilbert, H.S., Mendelsohn, N., Acs, G., Christman, J.K.: Evidence for the myelomonocytic character of uninduced HL-60 cells. Blood *54* (Suppl. 1), 170 a (1979)

Greenberger, J.S.: Phorbol myristate acetate inhibits normal hemopoiesis but stimulates leu-kemogenesis in long term bone marrow cultures. Blood *54* (Suppl. 1), 171 a (1979)

Gusella, J.F., Housman, D.: Induction of erythroid differentiation in vitro by purines and purine analogues. Cell *8*, 263–269 (1976)

Gusella, G., Geller, R., Clarke, B., Weeks, V., Housman, D.: Commitment to erythroid dif-ferentiation by Friend erythroid leukemia cells: a stochastic analysis. Cell *9*, 221–229 (1976)

Harrison, P.R.: The biology of the Friend cell. In: International review of biochemistry. Bio-chemistry of cell differentiation. Paul, J. (ed.), vol. 15, pp. 227–267. Baltimore: Univer-sity Park Press 1977

Harrison, P.R.: New human myeloid leukaemia cell line undergoes red shift. Nature *281*, 632–633 (1979)

Harrison, P.R., Rutherford, T., Conkie, D., Affara, N., Sommerville, J., Hissey, P., Paul, J.: Analysis of erythroid differentiation in Friend cells using non-induced variants. Cell *14*, 61–70 (1978)

Honma, Y., Kasukabe, T., Okabe, J., Hozumi, M.: Glucocorticoid binding and mechanisms of resistance in some clones of mouse myeloid leukemic cells resistant to induction of differentiation by dexamethasone. J. Cell. Physiol. *93*, 227–235 (1977 a)

Honma, Y., Kasukabe, T., Okabe, J., Hozumi, M.: Glucocorticoid-induced differentiation of cultured mouse myeloid leukemia cells. Gann *68*, 241–246 (1977 b)

Honma, Y., Kasukabe, T., Okabe, J., Hozumi, M.: Prolongation of survival time of mice inoculated with myeloid leukemia cells by inducers of normal differentiation. Cancer Res. *39*, 3167–3171 (1979)

Housman, D., Gusella, J., Geller, R., Levenson, R., Weil, S.: Differentiation of murine ery-throleukemia cells: the control role of the commitment event. In: Differentiation of nor-mal and neoplastic hematopoietic cells. Clarkson, B., Marks, P., Till, J. (eds.), pp. 193–207. New York: Cold Spring Harbor 1978

Huberman, E., Callahan, M.F.: Induction of terminal differentiation in human promyelo-cytic leukemia cells by tumor-promoting agents. Proc. Natl. Acad. Sci. U.S.A. *76*, 1293–1297 (1979)

Ichikawa, Y.: Differentiation of a cell line of mouse myeloid leukemia. J. Cell. Physiol. *74*, 223–234 (1969)

Ikawa, Y., Aida, M., Inoue, Y.: Isolation and characterization of high and low differentiation-inducible Friend leukemia lines. Gann *67*, 767–770 (1976a)

Ikawa, Y., Inoue, A., Aida, M., Kameji, R., Shibata, C., Sugano, H.: Phenotypic variants of differentiation-inducible Friend leukemia lines: Isolation and correlation between inducibility and virus release. Bibl. Haematol. *43*, 37–47 (1976b)

Keppel, F., Allet, B., Eisen, H.: Appearance of a chromatin protein during the erythroid differentiation of Friend virus transformed cells. Proc. Natl. Acad. Sci. U.S.A. *74*, 653–656 (1977)

Koeffler, H.P., Bar-Eli, M., Territo, M.: Heterogeneity of human myeloid leukemia cell response to phorbol diester. Blood *54*, (Suppl. 1), 174a (1979)

Krystosek, A., Sachs, L.: Control of lysozyme induction in the differentiation of myeloid leukemic cells. Cell *9*, 675–684 (1976)

Lacour, F., Harel, L., Friend, C., Huynh, T., Holland, J.G.: Induction of differentiation of murine erythroleukemia cells by aminonucleoside of puromycin and its inhibition by purines and purine derivatives. Proc. Natl. Acad. Sci. U.S.A. *77*, 2740–2742 (1980)

Leder, A., Leder, P.: Butyric acid: a potent inducer of erythroid differentiation in cultured erythroleukemia cells. Cell *5*, 319–322 (1975)

Leder, A., Orkin, S., Leder, P.: Differentiation of erythroleukemia cells in the presence of inhibitors of DNA synthesis. Science *190*, 893–894 (1975)

Levenson, R., Kernen, J., Mitrani, A., Housman, D.: DNA synthesis is not required for the commitment of murine erythroleukemia cells. Devel. Biol. *74*, 224–230 (1980)

Levy, J., Terada, M., Rifkind, R.A., Marks, P.A.: Induction of erythroid differentiation in Friend virus infected cells by dimethylsulfoxide: relationship to the cell cycle. Proc. Natl. Acad. Sci. U.S.A. *72*, 28–32 (1975)

Lo, S.C., Aft, R., Ross, F., Mueller, G.C.: Control of globin gene expression by steroid hormones in differentiating Friend leukemia cells. Cell *15*, 447–453 (1978)

Lotem, J., Sachs, L.: Different blocks in the differentiation of myeloid leukemic cells. Proc. Natl. Acad. Sci. U.S.A. *71*, 3507–3511 (1974)

Lotem, J., Sachs, L.: Induction of specific changes in the surface membrane of myeloid leukemic cells by steroid hormones. Int. J. Cancer *15*, 731–740 (1975a)

Lotem, J., Sachs, L.: Control of normal differentiation of myeloid leukemic cells. VI. Inhibition of cell multiplication and the formation of macrophages. J. Cell. Physiol. *85*, 587–596 (1975b)

Lotem, J., Sachs, L.: Regulation of normal differentiation in mouse and human myeloid leukemic cells by phorbol esters and the mechanism of tumor promotion. Proc. Natl. Acad. Sci. U.S.A. *76*, 5158–5162 (1979)

Lozzio, C.B., Lozzio, B.B.: Human chronic myelogenous leukemic cell line with positive Philadelphia chromosome. Blood *45*, 321–334 (1975)

Mager, D.L., MacDonald, M.E., Bernstein, A.: Growth in high-K$^+$ medium induces Friend cell differentiation. Devel. Biol. *70*, 268–273 (1979)

Marks, P.A., Rifkind, R.A.: Erythroleukemic differentiation. Ann. Rev. Biochem. *47*, 419–448 (1978)

Matarese, G.P., Rossi, G.B.: Effect of interferon on growth and division cycle of Friend erythroleukemic murine cells in vitro. J. Cell Biol. *75*, 344–354 (1977)

McClintock, P.R., Papaconstantinou, J.: Regulation of hemoglobin synthesis in a murine erythroblastic leukemic cell: the requirement for replication to induce hemoglobin synthesis. Proc. Natl. Acad. Sci. U.S.A. *71*, 4551–4555 (1974)

Metcalf, D.: Haemopoietic colonies. Recent Res. Cancer Res. *61*, 1–213 (1977)

Metcalf, D., Moore, M.A.S.: Haemopoietic cells. Amsterdam: North Holland 1971

Miao, R.M., Fieldsteel, A.H.: Regulation of erythroid differentiation: characterization of Rauscher and Friend virus-transformed proerythroid cells. Proc. Soc. Exp. Biol. Med. *160*, 24–27 (1979)

Miao, R.M., Fieldsteel, A.M., Fodge, D.W.: Opposing effects of tumor promoters on erythroid differentiation. Nature *274*, 271–272 (1978)

Moore, M.A.S.: Regulation of leukocyte differentiation and leukemia as a disorder of differentiation. In: Recent advances in cancer research: Cell biology and tumor virology. Gallo, R.C. (ed.). pp. 79–103. Cleveland: CRC Press 1977

Moore, M.A.S., Kurland, J., Broxmeyer, H.E.: Regulatory interactions in normal and leukemic myelopoiesis. In: Differentiation of normal and neoplastic hematopoietic cells. Clarkson, B., Marks, P. and Till, J. (eds.), pp. 393–404. New York: Cold Spring Harbor 1978

Nakayasu, M., Terada, M., Tamura, G., Sugimura, T.: Induction of differentiation of human and murine myeloid leukemia cells in culture by tunicamycin. Proc. Natl. Acad. Sci. U.S.A. 77, 400–413 (1980)

Ohta, Y., Tanaka, M., Terada, M., Miller, O.J., Bank, A., Marks, P.A., Rifkind, R.A.: Erythroid cell differentiation: murine erythroleukemia cell variant with unique pattern of induction by polar compounds. Proc. Natl. Acad. Sci. U.S.A. 73, 1232–1236 (1976)

Okabe, J., Honma, Y., Hozumi, M.: Differentiation of a resistant clone of mouse myeloid leukemia cells with dimethyl sulfoxide and ascitic fluid. Gann 68, 151–157 (1977)

Okabe, J., Honma, Y., Hayashi, M., Hozumi, M.: Actinomycin D restores in vivo sensitivity to differentiation induction of non-differentiating mouse myeloid leukemia cells. Int. J. Cancer 24, 87–91 (1979)

Papac, R.J., Browna, A.E., Sartorelli, A.C.: Induction of differentiation of human myeloid cells in vitro by 6-thioguanine. Proc. Am. Soc. Clin. Oncol. 20, 419 (1979)

Patuleia, M.C., Friend, C.: Tissue culture studies on murine virus induced leukemia cells: isolation of single cells in agar liquid medium. Cancer Res. 27, 726–730 (1967)

Paul, J., Hickey, I.: Hemoglobin synthesis in inducible, uninducible and hybrid Friend cell clones. Exp. Cell Res. 87, 20–30 (1974)

Pegoraro, L., Abrahm, J., Cooper, R.A., Levis, A., Lange, B., Meo, P., Rovera, G.: Differentiation of human leukemias in response to 12-O-tetradecanoyl-phorbol-13-acetate in vitro. Blood 55, 859–862 (1980)

Preisler, H.D., Giladi, M.: Differentiation of erythroleukemic cells in vitro: irreversible induction by dimethyl sulfoxide. J. Cell. Physiol. 85, 537–545 (1975)

Preisler, H.D., Lyman, G.: Differentiation of erythroleukemic cells in vitro: Properties of chemical inducers. Cell Differ. 4, 179–185 (1975)

Preisler, H.D., Housman, D., Scher, W., Friend, C.: Effects of 5-Bromo-2′-deoxyuridine on production of globin messenger RNA in dimethyl sulfoxide-stimulated Friend leukemia cells. Proc. Natl. Acad. Sci. U.S.A. 70, 2956–2959 (1973)

Preisler, H.D., Bjornsson, S., Mori, M., Lyman, G.H.: Inducers of Friend leukemia cell differentiation in vitro – effects of in vivo administration. Br. J. Cancer 33, 634–645 (1976a)

Preisler, H.D., Christoff, G., Taylor, E.: Cryoprotective agents as inducers of erythroleukemic cell differentiation in vitro. Blood 47, 363–368 (1976b)

Preisler, H.D., Shiraishi, Y., Mori, M., Sandberg, A.A.: Clones of Friend leukemia cells: differences in karyotypes and responsiveness to inducers of differentiation. Cell Differ. 5, 207–216 (1976c)

Quesenberry, P., Levitt, L.: Hematopoietic stem cells. New Engl. J. Med. 301, 755–760 (1979a)

Quesenberry, P., Levitt, L.: Hematopoietic stem cells. New Engl. J. Med. 301, 819–823 (1979b)

Quesenberry, P., Levitt, L.: Hematopoietic stem cells. New Engl. J. Med. 301, 868–872 (1979c)

Reuben, R.C., Wife, R.L., Breslow, R., Rifkind, R.A., Marks, P.A.: Identification of a new group of potent inducers of differentiation in murine erythroleukemia cells. Proc. Natl. Acad. Sci. U.S.A. 73, 862–866 (1976)

Reuben, R.C., Khanna, P.L., Gazitt, Y., Breslow, R., Rifkind, R.A., Marks, P.A.: Inducers of erythroleukemic differentiation. J. Biol. Chem. 253, 4214–4218 (1978)

Ross, J., Sautner, D.: Induction of globin mRNA accumulation by heme in cultured erythroleukemic cells. Cell 8, 513–520 (1976)

Rossi, G.B., Dolei, A., Cioe, L., Benedetto, A., Matarese, G.P., Belardelli, F.: Inhibition of transcription and translation of globin messenger RNA in dimethyl sulfoxide-stimulated Friend erythroleukemic cells treated with interferon. Proc. Natl. Acad. Sci. U.S.A. *74*, 2036–2040 (1977 a)

Rossi, G.B., Matarese, G.P., Grappelli, C., Belardelli, F.: Interferon inhibits dimethyl sulfoxide-induced erythroid differentiation of Friend leukaemia cells. Nature *267*, 50–52 (1977 b)

Rovera, G., Bonaiuto, J.: The phenotypes of variant clones of Friend mouse erythroleukemic cells resistant to dimethylsulfoxide. Cancer Res. *36*, 4057–4061 (1976)

Rovera, G., Surrey, S.: Isolation and characterization of a subline of Friend erythroleukemia cells that differentiate in tissue culture in the absence of inducers. Cancer Res. *37*, 4211–4219 (1977)

Rovera, G., Surrey, S.: Use of resistant or hypersensitive variant clones of Friend cells in analysis of the mode of action of inducers. Cancer Res. *38*, 3737–3744 (1978)

Rovera, G., O'Brien, T., Diamond, L.: Tumor promoters inhibit spontaneous differentiation of Friend erythroleukemia cells in culture. Proc. Natl. Acad. Sci. U.S.A. *74*, 2894–2898 (1977)

Rovera, G., O'Brien, T.G., Diamond, L.: Induction of differentiation in human promyelocytic leukemia cells by tumor promoters. Science *204*, 868–870 (1979 a)

Rovera, G., Santoli, D., Damsky, C.: Human promyelocytic leukemia cells in culture differentiate into macrophage-like cells when treated with phorbol diesters. Proc. Natl. Acad. Sci. U.S.A. *76*, 2779–2783 (1979 b)

Rovera, G., Olashaw, N., Meo, P.: Terminal differentiation in human promyelocytic leukemic cells in the absence of DNA synthesis. Nature *284*, 69–70 (1980)

Rutherford, T.R., Clegg, J.B., Weatherall, D.J.: K 562 human leukemic cells synthesize embryonic hemoglobin in response to hemin. Nature *280*, 164–165 (1979)

Sachs, L.: Control of normal cell differentiation and the phenotypic reversion of malignancy in myeloid leukemia. Nature *274*, 535–539 (1978)

Santoro, M.G., Benedetto, A., Jaffe, B.M.: Hydrocortisone inhibits DMSO-induced differentiation of Friend erythroleukemia cells. Biochem. Biophys. Res. Commun. *85*, 1510–1518 (1978)

Scher, W., Friend, C.: Breakage of DNA and alterations in folded genomes by inducers of differentiation in Friend erythroleukemic cells. Cancer Res. *38*, 841–849 (1978)

Scher, W., Preisler, H.D., Friend, C.: Hemoglobin synthesis in murine virus-induced leukemic cells in vitro. Effect of 5-bromodeoxyuridine, dimethyl formamide and dimethyl sulfoxide. J. Cell. Physiol. *81*, 63–79 (1973)

Scher, W., Tsuei, D., Sassa, S., Price, P., Gabelman, N., Friend, C.: Inhibition of DMSO-stimulated Friend cell erythrodifferentiation by hydrocortisone and other steroids. Proc. Natl. Acad. Sci. U.S.A. *75*, 3851–3855 (1978)

Selonick, S.E., Breitman, T.R., Collins, S.J.: Retinoic acid induces differentiation of the human promyelocytic leukemic cell line (HL 60). Proc. Am. Fed. Clin. Res. *180 a* (1980)

Silagi, S., Bruce, S.A.: Suppression of malignancy and differentiation in melanotic melanoma cells. Proc. Natl. Acad. Sci. U.S.A. *66*, 72–78 (1970)

Stockdale, F., Okazaki, K., Nameroff, M., Holtzer, H.: 5-bromodeoxy-uridine: effect on myogenesis in vitro. Science *146*, 533–535 (1964)

Stuart, R.K., Hamilton, J.: Tumor promoting phorbol esters stimulate hemopoietic colony formation in vitro. Blood *54* (Suppl. 1) 176 a (1979)

Sugano, H., Furusawa, M., Kawaguchi, T., Ikawa, Y.: Enhancement of erythrocytic maturation of Friend virus induced leukemic cells in vitro. Bibl. Hematol. *39*, 943–954 (1973)

Tabuse, Y., Furasawa, M., Eisen, H., Shibata, K.: Prostaglandin E_1, and inducer of erythroid differentiation of Friend erythroleukemia cells. Exp. Cell Res. *108*, 41–45 (1977)

Takahashi, E., Yamada, M., Saito, M., Kuboyama, M., Ogasa, K.: Differentiation of cultured Friend leukemic cells induced by short-chain fatty acids. Gann *66*, 577–580 (1975)

Tanaka, M., Levy, J., Terada, M., Breslow, R., Rifkind, R.A., Marks, P.A.: Induction of erythroid differentiation in murine virus infected erythroleukemia cells by highly polar compounds. Proc. Natl. Acad. Sci. U.S.A. *72*, 1003–1006 (1975)

Terada, M., Epner, E., Nudel, U., Salmon, J., Fibach, E., Rifkind, R.A., Marks, P.A.: Induction of murine erythroleukemia differentiation by Actinomycin D. Proc. Natl. Acad. Sci. U.S.A. *75*, 2795–2799 (1978a)

Terada, M., Marks, P.A., Rifkind, R.A.: Erythroid cell differentiation. Mol. Cell. Biochem. *21*, 33–41 (1978b)

Terada, M., Nudel, U., Fibach, E., Rifkind, R.A., Marks, P.A.: Changes in DNA associated with induction of erythroid differentiation by dimethyl sulfoxide in murine erythroleukemia cells. Cancer Res. *38*, 835–840 (1978c)

Terada, M., Fujiki, H., Marks, P.A., Sugimura, T.: Induction of erythroid differentiation of murine erythroleukemia cells by nicotinamide and related compounds. Proc. Natl. Acad. Sci. U.S.A. *76*, 6411–6414 (1979)

Till, J.E., McCulloch, E.A., Siminovitch, L.: A stochastic model of stem cell proliferation based on the growth of spleen colony-forming cells. Proc. Natl. Acad. Sci. U.S.A. *51*, 29–36 (1964)

Tomida, M., Yamamoto, Y., Hozumi, M.: Induction by synthetic polyribonucleotide Poly (I) of differentiation of cultured mouse myeloid leukemia cells. Cell Differ. 7, 305–312 (1978)

Tsiftsoglou, A.S., Gusella, J.F., Volloch, V., Housman, D.E.: Inhibition by dexamethasone of commitment to erythroid differentiation in murine erythroleukemia cells. Cancer Res. *39*, 3849–3855 (1979)

Weiss, B., Sachs, L.: Indirect induction of differentiation in myeloid leukemic cells by lipid A. Proc. Natl. Acad. Sci. U.S.A. *75*, 1374–1378 (1978)

Whang, J., Frei, E. III, Tjio, J.H., Carbone, P.C., Brecher, G.: The distribution of the Philadelphia chromosome in patients with chronic myeloid leukemia. Blood *22*, 664–673 (1963)

Wiens, A.W., McClintock, P.R., Papaconstantinou, J.: The dependence of erythroid differentiation on cell replication in dimethyl sulfoxide treated Friend leukemia virus infected cells. Biochem. Biophys. Res. Commun. *70*, 824–831 (1976)

Yamasaki, H., Fibach, E., Weinstein, I.B., Nudel, U., Rifkind, R.A., Marks, P.A.: Tumor promoters inhibit spontaneous and induced differentiation of murine erythroleukemia cells in culture. Proc. Natl. Acad. Sci. U.S.A. *74*, 3451–3455 (1977)

CHAPTER 15

Angiogenesis Factor(s)

P. M. GULLINO

A. Introduction

New formation of vessels occurs during embryonal development and throughout
the lifespan of an organism following repair processes or under certain pathologic
conditions. In particular, neoplastic cell populations can grow to form a clinically
evident tumor only if the host produces a vascular network sufficient to sustain tumor growth (FOLKMAN, 1974a, 1975; BREM et al., 1976). The hypothesis that angiogenesis is due to the production of a factor(s) by cells involved in the process of
neovascularization arose from several kinds of observations: During an inflammatory response, new formation of vessels appeared when certain types of cells invaded the area; a neoplastic cell population induced the formation of a vascular
network even when the tumor fragment was separated from the host by a Millipore
filter; (GREENBLATT and SHUBIK, 1968); and cell-free fluid sampled in vivo from the
extracellular compartment of a growing tumor elicited an angiogenic response
(GULLINO, unpublished observation), as does the media of cultured neoplastic cells
(FOLKMAN, 1974b).

   The objective of this chapter is to discuss evidence sustaining the hypothesis
that angiogenesis is induced by factor(s) controlling neovascularization. The importance of isolating and characterizing these factor(s) resides in the possibility to
interfere with the new formation of vessels, thus influencing several disease states,
including cancer.

B. Morphogenesis of Vascular Networks

In most vertebrates the earliest vascular plexus appears on the yolk sac and extends
into the embryo by sprouting. This process involves the migration and proliferation of endothelium followed by the formation of cords that later canalize. Therefore, capillaries are for a time bare endothelial tubes framed by a basement membrane; only later do the supporting structures develop from the surrounding mesenchyme. It is noteworthy that the peripheral vessels of the embryo still have a predominantly capillary structure by the time the heart already has completed its general structural plan. With the exception of the larger vessels, such as the dorsal seg-
mental branches of the aorta, definitive arteries and veins of vertebreates develop
from a temporary plexus of capillaries through a remodeling process. The conversion of the capillary plexus of a primitive anlage to the definitive vascular network
of the fully developed organ has been studied directly in transparent regions of the
living embryos and by timed injections (CLARK, 1918; BLOOM and BARTELMEZ,

1940). Remodeling involves atrophy of segments of the original plexus and expansion into major vascular pathways of other regions. Thus, arteries and veins of organs emerge from the primitive capillary plexus as service pipes for the supply of an expanding capillary bed. In the embryo, the general organizational plan of the established vascular network proceeds from a provisional capillary plexus toward the formation of major pathways, not vice versa.

The forces that govern the establishment of characteristic vascular patterns are poorly understood. Early investigations attributed major importance to the volume and pressure of blood and its rate of flow (THOMA, 1893). There are, indeed, convincing examples of the remodeling of vascular walls depending on a change in function. For instance, in an acardic twin parasite, the aorta shows the structure of a typical muscular artery because it functions as an ordinary peripheral artery. The modulation of vascular structures by blood circulation is a well-known event, but since the late 19th century, it has also been recognized that blood flow does not have a leading role in the mechanism of capillary induction. Studies on vessel development in embryos deprived of circulation convinced LOEB (1893) that induction of new capillaries must depend on stimuli exerted by a specific substance outside the vessels. This interpretation was shared by THOMA (1893) and by CLARK (1918) who, more than 20 years later, attributed the ultimate influence on the production of capillary sprouts to the metabolism and growth characteristics of the surrounding tissues. WILLIAMS (1959) considered hypoxia as the principal stimulus of endothelial proliferation and capillary formation.

The development of the transparent chamber technique brought a new experimental approach to the study of the mechanism of angiogenesis (SANDISON, 1924). The new formation of vessels could be followed in the ear chamber by direct observation. The migration of endothelial cells along the luminal surface of the vascular wall toward the sprout of a developing capillary was documented by time-lapse cinemicroscopy (CLIFF, 1965) and by the description of pseudopodia that were projected into adjacent tissue spaces by endothelial cells (SCHOEFL, 1964; AUSPRUNK and FOLKMAN, 1977). Since SCHOEFL (1963) and CLIFF (1963) observed mitotic figures more frequently at the proximal end of the sprout, they concluded that the sprout advanced by the migration of dividing endothelial cells at the proximal end of the tip. Early investigators (SANDISON, 1924; CLARK and CLARK, 1939) had observed a particular fragility of the newly formed vessels, and later electron microscopic studies revealed a lack of basement membrane (SCHOEFL, 1963; ERIKSSON and ZAREM, 1977) in the capillary wall proximal to the sprout. The absence of a basement membrane and "loose" cellular junctions are usually accepted as the two conditions mainly responsible for "leakage," a characteristic event observed in newly formed capillaries, in a variety of circumstances (SCHOEFL, 1964; HAUCK, 1971; GIMBRONE and GULLINO, 1976a). The basement membrane appears after the cord of endothelium has been canalized to form a tube and is most probably produced by the endothelial cells (JAFFEE et al., 1976). Indeed, ongoing experiments in our laboratory indicate that type IV collagen, a specific component of the basement membrane, favors growth of endothelium in culture (KIDWELL et al., 1980).

The transparent chamber technique permitted another observation of importance for the interpretation of the mechanism of angiogenesis. Neovascularization of the connective tissue included in the chamber was very slow when the chamber was empty, but the presence of tissue fragments accelerated the process considerably (MERWIN and ALGIRE, 1956). Again, stimulation of neovascularization induced by iso- or allografts, suggested that substances liberated by the implants could have an angiogenic influence.

C. Assays for Angiogenesis

I. Rabbit Eye

The material to be tested for angiogenic capacity may be implanted on the iris or into the corneal stroma. New formation of vessels is evaluated either by the appearance of capillaries in the cornea, normally devoid of them, or by the modification of the vascular pattern in the iris where a halo of capillaries appears around the implant in contrast with the straight and parallel position of the network in the background. Male or female New Zealand White rabbits, 4–6 months old, and weighing 2–3 kg, are routinely used in our laboratory. Intravenous anesthesia (pentobarbital sodium 25 mg/kg) may be supplemented by retrobulbar injection of 0.5 ml of 2% Xylocaine or by the application of a few drops of Xylocaine on the cornea. The eyeball is first secured in a proptosed position by clamping a fold in the upper lid, then flushed with sterile 0.9% NaCl solution. For a corneal transplant, a 1.5 mm incision is made just off center of the corneal dome to a depth of about one-half the thickness of the cornea, using a Bard-Parker blade N° 11. A pliable iris spatula 1.5 mm in width, is used to produce a pocket in the cornea sufficiently large to accommodate a 1 mm³ fragment of material to be assayed. The bottom of the pocket should be at 1.5 mm from the sclero-corneal margin (Fig. 1). This is a critical distance since transplants in avascular organs have shown that the angiogenic stimulus has a limited radius of action (FOLKMAN, 1974a; BREM et al., 1976; BEN EZRA, 1978a). If the tension of the aqueous humor makes it difficult to prepare the corneal pocket, a small amount of fluid can be drained from the anterior chamber through a 27-gauge needle. The top of the pocket seals spontaneously if the volume of the transplant is smaller than one-half the size of the pocket.

When a fluid or a cell suspension is to be tested, the material can be injected with a 23-gauge needle directly into the cornea. About 10 µl is the most suitable quantity to keep corneal lesions to a minimum. Rupture of the cornea and/or reflux of the injected material are two events to be carefully avoided. Sterility during the entire procedure is critical; no antibiotics or special eye care are necessary. A drop of a synthetic corticosteroid such as triamcinolone acetonide (Kenalog-40 suspension; Squibb & Sons, New York, N.Y.) may be useful once the transplant is completed. If a corneal perforation is produced during the procedure, the preparation must be discarded.

The iris transplant starts with a 2-mm incision through the cornea, about 1 mm from the limbus. With gentle pressure the aqueous humor is flushed out, then the explant is deposited on the iris using a blunt-tipped microdissecting fork. Up to three fragments can be implanted on one iris, and it is advisable to place them as far from the corneal incision as possible. If the amount of aqueous humor flushed out through the corneal incision is too small, the transplant will detach and float free in the anterior chamber and be lost. If insufficient fluid remains in the cham-

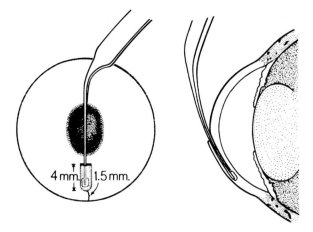

Fig. 1. Schematic presentation of the corneal implant procedure. Frontal and lateral views. Note the position and dimensions of the pouch

ber, the cornea tends to adhere to the iris and the positioning of the explant on the iris will be more difficult and hemorrhage may ensue. The operator must learn the proper balance.

Transplants can be examined at any time in unanaesthetized animals with a slit-lamp stereomicroscope at 10–40 X (Figs. 1 and 2). A green filter allows a clear definition of fine vessels growing into the cornea or in the iris around the explant. During the first 48 hr after the transplant, a slight opacity of the cornea usually appears due to edema around the wound. If the opacity does not diappear by the third day, it usually indicates that an inflammatory process is present. In our experience infection of the wound is very rare. In corneal explants, budding of vessels from the limbal plexus is observed at 4–5 days in most cases. Scoring is done at 5–7 days by counting the vessels and, if warranted, measuring their increment in length with an ocular micrometer reticule at 10 X. By the 8th day, the immunological rejection of the heterotransplant starts with infiltration of inflammatory-type cells. If the test substance was incorporated by a polymer, the observation period may be extended for 2 weeks. Expulsion of the pellet is a rare event.

In the iris, the newly formed vessels usually appear at 5–7 days with a radial distribution around the explant; this orientation is clearly different from that of the preexisting vessels which extend in a regular pattern radial to the pupil (GIMBRONE and GULLINO, 1976a; BREM et al., 1978). The newly formed vessels are permeable to fluorescein (0.5 ml/kg of 5% Fluorescite injected i.v.). By stereoscopic examination under cobalt blue illumination (Zeiss Photo-Slit Lamp, Carl Zeiss, Inc., New York, N.Y.), a green halo can be observed around newly formed capillaries more rapidly than around established vessels (GIMBRONE and GULLINO, 1976a).

After scoring the vascular response, the cornea or the iris is removed and several histologic sections are examined to evaluate the condition of the transplanted tissue. In tissue explants, the control is represented by a fragment of comparable tissue kept for 15 min at 100 °C before implantation.

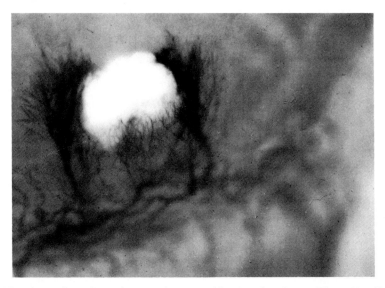

Fig. 2. New formation of vessels around a corneal implant bearing an Elvax 40 pellet containing lyophilized tumor interstitial fluid. Several large vessels from the limbus penetrate the cornea and constitute a dense array of capillaries that surround and occasionally penetrate the pellet. (Photographed in vivo with Zeiss slit-lamp stereomicroscope.) X 30

Fluids to be tested for angiogenic capacity are lyophilized, then incorporated into a polymer pellet. The most satisfactory polymer is Elvax 40 (E. I. DuPont de Nemours & Co., Plastic Products and Resins Dept., Wilmington, Del.) an ethylene-vinyl acetate copolymer (40% vinylacetate by weight) (LANGER and FOLKMAN, 1976). Before use it is important to wash the polymer in absolute ethanol at 37 °C for 10–15 days until the pellet fails to induce any inflammatory reaction in corneal implants. Under aseptic conditions 50–100 µl of a sterile 10% casting solution of Elvax 40 in methylene chloride is mixed with the dry substance to be tested. The mixture is placed on a flat surface and the solvent is evaporated under vacuum. The dry pellet is cut into 1-mm² pieces to be used as implants. Glass molds to produce corneal casts of the copolymer have been used by LANGER and FOLKMAN (1976). All manipulations should be done sterilely under a hood.

Polymers other than Elvax 40 can also be used. In our experience, polyvinylalcohol and Hydron, a polymer of hydroxymethylacrylate, (Type NCC, Hydron Labs., Inc., New Brunswick, N.J.) gave less satisfactory results than Elvax 40.

The cornea and iris of guinea pigs and rats could be used as transplant sites for either pellets or tissue fragments, but the size of the eye imposes technical limitations. Rats are more prone to keratitis than rabbits.

II. Chick Embryo Chorioallantoic Membrane

Fertilized white Leghorn eggs, 4–5 days old, are incubated horizontally for 3 days at 36.5 °C–37 °C and 65% relative humidity. At the end of day 3, the shell of the narrow end of the egg is sterilized with 70% isopropyl alcohol and punctured with a 20-gauge needle. An air bubble is produced between the chorioallantois and the shell membrane by slowly withdrawing with a syringe about 0.5 ml of albumin. The egg is returned to the incubator for about 24 h. Then, with the egg kept in a horizontal position, the shell is sterilized with isopropyl alcohol at its wide end and punctured. The air bubble will move to the upper surface causing the chorioallantoic membrane to separate from the egg shell. In this area, where the possibility of damaging the chorioallantoic membrane is minimal, a 1.5-cm² window is cut out using an abrasive disk (Dremel Mat Flex tool Model 323 fitted with abrasive disk $^1/_4$ shank, Emerson Electric Co., Racine, Wis.). The operator must be careful to cut only the shell and not the underlying shell membrane. The cut shell is removed with a sterile spatula and the shell membrane is lacerated and removed with sterile forceps. The chorioallantoic membrane appears underneath as a transparent jelly with a few large vessels. Care must be taken to avoid any fragments of shell or membrane from falling onto the chorioallantoic membrane because they could become foci of irritation and modify the local vascularization. After the chorioallantoic membrane has been exposed, the window is sealed with cellophane tape and the egg is returned to the incubator until day 9.

The test sample is dissolved in sterile distilled water and 5 µl are dropped onto the center of a 15-mm round plastic coverslip (Thermanox No. 1–½, Lux Scientific Corp., Newbury Park, Calif.). For future orientation, it is useful to mark the center of the reverse side of the coverslip. The sample is air dried in a sterile hood, then the cover slip is placed face down on the chorioallantoic membrane with the dried sample in contact with the membrane. Air bubbles between the cover slip and the membrane must be avoided. The window is resealed with the tape, and the egg is returned to the incubator. For the next 4 days, sample eggs are evaluated for an angiogenic response. A positive test is indicated by a halo of tiny, newly formed capillaries oriented toward the test sample, usually represented by a whitish spot in the center. The image has been described as a spoke-wheel pattern (Folkman, 1974 b). Scoring is done under a dissecting microscope. The chorioallantoic membrane is then either fixed in situ with formalin for histologic examination or removed and floated on saline for photographic examination of the halo of newly formed vessels.

The chorioallantoic membrane is extremely sensitive to a variety of materials that can initiate a local inflammatory reaction and evoke a vascular response (Jakob et al., 1978). In our laboratory, we use the chorioallantoic membrane assay to screen out negative samples. We have found the rabbit corneal assay to be the most reliable for testing material that was positive with the chorioallantoic membrane assay.

III. Hamster Cheek Pouch

The development of the transparent chamber (Sandison, 1924; Algire, 1943) was an important contribution to the study of vascularization and vascular physiology. In Shubik's laboratory, the transparent chamber technique was adapted to the hamster cheek pouch to study angiogenesis in tumors (Goodall et al., 1965; Greenblatt and Shubik, 1968; Greenblatt et al., 1969; Shubik et al., 1976). The preparation consists of distending the delicate pouch membrane over a transparent baseplate, then stripping off the nonvascular, muco-areolar tissue to enhance the transparency of the preparation. The transplant is placed on this membrane and sealed within a chamber molded from acrylic plastic. Details of the instrumentation and surgical procedure are reported by SEWELL (1966) and GREENBLATT and coworkers (1969). Observation of the transplant and changes in its vascular component is accomplished with an ordinary microscope equipped with long working distance objectives and from which the condenser assembly has been removed. The chamber is illuminated from beneath by a Lucite light-conducting rod inserted into the animal's mouth. Adaptation of this technique to other model systems has permitted an accurate study of tumor microcirculation (INTAGLIETTA et al., 1977; ENDRICH et al., 1979).

The major drawback of this procedure is the chronic irritation induced by the chamber which may interfere with the angiogenic response.

IV. Dorsal Air Sac and Intracutaneous Injections

The dorsal air sac technique was used by FOLKMAN and coworkers (1971) before transplantation in the cornea and iris were fully developed. Recently PHILLIPS and KUMAR (1979) used this technique in developing an antiserum to the angiogenesis factor. An aseptic injection of about 30 ml of air is used to produce a bubble in the subcutaneous tissue of the sacral region. The white triangle of the avascular fascia constitutes the background of the sac. The test material is placed in a container that insures a slow release over a period of a few days. Newly formed vessels appear on the white surface of the fascia as a dense, tortuous network and are usually scored on a 0–5+ scale.

The major limitations of this method are the need for a surgical procedure to examine the results and the need for a relatively large amount of test material.

A second method utilizing the response of skin vessels was utilized by SIDKY and AUERBACH (1975) to test for lymphocyte-induced angiogenesis. Lymphocytes in 0.1 ml saline were injected intracutaneously with a 27-gauge needle about 0.5 cm into the skin, as close to the epidermis as possible. Except for the mid-dorsal and mid-ventral regions, all other sites on the skin yielded comparable results. Newly formed vascular branches in the area of the injection were counted 2–3 days later at 7 × magnification. The vessels were readily detected because of their tortuosity and tendency to form loops. We have no direct experience with either the air sac or the intracutaneous methods.

V. Endothelial Cell Culture

Since endothelial cells of rat (SADE et al., 1972), human (GIMBRONE et al., 1974), or bovine origin (GOSPODAROWICZ et al., 1976) could be cultured, attempts have been made to evaluate the effect of preparations with angiogenic properties on the labeling index or growth stimulation of confluent monolayers. Reproducibility and specificity of the results have remained the two difficult problems to solve with these preparations. To my knowledge, there is no standardized procedure utilizing endothelial cell cultures to assess angiogenic capacity of test samples in vivo. MCAUSLAN and REILLY (unpublished observation) are proposing a new method that utilizes endothelial cells of bovine aorta cultivated on gold-coated surface (AL-BRECHT-BUEHLER, 1977). Addition of the angiogenic factor to the culture media increases cell motility; as the cells move, they phagocytose the gold particles, leaving a track clearly visible. Within 72 h the response is quite evident. Whether or not this ingenious method will have quantitative and routine applicability in assessing angiogenic activity is unknown at this writing.

VI. Renal Assay

HOFFMAN and coworkers (1976) and McAUSLAN and HOFFMAN (1979) injected subcutaneously Quackenbush mice, either newborn (20–40 µl/mouse/day) or adult (100 µl/mouse/day) with fractions of extracts from bovine parotid or mouse submaxillary glands. They observed an extensive increment in the number of capillary endothelial cells in a variety of organs, but particularly in the intertubular spaces of the kidney. They utilized this method for the initial fractionation of their angiogenesis factor and observed that the increment in the number of endothelial cells was not associated with an evident increase in their mitotic activity. This assay requires a relatively large amount of material and it is not quantitative. Later McAUSLAN (1980) utilized corneal implants and developed the endothelial motility test reported in the previous section.

D. Induction of Angiogenesis

In the adult organism new formation of vessels occurs under a variety of conditions and the morphogenesis is similar to that in the embryo. The ability to elicit new vessels formation has been evaluated for several tissues and cells as an approach to understanding the mechanism of angiogenesis. Using mostly transplantation techniques, it has been shown that, in general, most transplants of neoplastic tissues are able to induce angiogenesis while most normal, adult and embryonic tissues lack an appreciable angiogenic capacity. There are several exceptions to this generalization and they are mentioned below.

I. Angiogenesis by Normal Tissues

HUSEBY and coworkers (1975) transplanted fragments of testes derived from 1-day old mice into the subcutaneous area caudal to the axilla of castrated, histocompatible, young adult male mice. The transplants produced a strong neovascularization that was evident as early as 3 days after implantation. The newly formed capillaries proceeded from the surrounding stroma and extended through the interstitium of the explant. During angiogenesis the endothelial cells of the capillary walls incorporated extensively ^{3}H-thymidine. There was no evidence of inosculation between the newly formed capillaries and the original vessels of the implant. The study of testicular-induced angiogenesis has not progressed beyond the observation that neonatal testis is strongly angiogenic and may be a useful model to evaluate the mechanism of angiogenesis.

JAKOB and coworkers (1977) observed that fragments of corpus luteum removed from slaughtered cattle induced substantive angiogenesis in the chorioallantoic membrane of the chick embryo, in the subcutaneous tissue of the mouse, and in the hamster cheek pouch. The angiogenic capacity was retained by fragments heated to 56 °C but not 100 °C, and by the lyophilized extracts of corpus luteum in phosphate buffered saline. These facts support the hypothesis that an angiogenesis factor is present in this tissue (GOSPODAROWICZ and THAKRAL, 1978).

WARREN and coworkers (1972) implanted renal tissue removed from newborn hamsters on to the hamster cheek pouch membrane contained within a transparent chamber. An ultrathin type TH Millipore filter (0.01 μm pore size) was interposed between the pouch membrane and the kidney explant. Mitoses of the endothelial cells in capillaries and small veins were observed in the area underlying the Millipore filter where the explant was located. The results were interpreted to indicate production by the explant of a diffusible substance capable of inducing angiogenesis by acting on the cheek pouch membrane through the filter.

The influence of epithelium from the epidermis on blood vessel growth was studied by NISHIOKA and RYAN (1972) and WOLF and HARRISON (1973) with the cheek pouch technique. The skin from the trunk of 4-day old hamsters was removed and incubated in Eagle's Minimum Essential Medium containing 0.4% trypsin. The epidermis was peeled off after 3–4 h of incubation, washed and placed inside the fold of a type TH Millipore filter, 25 μ thick with 0.45 μ pore diameter. The filter and graft were placed on the vascular membrane of a hamster cheek pouch incorporated into a transparent chamber prepared according to GREEN-BLATT and SHUBIK (1968). Edema and widespread leukocyte infiltration occurred during the first 3 days after the operation, then the inflammatory reaction subsided and by the 6th–7th day, a large number of capillaries sprouted from the pre-existing vessels to form a 5-mm halo around the explant. A similar type of response was observed with buccal epithelium, but fragments of heart muscle, liver and spleen obtained from the same animal failed to produce angiogenesis, and the response of the lung was doubtful. The interpretation that the epidermis has a particular angiogenic effect is based on the specific localization and abundance of newly formed vessels around the explant as compared with other tissues, despite the inflammatory process in the background. WOLF and HARRISON (1973) confirmed the angiogenic capacity of the epidermis and attributed this property to a specific, heat-labile, diffusable but nondialyzable protein, the "epidermal angiogenic factor."

Crude extracts of salivary gland were found to produce angiogenesis (FOLKMAN and COTRAN, 1976). HOFFMAN and coworkers (1976) isolated from bovine parotid gland and mouse submaxillary glands substances of apparent mol. wt. 80,000–86,000 that stimulated endothelial proliferation when injected into newborn or adult mice. More details of this preparation will be reported in Section E.

Recently, GLASER and coworkers (1980) reported the induction of vasoproliferative activity by extracts of retinas removed from human, bovine and feline eyes. The retinas were simply extracted with a balanced salt solution (0.5 ml per retina) at room temperature for 2.5–3 h. The suspension was centrifuged, filter-sterilized and preserved at −20 °C. This crude extract was able to stimulate proliferation and thymidine uptake of bovine endothelium in culture and to induce angiogenesis in the chick chorioallantoic membrane. The activity persisted after dialysis, warming at 56 °C, but was lost at 100 °C. At this writing, purification and characterization of retinal substances with angiogenic and/or mitogenic activities have not been obtained. Angiogenic capacity was also demonstrated in both retinoblastomas and the aqueous humor of the tumor-bearing eye (TAPPER et al., 1979).

II. Angiogenesis by Normal Cells

Immunocompetent lymphocytes injected intracutaneously into irradiated, unimmunized mice were able to induce angiogenesis, and the number of newly formed vessels was found to correlate with the number of immunocompetent cells injected within a dose range of 2×10^5–4×10^6 lymphocytes (SIDKY and AUERBACH, 1975). Mouse lymph nodes and spleen cells as well as allogenic rabbit lymph nodes were also able to induce angiogenesis when grafted in the rabbit cornea (AUERBACH et al., 1976). A striking increase of the angiogenic response was observed after X-irradiation of the recipient with 700 R (KAMINSKI et al., 1978 a). Activation in vitro by concanavalin A stimulated the lymphocyte-induced neovascularization of the cornea (BEN EZRA, 1978 b). Mediators of angiogenesis appear to be released by certain lymphoid cells whenever a local graft-versus-host reaction is evoked. These soluble mediators may act either directly on endothelial cells or indirectly by activating macrophages which, in turn, induce the neovascular response (AUERBACH and SIDKY, 1979).

The role of macrophages in endothelial proliferation and angiogenesis was evaluated by POLVERINI and coworkers (1977 a). Macrophages were obtained from the peritoneal cavities of untreated guinea pigs or from guinea pigs and mice previously injected with paraffin oil or thioglycollate, respectively. Angiogenesis was tested by either injecting the macrophages into the cornea, or by incorporating media from macrophage cultures into pellets of Hydron and implanting them into the cornea of guinea pigs. Angiogenesis was elicited by the stimulated macrophages without an acute inflammatory response and in a completely syngeneic system. This indicates that induction of angiogenesis was a property of the macrophages and a local graft-versus-host reaction was not indispensable to obtain neovascularization (POLVERINI et al., 1977 b; AUERBACH and SIDKY, 1979). The fact that an "activation" of the macrophages was necessary suggests that the angiogenic capacity is an inducible rather than constitutive property. The culture media incorporated into the pellet also evoked neovascularization, indicating that secretion of an active molecule is involved in the macrophage-induced angiogenesis. A role for macrophages as stimulators of blood vessel growth was also found during wound healing (CLARK et al., 1976; GREENBURG and HUNT, 1978; THAKRAL et al., 1979).

From studies primarily concerned with corneal vascularization, the hypothesis was developed that leukocytes were necessary for neovascularization to occur. It is a common observation that in wounded corneas or in corneal explants, blood vessels invasion follows leukocyte infiltration (FROMER and KLINTWORTH, 1975 a, b, 1976). Wound repair, however, proceeds in neutropenic animals in manner similar to that observed in normal animals (STEIN and LEVENSON, 1966; SIMPSON and ROSS, 1972). In rats subjected to combined treatment with radiation (800 R) and repeated injections of antineutrophil serum, the vascular ingrowth of corneas cauterized with silver nitrate was still observed despite a leukocyte blood count of nearly zero and the absence of monocyte and neutrophil infiltration.

However, the mean length of newly formed vessels was reduced to 67% of the control at 3 days and 33% at 4 days. (SHOLLEY et al., 1978). It seems reasonable to conclude from these data that neutrophils are probably not indispensable to the induction of neovascularization although their presence may have an adjuvant effect which is presently difficult to characterize.

A similar supporting role has been suggested for mast cells. The basis for this hypothesis was a 40-fold increase in mast cell density around crude extracts of Walker 256 rat carcinoma implanted on the chick chorioallantoic membrane to elicit angiogenesis. Mast cells isolated from Sprague-Dawley rats, however, failed to induce neovascariztion when implanted on the chorioallantoic membrane (KESSLER et al., 1976).

III. Angiogenesis by Neoplastic Tissues and Cells

The importance of angiogenesis in the neoplastic process has been appreciated since the discovery that tumors could be transplanted. Tumor growth was dependent upon the promotion of a vascular network by the host, and the inability to do so could prevent tumor growth (GREEN, 1941 a, b; TANNOCK, 1970). This was confirmed experimentally during perfusion of the thyroid gland (FOLKMAN and GIMBRONE, 1972). A suspension of melanoma cells added to the perfusate produced a large number of small nodules clearly visible by their black color all over the gland. None of these nodules grew beyond 1–2 mm in diameter, but they grew readily when retransplanted into the original host. In every perfusion of isolated organs there is gradual degeneration of capillary endothelium. FOLKMAN attributed the lack of growth of melanomas in the perfused gland to the degeneration of the endothelium and the consequent inability to constitute functional vessels. Evidence that this interpretation was probably correct was derived from tumor implants into the vitreous body (BREM et al., 1976, 1977a; FINKELSTEIN et al., 1977). This tissue is avascular and the implanted neoplastic cells form 1–2 mm nodules that remain "dormant," i.e., unable to grow, unless the neoplastic cells are about 2 mm from the retina. In this case, vessels proliferate from the retinal network and a large tumor rapidly forms. The experiment suggested not only that a vascular network would be induced by neoplastic cells but also that induction was blocked by the interposition of a sufficient amount of vitreous tissue. If an angiogenesis factor was produced by the neoplastic cells, its action was limited to the immediate surroundings.

Evidence that neoplastic cells were secreting an angiogenesis factor in the environment was derived from two types of experiments. As mentioned before, tumor fragments contained in a folded Millipore filter induced angiogenesis in the cheek pouch of hamster (EHRMANN and KNOTH, 1968; GREENBLATT and SHUBIK, 1968; NISHIOKA and RYAN, 1972). The filter prevented direct contact between the explant and the host but did not prevent angiogenesis, suggesting that a stimulatory substance was crossing the filter.

The second experiment was based on the property of neoplastic tissue to grow around and incorporate extraneous material. A chamber with walls constituted by Millipore filters (0.45 μ pore diameter) was inserted subcutaneously and a few fragments of a transplantable tumor were seeded around it. Within a few days the chamber was incorporated by a solid tumor and the fluid surrounding the neoplastic cells penetrated the chamber while the filter excluded the cells (GULLINO, 1970). This fluid considered to be tumor interstitial fluid was lyophilized and found to have strong angiogenic activity. Chambers placed in the subcutaneous tissue of control rats also collected a fluid that is considered to be subcutaneous interstitial fluid. The lyophilisates of this fluid were also found to possess angiogenic activity but the minimal amount necessary to elicit a response was about 3-fold greater than the tumor fluid (GULLINO and coworkers, unpublished observation). Media from cultured neoplastic cells of rodent as well as human tumors have angiogenic activity (KLAGSBRUN et al., 1976), in particular human tumors of the nervous system (SUDDITH et al., 1975; KELLY et al., 1976). This indicates that the survival of the angiogenesis factor(s) in vivo as well as in vitro is sufficiently long to permit extraction from an acellular fluid. Moreover, culture medium conditioned by clonal glial or neuroblastoma cells specifically stimulated cultures of human endothelial cells derived from umbilical cords at term (SUDDITH et al., 1975).

In our experience, every tumor of human or rodent origin elicited an angiogenic response; this has also been the experience of other laboratories. This statement, however, should be qualified in the sense that every test for angiogenesis consists of implanting several tumor fragments (usually six, in three rabbits). In primary tumors of humans and rodents, about 70% of the fragments elicited a strong angiogenic response; in tumors transplanted for several generations, the response was usually seen in 100% of fragments (BREM et al., 1978; MAIORANA and GULLINO, 1978). The number of neoplastic cells that must be present in a fragment in order to induce angiogenesis is difficult to determine. For Walker 256 carcinoma, a mammary tumor of Sprague-Dawley rats, which was transplanted for many generations, we observed neovascularization of corneas bearing transplants of only 40 neoplastic cells counted in serial histologic sections. However, we also observed transplants of primary mammary carcinomas of both human and rodent origin, that were constituted by thousands of neoplastic cells which appeared viable and showed no evident morphologic alterations but were unable to induce an angiogenic response; while adjacent fragments from the same tumor were strongly angiogenic, often in the same rabbit. We could not attribute this difference to technical errors or immunologic diversity of the recipient and we have no convincing explanation of the observation. It is possible that some viable neoplastic cells acquire then lose angiogenic capacity while neighboring cells within the same tumor acquire and retain this capacity.

E. Isolation and Characterization of Angiogenesis Factor(s)

A vessel is an organ that may either acquire the complex structure of a large elastic or muscular artery or may have the simple structure of an endothelial tube delimited by a basement membrane. An angiogenesis factor should, by definition, be able to induce at least vessels with the simplest structure. As indicated in a previous section, new formation of vessels requires two events, migration and proliferation of the endothelium. The basement membrane necessary to complete the vessel can be synthesized by the endothelial cells themselves (HOWARD et al., 1976). The isolation of angiogenesis factors has focused on two approaches, namely, fractionation of fluids or tissue homogenates already known to induce angiogenesis or evaluation of the angiogenic capacity of substances having on the endothelium either a chemotactic or a growth-stimulating activity.

I. Fractionation of Fluids or Tissues with Angiogenic Capacity

FOLKMAN and coworkers (1971) were the first to fractionate homogenates of Walker 256 carcinomas with the objective of isolating the tumor angiogenesis factor, believed to be produced by the neoplastic cells. The fraction that showed the strongest angiogenic activity contained approximately 25% RNA, 10% protein, 50% carbohydrate and a residue probably constituted by lipids. The angiogenic activity of the fraction was destroyed by digestion with bovine pancreatic ribonuclease, or by heating for 1 h at 56 °C, or by incubation with subtilisin. The activity was not diminished after 3 months at 4 °C or by exposure to trypsin for up to 3 days (FOLKMAN et al., 1971).

A few years later TUAN et al. (1973) in FOLKMAN's laboratory, isolated from the nuclei of Walker 256 carcinoma a fraction containing nonhistone proteins able to induce angiogenesis in the rabbit cornea and mitosis in cultured endothelium. Chromatin was purified from nuclei isolated by differential centrifugation of cells disrupted by N_2 cavitation. DNA was separated from chromatin proteins by chromatography on Bio-Gel A 5 mm in the presence of 4 M guanidine-HCl. Histone and nonhistone proteins were further separated with carboxymethyl-Sephadex C-50, eluting alkylated nonhistones with 0.4 M guanidine · HCl in 0.01 M phosphate buffer and histone with 2 M guanidine · HCl in 0.1 M acetic acid. In the corneal bioassay, the histone fraction showed no activity but the nonhistone proteins were strongly angiogenic, and the authors concluded that the tumor angiogenesis factor was bound to the fraction of nonhistone nuclear proteins.

WEISS and coworkers (1979) recently isolated from Walker 256 carcinoma a low molecular weight (~ 200) nonprotein component with a strong angiogenic effect on the chick chorioallantoic membrane and a mitogenic effect on endothelial cells in culture (SCHOR et al., unpublished observation). Walter 256 carcinoma was extracted following the procedure of FOLKMAN and coworkers (1971) with the omission of the trypsin step. The crude preparation resulting from gel filtration on Sephadex G-100 was chromatographed on DEAE cellulose using a convex salt gradient between 0 and 0.3 M NaCl. Further purification was obtained by taking advantage of an antibody prepared by PHILLIPS and KUMAR (1979) against a crude preparation of angiogenesis factor derived from Walker 256 carcinoma. An affinity-chromatography column was prepared by coupling absorbed antibodies to CNBr Sepharose by conventional methods and each fraction from the DEAE column was individually applied. The bound material was eluted with 50 mM ammonium acetate buffer (pH 3.7) and applied to a Biogel P_2 column (exclusion limit 2,500 mol. wt.) with 10% isopropanol in water as packing and eluting solvent. The angiogenic peak emerged at a volume which corresponded to a mol. wt. of ~ 200. The chemical nature of the small molecule is presently being investigated and does not appear to be a prostaglandin or a polypeptide or a nucleic acid (WEISS et al., 1979).

HOFFMAN and coworkers (1976) isolated from extracts of bovine parotid glands and murine submandibular glands a fraction they called endothelium stimulating factor because it induced proliferation of renal capillaries following in vivo injection of both newborn and adult mice. A few years later, McAUSLAN and HOFFMAN (1979) observed the same effect with crude extracts of Walker 256 carcinomas. The renal assay also suggested that in the angiogenesis process, mitoses and mobilization of the endothelium were two separate events that could be influenced independently. McAUSLAN and coworkers proceeded to isolate fractions able to influence each of these events. They used Balb/c 3 T 3 cells, which have been shown to possess angiogenic activity (FOLKMAN, 1975). In these cultures they identified a polypeptide of mol. wt. 120,000 with mitogenic action on aortal bovine endothelium and a small molecular weight factor (3,000 or 210) without mitogenic action on cultured bovine endothelium but capable of stimulating angiogenesis in vivo. Further fractionation revealed that the 3,00 mol. wt. fraction was constituted by an inert carrier and a low molecular weight (210), heat stable endothelium stimulating factor. This heated factor was subjected to analysis by spark source mass spectrometry which revealed two elements at high concentration: zinc (2,000 ppm) and copper (8,000 ppm). Only copper salts were found to induce the same mobilization of renal endothelium as did the crude extracts of bovine parotid gland or the fractions from Walker 256 carcinoma or the Balb/c 3 T 3 cells. The data were confirmed with the corneal assay and the conclusion was reached that angiogenic activity was exerted by the cooper ions (McAUSLAN, 1980).

II. Angiogenic Capacity of Growth Factors and Prostaglandins

GOSPODAROWICZ and coworkers (1979) analyzed the angiogenic activities of the fibroblast and epidermal growth factors using the rabbit corneal assay. The first was isolated from bovine brain and pituitary glands, the second from the submaxillary glands of adult Swiss-Webster male mice injected with testosterone propionate. A positive angiogenic response was found for both, but the epidermal growth factor was about 2.5-fold more effective than the fibroblast growth factor. They also evaluated the effect of both factors on the growth of cultured endothelial cells (GOSPODAROWICZ et al., 1978). The endothelial cells were obained from human umbilical veins by collagenase treatment and cultivated in HEPES-buffered medium 199 supplemented with either the fibroblast or the epithelial growth factor in 10%–20% human serum. Strong potentiation of growth was observed when the cultures were supplemented with thrombin, and the growth increment was greater in cultures with the fibroblast growth factor. In contrast, growth of endothelium from bovine aorta was maximal in the presence of the fibroblast growth factor and was potentiated by neither the epidermal growth factor nor thrombin. This suggests that species-dependent factors may influence the proliferation of vascular endothelium. Moreover, the angiogenic effect may result from events not necessarily identical to those inducing proliferation of the endothelium.

An indication that this can be the case derives from the work of BEN EZRA (1978a) who observed neovascularization of corneas only with relatively high concentrations of epidermal or fibroblast growth factors. Strong proliferation of keratocytes around the implants was the major effect in vivo, followed by a moderate infiltrate of mononuclear cells. BEN EZRA hypothesized (1978b) that under these circumstances as well as a variety of others, cells may produce common metabolites that in large local concentration can influence the vascular response. In the quest for the identification of substances mediating neovascularization, BEN EZRA found a strong angiogenic activity produced by prostaglandin E_1. The dose necessary to induce corneal vascularization was at least 10-fold less than that of the epidermal growth factor. Prostaglandin E_2 stimulated a much weaker angiogenic response, prostaglandins of the F series were even less active, and PGA_1 and PGD_2 were unable to induce neovascularization at all. The results obtained by BEN EZRA (unpublished observation) with PGE_1 and PGE_2 were confirmed in our laboratory with the added observation of a lack of angiogenic capacity by prostacyclin. The possibility that PGE_1 is a mediator of angiogenesis is substantiated by the PGE production by cells normally involved in the neovascularization process such as macrophages (HUMES et al., 1977), endothelium (GIMBRONE and ALEXANDER, 1975; DUMONDE et al., 1977; JOYNER and STRAND, 1978), platelets (NALBANDIAN and HENRY, 1978) and neoplastic cells (JAFFEE, 1974; KIBBEY et al., 1979). KNAZEK (unpublished observation) in our laboratory found that the interstitial fluid obtained in vivo from MTW 9/A mammary carcinomas grown subcutaneously in the inguinal region of WF female rats contained PGE at 1.17 ± 0.3 ng/ml; in the normal subcutaneous fluid PGE was 0.19 ± 0.09 ng/ml. Both fluids are angiogenic in lyophilized form but the tumor fluid is 2- to 4-fold more effective. The concentration of $PGF_{2\alpha}$ of the tumor interstitial fluid was 153 ± 19 ng/ml versus 5.0 ± 1.1 ng/ml for the normal subcutaneous fluid. Whether these changes in prostaglandin concentration of tumor interstitial fluid are related to the stimulation of angiogenesis is a matter of speculation at this time.

The role of chemotactic agents as angiogenesis factors was also studied by BEN EZRA (1978a). Two formylated peptides, formyl-methyonyl-phenylalanine and formyl-methyonyl-leucine-phenylalanine, as well as two nonformylated peptides, methyonyl-leucine and methyonyl-phenylalanine, were tested in the cornea with the slow-release pellet and failed to induce any angiogenic response. These compounds are known to have a strong chemotactic effect on leukocytes in vitro (SCHIFFMANN et al., 1975; ASWANIKUMAR et al., 1976) but seldom induce infiltration of leukocytes in the cornea in vivo.

The results obtained with prostaglandins, chemoattractants and growth factors suggested to BEN EZRA (1978a) that PGE_1 could act as a mediator of the angiogenic response. If this were true, indomethacin could reduce or modify the new formation of capillaries in the corneal assay. Results of these experiments are incomplete at this writing (BEN EZRA, personal communication).

F. Physiologic Significance of Angiogenesis Factor(s)

When a preparation is tested for angiogenic capacity it is important to establish that new formation of vessels occurs without an inflammatory process. In fact, mobilization of some polymorphonuclear and mononuclear cells is evident in most cases of corneal implants but this is quite different from an active inflammatory reaction. In the corneal assay, the inflammatory response is represented by an opacity that does not disappear within about 48 h after implantation. Histologic control of the area with newly formed vessels is indispensable to assess the extent of the inflammatory process.

Another point to stress is the distinction between the angiogenic and the mitogenic properties of a substance. Epithelial and fibroblast growth factors have been found to possess both angiogenic and mitogenic activities (GOSPODAROWICZ et al., 1979), and several investigators tend to assume that mitogenic action on the endothelium in vitro is a marker of the angiogenic capacity in vivo. This is not necessarily true. MCAUSLAN (1980), for instance, observed that copper ions stimulated endothelial motility in vitro and induced angiogenesis in vivo without being mitogenic to endothelium in vitro. Present understanding of the angiogenesis process suggests that the angiogenic and mitogenic effects should be considered as two distinct events, even if they may be evoked by the same or similar factor(s).

The comparison of angiogenic capacity among different tissues or substances has been an approach followed to study the physiologic significance of the angiogenesis factor(s). Present methodology allows only a tentative forecast of this physiologic significance; the angiogenic capacity of prostaglandins may be taken as an example. PGE_1 at a dose of 1 µg incorporated into a 1-mm^2 Elvax pellet induces the formation of 10–40 vessels by the 5th day in practically all corneal implants. The same response can be obtained with PGE_2 with a dose of 10 µg while $PGF_{2\alpha}$ and prostacyclin do not induce new vessel formation at doses of up to 20 µg/pellet. Even less precise is the comparison of tissue fragments whose composition and cell population are unknown. More promising, however, appears to be the comparison between normal adult tissue and the proliferative lesions thereof, expecially when the angiogenic capacity of the normal tissue is practically nil and a correlation between onset of a proliferative activity and angiogenesis can be sought.

I. Angiogenesis as a Marker for Neoplastic Transformation

As mentioned before, studies on the morphogenesis of vascular networks in several organs suggested that the parenchyma should have some influence on the angiogenesis process. The possibility that this morphogenetic influence was exerted via an angiogenesis factor was emphasized by FOLKMAN and GIMBRONE (1972). We hypothesized that if tumor growth depends on angiogenesis and if the new formation of vessels is due to an angiogenesis factor(s), then angiogenic capacity should be acquired during neoplastic progression. Using the mammary gland of rats and mice as a model, we observed that whereas resting adult mammary gland has practically no angiogenic capacity, mammary carcinomas acquire the capacity almost consistently (GIMBRONE and GULLINO, 1976a, b; BREM et al., 1977b, 1978; MAIORANA and GULLINO, 1978). In both rats and mice (RIVERA, personal communication; DE OME, 1959), hyperplastic lesions have been found which are transplantable in the fat pad and grow in it to a limited extent. For some of these transplants one can predict fairly accurately the frequency of neoplastic transformation (MEDINA, 1973). We observed that lesions with high frequency of neoplastic transformation induced angiogenesis at a much higher rate than did lesions with low frequency of transformation. This elevated angiogenic capacity was observed long before any morphologic sign of neoplastic transformation was apparent (GIMBRONE and GULLINO, 1976a, b; MAIORANA and GULLINO, 1978). Hyperplastic lesions of the human mammary gland behaved like the rodent gland (BREM et al., 1977b). Since neoplastic transformation is believed to be a multistage process (FOULDS, 1975), acquisition of angiogenic capacity appears to indicate progression toward neoplastic transformation. If this interpretation is correct, one could utilize the increment of angiogenic capacity in hyperplastic-dysplastic tissues as a marker of neoplastic progression, i.e., a marker for cell populations at high risk of becoming neoplastic (GULLINO, 1978). Application of this approach to biopsy material is presently hampered by the complexity and imprecision of the angiogenesis assays.

II. Antiangiogenesis

The hypothesis that inhibition of angiogenesis might control tumor growth was an obvious consequence of the finding that neoplastic cells had angiogenic activity. Indirect evidence that the absence of vascularization prevented tumor formation was obtained in transplants of tumor cells into the vitreous body of the rabbit (BREM et al., 1976). If V_2 carcinoma cells are injected toward the center of this avascular body and care is taken to avoid any residual cells from trailing toward the sclera within the tunnel left by the needle, a nodule of 1–2 mm in diameter is formed but no further growth is observed. If this node is pushed toward the retina, rich in vessels, then rapid angiogenesis occurs and a large tumor is formed within a few days.

The search for substances that could prevent angiogenesis was guided by observations on avascular tissues, in particular cartilage. EISENSTEIN and coworkers (1973) called attention to the inability of capillaries to invade fragments of cartilage implanted in the chick chorioallantoic membrane. Factors able to inhibit invasion by vascular mesenchyme and growth of endothelium were extracted from cartilage (EISENSTEIN et al., 1975; SORGENTE et al., 1975). Arrest of vascular penetration into the cartilage was attributed to a collagenase inhibitor of low molecular weight ($\sim$11,000), synthesized and secreted by chondrocytes (KUETTNER et al., 1974, 1976, 1977). Interference with collagen metabolism, in particular type IV collagen of basement membranes, impairs cell proliferation. This observation appears to be applicable to several cell types, normal and neoplastic, not only to endothelium (KIDWELL et al., 1980).

Direct inhibition of angiogenesis by chondrocytes or cartilage extracts was reported by KAMINSKI and coworkers (1977, 1978b) and by LANGER and coworkers (1980). KAMINSKI and coworkers utilized the Sidky-Auerbach test based on the increment of the number of capillary branches following the injection of lymphocytes within the dermis. When chondrocytes were added to the lymphocyte suspension (10%), a marked inhibition of angiogenesis was observed. LANGER and coworkers (1980) infused cartilage extract continuously into the right carotid artery of rabbits. They used the rabbit corneal implant and showed by daily measurements an inhibition of the vascular proliferation elicited by a tumor explant. With the same procedure they also demonstrated a sharp inhibition of growth of B 16 melanoma implants under the conjunctive of C 57 Bl/6 J mice.

The antiangiogenesis effect is not specific for the cartilage extract; corneal cells (KAMINSKI and KAMINSKA, 1978) as well as extracts of the vitreous body (BREM et al., 1977a) have also been reported to decrease angiogenesis.

G. Concluding Remarks

The study of angiogenesis factor(s) is based on two working hypotheses: Firstly, the new formation of vessels is an integral part of a tissue, therefore factors controlling angiogenesis are determinants of tissue morphogenesis, normal or neoplastic. Secondly, neoplastic cells have angiogenic properties while the normal counterparts in the adult usually do not, therefore the acquisition of angiogenic capacity is a marker of progression toward a clinically evident tumor. An important corollary to the first hypothesis is that a block of angiogenesis should result in the arrest of tumor growth. A corollary to the second hypothesis is that presence of angiogenic capacity in biopsy tissues without morphologic signs of neoplasia may indicate a high risk for neoplastic transformation and thus justify preventive measures.

The new formation of a vessel requires endothelial mobilization, mitogenic activity, canalization, and formation of a basement membrane as the structural backbone of an endothelial wall. The original hypothesis of FOLKMAN and coworkers (1971) was that a factor could trigger this sequence of events. If one considers that the formation of even a simple capillary is an organogenetic event, then the angiogenesis factor should only be the initiator of the first in a cascade of events that the host tissue itself should be able to generate. Since mobilization of endothelium appears to be the first step of the angiogenesis process, the triggering event could be the concentration of an appropriate attractant for the endothelium. Present knowledge suggests that PGE_1 and copper ions could be these attractants (BEN EZRA, 1978a; McAUSLAN, 1980). Other substances, i.e., growth factors, may also have chemotactic effects on endothelium and act as triggers of angiogenesis (GOSPODAROWICZ et al., 1979).

If more than one chemoattractant for endothelium exists, one should also look for the specific conditions that permit and sustain the effect of a chemoattractant. Since PGE_1 or copper ions are normally present in tissue fluid, the simplest "activation" could be a specific increment of concentration at an appropriate location. This concentration can be obtained by direct production of the chemoattractant by the cells with angiogenic capacity or by secretion of a carrier that concentrates the chemoattractant. The fractionation procedures of tumor interstitial fluid or media of neoplastic cell cultures suggest this possibility. It is interesting that the antiangiogenic effect of cartilage extracts has been ascribed to a low molecular weight collagenase inhibitor synthesized and secreted by chondrocytes (KUETTNER et al., 1977), and interference with collagenogenesis, particularly type IV collagen, has been found to impair cell growth (KIDWELL et al., 1980).

The study of angiogenesis to date provides evidence to sustain the following statements: a) a tumor cannot develop or be transplanted unless the host provides a vascular network, and b) angiogenesis is induced by a factor(s) produced by the cells, normal or neoplastic. Moreover, evidence has been collected to sustain two working hypotheses: a) angiogenic capacity acquired by a tissue normally lacking it indicates progression toward neoplastic transformation, and b) arrest of angiogenesis may block tumor growth.

Fractionation and characterization of an angiogenesis factor(s) are progressing and appear to be indispensable steps in making both working hypotheses a practical reality.

Acknowledgements. I wish to acknowledge the help over the years of Drs. S. S. Brem, M. A. Gimbrone, Jr., A. Maiorana and M. Ziche; the expert technical assistance of Mrs. F. H. Grantham, Mrs. J. Higgs, Mrs. I. Losonczy and Mr. D. M. Hill; and the valuable contribution to the preparation of the manuscript by Ms. U. L. Walz.

References

Albrecht-Buehler, G.: The phagokinetic tracks of 3 T 3 cells. Cell *11*, 395–404 (1977)

Algire, G.H.: An adaptation of the transparent chamber technique to the mouse. J. Natl. Cancer Inst. *4*, 1–11 (1943)

Aswanikumar, S., Schiffmann, E., Corcoran, B.A., Wahl, S.M.: Role of a peptidase in phagocyte chemotaxis. Proc. Natl. Acad. Sci. U.S.A. *73*, 2439–2442 (1976)

Auerbach, R., Sidky, Y.A.: Nature of the stimulus leading to lymphocyte-induced angiogenesis. J. Immunol. *123*, 751–754 (1979)

Auerbach, R., Kubai, L., Sidky, Y.: Angiogenesis induction by tumors, embryonic tissues, and lymphocytes. Cancer Res. *36*, 3435–3440 (1976)

Ausprunk, D.H., Folkman, J.: Migration and proliferation of endothelial cells in preformed and newly formed blood vessels during tumor angiogenesis. Microvasc. Res. *14*, 53–65 (1977)

Ben Ezra, D.: Neovasculogenic ability of prostaglandins, growth factors, and synthetic chemoattractants. Am. J. Ophthalmol. *86*, 455–461 (1978 a)

Ben Ezra, D.: Mediators of immunological reactions: Function as inducers of neovascularisation. Metab. Ophthalmol. *2*, 2–4 (1978 b)

Bloom, W., Bartelmez, G.W.: Hematopoiesis in young human embryos. Am. J. Anat. *67*, 21–53 (1940)

Brem, S., Brem, H., Folkman, J., Finkelstein, D., Patz, A.: Prolonged tumor dormancy by prevention of neovascularization in the vitreous. Cancer Res. *36*, 2807–2812 (1976)

Brem, S., Preis, I., Langer, R., Brem, H., Folkman, J., Patz, A.: Inhibition of neovascularization by an extract derived from vitreous. Am. J. Ophthalmol. *84*, 323–328 (1977 a)

Brem, S.S., Gullino, P.M., Medina, D.: Angiogenesis: A marker for neoplastic transformation of mammary papillary hyperplasia. Science *195*, 880–882 (1977 b)

Brem, S.S., Jensen, H.M., Gullino, P.M.: Angiogenesis as a marker of preneoplastic lesions of the human breast. Cancer *41*, 239–244 (1978)

Clark, E.R.: Studies on the growth of blood vessels in the tail of the frog larva – by observation and experiment on the living animal. Am. J. Anat. *23*, 37–88 (1918)

Clark, E.R., Clark, E.L.: Microscopic observations on the growth of blood capillaries in the living mammal. Am. J. Anat. *64*, 251–301 (1939)

Clark, R.A., Stone, R.D., Leung, D.Y., Silver, I., Hohn, D.C., Hunt, T.K.: Role of macrophages in wound healing. Surg. Forum *27*, 16–18 (1976)

Cliff, W.J.: Observations on healing tissue: A combined light and electron microscopic investigation. Philos. Trans. R. Soc. Lond. (B) *246*, 305–325 (1963)

Cliff, W.J.: Kinetics of wound healing in rabbit ear chambers, a time lapse cinemicroscopic study. Q. J. Exp. Physiol. *50*, 79–89 (1965)

De Ome, K.B., Faulkin, L.J., jr., Bern, H.A., Blair, P.B.: Development of mammary tumors from hyperplastic alveolar nodules transplanted into gland-free mammary fat pads of female C 3 H mice. Cancer Res. *19*, 515–520 (1959)

Dumonde, D.C., Jose, P-J., Page, D.A., Williams, T.J.: Production of prostaglandins by porcine endothelial cells in culture. Br. J. Pharmacol. *61*, 504 P–505 P (1977)

Ehrmann, R.L., Knoth, M.: Choriocarcinoma: Transfilter stimulation of vasoproliferation in the hamster cheek pouch-studied by light and electron microscopy. J. Natl. Cancer Inst. *41*, 1329–1341 (1968)

Eisenstein, R., Sorgente, N., Soble, L.W., Miller, A., Kuettner, K.E.: The resistance of certain tissues to invasion: Penetrability of explanted tissues by vascularized mesenchyme. Am. J. Pathol. *73*, 765–774 (1973)

Eisenstein, R., Kuettner, K.E., Neapolitan, C., Soble, L.W., Sorgente, N.: The resistance of certain tissues to invasion. III. Cartilage extracts inhibit the growth of fibroblasts and endothelial cells in culture. Am. J. Pathol. *81*, 337–347 (1975)

Endrich, B., Intaglietta, M., Reinhold, H.S., Gross, J.F.: Hemodynamic characteristics in microcirculatory blood channels during early tumor growth. Cancer Res. *39*, 17–23 (1979)

Eriksson, E., Zarem, H.A.: Growth and differentiation of blood vessels. In: Microcirculation. Kaley, G., Altura, B.M. (eds), vol. 1, pp. 393–419. Baltimore: University Park 1977

Finkelstein, D., Brem, S., Patz, A., Folkman, J., Miller, S., Ho-Chen, C.: Experimental retinal neovascularization induced by intravitreal tumors. Am. J. Ophthalmol. *83*, 660–664 (1977)

Folkman, J.: Tumor angiogenesis. Advan. Cancer Res. *19*, 331–358 (1974a)

Folkman, J.: Tumor angiogenesis factor. Cancer Res. *34*, 2109–2113 (1974b)

Folkman, J.: Tumor angiogenesis. In: Biology of Tumors. Becker, F.F. (ed.), pp. 355–388. New York: Plenum Press 1975

Folkman, J., Cotran, R.S.: Relation of vascular proliferation to tumor growth. Int. Rev. Exp. Pathol. *16*, 207–248 (1976)

Folkman, J., Gimbrone, M.A., jr.: Perfusion of the thyroid. Acta Endocrinol. Suppl. *158*, 237–248 (1972)

Folkman, J., Merler, E., Abernathy, C., Williams, G.: Isolation of a tumor factor responsible for angiogenesis. J. Exp. Med. *133*, 275–288 (1971)

Foulds, L.: Neoplastic Development, Vol. 2. New York: Academic Press 1975

Fromer, C.H., Klintworth, G.K.: An evaluation of the role of leukocytes in the pathogenesis of experimentally induced corneal vascularization. I. Comparison of experimental models of corneal vascularization. Am. J. Pathol. *79*, 537–550 (1975a)

Fromer, C.H., Klintworth, G.K.: An evaluation of the role of leukocytes in the pathogenesis of experimentally induced corneal vascularization. II. Studies on the effect of leukocytic elimination on corneal vascularization. Am. J. Pathol. *81*, 531–544 (1975b)

Fromer, C.H., Klintworth, G.K.: An evaluation of the role of leukocytes in the pathogenesis of experimentally induced corneal vascularization. III. Studies related to the vasoproliferative capability of polymorphonuclear leukocytes and lymphocytes. Am. J. Pathol. *82*, 157–170 (1976)

Gimbrone, M.A., jr., Alexander, R.W.: Angiotensin II stimulation of prostaglandin production in cultured human vascular endothelium. Science *189*, 219–220 (1975)

Gimbrone, M.A., jr., Gullino, P.M.: Neovascularization induced by intraocular xenografts of normal, preneoplastic, and neoplastic mouse mammary tissues. J. Natl. Cancer Inst. *56*, 305–318 (1976a)

Gimbrone, M.A., jr., Gullino, P.M.: Angiogenic capacity of preneoplastic lesions of the murine mammary gland as a marker of neoplastic transformation. Cancer Res. *36*, 2611–2620 (1976b)

Gimbrone, M.A., jr., Cotran, R.S., Folkman, J.: Human vascular endothelial cells in culture. Growth and DNA synthesis. J. Cell Biol. *60*, 673–684 (1974)

Glaser, B.M., D'Amore, P.A., Michels, R.G., Patz, A., Fenseleau, A.: Identification of vasoproliferative activity from mammalian retina. J. Cell Biol. *84*, 298–304 (1980)

Goodall, C.M., Sanders, A.G., Shubik, P.: Studies of vascular patterns in living tumors with a transparent chamber inserted in hamster cheek pouch. J. Natl. Cancer Inst. *35*, 497–521 (1965)

Gospodarowicz, D., Thakral, K.K.: Production of a corpus luteum angiogenic factor responsible for proliferation of capillaries and neovascularization of the corpus luteum. Proc. Natl. Acad. Sci. U.S.A. *75*, 847–851 (1978)

Gospodarowicz, D., Moran, J., Braun, D., Birdwell, C.: Clonal growth of bovine vascular endothelial cells: Fibroblast growth factor as a survival agent. Proc. Natl. Acad. Sci. U.S.A. *73*, 4120–4124 (1976)

Gospodarowicz, D., Brown, K.D., Birdwell, C.R., Zetter, B.R.: Control of proliferation of human vascular endothelial cells. Characterization of the response of human umbilical vein endothelial cells to fibroblast growth factor, epidermal growth factor, and thrombin. J. Cell Biol. *77*, 774–788 (1978)

Gospodarowicz, D., Bialecki, H., Thakral, T.K.: The angiogenic activity of the fibroblast and epidermal growth factor. Exp. Eye Res. *28*, 501–514 (1979)

Greenblatt, M., Shubik, P.: Tumor angiogenesis: Transfilter diffusion studies in the hamster by the transparent chamber technique. J. Natl. Cancer Inst. *41*, 111–124 (1968)

Greenblatt, M., Choudari, K.V.R., Sanders, A.G., Shubik, P.: Mammalian microcirculation in the living animal: Methodologic considerations. Microvasc. Res. *1*, 420–432 (1969)

Greenburg, G.B., Hunt, T.K.: The proliferate response in vitro of vascular endothelial and smooth muscle cells exposed to wound fluids and macrophages. J. Cell Physiol. *97*, 353–360 (1978)

Greene, H.S.N.: Heterologous transplantation of mammalian tumors. I. The transfer of rabbit tumors to alien species. J. Exp. Med. *73*, 461–473 (1941 a)

Greene, H.S.N.: Heterologous transplantation of mammalian tumors. II. The transfer of human tumors to alien species. J. Exp. Med. *73*, 475–485 (1941 b)

Gullino, P.M.: Techniques for the study of tumor physiopathology. In: Methods in cancer research. Busch, H. (ed.), vol. 5, pp. 45–91. New York: Academic Press 1970

Gullino, P.M.: Angiogenesis and oncogenesis. J. Natl. Cancer Inst. *61*, 639–643 (1978)

Hauck, G.: Physiology of the microvascular system. Angiologica *8*, 236–260 (1971)

Hoffman, H., McAuslan, B., Robertson, D., Burnett, E.: An endothelial growth-stimulating factor from salivary glands. Exp. Cell Res. *102*, 269–275 (1976)

Howard, B.V., Macarak, E.J., Gunson, D., Kefalides, N.A.: Characterization of the collagen synthesized by endothelial cells in culture. Proc. Natl. Acad. Sci. U.S.A. *73*, 2361–2364 (1976)

Humes, J.L., Bonney, R.J., Pelus, L., Dahlgren, M.E., Sadowski, S.J., Kuehl, F.A., jr., Davies, P.: Macrophages synthesize and release prostaglandins in response to inflammatory stimuli. Nature *269*, 149–151 (1977)

Huseby, R.A., Currie, C., Lagerborg, V.A., Garb, S.: Angiogenesis about and within grafts of normal testicular tissue: A comparison with transplanted neoplastic tissue. Microvasc. Res. *10*, 396–413 (1975)

Intaglietta, M., Myers, R.R., Gross, J.F., Reinhold, H.S.: Dynamics of microvascular flow in implanted mouse mammary tumours. Bibl. Anat. *15*, 273–276 (1977)

Jaffee, B.M.: Prostaglandins and cancer: An update. Prostaglandins *6*, 453–460 (1974)

Jaffee, E.A., Minick, C.R., Adelman, B., Becker, C.G., Nachman, R.: Synthesis of basement membrane collagen by cultured human endothelial cells. J. Exp. Med. *144*, 209–225 (1976)

Jakob, W., Jentzsch, K.D., Mauersberger, B., Oehme, P.: Demonstration of angiogenesis activity in the corpus luteum of cattle. Exp. Pathol. *13*, 231–236 (1977)

Jakob, W., Jentzsch, K.D., Mauersberger, B., Heder, G.: The chick embryo chorioallantoic membrane as a bioassay for angiogenesis factors: Reactions induced by carried materials. Exp. Pathol. *15*, 241–249 (1978)

Joyner, W.L., Strand, J.C.: Differential release of prostaglandin E-like and F-like substances by endothelial cells cultured from human umbilical arteries and veins. Microvasc. Res. *16*, 119–131 (1978)

Kaminski, M., Kaminska, G.: Inhibition of lymphocyte-induced angiogenesis by enzymatically isolated rabbit cornea cells. Arch. Immunol. Ther. Exp. *26*, 1079–1082 (1978)

Kaminski, M., Kaminska, G., Jakobisiak, M., Brzezinski, W.: Inhibition of lymphocyte-induced angiogenesis by isolated chondrocytes. Nature *268*, 238–240 (1977)

Kaminski, M., Kaminska, G., Majewski, S.: Local graft-versus-host reaction in mice evoked by Peyer's patch and other lymphoid tissues cells tested in a lymphocyte-induced angiogenesis assay. Folia Biol. *24*, 104–109 (1978 a)

Kaminski, M., Kaminska, G., Majewski, S.: Inhibition of new blood vessel formation in mice by systemic administration of human rib cartilage extract. Experientia *34*, 490–491 (1978 b)

Kelly, P.J., Suddith, R.L., Hutchison, H.T., Werbach, K., Haber, B.: Endothelial growth factor present in tissue culture of CNS tumors. J. Neurosurg. *44*, 342–346 (1976)

Kessler, D.A., Langer, R.S., Pless, N.A., Folkman, J.: Mast cells and tumor angiogenesis. Int. J. Cancer *18*, 703–709 (1976)

Kibbey, W.E., Bronn, D.G., Minton, J.P.: Prostaglandin synthetase and prostaglandin E_2 levels in human breast carcinoma. Prostaglandins Med. *2*, 133–139 (1979)

Kidwell, W.R., Wicha, M.S., Salomon, D., Liotta, L.A.: Differential recognition of basement membrane collagen by normal and neoplastic mammary cells. In: Systematics of Mammary Cell Transformation. McGrath, C., Nandi, S. (eds.), pp. 17–32. New York: Academic 1980

Klagsbrun, M., Knighton, D., Folkman, J.: Tumor angiogenesis activity in cells grown in tissue culture. Cancer Res. *36*, 110–114 (1976)

Kuettner, K.E., Croxen, R.L., Eisenstein, R., Sorgente, N.: Proteinase inhibitor activity in connective tissue. Experientia *30*, 595–597 (1974)

Kuettner, K.E., Hiti, J., Eisenstein, R., Harper, E.: Collagenase inhibition by cationic proteins derived from cartilage and aorta. Biochem. Biophys. Res. Commun. *72*, 40–46 (1976)

Kuettner, K.E., Soble, N., Croxen, R.L., Marczynska, B., Hiti, J., Harper, E.: Tumor cell collagenase and its inhibition by a cartilage-derived protease inhibitor. Science *196*, 653–654 (1977)

Langer, R., Folkman, J.: Polymers for the sustained release of proteins and other macromolecules. Nature *263*, 797–800 (1976)

Langer, R., Conn, H., Vacanti, J., Haudenschild, C., Folkman, J.: Control of tumor growth in animals by infusion of an angiogenesis inhibitor. Proc. Natl. Acad. Sci. U.S.A. *77*, 4331–4335 (1980)

Loeb, J.: Über die Entwicklung von Fischembryonen ohne Kreislauf. Pflügers Arch. Ges. Physiol. *54*, 525–531 (1893)

Maiorana, A., Gullino, P.M.: Acquisition of angiogenic capacity and neoplastic transformation in the rat mammary gland. Cancer Res. *38*, 4409–4414 (1978)

McAuslan, B.R.: A new theory of neovascularisation based on identification of an angiogenic factor and its effect on cultured endothelial cells. In: Control Mechanisms in Animal Cells. Jimenez de Asua, L., Shields, R., Levi-Montalcini, R., Iacobelli, S. (eds.), pp. 285–292. New York: Raven Press 1980

McAuslan, B.R., Hoffman, H.: Endothelium stimulating factor from Walker carcinoma cells. Relation to tumor angiogenic factor. Exp. Cell Res. *119*, 181–190 (1979)

Medina, D.: Preneoplastic lesions in mouse mammary tumorigenesis. In: Methods in cancer research. Busch, H. (eds.), vol. 7, pp. 3–53. New York: Academic Press 1973

Merwin, R.M., Algire, G.H.: The role of graft and host vessels in the vascularization of grafts of normal and neoplastic tissue. J. Natl. Cancer Inst. *17*, 23–33 (1956)

Nalbandian, R.M., Henry, R.L.: Platelet-endothelial cell interactions. Metabolic maps of structures and actions of prostaglandins, prostacyclin, thromboxane and cyclic AMP. Semin. Thromb. Hemostas. *5*, 87–111 (1978)

Nishioka, K., Ryan, T.J.: The influence of the epidermis and other tissues on blood vessel growth in the hamster cheek pouch. J. Invest. Dermatol. *58*, 33–45 (1972)

Phillips, P., Kumar, S.: Tumour angiogenesis factor (TAF) and its neutralisation by a xenogeneic antiserum. Int. J. Cancer *23*, 82–88 (1979)

Polverini, P.J., Cotran, R.S., Gimbrone, M.A., jr., Unanue, E.R.: Activated macrophages induce vascular proliferation. Nature *269*, 804–806 (1977a)

Polverini, P.J., Cotran, R.S., Sholley, M.M.: Endothelial proliferation in the delayed hypersensitivity reaction: An autoradiography study. J. Immunol. *118*, 529–532 (1977b)

Sade, R.M., Folkman, J., Cotran, R.S.: DNA synthesis in endothelium of aortic segments in vitro. Exp. Cell Res. *74*, 297–306 (1972)

Sandison, J.C.: A new method for the microscopic study of living growing tissues by the introduction of a transparent chamber in the rabbit's ear. Anat. Rec. *28*, 281–287 (1924)

Schiffmann, E., Corcoran, B.A., Wahl, S.M.: N-Formylmethionyl peptides as chemoattractants for leucocytes. Proc. Natl. Acad. Sci. U.S.A. *72*, 1059–1062 (1975)

Schoefl, G.I.: Studies on inflammation. III. Growing capillaries: Their structure and permeability. Virchows Arch. Pathol. Anat. *337*, 97–141 (1963)

Schoefl, G.I.: Electron microscopic observations on the regeneration of blood vessels after injury. Ann. N.Y. Acad. Sci. *116*, 789–802 (1964)

Sewell, I.A.: Studies of the microcirculation using transparent tissue observation chambers inserted in the hamster cheek pouch. J. Anat. *100*, 839–856 (1966)

Sholley, M.M., Gimbrone, M.A., jr., Cotran, R.S.: The effects of leukocyte depletion on corneal neovascularization. Lab. Invest. *38*, 32–40 (1978)

Shubik, P., Feldman, R., Garcia, H., Warren, B.A.: Vascularization induced in the cheek pouch of the Syrian hamster by tumor and nontumor substances. J. Natl. Cancer Inst. *57*, 769–774 (1976)

Sidky, Y.A., Auerbach, R.: Lymphocyte-induced angiogenesis: A quantitative and sensitive assay of the graft-vs.-host reaction. J. Exp. Med. *141*, 1084–1100 (1975)

Simpson, D.M., Ross, R.: The neutrophilic leukocyte in wound repair: A study with antineutrophil serum. J. Clin. Invest. *51*, 2009–2023 (1972)

Sorgente, H., Kuettner, K.E., Soble, L.W., Eisenstein, R.: The resistance of certain tissues to invasion. II. Evidence for extractable factors in cartilage which inhibit invasion by vascularized mesenchyme. Lab. Invest. *32*, 217–222 (1975)

Stein, J.M., Levenson, S.M.: Effect of the inflammatory reaction on subsequent wound healing. Surg. Forum *17*, 484–485 (1966)

Suddith, R.L., Kelly, P.J., Hutchison, H.T., Murray, E.A., Haber, B.: In vitro demonstration of an endothelial proliferative factor produced by neural cell lines. Science *190*, 682–684 (1975)

Tannock, I.F.: Population kinetics of carcinoma cells, capillary endothelial cells, and fibroblasts in a transplanted mouse mammary tumor. Cancer Res. *30*, 2470–2476 (1970)

Tapper, D., Langer, R., Bellows, A.R., Folkman, J.: Angiogenesis capacity as a diagnostic marker for human eye tumors. Surgery *86*, 36–40 (1979)

Thakral, K.K., Goodson, W.H., III, Hunt, T.K.: Stimulation of wound blood vessel growth by wound macrophages. J. Surg. Res. *26*, 430–436 (1979)

Thoma, R.: Untersuchungen über die Histogenese und Histomechanik des Gefäßsystems. Stuttgart: Enke 1893

Tuan, D., Smith, S., Folkman, J., Merler, E.: Isolation of the nonhistone proteins of rat Walker carcinoma 256. Their association with tumor angiogenesis. Biochemistry *12*, 3159–3165 (1973)

Warren, B.A., Greenblatt, M., Kommineni, V.R.C.: Tumor angiogenesis: Ultrastructure of endothelial cells in mitosis. Br. J. Exp. Pathol. *53*, 216–224 (1972)

Weiss, J.B., Brown, R.A., Kumar, S., Phillips, P.: An angiogenic factor isolated from tumours: A potent low-molecular-weight compound. Br. J. Cancer *40*, 493–496 (1979)

Williams, R.G.: Experiments on the growth of blood vessels in thin tissue and in small autografts. Anat. Rec. *133*, 465–485 (1959)

Wolf, J.E., jr., Harrison, R.G.: Demonstration and characterization of an epidermal angiogenic factor. J. Invest. Dermatol. *61*, 130–141 (1973)

CHAPTER 16

Growth of Human Tumors in Culture

Helene S. Smith and Ch. M. Dollbaum

A. General Introduction

Development of malignancy is a complicated progressive process. Attempts to study cellular aspects in vivo have been plagued by the difficulty in separating cellular from other aspects (i.e. hormonal or immunological) in the whole organism. Furthermore cells in a tumor are in very different physiological states than their normal counterparts and these differences may obscure fundamental changes relating to malignancy. For example, the tumors contain growing, nongrowing and necrotic cells while their normal counterparts are usually in a more uniform state. In addition, studies comparing tumor and nonmalignant tissue are hampered by the fact that the tissues contain other cell types besides the one involved in the malignancy.

One way to circumvent these problems is to grow cells in culture where hormonal and other systemic influences can be examined in a controlled manner. In culture, tumor and normal cells are also in similar growth states, hence differences unique to malignancy may be magnified. A great deal of scientific effort has been directed toward developing the technology for culturing human tissues. While there has been much progress, particularly in the past decade, we are far from having developed the ideal culture system for studying malignancy.

I. The Ideal System

To define some of the problems encountered in culturing human tumor cells, it is useful to first delineate what we consider an ideal culture system:
1. The system must be capable of generating sufficient cells for biochemical studies.
2. The system must take into account the biology of malignancy.
    a) Cancer is a disease involving cell-cell interactions. The system must be able to isolate, culture, and identify each cell type within a given organ system.
    b) A given tumor is comprised of a heterogeneous population of cells. The system must culture each specimen with high efficiency.
    c) Each patient's tumor is unique. The system must culture every specimen irregardless of donor's age or pathology.
3. The cultured cells must remember their in vivo origins with regard to cell of origin and state in malignant progression.

Besides the obvious criterion that the system must be capable of generating sufficient cells for the desired studies, the ideal culture system should also take into

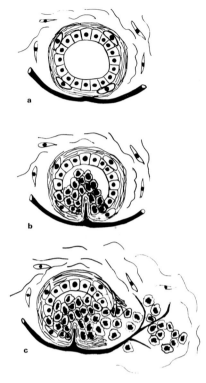

XBL793-3245

Fig. 1 a–c. Diagram of progression to malignancy of a typical mammary duct. **a** Normal mammary duct. Note uniform polar epithelial cells and myoepithelial cells surrounding the lumen separated by basement membrane from the blood vessels and stroma. **b** Atypical hyperplasia. Note the abnormal pleomorphic cells and intact basement membrane. **c** Frank carcinoma. Note the disruption of basement membrane and invasion of epithelial cells into stromal area. (SMITH et al., 1979a)

account the biology of malignancy. Cancer is a disease involving cell:cell interactions (PONTEN, 1976). Figure 1 illustrates this point by diagraming the progression to malignancy of a typical mammary duct. In the normal duct (Fig. 1 A) a uniform layer of epithelial cells with a regular and normal morphology is separated from the blood vessels and stroma by a basement membrane. As these cells progress to an atypical, premalignant state they lose growth control and proliferate extensively; morphologically, the atypical cells can become as irregular and pleomorphic as frank carcinoma cells (Fig. 1 B). The notable difference between this extreme atypia (sometimes described as "carcinoma in situ") and frank carcinoma is the relationship of the epithelial cells to the surrounding stromal cells. In contrast to the premalignant state, where the normal relationship of the various cellular components is maintained, the carcinoma cells invade the basement membrane and grow in the surrounding stromal matrix inducing blood vessel cells to proliferate and vascularize the malignant growth (Fig. 1 C). To understand the malignant process, clearly it will be necessary to decipher the contributions of all of the cell types that are involved. For example, one possibility is that in the process of becoming malignant the tumor cells themselves gain the ability to invade the stroma. Alternately, perhaps the stromal cells play an active role in this process by losing the ability to produce a signaling factor which inhibits migration and invasion of epithelial cells. Hence returning to the concept of the ideal culture system, it clearly should include the technology for isolating and separately growing all of the cell types within an organ system. In this way, the various cells can be recombined in a controllable manner. In addition, markers must be available that identify each cell type in culture.

A second relevant aspect of the biology of malignancy is that the progression from normal growth to carcinoma illustrated in figure 1 can follow alternate paths of development. It can occur by either gradual or abrupt changes, and it continues beyond the stage of primary carcinoma to metastases (FOULDS, 1969, 1975; MEDINA, 1975). Because of the nature of this malignant progression, the frank carcinoma is comprised of a heterogeneous population, only some cells of which may be invasive. Therefore, to study the biological properties of malignant cells in culture, one must develop the technology for culturing tumor cells with high efficiency. If only a small subpopulation of tumor cells within a specimen are successfully grown, the cultured cells may not be representative of the heterogeneous tumor cell populations.

Associated with each step in malignant progression, there must be changes in cellular physiology which could potentially be measured in vitro. According to the principles of malignant progression defined in vivo, a particular change might appear early on some pathways to malignancy and late on others, whereas another change may be present and important for some pathways to malignancy but not all pathways. Hence different patients with the same disease (i.e. breast cancer) may have tumors that are biologically and biochemically dissimilar. Therefore a third important aspect of an ideal culture system for studying malignancy is that every specimen should be cultured irregardless of the donor's age or pathology.

Finally, the ideal culture system must allow the tumor cells to be grown under conditions where they remember their state in vivo. One of the basic assumptions in all culture systems is that cells in culture are a valid representation of cells in vivo. It is clear that cells in culture are in a markedly different state than their in vivo counterpart; they are uniformly growing. lack interaction with various types of cells comprising a particular organ and lack a three dimensional architecture. There is much evidence that cells in culture do not behave like their in vivo counterpart. For example, cultured cells no longer express many differentiated characteristics (DAVIDSON, 1964; SCHWARZ and BISSELL, 1977). However, there is some evidence that differentiated cells in culture remember their tissue of origin and if placed under suitable conditions, will reexpress differentiated functions. For example, it has recently beeen shown that dedifferentiated cultured mouse mammary epithelial cells can be induced to express mammary specific properties by growing the cells on floating collagen gels (EMERMAN and PITELKA, 1977; EMERMAN et al., 1977). Similar results have been reported for liver (MICHALOPOULOS and PITOT, 1975). As another example, rat tracheal epithelial cells when placed in culture no longer have cilia; however, when the cells are allowed to repopulate a trachea from which the epithelium has been removed, the cultured cells again express cilia

(Steele et al., 1978). In these studies, the expression of differentiated properties depended on placing the cells in a different physical environment. In other experiments, merely manipulating the culture nutrients increased production of the differentiated product, collagen, by chick tendon cells (Schwarz and Bissell, 1977).

The concept that cells in culture remember their tissue of origin has important implications for in vitro tumor cell biology. It is also important to determine whether the tumor and normal cells remember their state in malignant progression. Richards and Nandi (1978) recently cultured mammary cells from normal and carcinomatous mouse mammary glands. The normal cells when reinoculated into cleared mammary fat pads regrew into normal mammary ducts while the carcinoma cells produced atypical and often invasive lesions. In contrast, there are numerous reports in the literature that normal cells after extensive subculture spontaneously transform and induce malignancies when reinoculated into mice.

Experiments to determine whether cells in culture remember their characteristics in vivo are particularly difficult with human cells. Obviously, one cannot reinoculate cultured tumor cells into patients. One approach to this problem has been reported by Salmon et al. (1978) who have examined the toxicity of various chemotherapy drugs to cultured human tumor cells. From the response of the cultured cells, they could accurately predict the patient's response suggesting that the cells did in fact retain their characteristic drug sensitivities in culture. There are many other examples where tumor cells in culture remember their tissue of origin and express various differentiated functions. Specific examples will be discussed under the appropriate organ system later in this chapter.

II. Current State of Technology for Culturing Tumor Cells

During the past 10 years a number of cell lines have been established in culture from a variety of human malignancies (for reviews, see Fogh and Trempe, 1975; Sinkoviks et al., 1978; Engel and Young, 1978; Rafferty, 1975). The major disadvantage of these tumor lines is that they represent small subpopulations within unusual tumor specimens that are capable of vigorous growth in culture. Typically, when tumor or normal tissue specimens are placed in culture, the normal stromal fibroblasts rapidly overgrow tumor cells or other normal cell types such as epithelial cells or melanocytes. Under these conditions, the tumor or normal epithelial cells initially form growing patches; however, after a rather low number of doublings, division ceases, the cells become vacuolated and eventually detach and die. In approximately 1%–10% of the specimens, a relatively few tumor patches continue proliferating vigorously so that they overgrow the normal fibroblasts and eventually become a cell line. Attempts have been made to remove fibroblasts by differential trypsinization (Owens et al., 1976; Smith et al., 1978). Under these conditions, carcinoma derived cell lines have been established which have not been selected for the ability to overgrow fibroblasts (Smith, 1979). Unfortunately, even when fibroblasts are removed, many of the tumor cell lines grow very slowly in culture (Owens et al., 1976; Smith et al., 1978).

The fact that currently existing cell lines represent a small subpopulation capable of vigorous growth in culture may severely limit the usefulness of cell lines in understanding the biology of malignancy. For example, growth without anchorage (i.e. in methocel or in agar) is thought to be a characteristic of tumor cells since most established tumor cell lines grow in agar. However, SMITH (1979) found that tumor cell lines which were not selected for the ability to overgrow normal fibroblasts often lacked this ability, and also grew poorly if at all in methocel. Therefore, it is possible that growth without anchorage is not necessarily a property of malignancy but rather correlates with increased growth capacity in culture. In other situations, selection for a small subpopulation capable of growth in culture may not affect the properties examined. For example, in recent studies on drug toxicity, tumor cells were plated in agar where only 1 in 10^3–10^4 cells grew into colonies. The response of these rare colony forming cells to drugs in culture predicted the patient's response. Therefore, those colonies selected for growth in culture were representative of the whole tumor population with regard to drug sensitivity (SALMON et al., 1978; HAMBURGER and SALMON, 1977).

Another problem associated with the use of cell lines is that the properties of the cells in culture may change as a function of passage. SMITH and colleagues (SMITH, 1979; SMITH et al., 1979a) found that a carcinoma derived line initially lacked plasminogen activator and could not grow on confluent monolayers. After 30–50 subcultures, the cell line gained these abilities. Studies of chromosome markers indicated that the changed growth properties were not the result of inadvertent cross contamination by another cell line. A change in growth properties as a function of passage has also been carefully documented for a melanoma line (CREASEY, 1978). This line reproducibly gained the ability to grow on contact inhibited monolayers after 20 subcultures. All of the single cell clones which were isolated at low passage from the line gained the ability to grow on monolayers. Therefore, the culture was not being overgrown by a small subpopulation capable of growth on monolayers; rather the whole population of melanoma cells was changing as a function of passage.

Despite the potential disadvantages that cell lines present, they have also been extremely useful and have contributed greatly to human tumor cell biology and biochemistry. Cell lines may be a valuable source of hormones and growth factors. In vivo, human tumors ectopically produce a variety of hormones. In culture, tumor lines have been found to produce various hormones and growth factors. For example, a human fibrosarcoma line produced a variety of growth factors related to multiplication stimulating activity (MSA) and presumably to the human somatomedins (DE LARCO and TODARO, 1978) while a melanoma line produced a factor related to nerve growth factor (TODARO et al., 1979). In other examples, hyperosmolar treatment of HeLa cells stimulated the ectopic production of gonadotropic hormones (FALLON et al., 1979) while concanavalin A stimulated chorionic gonadotropin secretion by human choriocarcinoma cells (BENVENISTE et al., 1978). Lung tumor cells ectopically produce various hormones (BERTAGNA et al., 1978; RABSON et al., 1973). Recently, STAMPFER et al. (1980) has shown that two normal human epithelial lines, one derived from fetal intestine and one from adult bladder both produce some factor(s) necessary for proliferation of primary human mammary epithelial cultures and GADJUSEK et al. (1980) have found that normal endothelial cells produce a growth factor stimulatory for smooth muscle cells in the absence of serum.

Tumor cell lines are also extremely important tools for the study of human tumor immunology. Cell lines often retain tumor specific antigens; hence, antisera made against cell lines recognize other primary tumors of a similar cell type (LIAO et al., 1979; BUBENICK et al., 1973; KENNETT and GILBERT, 1979; BRAATZ et al., 1978). Cell lines have also been extremely valuable for studies of potentially oncogenic viruses. The probability of detecting an oncogenic virus is greatly enhanced if a tumor can be grown in vitro since antibodies directed against viral antigens may bind to a virus and thus prevent its detection in vivo. Such antibodies are eliminated in the in vitro environment. Thus, Epstein-Barr virus has been detected in Burkett's lymphoma cell lines (TOOZE, 1973), C-type viruses in lymphoma lines (KAPLAN et al., 1979) and putative mammary tumor virus expression has been seen in a human mammary carcinoma line (YANG et al., 1978).

The fact that most tumor cells grow poorly in culture may simply be that most existing culture techniques were designed to optimize growth of normal fibroblasts. Tumor cells and/or other normal cell types may require different concentrations of basal salts (HAM and MCKEEHAN, 1978) different growth factors (BARNES and SATO, 1980) or substrates other than plastic or glass. An example illustrating this point is the system recently developed by GREEN and colleagues for culturing human epidermal cells. In previous attempts to grow epidermal cells, some multiplication took place but the cells could not be passaged and high initial inoculation densities were required (for a summary of references see GREEN et al., 1979). RHEINWALD and GREEN (1975) noted that cells could be serially cultivated from small inocula by supporting them with cocultivated lethally irradiated fibroblasts. The cultured epidermal cells showed differentiated behavior which imitated to a remarkable degree that of the epidermis (SUN and GREEN, 1978 a, 1978 b; RICE and GREEN, 1977). Higher plating efficiency, and long culture life-times were obtained by the addition of epidermal growth factor (RHEINWALD and GREEN, 1977) and agents known to increase cellular cyclic AMP levels such as cholera toxin (GREEN, 1978). Unfortunately, no studies on the growth properties of cultured malignant epidermal cells have been reported. However, the fact that the efficiency of normal epidermal cell culture was improved by both changes in substrate and growth factors, suggests that other cell types which previously could not be cultured might also respond to different culture conditions. Thus there have been a number of recent attempts to improve culture conditions for a variety of epithelial cell types including mammary gland, prostate, and bladder. These studies will be discussed under their specific sections.

III. Identifying Cell Types in Culture

As discussed above, one extremely important aspect of developing a culture system for studying malignancy is the ability to isolate and identify all of the different cell types within an organ. The cell types that need to be distinguished will of course vary with each organ system. Studies with some organs such as kidney or lung will be complicated by the presence of numerous differentiated cell types while other organs such as bladder will be relatively simpler. Typically, an organ specimen will contain blood vessels, smooth muscle, stromal fibroblasts, fat, blood, and lymphoid cells in addition to the epithelium. In general, fat cells usually float in medium and do not settle on the culture vessel; red blood cells lack nuclei and don't divide; macrophages stick to the plastic and will pinocytose particulate matter, but will not divide; other lymphoid cells grow in suspension and require various specialized growth factors to culture (see Chapters in this text book by CLINE and METCALF). Therefore, the cell types most commonly requiring identification include endothelial cells, smooth muscle cells, stromal fibroblasts, and epithelial cells. Generally, epithelial cells and endothelial cells have a cuboidal morphology and other similar characteristics which make identification difficult. Smooth muscle cells and fibroblasts are similar to each other, but are more readily distinguished from epithelial and endothelial cells by their elongated morphology and swirling growth pattern. Generally speaking, most reports on culturing tumor cells have been concerned with separating carcinomatous epithelial cells from elongated fibroblastoid cells. Relatively little work has been directed toward distinguishing whether these fibroblastoid cells originate from smooth muscle or stromal fibroblasts. Until recently, no one could culture capillary endothelium, therefore, there has also been very little concern with distinguishing epithelial from endothelial cells in culture. As the technology for culturing various cell types improves, however, it will become increasingly important to devise criteria for distinguishing the cell types in culture.

1. Endothelial Cells

It is now possible to obtain long term cultures of human vascular endothelial cells. Most success has been achieved with endothelial cells derived from large vessels such as the umbilical vein (for review see GIMBRONE, 1976). However, from the point of view of culturing all of the cell types within a tumor specimen, it is more important to culture capillary endothelial cells. Recently, FOLKMAN et al. (1979) reported the successful long term culture of capillary endothelial cells from a variety of mammalian organs including human adrenal gland, spleen, and foreskin. Successful culturing of capillary endothelial cells required conditioned media from cultured murine sarcoma cells, gelatin-coated culture vessels and seeding the endothelial cells with irradiated feeder cells. In addition, contaminating cell types were manually removed from the flasks.

The ability to culture human endothelial cells from various organs may be extremely important for understanding metastatic mechanisms. The metastatic process operates mainly at the level of capillary endothelium not aortic or large vein endothelium. Target site specificity of metastatic cells may be guided by information residing in the tumor cell itself (FIDLER, 1978; BRUNSON et al., 1978). Whether tumor cells in culture will adhere differently to capillary endothelial cells from different organs is now susceptible to test.

From these recent studies on culturing endothelial cells, a number of properties characterizing endothelial cells in culture have been defined. Human endothelial cells from both large vessels and capillaries grow in culture to form a monolayer of polygonal cells. Umbilical vein endothelial cells (HAUDENSCHILD et al., 1975) contain a unique ultrastructural marker, the Weibel-Palade body originally described in endothelial cells of small arteries in rats and man (WEIBEL and PALADE, 1964). These structures which are rod-shaped cytoplasmic inclusions limited by a unit membrane, have not been reported in any other cell type including fibroblasts and smooth muscle (HAUDENSCHILD et al., 1975; GIMBRONE et al., 1974; JAFFE et al., 1973a; GIMBRONE et al., 1974). It has been suggested that these bodies play a role in the release of procoagulant activity (BURRI and WEIBEL, 1968). Although Weibel-Palade bodies are rarely seen in small vessels, the structures were observed in human (but not bovine) capillary endothelial cells cultured by FOLKMAN et al. (1979). Endothelial cells share with epithelial cells the property of specialized junctions and desmosome-like structures (FRANKS and WILSON, 1977; HAUDENSCHILD et al., 1975). These structures are areas where membranes of adjacent cells come into immediate contact. Junctional complexes are not seen in either fibroblasts or smooth muscle cells.

Endothelial cells in vivo (BLOOM et al., 1973; HOYER et al., 1973; TUDDENHAM et al., 1974) and in culture (JAFFE et al., 1973b; FOLKMAN et al., 1979) also produce antihemophilic factor (factor VIII antigen) which can be detected by immunofluorescence microscopy. With the exception of platelets and megakaryocytes (HOYER et al., 1973), this localization appears to be selective for endothelial cells. Angiotensin converting enzyme has also been used as a marker for endothelial cells (FOLKMAN et al., 1979) but has also been found in brush border cells of the gut and in the proximal tubules of the kidney (HIAL et al., 1979). Normal endothelial cells in culture also have a nonthrombogenic cell surface and do not bind platelets (ZETTER et al., 1978), a property which has been used to distinguish them from epithelial cells (SMITH et al., 1979a).

2. Epithelial Cells

The textbook definition of epithelium is "a tissue composed of closely aggregated cells that are in apposition over a large part of their surface and which have very little intercellular substance" (BLOOM and FAWCETT, 1968). This definition would include endothelial cells as a type of epithelium; however, epithelial cells are generally considered to be those closely apposed cells specialized for the functions of adsorption, secretion transport, protection and sensory reception. From this all encompassing definition it is clear that epithelial cells can be very different and that the ideal way to identify them in culture would be to monitor a specific differentiated function. As discussed previously, unfortunately, many cells lose their differentiated properties when cultured on glass or plastic substrates. Attempts to define ways in which systems can be manipulated to allow cultured cells to regain differentiated properties have been severely limited by the inability to grow the cells at all. Hence one is faced with a "Catch 22" situation – we can't culture the cells because we don't know what cell type is growing and we can't develop markers for each cell type because we can't culture the cells! Operationally, most workers ignore the problem and concentrate on culturing cuboidal, closely apposed cells without concern for whether they are endothelial or epithelial in origin. The properties noted first are cuboid morphology and junctional complexes (FRANKS and WILSON, 1977). Since many epithelial cells are involved with secretion, evidence of secretory activity is used as an identifying criterion for epithelium in culture. Ultrastructural evidence of secretory activity includes well-developed Golgi apparatus, secretory droplets and dilated rough endoplasmic reticulum. In addition, many types of epithelial cells in culture produce a fluid-filled blister termed "dome" or "hemicyst" formed by localized detachment of the epithelial layer from the dish (LEIGHTON et al., 1970; AUERSPERG, 1969; LEVER, 1979). Domes result from fluid accumulation between the cell layer and the culture dish and are thought to be manifestations in culture of transepithelial transport phenomena (LEIGHTON et al., 1970).

Recently keratin has been found in various epithelial cells but not other cell types (SUN et al., 1979). Using antisera prepared against human stratum cornea, keratin has been detected by immunofluorescence microscopy in various epithelial tissues including all stratified squamous epithelia as well as epithelium of the intestinal, urinary, reproductive and respiratory tracts. Other epithelial cell types such as the parenchymal cells of the liver and pancreas did not stain. No staining was observed for endothelial cells, smooth muscle or stromal fibroblasts. Cells in culture behaved like their in vivo counterparts suggesting that keratin synthesis might be an excellent marker to identify epithelial cells in culture.

Another antiserum specific for epithelial cells has been prepared by absorbing antiserum raised against calf bladder with various nonepithelial tissues (NATHRATH et al., 1979). Whether or not this antiserum also recognizes keratins remains to be seen. By further absorbing the antiserum with other epithelial organs, NATHRATH et al. (1979) produced an antiserum specific for bladder. The antigen was present, although to a lesser extent in human bladder tumor cell lines. The presence of blood group antigens have also been used to distinguish normal epithelial cells in culture from fibroblasts (KATOH et al., 1979). Unfortunately, endothelial cells also synthesize blood group antigens in culture (GIMBRONE, 1976); therefore, this marker cannot be used to distinguish endothelial and epithelial cells.

Other workers have attempted to identify epithelial cells by developing antisera specific for a particular organ. For example, antigen prepared from milk fat globule has been used to prepare antisera specific for mammary cells (PETERSON et al., 1978; CERIANI et al., 1977, 1979). Breast stromal cells, myoepithelial cells and epithelial cells from other organs are negative for this antigen. Mammary epithelial cells in culture retain this antigen even though they are incapable of differentiating by other criteria.

3. Fibroblastoid Cells

Most studies on culturing human tumors are not concerned with the fibroblastoid cells except from the perspective of trying to eliminate them from the tumor cultures. As the techniques for culturing tumors improve, it is likely that the fibroblastoid cells will receive more attention. As discussed above, cancer is a disease of cell-to-cell interactions. Tumor cells invade stromal tissue; hence, studies on the interactions between tumor cells and stromal fibroblasts will become increasingly important. Fibroblastoid cells are the elongated, spindle-like cells which readily grow out of both tumor and nonmalignant specimens. They are likely to be a mixture of stromal fibroblasts and smooth muscle cells from capillary walls. These cells can easily be distinguished from epithelial and endothelial cells by morphology and the fact that they usually lack junctional complexes.

Distinguishing normal fibroblastoid cells from tumor cells of fibroblastic origin, sarcomas, is much more difficult since there are no uniformly accepted criteria for defining normal and malignant cells in culture (see discussion of sarcomas).

A recent review article by CHAMLEY-CAMPBELL et al. (1979) gives an excellent description of the current techniques for distinguishing smooth muscle cells from stromal fibroblasts. Under conditions where smooth muscle cells are plated densely, so that there is little proliferation, the cells will be packed with myosin containing myofilaments and will undergo spontaneous or drug induced contraction. If the smooth muscle cells are plated sparsely (the likely situation in cultures of tumor specimens) the cells grow readily in media containing 5% or 10% serum and modulate their phenotype so that they no longer contain a lot of myofilaments and cannot contract. Modulated smooth muscle cells can still be distinguished from stromal fibroblasts by a number of criteria. Antisera made against smooth muscle actin, tropomyosin, and the 55,000-dalton protein of 100-A filaments specifically stains cultured smooth muscle cells but not fibroblasts. These antisera used in an immunofluorescence test would be particularly valuable for tumor biologists since they would discriminate smooth muscle cells from fibroblasts in mixed culture, the situation most likely to be encounted when culturing tumor specimens. Morphologic differences between the two cell types are subtle but distinct; smooth muscle cells grow in overlapping monolayers or in "hills and valleys" while fibroblasts characteristically grow in concentric whorls. Subcultured smooth muscle cells produce relatively more type III collagen than cultures of fibroblasts and differ in the relative proportions of type I and type III procollagens synthesized, in the proteolytic activities involved in conversion of type I procollagen to collagen, and in the relative synthesis of the microfibrillar protein of elastic fibers (BURKE et al., 1977). Smooth muscle subcultures incorporate C^{14} lysine into desmosine and isodesmosine (the cross-linked amino acids unique to elastin), while fibroblast cultures show no desmosine formation (NARAYANAN et al., 1976). Finally, the principal glycosaminoglycans synthesized by primate arterial smooth muscle cells are dermatin sulfate and/or chondroitin sulfates A and C with little or no hyaluronic acid, whereas fibroblasts synthesize primarily hyaluronic acid with lesser amounts of dermatan sulfate and chondroitin sulfates (WIGHT and ROSS, 1975).

Another cell type present in sweat glands and mammary specimens is the myoepithelial cell. In the mammary gland, this cell type is thought to be derived from the same cuboidal stem cell as the mammary epithelial cell (BENNETT et al., 1978). Mammary myoepithelial cells function to contract the ductules for milk ejection. Morphologically, myoepithelial cells resemble smooth muscle cells and there are no completely valid criteria to distinguish myoepithelial cells from smooth muscle cells. Recently, a marker has been described which could potentially distinguish these two cell types. SUN and GREEN (1979) reported that sweat gland myoepithelial cells stained positively by immunofluorescence microscopy using antiserum to keratin. In the same assay, smooth muscle cells were negative.

IV. Identifying Tumor Vs Nonmalignant Cells in Culture

One extremely difficult problem in the field of culturing human tumors is distinguishing whether the cultured cells originate from the malignancy or are derived from normal peripheral tissue. With human cells, there are few systems where the normal counterparts of various tumor types are capable of being cultured; hence, few studies have been done comparing the properties of cultured normal and tumor cells. An additional problem is the fact that the tumors themselves are heterogeneous – they contain cells representing various states in the progression to malignancy. Within a primary tumor, the majority of cells may be capable of invading the stromal tissue of that organ, while only a few cells may be capable of all the steps involved in metastatic growth (i.e. invading blood vessels or lymph channels, surviving in the circulation, seeding and growing in another organ). Identifying tumor cells in culture requires that one have particular markers which define each stage of malignant progression.

One criterion commonly used to distinguish malignant cells is the formation of tumors when the cells are inoculated into immunosuppressed mice (PONTEN, 1976). While many human cancer derived lines produce tumors in immunosuppressed mice, other lines derived from human malignancies have not been successfully transplanted under these conditions. One explanation for the lack of success in transplanting these human tumor lines is that the cells are highly antigenic and that methods for making mice immunodeficient are imperfect. The most commonly used system for transplanting human tumor cells is the nude mouse (RYGAARD and POVLSEN, 1969; GIOVANELLA et al., 1974; GERSHWIN et al., 1978). The nude mouse has a congenital defect in thymus gland development; hence it lacks T cell mediated immunity. While tumor rejection is thought to be primarily mediated through T cells, there is a growing body of evidence suggesting that non-T-cell mediated immune pathways may be critical in graft rejection and/or in failure of tumor survival (STILES et al., 1976). Since original tumor explants do not uniformly grow in nude mice (STEEL, 1978; GIOVANELLA et al., 1978), it is not surprising that human tumor cell lines do not always grow in them either. Recently, GERSHWIN et al. (1978) reported increased tumorigenicity of various human tumor cell lines when they were inoculated into mice congenitally lacking both T and B cell immunity. Unfortunately, in these studies, normal cell lines were not included as controls.

A second difficulty with interpreting tumorigenicity data is related to the nature of the assay itself. The assay may actually be measuring metastatic ability rather than the malignant origin of tumor derived cells in culture. To perform the test, cells are trypsinized and inoculated as single cell suspensions into mouse subcutaneous tissue. Therefore, the cells must be capable of growth in the subcutaneous environment of the mouse either by growing in suspension or by anchoring into the subcutaneous tissue. In vivo only metastatic cells seed and grow as single cells or small clumps within a different tissue environment. The malignant cells within a primary tumor arise already anchored into the tissue of the originating organ; thus they are only required to grow within that organ. Within a primary tumor, only a small subpopulation may have progressed to complete metastatic ability.

Numerous other growth properties in vitro have been associated with malignancy. Most of these properties have been defined using model systems for quantitating the effects on normal cells of various oncogenic agents such as viruses and chemicals (for reviews see BARRETT and TS'O, 1978; KAKUNAGA, 1978; PONTEN, 1974; SANFORD, 1974; COLBURN et al., 1978; PASTAN and WILLINGHAM, 1978;

EASTY and EASTY, 1976; NICHOLSON, 1976). Growth properties most commonly studied include morphological changes, increased saturation density, decreased serum requirement, infinite life, and decreased anchorage requirement for growth. Generally, these in vitro measurements are compared to assays of tumorigenicity in nude mice. The property which correlates best with tumorigenicity is growth in suspension (KAHN and SHIN, 1979).

As discussed above, this correlation may be related to the fact that assays for tumorigenicity require that the cells grow in suspension. When primary human tumors are assayed for growth in agar suspension, they plate with very low efficiency (HAMBURGER and SALMON, 1977). Not surprisingly, SMITH (1979) found that human tumor cell lines derived from metastatic lesions were more likely to grow without anchorage than those derived from primary carcinomas.

Karyotype has also been used to distinguish normal from tumor cells; nearly all malignant tumors examined to date have shown chromosomal abnormalities (for reviews see NOWELL, 1975; BIEDLER, 1975; GERMAN, 1974). There has generally been a correlation between the extent of chromosome change and the stage of progression of the tumor with the most advanced tumors showing the most extensive cytogenetic alterations. The karyotypes of the metastases have generally been similar to those of the primary neoplasms, but frequently with additional superimposed cytogenetic alterations, tending toward higher ploidy and more variation in chromosome number. There is also a relationship between the chromosomal pattern and the particular organ involved. Tumors at some sites tend to have chromosome numbers with a diploid mode (e.g. cervix) while in other organs the tetraploid range is more common (e.g. colon, bladder), (NOWELL, 1975).

Karyotypic abnormalities have also been detected in premalignant lesions, suggesting that gross chromosomal rearrangements begin to appear relatively early in malignant progression. There is a rough correlation between degree of histological abnormality and the extent of cytogenetic changes; however, data is relatively sparse because of the scarcity of mitoses in slower growing benign lesions (AUERSPERG et al., 1967; KOLLER, 1972). Because premalignant lesions also show karyotypic changes, the presence of chromosomal changes in cultured tumor cells does not rule out the possibility that the cells were derived from premalignant tissue peripheral to the tumor.

In a few cases, specific chromosomal changes have been associated with certain cancers. Besides the well known Philadelphia chromosome in chronic myelogenous leukemias (NOWELL, 1975; ROWLEY, 1973; ISHIHARA et al., 1974; HAYATA et al., 1975) non-random chromosome changes have been found in human meningiomas (MARK et al., 1972), Burkett lymphomas (MANOLOV and MANOLOV, 1972) and retinoblastomas (WILSON et al., 1973). Small DNA containing spherical particles, called "double minutes," were observed in a number of different human tumor cells (BARKER and HSU, 1979) both in vitro and in vivo. These structures have not been detected in normal cells. It has also been reported that most breast carcinoma lines derived from metastatic lesions had marker translocations of chromosome number one (CRUCIGER et al., 1976). Not nearly enough studies have been done to determine whether specific chromosomal changes characterize different stages of malignant progression.

Malignancies are also characterized by expression of various oncofetal antigens, abnormal gene expressions having traits shared with fetal tissue (for an excellent review see IB SEN and FISHMAN, 1979). In vivo, these shifts in gene expression are also observed in proliferating normal tissue and in nonmalignant tissue recovering from stress. However, it is usually only in cancer that these shifts are irreversible. In addition, undifferentiated cancer cells may also exhibit a secondary amplification in gene expression which accounts for extremely high levels of expression typically associated with these tumors. One example of an oncodevelopmental protein, alpha fetoprotein, is associated with germ cell tumors of yolk sac origin (SAKASFITA et al., 1976); hepatomas (ABELEV, 1971; SELL et al., 1976) and some gastrointestinal malignancies (MASSEYEFF, 1972). A second common example is carcinoembryonic antigen (CEA) a glycoprotein which is more strongly expressed in fetal gut and gastrointestinal malignancies than in adult gastrointestinal tissue (RULE and KIRCH, 1976). Ectopic production of peptides is another type of abnormal gene expression found in tumor cells.

Expression of these oncofetal antigens are often used to characterize cells in culture. For example, many cell lines derived from carcinomas of the gastrointestinal tract produce CEA in culture (LAING et al., 1972; EGAN and TODD, 1972; TOMPKINS et al., 1974; McCOMBS et al., 1976; TOM et al., 1976; LEIBOVITZ et al., 1976). However, it is not clear whether these antigens can be used to distinguish normal from malignant cells of the same organ type. Since proliferating nonmalignant cells also express these antigens, it will be necessary to look for increased expression by cultured tumor cells, or for conditions in culture where normal cells reversibly stop synthesizing the antigens while tumor cells remain constitutive.

V. Identifying Cell-to-Cell Contamination in Culture

Another problem with culturing human tumor cells is the possibility of cross-contamination with other lines. Therefore, a number of techniques have been developed to establish the authenticity of various cell lines. One technique has been to examine the cells for allozyme phenotype. In the human population, there are many enzyme loci which have alleles whose products are easily distinguishable (HARRIS and HOPKINSON, 1976). The allozyme phenotype at any locus is fixed and except for somatic mutation, is shared by all of the cells, as well as cell lines derived from that individual. The allozyme phenotypes over a spectrum of loci for which variant alleles have been identified have been referred to as the genetic signature of the cell lines (O'BRIEN et al., 1977). Since these phenotypes have been shown to be stable irrespective of chromosome changes (AUERSPERG and GARTLER, 1970; POVEY et al., 1976), determining the genetic signature for a given line is a useful method for verifying its unique identity (O'BRIEN et al., 1977; POVEY et al., 1976; SICILIANO et al., 1979).

A second technique commonly used to monitor for cross-contamination is chromosome banding. NELSON-REES et al. (1974) have described a number of chromosome markers characteristic of the human cervical carcinoma line designated as HeLa (GEY et al., 1952). This line grows very rapidly in culture and is commonly used in laboratories throughout the world. NELSON-REES and his associates have found that many other commonly used human cell lines have marker chromosomes similar to HeLa and suggest that these lines are not bona fide unique tumor cell lines but represent cross-contamination by HeLa (NELSON-REES, 1978; NELSON-RESS and FLANDERMEYER, 1976; NELSON-REES et al., 1974). Further substantiating their claim are data on the allozyme type for the enzyme, glucose-6-phosphate dehydrogenase. This enzyme has an allozyme type (A) distinguished by electrophoretic mobility which is associated almost exclusively with a fraction of the world's Negro population (MOTULSKY, 1972). HeLa was established from a Negro donor and has this unusual type A glucose-6-phosphate dehydrogenase allozyme (GARTLER, 1967). Therefore, all lines, particularly those established from caucasian donors with the type A allozyme are suspect of being HeLa contaminants (GARTLER, 1967; PETERSON et al., 1968; MONTES DE OCA et al., 1969; POVEY et al., 1976).

While it is likely that the presence of glucose-6-phosphate dehydrogenase allozyme type A and marker chromosomes often indicate HeLa cell contamination, a recent report by PATHAK et al. (1979) indicate that the presence of these properties may not always be due to cross-contamination. PATHAK et al. (1979) document a case of malignant plural effusion in a Negro patient with breast cancer where the original uncultured tumor cells as well as the line derived from this effusion had a number of HeLa markers and type A glucose-6-phosphate dehydrogenase. The cultured effusions differed from HeLa in two other allozyme markers, acid phosphatase and phosphoglucomutase III. More recently, SATYA-PRAKASH et al. (1980), detected HeLa type marker chromosomes in many cell lines derived from metastatic breast cancers which had unique genetic signatures thereby excluding the possibility of cross contamination by HeLa cells (SICILIANO et al., 1979). These studies suggest that a common set of chromosomal rearrangements may occur (similar to the ones discussed above) which, in lines derived from Negro individuals, could be misconstrued as HeLa cell contamination. The authors suggest that extensive allozyme analysis should be undertaken before lines with HeLa type markers are discarded as contaminations. In summary, it is likely that allozyme profiles will become the method of choice for determining the unique identity of established cell lines. They are technically easier than chromosome banding and seem to be more sensitive, especially in distinguishing cross contamination by cell types other than HeLa.

B. Specific Systems

It is beyond the scope of any review article, to describe all of the studies establishing tumor cell lines or using human tumor cells in culture. In the following sections, we have attempted to stress examples illustrating the points discussed above using the recent literature not previously discussed in reviews. The book edited by FOGH (1975) is an excellent source of information on the field of human tumor cell biology prior to 1975. Especially relevant is an article by FOGH and TREMPE (1975) describing numerous cell lines developed in their laboratory as well as a summary of publications on other cell lines established from human tumors. An excellent source of information on short-term culture of human tumors is DENDY (1976).

Other pertinent reviews include ENGEL and YOUNG (1978), SINKOVIKS et al. (1978), and FOGH et al. (1977 a, b).

I. Carcinomas

1. Mammary Gland

There have been a number of attempts to culture normal human mammary epithelial cells. One source of normal mammary epithelial cells free of contaminating fibroblasts is lacteal secretions (BUEHRING, 1972; FURMANSKI et al., 1974; RUSSO et al., 1975; GAFFNEY et al., 1976; TAYLOR-PAPADIMITRIOU et al., 1977 a, b). Ductal epithelial cells are normally sloughed in milk (HOLMQUIST and PAPANICOLAOU, 1956), but the majority of adherant cells, termed "foam cells" do not divide, and show functional and morphological properties of macrophages (RUSSO et al., 1975; TAYLOR-PAPADIMITRIOU et al., 1977 a, b; LASCELLES et al., 1969). These foam cells also lack a membrane component specific for human mammary epithelial cells, the milk fat globule antigen (CERIANI et al., 1979). In addition to foam cells, cultured milk fluids also contain growing cells which have ultrastructural features of epithelium (RUSSO et al., 1975) and are positive for the milk fat globule antigen. In standard medium, these cells grow initially; but within 7–14 days, the cells stop dividing and can not be subcultured (BUEHRING, 1972). Many attempts have been made to improve culture conditions for these cells. TAYLOR-PAPADIMITRIOU et al. (1977 a, b) have obtained improved growth of mammary epithelial cells when epidermal growth factor (EGF) and cholera toxin or analogues of cyclic AMP were added to the medium (TAYLOR-PAPADIMITRIOU et al., 1980), and when the medium was buffered to pH of 6.8 (KIRKLAND et al., 1979).

However, none of the techniques described above result in extensive growth of the epithelial cells. One possible explanation is that the epithelial cells sloughed into the milk are terminally differentiated and therefore represent a population with less growth potential than a stem cell population. Recently, a technique has been developed to isolate mammary epithelial cells from reduction mammoplasties (HALLOWES et al., 1977a; STAMPFER et al., 1980). The technique involves treating dissected ductal elements with collagenase and hyaluronidase to hydrolyze the stromal matrix. Under these conditions, the ductal and alveolar elements remain as large clumps and are separated from the stromal fibroblasts, which are mostly single cells, by filtration through nylon mesh filters. Using this technique, STAMPFER et al. (1980) reported extensive growth of normal mammary epithelial cells in primary and secondary culture with doubling times of 24–48 h. The factors most stimulatory for epithelial cell growth were conditioned medium from certain human epithelial cell lines developed previously by OWENS et al. (1976). Conditioned medium from a human myoepithelial line (HACKETT et al., 1977) also contained some growth promoting activity. Using this complex medium, STAMPFER et al. (1980) found that the epithelial cells from reduction mammoplasties grew better than the epithelial cells from breast fluids. However, even under optimal conditions, the mammary epithelial cells could only be subcultured one to three times.

In all of the literature, there is only one report of an epithelial cell line developed from nonmalignant breast tissue. The line, HBL 100, was derived from breast fluids of an asymptomatic woman; however, the cells are probably not normal since they have an abnormal karyotype and will grow in soft agar. In the original report (POLANOWSKI et al., 1976), the cells were identified as being derived from mammary epithelium since they had junctional complexes and responded to lactogenic hormones by synthesizing increased amounts of isoelectrically precipitable casein. No other report of mammary specific properties has been published although the line still contains junctional complexes indicating either epithelial or endothelial origin. The problem of identifying the origins of HBL 100 illustrates some of the difficulties in working with cell lines. From the time of the initial report, the line may have undergone extensive subculture and have altered in expression of differentiation properties.

Much less work has been done to culture mammary carcinoma cells. Using the same techniques described for culturing normal cells, HALLOWES et al. (1977b) compared the properties of epithelial cultures from benign and malignant specimens. Cultures from benign specimens had healthy looking epithelial patches (termed E' cells) while cultures of primary carcinomas had both patches of E' cells and vacuolated, non-growing patches (termed E'). Cultures from metastatic lesions contained only E' cells. These observations led HALLOWES et al. (1977b) to postulate that culture conditions favored normal cell growth and that only the nonmalignant cells peripheral to tumor specimens were capable of growth in culture. Unfortunately, the conditions used permitted only minimal growth of the normal epithelial cells. BUEHRING and WILLIAMS (1976) examined the growth rates of normal and abnormal mammary epithelial cells in standard culture medium. They found that malignant cells divided at a slower rate than normal cells and that the population-doubling times of malignant cells were more heterogeneous than those of normal cells. KIRKLAND et al. (1979) reached similar conclusions and also reported that the tumor derived cells were less sensitive to pH changes than normal cells and were inhibited (rather than stimulated) by epidermal growth factor.

Because the tumor cells grow so poorly, there have been few attempts to look for markers that distinguish tumor from normal cells. For example, studies to examine chromosomal changes require rapidly dividing cells. One very interesting difference between the tumor and normal cells that has been described is that the human breast cancer cells lose selectivity in direct intercellular communication (FENTIMAN et al., 1979; FENTIMAN and TAYLOR-PAPADIMITRIOU, 1977; FENTIMAN et al., 1976). In these studies, communication is defined as a direct exchange of small molecules between touching cells and is measured by labelling one cell type with radioactive uridine, cocultivating these cells with a second, nonradioactive cell type and then measuring autoradiographically the amount of grains in unlabelled cells touching as compared to not touching the labelled donor. These studies have shown that metabolic cooperation normally is not an all-or-none phenomenon, but is selective. Mammary epithelial cells from benign tumors or milk do not communicate with breast fibroblasts, although either type can communicate well with homologous cells. Unlike normal mammary cells, the carcinoma derived epithelial cells all display an altered pattern of cell-to-cell communication. They either communicate with all cells capable of communication or they are unable to communicate with any cells.

Another difference in cell-to-cell interactions between mammary epithelial cells derived from normal and malignant tissue has recently been described by STOKER et al. (1978). By time lapse cinematography, they found that mammary fibroblast are unable to invade the cellular boundary at the edge of a growing normal epithelial patch. In contrast, the fibroblasts readily invade a boundary of malignant epithelial cells. One criticism of these very interesting studies on cell-to-cell interactions is that the abnormal social behavior of the malignant cells may be related to the fact that they are growing so poorly in culture. However, tumor derived cell lines which grow readily in culture also show similar abnormalities of interaction. These studies clearly illustrate the need to develop better culture conditions for the malignant cells. They also illustrate how cell lines can be powerful tools to substantiate studies done with primary cultures.

In contrast to normal epithelial cells, many carcinoma derived cell lines have been developed even though most specimens grow poorly in primary culture. It is not clear whether these lines represent an inherent difference between normal and tumor cells in the capacity to adapt to long-term culture or whether there have been fewer attempts to develop lines from normal tissues. Most of the tumor lines have been developed from metastatic tissue, suggesting that cells more readily adapt to culture as they progress in malignant potential. A series of nineteen long-term human breast carcinoma cell lines of metastatic origin have been established (YOUNG et al., 1971; CAILLEAU et al., 1974 a, b, 1978). As in other tumor cell systems, only approximately 10% of the specimens developed into cell lines. In most cases mesothelial cells settled first and overgrew the cultures. CAILLEU noted that the tumor cell clumps settled slowly under these conditions. By decanting the supernatant after letting the mesothelial cells settle, after an initial lag period she was able to establish the tumor cell lines. Each line was unique in culture-differing in growth rate, morphology and genetic signature. In addition, the lines varied in karyotype, with chromosomal modes ranging from 43 (hypodiploid) to 63 (hyperdiploid). Many of these lines show a common alteration in chromosome one (CRUCIGER et al., 1976). Recently, CAILLEAU and colleagues (SATYA-PRAKASH et al., 1980), have found similar markers in many, but not all other breast cancer cell lines derived from both primary and metastatic lesions. The fact that these markers are commonly seen in cancer cells suggests that they may be important in the development of the disease. Those lines which lack the marker may have lost it during establishment in culture and subsequent passage. The lines developed by CAILLEAU each have a unique genetic signature; hence the marker chromosomes are not the result of cross contamination (SICILIANO et al., 1979; PATHAK et al., 1979). Another interpretation of the data is that the marker is common to those specimens capable of growing well in culture. To test this possibility, it will be necessary to examine breast cancers in primary culture, another example of why it is increasingly important to develop culture conditions allowing growth and division of all carcinoma specimens.

Breast carcinoma derived cell lines have been used as models for studying the mechanisms of hormone action in tumors (for review see OSBORNE and LIPPMAN, 1978). HORWITZ et al. (1978) examined nine human breast cancer cell lines for various steroid hormone receptors. Receptor distribution varied considerably among the lines; however, the distribution of receptors differed from the expected distribution predicted from solid tumors. For example, only one line was positive for cytoplasmic estrogen receptors despite the fact that 50%–60% of human tumors are positive. This observation is not surprising since cell lines more readily develop from the most anaplastic tumors. KEYDAR et al. (1979) recently reported another cell line derived from the pleural effusions of a patient with breast cancer which contains cytoplasmic receptors for various steroid hormones. Another explanation for the discrepancies between data from human biopsies and those obtained from cultured cells may reflect adaption of cells to culture conditions or changes as a function of extensive subculture.

A number of studies have been done on the growth response of various mammary carcinoma lines to various hormones. One cell line in particular, MCF-7 (SOULE et al., 1973) possesses specific high affinity receptors for estradiol, progesterone, androgens, and glucocorticoids (LIPPMAN et al., 1975, 1976, 1977; OSBORNE et al., 1976; BROOKS et al., 1973). These cells respond to estrogen, androgen, and insulin with an increased rate of macromolecule synthesis. These studies with steroid-responsive cell lines strongly support the notion that a lack of receptors is uniformly associated with absence of relevant responses to steroid, but the converse is not considered true (LIPPMAN, 1976). However, one problem with many of the studies on hormone responsiveness is that some cells grown in the presence of hormones retain high intracellular levels for extensive periods after the hormone is removed from the medium (STROBL and LIPPMAN, 1979); therefore, lack of response may be artifactually caused by the cells retaining sufficient intracellular hormone levels after serum removal.

Studies on cell physiology are greatly hampered by the presence of serum in the medium. Therefore, there have been a number of recent attempts to develop serum free media for various lines (for an excellent review see BARNES and SATO, 1980; BOTTENSTEIN et al., 1979). Recently two breast carcinoma lines have been grown in hormone supplemented medium without serum. The factors required for one line, ZR-75-1, include estradiol, insulin, transferrin, dexamethasone, and triiodothyronine (ALLEGRA and LIPPMAN, 1978). These components could not replace serum for another line, MCF-7, which required insulin, transferrin, epidermal growth factor, prostaglandin F2α and fibronectin for serum free growth (BARNES and SATO, 1979). Thus, carcinoma cells from two tumors of the same organ differ in growth requirements; just as cancers differ in many other properties in vivo.

Recently, a number of new cell lines have been developed from primary carcinomas (NORQUIST et al., 1975; LASFARGUES et al., 1978; ENGEL et al., 1978; HACKETT et al., 1977; LANGLOIS et al., 1979). In one case (ENGEL et al., 1978) two lines were developed from the same person. These lines both had the same allozyme pattern giving further support to the idea that allozyme patterns are constant in culture and are good markers for cell identification. These new lines will be very interesting reagents to compare with lines derived from metastatic lesions for determining properties associated with later stages of malignant progression. Relating to malignant progression, there is one report by YUHAS et al. (1978) that breast cancer cells from pleural effusions differ from the bulk of tumor cells in solid breast cancer specimens. Cell lines from the effusion were unable to grow as organized spheroid aggregates while those from primary cancers could. Whether these studies will be confirmed with the newer lines remains to be seen.

2. Bladder and Kidney

Very few established cell lines of urinary tract origin have been described. RIGBY and FRANKS (1970) characterized the line RT-4 from a differentiated transitional cell carcinoma of the bladder. BUBENIK et al. (1973) described another line, T-24, that originated from an anaplastic bladder carcinoma. ELLIOT et al. (1974) characterized a line, 253 J, derived from a lymphnode metastasis of a patient with multiple transitional cell carcinomas of the urinary tract. In the last few years several other bladder cancer cell lines have been reported (ELLIOTT et al., 1976; O'TOOLE et al., 1976, 1978; NAYAK et al., 197 7; RASHEED et al., 1977). There are also a few lines of renal cell carcinoma (WILLIAMS et al., 1976; FOGH and TREMP, 1975; KATSUOKA, 1978).

Only one epithelial line derived from nonmalignant bladder tissue has been developed (OWENS et al., 1976) and extensively characterized. It has normal morphology and ultrastructure; the presence of junctional complexes suggest an epithelial origin (SMITH et al., 1979 a, b). The line does not grow in methocel, does not make tumors in immunosuppressed mice (SMITH, 1979), and is positive for fibronectin matrix (SMITH et al., 1979 c). The cells showed no evidence of being endothelial in origin since they were negative for Weibel-Palade bodies, lacked factor VIII antigen and possessed a thrombogenic surface (SMITH et al., 1979 a).

Bladder cancer cell lines have been used in immunological studies. The results have suggested that cells derived from transitional cell carcinomas retain tumor associated antigens (BUBENIK et al., 1973; O'TOOLE et al., 1974). Recently, O'TOOLE et al. (1978) found that lymphocytes from some patients with transitional cell carcinoma were cytotoxic to a bladder cell line in early passage; however, this selective toxicity was not detected after the cells were extensively subcultured, again illustrating how cell lines can change as a function of subculture.

The percentage of bladder tumor specimens that develop into cell lines is very low. Therefore, LEIGHTON and his associates have been concerned with developing techniques for studying bladder carcinomas in short-term culture. They found that most freshly prepared bladder tumor specimens could be kept in primary culture using three dimensional sponge matrices (ABAZA et al., 1978). Under these conditions, bladder cancers from different patients exhibit differences in histology and migration pattern. Recently, LEIGHTON et al. (1980) have developed a novel procedure for culturing stratified epithelium using histophysiologic gradients. In vivo, stratified epithelium is always constituted so that the plane of attachment to the underlying connective tissue is also the path of nutritional and oxygen exchange. In conventional tissue culture, however, the explant is attached to the glass or other support on its underside while culture medium and atmosphere provide nutrition from above. LEIGHTON et al. (1980) described a unique system for culturing rat bladder epithelial tissue using a cylindrical chamber where the epithelial cells are plated onto a collagen membrane diaphragm through which the cells are fed. Under these culture conditions bladder epithelial cells migrate out of explant pieces. After a few weeks the cells pile up and stratify in a manner similar to bladder in vivo. By autoradiography, after labelling with H^3 thymidine, DNA replication was concentrated in the zone of cells in contact with the diaphragm. To date no attempts have been made to subculture bladder cells grown in this manner (LEIGHTON, personal communication).

In contrast to bladder, WILLIAMS et al. (1976) report that most renal cell cultures grow well in primary culture and can be subcultured at least once. Very little work has been done to characterize and identify the cell types which grow out of these primary specimens. Only few of these cultures develop into lines. Using a serum free medium developed for a canine kidney cell line (TAUB et al., 1979 a), normal fetal kidney cells could also be grown in primary culture (TAUB and SATO, 1979 b).

3. Colon

Both tumor and normal colon cells can be maintained in short-term culture; however, the cells grow very poorly and rather quickly become overgrown by rapidly proliferating stromal cells (FRANKS, 1976). As with other systems, occasional tumor specimens grow well in culture and have been established into lines (FOGH and TREMPE, 1975; EGAN and TODD, 1972; GOLDENBERG et al., 1972; TOMPKINS et al., 1974; DREWINKO et al., 1976; TOM et al., 1976; LEIBOVITZ, 1975; McCOMBS et al., 1976; DEXTER et al., 1979). Many of these lines synthesize carcinoembryonic antigen (CEA), an antigen associated with gastrointestinal carcinomas. It had previously been proposed that tumor associated antigens such as CEA were not indigenous to the tumors, but were glycoproteins produced elsewhere in the body while coating the tumor cells secondarily. The fact that human colon cancer cells in long-term culture synthesized CEA proved that the antigen was actually produced by the colon carcinoma cells (GOLDENBERG et al., 1972).

There are no established cell lines derived from normal adult colon although a number of lines derived from fetal intestine have been described (OWENS et al., 1976). Three lines have normal karyology (SMITH et al., 1978), morphology, ultrastructure (SMITH et al., 1979) and growth properties (SMITH, 1979) and contain junctional complexes (SMITH et al., 1979 b) suggesting epithelial origin. They lack factor VIII antigen and bind platelets suggesting that they are not endothelial in origin. However, none of them have been examined for CEA synthesis nor do they exhibit any other specific differentiation property that would confirm their epithelial origin. The possibility that they are derived from undifferentiated mesothelial cells as suggested by FRANKS (1976) can not be ruled out. Six other normal fetal intestine lines have been reported by WHITEHEAD et al. (1979); however, no description of the origins nor attempts to identify the cell of origin were done.

WHITEHEAD et al. (1979) were concerned with comparing the fetal intestine lines with colon carcinoma lines for various aspects of membrane biosynthesis. Their results suggest that decreased glycosyltransferases may be characteristic of colon tumor cells. They examined glycosyltransferases and glycosidases in the various cell lines and also in biopsy specimens of colon carcinomas and normal peripheral tissue. In the biopsy specimens, there was no significant differences between tumor and peripheral tissue in glycosidase content while most of the glycosyltransferases were higher in normal tissue than in the tumor. In culture, the cells secreted these enzymes into the media. If one combines cell associated plus secreted enzymes as a measure of total enzyme synthesized in culture, then the cell lines were similar to the biopsy specimens. WHITEHEAD et al. (1979) also reported that the tumor cell lines synthesized a different isozyme of galactosyltransferase than the fetal intestine lines, suggesting the possibility of using this isozyme as a tumor marker. Since there are so few good tumor markers, these results could be very promising. Therefore, it will be extremely important to characterize the normal fetal intestine derived lines used as controls in the studies to identify their cell of origin.

Another comparison between tumor biopsies and cultured colon carcinoma cells has been described by HERBERMAN et al. (1975). These experiments illustrate how studies in vivo and in culture can compliment each other. By immunoelectron microscopy, they found that CEA was located closely adjacent to the plasma membrane in a colon tumor cell line while the antigen was primarily in the glycocalyx coat surrounding the plasma membrane in fresh colon cancer tissues (GOLD et al., 1970). The colon carcinoma line did not synthesize the mucoid glycocalyx present in the biopsy specimens. HERBERMAN et al. (1975) concluded that the close binding of anti-CEA antibody was rare on biopsy tumor cells because the reagent was unable to effectively penetrate the mucous coating, but that CEA, like other cell surface antigens was also found close to the surface of the plasma membrane.

A number of studies have been directed toward inducing differentiation of colon carcinoma cell lines. Increased production of CEA was found in a colon tumor cell line at stationary phase (DREWINKO and YANG, 1976). RUTSKY et al. (1979) found that a colon carcinoma cell line grown in capillary fibers organized into glandular elements which produced much more mucin and CEA than the cells did when grown under standard culture conditions. DEXTER et al. (1979) recently reported that polar solvents increased the synthesis of CEA and normal colonic mucoprotein antigens by two colon carcinoma lines. Polar solvent induction of maturation in cultured tumor cells has been shown with various murine tumors (BORENFREUND et al., 1975; DEXTER, 1977; FRIEND et al., 1971; KIMHI et al., 1976). This induction is associated with the loss of tumorigenicity and the appearance of differentiated cell products. The human colon carcinoma lines responded to the polar solvent with complete loss of clonogenicity in semisolid medium, and a marked reduction of tumorigenicity in nude mice. Removal of the polar solvent from the culture medium was accompanied by the reappearance of tumorigenicity and the original cell culture characteristics.

4. Prostate

For a number of years it has been possible to prepare primary explant cultures from prostate tissue (STONINGTON and HEMMINGSEN, 1971; BREHMER et al., 1972; SCHRODER et al., 1971; ROSE et al., 1975; KAIGHN and BABCOCK, 1975; WEBBER and STONINGTON, 1975). In most cases the epithelial cells grow poorly if at all (FRANKS, 1977) and are overgrown by fibroblasts (WEBBER, 1974; YANAGISAWA, 1977). In primary cultures, a slight improvement in growth has been noted with human serum (BOILEAU et al., 1978); however, the epithelial cells never grow well enough to be subcultured. SCHRODER and JELLINGHAUS (1976) studied 20 adenomas and

20 carcinomas by histologic examination at various times after explanting into culture. They concluded that the common feature of adenoma and carcinoma explants is that cells from benign acini become metaplastic, proliferate within the acini and epithelialize the surface of the explants. Cells derived from the actual malignancies (both primary carcinomas and metastatic tissues) died within ten days of culture. The carcinomatous and benign prostate explants were also identical in karyotype and morphology further indicating a common origin. These studies are reminiscent of the E and E′ cells present in mammary carcinoma specimens where E′ cells are throught to be the carcinoma cells, which do not proliferate in culture.

The presence of CEA is thought to be a good marker for prostate epithelial cells since primary cultures of normal, adenomatous, and carcinomatous prostate all released CEA into the medium while stromal cells from the same specimens were negative (WILLIAMS et al., 1977, 1978 a, b). There were no differences in levels of CEA released by normal and tumor tissue. This finding coupled with the fact that two specimens of metastatic human prostate carcinoma described in the same report did not release CEA may be further indication that the epithelial cells in primary cultures of prostate carcinoma are not the malignant cells.

There are only a few bona fide prostate carcinoma cell lines. Only two lines have been extensively characterized, one from a brain metastasis, DU 145 (STONE et al., 1978; MICKEY et al., 1977) and another from a bone metastasis, PC-3 (KAIGHN et al., 1978, 1979). Both lines are aneuploid, tumorigenic in nude mice and ultrastructurally similar in that they both have epithelial features and abnormal nuclei, nucleoli, and mitochondria. However, the karyotype of the PC-3 line differs significantly from that of DU 145; hence they are not cross-contaminants. Two other putative prostatic carcinoma cultures, EB-33 and MA 160 (FRALEY et al., 1970; OKADA and SCHRODER, 1974; OKADA et al., 1976) isolated in early 1970's are now suspected of being contaminated with HeLa cells inasmuch as they contain HeLa marker chromosomes and HeLa type glucose-6-phosphate dehydrogenase (NELSON-REES and FLANDERMEYER, 1976; WEBBER et al., 1977).

Recently an epithelial cell line has been developed from normal neonatal prostate tissue by KAIGHN and his associates (LECHNER et al., 1978, 1980). The authors attribute the ability to subculture the cells to the fact that a new nonenzymatic subculture procedure was used. The technique uses potassium ions to create high osmolality which caused expansion of the epithelial cells and altered their membranes to loosen cell-to-cell contacts. After high osmolality loosened these junctions, then ethylene-bis (oxyethylenenitrilo) tetracetic acid (EGTA) released the single cells from the dish. This nonenzymatic passing technique appeared to be most important for subculturing of primary and early passage cultures when cell-to-cell associations were most numerous. Another reason for successful culture related to the age of the donor. Prostate specimens from neonates grow much more readily than tissues from older donors (KAIGHN, personal communication). The cell line had a normal morphology, a normal male karyotype and when tested at early passage, was unable to grow in soft agar. The cells were defined as epithelial in origin by electron microscopy; they had junctional complexes, elements suggesting secretory activity and organized themselves into lumen-like regions similar to prostate in vivo. In addition, the cells were positive for keratin (LECHNER, personal communication) and lacked Weibel-Palade bodies suggesting epithelial rather than endothelial origin.

LECHNER and KAIGHN (1979 a, b) have compared the growth control by various nutritional factors in the normal prostate line and in one of the prostate carcinoma lines. They conclude that the carcinoma cells lose their ability to respond to growth regulatory signals present in the normal cells. The carcinoma line required considerably less fetal bovine serum protein for maximal growth than did the normal epithelial cells. The normal cells also recognized both epithelial growth factor and fibroblast growth factor as proliferation stimulating signals. Hydrocortisone, potentiated the mitogenic action of these factors. In contrast the growth rate of the malignant cell line was not increased by these factors. The normal cells also required 80 times more calcium than did the malignant cells; however, in the presence of epithelial growth factor, the calcium requirements of both normal and malignant cells were identical (LECHNER and KAIGHN, 1979 b).

The growth requirements of the normal prostate epithelial cells was also compared to normal foreskin fibroblasts. The epithelial cells and fibroblasts responded to epithelial growth factor and fibroblastic growth factor in quantitatively dissimilar ways. Epithelial growth factor was more potent than fibroblast growth factor for the epithelial cells whereas both growth factors were equally effective for fibroblasts.

II. Other Malignancies

Most human cancers are carcinomas, malignancies of epithelial cell origin. Malignancies of all other cell types combined (including connective tissue, endothelium, blood and lymphoid tissue, melanocytes, brain tissue, etc.) comprise only approximately 10% of human cancers. The biology of lymphoid and blood cells in culture will be covered in other chapters in this book.

1. Melanomas

Melanomas are malignant lesions of the melanocytes, specialized dendritic cells whose function is to synthesize and secrete the pigment, melanin. During embryonic development melanoblasts, precursor melanocytes (except those in the retinal pigment epithelium), arise in the neural crest and actively migrate to peripheral sites. The neural crest is a region between the neural plate and the remaining epidermal ectoderm at the neurula stage. It gives rise to other cell types which are therefore related to melanocytes including some cranial ganglia, the Schwann cell sheaths of peripheral nerves, leptomeninges, the branchial cartilages and the odontoblasts. In humans, melanocytes are found in skin, mucous membranes, nervous system, and the eye. All melanocytes except those of the retinal pigment epithelium can give rise to malignant melanoma (EBLING, 1970; JIMBOW et al., 1976).

Melanocytes in vivo function in intimate association with epidermal cells. In epidermis, each melanocyte is associated with approximately 30 viable keratinocytes. The melanocytes tend to be located at the dermal-ectodermal junction but produce long dendrite-like processes which extend into the epidermal layer.

Melanin pigment containing granules (melanosomes) are moved peripherally down the dendrites and are transferred to the keratinocytes by phagocytosis. Thus, melanocytes in vivo function in intimate association with epidermal cells. It is thought that the keratinocyte is a regulatory participant in melanocyte pigment synthesis, since ultraviolet light fails to stimulate melanogenesis in melanoma cells cultured in the absence of keratinocytes (Kitano and Hu, 1969). Since melanocytes, also proliferate in response to ultraviolet light, it is interesting to speculate that they may also respond to growth control signals by keratinocytes and in culture.

There have been a number of studies comparing the properties of melanomas in vivo. Histologically, melanomas are comprised of three major morphologic types – spindle, dendritic, cuboidal, or a combination of these morphologies. Morphology of the melanoma in vivo does not correlate with morphology in culture since cell lines derived from melanomas composed of either cuboidal or spindle shaped cells both produced a wide variety of morphologies in culture. However, if a single morphologic type was cloned in culture, the morphology remained stable even when grown as a tumor in nude mice (Kanzaki et al., 1979). Melanomas can contain mature melanosomes and be pigmented or be amelanotic with melanosomes of varying degrees of ultrastructural maturity. In contrast to cell morphology, tumor specimens were similar to the cell lines subsequently established from them when compared for various aspects of melanin production including presence or absence of mature melanin granules, premelanosome structures, and the relative proportion of melanosomes to mitochondria (Foa and Aubert, 1977). Melanoma cells in culture were also found to produce specific collagenases (Tane et al., 1978). The levels of collagenolytic activities of the lines in early cultures were comparable to those of human melanoma homogenates indicating that the tumor cells themselves rather than some other cell type in the tissue produced the collagenase. However, after a few subcultures, the collagenase activity drastically diminished, although a cell line which had lost collagenase activity regained it after transplantation into a nude mouse. The authors postulate that melanoma cells which were isolated from collagen ceased to produce collagenase, presumably an adaptive enzyme.

Numerous melanoma cell lines have been established and characterized as to morphology and karyotype (Fogh and Trempe, 1975; Pope et al., 1979; Zimmering et al., 1976; Qizilbash et al., 1977; Liao et al., 1975; Creasey et al., 1979; Giovanella et al., 1976; Seman et al., 1975; Quinn et al., 1977). Apparently, it is relatively easy to culture melanoma cells free of fibroblast contamination. The fact that melanomas are very cellular, rapidly proliferating tumors with little scirrhous response may explain why they grow readily in culture.

Numerous karyotypic changes have been found in the various melanoma cell lines. Most of the lines are hyperdiploid, although pseudodiploid and hypodiploid lines have been described. Cell lines derived from different individuals were identified by unique chromosomal markers while multiple cell lines derived from a single biopsy specimen or two lesions from the same patient had markers in common (Quinn et al., 1977). Even among the studies using newer chromosome banding techniques, no specific marker chromosomes common to all melanoma cell lines have been found (QUINN et al., 1977; POPE et al., 1979; ZIMMERING et al., 1976); however, GIOVANELLA et al. (1976) noted an excess of C chromosomes in the lines examined by them suggesting that hyperreduplication of one or more C chromosomes may be a specific characteristic of human melanomas.

Most of the melanoma lines grow rapidly in culture and have a high plating efficiency on plastic (COBB and WALKER, 1960; GERNER et al., 1975; GIOVANELLA et al., 1976; LIAO et al., 1975). Generally, the melanoma lines grow to relatively low saturation densities (5×10^4–20×10^4 cells/cm^2). In the few cases where it has been tested, the melanoma cell lines grow in soft agar (KANZAKI et al., 1977; CREASEY et al., 1979). Primary explants of melanoma also grow in soft agar (HAMBURGER and SALMON, 1977), and this system has been used to successfully predict response to chemotherapy drugs (SALMON et al., 1978). In at least one report (CHEN and SHAW, 1974), melanoma cells survived for an extended period in serum free medium; however, no attempts were made to subculture the cells without serum. There is only one study where normal melanocytes have been characterized extensively. GIOVANELLA et al. (1976) established four long-term cultures of normal human uveal embryo melanocytes. The cultured melanocytes synthesized melanin, had a normal karyotype, and could be subcultured a number of times, but were not permanent lines. They also differed from melanoma cells in morphology (they were flat, not refringent), and in growth properties (they were less dense at confluence and were unable to form colonies when plated sparsely on plastic). When melanoma cells were exposed to bromodeoxyuridine for long periods, GIOVANELLA et al. (1976) noted a reversible phenotypic change toward nonneoplastic characteristics. This reversible change is reminiscent of the effects of polar solvents on colon carcinoma cells described previously (DEXTER et al., 1979).

Melanoma cell lines have been used extensively for immunologic studies. This emphasis on immunology reflects the widespread belief that melanoma represents an immunogenic tumor in man. Many studies indicate that melanoma patients respond to their tumors by developing humoral and cellular immunity. The biology of the disease also suggests that the immune system may be involved. Melanomas are among the most unpredictable in oncology because some of them are highly malignant with the poorest prognosis, whereas others with a similar histologic picture may act in a relatively benign manner. In addition, they can sometimes spontaneously regress or remain dormant for extended periods. Studies on the antigenic composition of melanoma cells are complex and confusing (see GUPTA et al., 1979; LIAO et al., 1979, for a current summary of the literature and the problems in this area). Some antigens have been found in the cytoplasm whereas others have been demonstrated on the cell surface. These antigens have been defined as fetal, as common melanoma associated, as group-specific and as individual specific. There is conflicting data on the importance of common versus individually specific antigens, the relevance of tumor-associated or fetal antigens versus tumor-specific antigens as well as the effect of extent of disease on immune reactivity. Much of the confusion stems from the fact that most of the studies have not used purified antigen preparations. Therefore, the recent report describing monoclonal antibody synthesis using new hybridoma technology is especially promising (YEH et al., 1979).

2. Gliomas

Gliomas are those tumors within brain substance which arise from glia, the cells which form the myelin sheaths of the neural axons. The biology of normal and malignant glial cells has been extensively studied (for reviews see PONTEN, 1975; FRESHNEY, 1980). Normal glial cells proliferate readily in culture. They grow as diploid, elongated cells which die out usually after approximately 20 passages (PONTEN and MACINTYRE, 1968). There are no permanently established normal glial cell lines capable of infinite growth. The behavior of gliomas in culture is related to the histology of the lesion in vivo. Low grade malignancies (grade I and II) are always dominated by cells indistinguishable from normal glial cells and, like normal glia, they never develop into permanently established cell lines. The more aggressive gliomas (grade III and IV) are highly variable in culture; their behavior ranges from no attachment to rapid attachment with efficient migration and proliferation. All of the permanently established cell lines have been derived from these more aggressive specimens (PONTEN, 1975; BIGNER et al., 1977). Forty-three percent of those specimens derived from male parietotemporal tumors developed into permanent lines in contrast to twenty percent of unselected grade III–IV gliomas. The ability of any particular specimen to survive in culture was shown to be an intrinsic property of the specimen since multiple pieces from the same tumor carried in parallel behaved in the same fashion; i.e. either established lines developed from all primary cultures or cell degeneration occurred in all of them roughly simultaneously (WESTERMARK et al., 1973). All of the lines developed by PONTEN and associates were unique and uncontaminated as determined by chromosome composition and isozyme pattern. Most of the cell lines are hypotetraploid (PONTEN, 1975), although most gliomas are hypodiploid in early passages (FRESHNEY, 1980).

The surface antigenic characteristics of human glial cells have been examined. Antisera specifically recognizing only gliomas have been described (PONTEN, 1975; WAHLSTROM et al., 1974; WIKSTRAND et al., 1977) and these antisera recognize all of the human glioma lines tested. In addition the human glioma lines share antigenic activities with normal adult and fetal brain (WIKESTRAND and BIGNER, 1979).

When compared to normal glial cells, all of the glioma lines had decreased growth control in culture. The glioma lines – regardless of morphology and chromosome composition – all showed measurable deviations from the strict topoinhibition of normal glial cells (LINDGREN and WESTERMARK, 1977). In most cases, this deficient response to topoinhibition was of a quantitative rather than qualitative nature, with only a few glioma lines showing a complete absence of topoinhibition. All of the glioma lines required considerably less serum than normal glial cells to sustain growth and to reach comparable degrees of cell packing. In some, but not all cases, the gliomas grew to very high saturation densities and were able to grow on top of stationary normal glial cells. The glioma lines also exhibited less density dependent control of movement than normal glia. By electron microscopy, the glioma cell surface appeared hyperactive. Large segments of the glioma membranes were lost either in the process of endocytosis or by loss of cytoplasmic extrusions (PONTEN, 1975). Similar to various rodent systems, gliomas also have increased levels of plasminogen activator (VAHERI et al., 1976) and decreased fibronectin (TUCKER et al., 1978). Unlike the rodent systems, both normal and malignant glial cells agglutinate to the same extent with various plant lectins (GLI-

MELIUS et al., 1974). All of the characteristics described above relate to the more aggressive gliomas since they were the ones which developed into cell lines. How the cells progress to this aggressive state is still unknown.

3. Sarcomas

Sarcomas are malignancies of fibroblastic cells. All of the problems with culturing carcinomas also apply to sarcomas. As with carcinomas, the normal stromal tissue proliferates readily. Because the sarcoma cells are of fibroblastic origin, it is extremely difficult to distinguish the tumor cells from normal stroma in mixed cultures. Occasionally, cell lines develop from sarcomas (FOGH and TREMPE, 1975; RASHEED et al., 1974; MCALLISTER et al., 1975; SINKOVIKS et al., 1978). However, these lines probably represent the most aggressive and abnormal variants which are capable of overgrowing normal fibroblasts in culture.

Recently, there have been attempts to search for markers which might distinguish normal from malignant fibroblasts. For example, differences have been detected between cultured normal fibroblasts and osteosarcoma cells in levels of alkaline phosphatase (THORPE et al., 1979) and synthesis of various collagen types (STERN et al., 1980). These studies illustrate one important difficulty with studies on cultured human sarcoma cells. In both cases skin fibroblasts were used as normal controls while the malignancies were of bone origin. The assumption that all stromal fibroblastoid lines are identical has recently been challenged. Fibroblastoid lines derived from normal lung differed in culture from fibroblastoid cells derived from skin (DOLLBAUM, 1978; SCHNEIDER et al., 1977). Thus, it is extremely important to compare tumor cells with their normal fibroblastic cell of origin. For example, two sarcoma derived lines were considered to be malignant in origin because they were capable of growing on contact inhibited monolayers (SMITH et al., 1976). However, these lines were derived from lung metastases. Unlike other normal fibroblastoid cells, those derived from normal lung also grew on monolayers (DOLLBAUM, 1978).

Other recent studies with cultured human fibroblasts have focused attention on the role of interactions between normal stroma and cancer cells in the induction of malignancy. KOPELOVICH and colleagues found that the skin fibroblasts were abnormal in patients with familial adenomatosis of the colon and rectum, a congenital disease resulting in a high incidence of colon carcinomas. In culture, these fibroblasts had some properties associated with rodent cell transformation including growth in low serum, increased saturation density, disorganization of actin cables but not growth in agar (PFEFFER et al., 1976; KOPELOVICH et al., 1977). The fibroblasts of these patients also showed an increased frequency of transformation by oncogenic viruses when contrasted to fibroblasts from normal donors (PFEFFER and KOPELOVICH, 1976). Some of the asymptomatic offspring of these patients also had abnormal fibroblasts. These studies suggest the possibility than an abnormability of the stromal tissue (fibroblasts) may be a fundamental cause of an epithelial malignancy. SMITH et al. (1976) reported that normal skin fibroblasts from patients with osteogenic sarcoma showed some growth abnormalities in culture again raising the possibility that a defect in all stromal tissue might be involved in induction of this malignancy.

C. Summary

Numerous cell lines established from various human cancers have been invaluable for many types of studies. However, from the perspective of understanding the biology of malignancy, established lines are limited by the fact that only a small minority (1%–10%) of specimens develop into lines. Furthermore, with the exception of stromal fibroblasts or glial cells, there are very few lines derived from normal human tissues. Most tumor specimens grow to a limited degree in primary culture. Usually, the primary cultures contain a variety of cell types in addition to the tumor cells, including cells from vascular endothelium, smooth muscle, stromal fibroblasts and normal organ cells peripheral to the tumor. New techniques for identifying the various cell types will greatly aid attempts to develop culture conditions capable of reproducibly growing tumor and normal cells. Recent studies along these lines have focused on using altered nutrients and newly discovered growth factors as well as a variety of substrates other than standard plastics. These new culture systems will have a major impact on our understanding of human malignancy.

References

Abaza, N., Leighton, J., Zajac, B.: Clinical bladder cancer in sponge matrix tissue culture. Cancer 42, 1364–1374 (1978)

Abelev, G.I.: Alpha-fetoprotein in otogenesis and its association with malignant tumors. Adv. Cancer Res. 14, 295–358 (1971)

Allegra, J.C., Lippman, M.E.: Growth of a human breast cancer cell line in serum-free hormone supplemented medium. Cancer Res. 38, 3823–3829 (1978)

Auersperg, N., Corey, M.J., Worth, A.: Chromosomes in preinvasive lesions of the human uterine cervix. Cancer Res. 27, 1394–1408 (1967)

Auersperg, N.: Histogenetic behavior of tumors. I. Morphologic variation in vitro and in vivo of two related human carcinoma cell lines. J. Natl. Cancer Inst. 43, 151–173 (1969)

Auersperg, N., Gartler, S.M.: Isoenzyme stability in human heteroploid cell lines. Exp. Cell Res. 61, 465–469 (1970)

Barker, P.E., Hsu, T.C.: Double minutes in human carcinoma cell lines, with special reference to breast tumors. J. Natl. Cancer Inst. 62, 257–261 (1979)

Barnes, D., Sato, G.: Growth of a human mammary tumour cell line in a serum-free medium. Nature 5730, 388–389 (1979)

Barnes, D., Sato, G.: Method for growth of cultured celles in serum-free medium. Anal. Biochem. 102, 255–270 (1980)

Barrett, J.C., Ts'o, P.O.P: Mechanistic studies of neo-plastic transformation of cells in culture in polycyclic hydrocarbons and cancer, vol. 2, pp. 235–267. New York: Academic Press 1978

Bennett, D.C., Peachey, L.A., Durbin, H., Rudland, P.S.: A possible mammary stem cell line. Cell 15, 283–298 (1978)

Benveniste, R., Speeg, K.V., Long, A., Rabinowitz, D.: Concanavalin-A stimulates human chorionic gonadotropin (hCG) and hCG-alpha secretion by human choriocarcinoma cells. Biochem. Biophys. Res. Commun. 84, 1082–1087 (1978)

Bertagna, X.Y., Nicholson, W.E., Sorenson, G.D., Pettengill, O.S., Ortl, D.N.: Endorphin (END) corticotropin (ACTH) and lipotropin (LPH) production by a human nonpituitary tumor in tissue culture: Evidence for a common precursor. Clin. Res. 26, 489 A (1978)

Biedler, J.L.: Chromosome abnormalities in human tumor cells in culture. In: Human tumor cells in vitro. Fogh, J. (ed.), pp. 359–394. New York: Plenum Press 1975

Bigner, D.D., Markesberry, W.R., Pegram, C.N., Westermark, B., Porter, J.: Progressive neoplastic growth in nude mice of cultured cell lines derived from human gliomas. J. Neuropathol. Exp. Neurol. *36*, 593–593 (1977)

Bloom, A.L., Giddings, J.C., Wilks, C.J.: Factor VIII on the vascular intima: Possible importance in haemostasis and thrombosis. Nature New Biol. *241*, 217–219 (1973)

Bloom, W., Fawcett, D.W.: A Textbook of Histology. Philadelphia: Saunders Co. 1968

Boileau, M., Keenan, E., Kemp, E., Lawson, R., Hodges, C.V.: The effect of human serum 3 H-thymidine incorporation in human prostate tumors in tissue culture. J. Urol. *119*, 777–779 (1978)

Borenfreund, E., Steinglass, M., Korngold, G., Benedich, A.: Effect of dimethylsulfoxide and dimethylformamide on the growth and morphology of tumor cells. Ann. N.Y. Acad. Sci. *243*, 164–171 (1975)

Bottenstein, J., Hayashi, I., Hutchings, S., Masui, H., Mather, J., McClure, D.B., Ohasa, S., Rizzino, A., Sato, G., Serrero, G., Wolfe, R., Wu, R.: The growth of cells in serum-free hormone-supplemented media. Methods Enzymol. *58*, 94–109 (1979)

Braatz, J.A., McIntire, K.R., Princler, G.L., Kortright, K.H., Herbeman, R.B.: Purification and characterization of a human lung tumor-associated antigen. J. Natl. Cancer Inst. *61*, 1035–1045 (1978)

Brehmer, B., Marquardt, H., Madsen, P.O.: Growth and hormonal response of cells derived from carcinoma and hyperplasia of the prostate in monolayer cell culture. A possible in vitro model for clinical chemotherapy. J. Urol. *108*, 890–896 (1972)

Brooks, S.C., Locke, E.R., Soule, H.D.: Estrogen receptor in a human cell line (MCF-7) from breast carcinoma. J. Biol. Chem. *248*, 6251–6253 (1973)

Brunson, K.W., Beattie, G., Nicholson, G.L.: Selection and altered properties of brain-colonizing metastatic melanoma. Nature *272*, 543–545 (1978)

Bubenik, J., Baresova, M., Viklicky, V., Jakoubkova, J., Sainerova, H., Donner, J.: Established cell line of urinary bladder carcinoma (T24) containing tumor-specific antigen. Int. J. Cancer *11*, 765–773 (1973)

Buehring, G.C.: Culture of human mammary epithelial cells: Keeping abreast with a new method. J. Natl. Cancer Inst. *49*, 1433–1434 (1972)

Buehring, G.C., Williams, R.R.: Growth rates of normal and abnormal human mammary epithelia in cell culture. Cancer Res. *36*, 3742–3747 (1976)

Burke, J.M., Balian, G., Ross, R., Bornstein, P.: Synthesis of types I and III procollagen and collagen by monkey aortic smooth muscle cells in vitro. Biochemistry *16*, 3243–3249 (1977)

Burri, P.H., Weibel, E.R.: Beeinflussung einer spezifischen cytoplasmatischen Organelle von Endothelzellen durch Adrenalin. Zellforsch. Mikrosk. Anat. *88*, 426–440 (1968)

Cailleau, R., Young, R., Olive, M., Reeves, W.J.: Breast tumor cell lines from pleural effusions. J. Natl. Cancer Inst. *53*, 661–674 (1974a)

Cailleau, R., MacKay, B., Young, R.K., Reeves, W.M.: Tissue culture studies on pleural effusions from breast carcinoma patients. Cancer Res. *34*, 801–809 (1974b)

Cailleau, R., Olive, M., Cruciger, Q.V.J.: Longterm human breast carcinoma cell lines of metastatic origin: Preliminary characterization. In Vitro *14*, 911–915 (1978)

Ceriani, R.L., Thompson, K., Peterson, J.A., Abraham, S.: Surface differentiation of human mammary epithelial cells carried on the milk fat globule. Proc. Natl. Acad. Sci. U.S.A. *74*, 582–586 (1977)

Ceriani, R.L., Taylor-Papadimitriou, J., Peterson, J.A., Brown, P.: Characterization of cells cultured from early lactation milks. In Vitro *15*, 356–362 (1979)

Chamley-Campbell, J., Campbell, R., Ross, R.: The smooth muscle cell in culture. Physiol. Rev. *59*, 1–61 (1979)

Chen, T.R., Shaw, M.W.: Studies of a cell line derived from a human malignant melanoma. In Vitro *10*, 216–224 (1974)

Cobb, J.P., Walker, D.G.: Studies on human melanoma cells in tissue culture. I. Growth characteristics and cytology. Cancer Res. *20*, 858–867 (1960)

Colburn, N.H., Vorder Bruegge, W.F., Bates, J., Yuspa, S.H.: Epidermal cell transformation in vitro. In: Carcinogenesis. 2. Mechanisms of tumor promotion and cocarcinogenesis. Slaga, T.S., Sivak, A., Boutwell, R.K. (eds.), pp. 257–271. New York: Raven Press 1978

Creasy, A.A.: The biology of human malignant melanoma cells in tissue culture. Ph. Dissertation, U. of Calif. Berkeley (1978)

Creasy, A.A., Smith, H.S., Hackett, A.J., Fukuyama, K., Epstein, W.L., Madin, S.H.: Biological properties of human melanoma cells in culture. In Vitro *15*, 342–350 (1979)

Cruciger, Q.V., Pathak, S., Cailleau, R.: Human breast carcinomas: Marker chromosomes involving lq in seven cases. Cytogenet. Cell Genet. *17*, 231–235 (1976)

Davidson, E.H.: Differentiation in monolayer tissue culture cells. Adv. Genet. *12*, 143–280 (1964)

DeLarco, J.E., Todaro, G.J.: A human fibrosarcoma cell line producing multiplication stimulating activity (MSA)-related peptides. Nature *272*, 356–358 (1978)

Dendy, P.P.: Human tumor in short term culture: Techniques and clinical applications. New York: Academic Press 1976

Dexter, D.L.: N, N dimethylformamide-induced morphological differentiation and reduction of tumorigenicity in cultured mouse rhabdomyosarcoma cells. Cancer Res. *37*, 3136–3140 (1977)

Dexter, D.L., Barbosa, J.A., Calabresi, P.: N, N diamethylformamide-induced alteration of cell culture characteristics and loss of tumorigenicity in cultured human colon carcinoma cells. Cancer Res. *39*, 1020–1025 (1979)

Dollbaum, C.M.: In vivo and vitro studies of the metastatic process. Phd. Dissertation U. of Calif. Berkeley (1978)

Drewinko, B., Yang, L.Y.: Restriction of CEA synthesis to the stationary phase of growth of cultured human colon carcinoma cells. Exp. Cell Res. *101*, 414–416 (1976)

Drewinko, B., Romsdahl, M.M., Yang, L.Y., Ahearn, M.J., Trujillo, J.M.: Establishment of a human carcinoembryonic antigen-producing colon adenocarcinoma cell line. Cancer Res. *36*, 467–475 (1976)

Easty, G.C., Easty, D.M.: Tissue culture methods in cancer research in scientific foundations of oncology. Symington, T., Carter, R.L. (ed.), pp. 15–25. London: William Heinemann Medical Books Ltd. 1976

Ebling, F.J.G.: The embryology of skin. In: An introduction to the biology of the skin. Champion, R.H., Gillman, T., Rook, A.J., Sims, R.T. (eds.), pp. 23–24. Philadelphia, PA: F.A. Davos Co. 1970

Egan, M., Todd, C.W.: Carcinoembryonic antigen: Synthesis by a continuous line of adenocarcinoma cells. J. Natl. Cancer Inst. *49*, 887–889 (1972)

Elliott, A.Y., Cleveland, P., Cervenka, J., Castro, A.E., Stein, N., Hakala, T.R., Fraley, E.E.: Characterization of a cell line from human transitional cell carcinoma of the urinary tract. J. Natl. Cancer Inst. *53*, 1341–1349 (1974)

Elliott, A.Y., Bronson, D.L., Stein, N., Fraley, E.E.: In vitro cultivation of epithelial cells derived from tumors of the human urinary tract. Cancer Res. *36*, 365–369 (1976)

Emerman, J.T., Pitelka, D.R.: Maintenance and induction of morphological differentiation in dissociated mammary epithelium on floating collagen membranes. In Vitro *13*, 316–328 (1977)

Emerman, J.T., Enami, J., Pitelka, D.R., Nandi, S.: Hormonal effects on intracellular and secreted casein in cultures of mouse mammary epithelial cells on floating collagen membranes. Proc. Natl. Acad. Sci. *74*, 4466–4470 (1977)

Engel, L.W., Young, N.A.: Human breast carcinoma cells in continuous culture: A review. Cancer Res. *38*, 4327–4339 (1978)

Engel, L.W., Young, N.A., Tralka, T.S., Lippman, M.E., O'Brien, S.J., Joyce, M.J.: Establishment and characterization of three new continuous cell lines derived from human breast carcinomas. Cancer Res. *38*, 3352–3364 (1978)

Fallon, R.J., Cox, R.P., Ghosh, N.K.: Induction of human choriogonadotropin and follitropin in HeLa cell cultures by hyperosmolality. J. Cell. Physiol. *98*, 613–618 (1979)

Fentiman, I.S., Taylor-Papadimitriou, J.: Cultured human breast cancer cells lose selectivity in direct intercellular communication. Nature *269*, 156–157 (1977)

Fentiman, I.S., Taylor-Papadimitriou, J., Stoker, M.: Selective contact-dependent cell communication. Nature *264*, 760–762 (1976)

Fentiman, I.S., Hurst, J., Ceriani, R.L., Taylor-Papadimitrious, J.: Junctional intercellular communication pattern of cultured human breast cancer cells. Cancer Res. *39*, 4739–4743 (1979)

Fidler, I.J.: Tumor heterogeneity and the biology of cancer invasion and metastasis. Cancer Res. *38*, 2651–2660 (1978)

Foa, C., Aubert, C.: Ultrastructural comparison between cultured and tumor cells of human malignant melanoma. Cancer Res. *37*, 3957–3963 (1977)

Fogh, J. (ed.): Human tumor cells in vitro. New York: Plenum Press 1975

Fogh, J., Trempe, G.: New human tumor cell lines. In: Human cells in vitro. Fogh, J. (ed.), pp. 115–159. New York: Plenum Press 1975

Fogh, J., Fogh, J.M., Orfeo, T.: One hundred and twenty-seven cultured human tumor cell lines producing tumors in nude mice. J. Natl. Cancer Inst. *59*, 221–226 (1977a)

Fogh, J., Wright, W.C., Loveless, J.D.: Absence of HeLa cell contamination in 169 cell lines derived from human tumors. J. Natl. Cancer Inst. *58*, 209–214 (1977b)

Folkman, J., Haudenschild, C.C., Zetter, B.R.: Longterm culture of capillary endothelial cells. Proc. Natl. Acad. Sci. U.S.A. *76*, 5217–5221 (1979)

Foulds, L.: Neoplastic development I. New York: Academic Press 1969

Foulds, L.: Neoplastic development II. New York: Academic Press 1975

Fraley, E.E., Ecker, S., Vincent, M.M.: Spontaneous in vitro neoplastic transformation of adult human prostatic epithelium. Science *170*, 540–542 (1970)

Franks, L.M.: Cell and organ culture techniques applied to the study of carcinoma of colon and rectum. Path. Europ. *11*, 167–177 (1976)

Franks, L.M.: Research methods; tissue culture in the investigation of prostatic cancer. Urol. Res. *5*, 159–162 (1977)

Franks, L.M., Wilson, P.D.: Origin and ultrastructure of cells in vitro. Int. Rev. Cytol. *48*, 55–139 (1977)

Freshney, R.I.: Tissue culture of glioma of the brain. Brain tumors: Scientific basis, clinical investigation and current therapy. Thomas, D.G.T., Graham, D.I. (eds.). London: Butterworth and Comp. (in press) (1980)

Friend, C.W., Scher, W., Holland, J.G., Sato, T.: Hemoglobin synthesis in murine virus-induced leukemic cells in vitro: Stimulation of erythroid differentiation by dimethylsulfoxide. Proc. Natl. Acad. Sci. U.S.A. *68*, 378–382 (1971)

Furmanski, P., Longley, C., Fouchey, D.: Normal human mammary cells in culture: Evidence for ancornavirus-like particles. J. Natl. Cancer Inst. *52*, 975–977 (1974)

Gadjusek, C., Decorletto, P., Ross, R., Schwartz, S.: An endothelial cell derived growth factor. J. Cell Biol. *85*, 467–472 (1980)

Gaffney, E.V., Polanowski, R.P., Blackburn, S.E., Lambiase, J.P.: Origin, concentration and structural features of human mammary gland cells cultivated from breast secretions. Cell Tissue Res. *172*, 269–279 (1976)

Gartler, S.M.: Apparent HeLa cell contamination of human heteroploid cell lines. Nature *217*, 750–751 (1967)

German, J.: Chromosomes and cancer. New York: John Wiley & Sons Inc. 1974

Gerner, R., Kitamura, H., Moore, G.: Studies of tumor cell lines derived from patients with malignant melanoma. Oncology *31*, 31–43 (1975)

Gershwin, M.E., Ideka, R.M., Erickson, K., Owens, R.: Enhancement of heterotransplanted human tumor graft survival in nude mice treated with antilymphocyte serum and in congenitally athymic-asplenic (lasat) mice. J. Natl. Cancer Inst. *61*, 245–248 (1978)

Gey, G.O., Coffman, W.D., Kubicek, M.T.: Tissue culture studies of the proliferative capacity of cervical carcinoma and normal epithelium. Cancer Res. *12*, 264–265 (1952)

Gimbrone, M.A.: Culture of vascular endothelium. In: Progress in hemostasis and thrombosis. Spaet, T.H. (ed.), vol. III, pp. 1–28. New York: Grune & Stratton 1976

Gimbrone, M.A., Cotran, R.S., Folkman, J.: Human vascular endothelial cells in culture. Growth and DNA synthesis. J. Cell Biol. *60*, 673–684 (1974)

Giovanella, B.C., Stehlin, J.S., Williams, L.R.: Heterotransplantation of human malignant tumors in "nude" thymusless mice. II. Malignant tumors induced by injection of cell cultures derived from human solid tumors. J. Natl. Cancer Inst. *52*, 921–930 (1974)

Giovanella, B., Stehlin, J.S., Santamaria, C., Yim, S.O., Morgan, A.C.: Human neoplastic and normal cells in tissue culture. I. Cell lines derived from malignant melanomas and normal melanocytes. J. Natl. Cancer Inst. *56*, 1131–1142 (1976)

Giovanella, B.C., Stehlen, J.S., Williams, L.J., Lee, S., Shepard, B.C.: Heterotransplantation of human cancers into nude mice. Cancer 42, 2269–2281 (1978)

Glimelius, B., Westermark, B., Ponten, J.: Agglutination of normal and neoplastic human
cells by concamalin A and ricinus communis agglutinin. Int. J. Cancer 14, 314–326
(1974)

Gold, P., Krupey, J., Ansari, H.: Position of the carcinoembryonic antigen of the human
digestive system in ultrastructure of tumor cell surface. J. Natl. Cancer Inst. 45, 219–222
(1970)

Goldenberg, D.M., Pavia, R.A., Hansen, H.J., Vandevoorden, J.P.: Synthesis of carcinoembryonic antigen in vitro. Nature New Biol. 239, 189–191 (1972)

Green, H.: Cyclic AMP in relation to proliferation of the epidermal cell: A new view. Cell
15, 801–811 (1978)

Green, H., Kehinde, O., Thomas, J.: Growth of cultured human epidermal cells into multiple epithelia suitable for grafting. Proc. Natl. Acad. Sci. U.S.A. 76, 5665–5668 (1979)

Gupta, R.K., Irie, R.F., Chee, D.O., Kern, D.H., Morton, D.L.: Demonstration of two distinct antigens in spent tissue culture medium of a human malignant melanoma cell line.
J. Natl. Cancer Inst. 63, 347–356 (1979)

Hackett, A.J., Smith, H.S., Springer, E.L., Owens, R.B., Nelson-Rees, W.A., Riggs, J.L.,
Gardner, M.B.: Two syngeneic cell lines from human breast tissue: The aneuploid mam-
mary epithelial (Hs 578 T) and the disploid myoepithelial (Hs 578 Bst) cell lines. J. Natl.
Cancer Inst. 58, 1795–1806 (1977)

Hallowes, R.C., Rudland, P.S., Hawkins, R.A., Lewis, D.J., Bennett, D., Durbin, H.: Comparison of the effects of hormones on DNA synthesis in cell cultures of non-neoplastic
and neoplastic mammary epithelium from rats. Cancer Res. 37, 2492–2504 (1977a)

Hallowes, R.C., Millis, R., Piggot, D., Shearer, M., Stoker, M.G., Taylor-Papadimitriou,
J.: Results on a pilot study of cultures of human lacetal secretions and benign and malignant breast tumours. Clin. Oncol. 3, 81–90 (1977b)

Ham, R.G., McKeehan, W.L.: Development of improved media and culture conditions for
clonal growth of normal diploid cells. In Vitro 14, 11–22 (1978)

Hamburger, A.W., Salmon, S.E.: Primary bioassay of human tumor stem cells. Science 197,
461–463 (1977)

Harris, H., Hopkinson, D.A.: Handbook of enzyme electrophoresis in human genetics. Oxford: North Holland Publ. Co. 1976

Haudenschild, C.C., Cotran, R.S., Gimbrone, M.A., Folkman, J.: Fine structure of vascular
endothelium in culture. J. Ultrastruct. Res. 50, 22–32 (1975)

Hayata, I., Sukurai, M., Kakati, S., Sandberg, A.A.: Chromosomes and causations of human cancer and leukemia, XVI. Banding studies of chronic myelocytic leukemia, including five unusual Ph¹ translocations. Cancer 36, 1177–1191 (1975)

Herberman, R.B., Aoki, T., Cannon, G., Liu, M., Sturm, M.: Location by immunoelectron
microscopy of carcinoembryonic antigen on cultured adenocarcinoma cells. J. Natl.
Cancer Inst. 55, 797–799 (1975)

Hial, V., Gimbrone, M.A., Peyton, M.P., Wilcox, G.M., Pisano, J.J.: Angiotensin metabolism by cultured human vascular endothelial and smooth muscle cells. Microvasc. Res.
17, 314–329 (1979)

Holmquist, D.G., Papanicolaou, G.N.: Exfoliative cytology of mammary gland during
pregnancy and lactation. Acad. Sci. 63, 1422–1435 (1956)

Horwitz, K.B., Zava, D.T., Thilagar, A.K., Jensen, E.M., McGuire, W.L.: Steroid receptor
analysis of nine human breast cancer cell lines. Cancer Res. 38, 2434–2437 (1978)

Hoyer, L.W., Santos, R.P. de los, Hoyer, J.R.: Antihemophilic factor antigen. Localization
in endothelial cells by immunofluorescent microscopy. J. Clin. Invest. 52, 2737–2744
(1973)

Ibsen, K.H., Fishman, W.H.: Developmental gene expression in cancer. Biochim. Biophys.
Acta 560, 243–280 (1979)

Ishihara, T., Kohno, S.I., Kumatori, T.: Ph¹ translocation involving chromosomes 21 and
22. Brit. J. Cancer 29, 340–342 (1974)

Jaffee, E.A., Nachman, R.L., Becker, C.G.: Culture of human endothelial cells derived from
umbilical veins. Identification by morphological and immunologic criteria. J. Clin. Invest. 52, 2745–2756 (1973a)

Jaffe, E.A., Hoyer, L.W., Nachman, R.L.: Synthesis of antihemophilic factor antigen by cultured human endothelial cells. J. Clin. Invest. *52*, 2757–2764 (1973b)

Jimbow, K., Quevedo, W.C., Fitzpatrick, T.B., Szabo, G.: Some aspects of melanin biology 1950–1975. J. Invest. Dermatol. *67*, 72–89 (1976)

Kahn, P., Shin, S.: Cellular tumorigenicity in nude mice. J. Cell Biol. *82*, 1–16 (1979)

Kaighn, M.E., Babcock, M.S.: Monolayer cultures of human prostate cells. Cancer Chem. Rep. *59*, 59–63 (1975)

Kaighn, M.E., Lechner, J.F., Narayan, K.S., Jones, L.W.: Prostate carcinoma: Tissue culture cell lines. Natl. Cancer Inst. Monogr. *49*, 17–27 (1978)

Kaighn, M.E., Narayan, K.S., Ohnuki, Y., Lechner, J.F., Jones, L.W.: Establishment and characterization of a human prostatic carcinoma cell line (PC-3). Invest. Urol. *17*, 16–23 (1979)

Kakunaga, T.: Factors affecting polycyclic hydrocarbon-induced cell transformation. In: Polycyclic hydrocarbons and cancer 2. Gelvoin, H.V., Ts'o, P.O.P. (eds.), pp. 293–304. New York: Academic Press 1978

Kanzaki, T., Hashimoto, K., Bath, D.W.: Heterotransplantation of human malignant melanoma cell lines in athymic nude mice. J. Natl. Cancer Inst. *62*, 1151-1158 (1979)

Kanzaki, T., Hashimoto, K., Bath, D.W.: Human malignant melanoma in vivo and in vitro. J. Natl. Cancer Inst. *59*, 775–785 (1977)

Kaplan, H.S., Goodenow, R.S., Gartner, S., Bieber, M.M.: Biology and virology of the human malignant lymphomas. Cancer *43*, 1–24 (1979)

Katoh, Y., Stoner, G.D., McIntire, K.R., Hill, T.A., Anthony, R., McDowell, E.M., Trump, B.F., Harris, C.C.: Immunologic markers of human bronchial epithelial cells in tissue sections and in culture. J. Natl. Cancer Inst. *62*, 1177–1185 (1979)

Katsuoka, Y.: Functional and morphological studies of human renal cell carcinoma using a nude mouse and cell culture system. Jpn. J. Urol. *69*, 285–303 (1978)

Kennett, R.H., Gilbert, F.: Hybrid myelomas producing antibodies against a human neuroblastoma antigen present on fetal brain. Science *203*, 1120–1121 (1979)

Keydar, I., Chen, L., Karby, S., Weiss, F.R., Delarea, J., Radu, M., Chaitcik, S., Brenner, H.J.: Establishment and characterization of a cell line of human breast carcinoma origin. Europ. J. Cancer *15*, 659–670 (1979)

Kimhi, Y., Palfrey, C., Spector, I., Barak, Y., Littauer, U.Z.: Maturation of neurobastoma cells in the presence of dimethylsulfoxide. Proc. Natl. Acad. Sci. U.S.A. *73*, 462–466 (1976)

Kirkland, W.L., Yang, N.S., Jorgensen, T., Longley, C., Furmanski, P.: Growth of normal and malignant human mammary epithelial cells in culture. J. Natl. Cancer Inst. *63*, 29–39 (1979)

Kitano, Y., Hu, F.: The effects of ultraviolet light on mammalian pigment cells. J. Invest. Dermatol. *52*, 25–31 (1969)

Koller, P.C.: The role of chromosomes in cancer biology, vol. 38. Berlin, Heidelberg, New York: Springer 1972

Kopelovich, L., Conlon, S., Pollack, R.: Defective organization of actin in cultured skin fibroblasts from patients with inherited adenocarcinoma. Proc. Natl. Acad Sci. U.S.A. *74*, 3019–3022 (1977)

Laing, C.A., Heppner, G.H., Kopp, L.E., Calabresi, P.: Detection of carcino-embryonic antigen in the media of cultures of carcinomatous cells of digestive system origin. J. Natl. Cancer Inst. *48*, 1909–1911 (1972)

Langlois, A.J., Holder, W.D., Iglehart, J.D., Nelson-Rees, W.A., Wells, S.A., Bolognesi, D.P.: Morphological and biochemical properties of a new human breast cancer cell line. Cancer Res. *39*, 2604–2613 (1979)

Lascelles, A.K., Gurner, B.W., Coombs, R.R.A.: Some properties of human colostral cells. Aust. J. Exp. Biol. Med. Sci. *47*, 349–360 (1969)

Lasfargues, E.Y., Coutinho, W.G., Redfield, E.S.: Isolation of two human tumor epithelial cell lines from solid breast carcinomas. J. Natl. Cancer Inst. *61*, 967–973 (1978)

Lechner, J.F., Kaighn, M.E.: Application of the principles of enzyme kinetics to clonal growth rate assays: an approach for dilineating interactions among growth promoting agents. J. Cell. Physiol. *100*, 519–530 (1979a)

Lechner, J.F., Kaighn, M.E.: Reduction of the calcium requirement of normal epithelial cells by EGF. Exp. Cell Res. *121*, 432–435 (1979 b)

Lechner, J.F., Narayan, K.S., Ohnuki, Y., Babcock, M.S., Jones, L.W., Kaighn, M.E.: Replicative epithelial cell cultures from normal human prostate gland. J. Natl. Cancer Inst. *60*, 797–799 (1978)

Lechner, J.F., Babcock, M.S., Marnell, M., Narayan, K.S., Kaighn, M.E.: Normal human prostate epithelial cell cultures. Methods and perspectives in cell biology. II: Cultured human cells and tissues in biomedical research. Harris, C.C., Trump, B.F., Stoner, G.D. (eds.). New York: Academic Press (in press) 1980

Leibovitz, A.: Development of media for isolation and cultivation of human cancer cells. Fogh, J. (ed.), pp. 23–50. New York: Plenum Press 1975

Leibovitz, A., Stinson, J.C., McCombs, W.B., McCoy, C., Mazur, K., Mabry,N.: Classification of human colorectal adenocarcinoma cell lines. Cancer Res. *36*, 4562–4569 (1976)

Leighton, J., Estes, L.W., Mansukhani, S., Brada, Z.: A cell line derived from normal dog kidney (MDCK) exhibiting qualities of papillary adenocarcinoma and of renal tubular epithelium. Cancer *26*, 1022–1028 (1970)

Leighton, J., Tchao, R., Stein, R., Abaza, N.: Histophysiologic gradient culture of statified epithelium. Methods and perspectives in cell biology. II. Cultured human tissues and cells in biomedical research. Harris, C.C., Trump, B.F., Stoner, G.D. (eds.). (in press) 1980

Lever, J.E.: Inducers of mammalian cell differentiation stimulate dome formation in a differentiated kidney epithelial cell line (MDCK). Proc. Natl. Acad. Sci. U.S.A. *76*, 1323–1327 (1979)

Liao, S.K., Dent, P.B., McCulloch, P.B.: Characterization of human malignant melanoma cell lines. I. Morphology and growth characteristics in culture. J. Natl. Cancer Inst. *54*, 1037–1044 (1975)

Liao, S.K., Kwong, P.C., Thompson, J.C., Dent, P.B.: Spectrum of melanoma antigens on cultured human malignant melanoma cells as detected by monkey antibodies. Cancer Res. *39*, 183–192 (1979)

Lindgren, A., Westermark, B., Ponten, J.: Serum stimulation of stationary human glia and glioma cells in culture. Exp. Cell Res. *95*, 311–319 (1975)

Lippman, M.: Hormone-responsive human breast cancer in continuous tissue culture in breast cancer, trends in research and treatment. Heuson, J.C., Mattheiem, W.H., Rozencweig, M. (eds.), pp. 111–139. New York: Raven Press 1976

Lippman, M.E., Bolan, G., Huff, K.: Human breast cancer responsive to androgen in long term tissue culture. Nature *258*, 339–340 (1975)

Lippman, M., Huff, K., Bolan, G.: Progesterone and glucocorticoid interactions with receptor in breast cancer cells in long-term tissue culture. Ann. N.Y. Acad. Sci. *286*, 101–115 (1977)

Lippman, M., Bolan, G., Huff, K.: The effects of estrogens and antiestrogens on hormone-responsive human breast cancer in long-term tissue culture. Cancer Res. *36*, 4595–4601 (1976)

Manolov, G., Manolov, Y.: Marker band in one chromosome 14 from burkitt lymphomas. Nature *237*, 33–34 (1972)

Mark, J., Levan, G., Mitelman, F.: Identification by fluorescense of the G chromosome lost in human meningiomas. Hereditas 71, 163–168 (1972)

Masseyeff, R.: Human alpha feto protein. Pathol. Biol. (Paris) *20*, 703–725 (1972)

McAllister, R.M., Nelson-Rees, W.A., Peer, M., Lang, W.E., Isaacs, H., Gilden, R.V., Rongey, R.W., Gardner, M.B.: Childhood sarcomas on lymphomas: Characterization of new cell lines and search for type-C virus. Cancer *36*, 1804–1814 (1975)

McCombs, W.B., Leibovitz, A., McCoy, C.E., Stinson, J., Berlin, J.: Morphologic and immunologic studies of a human colon tumor cell line (SW-48). Cancer *38*, 2316–2327 (1976)

Medina, D.: Tumor progression. In: Cancer 3. Becker, F.F. (ed.), pp. 99–119. New York: Plenum Press (1975)

Michalopoulos, G., Pitot, H.C.: Primary cultures of parenchymal liver cells on collagen membranes. Exp. Cell Res. *94*, 70–78 (1975)

Mickey, D.D., Stone, K.R., Wunderli, H., Mickey, G.H., Vollmer, R.T., Paulson, D.F.: Heterotransplantation of a human prostatic adenocarcinoma cell line in nude mice. Cancer Res. *37*, 4049–4058 (1977)

Montes de Oca, F., Macy, M.L., Shannon, J.E.: Isoenzyme characterization of animal cell cultures. Proc. Soc. Exp. Biol. Med. *132*, 462–466 (1969)

Motulsky, A.G.: Hemolysis in glucose-6-phosphate dehydrogenase deficieny. Fed. Proc. *31*, 1286–1292 (1972)

Narayanan, A.S., Sandberg, L.B., Ross, R., Layman, D.L.: The smooth muscle cell. III. Elastin synthesis in arterial smooth muscle cell culture. J. Cell Biol. *68*, 411–419 (1976)

Nathrath, W.B.J., Detheridge, F., Franks, L.M.: Species cross-reading epithelial and urothelial specific antigens in human fetal, adult and neoplastic bladder epithelium. J. Natl. Cancer Inst. *6*, 1323–1330 (1979)

Nayak, S.K., O'Toole, C., Price, Z.H.: A cell line from an anaplastic transitional cell carcinoma of human urinary bladder. Br. J. Cancer *35*, 142–151 (1977)

Nelson-Rees, W.A.: The identification and monitoring of cell line specificity. In: Origin and natural history of the cell lines. Barigozzi, C. (ed.), pp. 25–79. New York: Alan R. Liss Inc. 1978

Nelson-Rees, W.A., Flandermeyer, R.R.: HeLa cultures defined. Science *191*, 96–98 (1976)

Nelson-Rees, W.A., Flandermeyer, R.R., Hawthorne, P.K.: Banded marker chromosomes as indicators of intraspecies cellular contamination. Science *184*, 1093–1096 (1974)

Nicholson, G.L.: Trans-membrane control of the receptors on normal and tumor cells. II. Surface changes associated with transformation and malignancy. Biochim. Biophys. Acta *458*, 1–72 (1976)

Norquist, R.E., Ishmael, D.R., Lovig, C.A., Hyder, D.M., Hoge, A.F.: The tissue culture and morphology of human breast tumor cell line BOT-2. Cancer Res. *35*, 3100–3105 (1975)

Nowell, P.C.: Cytogenetics. In: Cancer 1. A comprehensive treaty. Becker, F.F. (ed.), pp. 3–31. New York: Plenum Press 1975

O'Brien, S.J., Kleiner, G., Olsen, R., Shannon, P.E.: Enzyme polymorphisms as genetic signatures in human cell cultures. Science *195*, 1345–1348 (1977)

Okada, K., Schroder, F.H.: Human prostatic carcinoma in cell culture: Preliminary report on the development and characterization of an epithelial cell line. (EB 33). Urol. Res. *2*, 111–121 (1974)

Okada, K., Laudenbach, I., Schroeder, F.H.: Human prostatic epithelial cells in culture: Clonal selection and androgen dependence of cell line EB 33. J. Urol. *115*, 164–167 (1976)

Osborne, C.K., Bolan, G., Monaco, M.E., Lippman, M.E.: Hormone responsive human breast cancer in long-term tissue culture: Effect of insulin. Proc. Natl. Acad. Sci. U.S.A. *73*, 4536–4540 (1976)

Osborne, C.K., Lippman, M.E.: Human breast cancer in tissue culture: The effects of hormones. Experimental biology, 2: Breast cancer: Advances in research and treatment. McGuire, W.L. (ed.), pp. 103–154. New York: Plenum 1978

O'Toole, C., Stejskal, V., Perlmann, P., Karlsson, M.: Lymphoid cells mediating tumor-specific cytotoxicity to carcinoma of the urinary bladder. Separation of the effector population using a surface marker. J. Exp. Med. *139*, 457–466 (1974)

O'Toole, C., Nayak, S., Price, Z., Gilbert, W.H., Waisman, J.: A cell line (SCaBER) derived from squamous cell carcinoma of human urinary bladder. Int. J. Cancer *17*, 707–714 (1976)

O'Toole, C., Price, Z.H., Ohnuki, Y., Unsgaard, B.: Ultrastructure karyology and immunology of a cell line originated from a human transitional-cell carcinoma. Br. J. Cancer *38*, 64–76 (1978)

Owens, R., Smith, H.S., Nelson-Rees, W., Springer, E.L.: Epithelial cell cultures from normal and cancerous human tissues. J. Natl. Cancer Inst. *56*, 843–849 (1976)

Pastan, I., Willingham, M.: Cellular transformation and the morphologic phenotype of transformed cells. Nature *274*, 645–650 (1978)

Pathek, S., Siciliano, M.J., Cailleau, R., Wiseman, C.L., Hsu, T.C.: A human breast adeno
    carcinoma with chromosome and isoenzyme markers similar to those of the Hela line.
    J. Natl. Cancer Inst. *62*, 263–267 (1979)
Peterson, W.D., Stuhlberg, C.S., Swanborg, N.K., Robinson, A.R.: Glucose 6-phosphate
    dehydrogenase isoenzymes in human cell cultures determined by sucrose-agar gel and
    cellulose acetate zymograms. Proc. Soc. Exp. Biol. Med. *128*, 772–781 (1968)
Peterson, J.A., Buehring, G.C., Taylor-Papadimitriou, J., Ceriani, R.L.: Expression of hu-
    man mammary epithelial (HME) antigens in primary cultures of normal and abnormal
    breast tissue. Inst. J. Cancer *22*, 655–661 (1978)
Pfeffer, K.H., Lipkin, M., Stutman, O., Kopelovich, L.: Growth abnormalities of cultured
    human skin fibroblasts derived from individual with hereditary adenomatosis of the co-
    lon and rectum. J. Cell. Physiol. *89*, 29–38 (1976)
Pfeffer, L.M., Kopelovich, L.: Differential genetic susceptibility of cultured human skin fi-
    broblasts to transformation by kirsten murine sarcoma virus. Cell *10*, 313–319 (1976)
Polanowski, F.P., Gaffney, E.V., Burke, R.E.: HBL-100, a cell line established from human
    breast milk. In Vitro *12*, 328–328 (1976)
Ponten, J., Macintyre, E.H.: Long term culture of normal and neoplastic human glia. Acta
    Pathol. Microbiol. Scand. *74*, 465–486 (1968)
Ponten, J.: Carcinogenesis in vitro. In: Recent results in cancer research 44. Grundmann,
    E. (ed.), pp. 98–101. Berlin, Heidelberg, New York: Springer 1974
Ponten, J.: Neoplastic human glia cells in culture. In: Human tumor cells in vitro. Fogh, J.
    (ed.), pp. 175–206. New York: Plenum Press 1975
Ponten, J.: The relationship between in vitro transformation and tumor formation in vivo.
    Biochem. Biophys. Acta *458*, 397–422 (1976)
Pope, J.H., Morrison, L., Moss, D.J., Parsons, P.G., Mary, S.R.: Human malignant
    melanoma cell lines. Pathology *11*, 191–195 (1979)
Povey, S., Hopkinson, D.A., Harris, H., Franks, L.M.: Characterization of human cell lines
    and differentiation from HeLa by enzyme typing. Nature *264*, 60–63 (1976)
Qizilbash, A.H., Liao, S.K., Dent, P.B.: Characterization of human malignant melanoma
    cell lines. IV. Cytologic and histochemical characteristics. Acta Cytol. *21*, 147–150
    (1977)
Quinn, L.A., Woods, L.K., Merrick, S.B., Arabasz, N.M., Moore, G.E.: Cytogenetic anal-
    ysis of twelve human malignant melanoma cell lines. J. Natl. Cancer Inst. *59*, 301–307
    (1977)
Rabson, A.S., Rosen, S.W., Tashjian, A.H., Weintraub, B.D.: Production of human
    chorionic gonadotropin in vitro by a cell line derived from a carcinoma of the lung. J.
    Natl. Cancer Inst. *50*, 669–674 (1973)
Rafferty, K.A.: Epithelial cells: Growth in culture of normal and neoplastic forms. Adv.
    Cancer Res. *21*, 249–272 (1975)
Rasheed, S., Nelson-Rees, W., Toth, E.M., Arnstein, P., Gardner, M.B.: Characterization
    of a newly derived human sarcoma cell line (HT-1080). Cancer *33*, 1027–1033 (1974)
Rasheed, S., Gardner, M.B., Rongey, R.W., Nelson-Rees, W.A., Arnstein, P.: Human blad-
    der carcinoma: Characterization of two new tumor cell lines and search for tumor
    viruses. J. Natl. Cancer Inst. *58*, 881–890 (1977)
Rheinwald, J.G., Green, H.: Serial cultivation of strains of human epidermal keratinocytes:
    The formation of keratinizing colonies from single cells. Cell *6*, 331–343 (1975)
Rheinwald, J.G., Green, H.: Epidermal growth factor and the multiplication of cultured hu-
    man epidermal keratinocytes. Nature *265*, 421–424 (1977)
Rice, R.H., Green, H.: The cornified envelope of terminally differentiated human epidermal
    keratinocytes consists of cross-linked protein. Cell *11*, 417–422 (1977)
Richards, J., Nandi, S.: Neoplastic transformation of rat mammary cells exposed to 7,12-
    dimethylbenz(a)anthracene or N-nitrosomethylurea in cell culture. Proc. Natl. Acad.
    Sci. U.S.A. *75*, 3836–3840 (1978)
Rigby, C.C., Franks, L.M.: A human tissue culture cell line from a transitional cell tumor
    of the urinary bladder: Growth chromosome pattern and ultrastructure. Br. J. Cancer
    24, 746–748 (1970)

Rose, N.R., Choe, B.K., Pontes, J.E.: Cultivation of epithelial cells from the prostate. Cancer Chem. Rep. *59*, 147–163 (1975)

Rowley, J.D.: A new consistent chromosomal abnormality in chronic myelogenous leukemia identified by quinacrine fluorescence and giemsa staining. Nature *243*, 290–293 (1973)

Rule, A.H., Kirch, M.E.: Gene activation of molecules with carcinoembryonic antigen determinants in fetal development and in adenocarcinoma of the colon. Cancer Res. *36*, 3503–3509 (1976)

Russo, J., Furmanski, P., Rich, M.A.: An ultrastructural study of normal human mammary epithelial cells in culture. Am. J. Anat. *142*, 221–232 (1975)

Rutzky, L.P., Tomita, J.T., Calenoff, M.A., Kahan, B.D.: Human colon adenocarcinoma cells, III. In vitro organized expression and carcinoembryonic antigen kinetis in hollow fiber culture. J. Natl. Cancer Inst. *63*, 893–902 (1979)

Rygaard, J., Povlsen, C.O.: Heterotransplantation of a human malignant tumour to "nude" mice. Acta Pathol. Microbiol. Scand. *77*, 758–760 (1969)

Sakashita, S., Hirai, H., Nishi, S., Nakamura, K., Tsuji, I.: Fetoprotein synthesis in tissue culture of human testicular tumors and an examination of experimental yolk sac tumors in the rat. Cancer Res. *36*, 4232–4237 (1976)

Salmon, S.E., Hamburger, A.W., Sochnlen, B.S., Durie, B.G.M., Alberts, D.S., Moon, T.E.: Quantitation of differential sensitivity of human tumor stem cells to anticancer drugs. New Engl. J. Med. *298*, 1321–1327 (1978)

Sanford, K.K.: Biologic manifestations of oncogenesis in vitro: A critique. J. Natl. Cancer Inst. *53*, 1481–1485 (1974)

Satya-Prakash, K.L., Pathak, S., Hsu, T.C., Olive, M., Cailleau, R.: Cytogenetic analysis on eight human breast tumor cell lines: High frequencies of lq, llq and HeLa marker chromosomes[1]. Submitted. J. Natl. Cancer Inst. (1980)

Schneider, E.L., Mitsui, Y., Aw, K.S., Shorr, S.S.: Tissue-specific differences in cultured human diploid fibroblasts. Exp. Cell Res. *108*, 1–6 (1977)

Schroder, F.H., Jellinghaus, W.: Prostatic adenoma and carcinoma in cell culture and heterotransplantation in prostatic disease, pp. 301–312. New York: Liss Inc.

Schroder, F.H., Sato, G., Gittes, R.F.: Human prostatic adenocarcinoma: Growth in monolayer tissue culture. J. Urol. *106*, 734–739 (1971)

Schwarz, R.I., Bissell, M.J.: Dependence of the differentiated state on the cellular environment: Modulation of the collagen synthesis in tendon cells. Proc. Natl. Acad. Sci. U.S.A. *74*, 4453–4457 (1977)

Sell, S., Becker, F.F., Leffert, H.L., Watabe, H.: Expression of an oncodevelopmental gene product (fetoprotein) during fetal development and adult oncogenesis. Cancer Res. *36*, 4239–4249 (1976)

Seman, G., Hunter, S.J., Lukeman, J.M., Dmochowski, L.: Establishment of a cell line (SH-4) from pleural effusion of a patient with melanoma. In Vitro *11*, 205–211 (1975)

Siciliano, M.J., Barker, P.E., Cailleau, R.: Mutually exclusive genetic signatures of human breast tumor cell lines with a common chromosomal marker. Cancer Res. *39*, 919–922 (1979)

Sims, R.T.: Melanocytes. In: An introduction to the biology of the skin. Champion, R.H., Gillman, T., Rook, A.J., and Sims, R.T. (eds.), pp. 139–154

Sinkoviks, J.G., Gyorskey, F., Kusyk, C., Siciliano, M.J.: Growth of human tumor cells in established cultures. Methods Cancer Res. *14*, 243–323 (1978)

Smith, H.S.: In vitro properties of epithelial cell lines established from human carcinomas and nonmalignant tissue. J. Natl. Cancer Inst. *62*, 225–230 (1979)

Smith, H.S., Owens, R.B., Hiller, A.J., Nelson-Rees, A., Johnston, O.: The biology of human cells in tissue culture. I. Characterization of cells derived from osteogenic sarcomas. Int. J. Cancer *17*, 219–234 (1976)

Smith, H., Owens, R., Nelson-Rees, W., Springer, E., Dollbaum, C., Hackett, A.: Epithelial cell cultures from human carcinomas prevention and detection of cancer, pp. 1465–1478. Neiburgs, H.E. (ed.), New York: Marcel Dekker 1978

Smith, H.S., Hackett, A.J., Riggs, J.L., Mosesson, M.W., Walton, J.R., Stampfer, M.R.: Properties of epithelial cells cultured from human carcinomas and nonmalignant tissues. J. Supramol. Struct. *11*, 147–166 (1979a)

Smith, H.S., Springer, E.L., Hackett, A.J.: Nuclear ultrastructure of epithelial cell lines derived from human carcinomas and nonmalignant tissues. Cancer Res. *39*, 332–344 (1979b)

Smith, H.S., Riggs, J.L., Mosesson, M.W.: Production of fibronectin by human epithelial cells in culture. Cancer Res. *39*, 4138–4144 (1979c)

Soule, H.D., Vazquez, J., Long, A., Albert, S., Brennan, M.: A human cell line from a pleural effusion derived from a breast carcinoma. J. Natl. Cancer Inst. *51*, 1409–1416 (1973)

Stampfer, M., Hallowes, R.G., Hackett, A.J.: Growth of normal human mammary cells in culture. In Vitro *16*, 415–425 (1980)

Steel, G.G.: The growth and therapeutic response of human tumours in immune deficient mice. Bull. Cancer (Paris) *65*, 465–472 (1978)

Steele, V.E., Marchok, A.C., Nettesheim, P.: Establishment of epithelial cell lines following exposure of cultured tracheal epithelium to 12-O tetradecanoyl-phorbol-13-acetate. Cancer Res. *39*, 3563–3565 (1978)

Stern, R., Wilczek, J., Thorpe, W.P., Rosenberg, S.A., Cannon, G.: Procollagens as makers for the cell origin of human bone tumors. Cancer Res. *40*, 325–328 (1980)

Stiles, C.D., Desmond, W., Chuman, L.M., Sato, G., Saier, M.: Growth control of heterologous tissue culture cells in the congenitally athymic nude mouse. Cancer Res. *36*, 1353–1360 (1976)

Stoker, M.G.P., Piggott, D., Riddle, P.: Movement of human mammary tumour cells in culture: Exclusion of fibroblasts by epithelial territories. Int. J. Cancer *21*, 268–273 (1978)

Stone, K.R., Mickey, D.D., Wunderli, H., Mickey, G.H., Paulson, D.F.: Isolation of a human prostate carcinoma cell line (DU 145). Int. J. Cancer *21*, 274–281 (1978)

Stonington, O.G., Hemmingsen, H.: Culture of cells as a monoplayer derived from the epithelium of the human prostate: A new cell growth techniques. J. Urol. *106*, 393–400 (1971)

Strobl, J.S., Lippman, M.E.: Prolonged retention of estradiol by human breast cancer cells in tissue culture. Cancer Res. *39*, 3319–3327 (1979)

Sun, T.T., Green, H.: Immunofluorescent staining of keratin fibers in cultures cells. Cell *14*, 469–476 (1978a)

Sun, T.T., Green, H.: The keratin filaments of cultured human epidermal cells: Formation of inter-molecular disulfide bonds during terminal differentiation. J. Biol. Chem. *253*, 2053–2060 (1978b)

Sun, T.T., Shih, C., Green, H.: Keratin cytoskeletons in epithelial cells of internal organs. Proc. Natl. Acad. Sci. U.S.A. *76*, 2813–2917 (1979)

Tane, N., Hashimoto, K., Kanzaki, T., Ohyama, H.: Collagenolytic activities of cultured human malignant melanoma cells. J. Biochem. *84*, 1171–1176 (1978)

Taub, M., Sato, G.: Hormone regulation of growth and dome formation by kidney epithelial cells in serum free medium. J. Supramol. Struct. Suppl. *3*, 226–228 (1979b)

Taub, M., Chuman, L., Saier, M.H., Sato, G.: Growth of mandin-darby canine kidney epithelial cell (MDCK) line in hormone-supplemented serum-free medium. Proc. Natl. Acad. Sci. U.S.A. *76*, 3338–3342 (1979a)

Taylor-Papadimitrou, J., Shearer, M., Stoker, M.G.P.: Growth requirements of human mammary epithelial cells in culture. Int. J. Cancer *20*, 903–908 (1977a)

Taylor-Papadimitriou, J., Shearer, M., Tilly, R.: Some properties of cells from early lactation milks. J. Natl. Cancer Inst. *58*, 1563–1571 (1977b)

Taylor-Papadimitriou, J., Purkis, P., Fentiman, I.S.: Cholera toxin and analogues of cyclic AMP stimulate the growth of cultured human mammary epithelial cells. J. Cell. Physiol. *102*, 317–321 (1980)

Thorpe, W., Loeb, W., Hyatt, C., Rosenberg, S.A.: Alkaline phosphate measurements of paired normal and osteosarcoma tissue culture lines obtained from the same patient. Cancer Res. *39*, 277–279 (1979)

Todaro, G.J., DeLarco, J.E., Marquardt, H., Bryant, M.L., Sherwin, S.A., Sliski, A.J.: Polypeptide growth factors produced by tumor cells and virus-transformed cells: A possible growth advantage for the producer cells. Coldspring Harbor Conf. on Cell Prolif. Sato, G.H., Ross, G.R. (eds.), vol. 6, pp. 113–127. Hormones and Cell Culture, Long Island, N.Y.: Cold Spring Harbor Press (1979)

Tom, B.H., Rutsky, L.P., Jakstys, M.M., Oyasu, R., Kaye, C.I., Kahan, B.D.: Human colonic adenocarcinoma cells. I. Establishment and description of a new line. In Vitro 12, 180–191 (1976)

Tompkins, W.A., Watrach, A.M., Schmale, J.D., Schultz, R.M., Harris, J.A.: Cultural and antigenic properties of newly established cell strains derived from adenocarcinomas of the human colon and rectum. J. Natl. Cancer Inst. 52, 1101–1110 (1974)

Tooze, J.: The molecular biology of tumor viruses. New York: Cold Spring Harbor Press 1973

Tucker, W.S., Kirsch, W.M., Martinez-Hernandez, A., Fink, L.M.: In vitro plasminogen activator activity in human brain tumors. Cancer Res. 38, 297–302 (1978)

Tuddenham, E.G.D., Shearn, S.A.M., Peake, I.R.: Tissue localization and synthesis of factor VIII related antigen in the human foetus. Br. J. Haematol. 26, 669–677 (1974)

Vaheri, A., Ruoslaht, E., Westermark, B., Ponten, J.: A common cell type specific surface antigen in cultured human gliol cells and fibroblasts: Loss in malignant cells. J. Exp. Med. 143, 64–72 (1976)

Wahlstrom, T., Linder, E., Saksela, E., Westermark, B.: Tumor specific membrane antigens in established cell lines from gliomas. Cancer 34, 274–279 (1974)

Webber, M.M.: Effects of serum on the growth of prostatic cells in vitro. J. Urol. 112, 798–801 (1974)

Webber, M.M., Stonington, O.G.: Stromal hypercellularity and encapsulation in organ cultures of human prostate: Application in epithelial cell isolation. J. Urol. 114, 246–248 (1975)

Webber, M.M., Horan, P.K., Bouldin, T.R.: Present status of MA-160 cell line: Prostatic epithelium or HeLa cells? Invest. Urol. 14, 335–343 (1977)

Weibel, E.R., Palade, G.E.: New cytoplasmic components in arterial endothelia. J. Cell Biol. 23, 101–112 (1964)

Westermark, B., Ponten, J., Hergossen, R.: Determinants for the establishment of permanent tissue culture lines from human gliomas. Acta Pathol. Microbiol. Scand. 81, 791–805 (1973)

Whitehead, J.S., Fearney, F.J., Kim, Y.S.: Glycosyltransferase and glycosidase activities in cultured human fetal and colonic adenocarcinoma cell lines. Cancer Res. 39, 1259–1263 (1979)

Wight, T., Ross, R.: Proteoglycans in primate arteries. II. Synthesis and secretion of glycosaminoglycans by arterial smooth muscle cells in culture. J. Cell Biol. 67, 675–686 (1975)

Wikstrand, C.S., Mahaley, M.S., Bigner, D.D.: Surface antigenic characteristics of human glial brain tumor cells. Cancer Res. 37, 4267–4275 (1977)

Wikstrand, C.J., Bigner, D.D.: Surface antigens of human glioma cells shared with normal adult and fetal brain. Cancer Res. 39, 3235–3243 (1979)

Williams, R.D., Elliott, A.Y., Stein, N., Fraley, E.E.: In vitro cultivation of human renal cell cancer. I. Establishment of cells in culture. In Vitro 12, 623–627 (1976)

Williams, R.D., Bronson, D.L., Elliott, A.Y., Fraley, E.E.: Production of carcinoembryonic antigen by human prostate epithelial cells in vitro. J. Natl. Cancer Inst. 58, 1115–1116 (1977)

Williams, R.D., Bronson, D.L., Elliott, A.Y., Gehrke, C.W., Kuo, K., Fraley, E.E.: Biochemical markers of cultured human prostatic epithelium. J. Urology 119, 768–771 (1978a)

Williams, R.D., Elliott, A.Y., Stein, N., Fraley, E.E.: In vitro cultivation of human renal cell cancer. II. Characterization of cell lines. In Vitro 14, 779–786 (1978b)

Wilson, M.G., Towner, J.W., Fujimoto, A.: Retinoblastoma and D-chromosome deletions. Am. J. Hum. Genet. 25, 57–61 (1973)

Yanagisawa, Y.: Tissue culture of benign and malignant human prostatic tumors. I. Morphological observations on cultured cells. Jpn. J. Urol. *68*, 1036–1045 (1977)

Yang, N.S., McGrath, C.M., Furmanski, P.: Presence of a mouse mammary tumor virus-related antigen in human breast carcinoma cells and its absence from normal and mammary epithelial cells. J. Natl. Cancer Inst. *61*, 1205–1208 (1978)

Yeh, M., Hellstrom, I., Brown, J.P., Warner, G.A., Hansen, J.A., Hellstrom, K.E.: Cell surface antigens of human melanoma identified by monoclonal antibody. Proc. Natl. Acad. Sci. U.S.A. *76*, 2927–2931 (1979)

Young, R.K., Cailleau, R.M., MacKay, B., Reeves, W.J.: Establishment of epithelial cell line MDA-MB-157 from metastatic pleural effusion fo breast carcinoma. In Vitro *9*, 239–245 (1971)

Yuhas, J.M., Tarleton, A.E., Molzen, K.B.: Multicellular tumor spheroid formation by breast cancer cells isolated from different sites. Cancer Res. *38*, 2486–2491 (1978)

Zetter, B.R., Johnson, L.K., Shuman, M.A., Gospodarowicz, D.: The isolation of vascular cell lines with altered cell surface and platelet-binding properties. Cell *14*, 501–509 (1978)

Zimmering, A.K., Mansell, P.W.A., Dietrick, R.S., O'Neil, C.: Cytogenetic and ultrastructural observations on three lines of human melanoma cells kept in long-term culture. Pigment Cell *2*, 79–93 (1976)

CHAPTER 17

The Chalones

O. H. Iversen

> *Throughout the history of science one notices that the observers rush too quickly from the phenomenon to theory, and thereby become inadequate, hypothetical.*
> Johann Wolfgang Goethe

Everyone who has followed the development of our field during the past 20 years has to acknowledge, willingly, and with gratitude, how valuable, perhaps irreplaceable in this particular instance, the approach has been to connect the sparse fact by persistent hypothesizing and constructing a unifying vision, the true nature of which we may hope to know only in the distant future. Building theoretical castles in the air can provide the stimulus which is essential for carrying out painstaking experiments. However, in order to assess the true progress we have made, we have to realize just how far observation and experimentation by themselves are able to carry us today. To demarcate this is the purpose of my treatise. However, by confining myself to this task I do not want to deprive myself of hypotheses entirely, without which a mere body of factual data must remain lifeless.

> *Theodor Boveri* (1904), on the problems of chromosomal changes and malignancy.

A. Introduction

I. Definition and Properties

Although the concept of chalones has gained increasing acceptance in recent years, and the evidence in its favour has become substantial, much of it still remains circumstantial and falls short of direct scientific proof. Any description must therefore be provisional and operational only since a 'chalone' has not yet been completely purified and characterized. Most investigators feel, however, that the following covers the substances they are searching for:

Definitions

1. Chalones are naturally occurring, physiological, cell proliferation-inhibitory substances.

2. Chalones must be produced in and be present in the tissues (the cell lines) on which they selectively act. They are probably mainly produced by the more mature cells of the series.

3. Chalones are tissue- or cell line-specific (or show tissue or cell line preference or selectivity).

4. Chalone action is non-toxic and reversible, and hence chalones do not injure cells or cell membranes.

Observations

1. Chalones prolong the mean cell cycle time. Due to the methods available, this has been most often measured as inhibition of DNA synthesis or mitotic rate. It has become clear that one type of chalones (the G_1 chalone) seems to act mainly in the late G_1 and/or in the S phase by delaying the flux cells into DNA synthesis and through S; another type of chalones (the G_2 chalone) seems to act in the late G_2 phase by delaying the flux of cells into mitosis and prolonging the mitotic duration. (However, see also Sects. C.III and C.VI.)

2. Chalones are species non-specific.

3. Chalones are water soluble molecules.

II. Limitations of This Chapter

As clearly demonstrated in this book, there are many known growth stimulatory and inhibitory physiological substances. Growth regulation in vivo is a complex system, depending upon the balance between stimulatory, inhibitory, and modulatory influences.

The author of this chapter works within the paradigm (see, e.g., Kuhn, 1970; Naess, 1972) that chalones exist. There are many scientists, however, who dispute this (see, e.g. Allen and Smith, 1979; Cline and Fitchen's chapter in this book). They interpret and produce experimental data within another framework of basic assumptions (cf. "Data are created by theories –"; Scheffler, 1967). The existing credibility gap can only be bridged when the chalones and/or the non-chalone growth inhibitory substances responsible for the non-disputable biological effects are purified and characterized.

Only certain growth inhibitory substances will be discussed in this chapter. Growth stimulatory substances are treated elsewhere in this book and in many other reviews (see, e.g. Pardee and Dubrow, 1977; Rudland and De Asua, 1979). Alternative growth regulation models have also been proposed (see, e.g. reviews by Iversen, 1965; Iversen et al., 1974; Gelfant, 1977). Main emphasis will be on the epidermal, the granulocyte and the lymphocyte chalones, since these are the most thoroughly studied.

There are three monographs on chalones, i.e. (i) Forscher and Houck (eds.): Natl. Cancer Inst. Monogr. 38, 1973; (ii) J. Houck (ed.): Chalones, 1976 (a), and (iii) A. Balázs: Control of cell proliferation by endogenous inhibitors, 1979.

Personally, I am an expert only on the epidermal chalone. The rest is rumination on the literature. A recent review paper on epidermal growth regulation with emphasis on the chalones has been published by Prunieras et al. (1980). The editorial board required that "The complete world literature should be critically discussed." This was impossible within the space available. Especially as regards epidermal and blood cell chalones, a selection had to be made. Almost all references can be found in the review articles referred to in this chapter. The Reference list has been updated to March 1981.

III. The Name

The term "chalone" has always been controversial, but has recently (1979) been included in all important literature retrieval systems.

The term was probably first used for an inhibitory growth regulator at the Seventeenth International Congress of Medicine, 1913. Sir E. A. SCHÄFER (IVERSEN, 1976) suggested that a special name was needed for those substances which were inhibitory in action, and suggested the word "chalone" from the Greek *chalao*, meaning to loosen or to lower. In 1962, BULLOUGH adopted the word for tissue-specific anti-mitotic substances.

In 1970, the Growth Control Panel of the Medical Research Council held hearings in London and, according to HOUCK (1976a), it was more or less agreed that if a name were to be granted to a functional material, then this material should be completely characterized. I disagree. The term "hormone" functioned well long before any hormone was purified, and even today "hormone" continues to be a common and most useful name for a particular functional class of biological substances, even if they differ widely in chemical composition and mechanisms of action. The name interferon was used for twenty years before the substance was purified (see, e.g. MAURER, 1979).

The present situation in chalone research closely resembles that in hormone research in the early years, as described by LASNITSKY (1965): "... the fact that the hormones were not properly quantitated and often contained impurities, makes it hardly surprising that the results were often inconsistent or contradictory ..." Hence, the term chalone should be accepted as an operational term for tissue specific growth inhibitory substances acting according to the negative feedback principle. It seems to inspire active research and even fruitful, heated discussions. The number of scientific papers on chalones or related to chalones in my files in January 1980 amounts to more than 600. "What's in a name? That which we call a rose by any other name would smell as sweet!" (Romeo and Juliet, Act II, Sc. ii)".

IV. Theoretical and Biological Background

1. Theory

BERNARD (1878) was the first physiologist to formulate a specific idea of cybernetic control systems in biology. He stressed the importance of a constant "milieu interieur" as a necessary condition for the independent life of an organism: "La fixité du milieu suppose un perfectionnement de l'organisme tel que les variations externes soient à chaque instant compensées et equilibrées." In 1939 W. B. CANNON coined the term homeostasis to describe this phenomenon. In 1957 WEISS and KAVANAU formulated a general theory of growth regulation based on cybernetic principles (Fig. 1).

The "anti-templates" suggested by WEISS and KAVANAU correspond directly to what is presently called chalones.

In general and mathematical terms, growth regulation by inhibitors of cell proliferation is only a special case of general, biological cybernetics, as it was pioneered by WIENER (1948).

MAZIA (1961) postulated that stimulation of cell division in a multicellular organism is due to "the removal of a block."

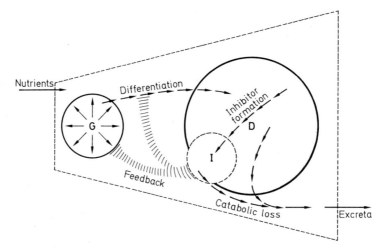

Fig. 1. The model of WEISS and KAVANAU (1957). *G*, generative mass; *D*, differentiated mass; *I*, inhibiting principle comparable to the present chalone concept. (Reproduced by permission from Rockefeller University and Dr. PAUL WEISS)

2. Biology

The organs and tissues of the adult body are constantly kept at a normal size. After cell injury and/or cell loss in renewing and stable cell populations (but not in permanent ones), regenerative reactions start with production of new cells, cell migration and cell maturation. Some of these reactions are local, like those in the epidermis after a superficial scratch; others are obviously dispersed, like those in the bone marrow after a haemorrhage.

It is an accepted hypothesis (i) that cells with the same type of differentiation have developed a mechanism to keep each other informed about their total number in the body or in a local area and about their degree of maturation, and (ii) that this system consists of certain specific signal molecules produced by the cells for the regulation of growth and differentiation. When such signals are inhibitory, they fit into the concept of chalones.

The idea of chalones came from the observations that (i) in the adult organism the cell number in any organ or tissue is kept at a "normal" level; (ii) all types of chronic, moderately increased rate of cell loss lead to a state of hyperplasia; (iii) the pattern of cell kinetics after acute cell loss point to inhibitory forces; (iv) an acute cell loss (a wound) in one tissue does not induce a mitotic response in other, neighbouring tissues and (v) the pattern of liver growth in parabiotic rats after partial hepatectomy in one of a pair.

There were scattered reports in the literature in the 1930s (e.g. MURPHY and STURM, 1931; SIMMS and STILLMAN, 1937) on growth inhibitory effects of tissue extracts, but the reports did not, strangely enough, attract much attention. This may partly be due to the fact that the history of medicine was, and still is, full of reports on scientifically unfounded cancer treatment regimens utilizing extracts of tumours and normal tissues (see, e.g. BAKER, 1933, 1935) and falsely claiming good results.

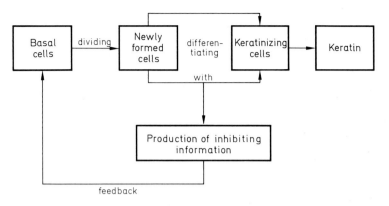

Fig. 2. Simple schematic representation of the model presented by IVERSEN at the First International Congress of Cybernetic Medicine, Naples, October 1960a. The production of inhibiting information (chalone) is thought to be dependent upon the process of cell maturation and thus linked to keratinization

In 1956, however, SAETREN studied kidney and liver regeneration and reported the tissue specific, inhibitory effect of extracts from the same organs. He interpreted his observation thus: "both the kidney and the liver maintain a certain blood level of a thermolabile nondialyzable substance which exerts a specific inhibitory action on the growth of its organ of origin." This was the starting point of specific chalone research.

Independently of each other, and at the same time, BULLOUGH and LAURENCE (1960) and IVERSEN (1960a, b) also proposed a similar theory, the first-mentioned authors from a study of skin wound healing on the mouse ear, and the latter from studies of the cell kinetics in epidermal carcinogenesis (Fig. 2).

B. Sources of Chalones: Methods to Extract, Purify and Measure Their Effects

I. Sources of Chalones

Chalones originate within cells and tissues. When tissues and organs are used for extraction, they must be homogenized and the active components extracted with water. Chalones coming from dispersed cells can also be found in, and extracted from, the fluid medium in which the cells have grown (so-called conditioned medium).

If one has a pure cell line, the chalones extracted therefrom may be homogeneous and monospecific. Tissues and organs, however, are always composite and all tissue extracts will contain putative chalones from all the cells present. MAURER et al. (1976) and KASTNER and MAURER (1980) tried to obtain a pure source by leucoadhesion of granulocytes to nylon and cotton wool and/or EDTA. Sophisticated work has been done on the epidermis to obtain G_1 chalones from the mature cells and G_2 chalone from the basal cells (ELGJO and HENNINGS, 1971b; ELGJO et al., 1971, 1972). OKULOV and KETLINSKY (1977) got rid of non-epidermal components by using immunoabsorbtion methods. In this field there is a caveat. If the procedure to obtain a pure source is too rigorous, one may also lose the chalone effect (see HERMAN et al., 1978; RYTÖMAA and KIVINIEMI, 1968 b).

The chalone concentration in the cells is probably extremely low. MAURER (1975) estimated that about 1,000 kg white blood cells would be necessary to prepare 1 mg chalone, assuming a hormone-like concentration of a putative chalone of 1,000 daltons.

This makes the question of the sources for chalones extremely important (see MASCHLER and MAURER, 1978) and emphasizes the necessity of purification and possible synthesis of such substances.

II. Methods to Assess Chalone-Mediated Growth Inhibition

The methods used to assess chalone effects are those of cell kinetics, based on our knowledge of the cell cycle. The methods generally measure a negative effect (an inhibition) on the incorporation of radioactivity labelled precursors into DNA, the fraction of cells in each cell cycle phase (cytophotometry), the mitotic activity (stathmokinetic methods) or the increase in cell number, organ size or weight. Most of these methods are extremely tedious and time-consuming and prone to potential artefacts, which explains the slow progress in chalone research.

Broad discussions of the methods have been given, e.g. by LORD (1976), MAURER and LAERUM (1976) and IVERSEN (1978 a). Many studies have been performed in vivo, others in organ culture (e.g. pieces of ear skin, BULLOUGH and LAURENCE, 1960; GRADWOHL, 1978 a, b), in cell cultures in vitro or in diffusion chambers (e.g. most works on granulocytes, lymphocytes and fibroblasts, see below), or in ascites tumours (see below). Long-term organ or cell cultures seem to be of limited value (personal experience; HANSTEEN et al., 1979). NAESER et al. (1978) showed that whereas chalones regulate the growth of ascites tumors in vivo, there was rather a stimulation from extracts of used media in vitro.

There are many artefactual problems in chalone assay methodology, the most important ones being: (i) reduced thymidine uptake does not eo ipso mean reduced proliferation (MAURER and LAERUM, 1976); (ii) general cytotoxicity leads to growth retardation in cell cultures (cf. the so-called Houck's Law (HOUCK, 1976 a): "Dead cells don't divide and dying cells divide damn slowly!" and (iii) general growth inhibition is one of the stress-hormone effects which may follow injection of too large doses of foreign proteins (often in strong, hypertonic salt solutions) in vivo. Problems of (iv) bacterial contamination of extracts with endo- or exotoxins is also a reality, as shown by MOHR et al. (1972 a, b).

One has to agree with MAURER and LAERUM (1976): "... the progress of ... chalone research is more dependent on reliable and rapid assay systems than on biochemical purification procedures."

To measure chalone effects in organs or tissues, one must preferably have a tissue with a normal, appreciable, proliferative activity (a "labile" cell population; BIZZOZERRO, 1894); or, a state of increased proliferation must be induced, e.g. the study of the liver chalones must be performed on the growing, the regenerating or the chemically stimulated liver because the normal liver has such a low proliferative activity. However, it must be kept in mind that at least the epidermal G_1-chalone is inoperative in neonatal epidermis (BERTSCH and MARKS, 1978 a, b; MARKS et al., 1978; ELGJO and CROMARTY, 1977); TPA-treated and rapidly regenerating epidermis is also relatively refractory to the epidermal G_1 chalones (BERTSCH et al., 1976; KRIEG et al., 1974). Rapidly regenerating tissue contains no or very little chalone, as shown, for example, for the epidermis by ROHRBACH et al. (1972, 1976a, b, 1977), for the stem cells of the red marrow by LORD et al. (1976), for the liver by PAERMENTIER and BARBASON (1979) and for the submandibular gland by BARKA (1973). From a relatively undifferentiated squamous cell carcinoma BERTSCH and MARKS (1979) could not extract any enrichable epidermal G_1 chalone.)

In this connection one may refer to the theory of OSGOOD (1957), who postulated different growth regulatory substances for exponential and linear growth situations. (See also below on epidermal G_1 and G_2 chalones.)

As regards the methods for selecting proper controls, see Sect. C.I.

III. Methods to Purify and Characterize Chalones

It was long held that chalones were glycoproteins, and the various glycoprotein separation methods were generally applied. The tissues and/or cells were homogenized and usually extracted with water. Tissue culture fluid or ascites fluid wherein the cells had been growing have been used directly. Often, slow alcohol precipitation in the cold was used, and the highest activity was found, for example, for the epidermal G_1 chalone in the alcohol precipitate between 55% and 71%, and for the epidermal G_2 chalone between 72% and 81%. Thereafter, fractionation and purification have been tried using the various chromatographic methods, molecular sieving and isoelectric focussing. A major problem has been aggregation and clumping of molecules. Aggregation was sometimes counteracted by treatment with 6 M urea. Since the chalones seemed to be unstable, most workers have done all procedures at low temperatures. From tissue homogenates, enzymes (e.g. TdR-kinase, hydrolases and complement factors) are certainly also extracted and to counteract their effects some have added enzyme blockers. Some workers have lost the tissue specific chalones during the purification procedure and ended up with non-specific growth inhibitory substances, e.g. the polyamines spermine or spermidine (DEWEY et al., 1977; ALLEN et al., 1977), or with specific ones, e.g. lactoferrin (BROXMEYER et al., 1978). See also Sect. G.IV.3.

C. Some Chalone Properties

I. Tissue or Cell Line Specificity or Preference

When we talk of chalones as being tissue- or cell line-specific, or showing such preference or selectivity, we mean that they mainly or preferably act on the same cell line or tissue as that in which they are produced, i.e. cells with the same set of gene expressions. The term "differentiation" is here used for the qualitative *type of cell function* depending on a specific set of gene expressions, and "maturation" is a term for the *degree* of maturity of the cells, e.g. the degree of keratin formed by squamous cells. The degree of maturation, then, usually depends on cell age.

Table 1a. Percentage reduction of mitotic rate after 5 mg extract injection. (NOME, 1975)

Tissue in which mitoses were counted	Tissues from which extracts were made							
	Skin	Fore-stomach	Glandular stomach	Lung	Liver	Kidney	Spleen	Striated muscle
Epidermis	70	59	27	5	14	13	9	8
Forestomach	25	44	30	—	10	—	—	3
Jejunum	2	12	1	—	—	—	11	—
Colon	8	10	2	—	7	3	13	9

Table 1b. Percentage maximum reduction of LI and DNA-specific H3Tdr incorporation after injection of 10 mg extract (NOME, 1975)

Tissue on which effect was measured		Tissue from which extracts were made			
		Skin	Forestomach	Glandular stomach	Small intestine
Epidermis	LI	41	50	—	12
	Incorp.	41	60	nd	nd
Forestomach	LI	36	55	—	9
	Incorp.	nd	nd	nd	nd
Jejunum	LI	11	7	—	3
	Incorp.		7	nd	nd
Colon	LI	19	15	—	6
	Incorp.	22	14	nd	nd

To prove cell line specificity, convincing control experiments must be carried out. Ideally a chalone from one cell line ought to be without effect on other cell lines or tissues, and aqueous extracts of other organs (with their own endogenous chalone effect) should have no effect on the cell line studied. NOME (1975), for example, showed (see, Table 1) that extracts of the epidermis inhibit proliferation in vivo in all keratinizing cells, but not in small intestine or colon. On the other hand, extracts of a variety of organs had no significant growth inhibitory effects on the epidermis or the forestomach.

VILPO (1979) recently reported that using the partially purified granulocyte chalone (Myleostat, Weddel, see Sect. G.IV.1) in in vitro short-term cultures of various hematopoietic cell lines, there was a granulocyte-specific effect of small doses (less than 10 µg/ml), whereas higher doses inhibited the incorporation of ^{125}I-UdR in many cell lines. VILPO ascribed the non-specific effect to possible impurities still residing in the preparations.

Similar types of cell line specificity have been demonstrated in more or less extensive assay systems for all substances called chalones, as will be discussed in Sect. G. The evidence available (see, e.g. BULLOUGH and LAURENCE, 1964b; IVERSEN, 1969; ELGJO and HENNINGS, 1971b; BICHEL, 1972; CHOPRA et al., 1972;

LAERUM and MAURER, 1973; KUO and YOO, 1977; BOLL et al., 1979; LAURENCE et al., 1979) supports the theory that the cells of each type of differentiation produce their own specific chalones. This, of course, needs to be confirmed by detailed experiments with pure chalone substances. HARRINGTON and GODMAN (1980) reported on a selective inhibitor of cell proliferation from normal serum. This had no cell line specificity, and acted on rat hepatocytes, human lymphoblasts and murine spleen cells, and to a small extent on human colon carcinoma cells.

II. Species Non-specificity

Epidermal extracts prepared from human, pig, mouse, and cod skin have been shown to inhibit mitosis in mouse skin in vitro and in vivo (BULLOUGH et al., 1967) and mouse and rat skin extracts inhibit mitotic activity and DNA synthesis in human skin in vitro (IVERSEN, 1968; DOSYCHEV et al., 1980). Bovine granulocyte chalone inhibits DNA synthesis in granulocyte precursors of other species, including man (see, e.g. BALAZS, 1979; RYTÖMAA et al., 1977). Thus, chalones seem to resemble some hormones in their target cell specificity and their lack of species specificity.

III. Sites of Attack in the Cell Cycle

Basic mechanisms are unknown although tentative hypotheses have been brought forth. The methods available in cell kinetics are mostly static and concentrate mainly on DNA synthesis and mitosis. What matters for proliferation is the speed at which the cells pass through the whole cylce and divide, the flux, and this can theoretically be regulated at many points. Only a combination of many methods can give a dynamic interpretation, which makes it possible to calculate *flux* through the different cell cycle compartments (IVERSEN, 1978a).

The first epidermal chalone concept concentrated on chalone as a mitotic inhibitor. Many biologists had difficulties in accepting a G_2 inhibitor as "a chalone", since mitosis is obviously consequential to DNA synthesis. However, some chalone systems investigated seem to contain chalones inhibiting cell entry into both the S phase and mitosis. The first type of chalone has been called the S factor or the G_1 chalone, and the second type the M factor or the G_2 chalone.

LAERUM (1970), using an ingenious method, found it possible to separate basal and maturing cells of hairless mouse epidermis, and ELGJO et al. (1971, 1972) showed that the G_2 inhibitor of epidermis was mainly present in the basal cells and the G_1 inhibitor in the maturing cell layer. Bleomycin provoked alterations of the G_2 chalone content in the epidermis, whereas the G_1 chalone was not much altered (IVERSEN et al., 1977).

In 1971 MARKS showed that crude pig skin extract contained both G_1 and G_2 chalone factors that could be separated from each other by various fractionation procedures. In 1971, BICHEL showed that the JBI ascites tumour cells produced two different inhibitors, one acting on cells in G_2, the other on cells in late G_1. These observations have been confirmed in many chalone systems (e.g., see LAERUM and MAURER, 1973; OKULOV et al., 1979) and in addition there is no cross-reaction between the two epidermal chalones (BERTSCH and MARKS, 1974; BERTSCH, 1975; THORNLEY and LAURENCE, 1975, 1976; OKULOV and KETLINSKI, 1977).

However, not only do chalones block the entry of cells into S and M, they also seem to delay the passage of the affected cells through these phases (e.g., LORD, 1975). In 1964 (a) BULLOUGH and LAURENCE found that the epidermal G_2 chalone (partly purified, alcohol-precipitated pig skin extract) also increased the mitotic duration. This was confirmed by RANDERS-HANSEN (1967). IVERSEN (1978 c) confirmed that epidermal extract prolonged the mitotic duration both in the epidermis and in a transplantable squamous cell carcinoma. The prolongation of the mitotic duration by G_2 chalone is parallel to the prolongation of the S phase caused by the epidermal G_1 chalone as reported by YAMAGUCHI et al. (1974) and by BORN and BICKHARDT (1968), and corresponds to a similar observation regarding the granulocytic G_1 chalone (LAERUM and MAURER, 1973).

Thus, in addition to preventing cells from entering DNA synthesis and mitotis, chalones also seem to prolong the cell passage through these compartments. Quite recently, ELGJO et al. (1980) have shown that epidermal chalones delay the flux of cells from S to G_2. LORD (1975) and RYTÖMAA (1978) concluded that granulocyte chalone prolong the cell cycle time. NAKAI (1976) demonstrated Ehrlich ascites chalone to inhibit DNA polymerase. Thus, the problem of site of attack in the cell cycle seems to be a complex one, and the last word has not yet been said. The recently developed flow cytometers with cell sorters will certainly throw more light on it.

IV. Reversibility and Turnover Time of Chalones

According to general biological reasoning, a growth regulatory substance that is continuously produced has to be short-lived in vivo. The problem is difficult because the assay systems make it difficult to distinguish between the possibilities that the putative chalone molecules have a short life span as a substance, or the short- (or long-) time persistence of a chalone effect. The question of reversibility has been thoroughly studied by BULLOUGH and LAURENCE (1964c), who proposed an "adrenaline wash" in an organ culture system as a "discriminative" test for reversibility. Reversibility of the granulocyte chalone was shown by LORD et al. (1974b).

ELGJO (1974a, b) found that epidermal cells which had been blocked in their progress through the cell cycle re-entered cycling when the chalone infusion was stopped. ELGJO and HENNINGS (1971 b) showed that the epidermal G_2 chalone has a short turnover time, whereas the G_1 inhibitor is more stable.

As regards the instability of crude skin extracts, ELGJO (1969) showed that at room temperature the mitosis inhibiting effect remains only for 30 min. This may be due to the presence of chalone-degrading enzymes in the solution or to aggregate or complex formation.

Since normal proliferation can always be resumed after the cessation of chalone effect, the growth inhibitory chalones are probably not toxic in the small doses that show growth inhibitory effect (see, e.g. BALÁZS, 1976; BENESTAD and RYTÖMAA, 1977; FOA et al., 1979). In cell culture the distinction between a true chalone effect and a non-specific inhibition of thymidine uptake (caused by, for example, spermidine) or by moderate cytotoxicity is extremely difficult.

V. Chalones and Stimulators of Cell Proliferation

Almost any dynamic biological system depends on a balance between opposing forces. Thus, there are physiological substances that stimulate growth as well as others that inhibit it. Some hormones have dual effects, and stimulate the target organs, but inhibit the production of trophic hormones from the pituitary. Many hormones, however, stimulate cell proliferation in target organs, and high doses over a long period lead to target organ hyperplasia. Specifically growth-stimulating substances have been isolated with great success and precision, e.g. the nerve growth factor, the epidermal growth factors and the platelet-derived growth factors (see other chapters in this volume).

Formally, all these ought to act the opposite way of chalones, but it is not known whether some of them ought to be classified as specific chalone antagonists. BJERKNES and IVERSEN (1974) published a theoretical study on the role of "antichalones" in growth control systems. RYTÖMAA and KIVINIEMI (1968a) suggested a balance between granulocytic chalone (MW around 4,000) and a chalone antagonist that he called antichalone (MW around 35,000). AARDAL et al. (1977) presented evidence for the antagonism between CSF and granulocyte chalone, and LEHMANN et al. (1979) showed such antagonism between insulin and Ehrlich ascites tumor chalone. A mesenchymal factor in corium seems to act as an epidermal "antichalone" (see review by BULLOUGH, 1975; see also BULLOUGH and DEOL, 1975; and BALÁZS and BARABÁS, 1979).

All growth stimulatory substances are thus formally chalone antagonists, but so far no particular substance is known to act by specifically blocking the chalone effect, and thus the term "antichalone" is at present rather vague.

VI. Dose-Response Relationship

This is the weakest link supporting chalones because no chalones are yet in a completely pure form. Dose-response studies are therefore difficult to evaluate for the time being.

It is unusual to obtain much more than 50% (30%–75%) inhibition with a chalone. The amount of chalone necessary to obtain this has sometimes been named "a chalone unit" (BULLOUGH and LAURENCE, 1964c; VERLY, 1973). However, it is perhaps correct to distinguish between cell cycle modulators (LORD et al., 1974b) and switch on/switch off mechanisms as indicated for the red marrow stem cell inhibitor (LORD, 1979). The reason for this may be that in most tissues only some cells are sensitive to the G_1 chalone and others to the G_2 chalone, whereas all stem cells may react to their chalone. MARKS et al. (1978) have given this idea a great deal of thought.

HONDIUS-BOLDINGH and LAURENCE (1968) showed shallow straight line dose-response effects with the purified epidermal G_2 chalone, and such relationships have also been published for the granulocyte chalone by PAUKOVITS (1973), BALÁZS et al. (1972) and RYTÖMAA and KIVINIEMI (1968b). MARKS (1973) showed dose-response curves for the partly purified epidermal G_1 chalone. The effect on DNA synthesis increased up to a dose of about 40 µg per mouse, then flattened off. A dose-response effect is also reported for the lymphocyte chalone (HOUCK, 1978) and for Ehrlich ascites chalone (NAKAI, 1976; NAKAI and GERGERLY, 1980). It is doubtful whether the dose-response curves presented for the epidermal chalones by IVERSEN (1978 b, c) are straight-line or biphasic. Biphasic effects of proliferation regulators have been demonstrated and discussed by AOYAMA et al. (1975). Quite recently, ANISIMOV et al. (1979) and OKULOV et al. (1979) published a method by which they claimed to quantitate the amount of epidermal G_2 chalone in cervical epithelium, utilizing the so-called radial immunodiffusion method. If this method works, better dose-response curves for this chalone may be obtained.

One can therefore conclude that reliable dose-response curves for chalones cannot be found until the pure chalones have been produced in sufficient quantities.

VII. Chemical Composition

The chemical composition of the various partially purified chalones will be discussed below. Most of the chalones have been assumed to be glycoproteins or glycopeptides, although PAUKOVITS and PAUKOVITS (1978 b) claims that the granulocyte chalone, HOUCK (1978) that the lymphocyte chalone and BARFOD and BICHEL (1976) that the JBI ascites tumor chalone are peptides.

There is certainly a possibility that chalones in vivo are composite molecules consisting of a small peptide molecule which is the general growth inhibitor (and which need not to be tissue specific) coupled to a tissue specific glycoprotein formed during differentiation and therefore specific for each type of differentiated cell. This theory seems to me so far the best explanation of most of the experimental facts.

D. Mechanisms of Action

I. General Considerations

Since no pure chalone substance is available, the molecular events in the mechanism of action cannot yet be described. Almost all evidence is of biological (not of biochemical) nature and speculations on the mechanism of action must be based on interpretations of the biological events following the action of chalone-containing extracts upon cells and tissues.

There is general agreement that the negative feedback principle is the basis for chalone action, although the system is a complex one involving stimulatory influences as well (BULLOUGH and MITRANI, 1976).

As regards the general mechanism of a chalone effect in cybernetic terms, the rate of cell proliferation may be directly proportional to the immediate concentration of chalones bathing the cells. Such a mechanism could be termed *proportional* feedback. It may be, however, that the signals to which the generative mass reacts are not the actual chalone concentration but the speed and the direction of alterations in concentration. This principle could be called *derivative* feedback. On the other hand, perhaps the cells "remember" the chalone concentrations during the previous time periods and react to this integral. This could then be called *integral* feedback. The immediate result of changing concentrations of chalone on proliferation would be different in these three situations, but the long-term effects would be similar. This needs to be investigated in more detail.

ELGJO (1968, 1974 a) presented results from epidermal carcinogenesis and from adhesive tape stripping indicating a *derivative* mechanism by which the state of proliferation was regulated by the "speed-of-the-variation" of the chalone concentration.

If epidermal chalones are diffusable in the skin, and if chalone antagonists diffuse upwards from the dermis, there ought to be a steep gradient in concentration at the level of the basal cells (BJERKNES and IVERSEN, 1974).

An extensive theoretical treatment of the mechanism of action of the chalones has been published by BJERKNES (1977).

II. Do Chalones Primarily Inhibit Proliferation, or Do They Promote Maturation?

It is generally agreed that in any tissue the mitotic activity and the degree of maturation are inversely correlated. The most mature (most highly differentiated) tissues (e.g. nerve cells) have little or no capacity for division, e.g. an epidermal cell or a blood cell cannot divide after it has passed a certain degree of maturity, a stage that BULLOUGH (1967) calls dichophase.

What, then, is the primary role of the chalones? If it were to promote maturation, then inhibition of cell proliferation might follow secondarily. In WEISS and KAVANAU'S model (1957, see Fig. 1) the antitemplates are assumed to influence both maturation and cell division. BULLOUGH (1967) and BULLOUGH and MITRANI (1976) maintain that the basic action of the epidermal chalone is to divert the cell from synthesis for mitosis to synthesis for aging and keratinization. But they also maintain that chalones inhibit the rate at which the cells age and die. This seems contradictory and his concept is based mainly on theoretical speculations.

BENESTAD and RYTÖMAA (1977) showed that the maturation rate of granulocytes could be increased by several experimental methods that inhibit cell proliferation, among them also treatment with chalone. If one assumes that the flux through the cell cycle is an ongoing process (RYTÖMAA, 1978) the speed of which is controlled by chalones, and if one also assumes that cells which are not allowed to divide are preprogrammed to pass the dichophase and start maturation, then it seems more easy to accept increased maturation as a consequence of growth inhibition. PAERMENTIER and BARBASON (1979) observed a marked chalone-induced morphological and functional differentiation in hepatocytes in vitro.

If one assumes that chalones are produced during maturation, and as a by-product of it (IVERSEN, 1969), and if the more mature cells produce more chalone per cell per time unit (the only way a feedback system can work, see IVERSEN and BJERKNES, 1963), then it becomes unnecessary and even contradictory to believe that chalones inhibit maturation rate.

III. Possible Relationship to Hormones and Cyclic AMP

Adrenalin has a dual effect on epidermal mitotic activity. It speeds up the passage of cells through mitosis and late G_2 (BULLOUGH and LAURENCE, 1966 b), empties the G_2 compartment and thereby secondarily reduces the mitotic rate (YOUNG et al., 1975).

In in vitro assays with short-term organ culture of mouse ear, adrenaline seems to be necessary cofactor to demonstrate G_2 chalone effect (Bullough and Laurence, 1961, 1964b, c, 1966a, 1968a). These authors also showed that glucocorticoid hormones enhanced G_2 chalone action. However, Laurence and Randers-Hansen (1971, 1972) later showed that adrenalectomized mice reacted to the epidermal G_2 chalone. Adrenalin acts through the cell membrane via the cyclic AMP system. A complete "second messenger system" is found in the epidermis (for reviews, see Marks, 1976, and Marks et al., 1978). There is thus reason to believe that the G_2 chalone effect is influenced by several hormones of the stress-hormone type and is mediated through the cyclic AMP-dependent second messenger system through membrane receptors (Elgjo, 1975). For the G_1 chalones no correlation with cyclic AMP has been found (see Marks, 1976).

IV. Do Chalones Act via the Cell Membranes?

Experimental evidence suggests a chalone effect via the cell membrane also for the G_1 chalone, which has no effect on the immature, neonatal epidermis. The tumour promoter TPA is known to affect cell membranes, and TPA-treated epidermis does not respond to G_1 chalone (Krieg et al., 1974).

Trypsinization of granulocytes and ascites tumour cells destroy their reaction to their respective chalones (Paukovits and Paukovits, 1975; Barfod and Bichel, 1976). Indications of membrane binding can also be found in Muller-Berat et al. (1973) and in Aardal et al. (1977), demonstrating lack of chalone effect after washing the cells and a temperature dependency of the assumed binding. Neustroev (1978), studying electrophoretic mobility of bone marrow cells, also found evidence of membrane binding of the erythrocytic chalone.

In tissue culture the cells often lose their sensitivity to chalone after some cell passages.

Thus, there are suggestions that chalones may act through receptors on the cell membrane, even if Nakai and Gergely (1980) could show that Ehrlich ascites chalone was effective on isolated nuclei.

E. Chalones and Malignancy

I. General Considerations

The possible role, if any, of chalones in carcinogenesis and neoplasia has unfortunately been the object of much naive, wishful thinking. Whatever the cause of the malignancy – chemical, physical, hormonal or viral carcinogens – and whatever mechanism of action may be suggested for these, there is general agreement that something is also wrong with the growth regulation. When a tumour grows the balance between cell gain and cell loss is disturbed in favour of cell gain. Chalones are concerned with the regulation of this balance, but tumours are not generally characterized by a rapid rate of cell proliferation; in fact many tumour cell populations have a lower rate of cell proliferation than that of the tissue of origin (see, e.g. Baserga, 1971). Hence, the rate of cell loss is reduced in tumours compared to normal tissue.

The capacity of infiltration and destructive growth and the formation of metastases is probably the most important property of a malignant neoplasm. Metastases in vital organs and often general toxicity is what kills the patient. Many benign tumours therefore attain a much larger size than malignant tumours.

The general conclusion must be that a priori nobody can know whether a change in the numerical balance between cell gain and cell loss is the basic or a secondary event in carcinogenesis. It is probably secondary, since this balance is also upset in benign hyperplasia and in benign tumours.

II. Chalones in Malignant Tumours

There is no doubt that many malignant tumours and tumour cells produce the chalone of their mother cell line, and that both the rate of DNA synthesis and mitosis can be reduced by injection of exogenous chalones produced from normal or from malignant cells. This has been shown repeatedly in transplanted squamous cell carcinomas (BULLOUGH and LAURENCE, 1968 b; BULLOUGH and DEOL, 1971; ELGJO and HENNINGS, 1971 a; LAURENCE and ELGJO, 1971; IVERSEN, 1978 b) and in melanomas (BULLOUGH and LAURENCE, 1968; FIEDLER et al., 1979). OKULOV et al. (1978) reported that extract of rat skin was effective on a transplanted highly differentiated squamous cell carcinoma of the cervix, but not on an undifferentiated skin carcinoma. Mouse epidermal chalone is effective on human squamous cell carcinoma cells (KORSGAARD et al., 1977). BERTSCH and MARKS (1979) showed that the epidermal G_2 chalone inhibited mitoses in a transplanted squamous cell carcinoma, but highly enriched G_1 chalone had no inhibitory effect on the DNA synthesis in the tumour.

RYTÖMAA and KIVINIEMI (1968 a, b, 1969, 1970) made similar observations in rats with Shay's chloroleukaemia in both the solid and the dispersed form, and a chalone effect has also been found in lymphomas (BULLOUGH and LAURENCE, 1970 a; BALÁZS et al., 1978). BICHEL (1976) demonstrated cell line specificity of G_1 and G_2 chalones of three different lines of ascites tumours. Similar observations have been made on Ehrlich ascites tumours (BURNS, 1969; LEHMANN et al., 1977).

BULLOUGH (1975) and RYTÖMAA (1976) believe that the chalone produced by tumour cells are readily released into the general circulation of the host (KARINIEMI and RYTÖMAA, 1976), and that tumour cells are less sensitive to the specific chalones than are their normal counterpart (RYTÖMAA and KIVINIEMI, 1968 b), suggesting a disturbed chalone receptor mechanism.

BURNS (1969) concluded that the growth rate of tumours is probably regulated by the production of a homologous specific mitotic inhibitor.

Thus, it is well documented that temporary retardation of tumour cell proliferation can be achieved by chalones. However, this does not mean that tumours can be *cured* by these substances (see below).

III. Chalones and Carcinogenesis

After a single dose of a chemical carcinogen or of radiation there is an initial block in DNA synthesis and mitosis, followed by a wave of proliferative activity (for a review see IVERSEN, 1973).

In the epidermis these periods of low and high proliferation are associated with corresponding alterations in the content of both G_1 and G_2 chalone, so that periods of high proliferative activity are associated with a low chalone content and vice versa. This has been observed after the strong complete carcinogen methylcholanthrene (Rohrbach et al., 1972, 1976a; Rohrbach and Laerum, 1974), after the tumour promoter (also a weak carcinogen) croton oil (Rohrbach et al., 1976b), after roentgen irradiation (Elgjo and Devik, 1978), but also after adhesive tape stripping (Rohrbach et al., 1977).

Unfortunately, not enough epidermal extract has been available to study the effect of chalone during a many weeks course of epidermal carcinogen application. The only information of this type comes from Modjanova et al. (1979), who showed that a concomitant dose of their lung chalone (called lung contactin) led to a 2.5-fold decrease in the number of lung adenomas per mouse after an injection of urethan.

Hence, the amount of factual information on the effect of chalones on carcinogenesis is sparse, which has left the field wide open to speculations.

F. Possible Practical Uses of Chalones

I. For Diagnostic Purposes

The cell line specificity of G_2-chalones may be used for diagnostic purposes to determine which cell line a tumour belongs to.

An example has been published by Korsgaard et al. (1978) in which the epidermal G_2 chalone showed a good response on human squamous cell carcinoma in short-term culture. However, an assumed epidermoid (histological diagnosis) human cancer did not respond to the epidermal G_2 chalone. Later autopsy showed that the tumour was a metastasis from an adenocarcinoma in the ovary.

II. For Therapeutic Purposes

1. Diseases with Benign Increased Cell Proliferation

The obvious candidate here is psoriasis, which is a skin lesion characterized by scaly plaques in which the rate of cell loss and the rate of cell proliferation are greatly increased.

Since all crude epidermal chalone preparations are unstable, and since the purified ones are produced only in minute quantities, there have been only abortive, not published, clinical trials with no favourable results yet.

2. Treatment of Cancer

Repeated attempts to use tissue extracts to cure cancer in humans have been made, so far without success. For example, Baker reported in 1933 that he had treated cancer patients with extracts of tissue prepared "from one area in an animal corresponding to that of the primary growth in the patient." In 1935 he gave a further report on "good results obtained in inoperable cases." The rest is silence. As already mentioned, in 1931, 1932, and 1933 Murphy and Sturm and later Macfay-

DEN and STURM (1936) reported on the presence of inhibiting factors which could retard the growth of transplanted and natural cancers in mice. The most interesting experiment was that of MURPHY and STURM in 1933, who injected mice intraperitoneally with embryo skin and placenta extracts. The mice had mammary carcinomas, and after weekly injections with embryonic skin extract, 28% of the tumours were stationary, 12% showed marked regression and 27% showed complete absorption. After placental extract injection, 24% were stationary, 25% showed marked regression and 19% showed complete absorption. No control groups were reported, but the authors conclude: "There seems to be little doubt that the extracts of the two tissues which have previously been shown to reduce the takes of transplanted cancer have an influence on natural or spontaneous cancer." The authors make, however, the following reservation:

"We do not consider that the results stated necessarily establish the hypothesis on which the experiment was based, for the complexity of the materials makes it quite possible that this explanation is not the correct one. The general relations between the factors which influence the origin and growth of spontaneous tumors, the balance in mechanism of normal tissues, and the inhibitor which has been isolated from the chicken sarcomas cannot be seriously discussed until further knowledge is available."

In 1965 PARSHLEY reported the inhibiting effects on the growth of a series of tumours in vitro of extracts from adult connective tissue, which seemed to be particularly effective against sarcomas.

In a series of papers in 1968 and 1969 (for a review see BULLOUGH, 1969), the mitosis-inhibiting and necrotizing effect of certain tissue extracts on four malignant transplantable tumours (the VX2 epithelial tumour, a chloroleukaemia and two melanomas) were published. These results were interpreted as being due to a chalone effect.

In 1970, I published a paper criticizing such statements, in which I objected to the acute necrosis of the melanomata being attributed to chalones and stated: "It seems more likely that the extensive necrosis in the tumour is not due to growth inhibition, but is mediated through the vascular bed, possibly as a local Shwartzman's reaction. Contaminating bacterial toxins might, for instance, be present in the tissue extracts, especially from tumors." And in fact in 1972 (a, b) MOHR et al. reported that the acute necrosis in the melanomas was not due to chalones, but to toxins from *Clostridium* spores present in the extracts.

FIEDLER et al. (1979) treated transplanted malignant melanomas in Syrian hamsters with i.p. injections of the Organon pig skin extract, which is shown to contain also a melanocyte-inhibiting factor (BULLOUGH and LAURENCE, 1968c). They observed tumour regressions, but no cures. The regressions were comparable to those seen after treatment schedules with some cytostatics.

In a systematic investigation of the presence of growth-inhibitory substances in animal tissues, BARDOS et al. (1968) tested over 1,000 fractions of extracts from 28 different bovine and porcine tissues. Anti-tumour activity was found in 14 fractions from different tissues and cell culture cytotoxicity was found in eight fractions from the liver, one from the lung and two from the pineal gland.

Experimental evidence shows that chalone treatment has an inhibitory effect on the proliferation of malignant cells. After i.p. injection of partly purified granulocyte chalone (Rytömaa and Kiviniemi, 1969, 1970) "... subcutaneous chloroma tumours (transplanted myeloid leukaemia in the rat) regressed; if the regression was complete, it was also permanent. The cell-line specific action of chalone was apparent. Fur renewal was activated, anaemia disappeared and body growth remained unimpaired (fast-growing young rats were used in these experiments)."

It is, however, well known that the tumour-host relationship in transplanted tumours is very different from the tumour-host relationship in tumours primarily induced by viruses, irradiation or chemical carcinogens, or occurring "spontaneously" in an animal or human being.

Tumour regressions and cures of transplanted experimental animal tumours are therefore not related to the cure of "spontaneous" or carcinogen-induced human cancers; however, such treatment has been reported.

Quoting Rytömaa and Toivonen (1979):

"When sufficiently large quantities of semi-purified granulocyte chalone (prepared by Weddel Pharmaceuticals, Ltd., London) became available the first attempts were made to test the effect of the material on myeloid leukaemia in man. It must be stressed at once that these experiments (Rytömaa et al., 1976, 1977) were not conducted as a real therapeutic trial, simply because there was not enough chalone for a long-term treatment and because the purity of the first samples was not sufficient for adequate dosing. Thus the main aim of these experiments was to test whether granulocytic chalone has any effect against myeloid leukaemia in man.

In view of all these difficulties the results obtained were surprisingly good. Thus, a distinct inhibition of leukaemic growth was observed in six of the seven patients; in five cases the inhibition was followed by actual regression of the leukaemia, lasting up to several months in the absence of any maintenance therapy, and in one case the treatment, quite unexpectedly, led to complete remission of the disease. In three of the patients the regression of leukaemia was followed by a dramatic improvement in the patients' general condition, and the survival time of the patients with acute myeloid leukaemia (5 cases) was significantly prolonged compared with appropriate historical controls." The results of this heroic experiment in treatment seem to be interesting, but not convincing.

Fiedler et al. (1979) treated a transplantable melanoma growing in hamsters with cytostatics and pig skin extract assumed to contain also the melanocyte chalone. They could not find any statistically significant effect of the epidermal extract on the growth of these tumours.

As a pathologist, working in contact with human cancer cases and having worked with chalones for more than 20 years, I feel that a word of caution is still very necessary as regards optimistic speculations that a direct use of chalones for the cure of human cancers is just around the corner. The evidence for such hopeful suggestions simply does not exist. From a theoretical point of view (Iversen, 1970), it seems unlikely that chalones can be used to cure cancer directly, although they may be of some help in protecting the normal cell during a course of chemotherapy (Houck, 1976a).

Those who believe in chalones as a treatment for cancer maintain that there is always a cell loss in a tumour. If chalones can reduce the birth rate of cells, the "natural" cell loss may kill enough cells to reduce the tumour load to such low values that the body's own immune defence can become operative and kill the remaining cells. A clear description of this view can be found in RYTÖMAA and TOIVONEN (1979). I do not believe that a numerical increase in cell number is the key to understand malignancy (IVERSEN, 1980). One can only hope that my sceptical view is wrong, but so far the scientific evidence is unfortunately on my side.

3. Immunosuppression

If pure B- and T-lymphocyte chalones can be produced, they will be of great value in immunosuppressive therapy (see KIGER et al., 1972b, 1973a, b; HOUCK et al., 1973; MATHÉ et al., 1973; BOREL et al., 1978 and Sect. G.IV.3). There are no immune reactions without initial cell proliferation, and a physiological growth inhibitor of T-lymphocytes administered for a period of time might well prove to be better than the presently known methods for immunosuppression, which all have many undesirable side effects.

4. A Male Antifertility Drug

There is good evidence for a spermatogonial chalone inhibiting cell proliferation in the male testis (see Sect. G.VIII.1). This is obviously a prospective candidate for a physiological drug reducing male fertility.

G. The Various Chalones

I. The Epidermal Chalones

1. The Epidermis

The epidermis with its cell product, a fibrous, disulfide-rich protein called *keratin* covers the external surface of the body and protects us from a multitude of external influences. The main bulk of the cells in the epidermis are the keratinocytes. They are created by mitotic division in the basal layer and migrate to the surface while they mature to keratin squames which are eventually shed.

The kinetic methods used to assess epidermal chalone activity have been applied both in vivo and in vitro. Long-term organ culture of skin seem to be unsuited for G_1 chalone measurements (GRADWOHL, 1978a, b; HANSTEEN et al., 1979). However, PRUNIERAS et al. (1980) feel that their cell culture system of epidermal basal cells has a great potential for the study of inhibitors of DNA synthesis and mitosis in epidermal cells. The G_2 chalone effect is measured by stathmokinetic methods, whereas the G_1 chalone requires cytometry and/or incorporation of tritiated thymidine (LI, grain counts, DNA specific uptake). Much evidence on the epidermal chalones is presented in the general part of this chapter and will not be repeated here.

2. The Epidermal G_2 Chalone

The first G_2 chalone effects on the epidermal mitotic rates were reported from mouse ear in in vitro organ culture with adrenalin added by BULLOUGH and LAURENCE (1964b) and from in vivo experiments by IVERSEN et al. (1965).

There are three reports on the partial purification of this regulatory substance.

HONDIUS-BOLDINGH and LAURENCE (1968) reported purification of G_2 chalone produced from pig skin. After in vivo i.p. injection of 30–180 µg the mitotic rate was reduced by about 50%; in in vitro conditions a dose of 1.5 µg/ml medium was sufficient. The source was mainly commercial rind (epidermis together with dermis) obtained from pigs which had been immersed in a water bath at 60 °C for about 15 min. The hair was then burned off, the skin removed and the rind separated. After lyophilization, the tissue was defatted with petroleum ether for 2 h at 45 °C and then ground to a powder. An aqueous extract was made at 0 °–4 °C in an atmosphere of nitrogen by means of a high-speed mixer. The extract was centrifuged and fractionated by the slow addition of ethanol at 0 °–4 °C. Usually most of the chalone activity was found in the 71%–82% ethanol precipitate; however, some activity was also collected with the 55%–72% ethanol precipitate as well as with the 81% ethanol supernatant. For further purification, preparative electrophoresis on cellulose columns at pH 3 (0.1 M acetic acid) proved to be superior to other techniques such as gel or ion exchange chromatography. A further separation from a large amount of inactive contaminants was achieved by dialysis at 4 °C against water. Finally the authors ended up with a preparation which was purified approximately 2,000 times over the crude skin extract.

The amino acid composition of the purified preparation showed a remarkably high content of proline and hydroxyproline, probably showing a great amount of connective tissue components from the corium. About 63% of the weight of the sample was accounted for by amino acids. Further analysis rendered possible a crude estimate of about 15% carbohydrate.

The chalone was resistant to pepsin but could be destroyed by trypsin; it was stable at pH 3–6 in aqueous solution for at least 3 days, but was inactivated at pH 9. The isoelectric point was estimated to be in the range of pH 5.2–6.0. The preparation was found to be almost homogeneous in the ultracentrifuge, and the sedimentation behavior as well as gel chromatography indicated an MW of 2.5–4×10^4 daltons.

The authors concluded that their most purified fraction was either a glycoprotein or a mixture of protein and polysaccharides. HONDIUS-BOLDINGH and LAURENCE (1968) did not claim to have isolated the pure epidermal chalone, but point to the heterogeneity of their preparation. THORNLEY and LAURENCE (1975) tested this material and showed that it contained only G_2 chalone effect and did not inhibit the influx of cells to the S phase or the DNA synthesis itself.

Another attempt at purifying the epidermal chalone has been published by the Lund-Oslo group (ISAKSSON-FORSÉN et al., 1977).

The procedure involved the sequential use of ammonium sulphate precipitation, affinity chromatography and gel filtration. Mitosis-inhibiting activity at each stage in the purification was tested in an in vitro assay system employing human epidermoid carcinoma cells in exponential growth phase and Colcemid (Ciba) for arresting mitoses. By the procedure described, a 10,000-fold material purification of the active component was obtained from the lyophilized powder. This corresponds to a 3,000-fold purification measured by protein content and a 300-fold increase in mitosis-inhibiting activity per unit weight. The active component was acidic and contained sugar residues and had gel chromatographic properties characteristic of a substance with molecular weight of approximately 20,000. On SDS polyacrylamide gel electrophoresis, however, three weak bands were found. The active component was resistant to trypsin and protease and was stable between pH 6.0 and 8.5. At the present stage of purification, several components other than the active one still remained in the material. Recently OKULOV et al. (1978) reported on an epidermal G_2 chalone extracted, MW about 35,000 daltons, isoelectric point 5.8 and containing no S-containing amino acids.

Even if purification procedures have ended with a glycoprotein (?) MW about 20,000 daltons, one cannot say that the epidermal G_2 chalone necessarily is a glycoprotein. There still remains the possibility that this chalone is a smaller molecule, perhaps attached to a larger one.

One must therefore conclude that epidermal G_2 chalone activity is found in a 20,000 dalton fraction partly purified from watery epidermal extract. This impure "soup" contains mainly one or more glycoprotein(s), but we do not know whether this is the G_2 chalone, or not. There may still be aggregates, and hence the G_2 chalone molecule may be a smaller one.

3. The Epidermal G_1 Chalone

We owe to FREDRICH MARKS in Heidelberg our biochemical knowledge on the purification of this substance (MARKS, 1975, 1976; MARKS et al., 1978). MARKS started out with the same pig skin crude extract as HONDIUS-BOLDINGH and LAURENCE (see above) and the effect was tested in vivo by the liquid scintillation technique measuring the DNA-specific uptake of ^{3}H-TdR. Disaggregation of the G_1 chalone-active components was achieved by incubating a 5%–10% solution of the ethanol precipitate in 0.1 M Tris-HCl buffer (pH 7.6) containing 0.02% sodium azide and 0.2% toluol with 1% sodium dodecyl sulphate and 0.8% dithiothreitol at room temperature for 60 h. Free sulfhydryl groups where then blocked with iodoacetamide (5% solution, pH 7.4, 3 h, room temperature). An excess of the reagent was destroyed with mercaptoethanol. The solution was extensively dialyzed against water (4 °C) or diafiltered through a Diaflo membrane filter PM-10 (exclusion limit 10^4 daltons, Amicon B.V., Oosterhout, Holland). A 1.5% solution of the residue was made in 0.125 M imidazol/HCl buffer (pH 7.5) containing 2.5×10^{-3} M CaCl$_2$. 0.5% Dodecylsulphate and a small amount of toluol and digested with three different concentrations of pronase (1, 0.5, and 0.2% in relation to the substrate) for 7 days.

After this procedure, most of the chalone activity was still retained by a PM-10 filter. It could be eluted from a DEAE-cellulose column (pH 7.4, 7 M urea) only with 0.3–0.5 M NaCl; this indicated a rather acidic compound such as a mucopolysaccharide or a nucleic acid.

After diafiltration through a PM-10 filter, the chalone could be quantitatively precipitated from the PM-10 residue with a 5% solution of cetylpyridinium chloride. The precipitate was dissolved in 1 M MgCl$_2$- and reprecipitated several times with 5 vol ice-cold ethanol/ether (4:1, 20 °C, 24 h). At this stage the chalone activity was still resistant to pronase, trypsin, collagenase, ribonuclease, deoxyribonuclease, hyaluronidase, β-neuraminidase, 5% dodecylsulfate, 7 M urea and heat (boiling water bath, 5 min), but could be destroyed by either 0.1 M NaOH or 1 M HCl (6 h, room temp.). After ultrafiltration in the presence of 2% dodexylsulfate, most of the activity was found in a fraction corresponding to 10^4–5×10^4

daltons; however, considerable activity was also present in the fractions corresponding to 10^3–10^4 daltons as well as 5×10^4 daltons.

From an aqueous solution of the cetyl pyridimium chloride-precipitated material, the chalone activity could be quantitatively extracted with 80% aqueous phenol. By this procedure it could be separated from the bulk of cetylpyridium chloride-precipitable material such as acidic mucopolysaccharides and nucleic acids.

An 800,000-fold purification over the defatted lyophilized skin poweder (2,000-fold over the ethanol precipiate) was achieved. Within the limits of the rather insensitive in vivo assay, the recovery of the factor seemed to be quantitative. This may be partially due to its extraordinary stability. A dose level as low as 0.05–0.1 µg of the phenol extract intraperitoneally injected into an adult female mouse was sufficient to reduce the DNA labelling in back skin epidermis by about 50%. This dose level corresponds to 2–4 µg/kg body weight. Therefore, the specific biological activity of the epidermal G_1 chalone seems to be in the range of that of a hormone.

The factor is probably a glycopeptide; its remarkable resistance to proteases may be due to its carbohydrate moiety. The fact that it is precipitated by cetyl-pyridinium chloride suggests that it contains a polyanionic residue (mucopolysaccharide or oligonucleotide).

The G_1 chalone has peculiar properties. Its molecular weight cannot be specified, and ranges from values below 10,000 daltons to more than 100,000. The lowest value is between 10^3 and 10^4, and MARKS (1975) believes this is the correct value.

The highly purified substance was also tested by THORNLEY and LAURENCE in London (1976) in vivo and they found a strictly tissue-specific depression of DNA labelling in mouse epidermis with no effect on the mitotic rate, and no effect on DNA synthesis in bone marrow or duodeum. Later (THORNLEY et al., 1977; LAURENCE et al., 1979) they showed that this epidermal G_1 chalone had no effect on the sebaceous gland. Thus, there is good reason to believe that this material – though not yet completely pure – contains the epidermal G_1 chalone.

As far as I know, a large-scale purification procedure is now under way.

4. Conclusions About Epidermal Chalones

It is now established beyond reasonable doubt that aqueous extracts of homogenized epidermis contain substances with specific growth inhibitory effect on keratinocytes. They are found in, and are effective on, both normal tissue and squamous cell carcinomas. The biological properties of these substances, which have been highly, but still not completely purified, make it reasonable to call them epidermal G_2 and G_1 chalones.

II. Chalones from Epidermal Derivatives

1. The Mammary Gland Chalone?

The mammary gland is an epidermal derivative, but the differentiated product is not keratin, but milk. Its epithelial cells are hormone sensitive, at least to oestrogens, progesterone, insulin, and prolactin. These hormones activate cell proliferation, but also function. Mammary growth control is therefore a very complicated one.

GONZALES and VERLY (1976, 1978) reported the purification of an inhibitor of DNA synthesis in mammary cells, and they called this inhibitor a chalone. The source of the chalone was bovine mammary gland and inhibitory activity was tested by measuring the uptake of ^{3}H-TdR and ^{14}C orotic acid by various cells with and without inhibitor added. The test cells were stimulated rat mammary cells from 11 to 12 days pregnant females and the mammary cancer derived cell lines of Ehrlich and Walker. Control cells were HeLa cells, Novikoff cells and chick embryo cells.

A 2,000–3,000 dalton molecule was found which inhibited DNA synthesis, but not RNA synthesis, in a cell line-specific way. The bovine mammary gland was homogenized, water extracted and separated by chromatography: Sephadex G-50, carboxymethyl-Sephadex and Sephadex G-25. After gel-electrophoresis there was a single absorbtion band. No real cell kinetic analysis has been done, and no in vivo experiments. A dose-response curve has not been produced.

My conclusion is that there is probably a specific chalone for the mammary gland, but this has hitherto been only indicated. Much further work remains to be done, especially in relation to the hormonal regulation of the proliferation of the mammary gland.

2. Sebaceous Gland, Sweat Gland and Hair Follicle Chalones?

a) Sebaceous Gland

BULLOUGH and LAURENCE (1970b) and THORNLEY et al. (1977) reported that the DNA synthesis and the mitotic rate in the sebaceous gland was not inhibited by epidermal G_1 or G_2 chalones, respectively, but both in vitro and in vivo by another skin extract fraction produced from skin with sebaceous glands. Extracts of skins without sebaceous glands (human gingiva, pig palate, and cod fish epidermis) were without inhibiting effect on the gland proliferation. No efforts were made to purify this putative sebaceous gland chalone, which has thus only been indirectly indicated.

b) Sweat Gland (Eccrine) Chalone

BULLOUGH and DEOL (1972) showed that the mitotic rate of sweat gland epithelium was not affected by epidermal G_2 chalone (partially purified) or by the fraction of skin extract that inhibited mitoses in the sebaceous gland. However, human sweat injected i.p. undiluted in doses of 0.1–2.0 ml inhibited mitotic activity in the eccrine gland duct epithelium, but not in the epidermis or the sebaceous glands. The inhibition was dose related, i.e. 0.1 ml sweat gave 8% inhibition, 0.5 ml 25% and 2.0 ml 42%.

No efforts have been made to purify the putative chalone in sweat, a task that will probably require blood, toil, tears, and sweat (W. Churchill, 13 May, 1940).

c) Hair Root Chalone

There is no experimental evidence in favour of a hair root chalone, but from biological reasoning BULLOUGH (1967) inferred its existence.

III. The Melanocyte Chalone (?)

Bullough and Laurence (1968 c) used two malignant, transplantable melanomas both as source and as assay system for a putative melanocytic G_2 chalone. They found an antimitotic factor in the tumour extracts and in the 71%–80% ethanol fraction of pig skin, whereas the more purified epidermal chalone had no such effect. The melanoma extracts had no effects on ear epidermal mitosis in vitro (Laurence, 1973), or on lymphomas (Bullough and Laurence, 1970a).

Seiji et al. (1974) reported that they had partly purified a melanoma cell cycle inhibitor, reducing both thymidine and leucine incorporation and the mitotic index in melanoma cells in vitro. The growth inhibitory, cell line specific activity was found in a fraction which was heat labile and contained both protein and RNA. No in vivo results were presented.

Dewey (1973) reported on the melanoma cell line-specific effect of a factor found in the extract of Harding-Passey transplantable melanomas. The effect was directly measurable as a reduced growth rate of these cells in culture, with no effect on cells derived from hamster lung, human embryo or liver cells. The factor was dialyzable and sensitive to heat, chymotrypsin, trypsin, and neuraminidase, and had a low MW of around 2,000. Later Dewey reported further purification. The factor inhibited 3 H TdR incorporation into DNA in shortterm cultures of melanoma cells, and now seemed to be heat stable and resistant to the enzymes. Finally, Dewey et al. (1977) reported that this factor was identified as spermidine. Thus, the search for a putative melanocyte chalone had led to sperimidine, just as Allen et al. (1977) found spermine in the lymphocyte chalone. It has repeatedly been shown that the lymphocyte chalone is not spermine (see below) and the possibility that a melanocyte chalone exists is still real.

Finally, the problem of a putative melanocyte chalone still suffers from the blow it got when it was wrongly reported that it could cure transplanted melanomas (Mohr et al., 1968). The injection of melanoma extract produced an acute local tumour necrosis, obviously related to a vascular block. Two years later the same authors (Mohr et al., 1972a, b) reported that the oncolysis was due to a toxin from *Clostridium* spores present in the extracts.

See also Fiedler et al. (1979), who treated transplanted melanomas with i.p. injections of melanocyte chalone in skin extract, and saw some tumour regressions, but no cures.

IV. Chalones in the Cells of the Red Bone Marrow, Blood and Lymphoid System

The red and white blood cells and the platelets are formed in the red marrow whence they are released into the circulation and the tissues. The lymphoid cell lines (B- and T-lymphocytes) later proliferate in the thymus, the spleen, the lymph nodes and the other lymph follicles in the body.

The functional capacity of the whole system depends on the cell number and the degree of maturation of each type of cell (i.e. granulocytes, lymphocytes, monocytes, platelets, erythrocytes). Most of these cells have a comparatively short life span and the demand for them varies greatly.

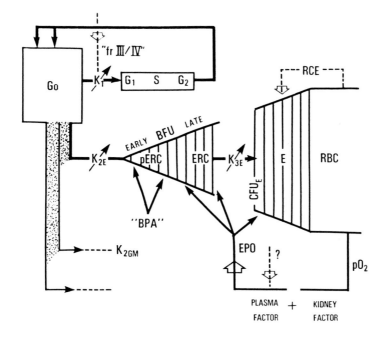

Fig. 3. Cell populations involved in erythropoiesis. The pluripotent stem cell, partly in G_0 state or in cell cycle, is controlled by a local feedback control system at rate K_1. K_2 is the rate at which such cells differentiate into the committed cell population, which has its own cycle length control system as well as a signal for differentiation from the pluripotent stem cell. The nature of this is not known, but evidence indicates a mechanism wider than local. Finally, erythropoietin induces second-step differentiation at rate K_3, to produce the erythron which has both a built-in suicide mechanism of maturation as well as a control of cell cycle length; the nature of the latter is not properly understood. (Reproduced by permission of Dr. L. LAJTHA and the publisher. Will appear in the updated version of "Blood and its Disorders" by Hardisty and Weatherall (ed), Blackwells Scientific Publ., Oxford 1980)

There are many systems for regulating the number of mature cells to maintain general homeostasis (see BROXMEYER and MOORE, 1978; and METCALF'S and CLINE'S chapters in this book). For most (if not all) of the blood cells the kinetic system of renewal and maturation is built up in the same general way (see, e.g. LAJTHA, 1979) Fig. 3 From a self-maintaining, pluripotent stem cell population, early committed precursor cells are formed and from these, series of more and more mature precursor cells arise until the functionally mature cells are released into the blood or lymph circulation and thence into the tissues (see, e.g. LAJTHA, 1979). Each round of cell division doubles the population, and thus the numerical and functional consequences of shortening or prolonging the mean cell cycle time, and of subtracting or adding one round of mitoses, are very great. In this chapter we are only concerned with the physiological growth regulators of a negative feedback

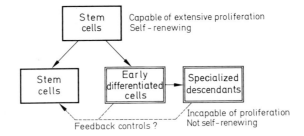

Fig. 4. A diagrammatic representation of the principal modes of proliferation of blood-forming cells, as suggested by TILL et al. (1964). (Reproduced by permission from the authors and Natl. Acad. of Sciences)

type, the chalones. A good and comprehensive review on chalones and blood cells was published by RYTÖMAA (1978).

1. The Granulocyte Chalone

Since most of the granulocytes are neutrophils and only a few are eosinophils and basophils, we shall only be concerned with the neutrophils. Their main function is the defence of the body against certain commonly occurring foreign substances (living or dead) and they are the cells seen early in the acute inflammatory reaction.

It has long been held that granulocytes probably release substances that inhibit their proliferation. A good early example of such a suggestion can be found in TILL et al. (1964) (Fig. 4). Many authors recommend the removal of inhibitory factors if serum is added to cultures of bone marrow cells (see, e.g. CHAN and METCALF, 1970; CHAN, 1971). Specific research directed at the putative granulocytic chalone was initiated by RYTÖMAA of Helsinki in 1964 (RYTÖMAA and KIVINIEMI, 1964), and followed by a series of papers up to the present (for a summary, see, e.g. RYTÖMAA, 1978). The main methods used to assess granulocytic chalone activity are based on the incorporation into the cell nucleus of tritiated precursors of DNA which is then measured by labelling indices and liquid scintillation countings. Colony formation techniques seem to be more valuable (MASCHLER and MAURER, 1978) and have been used both in the spleen and in agar. Real increase in cell number in diffusion chambers has also been measured (LAERUM and MAURER, 1973; BENESTAD and REIKVAM, 1975). Most research has been directed at the G_1 chalone. Mitotic rates have only rarely been studied (however, see LORD, 1975; VILPO, 1979), but the colony formation techniques and the cell counts certainly presuppose that mitoses have taken place.

The inhibiting effect of the granulocyte chalone on red marrow or granulocyte cultures has mainly been measured in vitro or in intraperitoneal diffusion chambers (BENESTAD et al., 1973; LAERUM and MAURER, 1973; LØVHAUG and BØYUM, 1977; McVITTIE and McCARTHY, 1975), but occasionally also in vivo on bone marrow proliferation (BALÁZS, 1976) and on transplanted leukaemias and chloroleukaemias. A series of tests on rat and human leukaemia has also been reported

(RYTÖMAA et al., 1976, 1977). Cell line specificity has been tested against erythrocytes, lymphocytes, monocytes (macrophages), L-929 fibroblasts and HeLa cells (see BALÁZS, 1979). Much evidence on the granulocyte chalone(s) has been discussed in the general part of this chapter, and will not be repeated here.

In summary, the results of all this search for a granulocyte chalone show that granulocyte conditioned media prepared in various ways with mature granulocytes or aqueous extracts of homogenized granulocytes contain inhibitor substance(s) that modulate the proliferation of immature granulocytes (see, e.g. AARDAL et al., 1977; BATEMAN, 1974; BENESTAD and RYTÖMAA, 1977, 1979; BLAZSÉK et al., 1976; BOLL et al., 1979; BØYUM et al., 1976; MACVITTIE and McCARTHY, 1975; MAIOLO et al., 1975; MAURER et al., 1976; MASCHLER and MAURER, 1978; PAUKOVITS, 1973; PAUKOVITS and HINTERBERGER, 1978a; RYTÖMAA and KIVINIEMI, 1968a, b; SCHUNK et al., 1978). Some of these inhibiting factors seem to be dialyzable, peptide-like substances with cell line specificity strictly for the myeloid cell family, thus deserving the name chalone.

The inhibition of cell proliferation is reversible, not due to polyamines, and these chalones are not cytotoxic (BALÁZS, 1976; FOA et al., 1979). PAUKOVITS has suggested that the cell line specific inhibitory effect is due to a single peptide. The inhibitory effect of this peptide can be observed on mitotically competent myeloid cells of the granulocytic line (PAUKOVITS and PAUKOVITS, 1978; PAUKOVITS and HINTERBERGER, 1978b). The inhibition can readily be reversed by washing the cells after incubation with granulocyte chalone, but prior to plating (AARDAL et al., 1977). Some experiments strongly indicate the presence of chalone-specific receptor sites on the surface of the target cells (PAUKOVITS and PAUKOVITS, 1975). The inhibitory effect of this peptide can only be observed when the inhibitor is present in reduced form. Prolonged exposure to atmospheric oxygen renders the substance inactive as an inhibitor, and it can then even stimulate colony formation in agar (AARDAL et al., 1977). The inhibitory activity can, however, be restored by treatment with 2-mercaptoethanol (see, e.g. LAERUM et al., 1980).

The substance responsible for all these effects has been isolated from leucocyte-conditioned media by sequential application of several steps of gel-chromatography, ion-exchange and partition chromatography. According to its chromatographic properties the granulocyte chalone is an acidic peptide with an MW of nearly 600 daltons. The presence of thiol groups is necessary for its action as an inhibitor, whereas formation of a disulfide bridge results in growth stimulatory properties (PAUKOVITS et al., 1980).

Not all research haematologists agree on the granulocyte chalones. Other more or less specific inhibitors have been found in granulocytes, among them lactoferrin (BROXMEYER et al., 1978; but, see also DELFORGE et al., 1979), prostaglandin E and leukaemia inhibitory activity (LIA) (BROXMEYER and MOORE, 1978). Some workers also doubt that the observed effects are actually derived from granulocytes (HERMAN et al., 1978). However, the method used by the latter authors to purify the granulocytes must be considered to be also a "repeated extraction procedure" which has a strong diluting effect on the chalone, as shown by RYTÖMAA and KIVINIEMI (1968a, b).

RYTÖMAA and the Weddel Pharmaceuticals, London, have never published in detail the chemical composition of their purified chalone factor *("Myelostat")*.

However, they report (Foa et al., 1979) the following: "... leucocytes were isolated from ox blood and extracted with Hanks' solution using intact cells; the conditioned medium was then collected and purified by a series of ultrafiltrations, followed by steps of gel filtration, chromatography, then desalted and lyophilized." The nominal MW is said to be 500–1,000 daltons. Foa et al. (1979) showed that this granulocyte chalone was not a polyamine complex. For investigations of tissue specificity of this preparation, see Vilpo (1979). Nobody knows whether *Myelostat* is the same as Paukovits' peptide. Maurer et al. (1976) reported on the purification of granulocyte chalone and came up with a thermostable mixture of peptides which met the criteria of a granulocytic chalone. A large-scale isolation of the granulospecific chalone fraction from granulocytes has been going on in Budapest, and Balázs et al. (1980) published the purification of an endogenous regulatory peptide which selectively inhibited the proliferation of normal and leukaemic myeloid cells. Leucocytes were isolated from calf spleen or horse blood, homogenized, extracted by acetone, chloroform and destilled water, ultrafiltrated by Amicon Diaflo XM 50 and PM 10 membranes. The lyophilized filtrate was chromatographed and the active fractions were submitted to paper electrophoresis at pH 6.5 and 1.9, and the peptides were re-tested. A ninhydrin negative and chlortolydine positive oligopeptide was identified as granulocyte-specific inhibitor, effective at 0.2–3.0 pmol/ml MED value in vitro. It is said that the oligopeptide nature and the exact chemical structure of this biologically verified leucocyte chalone will be documented in a fortcoming paper. For information on other inhibitory substances from granulocytes, see Balázs (1979) and Broxmeyer and Moore (1978).

Thus, in conclusion it may be said that highly convincing biological evidence for a granulocytic chalone has been produced from a series of centres all over the world. It has yet to be proven that the small peptide of Paukovits really is the granulocyte chalone. This problem is difficult to solve quickly, because so small quantities are available. If the synthesis of this peptide should be possible, all the above-mentioned effects have to be reproduced before one can definitely say that this is the granulocyte chalone.

2. The Erythrocyte Chalone

The regulation of growth and function of the erythrocyte cell line is complicated and involves many stimulatory and inhibitory factors (Fig. 3).

One of the advantages of this system is that it is relatively easy to get a fairly pure source for extraction of a possible chalone. The existence of an erythrocyte G_1 chalone was first suggested by Rytömaa and Kiviniemi (1967) and by Kivilaakso and Rytömaa (1971). The latter authors (1971) showed the cell line specificity, results which were confirmed by Bateman (1974), Bateman and Goodwin (1976), Bateman and Pollock (1977), Lord et al. (1974a, b, 1977) and Neustroev (1978). It seems well established that partially purified erythrocyte extract inhibits the incorporation of 3 HTdR in erythroid cells, and not in lymphoid or granuloid cells. Extract of granulocytes and lymphocytes have no such effect on erythrocytes. The erythrocytic G_1 chalone has no effect on the G_2 phase, but it inhibits iron incorporation into haemoglobin, perhaps because less cells are allowed to mature (?). The cellularity of erythroid spleen colonies became reduced by 50% after erythrochalone treatment (LORD et al., 1977) and there was an inhibition of erythrocyte release into the general circulation. The active molecule seems to have a molecular weight of between 500 and 1,000 daltons, probably around 1,000. A dose of 10 μg i.v. to a mouse gives a maximum depression of iron incorporation. The inhibitory factor is not haemoglobin (BATEMAN and GOODWIN, 1976) and probably not the factor EiF which LINDEMANN (1971, 1975) found in human urine, because this factor acts by interferring with the stimulator erythropoitin. However, EiF seems to be erythroid-cell specific (LINDEMANN and LAERUM, 1976). An erythrocyte chalone has also been demonstrated by NEUSTROEV (1978) and NEUSTROEV and KOROLEV (1979). Another erythropoietin inhibitor can be found in the kidney and is a complex lipid (ERSLEV et al., 1972). Another EiF was reported by LEWIS et al. (1969, 1973), not identical with LINDEMANN's EiF, and probably not the erythrocyte chalone either.

3. The Lymphocyte Chalone(s)

The regulation of lymphocyte proliferation is extremely complex. Antigens induce proliferation of B- and/or T-lymphocytes in a specific way through membrane receptors on memory cells, leading to functional clones of T- and/or B-lymphocytes (plasma cells) with high specificity against the provoking antigen. AOYAMA et al. (1975) reported that extracts of thymocyte suspensions had both stimulatory and inhibitory effects on the growth of mammalian cells in vitro. Corticosteroid hormones and many other factors inhibit lymphocyte proliferation (RANNEY, 1975). Recent reviews on factors inhibiting proliferation can be found in, for example, COOPERBAND et al. (1976), WAKSMANN and NAMBA (1976) and RIJKE (1979).

Specific research on lymphocyte chalone(s) started when MOORHEAD et al. (1969) reported that extracts of lymphnodes from pigs could inhibit incorporation of tritiated thymidine into DNA in stimulated human lymphocytes in vitro. In 1970 BULLOUGH and LAURENCE (1970a) reported a G_2 chalone effect of lymphocyte extracts on mouse lymphoma cells.

Cell line-specific endogenous inhibitors of lymphocyte DNA synthesis, mostly measured as reduced uptake of 3HTdR in stimulated cells, were then claimed to be found in a series of laboratories, see, e.g. JONES et al. (1970), LASALVIA et al. (1970), GARCIA-GIRALT et al. (1970), HOUCK et al. (1971), KIGER (1971), LORD et al. (1974a), MAIOLO et al. (1975), OLSSON and CLAËSSON (1975), GRUNDBOECK-JUSKO (1975, 1976, 1977), HEIDEMANN et al. (1976, 1979a, b), HIESTAND et al. (1977), BLAZSÉK et al. (1976), LENFANT et al. (1976, 1978), GARCIA-GIRALT et al. (1978), MASCHLER and MAURER (1977, 1979), RIJKE and BALLIEUX (1979), FEIJES et al. (1980). However, BOREL et al. (1978) reported negative results when testing thymus and spleen fractions in vivo.

Lymphocyte extracts with chalone activity have been prepared from lymph nodes, spleen, thymus, some established lymphoid cell lines and malignant lymphomas. No such activity has been found in similarly prepared extracts of granulocytes, epidermis, brain, muscle, lung, kidney or erythrocytes. Lymphocyte extracts, on the other hand, had no chalone effect on fibroblasts, HeLa cells, choriocarcinoma cells and erythroid or granulocytic precursor cells (see, e.g., ATTALLAH,

1976; Lord et al., 1974 a). Lymphocyte chalone activity seems to be non-toxic and reversible.

During assays to extract and purify lymphocyte chalone, Allen et al. (1977) suggested that lymphoid chalone might be a protein-spermine complex. The polyamines are by themselves not growth inhibitory, but seem to stimulate proliferation in some assay systems (Jänne et al., 1978; Raina and Jänne, 1975). However, they may interact with an amine oxidase in fetal calf serum (Byrd et al., 1977) and be converted to toxic aldehydes (Tabor et al., 1964). Allen et al. (1977) stress that their findings only point to the difficulties with the isolation of a lymphocyte chalone, and ... "by no means rule out its existence." Houck (1979, personal communication) has experienced that neutral extracts of thymus will contain only a trace of polyamine, whereas acid extracts contain a lot of spermidine. Spermine inhibits cell growth in vitro in the presence of fetal calf serum, but not when human serum is used. The lymphocyte chalone operates also in human serum. Recently Lenfant and Digiusto (1979) presented evidence that at least the immunosuppressive activity of lymphocyte chalone is not due to polyamines. The inhibitory activity of lymphocyte extracts on lymphocyte colony growth was shown by Maschler and Maurer (1979) not to result from spermine, a result also confirmed by Rijke and Ballieux (1978, 1979) and by Foa et al. (1979).

Since all immune reactions are initially also dependent upon lymphoid cell proliferation to produce the mature cells, immunosuppressive action has been used to assess lymphocyte chalone activity (evidence is found in the literature cited above). For a more comprehensive discussion on this problem, the reader should consult Rytömaa (1978) or Houck (1978).

There are indications that there might be two cell line-specific lymphocytic chalones, one for the B and one for the T cells. The plasma cell tumour JB-1 ascites tumour chalone (Bichel, 1976) is probably related to the B cell chalone. In this connection, one may also refer to the report by Danielsson and Van Alten (1974).

Kiger et al. (1972 a), Florentin et al. (1973 a, 1976) and Maschler and Maurer (1979) reported a T cell chalone effect and indicated that it might be an antigen-nonspecific suppressor lymphokine. See also Grundboeck-Jusko (1975, 1976, 1977). Rijke (1979) reported that his lymphocyte chalone inhibited both T- and B-lymphocyte proliferation.

Ranney (1975) published evidence on a chalone-like low-molecular-weight factor released from lymphocytes. Waksman's group has published interesting results clearly showing that lymphocytes produce an inhibitor of DNA synthesis. The factor has been extensively purified (Jegasothy and Battles, 1979) and appears to act by raising the intracellular cAMP level in late G_1 (thus in contrast to what has been described for the epidermal G_1 chalone). The possible relationship of these factors to the lymphocyte chalones is not at all clear.

Many laboratories have, as listed above, reported purification steps on lymphocyte chalone(s). Table 2 (from Allen and Smith, 1979) shows a list of the various molecular weights reported. Houck (1978) assumes to have found a cell specific and endogenous G_1 inhibitor of lymphocyte transformation and proliferation in vivo and in vitro. This appears to be a small cationic glycopeptide, often bound to RNA. The MW was in the range of 1,400 daltons. In a personal letter (1979) report he stated:

Table 2. The reported molecular weight of the active principles in lymphocyte chalone preparations (Modified from ALLEN and SMITH 1979)

Reported molecular weight	Reference
50,000–70,000	MOORHEAD et al. (1969)
50,000–70,000	JONES et al. (1970)
45,000	LASALVIA et al. (1970)
30,000–50,000	HOUCK et al. (1971)
20,000	KIGER et al. (1972)
45,000	GARCIA-GIRALT et al. (1972)
30,000–50,000 10,000–30,000	HOUCK et al. (1972)
30,000–50,000	ATTALLAH et al. (1975)
38,000 (B-cell-specific) 2,500 (T- and B-cell activity) 2,100 (T-cell-specific)	GRUNDBOECK-JUSKO (1976)
30,000–50,000	LENFANT et al. (1976)
10,000–20,000	KIGER et al. (1977)
1,000–10,000	HOUCK et al. (1977)
30,000–50,000	HEIDEMANN et al. (1976)
< 2,000	LENFANT et al. (1978)
1,400	J.C. HOUCK (1978)
1,400	MASCHLER and MAURER (1979)
10,000–50,000	HEIDEMANN (1980)
700	J.C. HOUCK (1980)

"Normally lymphocyte inhibitor does not pass through a 30,000 dalton filter very effectively, but after treatment with ribonuclease or increasing the concentration of the aqueous extracts to 18 mg/ml with streptomycin sulfate, essentially all of the inhibitory activity passed through the filter. The resulting solution could then be concentrated on a 500 dalton filter and subjected to Sephadex G-25 chromatography where the inhibitor was eluted at a position equivalent to 1,400 daltons as marked by bacitracin standard. This inhibitor operated equally well in human serum or in calf serum to inhibit the two-way MLC (Mixed Lymphocyte Culture) transformation of murine or human lymphocytes in vitro. The bulk of this fraction was actually streptomycin sulfate, since it too has a similar molecular weight and this could be partially (but not completely) desalted by chromatography on G-15, using ammonium bicarbonate as a buffer. It was imperative that some salt or buffer was in the solution in which Sephadex was used, because the inhibitor binds powerfully to Sephadex polysaccharides as well as to cellulose.

Streptomycin sulfate is a well known reagent from bacterial chemistry for the precipitation of RNA, including relatively small molecular weight RNA. In my judgement it displaces the small molecular weight cationic chalone from the large molecular weight anionic RNA. This complex can also be separated by exclusion chromatography on G-200. The biophysics of exclusion chromatography is such that the complex is not that stable and resolution of the material held back on G-200 by G-25 indicates the same amount of activity at the same molecular size can be produced as with ultrafiltration."

GRUNDBOECK-JUSKO (1975, 1976, 1977) claims to have isolated from bovine spleen specific T and B lymphocyte G_1 chalones with MWs of 2,100 and 38,000, respectively. HOUCK (November 1980, personal communication) reports that his lymphocyte inhibitory material now seems to contain 85% spermin, 10% thymidine nucleotides, and only about 5% seems to be what might be called a true lymphocytic chalone. This substance has a molecular weight of 700 daltons. The activity is destroyed by acid hydrolysis, acetylation and by pronase.

The material partly purified by KIGER et al. (1975) was sensitive to trypsin and pronase, and had an MW of 10,000–15,000 daltons, but they also found a low-molecular-weight inhibitor probably released from a larger carrier.

LENFANT et al. (1976, 1978) showed that in spleen extract the chalone activity was due to a low-molecular-weight factor non-covalently associated to high-molecular-weight carriers. This factor has been further purified, LENFANT et al. (1979). Through ultrafiltration (Amicon PM 10) molecular weight sieving (Biogel P-2), partition chromatographies (Alumina and Sephadex LH-20) and cellulose thin layer chromatography a 10^7 times purification factor could be achieved. The purified product was active in vivo at a 1 ng level per mice and in vitro at less than 1 pg/ml. The factor was characterized by high pressure liquid chromatography. Enzymatic degradations allowed the conclusion that the active substance had a peptide nature. The chalone purified by GARCIA-GIRALT et al. (1978) and by LENFANT et al. (1979) seems to affect both the proliferation and the function of lymphocytes.

MASCHLER and MAURER (1979) found T-lymphocyte chalone activity in two fractions extracted from calf thymus. Fraction I had an MW of 1,000–10,000 daltons, was stable upon heating, whereas Fraction II (MW 10,000–30,000 daltons) was unstable. Further Biogel P 6 or DEAE ion-exchange chromatography of Fraction I gave a specific chalone effect of a substance with an MW around 1,400 daltons. This effect was reversible, and probably not due to spermine.

HEIDEMANN et al. (1979 a, b) reported G_1-chalone activity in spleen extract. The crude extract was separated into 16 different fractions, and one of these contained the chalone properties, assumed to be 10,000 times concentrated. They think their fraction is similar to that of HOUCK (1978), and report an MW of 10,000–50,000 daltons (HEIDEMANN, 1980 personal communication).

It must be remembered that MATHÉ (1972) and MATHÉ et al. (1973) reported on the immunosuppressive action of semipurified extracts of the spleen and thymus (see also Sect. F.II.3).

A more detailed account of many substances claimed to have shown lymphocytic chalone activity can be found in BALÁZS (1979). Some are proteins or glycoproteins, others are peptides or peptide-sugar-RNA complexes, or spermine-protein complexes.

My own conclusion is that there is increasing evidence that lymphocytes contain factor(s) with chalone properties, inhibiting lymphoblastic transformation and proliferation in a cell line-specific manner. The weak point is that most of the studies are based on 3 HTdR uptake assays alone (see Sect. B.II). There is thus still work to be done before the lymphocyte chalone(s) is/are purified and their exact role in the complex set of growth regulatory mechanisms in lymphocytes elucidated.

4. The Monocyte (Macrophage) Chalone (?)

Monocytes are formed in the red marrow and differentiate from the same early differentiated cells as do the granulocytes. Monocyte proliferation is stimulated by the chemical inducers of inflammation, notably by prostaglandin E, which is also secreted by macrophages (positive feedback?) (see BROXMEYER and MOORE, 1978).

The presence of a specific macrophage chalone has been indicated by ISCHI-KARA et al. (1967) and PARAN et al. (1969) and WAARDE et al. (1978). LAERUM and MAURER (1973) and BENESTAD et al. (1973) could confirm this and it was shown that macrophages were not inhibited by granulocyte extracts. Partial purification included removal of macrophages by cell separation, and the specific activity rested in a factor between 500 and 10,000 daltons.

5. The Platelet Chalone (?)

The platelets (thrombocytes) are produced by the megakaryocytes in the red marrow. They have a variety of functions and secrete many biologically active substances into the circulation (see ROSS' chapter in this book, and ROSS and VOGEL, 1978).

A review on the possible humoral growth control substances for thrombocytopoiesis was given by LOZZIO (1973) and by ODELL (1973). He mainly commented upon a stimulative factor, thrombopoietin. However, negative feedback growth regulation has also been considered, and McDONALD (1973) reviewed the evidence from experiments with platelet infusions and concluded that the data were consistent with the existence of a chalone principle (he did not use the word chalone). ODELL (1973) and PAULUS (1967) suggested that this factor could be thrombostenin.

Thrombocytopoiesis can also be suppressed by extracts from injured platelets (KRIZSA et al., 1977). It must be concluded that the evidence for a putative platelet chalone is only circumstantial.

6. The Stem Cell Chalone (?)

The proliferation of pluripotent stem cells in the red marrow can be assessed by the CFU-S technique (spleen colony forming unit). The growth of such colonies is not inhibited by the erythrocytic or the granulocytic chalones (see RYTÖMAA, 1978). However, LORD et al. (1976) and WRIGHT and LORD (1978) indicated that extracts from normal (but not from rapidly proliferating) bone marrow inhibited stem cell proliferation in a specific way. This effect did not seem to be a modulation, but a switch on/switch off mechanism. Partial purification of this factor was achieved by ultrafiltration, and found to have an MW of 50,000–100,000 daltons. The activity was destroyed by trypsin, and thus their stem cell chalone may be a protein or bound to a protein. FRINDEL and GUIGON (1977), GUIGON and FRINDEL (1978, 1979), and GUIGON et al. (1979) have reported an endogenous stem cell inhibitor with a low molecular weight (700 daltons) obtained from fetal calf red marrow (GUIGON, 1979, personal communication).

7. Conclusion About Blood Cell Chalones

In 1973 LOZZIO wrote: "The functional interrelationship of stimulators – and inhibitors of haemopoietic cell production and differentiation is obscure." This is still true, but according to my opinion convincing evidence has been published on chalones for granulocytes and fair evidence exists for chalones in erythrocytes, lymphocytes, and stem cells. The existence of a putative monocyte chalone is also a probability. The possible platelet chalone is more in the dark. The purification of the lymphocyte and the granulocyte G_1-chalones is far advanced, and their chemical nature (probably peptides) may be clear in the not too distant future.

V. Chalones in the Gastro-intestinal Tract

1. Salivary Gland Chalone (?)

BARKA (1973) reported that isoproterenol-induced stimulation of DNA synthesis in the submandibular gland of rats and mice could be prevented by i.p. injection of a cell-free extract of the non-stimulated gland, but not if the gland from which the extract was made was already stimulated by isoproterenol. The effect was organ specific, but unusually enough, species specific. No effect was seen after similarly prepared liver homogenates. The inhibiting factor could be partly concentrated by Sephadex G 75 and DEAE cellulose chromatography, but the purified fraction contained at least eight protein bands when studied by polyacrylamide gel electrophoresis.

It is difficult to know whether this factor is a chalone.

2. Oral Mucosa, Oesophageal and Forestomach Chalone

The oral mucosa and the oesophagus, and also the forestomach in rodents, are covered by squamous cell epithelium. TEEL (1972) and NOME (1975) showed that these organs contain and react to the epidermal chalones.

3. Gastric Chalone (?)

In 1971 PHILPOTT reported that aqueous extracts of hatching chick stomach mucosa selectively suppressed mitotic activity in younger chick stomach epithelial cells in vitro, but had no such effect on intestinal, skin or mesenchymal cells. He concluded that the results pointed to "the possibility of a tissue-specific inhibitory feedback control of cell proliferation."

In 1972 WILLEMS reviewed the field of cell proliferation in the gastric mucosa, and concluded that the existence of a gastric chalone was a possibility. THOMPSON (1974) did not agree. No attempt at purification of a putative gastric chalone has been published. A recent review was published by GLASS (1980).

4. Small Intestinal Chalone

WRIGHT et al. (1979) presented evidence from cell kinetic studies which led them to conclude that the control of cell proliferation in the small intestine was mediated through a negative feedback mechanism.

BISCHOFF (1964) demonstrated that saline extracts of duodenal mucosa of 19-day-old animals injected i.v. into embryos inhibited mitotic activity in the duodenum and reduced the villus length. Saline or liver extract had no such effect.

A G_2 chalone effect for the small intestine was later found also in intestinal extracts from mature animals, and the effect was confirmed in embryos (PHILPOTT, 1971; GALJAARD et al., 1972; BRUGAL, 1973; TUTTON, 1973; BRUGAL and PELMONT, 1974, 1975). This inhibitor was been partly purified and is water soluble, thermolabile, dialysable and with an MW of 2,000 daltons. Adrenaline is not necessary for its effect. ANDERSSON (1973) found only a weak, non-significant intestinal G_2 chalone effect. A G_1 small intestinal chalone effect was demonstrated by BRUGAL and PELMONT (1974). This factor was non-dialyzable with an MW of 120,000–150,000 daltons.

SASSIER and BERGERON (1977) found a similar intestinal G_1 chalone effect in two of their fractions called F_1. The effect was reversible and not found in kidney or testis extracts, but a weak effect was observed in colon extracts. CLARKE (1974) was not able to find evidence of a blood-borne inhibitory factor.

There is thus good biological evidence for locally acting small intestinal G_1 and G_2 chalones. A detailed review was published by BRUGAL (1976).

5. A Colon Chalone?

KANAGLINGAM and HOUCK (1976) partly purified a factor from dog and pig colon and called it "a putative colon chalone." The factor had an MW of 10,000–50,000 daltons, and was sensitive to trypsin and chymotrypsin but resistant to RNase. It seemed to be a G_1 chalone, was effective on colon carcinoma cells and showed no inhibitory effect on a bronchial carcinoma, HeLa cells or stimulated lymphocytes. No further studies have been published.

VI. The Liver Chalones

The growth regulation of the liver presents a formidable problem, and there is a vast literature on the subject. It seems clear, however, that partial hepatectomy leads to a wave of DNA synthesis in the remaining liver after about 24 h, followed by a peak of mitosis about 10 h later. Then the liver grows up to about normal size, whereafter growth is arrested. Similar proliferative peaks can be seen in non-operated livers in animals which are connected to another hepatectomized animal by parabiosis or by cross-circulation, and also in liver autotransplants in other sites of the body. To explain such phenomena, the existence of inhibitory (chalones) and stimulatory factors have been proposed.

Good reviews can be found, e.g. GRUNDMANN and SEIDEL (1969), MORLEY (1974), NADAL (1979). For the chalone problem, NADAL's review (1979) is especially relevant. There are many specific problems in the study of the putative liver chalones. First, the normal liver has a very low proliferative activity and thus a stimulation is necessary for cell kinetic studies. But, as mentioned above, a rapidly proliferating tissue with many newly formed cells seems to be less reactive to chalones.

The source of the hepatocyte chalone is also a problem, since only 40% of the cells in a liver are hepatocytes (IYPE et al., 1965). Many different methods have been applied both in vivo and in vitro (see NADAL, 1979).

SAETREN (1956) was the first to show directly the existence of a liver chalone mechanism, based on in vivo studies. Since then a variety of methods have demonstrated both G_1 and G_2 chalone effects on the liver, but the results could not always be confirmed by others.

As an example of a series of studies, one may mention those of VERLY et al. (1971) and DESCHAMP and VERLY (1975), who found a liver G_1 chalone in a low-molecular-weight fraction of liver extract (about 2,000 daltons), probably a peptide, sensitive to trypsin and pronase but not affected by neuraminidase, DNase, RNase or heating. SIMARD et al. (1974, 1976) confirmed this result.

VINET and VERLY (1976) also found a protein with an MW of 40,000 daltons having growth inhibitory effect on hepatoma cells, but not on liver slices, in vitro. Such high-molecular-weight fractions with chalone-like effects were also found by CHANY and FRAYSSINET (1971), AUJARD et al. (1973); HIGUERET et al. (1975); and MOLIMARD et al. (1975). Liver G_1 chalones have been studied by MIYAMOTO and TERAYAMA (1971), BARBASON et al. (1977), ECHAVE LLANOS et al. (1970), MAHARA-JAN and BATRA (1976), SEKAS and COOK (1976), OKULOV and CHEKULAEVA (1976), LOGINOV et al. (1976), LAVIGUE et al. (1977), KUO and YOO (1977), MALENKOV et al. (1977), CHEKULAEVA et al. (1978), PIETU et al. (1979), PAERMENTIER and BAR-BASON (1979).

Liver G_2 chalones were indicated by BRUGAL (1973), MENZIES and KERRIGAN (1974), CANELLA (1977), LOGINOV et al. (1976). Inhibitory factors in blood and serum have also been studied NADAL (1979), SEKAS et al. (1979), McMAHON and IYPE (1980).

In summary, two groups of factors have been found with liver-specific chalone-like effect. One consists of peptide(s) with MW less than 10,000 daltons, the other of proteins with MW more than 40,000 daltons. It is not clear whether these are two different factors, or whether the high MW is due to aggregations, or to a combination of a small molecular inhibitory factor sometimes bound to a larger carrier molecule. Nor is it clear whether there is a difference between a G_1 and a G_2 liver chalone.

In conclusion, one can echo NADAL (1979): "Much work remains to be done before it will be possible to determine how many physiological inhibitor factors there are, whether they belong to the chalone family, and the part played by the inhibitory mechanism in the overall regulation of liver growth."

I find the situation exciting, and it ought to represent a great challenge to scientists interested in growth regulatory mechanisms. DIGERNES and BOLUND (1979) have recently shown that a pure population of hepatocytes can be studied by flow cytometry, a technique that may open up new possibilities for the study of liver chalones.

VII. Kidney Chalones (?)

SAETREN (1956, 1963, 1970) showed the presence of tissue specific proliferation inhibitors in kidney extracts. After subtotal nephrectomy kidney macerate was injected i.p. 18 h before the expected rise in mitotic rate in the remaining kidney. This procedure reduced the mitotic count by a factor of 7–8.

Liver, spleen, testis, and brain extracts had no such effect. The first wave of mitoses in resected livers in other animals was not depressed by kidney homogenate. The chalone-factor was undialyzable, heat and freezing stable.

CHOPRA and SIMNETT (1969, 1971) and SIMNETT and CHOPRA (1969) treated embryonic kidneys with the supernatant of kidney homogenate. The mitotic rate was reduced by 74%, and similarly prepared extracts of liver or lung had no effect. Kidney extract had no effect on epidermal mitotic rate. Adrenaline and hydrocortisol augmented the effect of the renal chalone, and the authors believe they demonstrated a G_2 chalone for the kidney tubular cells.

Recently KLEIN et al. (1979) showed that a factor from kidney KCl extract inhibited in a reversible manner 3 HTdR incorporation into Kirsten murine sarcoma virus-transformed cells. Weaker, similar effects were found of extracts from liver. However, the kidney factor is heat stable, resistant to pronase, DNase and RNase and possibly a mucoprotein. The inhibition of phosphorylation of thymidine appeared to be its mechanism of action. No cell kinetic studies were made and tissue specificity was not proven. It is difficult to know whether this factor may be related to the renal G_1 chalone. CAIN et al. (1976) showed increased mitotic count (no stathmokinetic method) following direct injection into the organ of homogenate from infarcted or regenerating kidney. This increase in the mitotic count is probably due to a prolonged mitotic duration.

Biological evidence of the same type as that obtained from the liver, pointing to a circulating factor regulating cell proliferation in the kidney (stimulator or inhibitor), has been brought forward from experiments with unilateral nephrectomy (see, e.g. REITER, 1965) and from parabiosis (see, e.g. KURNICK and LINDSAY, 1968).

MARTEL-PELLETIER and BERGERON (1975) found a factor in kidney homogenate with renal G_1 chalone properties, inhibiting 3 HTdR incorporation in tissue culture slices of renal cortex. This factor had an MW of 4,000 daltons after dialysis. With Sephadex G 25 chromatography they found similar effects in a factor with MW around 5,000 daltons.

Similar renal G_1 effects were demonstrated by GOLDIN and FABIAN (1978); their active factor was found in the ultrafiltrate with an MW of less than 10,000 daltons.

However, it must be mentioned that Goss (1963) found mitosis inhibitory effect on the remaining kidney after one-sided nephrectomy, not only in kidney extracts, but also in extracts from liver, testis, spleen, and blood cells.

In conclusion, it must be said that fairly convincing biological evidence indicates the presence of factors that may be kidney tubular cell G_1 and G_2 chalones. Much work remains to be done, however, before the situation is cleared up and the true nature of these putative chalones are demonstrated. However, the partly purified fractions are of the same general MW as chalone-containing fractions from other organs.

VIII. Chalones in the Male Reproductive Organs (?)

1. Testicular Chalone

The testis has a complicated anatomy and in addition to the spermatogenic cells the testis contains stromal cells, some Leydig cells and Sertoli cells. The main component numerically, however, is that of the spermatogenic cells, which divide several times by mitosis and once by meiosis (including a reduction division), and then the cells mature to sperm cells with a haploid DNA content. The precursor cells are sometimes subdivided into early A_0–A_4 spermatogonia, then follow other classes of spermatogonial cells (intermediary and B-type) and finally the preleptogenic spermatocytes that mature to sperm cells.

Theoretically, it has been suggested that the growth control of this cell line might be similar to those of the red marrow in that the more advanced types of A spermatogonia regulate proliferation of early type A spermatogonia via a negative feedback cycle based on a chalone. The proliferation of the spermatogonial cell line and its maturation to spermatozoa may be stimulated by hormones and modulated by many other factors. However, negative feedback inhibition of proliferation may still be mediated through a chalone. CLERMONT and GIRARD (1973) and CLERMONT and MAUGER (1974) reported experiments in which early spermatogonia (A_1–A_4) were injured by local radiation (300 rad). As a consequence of this, 11 days after the injury, the subsequent cells in the line, i.e. the intermediary, the B-type spermatogonia and the more mature cells fell to low levels. At this time, a population of obviously non-injured A_0 spermatogonia started to proliferate rapidly. Incorporation of H 3 TdR into these cells could be reduced by 58% by i.p. injection of testicle extract into the animals. The reduction was seen only in the A_0–A_4 cells, the later stages were not inhibited. The authors claimed that this effect was non-toxic and cell-specific. Liver extracts had no similar effect.

These results have been confirmed by CLERMONT and MAUGER (1976) and by THUMANN and BUSTOS-OBREGON (1978) who confirmed them both in vivo and in vitro; and finally, and convincingly, by DE ROOIJ (1980) showing reduced cell number of the early stages of spermatogonia after two injections of testicular extracts. Finally, IRONS and CLERMONT (1979) reported the existence of both G_1 and G_2 spermatogonial chalones.

CUNNINGHAM and HUCKINS (1979) could not confirm the results of CLERMONT'S group. They used whole body irradiation and thus may have induced general influences not present in the locally irradiated animals. CATTANACH et al. (1979), performing part-body irradiation and attempting to analyze the results in terms of a genetic response, were also unable to find signs of a chalone effect. However, they admit that their method may not have been sensitive enough.

My own conclusion is that although the assay system is complex, there is convincing biological evidence for spermatogonial chalones both of the G_1 and G_2 type, but nothing is known about the chemical nature of these putative chalones.

2. A Chalone in the Seminal Vesicle?

SÖDERSTRÖM and TUOHIMAA (1976) indicated that crude supernatant of seminal vesicle homogenate inhibited the H 3 TdR labelling index in testosterone-stimulated seminal vesicle epithelium. Liver extract had no such effect. Most activity was found in a fraction with MW higher than 3,000 daltons. Seminal vesicle extract had no effect on epidermal DNA synthesis. No further reports have been published on this subject.

3. A Prostatic Chalone?

In addition to the hormonal influences that regulate cell proliferation in the prostatic gland (see, e.g. LESSER and BRUCHOVSKY, 1973). BRUCHOVSKY et al. (1975) considered the presence of a prostatic chalone an attractive hypothesis.

However, when MÜNTZING and MURPHY (1977) and MÜNTZING et al. (1979) searched for a possible prostatic G_1 chalone, they found none. They assessed the cell number by measuring total DNA content and weight of organ (including stroma) only. No study of labelling index or mitotic rate in the epithelial cells were done. Thus, the question is still open.

IX. Ascites Tumour Cell Chalones

1. JBI Ascites Tumour Cell Chalone

Most work on this tumour (and two others for control) has been done by BICHEL in Denmark and his school, see BICHEL (1973) for a review up to that year; and, BARFOD and BICHEL (1976); BICHEL and BARFOD (1977); NAESER et al. (1978), and BARFOD and MARCKER (1979).

These authors have used cell kinetic studies of these ascites tumours for a comprehensive analysis of growth inhibitory substances found in such cells, excluding factors such as lack of oxygen or nutrients, presence of non-specific cytotoxicity or the effects of immune mechanisms. They have employed cell counts, incorporation of 3 HTdR and labelling index, mitotic rate studies and flow cytometry. They have provided evidence for the existence of cell line specific G_1 and G_2 chalones in these cell lines, and shown that these factors operate only in vivo, not in vitro. In cell culture even a stimulatory effect may appear.

In 1976 BICHEL reported a G_1 JBI ascites tumour chalone to be in a fraction with an MW of 10,000–50,000 daltons, and the G_2 chalone was found in a fraction between 10,000 and 1,000 daltons.

Later, they provided evidence that the G_1 chalone was of basic or polycationic nature and probably bound to anionic receptors on the cell surface (BARFOD and BICHEL, 1976). Finally, in 1979, BARFOD and MARCKER reported further purification of the G_1 chalone by boiling ascites fluid, combined with various chromatographic procedures, and now found that the active chalone with G_1 activity seemed to be an acidic and hydrophobic peptide (MW 400–700 daltons). Thus, the JBI G_1 chalone may be of similar structure to the granulocytic G_1-chalone.

Hence, the JBI ascites tumour chalone system is one of the most thoroughly investigated ones.

2. Ehrlich Ascites Tumour Chalone

This ascites tumours originated from a mammary carcinoma and it has been used to study the chalones (see LEHMANN et al., 1979).

In 1972 LALA reported that the growth of exponentially proliferating Ehrlich ascites tumour could be transiently delayed by the reintroduction of the cell-free ascites.

In 1973 COOPER and SMITH reported on a similar substance and referred to it as a chalone, and related the putative chalone effect to the cyclic AMP system. NAKAI (1976; NAKAI and GERGELY, 1980) reported Ehrlich ascites tumour G_1 chalone effect, and showed in his system that the chalone inhibited both α and β DNA – polymerases, but that DNA ligase was not inhibited. LEHMANN et al. (1977) also presented evidence for a chalone factor from these cells, and this factor had no effect on a lymphocytic leukaemia cell line L 1210. They used the increase in cell number as a parameter of growth. Later, LEHMANN et al. (1979, 1980) reported that the chalone effect could be prevented by preincubating the cells in high serum concentration, and that chalone and insulin (which is a known growth factor) were antagonistic. A G_2 chalone for Ehrlich ascites tumour cells was also indicated by ROMANOV and SEMENOVA (1980).

There seems to be a fact, also for this chalone, that it is specific only when the tumour grows as ascites tumour in vivo. In vitro, the cells can be inhibited also by factors from liver and skeletal muscle homogenate (TONG et al., 1978). Similar non-specificity was found by ROTHBARTH et al. (1977) and by MAIER and ROTHBARTH (1979).

X. Connective Tissue and Fibroblast Chalones (?)

Already in 1965 PARSHLEY reported that "mild tryptic digestion" of the constituents of normal adult connective tissue contained fractions with growth inhibitory activity on fibroblasts in tissue culture. She described these as being "highly active biologically" and assumed that they were peptides or conjugated peptides. She also found some growth inhibition in cultures of some human tumour cells.

In 1971 a, b NJEUMA studied the growth of embryonic chick and mouse fibroblasts using the Colcemid technique. She discussed the possibility of "the accumulation of inhibitory substances in the medium" to explain the growth retardation. HOUCK et al. (1972) reported on a fibroblast G_1 chalone with MW at 30,000–50,000 daltons and isoelectric point at about pH 3.5–4.3, probably of protein nature.

RICHTER (1978) demonstrated a fibroblast proliferation inhibitor of chalone nature, with an MW between 10,000 and 100,000 daltons. The active part seemed to be of peptide nature.

Later HOUCK (1979, personal communication) assued that the fibroblast chalone may be a peptide as small as 2,100 daltons. It is interesting to note that NATRAJ and DATTA (1978) and DATTA and NATRAJ (1980) reported on "a cell surface component" that inhibited DNA synthesis and cell multiplication in the mouse fibroblast BALB/C 3 T 3. This factor can be extracted from quiescent cell cultures in plateau phase, but not from growing cells. The factor is protein in nature and heat labile and sensitive to trypsin. Strangely enough, these authors do not mention the word chalone in their papers, but name it a "fibroblast growth regulatory factor."

HANKS and SMITH (1978) reported on a water soluble factor extracted from the dermis, and this factor inhibited DNA synthesis in regenerating connective tissue cells. No similar inhibition was found in extracts from epithelial cells from wound margins. The authors conclude that chalone-like negative feedback mechanisms may be partially responsible for in vivo control of fibroblast proliferation in wound healing.

There is interesting evidence in favour of a fibroblast chalone, but much work remains to be done before its existence and its chemical nature is clarified.

STECK et al. (1979) demonstrated growth inhibitory activity on fibroblast target cell by a factor in tissue culture medium which had been conditioned by exposure to density-inhibited fibroblast culture. The effect was reversible.

XI. Chalones of the Lung?

In 1969 SIMNETT et al. reported that when adding a lung extract (2–8 mg dry weight/ml medium) to mouse lung short-term organ culture, a 20%–36% depression of the mitotic rate occurred in the alveolar cells.

HOUCK (1976 b) prepared an ultrafiltrate from lung homogenate extract, MW $1-5 \times 10^5$ daltons, and tested it on human bronchial carcinoma cell cultures. After 96 h incubation, there was a 60%–83% inhibition in cell count. The effect was reversible and non-toxic, and the factor was sensitive to trypsin. MODJANOVA and MALENKOV (1975) and MALENKOV et al. (1977) also demonstrated chalones for the rat lung. Since their inhibitory substance also seemed to enhance contact between cells, they called their partly purified factor *lung contactin*. Finally, MENYHÁRT and MARCSEK (see BALÁZS, 1979) isolated a factor of MW about 40,000 daltons with the properties of a lung chalone. Thus, there is some evidence for the existence of chalone(s) for the lung epithelial cells. However, one must keep in mind that the bronchial epithelium has a great propensity for squamous metaplasia and therefore may share the epidermal chalones, whereas the proliferating lung alveolar epithelium (pneumocytes type II) belong to another cell line. The situation is therefore far from clear.

XII. Other Chalones Indicated

1. Lens of the Eye

Some work has been published on this by SMITH (1965), VOADEN (1968), VOADEN and LEESON (1970), and VOADEN (1971).

VOADEN observed a chalone effect of the partly purified pig skin epidermal chalone on the lens's mitotic rate (61% inhibition), and she thereafter also extracted an inhibiting factor from the crystalline lens. This was a crystalline protein with an MW of around 25,000 daltons. The factor was extracted both from rat, rabbit, and mouse lenses, and VOADEN (1971) assumed that it corresponded to a chalone.

2. Endothelial Chalone?

The presence of negative feedback regulation of endothelial cell regeneration is a possibility that may be inferred from the quantitative analysis of initial stages of endothelial regeneration published by SCHWARTZ and HAUDENSCHILD (1978).

3. Smooth Muscle Chalone?

R. A. FLORENTIN et al. (1973) and THOMAS et al. (1976) indicated that an extract of pig aorta precipitated with 80% ethanol injected i.p. reduced the mitotic rate in the smooth muscle cells in the carotid artery by 55%. No further studies have been reported.

4. Heart Muscle Chalone?

LEZHAVA et al. (1979) reported that injection of extract of left ventricle heart walls from rat or hen into 11-day-old chick embryos inhibited the cell proliferation in the embryonal heart. No further studies have been published.

5. Placental Chalone?

BADEN (1973) and KUBILUS and BADEN (1975) reported on a growth inhibitory factor extracted from placenta. It inhibited the uptake of 3 HTdR by fibroblasts, epidermis and lymphocytes. The factor had an MW of 150,000 daltons and differed from the other "chalones" by a broader specificity. Even if the authors called this a chalone, it is highly doubtful whether this factor belongs to the chalones.

Acknowledgements. I am, of course, alone responsible for this chapter, my opinions humbly expressed, possible errors, obvious shortcomings and misinterpretations. However, I have circulated a draft of the manuscript among many of my colleagues, and have received many valuable comments, and criticism – positive and negative. I have not always been able to follow their advice, but I wish to thank the following who have helped: John C. Allen, Clwyd; Andras Balázs, Budapest; Renato Baserga, Philadelphia; Haakon B. Benestad, Oslo; Roswell K. Boutwell, Madison, WI; Hal E. Broxmeyer, New York; William S. Bullough, London; Arne Bøyum, Oslo; Kjell M. Elgjo, Oslo; Jadwiga Grundboek-Jusko, Pulawi; Martine Guigon, Villejuif; John C. Houck, Seattle; Gunilla Isaksson-Forsén, Lund; Nicole Kiger, Villejuif; Ole D. Laerum, Bergen; Laszlo G. Lajtha, Manchester; Peter Langen, Berlin-Buch; Edna B. Laurence, London; Maryse Lenfant, Poitiers; Brian I. Lord, Manchester; H. Rainer Maurer, Berlin; Janos Menyhárt, Budapest; George S. Nakai, Long Beach; Valeri B. Okulov, Leningrad; Walter R. Paukovits, Vienna; Karl L. Reichelt, Oslo; Eric O. Rijke, Utrecht; Dirk G. de Rooij, Utrecht; Tapio Rytömaa, Helsinki; and James E. Till, Toronto.

The work was undertaken during the tenure of an American Cancer Society – Eleanor Roosevelt International Cancer Fellowship awarded by the International Union Against Cancer. I enjoyed the benefits of this grant during a sabbatical 1 year leave from the Institute of Pathology, University of Oslo, Norway, to work with experimental skin carcinogenesis in the laboratory of Professor Roswell K. Boutwell at the McArdle Laboratory for Cancer Research, Madison, Wisconsin, U.S.A.

Last, but not least, thanks to Karen Denk who has patiently performed most of the typing, and retyping, $\rightarrow \infty$. *"When the dust falleth, Thou wilst see whether Thou ridest a horse or an ass!"* Arab proverb

References

Aardal, N.P., Laerum, O.D., Paukovits, W.R., Maurer, H.R.: Inhibition of agar colony formation by partially purified granulocyte extracts (chalone). Virchows Archiv [Cell Path.] *24*, 27–39 (1977)

Allen, J.C., Smith, Ch.J.: Chalones: a reappraisal. Biochem. Soc. Trans. *7*, 584–592 (1979)

Allen, J.C., Smith, C.J., Curry, M.C., Gaugas, J.M.: Identification of a thymic inhibitor ("chalone") of lymphocyte transformation as a spermine complex. Nature *267*, 623–625 (1977)

Andersson, P.: Effect of an aqueous extract of small intestinal mucosa on the G-2 phase of jejunal proliferative cells in mice. Virchows Archiv [Cell Path.] *13*, 233–246 (1973)

Anisimov, V.N., Okulov, V.B.: Effect of ageing on concentration of estradiol in serum and the epidermal G_2 chalone in vaginal mucosa of rats. Exp. Geront. *15*, 87–91 (1980)

Anisimov, V.N., Ivanov, M.N., Okulov, V.B.: Changes in the level of epidermal G_2 chalone and mitotic activity in the vaginal epithelium of rats. Bull. Exp. Biol. Med. *87*, 357–360 (1979)

Aoyama, T., Khara, K., Nishiguchi, K.: Dual effects of thymic substance(s) on growth of cultured mammalian cells. Exp. Cell Res. *93*, 427–437 (1975)

Attallah, A.M.: Regulation of cell growth in vitro and in vivo: point/counterpoint. In: Chalones. Houck, J.C. (ed.), pp. 141–172. Amsterdam, Oxford, New York: North-Holland and American Elsevier 1976

Aujard, C., Chany, E., Frayssinet, C.: Inhibition of DNA synthesis of synchronized liver cells by liver extracts acting in G_1 phase. Exp. Cell Res. *78*, 476–478 (1973)

Baden, H.P.: Inhibition of DNA synthesis by an extract from human placenta. J. Natl. Cancer Inst. *50*, 43–48 (1973)

Baker, H.S.: The treatment of cancer with connective tissue extract. Lancet *1933 II*, 643–645

Baker, H.S.: Tissue extracts in the treatment of cancer. Lancet *1935 II*, 583–584

Balázs, A.: Acute effect of endogenous inhibitors and exogenous cytostatics on the ultrastructure of bone marrow cells. II. Single dose of granulocyte crude extract (GCE). Haematologia *10*, 445–453 (1976)

Balázs, A.: Control of cell proliferation by endogenous inhibitors. Budapest: Akademiai Kiado 1979

Balázs, A., Barabás, K.: Control of CFU_c proliferation by selective endogenous inhibitors. Virchows Archiv [Cell Path.] *31*, 125–134 (1979)

Balázs, A., Fazekas, I., Bukulya, B., Blazsek, I., Rappay, G.Y.: An intracellular factor (CDI) controlling differentiation and cell division. Mech. Ageing Dev. *1*, 175–182 (1972)

Balázs, A., Gál, F., Klupp, T., Blazsek, I.: In vitro sensitivity of transplantable leukemias to endogenous granuloid (GCE, GI 2) and lymphoid (T_4, T_4-1) inhibitors of proliferation. Oncology *35*, 8–14 (1978)

Balász, A., Sajgo, M., Klupp, T., Kemeny, A.: Purification of an endopeptide to homogenity and the verification of its selective inhibitory action on myeloid cell proliferation. Cell Biol. Int. Rep. *4*, 337–345 (1980)

Barbason, H., Fridman-Manduzio, A., Lelievre, P., Betz, E.H.: Variations of liver cell control during diethylnitrosamine carcinogenesis. Eur. J. Cancer *13*, 13–18 (1977)

Bardos, T.J., Gordon, H.L., Chmielewicz, Z.F., Kutz, R.L., Nadkarni, M.V.: A systematic investigation of the presence of growth-inhibitory substances in animal tissues. Cancer Res. *28*, 1620–1630 (1968)

Barfod, N.M., Bichel, P.: Characterization of a G_1 inhibitor from old JB-1 ascites tumor fluid. Interaction with polyions and ion exchangers. Virchows Archiv [Cell Pathol.] *21*, 249–259 (1976)

Barfod, N., Marcker, K.: Purification of a G_1 inhibitor of proliferation from JB-I ascites tumours. Abstracts Xth Meeting Europ. Study Group Cell Prolif., p. 2. Oxford: Blackwell Sci. Publ. 1979

Barka, T.: Partial purification of a mitotic suppressor from the salivary gland. Exp. Mol. Path. *18*, 225–233 (1973)

Baserga, R.: The cell cycle and cancer. New York: Marcel Dekker 1971

Bateman, A.E.: Cell specificity of chalone-type inhibitors of DNA synthesis released by blood leukocytes and erythrocytes. Cell Tissue Kinet. *7*, 451–461 (1974)

Bateman, A.E., Goodwin, B.C.: An inhibitor of DNA synthesis in erythrocyte-conditioned medium and its separation from haemoglobin. Biomedicine *25*, 77–78 (1976)

Bateman, A.E., Pollock, K.: Effects of erythrocyte lysate and erythrocyte-conditioned medium on erythroid cells in vitro. Virchows Archiv [Cell Pathol.] *25*, 171–177 (1977)

Benestad, H.B., Reikvam, Å.: Diffusion chamber culturing of haematopoietic cells. Methodological investigations and improvement of technique. Exp. Hemat. *3*, 249–260 (1975)

Benestad, H.B., Rytömaa, T.: Regulation of maturation rate of mouse granulocytes. Cell Tissue Kinet. *10*, 461–468 (1977)

Benestad, H.B., Rytömaa, T.: Granulocytic chalone inhibits rapidly proliferating committed murine progenitor cells (CFU-C). Biomedicine *31*, 33–37 (1979)

Benestad, H.B., Rytömaa, T., Kiviniemi, K.: The cell specific effect of the granulocyte chalone demonstrated with the diffusion chamber technique. Cell Tissue Kinet. *6*, 147–154 (1973)

Bernard, C.: Lecons sur les Phenomenes de la Vie. Paris: J.B. Ballière et fils 1879

Bertsch, S.: Untersuchungen zur Wachstumskontrolle in embryonaler, neonataler und regenerierender Epidermis. Ph. D. Thesis, University of Heidelberg, 1975

Bertsch, S., Marks, F.: Lack of an effect of tumor-promoting phorbol esters and of epidermal G_1 chalone on DNA synthesis in the epidermis of newborn mice. Cancer REs. *34*, 3283–3288 (1974)

Bertsch, S., Marks, F.: Removal of the horny layer does not stimulate cell proliferation in neonatal mouse epidermis. Cell Tissue Kinet. *11*, 657–658 (1978)

Bertsch, S., Marks, F.: Appearance of growth-inhibiting activity (G$_1$ chalone) during onto-genetic development of rat and chick epidermis in vivo and in vitro. Experientia 35, 897–898 (1978b)

Bertsch, S., Marks, F.: Effects of epidermal chalone and epidermal growth factor on a trans-plantable epidermal carcinoma (Hewitt) of the mouse in vivo. Cancer Res. 39, 239–243 (1979)

Bertsch, S., Csontos, K., Schweizer, J., Marks, F.: Effect of mechanical stimulation on cell proliferation in mouse epidermis and on growth regulation by endogenous factors (cha-lones). Cell Tissue Kinet. 9, 445–457 (1976)

Bichel, P.: Autoregulation of ascites tumour growth by inhibition of the G$_1$ and G$_2$ phase. Europ. J. Cancer 7, 349–355 (1971)

Bichel, P.: Specific growth regulation in three ascites tumours. Europ. J. Cancer 8, 167–173 (1972)

Bichel, P.: Self-limitation of ascites tumor growth: a possible chalone regulation. Natl. Can-cer Inst. Monogr. 38, 197–203 (1973)

Bichel, P.: Ascites tumors and chalones. In: Chalones. Houck, J.C. (ed.), pp. 429–499. Am-sterdam: North-Holland 1976

Bichel, P., Barfod, N.M.: Specific chalone inhibition of the regeneration of the JB-1 ascites tumour studied by flow microfluorometry. Cell Tissue Kinet. 10, 183–193 (1977)

Bischoff, F.: Inhibition of mitosis by homologous tissue extracts. J. Cell Biol. 23, 16 (1964)

Bizzozerro, G.: Wachstum und Regeneration im Organismus. Wien Med. Wochenschr. 16, 699, and 17, 744 (1894)

Bjerknes, R.: Model studies of epidermal growth regulation and epidermal carcinogenesis. Thesis. Oslo Univ. Inst. of Pathol., Oslo, 1977

Bjerknes, R., Iversen, O.H.: Antichalone: a theoretical treatment of the possible role of an-tichalone in the growth control system. Acta Pathol. Microbiol. Scand. [A] Suppl. 248, 33–42 (1974)

Blazśek, I., Balázs, A., Gáal, D., Holczinger, L.: A multifactorial system controlling myeloid cell differentiation and division. Mech. Ageing Devel. 5, 57–65 (1976)

Bøyum, A., Løvhaug, L., Boecker, W.R.: Regulation of bone marrow cell growth in diffu-sion chambers: The effect of adding normal and leukemic (CML) polymorphonuclear granulocytes. Blood 48, 373–383 (1976)

Boll, I.T.M., Sterry, K., Maurer, H.R.: Evidence for a rat granulocyte chalone effect on the proliferation of normal human bone marrow and of myeloid leukemias. Acta Haemat. 61, 130–137 (1979)

Borel, J.F., Feurer, C., Hiestand, P.C., Stähelin, H.: The effects of fractions (chalones) ob-tained from lymphoid organs on the immune response in vivo. Agents Actions 5, 523–531 (1978)

Born, W., Bickhardt, R.: Zur Regelung des Zellnachschubs in der Epidermis. Klin. Wo-chenschr. 46, 1312–1314 (1968)

Boveri, Th.: Ergebnisse über die Konstitution der chromatischen Substanz des Zellkerns. Jena: G. Fischer 1904

Broxmeyer, H.E., Moore, M.A.S.: Communication between white cells and the abnor-malities of this in leukemia. Biochim. Biophys. Acta 516, 129–166 (1978)

Broxmeyer, H.E., Smithyman, A., Eger, R.R., Meyers, P.A., De Sousa, M.: Identification of lactoferrin as the granulocyte-derived inhibitor of colony-stimulating activity pro-duction. J. Exp. Med. 148, 1052–1067 (1978)

Bruchovsky, N., Lesser, B., Van Doorn, E., Craven, S.: Hormonal effects on cell prolifer-ation in rat prostate. Vitam. Horm. 33, 61–102 (1975)

Brugal, G.: Effects of adult intestine and liver extracts on the mitotic activity of correspond-ing embryonic tissues of pleurodeles waltlii Michah. (Amphibia, Urodela). Cell Tissue Kinet. 6, 519–524 (1973)

Brugal, G.: Presence of intestinal chalones. In: Stem cells of renewing cell populations. Cair-nie, A.B., Lala, P.K., and Osmond, D.G. (eds.), pp. 41–50. New York: Academic Press 1976

Brugal, G., Pelmont, J.: Presence, dans l'intestin du Triton adulte pleurodeles waltlii Mi-chah, de deux facteurs antimitotiques naturels (chalones) actifs sur la proliferation cel-lulaire dans l'intestin embryonnaire. C.R. Acad. Sci. Paris 278 D, 2831–2834 (1974)

Brugal, G., Pelmont, J.: Existence of two chalone-like substances in intestinal extracts from the adult newt, inhibiting embryonic intestinal cell proliferation. Cell Tissue Kinet. *8*, 171–187 (1975)

Bullough, W.S.: The control of mitotic activity in adult mammalian tissues. Biol. Rev. *37*, 307–342 (1962)

Bullough, W.S.: The evolution of differentiation. London, New York: Academic Press 1967

Bullough, W.S.: The use of tissue extracts in the treatment of cancer. In: Fortschritte der Krebsforschung. Schmidt, C.G. and Wetter, O. (eds.), pp. 315–319. Stuttgart, New York: Schattauer 1969

Bullough, W.S.: Chalone control mechanisms. Life Sci. *16*, 323–330 (1975)

Bullough, W.S., Deol, J.U.R.: Chalone-induced mitotic inhibition in the Hewitt keratinising epidermal carcinoma of the mouse. Europ. J. Cancer *7*, 425–431 (1971)

Bullough, W.S., Deol, J.U.R.: Chalone control of mitotic activity in the eccrine sweat glands. Br. J. Derm. *86*, 586–592 (1972)

Bullough, W.S., Deol, J.U.R.: Dermo-epidermal adhesion and its effect on epidermal structure in the mouse. Br. J. Derm. *92*, 417–424 (1975)

Bullough, W.S., Laurence, E.B.: The control of epidermal mitotic activity in the mouse. Proc. Roy. Soc. B. *151*, 517–536 (1960)

Bullough, W.S., Laurence, E.B.: Stress and adrenaline in relation to the diurnal cycle of epidermal mitotic activity in adult male mice. Proc. Roy. Soc. B *154*, 540–556 (1961)

Bullough, W.S., Laurence, E.B.: Duration of epidermal mitosis in vitro. Exp. Cell Res. *35*, 629–641 (1964a)

Bullough, W.S., Laurence, E.B.: The production of epidermal cells. Symp. Zool. Soc. Lond. *12*, 1–23 (1964b)

Bullough, W.S., Laurence, E.B.: Mitotic control by internal secretion: the role of the chalone-adrenalin complex. Exp. Cell Res. *33*, 176–194 (1964c)

Bullough, W.S., Laurence, E.B.: The diurnal cycle in epidermal mitotic duration and its relation to chalone and adrenalin. Exp. Cell Res. *43*, 343–350 (1966a)

Bullough, W.S., Laurence, E.B.: Accelerating and decelerating actions of adrenalin on epidermal mitotic activity. Nature *210*, 715–716 (1966b)

Bullough, W.S., Laurence, E.B.: The role of glucocorticoid hormones in the control of epidermal mitosis. Cell Tissue Kinet. *1*, 5–10 (1968a)

Bullough, W.S., Laurence, E.B.: Epidermal chalone and mitotic control in the V X 2 epidermal tumor. Nature *220*, 134–135 (1968b)

Bullough, W.S., Laurence, E.B.: Control of mitosis in mouse and hamster melanomata by means of the melanocyte chalone. Europ. J. Cancer *4*, 607–615 (1968c)

Bullough, W.S., Laurence, E.B.: The lymphocyte chalone and its antimitotic action on a mouse lymphoma in vitro. Europ. J. Cancer *6*, 525–531 (1970a)

Bullough, W.S., Laurence, E.B.: Chalone control of mitotic activity in sebaceous glands. Cell Tissue Kinet. *3*, 291–300 (1970b)

Bullough, W.S., Mitrani, E.: An analysis of the epidermal chalone control mechanism. In: Chalones. Houck, J.C. (ed.), pp. 7–36. New York: American Elsevier Publ. 1976

Bullough, W.S., Laurence, E.B., Iversen, O.H., Elgjo, K.: The vertebrate epidermal chalone. Nature *214*, 578–580 (1967)

Burns, E.R.: On the failure of self-inhibition of growth in tumors. Growth *33*, 25–45 (1969)

Byrd, W.J., Jacobs, D.M., Amoss, M.S.: Synthetic polyamines added to cultures containing bovine sera reversibly inhibit in vitro parameters of immunity. Nature *267*, 621–623 (1977)

Cain, H., Egner, E., Redenbacher, M.: Increase of mitosis in the tubular epithelium following intrarenal doses of various kidney homogenates and homogenate fractions in the rat. Virchows Arch. [Cell Pathol.] *22*, 55–72, 1976

Canella, N.K.: A difference in the effect of rat liver extract (chalone) on the mitotic index of chick embryo liver when extracted at different times of a day. Life Sci. *20*, 155–158 (1977)

Cannon, W.B.: The wisdom of the body. New York: Norton Publ. Co. 1939

Cattanach, B.M., Jones, J.T., Andrews, S.J., Crocker, M.: An attempt to distinguish a modified genetic response of the mouse testis to X-ray exposure by the action of a spermatogonial chalone. Mutat. Res. *62*, 197–201 (1979)

Chan, S.H.: Influence of serum inhibitors on colony development in vitro by bone marrow cells. Aust. J. Exp. Biol. Med. Sci. *49*, 553–564 (1971)

Chan, S.H., Metcalf, D.: Inhibition of bone marrow colony formation by normal and leukaemic human serum. Nature *227*, 845–846 (1970)

Chany, E., Frayssinet, C.: Presence dans le foie normal de substances inhibant precocement la croissance de cultures de cellules cancereuses. C.R. Acad. Sci. Paris *272*, 2644–2647 (1971)

Chekulaeva, L.I., Ketlinsky, S.A., Okulov, V.B.: Some pecularities of the rat's liver chalone influence on hepatocyte proliferation at various periods after a partial hepatectomy. Tsitologica *20*, 436–446 (1978)

Chopra, D.P., Simnett, J.D.: Demonstration of an organ-specific mitotic inhibitor in amphibian kidney. The effect of adult xenopus tissue extracts on the mitotic rate of embryonic tissue (in vitro). Exp. Cell Res. *58*, 319–322 (1969)

Chopra, D.P., Simnett, J.D.: Tissue-specific mitotic inhibition in the kidneys of embryonic grafts and partially nephrectomized host xenopus laevis. J. Embryol. Exp. Morphol. *25*, 321–329 (1971)

Chopra, D.P., Rucy, J.Y., Flaxmann, B.A.: Demonstration of a tissue-specific inhibitor of mitosis of human epidermal cells in vitro. J. Invest. Dermatol. *59*, 207–210 (1972)

Clarke, R.M.: Control of intestinal epithelial replacement: lack of evidence for a tissue-specific blood-borne factor. Cell Tissue Kinet. *7*, 241–250 (1974)

Clermont, Y., Girard, A.: Existence of a spermatogonial chalone in the rat testis. Anat. Rec. *175*, 294 (1973)

Clermont, Y., Mauger, A.: Existence of a spermatogonial chalone in the rat testis. Cell Tissue Kinet. *7*, 165–172 (1974)

Clermont, Y., Mauger, A.: Effect of spermatogonial chalone in the growing rat testis. Cell Tissue Kinet. *9*, 99–104 (1976)

Cooper, P.R., Smith, H.: Influence of cell-free ascites fluid and adenosine $3',5'$-cyclic monophosphate upon the cell kinetics of Ehrlich's ascites carcinoma. Nature *241*, 457–458 (1973)

Cooperband, S.R., Nimberg, R., Schmid, K., Mannick, J.A.: Humoral immunosuppressor factors. Transplant. Proc. *8*, 225–242 (1976)

Cunningham, G.R., Huckins, C.: Failure to identify a spermatogonial chalone in adult irradiated testes. Cell Tissue Kinet. *12*, 87–89 (1979)

Danielsson, J.R., Van Alten, P.J.: Lymphocyte proliferation inhibited by cells and by effector substances obtained from bursal lymphocytes. In: Prog. Exp. Tumor Res. *19*, 194–202 (1974)

Datta, P., Natraj, C.V.: Fibroblast growth regulatory factor inhibits DNA synthesis in Balb/C 3T3 cells by arresting in G_1. Exp. Cell Res. *125*, 431–439 (1980)

Delforge, A., Ronge-Collard, E., Schlusselbergh, J., Prieels, J.P., Matterlaer, M.A., Stryckmans, P.: Role of lactoferrin in regulation of human granulopoiesis. Abstract Xth Meeting of the European Study Group for Cell Proliferation, pp. 7–8. London: Blackwell Sci. Publ. 1979

de Rooij, D.G.: Effect of testicular extracts on proliferation of spermatogonia in the mouse. Virchows Archiv (Cell Pathol.) *33*, 67–75 (1980)

Deschamp, Y., Verly, W.G.: The hepatic chalone II Chemial and biological properties of the rabbit liver chalone. Biomedicine *22*, 195–208 (1975)

Dewey, D.L.: The melanocyte chalone. Natl. Cancer Inst. Monogr. *38*, 213–216 (1973)

Dewey, D.L., Butcher, F.W., Galpine, A.R.: Control of melanoma cells. In: Annual Report, Gray Laboratories, Mount Vernon Hospital, pp. 95–100. Middlesex, England: Northwood 1977

Digernes, V., Bolund, L.: The ploidy classes of adult mouse liver cells. A methodological study with flow cytometry and cell sorting. Virchows Archiv [Cell Pathol.] *32*, 1–10 (1979)

Dosychev, E.A., Topolnitsky, N.F., Ketlinsky, S.A.: Inhibition of cell proliferation in normal human epidermis by epidermal chalones in vitro. Vestn. Dermatol. Venerol. *2*, 4–8 (1980)

Echave-Llanos, J.M., Balduzzi, R., Surur, J.M.: Factores tisulares hepaticos inhibidores des crecimiento del higado. Effecto de la technica empleada en la preparacion de los extractos. Rev. Soc. Argent. Biol. *46*, 42–50 (1970)

Elgjo, K.: The stability of the epidermal mitosis inhibiting factor (chalone) in water solution. Acta Path. Microbiol. Scand. *76*, 31–34 (1969)

Elgjo, K.: Reversible inhibition of epidermal G-1 cells by repeated injections of aqueous skin extract (chalone). Virchows Archiv [Cell Pathol.] *15*, 157–163 (1974a)

Elgjo, K.: Evidence for presence of the epidermal G_2-inhibitor („epidermal chalone") in dermis. Virchows Archiv [Cell Pathol.] *16*, 243–247 (1974b)

Elgjo, K.: Epidermal chalone and cyclic AMP: An in vivo study. J. Invest. Dermat. *64*, 14–18 (1975)

Elgjo, K., Cromarty, A.: Growth kinetics in newborn mouse epidermis: response to epidermal chalone. Virchows Archiv [Cell Pathol.] *24*, 101–108, 1977

Elgjo, K., Devik, F.: Growth regulation in X-irradiated mouse skin; the possible role of chalones. Int. J. Radiat. Biol. *34*, 119–126 (1978)

Elgjo, K., Hennings, H.: Epidermal chalone and cell proliferation in a transplantable squamous cell carcinoma in hamsters. I. In vivo results. Virchows Arch. [Cell Pathol.] *7*, 1–7 (1971a)

Elgjo, K., Hennings, H.: Epidermal mitotic rate and DNA synthesis after injection of water extracts made from mouse skin treated with actinomycin D: Two or more growth-regulating substances? Virchows Archiv [Cell Pathol.] *7*, 342–347 (1971b)

Elgjo, K., Laerum, O.D., Edgehill, W.: Growth regulation of mouse epidermis. I. G-2 inhibitor present in the basal cell layer. Virchows Archiv [Cell Pathol.] *8*, 277–283 (1971)

Elgjo, K., Laerum, O.D., Edgehill, W.: Growth regulation of mouse epidermis. II. G-1 inhibitor present in the differentiating cell layer. Virchows Archiv [Cell Pathol.] *10*, 229–236 (1972)

Elgjo, K., Clausen, O.P.F., Thorud, E.: Epidermis extracts (chalone) inhibit cell flux at the G_1/S, S/G_2, and G_2/M transitions in mouse epidermis. Cell Tissue Kinet. *14*, 21–29 (1980)

Erslev, A.J., Kazal, L.A., Miller, O.P., Abaidoo, K.-J.R.: The renal erythropoietin inhibitor. In: Regulation of erythropoiesis. Gordon, A.S., Condorelli, M., and Peschle, C. (eds.), pp. 217–222. Milano: Il Ponte 1972

Fejes, M., Pasqualeni, Ch.D., Braun, M.: Normal murine endogenous lymphoid factor (S) inhibiting lymphocyte functions. I. In vivo effect on allogeneic tumor growth. Oncology *37*, 96–100 (1980)

Fiedler, H., Wohlrab, W., Zaumseil, R.-P.: Zur Frage der Beeinflussung der Wirksamkeit von Chemotherapeutika durch Chalone beim malignen Melanom. Dermatol. Monatsschr. *165*, 198–201 (1979)

Florentin, I., Kiger, N., Mathe, G.: T lymphocyt specificity of a lymphocyte inhibiting factor extracted from the thymus. Eur. J. Immunol. *3*, 624 (1973a)

Florentin, R.A., Nam, S.C., Janakidevi, K., Lee, K.T., Reiner, J.M., Thomas, W.A.: Population dynamics of arterial smooth muscle cells. II. In vivo inhibition of entry into mitosis of swine arterial smooth muscle cells by aortic tissue extracts. Arch. Pathol. *95*, 317–320 (1973b)

Florentin, I., Kiger, N., Mathe, G.: The lymphocyte inhibiting factor extracted from the thymus (LIFT): Inhibition of the DNA synthesis in vitro. Cell. Immunol. *23*, 1–10 (1976)

Foa, P., Paile, W., Yuen, T.L.S.T.H., Jones, W.A., Jänne, J., Rytömaa, T.: Granulocytic chalone is not a polyamine complex. Biomedicine *31*, 163–166 (1979)

Forscher, B.K., Houck, J.C.: Chalones: concepts and current researchers. Nat. Cancer Inst. Monographs 38. Washington: Superintendent of documents, 1973

Frindel, E., Guigon, M.: Inhibition of CFU entry into cycle by a bone marrow extract. Exp. Hemat. *5*, 74–76 (1977)

Galjaard, H., Meier-Fieggen, W., Giesen, J. van der: Feedback control by functional epithelium. Exp. Cell Res. *73*, 197–207 (1972)

Garcia-Giralt, E., Lasalvia, E., Florentin, I., Mathe, G.: Evidence for a lymphocyte chalone. Eur. J. Clin. Biol. Res. *15*, 1012–1015 (1970)

Garcia-Giralt, E., Lenfant, M., De Garilhe, M.P., Mayadoux, E., Motta, R.: Purification of immunosuppressive factors extracted from bovine spleen (lymphoid chalone). II. Biological activity. Cell Tissue Kinet. *11*, 465–476 (1978)

Gelfant, S.: A new concept of tissue and tumor cell proliferation. Cancer Res. *37*, 3845–3862 (1977)

Glass, G.B.J.: Antral chalone and gastrones. In: Gastrointestinal hormones. Glass, J.B.G. (ed.), chap. 42, pp. 929–970. New York: Raven Press 1980

Goldin, G., Fabian, B.: The regulation of growth in the mesonephric kidney of adult *xenopus laevis* by an endogenous inhibitor of proliferation. Develop. Biol. *66*, 529–538 (1978)

Gonzalez, R., Verly, W.G.: Isolation of an inhibitor of DNA synthesis specific for normal and malignant mammary cells. Proc. nat. Acad. Sci. (Wash.) *73*, 2196–2200 (1976)

Gonzalez, R., Verly, W.G.: Purification of an inhibitor of DNA synthesis in mammary cells. Europ. J. Cancer *14*, 689–697 (1978)

Goss, R.J.: Mitotic responses of the compensating rat kidney to injections of tissue homogenates. Cancer Res. *23*, 1031–1035 (1963)

Gradwohl, P.R.: The proliferation of epidermal cells in mouse ear organ culture. Arch. Dermatol. Res. *263*, 273–281 (1978 a)

Gradwohl, P.R.: Mouse epidermal and skin extracts tested for cytostatic activity ("chalones"). Effects of organ and cell cultures. Arch. Dermatol. Res. *263*, 283–295 (1978 b)

Grundboeck-Jusko, J.: Purification of chalone from the spleen of cattle. Bull. Vet. Inst. Pulawy *19*, 41–48 (1975)

Grundboeck-Jusko, J.: Chemical characteristics of chalones isolated from bovine-spleen. Acta Biochim. Polon. *23*, 165–170 (1976)

Grundboeck-Jusko, J.: Biological activity of spleen chalones determined in the culture of lymphocytes. Bull. Vet. Inst. Pulawi *21*, 94–100 (1977)

Grundmann, E., Seidel, H.J.: Die reparative Parenchymregeneration am Beispiel der Leber nach Teilhepatektomie. In: Handbuch der Allgemeinen Pathologie. Büchner, F. (ed.), vol. 6,II, pp. 129–243. Berlin, Heidelberg, New York: Springer 1969

Guigon, M., Frindel, E.: Inhibition of CFU-S entry into cell cycle after irradiation and drug treatment. Biomedicine *29*, 176–178 (1978)

Guigon, M., Frindel, E.: Inhibition of bone marrow stem cell kinetics. Exp. Hematol. *7*, 34 (1979)

Guigon, M., Enouf, J., Frindel, E.: Inhibition of bone marrow stem cell proliferation. Abstracts Xth Meeting of the European Study Group Cell Prolif, p. 15. Oxford: Blackwell Sci. Publ. 1979

Hanks, C.T., Smith, E.O.: Inhibition of connective tissue proliferation by dermal extracts. J. Invest. Dermatol. *71*, 172–176 (1978)

Hansteen, I.L., Iversen, O.H., Refsum, S.B.: Epidermal DNA synthesis in organ culture explants. Virchows Archiv [Cell Pathol.] *31*, 187–191 (1979)

Harrington, W.N., Godman, G.C.: A selective inhibitor of cell proliferation from normal serum. Proc. Natl. Acad. Sci. U.S.A. *77*, 423–427 (1980)

Heideman, E., Jung, A., Wilms, K.: Gewebsspezifische Hemmung der Lymphocytenproliferation durch Milzextrakt (Lymphocytenchalon). Klin. Wochenschr. *54*, 221 (1976)

Heideman, E., Jung, A., Masurczak, J., Podgornik, N., Schmidt, H., Wilms, K.: Weitere Anreicherung eines gewebsspezifischen Inhibitors der Lymphopoese (Lymphozytenchalon?). Blut *39*, 99–106 (1979 a)

Heidemann, E., Podgornik, N., Wilms, K.: Tissue-specific inhibitor of lymphocyte proliferation extracted and purified from calf spleen. Biological and chemical properties. Blut *39*, 271–279 (1979 b)

Herman, S.P., Golde, D.W., Cline, M.I.: Neutrophil products that inhibit cell proliferation: Relation to granulocytic "chalone". Blood *57*, 207–220 (1978)

Hiestand, P.C., Borel, J.F., Bauer, W., Kis, Z.L., Magnée, C., Stähelin, H.: The effects of fractions (Chalones) obtained from lymphoid organs on lymphocyte proliferation in vitro. Agents Actions *7*, 327–335 (1977)

Higueret, P., Chany, E., Mousset, S., Gardes, M., Frayssinet, C.: Activité inhibitrice de fractions protéiques isolées du foie de rat sur l'hypertrophie compensatrice après hépatectomie partielle. C.R. Acad. Sci. Paris *281*, 49–52 (1975)

Hondius-Boldingh, W.H., Laurence, E.B.: Extraction, purification, and preliminary characterization of the epidermal chalone: a tissue-specific mitotic inhibitor obtained from vertebrate skin. Europ. J. Biochem. *5*, 191–198 (1968)

Houck, J.C. (ed.): Chalones. Amsterdam, Oxford, New York: North-Holland Publ. Co. 1976a

Houck, J.C.: Putative bronchial chalone. In: Chalones. Houck, J.C. (ed.), pp. 394–400. Amsterdam-Oxford-New York: North-Holland and American Elsevier 1976b

Houck, J.C.: Lymphocyte chalone. J. Reticuloendothel. Soc. *24*, 571–581 (1978)

Houck, J.C., Irausquin, H., Leikin, S.: Lymphocyte DNA synthesis inhibition. Science *173*, 1139–1141 (1971)

Houck, J.C., Weil, R.L., Sharma, V.K.: Evidence for a fibroblast chalone. Nature New Biol. *240*, 210–211 (1972)

Houck, J.C., Attalah, A.M., Lilly, J.R.: Immunosuppressive properties of the lymphocyte chalone. Nature *245*, 148–149 (1973)

Ichikawa, Y., Pluznik, D.H., Sachs, L.: Feedback inhibition of the development of macrophage and granulocyte colonies. In: I. Inhibition by macrophages. Proc. Natl. Acad. Sci. U.S.A. *58*, 1480–1486 (1967)

Irons, M.J., Clermont, Y.: Spermatogonial chalone(s): effect on the phases of the cell cycle of type A spermatogonia in the rat. Cell Tissue Kinet. *12*, 425–433 (1979)

Isaksson-Forsén, G., Burton, D.R., Korsgaard, R., Elgjo, K., Iversen, O.H.: Partial purification of epidermal G2 chalone. Virchows Archiv [Cell Pathol.] *26*, 97–103 (1977)

Isaksson-Forsén, G., Elgjo, K., Burton, D., Iversen, O.H.: Partial purification of the epidermal G_2 chalone based on an in vivo assay system. Cell Biol. Int. Rep. *5*, 195–199 (1981)

Iversen, O.H.: A homeostatic mechanism regulating the cell number in epidermis. In: Medicine Cibernetica, Proc. 1st Congr. Int. Cybernetic Med. Masturzo, A. (ed.), pp. 420–430. Naples: Gianno 1960a

Iversen, O.H.: Cell metabolism in experimental skin carcinogenesis. Acta Pathol. Microbiol. Scand. *50*, 17–24 (1960b)

Iversen, O.H.: Cybernetic aspects of the cancer problem. In: Progress in biocybernetics. Wiener, N. and Schade, J.P. (ed.), vol. 2, pp. 76–110. Amsterdam, London, New York: Elsevier Publ. Co. 1965

Iversen, O.H.: Effect of epidermal chalone on human epidermal mitotic activity in vitro. Nature *219*, 75 (1968)

Iversen, O.H.: Chalones of the skin. In: CIBA Foundation Symposium on Homeostatic Regulators. Wolstenholme, G.E.W., Knight, J. (eds.), pp. 29–56. London: Churchill 1969

Iversen, O.H.: Some theoretical considerations on chalones and the treatment of cancer. A review. Cancer Res. *30*, 1481–1484 (1970)

Iversen, O.H.: Cell proliferation kinetics and carcinogenesis: a review. In: 1973. Proc. Vth Int. Symp. Biol. Charact. Human Tumours. Davis, W. and Malton, C. (eds.), pp. 21–29. Amsterdam: Exerpta Med. 1973

Iversen, O.H.: The role of cytofluorometry in relation to all the other methods used in the study of cell kinetics. In Lutz, D. and Paukovits, W.R. (ed.), pp. 539–544. Ghent: Pulse-Cytophotometry, Part III, European Press 1978a

Iversen, O.H.: Epidermal chalones and squamous cell carcinomas. The growth inhibitor effects of aqueous epidermal extracts (G1 and G2 chalones) on the epidermis and on a transplantable keratinizing carcinoma in nude mice. Virchows Archiv [Cell Pathol.] *27*, 229–235 (1978b)

Iversen, O.H.: The effect of the epidermal G2 chalone on the mitotic duration in nude mouse epidermis and in a transplanted squamous cell carcinoma. Virchows Archiv [Cell Pathol.] *28*, 271–277 (1978c)

Iversen, O.H.: Comments on "chalones and cancer". Mech. Ageing Develop. *12*, 211–212 (1980)

Iversen, O.H., Bjerknes, R.: Kinetics of epidermal reaction to carcinogens. Acta Pathol. Microbiol. Scand. Suppl. *165* (1963)

Iversen, O.H., Aandahl, E., Elgjo, K.: The effect of an epidermis specific mitotic inhibitor extracted from epidermal cells. Acta Pathol. Microbiol. Scand. *64*, 506–510 (1965)

Iversen, O.H.: The History of Chalones. In: Chalones. Houck, J.C. (ed.), pp. 37–70. Amsterdam, Oxford, New York: North-Holland and American Elsevier 1976

Iversen, O.H., Bhangoo, K.S., Hansen, K.: Control of epidermal cell renewal in the bat web. Virchows Archiv [Cell Pathol.] *16*, 157–179 (1974)

Iversen, O.H., Clausen, O.P., Elgjo, K., Iversen, U.M., Rohrbach, R.: Effects of bleomycin on the epidermal content of growth-regulatory substances (chalones). Cell Tissue Kinet. *10*, 71–79 (1977)

Type, P.T., Bhargava, P.M., Tasker, A.D.: Some aspects of the chemical and cellular composition of adult rat liver. Exp. Cell Res. *40*, 233–241 (1965)

Jänne, J., Poso, H., Raina, A.: Polyamines in rapid growth and cancer. Biochim. Biophys. Acta *473*, 241–293 (1978)

Jegasothy, B.V., Battles, D.R.: Immuno-suppressive lymphocyte factors. I. Purification of inhibitor of DNA synthesis to homogeneity. J. Exp. Med. *150*, 622–631 (1979)

Jones, J.W.: A system of growth control in Ehrlich ascites tumor. Dissertation Abstr. Int. B. Sciences and engineering *40*, 5103–5104 (1980)

Jones, J., Paraskova-Tchernozenska, E., Moorhead, J.: In vitro inhibition of DNA synthesis in human leukemic cells by a lymphoid cell extract. Lancet, 654–655 (1970)

Kanagalingam, K., Houck, J.C.: Colon carcinoma: its genesis and chalone control. In: Chalones. Houck, J.C. (ed.), pp. 459–482. Amsterdam, Oxford, New York: North-Holland and American Elsevier 1976

Kariniemi, A.-L., Rytömaa, T.: Effect of the Hewitt keratinizing epidermal carcinoma on cell proliferation in different organs of the host mouse and in human psoriatic skin cultured in diffusion chambers. Br. J. Dermatol. *94*, 515–522 (1976)

Kastner, M., Maurer, H.R.: Pure bovine granulocytes as a source of granulopoiesis inhibitor (chalone). Hoppe-Seylers Z. Physiol. Chem. *361*, 197–200 (1980)

Ketlinski, S.A.: Chalones as factor of the tissue homeostasis. Arch. Anatom. Gist. embriol. *78*, 29–49 (1980)

Kiger, N.: Isolation and immunological study of thymic lymphocytic inhibitory factors. Eur. Etud. Clin. Biol. *16*, 566–572 (1971)

Kiger, N., Florentin, I., Mathé, G.: Further purification of the lymphocyte inhibiting extract from the thymus. Boll. 1st. Sieroter. Milan *54*, 244–249 (1975)

Kiger, N., Florentin, I., Mathé, G.: Some effects of a partially purified lymphocyte-inhibiting factor from calf thymus. Transplantation *14*, 448–454 (1972a)

Kiger, N., Florentin, I., Garcia-Giralt, E., Mathé, G.: Lymphocyte inhibitory factors (chalones) extracted from lymphoid organs. Extraction, partial purification and immuno-suppressive properties. Transplant. Proc. *4*, 531 (1972b)

Kiger, N., Florentin, I., Garcia-Giralt, E., Mathé, G.: Inhibition of graft-versus-host reaction (GVHR) by in vitro incubation of donor lymphocyte with thymic or splenic chalone(s). Exp. Hematol. *1*, 22 (1973a)

Kiger, N., Florentin, I., Mathé, G.: Inhibition of graft-versus-host reaction by preincubation of the graft with a thymic extract (lymphocyte chalone). Transplantation *16*, 393 (1973b)

Kivilaakso, E., Rytömaa, T.: Erythrocyte chalone, a tissue-specific inhibitor of cell proliferation in the erythron. Cell Tissue Kinet. *4*, 1–9 (1971)

Klein, K., Coetzee, M.L., Madhau, R., Ove, P.: Inhibition of tritiated thymidine incorporation in cultured cells by rat kidney extract. J. Natl. Cancer Inst. *62*, 1557–1564 (1979)

Korsgaard, R., Iversen, O.H., Burton, D.R., Isaksson-Forsén, G.: Artunspezifische und reversible Wachstumshemmung durch Chalone in Plattenepithelkarzinomen des Menschen in vitro. Z. Krebsforsch. *88*, 217–221 (1977)

Korsgaard, R., Iversen, O.H., Isaksson-Forsen, G., Burton, D.R.: Human epidermoid lung carcinomas in vitro seem to react in a specific way to epidermal G 2 chalone. Cell Tissue Kinet. *11*, 441–443 (1978)

Krieg, L., Kühlmann, I., Marks, F.: Effect of tumor-promoting phorbol esters and of acetic acid on mechanisms controlling DNA synthesis and mitosis (Chalones) and on the biosynthesis of histidine-rich protein in mouse epidermis. Cancer Res. *34*, 3135–3146 (1974)

Krizsa, F., Kelemen, E., Cserháti, I., Lajtha, L.G.: Specific thrombopoietic inhibition by syngeneic platelet homogenates. Biomedicine *27*, 145–148 (1977)

Kubilus, J., Baden, H.P.: A mitotic inhibitor from human placental membrane. Fed. Proc. *34*, 3463 (1975)

Kuhn, T.S.: The structure of scientific revolutions, 2nd ed. Chicago: University of Chicago Press 1970

Kuo, C.Y., Yoo, T.J.: In vitro inhibition of tritiated thymidine uptake in Morris hepatoma cells by normal rat liver extract: A possible liver chalone. J. Natl. Cancer Inst. *59*, 1691–1695 (1977)

Kurnick, N.B., Lindsay, P.A.: Compensatory renal hypertrophy in parabiotic mice. Lab. Invest. *19*, 45–48 (1968)

Laerum, O.D.: The separation and cultivation of basal and differentiating cells from hairless mouse epidermis. J. Invest. Dermatol. *54*, 279–287 (1970)

Laerum, O.D., Maurer, H.R.: Proliferation kinetics of myelopoietic cells and macrophages in diffusion chambers after treatment with granulocyte extracts (chalone). Virchows Archiv [Cell Pathol.] *14*, 293–305 (1973)

Laerum, O.D., Paukovits, W.R., Aardal, N.P., Morild, I.: Modification of myelopoiesis by granulocyte chalone: possible implications for chronobiology. In: Proceed. XIV Int. Conf. on Chronobiology, Hannover 1980. Mayersbach, H. von (ed.). Milano: Il Ponte 1980

Lajtha, L.G.: Review of leukocytes. Natl. Cancer Inst. Monogr. *38*, 111–116 (1973)

Lajtha, L.G.: Stem cell concepts. Differentiation *14*, 23–34 (1979)

Lajtha, L.G.: In: Cellular Kinetics of Haemopoiesis, Blood and its disorders. Hardisty, R.M., Weatherall, D. (eds.). Oxford: Blackwell Sci. Publ. 1980

Lala, P.K.: Influence of local environment on the growth parameters of the Ehrlich ascites tumor. Eur. J. Cancer *8*, 197–204 (1972)

Lasalvia, E., Garcia-Giralt, E., Macieria-Coelho, A.: Extraction of an inhibitor of DNA synthesis from human peripheral blood lymphocytes and bovine spleen. Rev. Eur. Etud. Clin. Biol. *15*, 789–792 (1970)

Lasnitzki, I.: The action of hormones on cell and organ cultures. In: Cell and tissue in culture. Wilmer, E.N. (ed.), vol. I, pp. 591–658. London: Academic Press 1965

Laurence, E.B.: Experimental approach to the epidermal chalone. Natl. Cancer Inst. Monogr. *38*, 37–46 (1973)

Laurence, E.B.: The significance of chalones in epidermal growth. In: The skin of vertebrates. Spearman, R.I.C., Riley, P.A. (eds.), pp. 139–150, Linnean Soc. Symp. Ser. 9. London (1980)

Laurence, E.B., Elgjo, K.: Epidermal chalone and cell proliferation in a transplantable squamous cell carcinoma in hamsters. II. In vitro results. Virchows Archiv [Cell Pathol.] *7*, 8–15 (1971)

Laurence, E.B., Randers-Hansen, E.: An in vitro study of epidermal chalone and stress hormones on mitosis in tongue epithelium and ear epidermis of the mouse. Virchows Archiv [Cell Pathol.] *9*, 271–279 (1971)

Laurence, E.B., Randers-Hansen, E.: Regional specificity of the epidermal chalone extracted from two different body sites. Virchows Archiv [Cell Pathol.] *11*, 34–42 (1972)

Laurence, E.B., Spargo, D.J., Thornley, A.L.: Cell proliferation kinetics of epidermis and sebaceous glands in relation to chalone action. Cell Tissue Kinet. *12*, 615–633 (1979)

Lehmann, W., Graetz, H., Schütt, M., Langen, P.: Chalone-like inhibition of Ehrlich ascites cell proliferation in vitro by an ultrafiltrate obtained from the ascitic fluid. Acta Biol. Med. Germ. *36*, 43–52 (1977)

Lehmann, W., Graetz, H., Schütt, M., Langen, P.: Antagonistic effects of insulin and a negative growth regulator from ascites fluid on the growth of Ehrlich ascites carcinoma cells in vitro. Exp. Cell Res. *119*, 396–399 (1979)

Lehmann, W., Graetz, H., Samtleben, R., Schütt, M., Langen, P.: On a „chalone"-like factor for the Ehrlich ascites mammary carcinomas. Acta Biol. Med. Germ. *39*, 93–105 (1980)

Lenfant, M., Digiusto, L.: Is lymphocytic chalone activity restricted to a spermine-protein complex? Nature *277*, 154 (1979)

Lenfant, M., Garilhe, M.P. de, Garcia-Giralt, E., Tempéte, C.: Further purification of bovine spleen inhibitors of lymphocyte DNA synthesis (chalones). Biochim. Biophys. Acta *451*, 106–117 (1976)

Lenfant, M., Garcia-Giralt, E., Thomas, M., Di Giusto, L.: Purification of immunosuppressive factors extracted from bovine spleen (lymphoid chalone). Cell Tissue Kinet. *11*, 455–463 (1978)

Lenfant, M., Garcia-Giralt, E., Digiusto, L., Thomas, M.: The purification of an immunosuppressive factor extracted from bovine spleen. III. Purification process. Mol. Immunol. *17*, 119–126 (1980)

Lesser, B., Bruchovsky, N.: The effects of testosterone, 5α-dihydrotestosterone and adenosine 3',5'-monophosphate on cell proliferation and differentiation in rat prostates. Biochem. Biophys. Acta *308*, 426–437 (1973)

Lewis, J.P., Neal, W.A., Moores, R.R., Gardner, E., Alford, A., Smith, L.L., Wright, C.-S., Welch, E.T.: A protein inhibitor of erythropoiesis. J. Lab. Clin. Med. *74*, 608–613 (1969)

Lewis, J.P., Neal, W.A., Welch, E.T.: The isolation of erythropoiesis regulatory factor by an electrofractionation technique combined with selective membrane permeability. Proc. Soc. Exp. Biol. Med. *142*, 293–298 (1973)

Lezhava, R.A., Tumanishvili, G.D., Gogsadse, L.A.: Growth inhibiting factors (chalones) in the heart tissue of adult hens and rats. Bull. Georgian Acad. Sci. (USSR) *3*, 435–443 (1977)

Lindemann, R.: Erythropoiesis inhibiting factor (EIF). I. Fractionation and demonstration of human urinary EIF. Brit. J. Haemat. *21*, 623–631 (1971)

Lindemann, R.: Erythropoiesis inhibiting factor(s) in intact and haemolyzed red blood cells. Scand. J. Haematol. *14*, 216–225 (1975)

Lindemann, R., Laerum, O.D.: Erythropoiesis inhibiting factor(s) (EIF). The specificity and toxicity of urinary EIF studied in vivo and in vitro. Scand. J. Haematol. *17*, 293–299 (1976)

Loginov, A.S., Speransky, M.D., Arwin, L.I., Matyushina, E.D., Magnitsky, G.S., Keilonakh, P.: Concerning the Liver Chalones. Biull. Eksp. Biol. Med. *82*, 1482–1484 (1976)

Lord, B.I.: Modification of granulocytopoietic cell proliferation by granulocyte extracts. Boll. Ist. Sieroter. Milan. *54*, 187–194 (1975)

Lord, B.I.: The assay of cell proliferation inhibitors. In: Chalones. Houck, J.C. (ed.), pp. 97–139. Amsterdam, Oxford, New York: North-Holland and American Elsevier 1976

Lord, B.I.: Cellular dynamics of haemopoiesis. In: Clin. Haematol. *8*, 435–445 (1979)

Lord, B., Cercek, L., Cercek, B., Shah, G., Dexter, T., Lajtha, L.: Inhibitors of hemopoietic cell proliferation? Specificity of action within the hemopoietic system. Br. J. Cancer *29*, 168–175 (1974a)

Lord, B.I., Cercek, L., Cercek, B., Shah, G.P., Lajtha, L.G.: Inhibitors of haemopoietic cell proliferation. Reversibility of action. Br. J. Cancer *29*, 407–409 (1974b)

Lord, B.I., Mori, E.G., Wright, E.G., Lajtha, L.G.: An inhibitor of stem cell proliferation in normal bone marrow. Brit. J. Haemat. *34*, 441–445 (1976)

Lord, B.I., Shah, G.P., Lajtha, I.G.: The effects of red blood cell extracts on the proliferation of erythrocyte precursor cells in vivo. Cell Tissue Kinet. *10*, 215–222 (1977)

Lozzio, B.B.: Regulators of cell division. A review. I. Endogenous mitotic inhibitors of hematopoietic cells. Exp. Hematol. *1*, 309–339 (1973)

Løvhaug, D., Bøyum, A.: Regulation of bone marrow cell growth in diffusion chambers: The effect of granulocyte extracts. Cell Tissue Kinet. *10*, 137–146 (1977)

MacFayden, D.A., Sturm, E.: Further observations on factors from normal tissues influencing the growth of transplanted cancer. Science *84*, 67–68 (1936)

MacVittie, T.J., McCarthy, K.F.: The influence of a granulocytic inhibitor(s) on hematopoiesis in an in vivo culture system. Cell Tissue Kinet. *8*, 553–559 (1975)

Maharajan, V., Batra, B.K.: Chalone: A possible problem for embryonic differentiation. Indian J. Exp. Biol. *14*, 324–325 (1976)

Maier, G., Rothbarth, K.: Factors involved in the growth regulation of Ehrlich ascites tumor cells in vitro. Abstracts Xth Meeting Europ. Study Group Cell Prolif., p. 22. Oxford: Blackwell Sci. Publ. 1979

Maiolo, A.T., Cazzangia, E., Depanger, V., Foa, P., Lombardi, L., Mozzana, R., Polli, E.E.: In vitro production of lymphocyte and granulocyte proliferation inhibitors (chalones?) from living cells. Boll. 1st Sieroter. Milan *54*, 235–243 (1975)

Malenkov, A.G., Modjanova, E.A., Jamskova, V.P.: Tissue specific adhesive factors – G-1-chalones. Biofizika *22*, 156–157 (1977)

Marks, F.: Direct evidence of two tissue-specific chalone-like factor regulating mitosis and DNA synthesis in mouse epidermis. Hoppe-Seylers Z. Physiol. Chem. *352*, 1273–1274 (1971)

Marks, F.: A tissue-specific factor inhibiting DNA synthesis in mouse epidermis. Natl. Cancer Inst. Monograph *38*, 79–90 (1973)

Marks, F.: Isolation of an endogenous inhibitor of epidermal DNA synthesis (G_1 chalone) from pig skin. Hoppe Seylers Z. Physiol. Chem. *356*, 1989–1992 (1975)

Marks, F.: The epidermal chalones. In: Chalones. Houck, J.C. (ed.), pp. 173–227. Amsterdam, Oxford, New York: North-Holland and American Elsevier 1976

Marks, F., Bertsch, S., Grimm, W., Schweizer, J.: Hyperplastic transformation and tumor promotion in mouse epidermis: Possible consequences of disturbances of endogenous mechanisms controlling proliferation and differentiation. In: Carcinogenesis. Slaga, T.J., Sivak, A., and Boutwell, R.K. (eds.), vol. 2, pp. 97–116. Mechanisms of tumor promotion and cocarcinogenesis. New York: Raven Press 1978

Martel-Pelletier, J., Bergeron, M.: Mise en evidence d'un facteur regulateur (chalone) de la masse renale. J. Urol. Nephrol. (Paris) *81*, 716 (1975)

Maschler, R., Maurer, H.R.: A lymphocyte chalone assay using lymphocyte colony growth in agar capillaries. Virchows Archiv [Cell Pathol.] *25*, 345–353 (1977)

Maschler, R., Maurer, H.R.: Screening for granulopoiesis inhibitors (chalones) by different assays. Hoppe-Seylers Z. Physiol. Chem. *359*, 825–834 (1978)

Maschler, R., Maurer, H.R.: Screening for specific calf thymus inhibitors (chalones) of T-lymphocyte proliferation. Hoppe-Seylers Z. Physiol. Chem. *360*, 735–745 (1979)

Mathé, G.: Lymphocyte inhibitors fulfilling the definition of chalones and immunosuppression. Eur. J. Clin. Biol. Res. *17*, 548–555 (1972)

Mathé, G., Kiger, N., Florentin, I., Garcia-Giralt, E., Martyre, M.-Cl., Hanne-Pannenko, O., Schwartzenberg, L.: Progress in the prevention of GVH: Bone marrow grafts after ALG conditioning, with lymphocyte split chimerism, use of a lymphocyte "chalone T", and soluble histocompatibility antigens. Transplant. Proc. *5*, 933–939 (1973)

Mattern, I., Wayss, K., Volm, M.: Studies on the specificity of tissue supernatants on the proliferation of liver in baby-rats. Exp. Path. *18*, 302–309 (1980)

Maurer, H.R.: Chalones: specific regulators of eukaryotic tissue growth. In: Regulation of growth and differentiated function in eukaryotic cells. Talwar, G.P. (ed.), pp. 69–77. New York: Raven Press 1975

Maurer, H.R.: Interferon-brauchbare Antivirus- und Antikrebsmittel? Dtsch. Apotheker Z. *41*, 1611–1616 (1979)

Maurer, H.R., Laerum, O.D.: Granulocyte testing: a critical review. In: Chalones. Houck, J.C. (ed.), pp. 331–354. Amsterdam, Oxford, New York: North-Holland/American Elsevier 1976

Maurer, H.R., Weiss, G., Laerum, O.D.: Starting procedures for the isolation and purification of granulocyte chalone activities. Blut *33*, 161–170 (1976)

Mazia, D.: Mitosis and the physiology of cell division. In: The cell. Brachet, J., Mirsky, A.E. (eds.), vol. III, chap. 2. New York: Academic Press 1961

McDonald, T.P.: The hemagluttination-inhibition assay for thrombopoitin. Blood *41*, 219–233 (1973)

McMahon, J.B., Iype, P.T.: Specific inhibition of proliferation of non-malignant rat hepatic cells by a factor from rat liver. Cancer Res. *40*, 1249–1254 (1980)

Menzies, R.A., Kerrigan, J.M.: Liver extract (chalone) anti-mitotic activity assayed with chick embryos. Exp. Cell Res. *86*, 430–433 (1974)

Miyamoto, M., Terayama, H.: Nature of rat liver cell sap factors inhibiting the DNA synthesis in tumor cells. Biochim. Biophys. Acta *228*, 324–329 (1971)

Modjanova, A.A., Malenkov, A.G.: Tissue-specific inhibition of DNA synthesis by lung and liver Ca-free salt extracts able to increase the mutual cell-tissue adhesion. Citologia *17*, 1155–1159 (1975)

Modjanova, A.A., Malenkov, A.G., Bocharova, O.A.: The long-term effects of lung contactin on urethane-induced adenomatosis. Abstr. Xth meeting Eur. Study Group Cell Prolif., p.24. Oxford: Blackwell Sci. Publ. 1979

Mohr, U., Althoff, J., Kinzel, V., Süss, R., Volm, M.: Melanoma regression induced by "chalone": a new tumour inhibiting principle acting in vivo. Nature 220, 138–139 (1968)

Mohr, U., Hondius Boldingh, W., Althoff, J.: Identification of contaminating Clostridium spores as the oncolytic agent in some chalone preparations. Cancer Res. 32, 1117–1121 (1972a)

Mohr, U., Hondius-Boldingh, W., Emminger, A., Behagel, H.S.: Oncolysis by a new strain of Clostridium. Cancer Res. 32, 1122–1128 (1972b)

Moltimard, R., Pietu, G., Chany, E., Trincal, G., Frayssinet, C.: An inhibitor of hepatoma cells multiplication in the efferent fluid from isolated perfused rat liver. Biomedicine 23, 434–437 (1975)

Moorhead, J.F., Paraskova-Tchernozenska, E., Pirrie, A.J., Hayes, C.: Lymphoid inhibitor of human lymphocyte DNA synthesis and mitosis in vitro. Nature 224, 1207–1208 (1969)

Morley, C.G.D.: Humoral regulation of liver regeneration and tissue growth. Perspect. Biol. Med. 17, 411–428 (1974)

Müntzing, J., Murphy, G.P.: Study on the growth limiting mechanism in the rat ventral prostate. Proc. Soc. Exp. Biol. Med. 154, 331–336 (1977)

Müntzing, J., Liljekvist, J., Murphy, G.P.: Chalones and stroma as possible growth-limiting factors in the rat ventral prostate. Invest. Urol. 16, 399–402 (1979)

Muller-Bérat, C.N., Laerum, O.D., Maurer, H.R.: Chalone inhibition of the committed stem cell to granulopoiesis in vitro. Abstracts 6th meeting European Study Group Cell Prolif., p.41. Moscow 1973

Murphy, J.B., Sturm, E.: Further observations on an inhibitor principle associated with the causative agent of a chicken tumour. Science 74, 180–181 (1931)

Murphy, J.B., Sturm, E.: Normal tissues as a possible source of inhibitor for tumors. Science 75, 540–541 (1932)

Murphy, J.B., Sturm, E.: Effect of inhibiting factor from normal tissues on spontaneous tumors of mice. Science 77, 631–633 (1933)

Nadal, G.: Control of liver growth by growth inhibitors (chalones). Arch. Toxicol. Suppl. 2, 131–142 (1979)

Naeser, F.K., Barfod, N.M., Bichel, P.: Lack of evidence of chalone activity in used medium and extract of JB-1 tumor cells in vitro. Virchows Archiv B [Cell Pathol.] 26, 321–329 (1978)

Naess, A.: The pluralist and possibilist aspect of the scientific enterprise. London: Allen & Unvin Ltd. 1972

Nakai, G.S.: Ehrlich ascites tumor (EAT) chalone effect on nascent DNA synthesis and DNA polymerase alpha and beta. Cell Tissue Kinet. 9, 553–563 (1976)

Nakai, G.S., Gergely, H.: Effect of Ehrlich ascites cell chalone on nascent DNA synthesis in isolated nuclei. Cell Tissue Kinet. 13, 65–73 (1980)

Natraj, C.V., Datta, P.: Control of DNA synthesis in growing BALB/c 3T3 mouse cells by a fibroblast growth regulatory factor. Proc. Natl. Acad. Sci. U.S.A. 75, 6115–6119 (1978)

Neustroev, G.V.: Effect of erythrocytic chalone on electrophoretic mobility of mouse bone marrow. Bull. Exp. Biol. Med. 85, 8–11 (1978)

Neustroev, G.V.: Mechanism of anemia induced by long-term administration of erythrocytic chalone. Probl. Gematol. Perelivaniia Krovi (Moscow) 25, 32–35 (1980a)

Neustroev, G.V.: The effect of clean erythrocytic chalone on the proliferative ability of mice marrow cells. Fiziol. J. of the USSR 66, 846–851 (1980b)

Neustroev, G.V.: Effect of erythrocytic chalone on haemopoiesis in the spleen and liver induced by cyclophosphamide and carboxymethylcellulose. Patol. Fiziol. eksp. Terapi 4, 69–73 (1980c)

Neustroev, G.V., Korolev, N.P.: Surface origin of erythrocytic chalone in vitro in experimental polycythemia. Bull. Exp. Biol. Med. (Moscow) 88, 1065–1069 (1979)

Njeuma, D.L.: Mitosis and population density in cultures of embryonic chick and mouse fibroblasts. Exp. Cell Res. *66*, 237–243 (1971 a)

Njeuma, D.L.: Non-reciprocal density-dependent mitotic inhibition in mixed cultures of embryonic chick and mouse fibroblasts. Exp. Cell Res. *66*, 244–250 (1971 b)

Nome, O.: Tissue specificity of the epidermal chalones. Virchows Arhiv [Cell Pathol.] *19*, 1–25 (1975)

Odell, T.T.: Humoral regulation of thrombocytopoiesis. In: Humoral control of growth and differentiation. Lo Bue, J., Gordon, A.S. (eds.), pp. 120–139. New York: Academic Press 1973

Okulov, V.B., Chekulaeva, L.F.: Contribution to the method of isolation of inhibitors of the liver hepatocyte proliferation activity. Arkh. Anat. Gistol. Embriol. *770*, 106–109 (1976)

Okulov, V.B., Ketlinski, S.A.: Immunological histospecificity of epidermal chalones and its possible application to cancer research. Cancer Letters *3*, 215–220 (1977)

Okulov, V.B., Ketlinsky, E.A., Ratovitsky, E.A., Kalinovsky, V.P.: Purification and characteristics of epidermal G_2 and G_1 chalones. Biochimia *43*, 971–978 (1978)

Okulov, V.B., Anisimov, V.N., Azarova, M.A.: Study of antiblastomogenic action of epidermal chalones. I. The effect of epidermal chalones on some transplantable mouse tumours. Exp. Pathol. (Jena) *15*, 178–181 (1978)

Okulov, V.B., Ivanov, M.N., Anisimov, V.N.: Regulation of mitotic activity in rat vaginal epithelium: Relationship between the level of its inhibitor (G_2-chalone) and estrogens. Endokrinologie *74*, 20–26 (1979)

Okulov, V.B., Anisimov, V.N., Chepik, O.F., Azarova, M.A.: Effect of epidermal chalones on induction and growth of tumors of the cervicovaginal epithelium in mice. Byull. Éksp. Biol. Med. *89*, 335–337 (1980)

Olsson, L., Claësson, M.H.: Studies on the regulation of lymphocyte production in the murine thymus and some effects of a crude thymus extract. Cell Tissue Kinet. *8*, 491–502 (1975)

Osgood, E.E.: A unifying concept of the etiology of the leukemias, lymphomas, and cancers. J. Natl. Cancer Inst. *18*, 155–166 (1957)

Paermentier, F. de, Barbason, H.: Mitotic homeostatic control by liver extract in hepatocytes in vivo or cultivated in vivo. Abstract Xth meeting of the European Study Group Cell Prolif., pp. 8–9. London: Blackwell Sci. Publ. 1979

Paran, M., Ichikawa, Y., Sachs, L.: Feedback inhibition of the development of macrophage and granulocyte colonies. In: II. Inhibition by granulocytes. Proc. Natl. Acad. Sci. U.S.A. *62*, 81–87 (1969)

Pardee, A.B., Dubrow, R.: Control of cell proliferation. Cancer *39*, 2747–2754 (1977)

Parshley, M.S.: Effect of inhibitors from adult connective tissue on growth of a series of human tumours in vivo. Cancer Res. *25*, 387–401 (1965)

Patt, L.M., Houck, J.C.: The incredibly shrinking chalone. FEBS lett. *120*, 163–170 (1980)

Paukovits, W.R.: Granulopoiesis-inhibiting factor: demonstration and preliminary chemical and biological characterization of a specific polypeptide (chalone). Nat. Cancer Inst. Monogr. *38*, 147–155 (1973)

Paukovits, W.R., Hinterberger, W.: Molecular weight and some chemical properties of the granulocytic chalone. Blut *37*, 7–18 (1978 a)

Paukovits, W., Hinterberger, W.: Biochemical characterization of humoral factors regulating myelopoiesis. In: Cell-separation and cryobiology, pp. 75–78. Stuttgart, New York: Schattauer 1978 b

Paukovits, W.R., Paukovits, J.B.: Mechanism of action of granulopoiesis inhibiting factor (chalone). I.: Evidence for a receptor protein on bone marrow cells. Exp. Pathol. *10*, 348–352 (1975)

Paukovits, W.R., Paukovits, J.B.: Peptide nature and proteolytic sensitivity of granulopoiesis inhibiting factor. IRCR Med. Sci. *6*, 175 (1978)

Paukovits, W.R., Paukovits, J.B., Laerum, O.D., Hinterberger, V.: Granulopoiesis inhibiting factor (chalone): Purification and chemical composition. IRCS Med. Sci. *8*, 305–306 (1980)

Paulus, J.M.: Multiple differentiation in megakaryocytes and platelets. Blood 29, 407–416 (1967)

Philpott, G.W.: Tissue-specific inhibition of cell proliferation in embryonic stomach epithelium in vitro. Gastroenterology 61, 25–34 (1971)

Pietu, G., Munsch, N., Mousset, S., Frayssinet, C.: Effect of an inhibiting factor isolated from rat liver on DNA polymerase in regenerating rat liver. Cell Tissue Kinet. 12, 153–160 (1979)

Privat de Garilhe, M., Vedel, M., Tempête, Ch., Federici, Ch., Garcia-Giralt, E.: Endogenous factors possessing lymphoid chalone properties, purified by chromatography. Biomedicine 32, 148–154 (1980)

Prunieras, M., Delescluse, C., Regnier, M.: A cell culture model for the study of epidermal (chalone) homeostasis. Pharmacol. Ther. [B] 9, 271–295 (1980)

Raina, A., Jänne, J.: Physiology of the natural polyamines putrescine, spermidine and spermine. Med. Biol. 53, 121–147 (1975)

Randers-Hansen, E.: Mitotic activity and mitotic duration in tongue and gingival epithelium of mice. Odontolog. Tidskr. (Denmark) 75, 480–487 (1967)

Ranney, D.F.: Biological inhibitors of lymphoid cell division. In: Advances in Pharmacology and Chemotherapy. Garattini, S., Goldin, A., Hawking, F., and Kopin, I.J. (eds.), vol.13, pp.359–408. New York: Academic Press 1975

Reiter, R.J.: Cellular proliferation deoxyribonucleic acid synthesis in compensating kidneys of mice and the effect of food and water restriction. Lab. Invest. 14, 1636–1643 (1965)

Richter, K.H.: Untersuchungen zum Chalon aus Fibroblasten. Hoppe-Seylers Z. Physiol. Chem. 359, 1435–1439 (1978)

Rijke, E.O.: Lymphocyte inhibitory factors. Thesis at the University of Utrecht, 1979

Rijke, E.O., Ballieux, R.E.: Is thymus derived lymphocyte inhibitor a polyamine? Nature 274, 804–805 (1978)

Rijke, E.O., Ballieux, R.E.: Thymus derived inhibitor of lymphocyte proliferation. I. Isolation and assessment of tissue specificity. Cell Tissue Kinet. 12, 435–444 (1979)

Rijke, E.O., Lempers, H.C.M., Ballieux, R.E.: Thymus-derived inhibitor of lymphocyte proliferation. II. Tissue specificity in vitro and in vivo. Cell Tissue Kinet. 14, 31–38 (1981)

Rohrbach, R., Laerum, O.D.: Variations of mitosis-inhibiting chalone activity in epidermis and dermis after carcinogen treatment. Cell Tissue Kinet. 7, 251–257 (1974)

Rohrbach, R., Elgjo, K., Iversen, O.H., Sandritter, W.: Effects of methyl-cholanthrene on epidermal growth regulators. I. Variations in the M-factor. Beitr. Path. Bd. 147, 21–27 (1972)

Rohrbach, R., Iversen, O.H., Elgjo, K., Sandritter, W.: Effects of methylcholanthrene on epidermal growth regulators. II. Variations in the S-factor. Beitr. Pathol. 158, 145–158 (1976a)

Rohrbach, R., Iversen, O.H., Elgjo, K., Riede, U.N., Sandritter, W.: Effects of croton oil on epidermal growth regulators (chalones). Beitr. Pathol. 159, 143–156 (1976b)

Rohrbach, R., Iversen, O.H., Riede, U.N., Sandritter, W.: Effects of cellophane tape stripping of mouse skin on epidermal growth regulators (chalones). Beitr. Pathol. 160, 175–186 (1977)

Romanov, Y.A., Semenova, M.V.: Effect of chalone from mice Erlich ascites tumor on mitotic activity of this tumor after single and repeated administration. Biull. Exp. Biol. Med. (Moscow) 89 (2), 189–191 (1980)

Ross, R., Vogel, A.: The platelet-derived growth factor. Cell 14, 203–210 (1978)

Rothbarth, K., Maier, G., Schöpf, E., Werner, D.: Inhibition of DNA synthesis by a factor from ascites tumor cells. Eur. J. Cancer 13, 1195–1196 (1977)

Rudland, Ph.S., DeAsua, L.J.: Action of growth factors in the cell cycle. Biochim. Biophys. Acta 560, 91–133 (1979)

Rytömaa, T.: Biology of the granulocyte chalone. In: Chalones. Houck, J.C. (ed.), pp.238–309. Amsterdam, Oxford, New York: North-Holland and American Elsevier 1976

Rytömaa, T.: Chalones and blood cells. In: The year in hematology. Gordon, A.S., Silber, P., and LoBue, I. (eds.), vol.2, pp.321–373. New York: Plenum 1978

Rytömaa, T., Kiviniemi, K.: In vitro experiments for the demonstration of specific feedback factors in rat serum. In: Proceedings of the XIV Scandinavian congress of pathology and microbiology. Kreyberg, L., Lerche, Ch., Iversen, D.H. (eds.), p. 169. Oslo: University Press 1964

Rytömaa, T., Kiviniemi, K.: Regulation system of blood cell production. In: Control of cellular growth in adult organisms. Teir, H. and Rytömaa, T. (eds.), pp. 106–137. London: Academic Press 1967

Rytömaa, T., Kiviniemi, K.: Control of granulocytic production. I. Chalone and antichalone, two specific humoral regulators. Cell Tissue Kinet. *1*, 329–340 (1968 a)

Rytömaa, T. Kiviniemi, K.: Control of granulocytic production. II. Mode of action of chalone and antichalone. Cell Tissue Kinet. *1*, 341–350 (1968 b)

Rytömaa, T., Kiviniemi, K.: Control of DNA duplication in rat chloroleukaemia by means of the granulocytic chalone. Eur. J. cancer *4*, 595–606 (1968 c)

Rytömaa, T., Kiviniemi, K.: Chloroma regression induced by the granulocytic chalone. Nature *222*, 995–996 (1969)

Rytömaa, T., Kiviniemi, K.: Regression of generalized leukemia in rat induced by the granulocyte chalone. Eur. J. Cancer *6*, 401–410 (1970)

Rytömaa, T., Toivonen, H.: Chalones: concepts and results. Mech. Ageing Dev. *9*, 471–480 (1979)

Rytömaa, T., Vilpo, J.A., Levanto, A., Jones, W.A.: Effect of granulocyte chalone on acute and chronic granulocytic leukaemia in man. Report of seven cases. Scand. J. Haematol. Suppl. *27*, 5–28 (1976)

Rytömaa, T., Vilpo, J.A., Levanto, A., Jones, W.A.: Effect of granulocytic chalone on acute myeloid leukaemia in man. A follow-up study. Lancet *1977 I*, 771

Saetren, H.A.: A principle of auto-regulation of growth. Production of organ specific mitosis-inhibitors in kidney and liver. Exp. Cell Res. *11*, 229–232 (1956)

Saetren, H.: The organ specific growth inhibition of the tubule cells of the rat's kidney. Acta Chem. Scand. *17*, 889 (1963)

Saetren, H.: The first mitotic wave released among rat kidney tubule cells by partial nephrectomy. Acta Pathol. Microbiol. Scand. [A] *78*, 55–65 (1970)

Sassier, P., Bergeron, M.: Specific inhibition of cell proliferation in the mouse intestine by an aqueous extract of rabbit small intestine. Cell Tissue Kinet. *10*, 223–231 (1977)

Sassier, P., Bergeron, M.: Existence of an endogenous inhibitor of DNA synthesis in rabbit small intestine specifically effective on cell proliferation in adult mouse intestine. Cell Tissue Kinet. *13*, 251–261 (1980)

Scheffler, I.: Science and subjectivity. Indianapolis: Bobbs-Merrill 1967

Schunk, H., Schütt, M., Langen, P.: Granulozytenchalon: Gewebespezifität der Wirkung in Kurzzeitkulturen. Acta Biol. Med. Germ. *37*, 593–600 (1978)

Schwartz, S.M., Haudenschild, Ch.E.E.M.: Endothelial regeneration. 1. Quantitative analysis of initial stages of endothelial regeneration in rat aortic intima. Lab. Invest. *38*, 568–580 (1978)

Sekas, G., Cook, R.T.: The isolation of a low molecular weight inhibitor of ^{3}H-TdR incorporation into hepatic DNA. Exp. Cell Res. *102*, 422–425 (1976)

Sekas, G., Owen, W.G., Cook, R.T.: Fractionation and preliminary characterization of a low molecular weight bovine hepatic inhibitor of DNA synthesis in regenerating rat liver. Exp. Cell Res. *122*, 47–54 (1979)

Seiji, M., Nakano, H., Akiba, H., Kato, T.: Inhibition of DNA protein synthesis in melanocytes by a melanoma extract. J. Invest. Derm. *62*, 11–19 (1974)

Simard, A., Corneille, L., Deschamps, Y., Verly, W.: Inhibition of cell proliferation in the livers of hepatectomized rats by a rabbit hepatic chalone. Proc. Nat. Acad. Sci. U.S.A. *71*, 1763–1766 (1974)

Simard, A., Lavigne, J., Boileau, G., Lalanne, M., Riendeau, G.: Recherches sur les inhibiteurs de la proliferation cellulaire de type „chalones" dans le foie de rat a l'Institut du Cancer de Montreal. Union Med. Canada *105*, 1052–1057 (1976)

Simms, H.S., Stillman, N.P.: Substances affecting adult tissue in vitro. II. A growth inhibitor in adult tissue. J. Gen. Physiol. *20*, 621–629 (1937)

Simnett, J.D., Chopra, D.P.: Organ-specific inhibitor of mitoses in the amphibian kidney. Nature *222*, 1189–1190 (1969)

Simnett, J.D., Fisher, J.M., Heppleston, A.G.: Tissue-specific inhibition of lung alvolar cell mitosis in organ culture. Nature *223*, 944–946 (1969)

Smith, S.D.: The effects of electrophoretically separated lens proteins on lens regeneration in *Diemycylus viridescens*. J. Exp. Zool. *159*, 149–166 (1965)

Söderström, K.O., Tuohimaa, P.: Control of mitotic activity in the rat seminal vesicle by "chalone". Acta Endocrinol. (Kbh.) *81*, 668–672 (1976)

Spargo, D.J.: The evaluation of isolated endogenous factors regulating cell proliferation in mammalian skin. Ph D dissertation, University of London (1980)

Steck, P.A., Voss, P.G., Wang, J.L.: Growth control in cultured 3 T 3 fibroblasts. Assays of cell proliferation and demonstration of a growth inhibitory activity. J. Cell Biol. *83*, 562–575 (1979)

Tabor, C.W., Tabor, H., Bachrach, V.: Identification of the amino aldehydes produced by the oxidation of spermine and spermidine with purified plasma amine oxidase. J. Biol. Chem. *239*, 2194–2203 (1964)

Teel, R.W.: Inhibition of DNA synthesis in hamster cheek pouch tissue in organ culture by dibutyryl cyclic AMP and a homologous extract. Biochem. Biophys. Res. Commun. *47*, 1010–1014 (1972)

Thomas, W.E.G.: Somatostatin – the long lost antral chalone. Medical Hypotheses *6*, 919–928 (1980)

Thomas, W.A., Janakidevi, K., Florentin, R.A., Lee, K.T., Reiner, J.M.: Search for arterial smooth muscle cell chalone(s). In: Chalones. Houck, J.C. (ed.), pp. 451–457. Amsterdam, Oxford, New York: North-Holland and American Elsevier 1976

Thompson, J.C.: Antral chalone. Gastroenterology *67*, 752–754 (1974)

Thornley, A.I., Laurence, E.D.: Chalone regulation of the epidermal cell cycle. Experientia (Basel) *31*, 1024–1026 (1975)

Thornley, A.L., Laurence, E.D.: The specificity of epidermal chalone action: the results of in vivo experimentation with two purified skin extracts. Develop. Biol. *51*, 10–12 (1976)

Thornley, A.L., Spargo, D.J., Laurence, E.B.: A simple method for measuring DNA synthesis in epidermis and sebaceous glands and its application in chalone assays. Br. J. Dermatol. *97*, 11–23 (1977)

Thumann, A., Bustos-Obregon, E.: An "in vitro" system for the study of rat spermatogonial proliferative control. Andrologia *10*, 22–25 (1978)

Till, J.E., McCulloch, E.A., Siminovitch, L.: A stochastic model of stem cell proliferation based on the growth of spleen colony-forming cells. Proc. Natl. Acad. Sci. U.S.A. *51*, 29–36 (1964)

Tong, Ch., Bergevin, P., Appelbaum, L., Dinesh, P., David, G.: Inhibition of growth of mouse Ehrlich ascites tumor by normal tissue extracts. Chemotherapy *24*, 34–38 (1978)

Tutton, P.J.M.: Control of epithelial cell proliferation in the small intestinal crypt. Cell Tissue Kinet. *6*, 211–216 (1973)

Verly, W.Y.: The hepatic chalone. Natl. Cancer Inst. Monogr. *38*, 175–184 (1973)

Verly, W.G., Deschamp, Y., Pushpathadam, J., Desrosiers, M.: The hepatic chalone. I assay method for the hormone and purification of the rabbit liver chalone. Can. J. Biochem. *49*, 1376–1383 (1971)

Vilpo, J.A.: Non-specific and specific effects of a granulocytic bio-inhibitor on different haematopoietic cell populations. Scand. J. Haematol. *22*, 433–441 (1979)

Vinet, B., Verly, W.G.: Purification of a protein inhibitor of DNA synthesis in cells of hepatic origin. Eur. J. Cancer *12*, 211–218 (1976)

Voaden, M.J.: A chalone in the rabbit lens? Exp. Eye Res. *7*, 326–331 (1968)

Voaden, M.J.: Chalones in the lens: a progress report. Exp. Eye Res. *2*, 141–142 (1971)

Voaden, M.J., Leeson, S.J.: A chalone in the mammalian lens. I. Effect of bilateral adrenalectomy on the mitotic activity of the adult mouse lens. Exp. Eye Res. *9*, 57–66 (1970)

Van Waarde, D., Hulsin-Hesselink, E., Van Furth, R.: Humoral control of monocytopoiesis by an activator and an inhibitor Agents Actions *4*, 432–437 (1978)

Waksmann, B.H., Namba, Y.: Commentary on soluble mediators of immunological regulation. Cell. Immunol. *21*, 161–176 (1976)

Weiss, P., Kavanau, J.L.: A model of growth and growth control in mathematical terms. J. Gen. Physiol. *41*, 1–47(1957)

Wiener, N., Cybernetics. New York: John Wiley & Sons 1948

Willems, G.: Cell renewal in the gastric mucosa. Digestion *6*, 46–63 (1972)

Wright, E.G., Lord, B.I.: Production of stem cell proliferation regulators by fractionated cell suspensions. Leukemia Res. *3*, 15–22 (1978)

Wright, N.A., Al-Nafussi, A., Britton, N.: Negative feedback control of cell proliferation in the small intestine after death of proliferative cells. Abstracts Xth meeting european study group cell prolif., p.39. Oxford: Blackwell Sci. Publ. 1979

Wright, E.G., Sheridan, P., Moore, M.A.S.: An inhibitor of murine stem cell proliferation produced by normal human bone marrow. Leukemia Res. *4*, 309–314 (1980)

Yamaguchi, T., Hirobe, T., Kinjo, Y., Manaka, T.: The effect of chalone on the cell cyle in the epidermis during wound healing. Exp. Cell Res. *89*, 247–254 (1974)

Young, J.M., Lawrence, H.S., Cordell, S.L.: In vitro epidermal cell proliferation in rat skin plugs. J. Invest. Dermatol. *64*, 23–29 (1975)

Zakharov, V.B.: Diurnal rhythm of cell division of hepatoma 22a under the influence of hepatic chalone. Byull. Éksp. Biol. Med. (Moscow) *89*, 594–596 (1980)

Addendum

Since this chapter was written, a number of new articles on chalones have appeared. Some of them have been included in the list of references, without making reference to them in the text, others have appeared after the proof-reading, and the most important of them are added as a reference list here.

Bala, I.M.: Chalones and antichalones in hematology. Probl. Gematol. Perelivan. Krovi (Moskva) *25* (8), 50–54 (1980)

Barfod, N.M.: Partial purification and characterization of a cell specific G_1-inhibitor (chalone) from JB-1 ascites tumors. Europ. J. Cancer *17*, 421–431 (1981)

Bassi, A.M., Burlando, F., Romano, S., Ferro, M.: Sul calone epidermico della pelle umana. Pathologica (Genova) *72*, 201–218 (1980)

Benestad, H.B., Rytömaa, T., Svensson, T.: Some methods of assaying inhibitors of cell proliferation. Exp. Hematol. *8* (8), 961–970 (1980)

Chow, O.: Regulation factors in hyperplasia and tumors. Progr. Physiol. Sci. (China) *11*, 41–47 (1980)

Cox, E.B., Woodbury, M.A., Myers, L.E.: A new model for tumor growth analysis based on a postulated inhibitory substance. Computers Biomed. Res. *13*, 437–445 (1980)

Guigon, M., Enouf, J., Frindel, E.: Effects of CFU-S inhibitors on murine bone marrow during ARA-C treatment. I. Effects on stem cells. Leukemia Res. *4*, 385–392 (1980)

Heidemann, E.: Humorale Regulatoren der Zellproliferation in der Hämatopoese. I. Granulopoese und Lymphopoese. Klin. Wochenschr. *58*, 1117–1133 (1980)

Hess, A.D., Gall, S.A., Dawson, J.R.: Partial purification and characterization of a lymphocyte-inhibitoy factor(s) in ascitic fluids from ovarian cancer patients. Cancer Res. *40*, 1842–1851 (1980)

Holley, R.W., Böhlen, P., Fava, R., Baldwin, J.H., Kleeman, G., Armour, R.: Purification of kidney epithelial cell growth inhibitors. Proc. Natl. Acad. Sci. USA 77: 5989–5992 (1980)

Lenfant, M., Millerioux, L., Duchange, N., Tanzer, J.: De la recherche d'une chalone lymphocytaire a la redecouverte du facteur thymique serique (FTS). C. R. Acad. Sc. Paris *292*, 403–406 (1981)

Maharajan, V., Menon, M.M.: An inhibitor of cell proliferation & HeLa specific antigen expression isolated from HeLa cells. Indian J. Exp. Biol. *18*, 576–579 (1980)

Mikhailov, V.V., Neustroev, G.V., Gerina, L.S.: Role of the sympathetic nervous system in the regulation of erythrocyte chalone production. Fiziol. Zhurnal SSSR imeni I.M. Secenova *66*, 1344–1349 (1980)

Nakai, G.S., Gergely, H.: Ehrlich ascites cell (EAC) chalone assay using purified calf thymus-DNA polymerase-Alpha. Biomedicine Express *33*, 208–209 (1980)

Neustroev, G.V.: Effect of the erythrocytic chalone on the postsynaptic period of mitotic cycle in erythroblastic cells of the mouse bone marrow. Arch. Anat. Gistol. Embriol. *79* (7), 69–73 (1980)

Okulov, V.B., Zabezhinsky, M.A.: Immunologic testing of epidermal G_2 chalone as a marker of squamous cell tumors of the rat lungs. Biull. Eksp. Biol. Med. *90* (8), 249–249 (1980)

Onda, H.: A theoretical consideration of fundamental biological phenomena on cell-specific mitosis-inhibiting protein excretion hypothesis. J. theor. Biol. *86*, 771–787 (1980)

Perrins, D.J.D., Wiernik, G., Jones, W.A.: Granulocyte chalone assayed in vivo in the mouse. Acta haemat. *64*, 72–78 (1980)

Author Index

Subject Index

Handbook of Experimental Pharmacology

Continuation of "Handbuch der experimentellen Pharmakologie"

Springer-Verlag
Berlin
Heidelberg
New York

Handbook of Experimental Pharmacology

Continuation of "Handbuch der experimentellen Pharmakologie"

Editorial Board
G. V. R. Born, A. Farah,
H. Herken, A. D. Welch

Springer-Verlag
Berlin
Heidelberg
New York